Topics in Applied Physics

Volume 134

Topics in Applied Physics is a well-established series of review books, each of which presents a comprehensive survey of a selected topic within the area of applied physics. Edited and written by leading research scientists in the field concerned, each volume contains review contributions covering the various aspects of the topic. Together these provide an overview of the state of the art in the respective field, extending from an introduction to the subject right up to the frontiers of contemporary research.

Topics in Applied Physics is addressed to all scientists at universities and in industry who wish to obtain an overview and to keep abreast of advances in applied physics. The series also provides easy but comprehensive access to the fields for newcomers starting research.

Contributions are specially commissioned. The Managing Editors are open to any suggestions for topics coming from the community of applied physicists no matter what the field and encourage prospective book editors to approach them with ideas.

2018 Impact Factor: **0.746**

More information about this series at http://www.springer.com/series/560

Tim Salditt · Alexander Egner · D. Russell Luke
Editors

Nanoscale Photonic Imaging

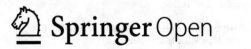

Editors
Tim Salditt
Institut für Röntgenphysik
Universität Göttingen
Göttingen, Germany

Alexander Egner
Laser Laboratorium
University of Göttingen
Göttingen, Germany

D. Russell Luke
Institut für Numerische
und Angewandte Mathematik
Universität Göttingen
Göttingen, Germany

ISSN 0303-4216 ISSN 1437-0859 (electronic)
Topics in Applied Physics
ISBN 978-3-030-34415-3 ISBN 978-3-030-34413-9 (eBook)
https://doi.org/10.1007/978-3-030-34413-9

This Springer imprint is published by the registered company Springer Nature Switzerland AG
The registered company address is: Gewerbestrasse 11, 6330 Cham, Switzerland

Preface

The word 'Nano' has been around for a long time. It became a topic of significant interest in the eighties of the last century, after instruments such as the scanning tunneling microscope and the atomic force microscope had been invented. The 'nanoscale' was probed based on electric currents through a tunneling tip or by measuring the forces with a cantilever. In other words, the 'room at the bottom' was conquered not by 'seeing', but rather by 'feeling'. Too strong was the belief that optical imaging was limited to the microscale due to the diffraction barrier. But the insight that photonics and nanoscale also make a perfect match followed only shortly after the advent of the scanning tunneling and atomic force microscopes. Around the turn of the millennium it became broadly accepted that plenty of 'nano' can be done with photons: Single molecule spectroscopy had been established, fluorescence correlation spectroscopy was emerging, and above all there was a new way to turn microscopes into nanoscopes based on optical switching, as pioneered by Stefan Hell here in Göttingen. While very few physicists cared about optical microscopes before, a time of rapid development had now set in. At the same time, a long-standing dream to realize X-ray microscopy was empowered by coherent optics and computational phase retrieval.

Pairing up optical and short wavelength to extend the scales of 'imaging', research teams in Göttingen set out for new discoveries. But how to empower their vessels? The solution was found by mathematics. Using results from inverse problems, stochastics, and optimization theory, new and bountiful shores were discovered, and photonic data was turned into useful information....

As we now come back from our expeditions funded for the last 12 years by the German Science Foundation (DFG) through SFB755 *Nanoscale Photonic Imaging*, we do not want to keep all the treasures for ourselves. The current book is a compilation of tutorials, experiments and experiences, and a compendium for further reading. In addition to the contributing authors and Angela Lehee at Springer, we are grateful to Leon Lohse, Shahroz Shahjahan for helping to keep this project on track. Above all we would like to express our deepest gratitude to Eva Hetzel who has been with this collaborative research center for the duration and has been essential to keeping the expedition on track, on budget and on time—all with grace and joyful optimism.

Now, let us dive deep into the nanoscale, and not just scratch at its surface!

Göttingen, Germany   Tim Salditt
                                                                Alexander Egner
                                                                D. Russell Luke

Contents

Contributors

Sadia Bari FS-Strukturdynamik (bio)chemischer Systeme, Deutsches Elektronen-Synchrotron DESY, Hamburg, Germany

Robert Beinert Institute for Mathematics and Scientific Computing, University of Graz, Graz, Austria

C. Charitha Indian Institute of Technology Indore, Indore, India

Alexey I. Chizhik Third Institute of Physics - Biophysics, Universität Göttingen, Göttingen, Germany

Alexander Egner Laser-Laboratorium Göttingen, Göttingen, Germany

Benjamin Eltzner Felix-Bernstein-Institute for Mathematical Statistics in the Biosciences, Universität Göttingen, Göttingen, Germany

Jörg Enderlein Third Institute of Physics - Biophysics, Universität Göttingen, Göttingen, Germany

Bernhard Flöter Laser-Laboratorium Göttingen eV., Göttingen, Germany

Claudia Geisler Laser-Laboratorium Göttingen, Göttingen, Germany

Helmut Grubmüller Department of Theoretical and Computational Biophysics, Max Planck Institute for Biophysical Chemistry Göttingen, Göttingen, Germany

Lara Hauke Third Institut of Physics - Biophysics, Universität Göttingen, Göttingen, Germany

Stefan W. Hell Department of NanoBiophotonics, Max Planck Institute for Biophysical Chemistry, Göttingen, Germany;
Department of Optical Nanoscopy, Max Planck Institute for Medical Research, Heidelberg, Germany

Thorsten Hohage   Institute for Numerical and Applied Mathematics, Universität Göttingen, Göttingen, Germany

Stephan Huckemann   Felix-Bernstein-Institute for Mathematical Statistics in the Biosciences, Universität Göttingen, Göttingen, Germany

Stefan Jakobs   Department of NanoBiophotonics, Max Planck Institute for Biophysical Chemistry, Göttingen, Germany

Isabelle Jansen   Department of NanoBiophotonics, Max Planck Institute for Biophysical Chemistry, Göttingen, Germany

Nickels A. Jensen   Department of NanoBiophotonics, Max Planck Institute for Biophysical Chemistry, Göttingen, Germany

Maria Kamper   Department of NanoBiophotonics, Max Planck Institute for Biophysical Chemistry, Göttingen, Germany

Jan Keller-Findeisen   Department of NanoBiophotonics, Max Planck Institute for Biophysical Chemistry, Göttingen, Germany

Ofer Kfir   IV. Physical Institute - Solids and Nanostructures, Universität Göttingen, Göttingen, Germany

Sarah Köster   Institute for X-ray Physics, Universität Göttingen, Göttingen, Germany

Hans-Ulrich Krebs   Institute for Material Physics, Universität Göttingen, Göttingen, Germany

Housen Li   Institute for Mathematical Stochastics, Universität Göttingen, Göttingen, Germany

D. Russell Luke   Institute for Numerical and Applied Mathematics, Universität Göttingen, Göttingen, Germany

Yura Malitsky   Institute for Numerical and Applied Mathematics, Universität Göttingen, Göttingen, Germany

Klaus Mann   Laser-Laboratorium Göttingen e.V., Göttingen, Germany

Simon Maretzke   Institute for Numerical and Applied Mathematics, Universität Göttingen, Göttingen, Germany

Anna-Lena Martins   Institute for Numerical and Applied Mathematics, Universität Göttingen, Göttingen, Germany

Tobias Mey   Laser-Laboratorium Göttingen eV., Göttingen, Germany

Matthias Müller   Laser-Laboratorium Göttingen e.V., Göttingen, Germany

Axel Munk Institute for Mathematical Stochastics, Universität Göttingen, Göttingen, Germany;
Max Planck Institute for Biophysical Chemistry, Göttingen, Germany

Markus Osterhoff Institute for X-ray Physics, Universität Göttingen, Göttingen, Germany

Gerlind Plonka Institute for Numerical and Applied Mathematics, Universität Göttingen, Göttingen, Germany

Katharina Proksch Institute for Mathematical Stochastics, Universität Göttingen, Göttingen, Germany

Florian Rehfeldt Third Institut of Physics - Biophysics, Universität Göttingen, Göttingen, Germany

Anna-Lena Robisch Institute for X-ray Physics, Universität Göttingen, Göttingen, Germany

Claus Ropers IV. Physical Institute - Solids and Nanostructures, Universität Göttingen, Göttingen, Germany

Steffen J. Sahl Department of NanoBiophotonics, Max Planck Institute for Biophysical Chemistry, Göttingen, Germany

Tim Salditt Institute for X-ray Physics, Universität Göttingen, Göttingen, Germany

Bernd Schäfer Laser-Laboratorium Göttingen eV., Göttingen, Germany

Ron Shefi Institute for Numerical and Applied Mathematics, Universität Göttingen, Göttingen, Germany

René Siegmund Laser-Laboratorium Göttingen, Göttingen, Germany

Murat Sivis IV. Physical Institute - Solids and Nanostructures, Universität Göttingen, Göttingen, Germany

Benjamin Sprung Institute for Numerical and Applied Mathematics, Universität Göttingen, Göttingen, Germany

Thomas Staudt Institute for Mathematical Stochastics, Universität Göttingen, Göttingen, Germany

Simone Techert FS-Strukturdynamik (bio)chemischer Systeme, Deutsches Elektronen-Synchrotron DESY, Hamburg, Germany

Sreevidya Thekku Veedu FS-Strukturdynamik (bio)chemischer Systeme, Deutsches Elektronen-Synchrotron DESY, Hamburg, Germany

Mareike Töpperwien Institute for X-ray Physics, Universität Göttingen, Göttingen, Germany

Benjamin von Ardenne Department of Theoretical and Computational Biophysics, Max Planck Institute for Biophysical Chemistry Göttingen, Göttingen, Germany

Frederic Weidling Institute for Numerical and Applied Mathematics, Universität Göttingen, Göttingen, Germany

Frank Werner Institute for Mathematical Stochastics, Universität Göttingen, Göttingen, Germany;
Max Planck Institute for Biophysical Chemistry, Göttingen, Germany

Carina Wollnik Third Institut of Physics - Biophysics, Universität Göttingen, Göttingen, Germany

Sergey Zayko IV. Physical Institute - Solids and Nanostructures, Universität Göttingen, Göttingen, Germany

Part I
Fundamentals and Tutorials

Chapter 1
STED Nanoscopy

Alexander Egner, Claudia Geisler and René Siegmund

> *The very first step of every understanding of the Microscope is ...*
> *to become familiar with the idea that it is a thing sui generis*
> – Ernst Abbe [1]

1.1 Fundamentals of Fluorescence Microscopy

This section will present the basics of fluorescence microscopy. Starting from the intensity distribution within the focal spot of an objective lens, we will discuss the image formation and derive the classical formula of the resolution limit. Furthermore, we will introduce the principle of confocal detection.

1.1.1 Vectorial Diffraction Theory and Intensity Distribution Within the Focal Spot

The complex electric vector field $\mathscr{E}(\mathbf{r})$ in the focal region of an optical system can be expressed in terms of a modified Huygens-Fresnel principle as the coherent superposition of secondary plane waves at the exit pupil [2, 3]

A. Egner (✉) · C. Geisler · R. Siegmund
Laser-Laboratorium Göttingen, Hans-Adolf-Krebs-Weg 1, 37077 Göttingen, Germany
e-mail: alexander.egner@llg-ev.de

C. Geisler
e-mail: claudia.geisler@llg-ev.de

R. Siegmund
e-mail: r.siegmund@llg-ev.de

© The Author(s) 2020
T. Salditt et al. (eds.), *Nanoscale Photonic Imaging*, Topics in Applied Physics 134,
https://doi.org/10.1007/978-3-030-34413-9_1

3

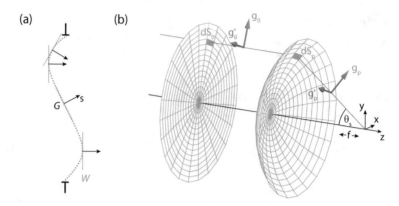

Fig. 1.1 Secondary plane waves as well as strength and polarization of a focused wavefront. **a** A secondary plane wave can be defined for each point G of a wave front W leaving the pupil of an optical imaging system. The wave front of the secondary plane wave is tangential to W in G. **b** When focusing through a lens, the projection of the electric vector field onto the meridional plane changes its direction of polarization (vectors $\mathbf{g}_0$ and $\mathbf{g}_p$). The part of the vector field orthogonal to the meridional plane retains its polarization direction (vectors $\mathbf{g}_0^*$ and $\mathbf{g}_p^*$). Due to the transition from a plane to a spherical wavefront, the strength of the electric field within associated surface segments (dS_0 and dS_p) changes such that the energy passing through the surface elements remains constant

$$\mathscr{E}(\mathbf{r}) = -\frac{ik}{2\pi} \iint_{\Omega} \sqrt{R_1 R_2} \mathbf{E}_p(\mathbf{s}) e^{ik\mathbf{sr}} d\Omega, \qquad (1.1)$$

where Ω is the solid angle, $\mathbf{E}_p$ is the complex amplitude of the secondary plane waves, R_1 and R_2 are the principle radii of curvature of the wavefront at the exit pupil and $\mathbf{s}$ is the unity vector in the respective direction of propagation, see Fig. 1.1a. The wave number k is given by $k = 2\pi n/\lambda_0$, with n being the refractive index and λ_0 the vacuum wavelength. Note that the geometric focus is at position $\mathbf{r} = (0, 0, 0)$.

In order to derive $\mathscr{E}(\mathbf{r})$ for an aplanatic, i.e. axially stigmatic and obeying the sine condition, imaging system such as the objective lens of a microscope, $\mathbf{E}_p(\mathbf{s})$, R_1 and R_2 have to be determined [4].

The typical scenario when focusing with an infinitely corrected lens is shown in Fig. 1.1b. Without loss of generality, we first assume that the light at the entrance pupil of the objective is linear polarized in the x-direction

$$\mathbf{E}_0(x_0, y_0) = E_0 \mathbf{e}_x e^{iP(x_0, y_0)}, \qquad (1.2)$$

where E_0 is the (real-valued) amplitude of the electric field, $\mathbf{e}_x$ is the unit vector in x-direction and P is the pupil function which encodes the phase distribution of the electric field [5]. For arbitrary polarization states, $\mathscr{E}(\mathbf{r})$ can then be calculated by the coherent superposition of several solutions of $\mathscr{E}(\mathbf{r})$ for correspondingly linear polarized vector fields at the entrance pupil. The case of unpolarized light is obtained by averaging over all possible polarization states.

According to Fig. 1.1b, the surface segment at the entrance pupil

$$dS_0 = r_0 d\phi_0 dr_0, \tag{1.3}$$

will be transformed by the objective lens to a surface segment at the exit pupil

$$dS_p = f^2 \sin\theta_s d\phi_s d\theta_s, \tag{1.4}$$

where r_0 is the distance of the surface segment from the optical axis and f is the focal length of the lens. As the lens obeys the sine condition

$$r_0 = f \sin\theta_s \tag{1.5}$$

and the intensity law of geometrical optics [6]

$$E_0^2 dS_0 = E_p^2 dS_p \tag{1.6}$$

has to be fulfilled, the amplitude of the electric field at the exit pupil is given by

$$E_p = \sqrt{\frac{r_0}{f\sin\theta_s}\frac{dr_0}{f d\theta_s}}\, E_0 = \sqrt{\cos\theta_s}\, E_0. \tag{1.7}$$

To determine the polarization of the electric field at the exit pupil, it is advisable to introduce two unit vectors for each light ray passing through the objective lens

$$\mathbf{g}_0 = \begin{pmatrix} \cos\phi_0 \\ \sin\phi_0 \\ 0 \end{pmatrix} \text{ and } \mathbf{g}_p = \begin{pmatrix} \cos\theta_s \cos\phi_s \\ \cos\theta_s \sin\phi_s \\ \sin\theta_s \end{pmatrix} \tag{1.8}$$

in the corresponding meridional plane and two unit vectors

$$\mathbf{g}_0^* = \begin{pmatrix} -\sin\phi_0 \\ \cos\phi_0 \\ 0 \end{pmatrix} \text{ and } \mathbf{g}_p^* = \begin{pmatrix} -\sin\phi_s \\ \cos\phi_s \\ 0 \end{pmatrix} \tag{1.9}$$

which are orthogonal to the meridional plane, Fig. 1.1b. Note that $\phi_0 = \phi_s$. When the light rays pass through the lens, the portion of their electric fields originally pointing in the $\mathbf{g}_0$ directions are re-polarized in the $\mathbf{g}_p$ directions and the portion pointing in the $\mathbf{g}_0^*$ directions do not change their polarization. Hence, using (1.2), (1.7), (1.8) and (1.9), we can write for the amplitude of the secondary plane waves

$$\mathbf{E}_p = \sqrt{\cos\theta_s}\left((\mathbf{E}_0 \cdot \mathbf{g}_0)\mathbf{g}_p + (\mathbf{E}_0 \cdot \mathbf{g}_0^*)\mathbf{g}_p^*\right), \tag{1.10}$$

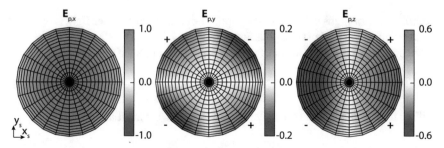

Fig. 1.2 Electrical field components pointing in x-, y- and z-direction for an initially x-polarized planar wave front after passing a lens

which after expansion results in

$$
\mathbf{E}_p = \sqrt{\cos\theta_s}\, E_0
\begin{pmatrix}
\cos\theta_s + (1 - \cos\theta_s)\sin^2\phi_s \\
(\cos\theta_s - 1)\cos\phi_s\sin\phi_s \\
\sin\theta_s\cos\phi_s
\end{pmatrix}
e^{i P(x_0, y_0)}. \tag{1.11}
$$

Note that the influence of the lens on the phase of the electric field has already been fully accounted for in the presented geometry and that $x_0 = f\sin(\theta_s)\cos(\phi_s)$ and $y_0 = f\sin(\theta_s)\sin(\phi_s)$. The electrical field components in all three directions are depicted in Fig. 1.2. Note that the maximum strength of $\mathbf{E}_{p,z}$ is about half of the maximum strength of $\mathbf{E}_{p,x}$ and that $\mathbf{E}_{p,y}$ is again a factor of three lower. While $\mathbf{E}_{p,x}$ points into the same direction over the entire aperture, both the y- and the z-component change their signs. As we will see later, this has a direct influence on the distribution of the polarization within the focus. In particular, the y- and z-component interfere destructively on the optical axes ($x = y = 0$).

If we assume that the entrance pupil of the objective is homogeneously illuminated and if we use that $R_1 = R_2 = f$ applies to the principle radii in (1.1), we derive for the electric vector field in the focal region

$$
\mathscr{E}_x(\mathbf{r}) = -\frac{iA}{\pi} \int_0^\alpha \int_0^{2\pi} \sqrt{\cos\theta_s}\,\sin\theta_s \left\{\cos\theta_s + (1 - \cos\theta_s)\sin^2\phi_s\right\} e^{i(k\mathbf{r}\cdot\mathbf{s} + P(\theta_s,\phi_s))}\,\mathrm{d}\phi_s\,\mathrm{d}\theta_s
$$

$$
\mathscr{E}_y(\mathbf{r}) = +\frac{iA}{\pi} \int_0^\alpha \int_0^{2\pi} \sqrt{\cos\theta_s}\,\sin\theta_s \left\{(1 - \cos\theta_s)\cos\phi_s\sin\phi_s\right\} e^{i(k\mathbf{r}\cdot\mathbf{s} + P(\theta_s,\phi_s))}\,\mathrm{d}\phi_s\,\mathrm{d}\theta_s
$$

$$
\mathscr{E}_z(\mathbf{r}) = +\frac{iA}{\pi} \int_0^\alpha \int_0^{2\pi} \sqrt{\cos\theta_s}\,\sin\theta_s \left\{\sin\theta_s\cos\phi_s\right\} e^{i(k\mathbf{r}\cdot\mathbf{s} + P(\theta_s,\phi_s))}\,\mathrm{d}\phi_s\,\mathrm{d}\theta_s
$$

$$
\tag{1.12}
$$

where

$$A = \frac{kf E_0}{2} \tag{1.13}$$

is a constant and α is the semi-aperture angle of the objective lens. Please note that θ_s is defined in negative z-direction, therefore the sign of the $\mathscr{E}_z$ component has to be changed. Usually, the numerical aperture

$$NA = n \cdot \sin(\alpha) \tag{1.14}$$

is specified instead of α to indicate the aperture angle of an objective lens, in which n is the refractive index of the immersion medium.

For an incident plane wave $P(\theta_s, \phi_s)$ is constant (e.g. 0) and the integration with respect to ϕ_s can be carried out and the analytic solution of $\mathscr{E}(\mathbf{r})$ is [4]

$$\begin{aligned}
\mathscr{E}_x(\mathbf{r}) &= -i A (I_0 + I_2 \cos 2\phi_r) \\
\mathscr{E}_y(\mathbf{r}) &= -i A I_2 \sin 2\phi_r \\
\mathscr{E}_z(\mathbf{r}) &= -2 A I_1 \cos \phi_r,
\end{aligned} \tag{1.15}$$

where the field is expressed in spherical coordinates $\mathbf{r} = (r, \theta_r, \phi_r)$ and the diffraction integrals are defined as

$$I_0(\mathbf{r}) = \int_0^\alpha \sqrt{\cos \theta_s} \, \sin \theta_s (1 + \cos \theta_s) J_0(kr \sin \theta_r \sin \theta_s) e^{ikr \cos \theta_r \cos \theta_s} d\theta_s$$

$$I_1(\mathbf{r}) = \int_0^\alpha \sqrt{\cos \theta_s} \, \sin^2 \theta_s J_1(kr \sin \theta_r \sin \theta_s) e^{ikr \cos \theta_r \cos \theta_s} d\theta_s \tag{1.16}$$

$$I_2(\mathbf{r}) = \int_0^\alpha \sqrt{\cos \theta_s} \, \sin \theta_s (1 - \cos \theta_s) J_2(kr \sin \theta_r \sin \theta_s) e^{ikr \cos \theta_r \cos \theta_s} d\theta_s,$$

and J_n are the Bessel functions of the first kind and order n. The overall intensity in the vicinity of the focal spot is given by

$$I(\mathbf{r}) = \mathscr{E}_x^2(\mathbf{r}) + \mathscr{E}_y^2(\mathbf{r}) + \mathscr{E}_z^2(\mathbf{r}). \tag{1.17}$$

The contributions of the electric fields of individual polarization directions to the intensity in the focal plane as well as the overall intensity is shown in Fig. 1.3a. It can be clearly seen that the symmetry of the polarization direction distribution on the exit pupil is transferred to the focal plane. As a consequence, the intensity distribution is not rotational symmetric. For example, the focal spot is narrower in the direction orthogonal to the polarization direction of the incident field. However, the focus can

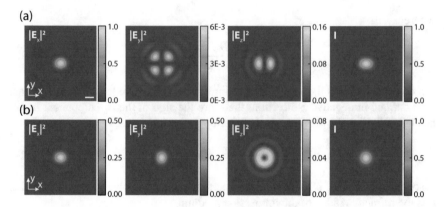

Fig. 1.3 Contributions of the electric fields of individual polarization directions and overall intensity in the focal plane for **a** linear and **b** circular polarized light in the entrance pupil of the objective lens. Calculations were performed for an NA 1.4 oil immersion objective lens ($\lambda = 640$ nm, $n = 1.518$). Scale bar 250 nm

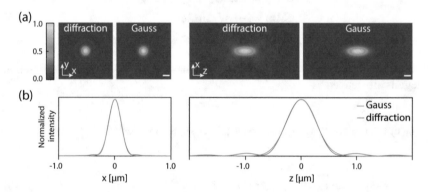

Fig. 1.4 Simulated intensity distribution and Gaussian approximation. The intensity distributions in the x-y (left) and x-z (right) plane through the geometric focus **a** and the corresponding intensity profiles along the x-(left) and the z-direction (right) **b** show the good agreement of the Gaussian approximation in the central area. Calculations were performed for an NA 1.4 oil immersion objective lens ($\lambda = 640$ nm, $n = 1.518$) and circular polarized ligth. Scale bars 250 nm

be made symmetrical by the use of circular polarized light, Fig. 1.3b. In many cases it is not necessary to know the intensity distribution in the focus down to the last detail. In this case it is useful to approximate it by a Gaussian function with a corresponding full width at half maximum (FWHM). As you can see in Fig. 1.4 this approximation is reasonably good in the central area.

1.1.2 Incoherent Image Formation

Far-field fluorescence microscopy has proven to be a powerful and versatile tool in the life sciences and beyond [7–11]. Since it allows to non-invasively image the interior of sufficiently translucent samples in three dimensions, it is well suited for imaging biological samples, even under living conditions [12, 13]. Further, tagging of target proteins or epitopes with fluorescent markers, e.g. by immunolabeling with organic fluorophores or by expression of fluorescent fusion proteins, lends an exceptional molecular specificity to the method [14–16]. In order to understand the implementation of a fluorescence microscope, it is instructive to first consider the fluorescence process on the molecular level.

Figure 1.5a illustrates the relevant molecular energy levels and transitions within the singlet state in a Jablonski diagram. Here, S_0 denotes the electronic ground state and S_1 the first excited electronic state. Please note that higher excited states as well as the triplet states are neglected here because they are not necessary for a basic understanding. The thick lines indicate the lowest vibrational energy level, whereas the thin lines indicate levels with higher vibrational energy. At ambient temperatures, a molecule typically resides in the lowest vibrational level of S_0 according to the Boltzmann distribution [17]. By absorption of a photon of suitable energy, the molecule can be excited to higher vibrational levels of S_1. From there, it relaxes radiation-less to the lowest vibrational level, which typically takes place within one picosecond or less [18]. The emission of a fluorescence photon takes place, as the molecule spontaneously returns to higher vibrational levels of S_0. This transition may also occur radiation-less via internal conversion. However, for fluorescent molecules, which are described here, this process is of minor importance. The time interval which the molecule spends in S_1 is known as the fluorescence lifetime and depends on the molecule itself and on its environment. It is typically several nanoseconds and therefore three to four orders of magnitudes longer than the characteristic time for vibrational relaxation [18]. The cycle is completed by vibrational relaxation back to the lowest vibrational level of S_0.

For the design of a fluorescence microscope, two consequences of the described excitation and emission process are of immediate relevance. First, due to the extended spectrum of vibrational levels, photons within a range of energies may excite the molecule. Likewise, fluorescence photons have a spectrum of energies. Second, due to the dissipation of energy by the vibrational relaxation after excitation, the emitted photon's energy is always lower than that of the absorbed photon. Figure 1.5b illustrates these aspects in terms of wavelength instead of energy, and shows the absorption and the emission spectrum of a typical fluorescent molecule. Since the photon wavelength scales inversely with the photon energy, the emission spectrum is at longer wavelengths than the absorption spectrum. This red-shift is called Stokes shift [17] and can be harnessed in the implementation of a fluorescence microscope.

Figure 1.5c shows a simple epi-illumination design of such a microscope. A broadband light source, e. g. a metal-halide lamp or a light emitting diode, is spectrally filtered by a bandpass filter (BP), such that the selected wavelength range lies within

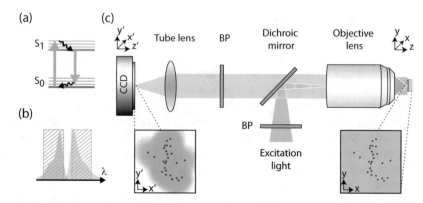

Fig. 1.5 Principle of fluorescence microscopy. **a** Jablonski diagram of a fluorescent molecule. By absorption of a photon the molecule can be excited from the electronic ground state, S_0, into any vibrational level of S_1. Fast nonradiative relaxation into the lowest level of S_1 takes place within a picosecond. The molecule can return into any vibrational levels of S_0 by the spontaneous emission of a photon (fluorescence). From there it relaxes non-radiatively into its lowest vibrational level. **b** Absorption (blue) and emission (green) spectrum of a fluorescent molecule. The hatched areas indicate the transmission range of the respective bandpass filters. **c** In a typical experimental implementation the excitation light is focused into the back aperture of the objective lens in order to generate a homogeneous light distribution within the sample. All fluorophores in the sample are equally excited (right inset). The fluorescence signal is collected by the objective lens, separated from the excitation light by a dichroic mirror and a detection bandpass (BP), and imaged onto an area detector such as a CCD camera. The inset on the left depicts the image on the camera (green). Note that the positions of the fluorophores are indicated only for illustration purposes

the absorption spectrum of the respective fluorescent molecule. This excitation light is reflected by a dichroic mirror through the objective lens into the sample. In a typical implementation it is focused into the back aperture of the objective lens in order to illuminate the sample over an extended area (wide-field illumination). All fluorescent molecules inside the sample are equally excited and their fluorescence is collected by the same objective lens. Since the fluorescence is red-shifted with respect to the excitation light, it is transmitted by the dichroic mirror and thus efficiently separated from the excitation light. After passing through a bandpass filter, which blocks residual excitation light and suppresses unwanted room light outside of the desired spectral detection window, the fluorescence is imaged by a tube lens onto a camera. The transmission ranges of the excitation and detection BP are illustrated by hatched regions in Fig. 1.5b.

The fluorescence light emitted in the sample plane is imaged to the detection plane by lenses and is therefore subject to diffraction. The image of a fluorescent molecule, which can be seen as a point emitter of electromagnetic radiation, is therefore not a point, but spread to an extended intensity distribution (cf. Sect. 1.1.1). The projection of this pattern into the sample is called the point spread function (PSF) and it is an important characteristic of a microscope, as it determines its resolution capability (cf. Sect. 1.1.3). Since fluorescence emission is a spontaneous process, emitted light of different sources has no constant phase relation. The light is incoherent and thus the image of several point sources (e.g. a fluorophore distribution in the sample) is composed of single overlapping PSFs. In mathematical terms, the image I_{image}, which is back-projected into the sample plane, is the convolution of the true object O and the PSF h:

$$I_{\text{image}}(\mathbf{r}) = O(\mathbf{r}) * h(\mathbf{r}) \tag{1.18}$$

1.1.3 Classical Resolution Limit

The resolution of an optical system, e.g. a microscope, describes its ability to distinguish two objects. Therefore, the spatial resolution of a microscope is given by the minimum distance of two structures at which their images can be discerned.

The Abbe Limit

By investigating line gratings, Ernst Abbe discovered in 1873 that the lateral resolution of a light microscope solely depends on the wavelength of the light and the numerical aperture of the objective lens used [19]. In order to resolve two adjacent lines they have to be separated by at least:

$$d_{\text{min}} = \frac{\lambda_0}{2\,\text{NA}}, \tag{1.19}$$

with λ_0 being the vacuum wavelength of the light used. This fundamental limit is often referred to as the diffraction limit. However, Abbe's considerations do not allow conclusions on light-emitting objects or the axial resolution of the microscope.

The Rayleigh Criterion

As already described in Sect. 1.1.1, the image of a point is not a point but a blurred spot. If we assume an incident plane wave and neglect the direction change of the polarization of the electric field by the focusing process, formula 1.12 simplifies to

$$\mathscr{E}(\mathbf{r}) = \mathscr{E}_x(\mathbf{r}) = -\frac{iA}{\pi} \int_0^\alpha \int_0^{2\pi} \sqrt{\cos\theta_s} \, \sin\theta_s e^{i(k\mathbf{r}\cdot\mathbf{s})} d\theta_s d\phi_s. \tag{1.20}$$

In this case the integration with respect to ϕ_s can be readily performed and we derive (compare (1.16))

$$\mathscr{E}(\mathbf{r}) = -iA \int_0^\alpha \sqrt{\cos\theta_s} \, \sin\theta_s J_0(kr\sin\theta_r\sin\theta_s) e^{ikr\cos\theta_r\cos\theta_s} d\theta_s. \tag{1.21}$$

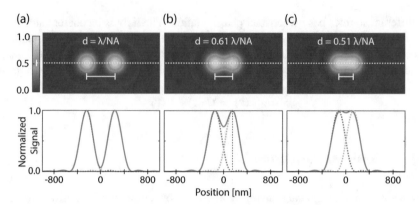

Fig. 1.6 Images of two point emitters with different distances and corresponding intensity profiles along the dashed white lines. **a** For large separations both emitters can be easily identified. **b** According to the Rayleigh criterion, the minimum distance to resolve both emitters is reached, when the maximum of one emitter coincides with the minimum of the other. **c** Emitters that are closer than this distance cannot be resolved in the image

Note that $r \sin(\theta_r)$ is the distance of $\mathbf{r}$ from the optical axes and $r \cos(\theta_r)$ corresponds to the z-coordinate of $\mathbf{r}$. If we concentrate on the focal plane ($z = 0$) or the optical axis ($x = y = 0$) and assume that the aperture angle is relatively low (paraxial approximation), the integration with respect to θ_s can also be performed and we obtain

$$\mathscr{E}(x, y, 0) = -iA \frac{J_1(k\sqrt{x^2 + y^2} \sin \alpha)}{k\sqrt{x^2 + y^2} \sin \alpha}$$

$$\mathscr{E}(0, 0, z) = -iA \frac{sin(\frac{k}{4}z \sin^2 \alpha)}{\frac{k}{4}z \sin^2 \alpha}. \tag{1.22}$$

The absolute square of $\mathscr{E}(x, y, 0)$ results in the so-called Airy pattern in the focal plane. Figure 1.6 shows the images of two point-like emitters for three different distances and the corresponding intensity profiles along the white dashed lines. Dashed blue and yellow curves indicate the profiles for the individual emitters, whereas the red lines show the profile when both emitters are radiating at the same time. When the distance of both emitters is sufficiently large, they can easily be identified as individual emitters. According to the Rayleigh criterion, two spatially separated point sources can be discerned, when the maximum of the diffraction pattern of one point emitter coincides with the first minimum of the other [6, 20]. This case is illustrated in Fig. 1.6b. Note that the Rayleigh criterion only holds true for incoherently radiating sources. When the distance between the emitters gets smaller they cannot be distinguished any more, Fig. 1.6c.

The distance between the main maximum and the first minimum of the Airy pattern is often used as a measure for the lateral resolution and is given by:

$$d_{x,y} = 0.61 \frac{\lambda_0}{\text{NA}}. \tag{1.23}$$

Likewise, the resolution in axial direction (z) can be defined as the distance between the main maximum and the first minimum in z-direction as:

$$d_z = 2.00 \frac{n\,\lambda_0}{(\text{NA})^2}. \tag{1.24}$$

The Full Width at Half Maximum Criterion

As you can see from Fig. 1.6b, two points with a distance corresponding to the Rayleigh criterion can still be resolved if the signal to noise ratio is sufficiently high. This is no longer the case when their distance corresponds to the FWHM of the Airy pattern, Fig. 1.6c. If the resolution is defined in such a way we get

$$d_{x,y} = 0.51 \frac{\lambda_0}{\text{NA}} \text{ and} \tag{1.25}$$

$$d_z = 1.77 \frac{n\,\lambda_0}{(\text{NA})^2}. \tag{1.26}$$

Thus, the resolution of a microscope is according to this criterion limited to approximately half the wavelength in lateral and twice the wavelength in the axial direction. The FWHM definition of the resolution is particularly well suited if the PSF is approximated by a Gaussian function which, as is well known, has no minima. In this case the resolution then can either be expressed by its standard deviation, σ, or its full width at half maximum (FWHM $= 2\sqrt{2\ln 2}\sigma$).

1.1.4 Confocal Microscopy

Wide-field fluorescence microscopy offers the possibility to image the entire field of view of the objective lens at once. However, wide-field illumination also has a decisive disadvantage as it not only excites dye molecules in the focal plane, but simultaneously in the entire sample volume. The light emitted by axially distant molecules is detected in addition to the signal from the fluorophores in the focal plane and generates a bright background in the image. This makes it difficult to acquire high quality data, especially in axially extended samples.

In confocal microscopy [21], the axial extent of the sample region from which the signal impinges on the detector can be narrowed down. For excitation, a point-like light source is imaged into the sample plane. Since conventional fluorescent lamps are spatially extended, their light has to be focused onto a pinhole and afterwards

(a)

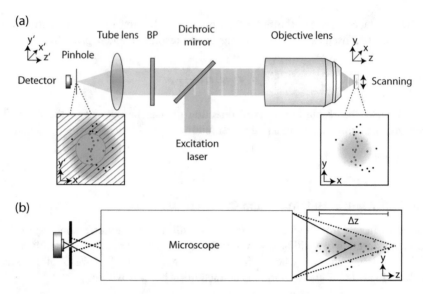

Fig. 1.7 Principle of a confocal microscope. **a** In a typical experimental implementation a collimated excitation laser is focused into the sample by the objective lens, creating a diffraction-limited excitation spot. Fluorophores within this spot will be excited with a probability proportional to the excitation light intensity. The fluorescence signal is collected by the objective lens, separated from the excitation light by a dichroic mirror and a detection BP, and imaged onto a detector. A confocal pinhole in front of the detector ensures that only fluorescence from a certain region is detected. The inset on the right depicts the excitation spot and the accordingly excited fluorophores in the sample plane, while the inset on the left depicts the image of the excited molecules in the pinhole plane. The gray circle indicates the detection pinhole. Only fluorescence within the circle is detected. **b** Additionally, fluorescence from planes distant to the focal plane is blocked by the detection pinhole. This allows to analyze signal from thin optical sections. Note that the positions of the fluorophores are indicated only for illustration purposes

collimated with a lens. The invention of lasers as bright and intense point-like light sources rendered the use of an excitation pinhole obsolete and quickly led to the first confocal laser scanning microscopes [22, 23].

The main components of a typical confocal fluorescence microscope are illustrated in Fig. 1.7. A collimated excitation beam is focused into the sample by an objective lens and generates a diffraction-limited excitation PSF, $h_{ex}(\mathbf{r})$. The inset on the right side in Fig. 1.7a depicts the focal plane and the excitation spot. Fluorescent markers within $h_{ex}(\mathbf{r})$ will be excited with a probability proportional to the light intensity and can therefore emit fluorescence. In order to acquire an image, either the sample must be scanned through the focus, or vice versa.

Each excited fluorophore can emit fluorescence, which is collected by the objective lens, separated from the excitation light by a dichroic mirror and a detection BP and imaged onto a pinhole. The pinhole ensures that only fluorescence from the direct vicinity of the geometrical focal point is detected. As the light path is invertible this can be interpreted as imaging the pinhole into the focal plane. This image is called the detection PSF, $h_{\text{det}}(\mathbf{r})$, and describes the probability to detect a photon emitted at position $\mathbf{r}$. The gray circle in the left inset in Fig. 1.7a indicates the pinhole. Only fluorescence originating from inside this circle is detected with high probability. Note that each fluorescent molecule is imaged diffraction-limited.

Another advantage of the detection pinhole is illustrated in Fig. 1.7b. The fluorescence from axial distant planes (with respect to the focal plane) is blocked by the detection pinhole. Therefore, the key feature of the confocal microscope, other than conventional microscopes, is that it efficiently (and sharply) images only those regions of a volume sample that lie within a thin section around the focal plane of the microscope. In other words, it is able to reject (effectively attenuate) light from out-of-focus regions of the sample [24–28].

The PSF of the confocal microscope is given by the probability that a fluorophore is excited multiplied with the probability that its fluorescence is detected:

$$h_{\text{conf}}(\mathbf{r}) = h_{\text{ex}}(\mathbf{r}) \cdot h_{\text{det}}(\mathbf{r}). \tag{1.27}$$

In the theoretical limit of an infinitesimally small detection pinhole and identical wavelengths for illumination λ_{ex} and detection λ_{det}, a confocal microscope improves the resolution by a factor of $\sqrt{2}$ [24].

The influence of the size of the detection pinhole on the lateral (black line) and axial (blue line) resolution, as well as on the detected signal (red line) is shown in Fig. 1.8. The graphs are obtained by calculating and analyzing images of a point emitter imaged with an oil-immersion objective lens (NA = 1.4, n = 1.518, $\lambda_{\text{ex}} = 640$ nm, $\lambda_{\text{det}} = 680$ nm) using (1.17) and varying pinhole diameters. The pinhole diameter is measured in Airy units (AU), with one AU corresponding to the diameter of the Airy disc in the focal plane (1 AU $= 1.22\,\lambda/\text{NA}$). It is clearly visible that the best achievable resolution in all directions is achieved with an infinitesimally small pinhole. With increasing pinhole size, the resolution of the confocal microscope decreases. The detected signal, however, grows with increasing pinhole diameter [29]. For experimental purposes, a finite pinhole size is necessary to collect sufficient signal. Often a pinhole size in the range of 1 AU is chosen as a tradeoff between collected signal and resolution. Even though the resolution increase in the lateral direction is almost negligible in this regime, the advantage of optical sectioning remains.

For a circular detection pinhole, the pinhole function is given by

$$p(\mathbf{r}) = p(x, y, z = 0) = \begin{cases} 1 & \text{for}\sqrt{x^2 + y^2} \leq p_0 \\ 0 & \text{otherwise} \end{cases}, \tag{1.28}$$

with p_0 being the pinhole radius. The real detection PSF, $h_{\text{det, real}}(\mathbf{r})$, is then given by the convolution of $h_{\text{det}}(\mathbf{r})$ with the pinhole function:

$$h_{\text{det, real}}(\mathbf{r}) = h_{\text{det}}(\mathbf{r}) * p(\mathbf{r}). \tag{1.29}$$

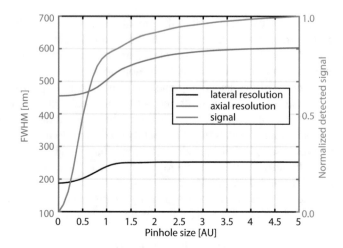

Fig. 1.8 Influence of the pinhole diameter on the lateral (black) and the axial (blue) resolution as well as the detected signal (red). Calculations are performed for an NA 1.4 oil-immersion objective lens ($\lambda_{ex} = 640$ nm, $\lambda_{det} = 680$ nm, $n = 1.518$). Increasing the pinhole size increases the detected signal, but also lowers the achievable resolution. Often pinholes with a size of 1 AU are utilized in a confocal microscope, as at this size sufficient signal is collected while the optical sectioning capability is mainly maintained

1.2 Fundamentals of STED Microscopy

For a long time, the resolution of a microscope was considered to be limited by diffraction. But during the last decades, physico-optical methods that circumvent the diffraction barrier emerged in far-field fluorescence microscopy [30]. These new super-resolution microscopy - in short 'nanoscopy'—methods have been awarded the Nobel prize in Chemistry in 2014 and allow a resolution improvement of at least one order of magnitude. The first method of this kind was stimulated emission depletion (STED) microscopy, proposed in 1994 by Hell and Wichmann [31] and demonstrated by Klar and Hell in 1999 [32].

Ever since their advent, super-resolution microscopy techniques are versatile tools for non-invasive investigations of structures. STED microscopes offer for example the possibility to measure intracellular structures in fixed [33, 34] and living cells [35, 36] with, in principle, unlimited resolution [31]. A lateral resolution of 15 nm was demonstrated by imaging single fluorescent molecules [37], and a resolution of 5.8 nm [38] resp. 2.4 nm [39] have been demonstrated on single nitrogen vacancy centres in diamonds. Furthermore, STED microscopes have been used to measure e.g. colloidal structures [40], and block copolymers [41, 42] and the underlying principle has been used for STED lithography [43, 44]. As STED microscopy is cutting edge technology, new, improved acquisition schemes are continuously being developed and integrated (e.g. RESCue-STED [45] or DyMIN [46]).

1.2.1   Basic Idea

A fundamental breakthrough in the achievable resolution of light microscopes was realized when fluorescent markers were not only considered as contrast agents, but the molecular transitions of the markers were additionally used to specifically switch on and off the ability of a subset of markers to fluoresce. Hereby, the fluorescence from markers within a diffraction-limited spot can be temporally separated and thus be read out sequentially.

As detailed above, confocal microscopy employs a targeted readout scheme. The diffraction-limited focus is scanned through the sample and the detected fluorescence is computationally assigned to the known position, thereby generating an image pixel by pixel. In this mode, increasing the resolution is synonymous to decreasing the spatial extent of the region from where the fluorescence is detected. STED microscopy realizes this by employing the process of stimulated emission to actively switch off fluorescent markers by forcing them to the electronic ground state S_0 without emission of a fluorescence photon (Fig. 1.9a). This can be achieved by overlapping the excitation spot with a spatially extended intensity distribution, $I(\mathbf{r}, t)$, featuring at least one zero-intensity region as off-switching requires $I(\mathbf{r}, t) > 0$ and is absent for $I(\mathbf{r}, t) = 0$. If the STED focus has a ring shape (doughnut shape) with a central intensity zero, molecules at its rim are switched off, while molecules in the center are not. This results in a spatial narrowing of the fluorescent spot, whose extent then defines the resolution of the microscope. The resolution, which theoretically can get arbitrarily good, depends not only on the applied STED intensity, but also on the photophysical properties of the fluorophores. A detailed discussion of the photophysics of dye molecules is presented in Sect. 1.2.2.

The key components of a STED microscope are illustrated in Fig. 1.9c. The setup is based on a confocal microscope (cf. Figure 1.7). Additionally, a STED laser, whose wavelength is at the red end of the fluorescence spectrum (cf. Fig. 1.9b), e.g. $\lambda_{em}^{max} = 654$ nm and $\lambda_{STED} = 775$ nm for Abberior STAR 635P, is phase-modulated and superimposed with the excitation laser. Further detail of how to shape the STED beam is given in Sect. 1.2.3. The emitted fluorescence is spectrally separated from the laser beams and detected by a point detector (e. g. a single photon counting module).

The right inset in Fig. 1.9c depicts the overlap of the excitation and STED beams in the sample plane. Only fluorescent molecules in the central region of the depletion pattern are allowed to remain in the excited state and can therefore emit fluorescence and contribute to the detected signal. The inset on the left side depicts the image plane with the gray circle indicating the detection pinhole. Usually, pulsed lasers are used as light sources for excitation and depletion in STED microscopy. The central inset indicates that a temporal delay between the excitation and depletion pulses is needed for an efficient fluorescence suppression (cf. section 1.2.2). Additionally, a helical phase mask, that is used to create the doughnut-shaped depletion pattern by imprinting a phase retardation from $0 - 2\pi$ onto the STED beam, is depicted.

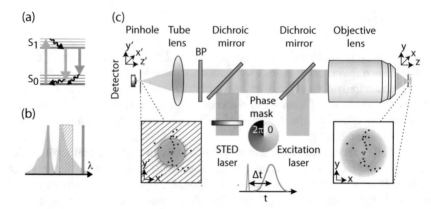

Fig. 1.9 Principle of a STED microscope. **a** Jablonski diagram of a fluorescent molecule. In addition to the processes of excitation and spontaneous emission, stimulated emission is now used to switch off excited molecules in a targeted way. **b** Absorption and emission spectrum of a fluorescent molecule. The depletion laser is shifted to the far right of the emission spectrum of the fluorophore. **c** In comparison to the confocal microscope, an additional depletion laser is now superimposed with the excitation beam. A helical phase mask imprints a phase retardation from $0 - 2\pi$ onto the STED beam, that when imaged into the sample plane creates a doughnut-shaped depletion pattern. The right inset shows the overlap of excitation and STED beam in the focal plane. Wherever the STED intensity is sufficiently high, excited fluorophores are driven into their off-state. Therefore, fluorescence is only emitted from sample regions where the STED intensity is negligible. This fluorescence is separated from the laser light and imaged onto a point-detector. Most STED microscopes utilize pulsed lasers for excitation and depletion. The central inset illustrates, that a time delay between the excitation and STED pulses is needed for an effective suppression of the fluorescence

1.2.2 Basic Photophysics of Dye Molecules

As described in the previous section, the key principle of STED microscopy is the inhibition of fluorescence emission by stimulated emission. The efficiency of this fluorescence depletion is a crucial parameter for the performance of a STED microscope and it depends on the interplay of the excitation light and the STED light with the fluorescent molecules. In the following, this will be discussed in detail with special attention to the timing between excitation and STED light and to the required STED power. Following [47], rate equations for the population of electronic states of fluorescent molecules will be formulated and their implications will be discussed.

In the context of STED microscopy, a fluorophore can be modelled as a simple four level system, in which photo-bleaching, intermediate dark states and radiation-less decay from S_1 to S_0 are neglected (cf. Fig. 1.10a). Note that in comparison to Fig. 1.9a, the spectrum of higher vibrational levels is merged to one level each and transition rates k and population probabilities N have been introduced. Specifically, N_1 and N_3 correspond to the population of the lowest vibrational level of S_0 and S_1, respectively. N_4 and N_2 represent the population of the higher vibrational states $S_{1,\text{vib}}$ and $S_{0,\text{vib}}$

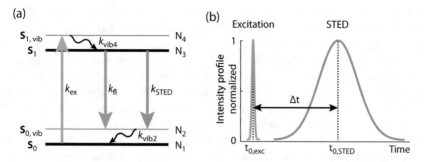

Fig. 1.10 Four level system of a fluorophore and laser pulse timing used in numeric calculations. **a** A fluorescent molecule can be described as a four level system with S_0 and $S_{0,vib}$ being the lowest, respectively a higher vibrational level of the electronic ground state. S_1 and $S_{1,vib}$ are the corresponding levels in the first excited state. N_1 to N_4 denote the respective population probabilities. The straight arrows indicate the excitation (blue) with its rate constant k_{ex}, fluorescence emission (green, k_{fl}) and stimulated emission (red, k_{STED}), while the wiggly arrows denote vibrational relaxation (k_{vib2} and k_{vib4}). **b** Gaussian-shaped excitation pulse (blue) and STED pulse (red) which exhibit their maximum intensities at time points $t_{0,ex}$ and $t_{0,STED}$. The delay between the pulses is Δt

after excitation and fluorescence emission, respectively. Since $N_i, i = 1, 2, 3, 4$ are probabilites, $\sum_i N_i = 1$.

The temporal evolution of the population probabilities N_1 to N_4 can be described by a set of coupled rate equations:

$$\frac{\partial N_1(t)}{\partial t} = k_{ex}\left[N_4(t) - N_1(t)\right] + k_{vib2} N_2(t)$$

$$\frac{\partial N_2(t)}{\partial t} = k_{fl} N_3(t) - k_{STED}\left[N_2(t) - N_3(t)\right] - k_{vib2} N_2(t)$$

$$\frac{\partial N_3(t)}{\partial t} = -k_{fl} N_3(t) + k_{STED}\left[N_2(t) - N_3(t)\right] + k_{vib4} N_4(t) \tag{1.30}$$

$$\frac{\partial N_4(t)}{\partial t} = -k_{ex}\left[N_4(t) - N_1(t)\right] - k_{vib4} N_4(t)$$

Here, k_{fl}, k_{vib2} and k_{vib4} are the rate constants for fluorescence decay from S_1 and vibrational decay from $S_{0,vib}$ and $S_{1,vib}$, respectively. The rates for these spontaneous processes are given by the inverse of the lifetimes of the starting states, with $k_{fl}^{-1} = \tau_{fl}$ in the range of several nanoseconds and $k_{vib}^{-1} = \tau_{vib}$ on the order of one picosecond or less [18]. Note that excitation from S_0 to $S_{1,vib}$ by the STED light has been neglected.

The rate constants for excitation k_{ex} and stimulated emission k_{STED}, however, depend on the intensity of the excitation and the STED light. They are given by the product of the molecular cross-section σ for the respective transition and the light intensity I divided by the photon energy hc/λ_0:

$$k = \frac{\sigma I}{hc/\lambda_0} \tag{1.31}$$

with the Planck constant h, the speed of light c and the vacuum wavelength λ_0. Please note that in order to make the notation easier to read, the indices $_{ex}$ and $_{STED}$ are omitted here and in the following and are introduced again later.

When considering the third line in (1.30), it becomes obvious that for an efficient depletion of fluorescence, the depopulation of S_1 by stimulated emission must not only dominate over the spontaneous fluorescence emission, but also over the refilling of S_1 from $S_{1,vib}$ caused by vibrational relaxation after excitation. This suggests that a pulsed scheme, which has already been implied in Fig. 1.9b, is beneficial [48]. An excitation pulse is followed by a STED pulse. This separates excitation and stimulated emission temporally, such that S_1 is not refilled during fluorescence depletion. Further, pulsed lasers typically provide a high peak intensity, while the average laser power and thus the light dose in the sample is kept rather low.

For modelling the pulsed scheme, the intensity-dependent rate constants k_{ex} and k_{STED} in the rate equations (1.30) need to be formulated time-dependently. For this, the laser pulses are assumed to have a Gaussian shape in time (cf. Fig. 1.10b) and the time-dependent intensity $I(t)$ is

$$I(t) = J \frac{hc}{\lambda_0} \sqrt{\frac{4\ln 2}{\pi \tau^2}} e^{\frac{-4\ln 2(t-t_0)^2}{\tau^2}} \tag{1.32}$$

with the photon fluence per pulse J (measured in number of photons per area per pulse), the temporal FWHM τ and pulse center position t_0.

Usually, in the experiment, the fluence per pulse in the focal plane cannot be measured directly. Instead, the laser power P is readily accessible, which is why the fluence J will now be expressed in terms of power P. The total number of photons per laser pulse n is given by

$$n = \frac{P}{r_{rep}hc/\lambda_0} \tag{1.33}$$

with the repetition rate of the laser pulses r_{rep} and the photon energy in the denominator. The distribution of photon fluences in the focal plane $J(x, y)$ is then given by

$$J(x, y) = nh(x, y) \tag{1.34}$$

with the focal probability distribution of a single photon h. Please note that in contrast to the previous notation, here the PSF h is not interpreted as an intensity distribution, but as the probability for a photon to be found at a certain position. Therefore, h is normalized such that $\iint_{-\infty}^{\infty} h(x, y)dxdy = 1$.

Combining (1.31), (1.32), (1.33) and (1.34) gives the time and position dependent rate constant

$$k(x, y, t) = \sigma \frac{P}{r_{rep}hc/\lambda_0} \sqrt{\frac{4\ln 2}{\pi \tau^2}} e^{\frac{-4\ln 2(t-t_0)^2}{\tau^2}} h(x, y) \tag{1.35}$$

with a molecule dependent, a laser light dependent and a microscope dependent part. Due to practical reasons, we simplify this expression further by approximating the PSF with a Gaussian function with FWHM $d_{x,y} \simeq \frac{\lambda_0}{2NA}$ (see Sect. 1.1.3 and (1.25))

$$h(x, y) \simeq \frac{4 \ln 2}{\pi d_{x,y}^2} e^{-\frac{4 \ln 2 \left(x^2 + y^2\right)}{d_{x,y}^2}} \tag{1.36}$$

and evaluate it at the geometric focus position

$$k(0, 0, t)_i \simeq 3.32 \, \sigma_i \frac{P_i}{r_{rep} h c \lambda_i \tau_i} e^{\frac{-4 \ln 2 (t - t_{0,i})^2}{\tau_i^2}} NA^2 \tag{1.37}$$

where $i \in \{ex, STED\}$. Note that here the indices $_{ex}$ and $_{STED}$ are introduced again. This equation depends on experimental parameters, which are easy to obtain, either by direct measurements or by consulting data sheets. Substituting this expression into the rate equations (1.30), we obtain the means to analyze the time-dependent state population of a fluorescent molecule in the pulsed STED scheme. A quantity of particular interest is the overall emitted fluorescence

$$F = \int_0^\infty k_{fl} N_3(t) dt \tag{1.38}$$

and its dependence on experimental parameters, since the STED microscope's performance is directly influenced by the efficiency of fluorescence depletion.

Influence of Laser Parameters on Fluorescence Depletion

For successful STED imaging in a pulsed scheme, it is particularly important to consider the influences of the relative timing between the laser pulses and the STED laser power on the efficiency of fluorescence depletion, since these two parameters need to be routinely set by the microscopist. Therefore, the overall emitted fluorescence (1.38) is simulated by numerically solving the rate equations (1.30). The rate constants k_{ex} and k_{STED} are assumed to be time-dependent and are analyzed at position (x,y) = (0,0) according to (1.37). From an experimental point of view, this corresponds to measuring the fluorescence from a very small bead which is located in the very center of the superimposed focal spots of the excitation and the (not spatially shaped) STED light.

For the simulations, fluorophore parameters are set to mimic a typical STED fluorophore: $\tau_{fl} = 3.3$ ns, $\tau_{vib2} = \tau_{vib4} = 1$ ps, $\sigma_{ex} = 4.6 \cdot 10^{-16}$ cm^2, $\sigma_{STED} = 4.6 \cdot 10^{-17}$ cm^2. Note that effects due to the polarization and the orientation of the transition molecular dipole are neglected. The NA of the objective lens is assumed to be 1.4. The laser wavelengths are set to $\lambda_{ex} = 640$ nm and $\lambda_{STED} = 775$ nm, which are typical for STED imaging of red fluorophores, and the laser repetition rate is assumed to be $r_{rep} = 20$ MHz.

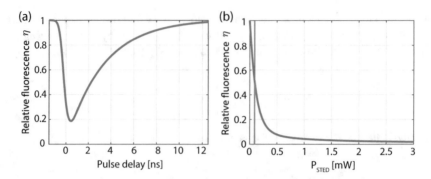

Fig. 1.11 Influence of pulse delay Δt **a** and STED power P_{STED} **b** on the relative fluorescence. **a** The relative fluorescence shows a pronounced minimum at $\Delta t = 440$ ps. Calculations were performed for $P_{STED} = 0.25$ mW. **b** For optimized pulse delay $\Delta t = 440$ ps, the relative fluorescence drops to half at $P_{STED} = 0.09$ mW, which is indicated by the red line. All other parameters are as mentioned in the main text

The question of suitable laser pulse lenghts deserves a short comment: While a very short excitation pulse in the range of a picosecond is beneficial, because fluorescence decay during excitation can be neglected in this case, there is a clear constraint on the shortest feasible pulse length of the STED laser. The rate for stimulated emission from the S_1 to $S_{0,vib}$ is equal to the rate for re-excitation from $S_{0,vib}$ back to S_1. Therefore, at best, an equal population of both states can be achieved, unless there is sufficient time for vibrational relaxation from $S_{0,vib}$ to S_0. Only due to this drain of $S_{0,vib}$, the state S_1 can be efficiently depleted. The STED pulse length should therefore be much longer than the vibrational lifetime [48]. On the other hand, it should be shorter than the fluorescence lifetime since STED photons arriving after the molecule has already fluoresced do not have any effect and are therefore wasted. Considering these aspects as well as specifications of commercially available laser systems, excitation and STED pulse lengths are set to $\tau_{ex} = 50$ ps and $\tau_{STED} = 800$ ps.

The results of the simulations are shown in Fig. 1.11. It illustrates the relative fluorescence η, which depicts the amount of remaining fluorescence, when STED light is applied compared to the case without applying any STED light. The STED power is $P_{STED} = 0.25$ mW and $P_{ex} = 10\,\mu$W is chosen such that no saturation effects occur during excitation.

In Fig. 1.11a the relative timing $\Delta t = t_{0,STED} - t_{0,ex}$ of the excitation and STED pulse is varied. This so called pulse delay spans a range from -1.5 ns to 12.5 ns, where a positive value corresponds to the situation where the STED pulse peak reaches the sample after the peak of the excitation pulse. If Δt is too short, the STED efficiency is low, either because STED photons reach the sample even before the molecules have been excited or because they have not yet vibrationally relaxed to the lowest level of S_1. This effect accounts for the steep slope on the left hand side, whose gradient is determined by τ_{STED}. If, however, the pulse delay is too long and the STED pulse reaches the sample too late, some of the molecules will have already fluoresced. The gradient of the right slope therefore depends on τ_{fl}. At optimal time delay the relative fluorescence is minimal, which is the case for $\Delta t = 440$ ps in this example.

Figure 1.11b shows the relative fluorescence at optimal time delay as a function of P_{STED}. The STED power at which the fluorescence drops to half is called saturation power P_{sat}. The shape of the curve has strong similarity to an exponential decay and, indeed, if re-excitation of the dye by the STED light is neglected (simple two-level system), η is given by [49]

$$\eta = e^{-\sigma_{STED} J_{STED}}. \tag{1.39}$$

Substituting J_{STED} using (1.34) and (1.33) gives the focal shape of the fluorescence suppression induced by the applied STED light

$$\eta(x, y) = e^{-\sigma_{STED} \frac{P_{STED}}{r_{rep} hc/\lambda_{STED}} h_{STED}(x,y)} \tag{1.40}$$

again with a dye dependent, a laser light dependent and a microscope dependent part.

Evaluating the relative fluorescence η at the center of the PSF and setting it to $1/2$

$$\eta(0, 0) = e^{-\sigma_{STED} \frac{P_{STED}}{r_{rep} hc/\lambda_S TED} h_{cal}(0,0)} \overset{!}{=} 1/2 \tag{1.41}$$

results in an analytical expression for the saturation power P_{sat}

$$P_{sat} = \frac{\ln 2/\sigma_{STED}}{h_{cal}(0, 0)} r_{rep} hc/\lambda_{STED}. \tag{1.42}$$

Note that a calibration PSF h_{cal} is introduced here in order to make the expression also applicable for more elaborated shapes of h_{sted}, e. g. exhibiting a central intensity zero. In practice, the thus defined P_{sat} defines the power of the STED light which is needed to suppress the fluorescence at the center of a Gaussian-shaped STED PSF by half. It depends on the optical properties of the microscope, photophysical properties of the dye and parameters of the STED laser and allows to write the relative fluorescence (cf. 1.40) in a particularly simple form

$$\eta(x, y) = e^{-\ln 2\zeta \frac{h_{STED}(x,y)}{h_{cal}(0,0)}}, \tag{1.43}$$

with the saturation factor ζ defined as P_{STED}/P_{sat}.

1.2.3 Shaping the STED Beam

As already mentioned in Sect. 1.2.1, STED nanoscopy is based on the idea of limiting the ability of molecules to fluoresce in the immediate vicinity of the geometric focus. Since the fluorescence is inhibited by the process of stimulated emission, it is necessary to shape the intensity distribution of the STED light such that it is zero in the geometric focus. In order to utilize the maximum power of the available STED light, the phase distribution of the electric field at the entrance pupil of the objective lens $P(x_0, y_0)$ has to be designed such that the contributions of all secondary plane waves $\mathbf{E}_p$ interfere destructively at the focus.

Central Retardation

Among the simplest ways to generate a central zero intensity is the generation of a phase delay of π in a central circular region of the aperture [50]

$$P(x_0, y_0) = \begin{cases} \pi & \sqrt{x_0^2 + y_0^2} \leq r_0 \\ 0 & \text{elsewhere.} \end{cases} \tag{1.44}$$

If the effect of focusing on the polarization direction of $\mathbf{E}_p$ is neglected, r_0 is exactly given by the diameter of the entrance pupil divided by $\sqrt{2}$. For the usually utilized high NA lenses, however, it is slightly smaller (compare (1.11) and Fig. 1.2). As shown in Fig. 1.12a, the polarization contributions of the secondary plane waves in the direction of the original polarization cancel each other out. The orthogonal polarization directions also interfere destructively, since the phase mask does not change the rotary symmetry of the corresponding field components on the exit aperture (compare Figs. 1.2 and 1.3). The intensity distribution generated by illuminating the phase mask with circular polarized light is shown in Fig. 1.12b. Since the phase of secondary plane waves, originating from the central region of the aperture, changes significantly faster as a function of z than that of secondary plane waves from the boundary region, spots of high constructive interference occur above and below the focal plane. Therefore, this phase mask is usually used to increase the resolution in the axial direction.

Helical Retardation

Another way to create a depletion pattern is to helicaly phase retard the STED beam [51]

$$P(x_0, y_0) = \phi \tag{1.45}$$

where ϕ is the angle between the vector (x_0, y_0) and the x-axis. The operation principle of this phase mask is based on the same effect, which ensures that when focusing a plane x-polarized wavefront, the y- and z-components of the electric field vanish on the optical axis (compare Figs. 1.2 and 1.3). Since two mirror-symmetrical points with respect to the optical axis always exhibit a phase difference of π, their x- and y-components of the electric field cancel each other out at the geometric focus (Fig. 1.13a). However, this also creates the effect that the z-components of $\mathbf{E}_p$ for these points face in the same direction, which means that they interfere constructively at the focal spot. However, this can be avoided by using circular polarized light. For the originally x-polarized part of the illuminating field, the effect still exists, but now the z-components of the originally y-polarized part for two points which are

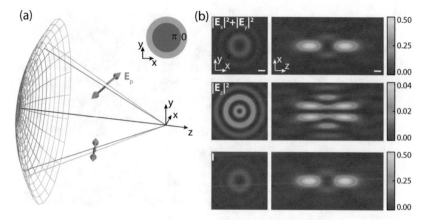

Fig. 1.12 Central phase retardation. **a** A phase delay of the central region of the aperture by π (see inset in the top right corner) causes the x-components of the secondary plane waves of the central and outer regions to cancel each other out. The illustration shows $\mathbf{E}_p$ for two opposing points in the inner and outer region when illuminated with x-polarized light. **b** Strength of the lateral and axial electric field components and overall intensity in the vicinity of the focal spot for illumination with circular polarized light. Calculations were performed for an NA 1.4 oil immersion objective lens ($\lambda = 775$ nm, $n = 1.518$). Scale bars 250 nm

rotated by $\phi = 90°$ with respect to the originally considered points face in the opposite direction. This causes the z-components of the electric field of the two point pairs to cancel each other out (Fig. 1.13a). Note that this effect is only achieved if the circularity of the light matches the rotation direction of the helical phase mask. If this is not the case, the described effect contradicts and the field distribution has maximum z-component in the geometrical focus. The intensity distribution for a correct circularity of the illuminating light field is depicted in Fig. 1.13b and forms a hollow cylinder around the optical axes. It has been shown that helical phase retardation generates the optimal inhibition pattern for isotropic resolution enhancement in the focal plane [51].

1.2.4 Resolution

In this section the effective PSF of a STED microscope is derived. It describes the volume in which fluorescence is still allowed, and whose spatial extent is a measure for the resolution. As an example, a 2D STED microscope utilizing a helical phase mask is considered.

We assume that the excitation and STED light is applied as temporally separated pulses with a pulse duration much shorter than the fluorescence lifetime. Photobleaching, intermediate dark-states or re-excitation of the dye by the STED light are neglected (simple two-level model) and dye molecules are assumed to rotate

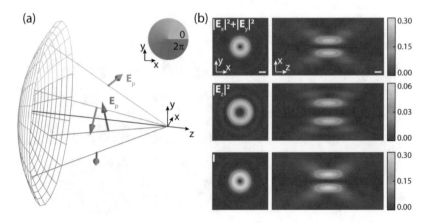

Fig. 1.13 Helical phase retardation. **a** A helical phase retardation (see inset in the top right corner) causes the lateral field components of the secondary plane waves of opposing points to cancel each other out. The illustration shows $\mathbf{E}_p$ for two opposing points when illuminated with x-polarized (red) and y-polarized (blue) light. The two point pairs are rotated by 90° with respect to each other. The phase delay between the x-polarized and y-polarized light was set to $\pi/2$. **b** Strength of the lateral and axial electric field components and overall intensity in the vicinity of the focal spot for illumination with circular polarized light. Calculations were performed for an NA 1.4 oil immersion objective lens ($\lambda = 775$ nm, $n = 1.518$). Scale bars 250 nm

fast enough to average the orientation of their molecular transition dipole relative to the polarization of the excitation and STED light. Under these conditions, the effective PSF of the STED microscope h_{eff} is the product of the excitation PSF and the remaining fluorescence in the presence of the STED light [49]

$$h_{\text{eff}}(x, y) = h_{\text{ex}}(x, y)\eta(x, y). \qquad (1.46)$$

According to Sect. 1.1.3, the excitation PSF can well be approximated by a symmetrical 2D Gaussian peak in the focal plane with a FWHM of $d_{\text{x,y}} \simeq \frac{\lambda_0}{2\text{NA}}$

$$h_{\text{ex}}(x, y) \propto e^{-\frac{4\ln 2\left(x^2 + y^2\right)}{d_{\text{x,y}}^2}} \qquad (1.47)$$

where a normalization constant has been neglected.

For sufficiently large saturation factors, the FWHM of the effective central spot of the STED microscope is much smaller than the wavelengths used and only the shape of the STED intensity distribution in the vicinity of the focal spot determines the shape of the central spot. In this region, the focal distribution h_{STED}, which is generated via helical phase retardation (cf. Sect. 1.2.3), can be well approximated by a 2D parabola [52]

$$\frac{h_{\text{STED}}(x, y)}{h_{\text{cal}}(0, 0)} \simeq 4a(x^2 + y^2). \qquad (1.48)$$

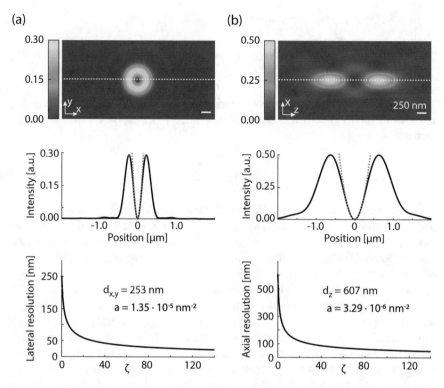

Fig. 1.14 Pattern steepness and resolution in the case of helical **a** and central **b** phase retardation. Top: Focal intensity distribution in the x-y-plane and the x-z-plane, respectively, through the geometric focus. Center: The intensity profiles (black) along the dotted white lines can be well fitted with a parabola (red) in the vicinity of the minimum in both cases. Bottom: With the fitted pattern steepnesses and the indicated FWHM of the excitation PSF in the respective direction, the lateral and axial resolution can be calculated according to (1.50). Calculations were performed for a 1.4 NA oil-immersion objective lens ($n = 1.518$), $\lambda_{\text{ex}} = 640\,\text{nm}$ and $\lambda_{\text{STED}} = 775\,\text{nm}$

Here, $h_{\text{cal}}(0, 0)$ is the calibration factor already known from Sect. 1.2.2 and a is the so called pattern steepness, which is proportional to the curvature of h_{STED} in the geometrical focus. Figure 1.14a shows the 2D STED intensity distribution (top) (cf. Fig. 1.13b) and the good agreement of the parabolic fit (center). Please note that the definition of the pattern steepness differs from a prior definition. Here, it is normalized to $h_{\text{cal}}(0, 0)$, while Harke et al. normalized a to the maximal intensity of $h_{\text{STED}}(x, y)$ in the focal plane [52].

Combining (1.46), (1.47), (1.48) with (1.43) from Sect. 1.2.2 for $\eta(x, y)$ gives a relatively simple expression for the effective STED PSF h_{eff}, which represents a 2D Gaussian peak shape

$$h_{\text{eff}}(x, y) = e^{-4\ln 2(x^2+y^2)(d_{x,y}^{-2}+a\zeta)}. \tag{1.49}$$

Its FWHM along the lateral direction is

$$d_{\text{STED}} = \frac{d_{\text{x,y}}}{\sqrt{1 + d_{\text{x,y}}^2 a \zeta}}. \tag{1.50}$$

For sufficiently large saturation factors, the attainable resolution of the STED microscope is only governed by the product of the pattern steepness and the saturation factor

$$d_{\text{STED}} = \frac{1}{\sqrt{a\zeta}}. \tag{1.51}$$

The dependence of the lateral STED resolution on the saturation factor ζ according to (1.50) is shown in Fig. 1.14a (bottom). In this example, which was calculated for an NA 1.4 oil immersion objective lens ($n = 1.518$, $\lambda_{\text{ex}} = 640$ nm, $\lambda_{\text{STED}} = 775$ nm), a resolution of 50 nm, which corresponds to a resolution improvement of factor 5, is achieved for a saturation factor $\zeta \simeq 28$.

The resolution formula is not limited to the 2D STED pattern considered here, but is applicable whenever a parabolic fit can be reasonably applied in the vicinity of the zero intensity spot. This specifically also applies to the STED pattern, which is usually used for axial resolution increase (cf. Sect. 1.2.3 and Fig. 1.12b). Figure 1.14b shows the good agreement of the fit to the corresponding focal intensity distribution along the axial direction and presents the attainable axial resolution. Again an oil immersion objective lens with NA = 1.4 was assumed ($\lambda_{\text{ex}} = 640$ nm, $\lambda_{\text{STED}} = 775$ nm). Because of the larger FWHM and the smaller pattern steepness, a saturation factor of $\zeta = 28$ yields a resolution of only 103 nm in this case. Still this corresponds to a resolution increase by a factor of 6.

For imaging three-dimensional structures, a resolution increase in all three dimensions is often desired. This can be achieved by an incoherent superposition of both STED patterns. It was shown that a distribution of the total available power of 30% in the 2D and 70% in the axial pattern is favorable in terms of focal volume size and axial resolution [40].

1.3 Imaging Examples

STED microscopy has become an indispensable tool in the life sciences, as it allows non-invasive uncovering of details hidden to conventional light microscopes. By now, nanoscopy has been successfully applied to various fields such as immunology, signaling, virology, bacteriology and cancer biology [53]. Particularly, the possibility to label different types of proteins simultaneously and to record their relative spatial distribution at super-resolution offers important insight into protein co-localization and interaction. In order to demonstrate the current performance of STED microscopy, some selected examples of cell imaging are presented in the following.

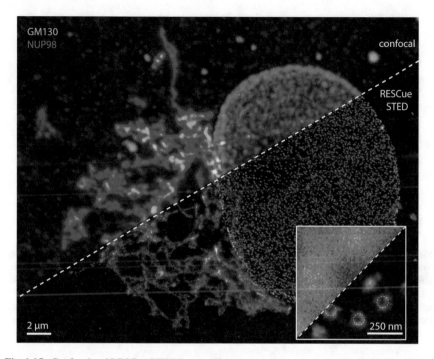

Fig. 1.15 Confocal and RESCue STED images of nuclear pore complex subunits (NUP98, red) and Golgi apparatus (GM130, blue) in Vero cells. Samples were prepared by indirect immunolabeling using Abberior STAR RED and Abberior STAR ORANGE. Acquisition was performed using an Abberior Instruments Facility Line STED microscope. Shown is a maximum projection of a raw image stack. The inset shows that RESCue STED microscopy can resolve the ring-like organization of the nuclear pore complex proteins (as highlighted by the dotted white circles). Please note that the diameter of individual NUP98 rings is ~70 nm. For a better visualization, the STED image in the inset is smoothed. Data are courtesy of Abberior Instruments, Germany

Example 1: Golgi Apparatus and Nuclear Pore Complex

Figure 1.15 shows the complexly structured Golgi apparatus (blue) in a Vero cell. This cell organelle is known to be a collection and dispatch station of protein products from the endoplasmatic reticulum. It synthesizes and modifies elements of the plasma membrane and generates primary lysosomes. Next to the Golgi apparatus, the easily recognizable oval shaped cell nucleus is visible in red. More precisely, the image depicts the nuclear pore complex, a part of the nuclear envelope surrounding the cell nucleus, that allows transportation across the envelope.

For confocal and STED imaging, the proteins GM130 (Golgi apparatus) and NUP98 (nuclear pore complex) have been immunolabeled with primary antibodies targeting the respective proteins and dye labeled secondary antibodies (Abberior STAR ORANGE, Abberior STAR RED) binding to the latter. It is evident, that the structures are resolved with much more detail in the RESCue STED image. Especially the ring-like arrangement of the nuclear pore complex proteins can be discerned (cf. dotted circles in the inset of Fig. 1.15).

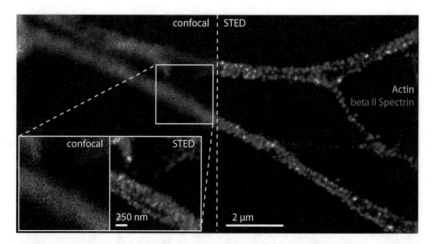

Fig. 1.16 Confocal and STED images of spectrin periodicity in primary rat hippocampal neurons. Please note the characteristic ~190 nm beta II spectrin periodicity along distal axons (red, blue) which is only visible in the STED image. Labelled structures: beta II spectrin (red, Abberior STAR635P), actin (blue, Abberior STAR 580). Acquisition was performed using an Abberior Instruments STEDYCON microscope. Shown are raw data. Data are courtesy of Abberior Instruments, Germany

Example 2: Nanoscopy of Neurons

Neurons are highly specialized cells, which are the basic building blocks of the nervous system and transmit information throughout the body. STED nanoscopy revealed that short actin filaments in neuronal axons, dendrites and spine necks are bridged by spectrin tetramers to form an ~190 nm periodic structure [53].

An exemplary measurement of the actin (blue) and beta II spectrin (red) distribution in the axons of a primary rat hippocampal neuron is shown in Fig. 1.16. While in the confocal image only little information on the co-localization can be obtained, the characteristic periodicity of the beta II spectrin as well as the actin is easily seen in the STED image. The inset emphasizes previously concealed details, that are clearly visible in the STED image.

Example 3: Nanoscopy of Mitochondria

Although mitochondria are best known for their role as the 'power houses' of the cell, they are also key players in executing apoptosis, a tightly regulated suicide program in eukaryotic cells [53]. Moreover, damage and subsequent dysfunction of mitochondria is known as an important factor for several human diseases. With a diameter of approximately 300–500 nm in cultured mammalian cells, their structure is not accessible to conventional light microscopy.

STED nanoscopy revealed that Tom20, a membrane-spanning receptor protein of the translocase of the outer membrane complex, is found in clusters on the surfaces of mitochondria [53]. Super-resolution studies also showed that the nucleoids in mitochondria have a diameter of 70–110 nm and allowed conclusions on the number of copies of mitochondrial DNA (mtDNA) per nucleoid [53].

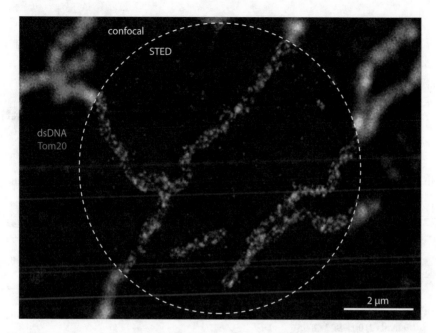

Fig. 1.17 Confocal and STED images of mitochondrial protein (Tom20, red) and the mitochondrial genome (dsDNA, green) in Vero cells. Samples were prepared by indirect immunolabeling using Abberior STAR RED and Abberior STAR ORANGE. Acquisition was performed using an Abberior Instruments Facility Line STED microscope. Shown are raw data. Data are courtesy of Abberior Instruments, Germany

Figure 1.17 depicts an image of the mitochondria in a Vero cell. The Tom20 proteins are illustrated in red and the mtDNA in green. Again in the STED image the clustering and co-localization of both proteins is evident, whereas only few conclusions can be drawn from the confocal image.

Acknowledgements We thank Andreas Schönle and Jan Keller-Findeisen for the with software for calculating electric fields according to vectorial diffraction theory, Jan Keller-Findeisen for his definition of the saturation power and Abberior Instruments for the application images. Finally, we also thank Jaydev Jethwa for proofreading.

References

1. Abbe, E.: On the Estimation of aperture in the microscope. J. Royal Microsc. Soc. **1**(3), 388–423 (1881) (London, UK: Williams & Norgate)
2. Wolf, E.: Electromagnetic diffraction in optical systems. I. An Integral Representation of the Image Field. Proc. R. Soc. London, Ser. A **253**, 349–357 (1959)
3. Egner, A., Hell, S.W.: Equivalence of the Huygens-Fresnel and Debye approach for the calculation of high aperture point-spread functions in the presence of refractive index mismatch. J.

Microsc. **193**, 244–249 (1999)
4. Richards, B., Wolf, E.: Electromagnetic diffraction in optical systems. II. Structure of the image field in an aplanatic system. Proc. R. Soc. London, Ser. A **253**, 358–379 (1959)
5. Egner, A., Schrader, M., Hell, S.W.: Refractive index mismatch induced intensity and phase variations in fluorescence confocal, multiphoton and 4Pi-microscopy. Opt. Commun. **153**, 211–217 (1998)
6. Born, M., Wolf, E.: Principles of Optics. Pergamon Press Ltd., Oxford (1999)
7. Haustein, E., Schwille, P.: Trends in fluorescence imaging and related techniques to unravel biological information. HFSP J. **1**, 169–180 (2007)
8. Lichtman, J.W., Conchello, J.-A.: Fluorescence microscopy. Nat. Methods **2**, 910–919 (2005)
9. Rost, F.W.D.: Fluorescence Microscopy, vol. 2. Cambridge University Press, Cambridge, New York (1992)
10. Wöll, D., Flors, C.: Super-resolution fluorescence imaging for materials Science. Small Methods **1**, 1700191 (2017)
11. Yuste, R.: Fluorescence microscopy today. Nat. Methods **2**, 902–904 (2005)
12. Ettinger, A., Wittmann, T.: Fluorescence live cell imaging. In: Waters, J.C., Wittman, T. (eds.), Quantitative Imaging in Cell Biology. Methods in Cell Biology, vol. 123, pp. 77–94. Elsevier (2014)
13. Stephens, D.J., Allan, V.J.: Light microscopy techniques for live cell imaging. Science **300**, 82–86 (2003)
14. Drummen, G.P.C.: Fluorescent probes and fluorescence (microscopy) techniques—illuminating biological and biomedical research. Molecules **17**, 14067–14090 (2012)
15. Shaner, N.C., Steinbach, P.A., Tsien, R.Y.: A guide to choosing fluorescent proteins. Nat. Methods **2**, 905–909 (2005)
16. Suzuki, T., Matsuzaki, T., Hagiwara, H., Aoki, T., Takata, K.: Recent advances in fluorescent labeling techniques for fluorescence microscopy. Acta Histochem. Cytochem. **40**, 131–137 (2007)
17. Valeur, B.: Molecular Fluorescence Principles and Applications. Wiley-VCH Verlag GmbH, Weinheim, New York, Chichester, Brisbane, Singapor, Toronto (2001)
18. Lakowicz, J.R.: Principles of Fluorescence Spectroscopy, 3rd edn. Springer, New York (2006)
19. Abbe, E.: Beiträge zur Theorie des Mikroskops und der mikroskopischen Wahrnehmung. Archiv f. mikrosk. Anatomie **9**, 413–468 (1873)
20. Lord Rayleigh Sec. R.S.: XV. On the theory of optical images, with special reference to the microscope. London Edinb. Dublin Philos. Mag. J. Sci. **42**, 167–195 (1896)
21. Minsky, M.: Microscopy apparatus, US Patent US3013467A, 19 Dec 1961
22. Davidovits, P., Egger, M.D.: Scanning laser microscope. Nature **223**, 831–831 (1969)
23. Davidovits, P., Egger, M.D.: Scanning laser microscope for biological investigations. Appl. Opt. **7**, 1615 (1971)
24. Sheppard, C.J.R., Choudhury, A.: Image formation in the scanning microscope. Opt. Acta **24**, 1051–1073 (1977)
25. Wilson, T.: Optical sectioning in confocal fluorescent microscopes. J. Microsc. **154**, 143–156 (1989)
26. Wilson, T.: Optical sectioning in fluorescence microscopy. J. Microsc. **242**, 111–116 (2010)
27. Wilson, T.: Resolution and optical sectioning in the confocal microscope. J. Microsc. **244**, 113–121 (2011)
28. Wilson, T., Sheppard, C.: Theory and Practice of Scanning Optical Microscopy. Academic Press Inc., London (1984)
29. Cox, I.J., Sheppard, C.J.R.: Information capacity and resolution in an optical system. J. Opt. Soc. Am. A **3**, 1152 (1986)
30. Hell, S.W.: Far-field optical nanoscopy. Science **316**, 1153–1158 (2007)
31. Hell, S.W., Wichmann, J.: Breaking the diffraction resolution limit by stimulated emission: stimulated-emission-depletion fluorescence microscopy. Opt. Lett. **19**, 780–782 (1994)
32. Klar, T.A., Hell, S.W.: Subdiffraction resolution in far-field fluorescence microscopy. Opt. Lett. **24**, 954–956 (1999)

33. Willig, K.I., Rizzoli, S.O., Westphal, V., Jahn, R., Hell, S.W.: STED microscopy reveals that synaptotagmin remains clustered after synaptic vesicle exocytosis. Nature **440**, 935–939 (2006)
34. Wurm, C.A., Neumann, D., Lauterbach, M.A., Harke, B., Egner, A., Hell, S.W., Jakobs, S.: Nanoscale distribution of mitochondrial import receptor Tom20 is adjusted to cellular conditions and exhibits an inner-cellular gradient. Proc. Natl. Acad. Sci. USA **108**, 13546–13551 (2011)
35. Takasaki, K.T., Ding, J.B., Sabatini, B.L.: Live-cell superresolution imaging by pulsed STED two-photon excitation microscopy. Biophys. J. **104**, 770–777 (2013)
36. Westphal, V., Rizzoli, S.O., Lauterbach, M.A., Kamin, D., Jahn, R., Hell, S.W.: Video-rate far-field optical nanoscopy dissects synaptic vesicle movement. Science **320**, 246–249 (2008)
37. Westphal, V., Hell, S.W.: Nanoscale resolution in the focal plane of an optical microscope. Phys. Rev. Lett. **94**, 143903 (2005)
38. Rittweger, E., Han, K.Y., Irvine, S.E., Eggeling, C., Hell, S.W.: STED microscopy reveals crystal colour centres with nanometric resolution. Nat. Photonics **3**, 144–147 (2009)
39. Wildanger, D., Patton, B.R., Schill, H., Marseglia, L., Hadden, J.P., Knauer, S., Schönle, A., Rarity, J.G., O'Brien, J.L., Hell, S., W., Smith, J.M.: Solid immersion facilitates fluorescence microscopy with nanometer resolution and sub-angström emitter localization. Adv. Opt. Mater. **24**, OP309–OP313 (2012)
40. Harke, B., Ullal, C.K., Keller, J., Hell, S.W.: Three-dimensional nanoscopy of colloidal crystals. Nano Lett. **8**, 1309–1313 (2008)
41. Ullal, C.K., Schmidt, R., Hell, S.W., Egner, A.: Block copolymer nanostructures mapped by far-field optics. Nano Lett. **9**, 2497–2500 (2009)
42. Ullal, C.K., Primpke, S., Schmidt, R., Böhm, U., Egner, A., Vana, P., Hell, S.W.: Flexible microdomain specific staining of block copolymers for 3D optical nanoscopy. Macromolecules **44**, 7508–7510 (2011)
43. Klar, T.A., Wollhofen, R., Jacak, J.: Sub-Abbe resolution: from STED microscopy to STED lithography. Phys. Scr. **2014**, 014049 (2014)
44. Wollhofen, R., Katzmann, J., Hrelescu, C., Jacak, J., Klar, T.A.: 120 nm resolution and 55 nm structure size in STED-lithography. Opt. Express **21**, 10831 (2013)
45. Staudt, T., Engler, A., Rittweger, E., Harke, B., Engelhardt, J., Hell, S.W.: Far-field optical nanoscopy with reduced number of state transition cycles. Opt. Express **19**, 5644–5657 (2011)
46. Heine, J., Reuss, M., Harke, B., D'Este, E., Sahl, S.J., Hell, S.W.: Adaptive-illumination STED nanoscopy. Proc. Natl. Acad. Sci. **114**, 9797–9802 (2017)
47. Klar, T.A.: Progress in Stimulated Emission Depletion Microscopy. (Berichte aus der Physik). Shaker Verlag, Aachen (2001)
48. Hell, S.W.: Increasing the resolution of far-field fluorescence light microscopy by point-spread-function engineering. In: Lakowicz, J.R. (ed.) Topics in Fluorescence Spectroscopy, vol. 5, pp. 361–426. Kluwer Academic Publishers, New York, Boston, Dordrecht, London, Moscow (2002)
49. Dyba, M., Keller, J., Hell, S.W.: Phase filter enhanced STED-4Pi fluorescence microscopy: theory and experiment. New J. Phys. **7**, 134 (2005)
50. Klar, T.A., Jakobs, S., Dyba, M., Egner, A., Hell, S.W.: Fluorescence microscopy with diffraction resolution barrier broken by stimulated emission. Proc. Natl. Acad. Sci. USA **97**, 8206–8210 (2000)
51. Keller, J., Schönle, A., Hell, S.W.: Efficient fluorescence inhibition patterns for RESOLFT microscopy. Opt. Express **15**, 3361–3371 (2007)
52. Harke, B., Keller, J., Ullal, C.K., Westphal, V., Schönle, A., Hell, S.W.: Resolution scaling in STED microscopy. Opt. Express **16**, 4154–4162 (2008)
53. Sahl, S.J., Hell, S.W., Jakobs, S.: Fluorescence nanoscopy in cell biology. Nat. Rev. Mol. Cell Biol. **18**, 685–701 (2017)

Chapter 2
Coherent X-ray Imaging

Tim Salditt and Anna-Lena Robisch

> *Science, for me, gives a partial explanation for life. In so far as it goes, it is based on fact, experience and experiment.*
> – Rosalind Franklin

2.1 X-ray Propagation

Coherent X-ray imaging is based on wave-optical propagation of electromagnetic waves, including free-space propagation and the interaction of short wavelength light with matter. Here we present an overview of fundamental principles of X-ray imaging and field propagation, with references to relevant literature. We first justify the use of scalar wave theory and approximations of paraxial (parabolic) wave equations. Then we show how to compute the wavefield at a distance d along the optical axis z with respect to a known field distribution in a plane at $z = 0$, assuming free space between planes $z = 0$ and $z = d$. Next, we address the projection approximation which is ubiquitous in X-ray imaging to describe the complex transmission function of an optically thin object. Finally, we present finite difference equations as a more general tool to treat X-ray propagation in matter and objects which cannot be approximated as thin.

T. Salditt (✉) · A.-L. Robisch
Institute for X-ray Physics, Universität Göttingen, Friedrich-Hund-Platz 1, 37077 Göttingen, Germany
e-mail: tsalditt@gwdg.de

A.-L. Robisch
e-mail: anna-lena.robisch@uni-goettingen.de

© The Author(s) 2020
T. Salditt et al. (eds.), *Nanoscale Photonic Imaging*, Topics in Applied Physics 134,
https://doi.org/10.1007/978-3-030-34413-9_2

2.1.1 Scalar Diffraction Theory and Wave Equations

Propagation of stationary X-ray fields in matter can be described by the well-known *Helmholtz equation* (HE)

$$\Delta \mathbf{E}(\mathbf{r}, \omega) + k^2 n^2 \mathbf{E}(\mathbf{r}, \omega) = 0, \tag{2.1}$$

with vacuum wave number $k = \omega/c$, vacuum speed of light c, angular frequency ω, and the complex refractive index n of the propagation medium. Here, $\mathbf{E}(\mathbf{r}, \omega)$ denotes the time-domain Fourier transform of the electric vector field $\mathcal{E}(\mathbf{r}, t)$

$$\mathbf{E}(\mathbf{r}, \omega) = \mathcal{F}_t[\mathcal{E}(\mathbf{r}, t)](\mathbf{r}, \omega) = \frac{1}{\sqrt{2\pi}} \int_{\mathbb{R}} \mathcal{E}(\mathbf{r}, t) \, e^{-i\omega t} \, dt. \tag{2.2}$$

The HE in the homogeneous form of (2.1) is derived from Maxwell's equations for stationary fields in media, which are homogeneous, isotropic, non-magnetic, non-conductive, and do not contain free charges. Further, the field intensity has to be sufficiently small to neglect the non-linear response of matter. While this derivation assumes a constant or at least piecewise constant index of refraction, the HE is still an excellent description even for $n \to n(\mathbf{r})$, i.e. spatially varying distributions of matter as in an object to be imaged or in an optical device (refractive lens, zone plate, waveguide). Also in a crystal, where the continuum approximation of the index of refraction $n(\mathbf{r})$ must certainly break down, Fourier expansion with respect to the lattice vectors, still allows using (2.1). Indeed, propagation in a source-less but inhomogeneous medium with spatially varying dielectric function $\epsilon(\mathbf{r})$ (equivalently magnetic permeability $\mu(\mathbf{r})$) is well described by (2.1). This is surprising, since inhomogeneous media result in an inhomogeneous wave equation with corresponding source terms on the right hand side of (2.1). Certainly, in an inhomogeneous medium $\epsilon(\mathbf{r})$ is not a slowly varying function on scales of the X-ray wavelength. However, the approximations used to derive the HE are rescued by the simple fact that the X-ray index of refraction in matter is very close to the vacuum index of refraction, i.e. $\epsilon \simeq \epsilon_0$ and $\mu \simeq \mu_0$ for X-rays. This is both a curse and a blessing. It is a blessing, because the weak interactions result in the bulk penetration capability for which hard X-rays are famous, as well as the beneficial approximation of kinematic diffraction, which often warrants a quantitative reconstruction. Note that multiple scattering events can be safely neglected in most X-ray imaging applications, contrary to electron and visible light optics. Yet, at the same time, weak interaction is a curse, because it severely limits our ability to create efficient optical elements such as focusing devices.

Let us therefore briefly consider the (continuous) index of refraction $n(\mathbf{r}) = 1 - \delta(\mathbf{r}) + i\beta(\mathbf{r})$. Compared to other regions in the electromagnetic spectrum, the high frequency of X-rays manifests itself in extremely small dispersion $\delta \ll 1$ and absorption decrements $\beta \ll 1$, describing the phase shifts and absorption in matter, respectively. For a given element and atom density ρ_a, the refractive index is given in terms of the atomic form factor

$$n = 1 - \delta + i\beta = 1 - \frac{r_e \lambda^2}{2\pi} \rho_a(\mathbf{r})[Z + f'(\omega)] + i \frac{r_e \lambda^2}{2\pi} \rho_a(\mathbf{r}) f''(\omega), \quad (2.3)$$

where Z is the atomic number, $r_e = 2.82 \cdot 10^{-15}$ m the Thomson scattering length; $f'(\omega) \ll Z$ and $f''(\omega) \ll Z$ are the dispersion and absorption corrections. For mixed elemental composition, the indices are weighted averages according to the local stoichiometry of elements.

Next, we want to justify scalar wave theory for X-rays. Strictly speaking, we have to solve the HE for all components of the electric and magnetic field: $\forall \psi \in \{E_x, E_y, E_z, B_x, B_y, B_z\}$. In general, the field components are coupled (since they have to obey the full Maxwell system and/or different boundary conditions). For example, given the HE for the electric field in (2.1), the solution must also fulfill

$$\mathbf{\nabla} \cdot \mathbf{E} = 0 \qquad \text{and} \qquad \mathbf{B} = \frac{1}{i\omega} \mathbf{\nabla} \times \mathbf{E}. \quad (2.4)$$

Instead, in scalar wave theory, one often treats only a single component

$$\triangle \psi + k^2 n^2 \psi = 0. \quad (2.5)$$

Note that this form can be further simplified to a second-order ordinary differential equation, if one takes a two-dimensional Fourier transform with respect to the perpendicular space directions $\mathbf{r}_\perp := (x, \ y)^T$

$$\frac{\partial^2 \tilde{\psi}}{\partial z^2} + (k^2 n^2 - k_\perp^2) \, \tilde{\psi} = \left(\frac{\partial^2}{\partial z^2} + \beta^2 \right) \tilde{\psi} = 0, \quad (2.6)$$

with $\tilde{\psi} = \mathcal{F}_{\mathbf{r}_\perp}[\psi](\mathbf{k}_\perp)$ and $\beta := \sqrt{k^2 n^2 - k_\perp^2}$ [1]. Scalar wave theory is ubiquitous in X-ray optics and X-ray imaging, but permissible only if polarization effects can be neglected and field propagation of different polarization states is equivalent. This is the case for many applications (apart from propagation in crystals for example), since the relevant diffraction angles are much smaller than the Brewster angle.

Given a solution ψ of the scalar HE in (2.5), how do we obtain meaningful solutions and permissible polarisation states in terms of $\mathcal{E}$ and $\mathcal{B}$? As shown in [1], one can construct solutions of the Maxwell system by setting $\psi = \psi \, \mathbf{e}_p$ for any unit vector $\mathbf{e}_p$, and then compute

$$\mathbf{E} = \frac{i}{k} (\mathbf{\nabla} \psi) \times \mathbf{e}_p \qquad \text{and} \qquad \mathbf{B} = \frac{1}{c} \left(\frac{1}{k^2} \mathbf{\nabla} (\mathbf{e}_p \cdot \mathbf{\nabla} \psi) + n^2 \psi \, \mathbf{e}_p \right). \quad (2.7)$$

General solutions can be constructed from linear combinations of three independent scalar potentials ψ_x, ψ_y, ψ_z so that $\psi = \psi_x \mathbf{e}_x + \psi_y \mathbf{e}_y + \psi_z \mathbf{e}_z$. For the special case of paraxial waves which can be written as the product of a slowly varying envelope $u(\mathbf{r})$ and a fast oscillating term (see Fig. 2.1) with the wavevector pointing in direction of the optical axis $\mathbf{e}_z$, i.e. $\mathbf{k} = k\mathbf{e}_z$, i.e. propagation axis along z, we have

$$\psi(\mathbf{r}) = u(\mathbf{r})\, e^{-ink\,\mathbf{e}_z \cdot \mathbf{r}}, \tag{2.8}$$

and the corresponding electromagnetic field vectors are $\mathbf{E} \approx n\psi\, \mathbf{e}_x$ and $\mathbf{B} \approx \frac{n^2}{c}\psi\, \mathbf{e}_y$ [1]. Given the numerically computed scalar field, the local (time-averaged) energy flux density also follows from the full electromagnetic field [1] via the Poynting vector

$$\mathbf{S} = \frac{1}{\mu_0}\, \boldsymbol{\mathcal{E}} \times \boldsymbol{\mathcal{B}}, \tag{2.9}$$

where μ_0 is the free-space permeability. The energy flux density (averaged or integrated over the exposure time) is often denoted as the *optical intensity*. For time-harmonic fields and requiring Hermitian symmetry $\mathbf{E}(-\omega) = \mathbf{E}^*(\omega)$, one can write the field based on a discrete sum of frequencies ω_i for $i \in \mathbb{N}$

$$\begin{aligned}
\boldsymbol{\mathcal{E}}(\mathbf{r}, t) &= \mathcal{F}_\omega^{-1}[\mathbf{E}(\mathbf{r}, \omega)](\mathbf{r}, t) \\
&= \sqrt{\frac{2}{\pi}} \sum_{i \in \mathbb{N}} \Re\left[\mathbf{E}(\mathbf{r}, \omega_i))\right] \cos(\omega_i t) + \Im\left[\mathbf{E}(\mathbf{r}, \omega_i)\right] \sin(\omega_i t),
\end{aligned}$$

and the equivalent expression for $\boldsymbol{\mathcal{B}}$, which is a starting point to compute $\mathbf{S}$. When the fields are constructed from a single scalar potential ψ, which is slowly varying in direction of $\mathbf{e}_p$ such that $\|\nabla(\mathbf{e}_p \cdot \nabla\psi)\| \ll |n^2 k^2 \psi|$, the time-averaged Poynting vector can be approximated as [1]

$$\langle \mathbf{S} \rangle \approx \frac{c\epsilon_0}{\pi} \sum_{i \in \mathbb{N}} \Re[n(\omega_i)]^2\, |\psi(\omega_i)|^2\, \frac{\nabla \arg(\psi(\omega_i))}{k_i}, \tag{2.10}$$

with $k_i = \omega_i/c$. For paraxial beams it then follows

$$\langle \mathbf{S} \rangle \approx \frac{c\epsilon_0}{\pi} \sum_{i \in \mathbb{N}} \Re[n(\omega_i)]^3\, |\psi(\omega_i)|^2\, \frac{\nabla \arg(\psi(\omega_i))}{\|\nabla \arg(\psi(\omega_i))\|}. \tag{2.11}$$

Hence, for a monochromatic paraxial beam, the magnitude of the time-averaged Poynting vector, i.e. the optical intensity I, can be written as [1]

$$I = \|\langle \mathbf{S} \rangle\| \approx \frac{c\epsilon_0}{\pi} \Re(n)^3\, |\psi|^2, \tag{2.12}$$

with an energy flow oriented in direction of the phase gradient.

The above approximation of the Poynting vector was derived under the assumption of paraxial waves. Indeed, many X-ray optical problems are well described by

Fig. 2.1 Separation ansatz
for solutions of the
Helmholtz equation: A
wavefield can be written as
the product of a slowly
varying envelope and a
rapidly oscillating term
changing sign on atomic
length scales. In general, one
is interested in finding the
envelope function

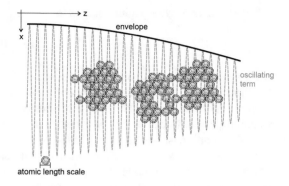

forward directed paraxial diffraction. Thus, we can further approximate the scalar
HE and arrive at the parabolic wave equation. To this end, we start with a separation
ansatz for solutions of the HE in case of waves $\psi(\mathbf{r})$ propagating along the optical
axis z

$$\psi(\mathbf{r}) = u(\mathbf{r})e^{ikz} , \qquad (2.13)$$

where $u(\mathbf{r})$ is the (slowly varying) envelope of the wave field $\psi(\mathbf{r})$. For X-rays in
particular, we are often not interested in the rapidly oscillating term which changes
sign on atomic length scales, and which is as irrelevant as the time-harmonic term.
Instead, in numerical computation and plots, we are only interested in phases and
amplitudes of $u(\mathbf{r})$, which are well suited to monitor the small phase and intensity changes building up over many hundreds and thousands of atoms, see Fig. 2.1.
Inserting (2.13) into the HE yields a differential equation for the envelope u

$$\nabla^2 \left(u(\mathbf{r})e^{ikz}\right) + n^2(\mathbf{r})k^2 u(\mathbf{r})e^{ikz} = 0 . \qquad (2.14)$$

Working out the differential operators of the left-hand term

$$\nabla_\perp^2 u(\mathbf{r})e^{ikz} + \partial_z \left(\partial_z u(\mathbf{r})e^{ikz} + ike^{ikz}u(\mathbf{r})\right)$$
$$= \nabla_\perp^2 u(\mathbf{r})e^{ikz} + \left(\partial_z^2 u(\mathbf{r}) + 2(\partial_z u(\mathbf{r}))ik\right)e^{ikz} - k^2 e^{ikz}u(\mathbf{r}) , \qquad (2.15)$$

and dividing by e^{ikz}, we obtain

$$\left[\nabla_\perp^2 + \partial_z^2 + 2ik\partial_z + k^2(n^2(\mathbf{r}) - 1)\right]u(\mathbf{r}) = 0 . \qquad (2.16)$$

For paraxial beams, the second order derivative in z can be neglected, given
$\left|\frac{\partial^2 u}{\partial z^2}\right| \ll \left|k\frac{\partial u}{\partial z}\right|$ since $\partial_z^2 u \ll k^2 u$, leading to the paraxial (or parabolic) wave equation

$$\left[\nabla_\perp^2 + 2ik\partial_z + k^2(n^2(\mathbf{r}) - 1)\right]u(\mathbf{r}) = 0 . \qquad (2.17)$$

The parabolic wave equation reduces to

$$\left[\nabla_\perp^2 + 2ik\partial_z\right]u(\mathbf{r}) = 0 \tag{2.18}$$

in free space with Gaussian beams as a well known family of solutions. Compared to the elliptic Helmholtz equation, its parabolic approximations offer higher numerical stability. Furthermore, solutions are more easily accessible in terms of boundary conditions. Initial values have to be specified in a plane at small z along with lateral boundary conditions, but no values at the high z boundary of the computational domain are required. For these reasons, paraxial approximations have become an important tool in X-ray optics [2–4], including generalizations to time-dependent propagation problems via the spectral approach [1]. As in Schrödinger's equation, the parabolic wave equation can be rearranged to

$$\frac{\partial u}{\partial z} = \frac{1}{2ik}\,\triangle_\perp u - ik\,\frac{n^2 - 1}{2}\,u. \tag{2.19}$$

This form of the paraxial wave equation is typically used in X-ray optics.

Turning back to the Helmholtz equation formulated as in (2.6), we complete this section by presenting a slightly different form of the parabolic wave equation, put forward in [1] and based on the approach in [5]. Due to the less restrictive approximations necessary for its derivation, we expect this form to have a larger range of validity. In particular, the assumption $\left|\frac{\partial^2 u}{\partial z^2}\right| \ll \left|k\frac{\partial u}{\partial z}\right|$ (or correspondingly $\partial_z^2 u \ll k^2 u$) is not required anymore. The differential operator in (2.6) can be written as a product of two operators

$$\left(\frac{\partial}{\partial z} + i\beta\right)\left(\frac{\partial}{\partial z} - i\beta\right)\tilde\psi = 0.$$

The scalar HE is hence solved by a solution $\tilde\psi$ of either of the differential equations (forward and backward HE)

$$\frac{\partial\tilde\psi}{\partial z} = i\beta\,\tilde\psi \quad \text{and} \quad \frac{\partial\tilde\psi}{\partial z} = -i\beta\,\tilde\psi. \tag{2.20}$$

In the context of paraxial beams, we are interested in the right equation describing a wave vector oriented in $\mathbf{e}_z$ direction. Further, for paraxial beams, the support of $\tilde\psi$ is constrained to low spatial frequencies with $k_\perp^2 \ll k^2$. Therefore, one can approximate the square root in β through its first-order Taylor series around $k_\perp^2 = 0$ [1]

$$\beta = \sqrt{k^2 n^2 - k_\perp^2} \approx kn - \frac{k_\perp^2}{2kn}.$$

Substitution into (2.20) (right) and transforming back to real space results in [1]

$$\frac{\partial \psi}{\partial z} = \frac{1}{2ikn} \Delta_\perp \psi - ikn\,\psi. \tag{2.21}$$

Equation (2.21) can again be formulated for the envelope u, defined by $\psi = u \exp(-ikz)$ [1]

$$\frac{\partial u}{\partial z} = \frac{1}{2ikn} \Delta_\perp u - ik(n-1)\,u. \tag{2.22}$$

For $n = 1$, (2.22) and (2.19) become identical. As a consequence in the regime of hard X-rays, solution of equations (2.22) and (2.19) will not differ. Yet, for soft X-rays, the difference could become relevant.

2.1.2 Propagation in Free Space

We first address the propagation in free space, following the *angular spectrum approach* as presented in the textbook of Paganin [6]. Again, we assume a time independent, monochromatic wave, i.e. we treat a single component $\psi_\omega(\mathbf{r})$ of the spectrum with angular frequency ω and corresponding wavelength λ. A general time-dependent field of finite bandwidth is then computed as superposition of its monochromatic components by

$$\Psi(\mathbf{r}, t) = \frac{1}{\sqrt{2\pi}} \int \psi_\omega(\mathbf{r}) \exp[i\omega t]\,d\omega. \tag{2.23}$$

As discussed above, the single spectral component $\psi_\omega(\mathbf{r})$ must obey the free-space Helmholtz equation

$$\left(\Delta + \left(\frac{2\pi}{\lambda}\right)^2\right) \psi(\mathbf{r}) = 0, \tag{2.24}$$

where we have dropped the subscript $\psi_\omega \to \psi$ for simplicity of notation. Particular solutions of the Helmholtz equation are plane waves

$$\psi_P(\mathbf{r}) = \exp[i\mathbf{k} \cdot \mathbf{r}] = \exp\left[i(k_x x + k_y y + k_z z)\right], \tag{2.25}$$

where $k^2 = k_x^2 + k_y^2 + k_z^2 = \frac{4\pi^2}{\lambda^2}$. The z-dependent part of the plane wave can be separated by

$$\psi_P(x, y, z) = \exp\left[i(k_x x + k_y y)\right] \exp\left[iz\sqrt{k^2 - k_x^2 - k_y^2}\right]. \tag{2.26}$$

The last equation entails an important message: Knowing the plane wave in the source plane $\psi_P(x, y, z = 0) = \exp\left[i(k_x x + k_y y)\right]$, the electromagnetic field at any distance z can be calculated by a simple multiplication with the so-called "free-space propagator" $\exp\left[iz\sqrt{k^2 - k_x^2 - k_y^2}\right]$. To propagate an arbitrary wave field ψ, we express ψ given in plane $z = 0$ by its Fourier transform $\hat{\psi}(k_x, k_y, z = 0)$

$$\psi(x, y, z = 0) = \frac{1}{2\pi} \int \int \hat{\psi}(k_x, k_y, z = 0) \exp\left[i\left(k_x x + k_y y\right)\right] dk_x dk_y. \quad (2.27)$$

This Fourier transform can be read as a superposition of plane waves. Since we can expand (almost) any wave field of interest in such a Fourier integral in the source plane and since we know how to propagate plane waves, we also know how to propagate general wave fields. It is thus possible to compute the field at any distance z from the given field in the xy-plane at $z = 0$. This allows interpreting an electromagnetic disturbance in a plane at $z = 0$ as a superposition of plane waves of fixed modulus of the wavevector leaving the plane of interest under different angles

$$\theta = \arcsin\left(\frac{\sqrt{k_x^2 + k_y^2}}{k}\right), \quad (2.28)$$

see also Fig. 2.2 and [7]. Each of these plane waves can be propagated from $z = 0$ to any distance $z > 0$ by multiplication with the free-space propagator

$$\psi(\mathbf{r}) = \frac{1}{2\pi} \int \int \hat{\psi}(k_x, k_y, z = 0) \cdot \exp\left[iz\sqrt{k^2 - k_x^2 - k_y^2}\right]$$
$$\cdot \exp\left[i\left(k_x x + k_y y\right)\right] dk_x dk_y. \quad (2.29)$$

Next, we restrict the wave fields of interest to paraxial waves, i.e. those which propagate at small angles with respect to the optical axis z. In this case, k_x and k_y are

Fig. 2.2 Angular spectrum approach for propagation of arbitrary wavefields. A wavefield in the source plane is decomposed into plane waves propagating under different angles (2.28) with respect to the optical axis. Each plane wave can be propagated by application of the free-space propagator

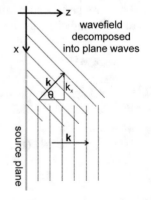

much smaller compared to k_z. As a consequence, we can approximate the free-space propagator by

$$\exp\left[iz\sqrt{k^2 - k_x^2 - k_y^2}\right] \approx \exp[ikz]\, G_k(k_x, k_y; z), \qquad (2.30)$$

where

$$G_k(k_x, k_y; z) := \exp\left[\frac{-iz(k_x^2 + k_y^2)}{2k}\right]. \qquad (2.31)$$

Using (2.30) and formulating (2.29) as an operator-equation the propagated field at distance z is

$$\mathcal{D}_z\left[\psi(x, y, z = 0)\right](x, y, z) := \\ \exp(ikz)\,\mathcal{F}^{-1}\left\{G_k(k_x, k_y; z)\mathcal{F}\left[\psi(x, y, z = 0)\right]\right\}(x, y, z), \qquad (2.32)$$

where $\mathcal{F}$ is the Fourier transform with respect to the xy-plane. Application of the Fourier convolution theorem to (2.32) and expansion of the exponent in the integrand leads to the Fresnel diffraction integral

$$\psi_\omega(\mathbf{r}) = \frac{k}{i2\pi z}\exp(ikz)\exp\left[\frac{ik}{2z}(x^2 + y^2)\right]\cdot\int\int\psi_\omega(x', y', 0)\cdot \\ \exp\left[\frac{ik}{2z}(x'^2 + y'^2)\right]\exp\left[\frac{-ik}{z}(xx' + yy')\right]dx'dy'. \qquad (2.33)$$

The chirp function $\exp\left[\frac{ik}{2z}(x'^2 + y'^2)\right]$ merits a closer look. Reformulating the argument of the exponential function results in

$$\frac{k}{2z}(x'^2 + y'^2) = \frac{\pi(x'^2 + y'^2)}{\lambda z}. \qquad (2.34)$$

The term $(x'^2 + y'^2)$ 'measures' the squared spatial cross-section of features in the input wavefield. Let a be the smallest such structure of interest, so that the argument of the exponential, the so-called chirp, is governed by the dimensionless ratio $\frac{a^2}{\lambda z} =: F$ called the Fresnel number. For large propagation distances (compared to wavelength and a), the Fresnel number approaches zero and hence the chirp function is close to unity: The Fresnel diffraction integral becomes the Fourier transform of the wavefield. This limiting case is known as the Fraunhofer far field approximation. Fresnel numbers close to one indicate the optical near field: The chirp function cannot be neglected. Indeed, as we will see below, it is the chirp function that makes numerical free-space propagation challenging. Importantly, paraxial propagation is governed only by a single parameter F, so that all simulations can be carried out in natural units of pixel size and with a single unitless parameter F.

Before we turn to numerical implementation, however, a brief comment on the choice of relevant feature size a is reasonable. A natural choice in the discrete setting of numerical calculations is the pixel size $a = \Delta x$. The Fraunhofer regime $F \ll 1$ is then quickly reached with increasing z. These 'far field' holograms, however, do not look like far field diffraction patterns (squared Fourier transform of the object), which is usually associated with the Fraunhofer approximation. The reason is that there is always another length scale a to be considered, namely the beam size. Only when the Fresnel number—computed for all object length scales *and* the length scale of the beam—is much smaller than one, we get the conventional Fraunhofer diffraction pattern without mixing of object wave and primary wave.

2.1.3 The Fresnel Scaling Theorem

Consider a setting where an object is illuminated by the incident wavefield ψ_i. Two limiting cases of the illumination geometry are of particular importance: plane

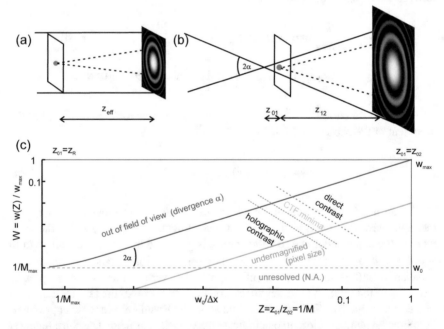

Fig. 2.3 **a** Illustration of the hologram created by a single point in plane wave illumination (parallel beam), and **b** the equivalent for spherical wave illumination (cone beam). According to the Fresnel scaling theorem, the holograms are identical up to a simple variable transformation, including the geometric magnification M and an effective defocus distance z_{eff}. **c** Cone beam geometry in unitless variables, taking a Gaussian beam as an example for a coherent and divergent beam, see text for explanations

wave illumination and spherical wave illumination. The spherical beam geometry (also denoted as cone-beam geometry) can be used to implement radiography at the nanoscale, well below the resolution limit of the detector pixel size Δx. As sketched in Fig. 2.3a, b, this requires X-ray nano-focusing optics [8, 9], as presented in Chap. 3 and a defocus geometry with object and detector positioned behind the focus at distances z_{01} and z_{02}, respectively, see also Chap. 13. Focusing serves two purposes: Firstly, the photon density is increased in the object plane, as long as the defocus position is smaller than the focal length of the optic $z_{01} \leq f$. Secondly, it magnifies the near field pattern (hologram) and thus enables a resolution below the detector pixel size Δx. To discuss the effect of the spherical illumination function ψ_i and following the projection approximation, we slightly rewrite the input wavefield as a product of illumination and a complex-valued object transfer function $\psi_\omega(x', y', 0) = \psi_i(x', y', 0)\tau(x', y')$ and insert it in (2.33) yielding

$$\psi(\mathbf{r}) = A(\mathbf{r}) \int \int dx'dy' \, \psi_i(x', y') \, \tau(x, y)$$
$$\cdot \exp\left[\frac{ik}{2z_{12}}\left(x'^2 + y'^2\right)\right] \exp\left[\frac{-ik}{z_{12}}\left(xx' + yy'\right)\right], \qquad (2.35)$$

where $z_{12} := z_{02} - z_{01}$ denotes the distance between object and detector. For simplicity of notation, the subscript ω was skipped, and the prefactors were abbreviated by the complex valued $A(\mathbf{r})$. The factorization underlying the projection approximation will be justified further below. Equation (2.33) essentially describes the case of a plane wave incident illumination $\psi_i = e^{ikz}$ with $\tau(x', y', 0) = \psi(x', y', 0)$, i.e. no sample in the beam path. However, for point source illumination with

$$\psi_i = \exp\left[ik\sqrt{x'^2 + y'^2 + z_{01}^2}\right] \approx \exp\left(ikz_{01}\right) \exp\left[\frac{ik}{2z_{01}}\left(x'^2 + y'^2\right)\right], \qquad (2.36)$$

the integral becomes

$$\psi(\mathbf{r}) = A(\mathbf{r}) \int \int dx'dy' \, \tau(x', y') \, \exp\left(ikz_{01}\right) \exp\left[\frac{-ik}{z_{12}}(xx' + yy')\right]$$
$$\cdot \underbrace{\exp\left[\frac{ik}{2z_{01}}(x'^2 + y'^2)\right] \exp\left[\frac{ik}{2z_{12}}(x'^2 + y'^2)\right]}_{\exp\left[\frac{ik}{2}\left(\frac{1}{z_{01}} + \frac{1}{z_{12}}\right)(x'^2 + y'^2)\right]}, \qquad (2.37)$$

where $A(\mathbf{r})$ may also account for the amplitude decrease of the spherical wave. With the geometric magnification given by

$$M = \frac{z_{01} + z_{12}}{z_{01}}, \qquad (2.38)$$

and the definition of an effective propagation distance z_{eff}

$$z_{\text{eff}} = \frac{z_{01} z_{12}}{z_{01} + z_{12}} = \left(\frac{1}{z_{01}} + \frac{1}{z_{12}}\right)^{-1} = \frac{z_{12}}{M}, \tag{2.39}$$

we obtain

$$\psi(\mathbf{r}) = A(\mathbf{r}) \, \exp\left(ikz_{01}\right) \int \int dx' dy' \, \tau(x', y') \, \exp\left[\frac{ik}{2z_{\text{eff}}} \left(x'^2 + y'^2\right)\right]$$
$$\cdot \exp\left\{-2\pi i \left[\left(\frac{x}{M}\right)\left(\frac{x'}{\lambda z_{\text{eff}}}\right) + \left(\frac{y}{M}\right)\left(\frac{y'}{\lambda z_{\text{eff}}}\right)\right]\right\}, \tag{2.40}$$

i.e. wave propagation is equivalent in plane and spherical geometry, if a simple coordinate transformation is performed. This result is known as the *Fresnel Scaling Theorem* (FST). Following the FST, numerical propagation which will be discussed in the next section, is always performed in the effective parallel beam coordinate system. But before doing so, we note down an immediate consequence of spherical beam propagation, which is also illustrated in Fig. 2.3. While the Fresnel number F decreases with increasing distance between object and detector, the opposite is true for the divergent beam setting. As z_{01} is reduced (at constant z_{02}), the imaging regime becomes more and more holographic (decrease in F) since the decrease in effective pixels size $\Delta x_{\text{eff}} = \Delta x / M$ enters quadratically and outweighs the effect of z_{eff}. For the minima of the contrast transfer function, which will be discussed further below, this means that their number increases as z_{01} is decreased (or equivalently M is increased), see the dashed lines in Fig. 2.3c. In order to plot the divergent beam geometry in unitless variables, the intensity of a Gaussian illumination can serve as a model for the diffraction limited beam

$$I(\mathbf{r}) = I_0 \left(\frac{w_0}{w(z)}\right)^2 \exp\left[-\frac{2 \cdot (x^2 + y^2)}{w(z)^2}\right], \tag{2.41}$$

with waist w_0, $w(z) = w_0 \sqrt{1 + (z/z_R)^2}$ and Rayleigh length $z_R = w_0^2 \pi / \lambda$. Next, we parameterize the position of the object along the optical axis by $Z = z_{01}/z_{02} = 1/M$, such that this inverse magnification varies between $1/M_{\text{max}} = z_R/z_{02}$ and one. The corresponding lateral width of the beam varies between w_0 and $w_{\text{max}} = \alpha z_{02}$, with divergence α defining the numerical aperture. In unitless variables we designate the beam width by $W = w(Z)/w_{\text{max}}$, and plot W as a function of $1/M$ on double-logarithmic scale, see Fig. 2.3c. Accordingly, the blue line shows the increase in beam width, while the orange line designates the effective pixel size $\Delta x_{\text{eff}} = (\Delta x / z_{02}) z_{01}$, i.e. the demagnification size of a single detector pixel. In unitless variables this line reaches $W = \Delta x / w_{\text{max}}$ for $Z = 1$ and crosses the dashed red horizontal line at $Z = w_0 / \Delta x$. The dashed red line separates lateral length scales which can be resolved (above) from those which are unresolved (below), based on the limited numerical aperture (N.A.).

2.1.4 Numerical Implementation of Free-Space Propagation

The numerical implementation of free-space propagation is not trivial. In particular, sufficient sampling of chirp functions has to be guaranteed. Extensive literature can be found at [7, 10, 11]. There are two main options to implement propagation. The first uses two Fourier transformations and has equal coordinate systems in source and destination plane, the second is based on a single transformation and can work with differently sized and sampled planes. We start with the first option, which is based on (2.29), according to which propagation can be designed as a filter operation of the wavefield in Fourier space. The corresponding filter is the free-space propagator, which is given by (2.31) in paraxial approximation $G_k(k_x, k_y; z)$ as a function in reciprocal space. The propagator function can also be written down analytically in real space (called the impulse response function), and then be numerically Fourier transformed. Importantly, the coordinate system of input and output field are identical, and the propagation is based on two fast Fourier transform (FFT) operations. For the second approach, the propagation can be directly calculated based on (2.33), involving a single Fourier transform, either by a single FFT or by other numerical solutions of the Fourier integral, e.g. for non-equidistant sampling. This approach is well suited for cases where the pixel sizes between input and output must vary, e.g. to cover the field of view in a detector after diffraction broadening.

Next, we consider the sampling criteria for the first method. In order to correctly sample the free-space propagator or reciprocal space chirp, the real space sampling interval Δx has to be [11]

$$\Delta x \geq \frac{\lambda z}{L}, \tag{2.42}$$

where L is the field of view consisting of N pixels: $L = N\Delta x$. Hence, only a large number N of pixels or a short propagation distance result in aliasing-free sampling of the propagated field. To find a remedy, one can artificially increase the number of pixels. Yet, this drastically decreases the computational speed. Alternatively, one can also write down the impulse response function, i.e. the real space counterpart of the reciprocal space chirp:

$$\mathcal{F}^{-1}\left[G_k(\cdot, \cdot; z)\right](x, y, z) = \frac{k}{iz} \exp\left[\frac{ik(x^2 + y^2)}{2z}\right]. \tag{2.43}$$

The corresponding sampling criterion is [11]

$$\Delta x \leq \frac{\lambda z}{L}. \tag{2.44}$$

Violation of the last equation results in periodic copies of the chirp interfering with each other. A way out is to use a suitable window function to limit the chirp [7, 12]. Following (2.28) and [7], the highest angle under which plane waves can be emitted is given by

$$\theta_{max} = \arcsin\left(\frac{\Delta k_{max}}{k}\right). \tag{2.45}$$

In the discrete version of the Fourier transform, the Shannon sampling theorem is fulfilled by the relation

$$\Delta k \Delta x = \frac{2\pi}{N}. \tag{2.46}$$

The highest resolvable spatial frequency corresponds to an oscillation that extends over two real space pixels, or equivalently

$$\Delta k_{max} = \frac{N}{2} \cdot \Delta k = \frac{N}{2} \frac{2\pi}{\Delta x N} = \frac{\pi}{\Delta x}. \tag{2.47}$$

Hence, the largest properly sampled angle θ_{max} under which radiation is emitted is given by

$$\theta_{max} = \arcsin\left(\frac{\lambda}{2\Delta x}\right). \tag{2.48}$$

A useful window function to be multiplied with the real space chirp is [7]

$$W(\theta) = \begin{cases} \cos\left(\frac{\pi}{2}\frac{\theta}{\theta_{max}}\right) & \text{if } \theta \leq \theta_{max} \\ 0 & \text{else} . \end{cases} \tag{2.49}$$

Finally, we consider the sampling requirement for the second propagation approach, based directly on the Fresnel diffraction integral. This requires sampling of chirp functions in the source plane.

$$\Delta x \leq \frac{\lambda z}{L}. \tag{2.50}$$

In fact, if only intensity in the destination plane is of interest, the observation chirp does not need to be sampled. Further, instead of a destination *plane*, the output field should be regarded as a function of spatial frequencies, rather than spatial coordinates of the detector. In this way one can dispense of the observation chirp altogether.

2.1.5 X-ray Propagation in Matter

After considering propagation in free space, the next challenge is to conceptualize the interaction of X-rays and matter as well as to give suitable approximations to compute stationary wave propagation in matter. For wave propagation, we are not interested in incoherent and inelastic interaction processes such as Compton scattering, nor in the cascade of processes following photo absorption of X-ray photons [13, 14], but only in the elastic scattering events of atoms which collectively result in diffraction and refraction effects. For X-ray imaging, matter mainly has two functions: either optical component or sample/object of interest. If objects and optical components are sufficiently thin so that propagation effects (also called *volume effects*) can be neglected, the propagation effects can be described by a simple multiplication, as shown below. For a more general case, (2.16) or (2.17) have to be solved for the paraxial case. This is a formidable task and analytical solutions only exist for a few very special cases (sphere, slab waveguide). Neglecting atomic-scale variations which would become important only for perfect crystals, we can write the index of refraction in continuum approximation as in (2.3) with the real-valued decrement or dispersion term δ given by [13]

$$\delta(\mathbf{r}) = \frac{r_0 \lambda^2}{2\pi} \rho_a(\mathbf{r})[Z + f'(E)], \tag{2.51}$$

where ρ_a is the number density of atoms, $r_0 = 2.82 \cdot 10^{-15}$ m the Thomson scattering length, Z the number of electrons in the atom and $f'(E)$ the real part of the atomic form factor correction at energy E. For the imaginary part (absorption term) β we have

$$\beta(\mathbf{r}) = \frac{r_0 \lambda^2}{2\pi} \rho_a(\mathbf{r}) f''(\omega), \tag{2.52}$$

with $f''(E)$ being the imaginary part of the atomic form factor. Note that form factors and their dependence on photon energy E are tabulated for each element in the International Tables for Crystallography and are also available online.[1] Away from absorption edges, the real part of $n(\mathbf{r})$ only depends on the total electron density (summed over all elements)

$$\delta = \frac{r_0 \lambda^2}{2\pi} \rho(\mathbf{r}). \tag{2.53}$$

Due to the small value of the Thomson scattering length resulting in $\delta \ll 1$, diffraction and refraction effects in the X-ray spectral range are weak, and hence there is ample room for approximations, notably the ansatz by Born or the ansatz by Rytov, which are discussed in view of X-ray propagation in [15–17].

Within the first order of the Born approximation, we can neglect multiple scattering if the sample is sufficiently thin and consider the scattering $\psi_s(\mathbf{r})$ as a small additive correction to the primary wave $\psi_0(\mathbf{r})$, i.e.

$$\psi(\mathbf{r}) = \psi_0(\mathbf{r}) + \psi_s(\mathbf{r}). \tag{2.54}$$

For a far field observation point, the scattered wave is given as a Fourier transform of the scattering length density, i.e. $r_0 \rho$, with ρ being the electron density and r_0 the Thomson scattering length

---

[1] http://it.iucr.org/Cb/ch4o2v0001/sec4o2o6/ and http://henke.lbl.gov/optical_constants/.

$$\psi_s(\mathbf{r}) \propto r_0 \int d\mathbf{r}' \, \rho(\mathbf{r}') \exp(i\mathbf{q}\mathbf{r}'), \tag{2.55}$$

where $\mathbf{q}$ is the scattering vector. Alternatively, according to the Rytov ansatz, we can also write the solution in a multiplicative form as

$$\psi(\mathbf{r}) = \psi_0(\mathbf{r}) \, \exp(\varphi_s(\mathbf{r})), \tag{2.56}$$

where $\varphi_s(\mathbf{r})$ is the complex-valued phase of the scattered field and $\psi_0(\mathbf{r})$ again the primary wave amplitude without perturbation by the object. Rytov's approximation then requires $|(\nabla \varphi_s)^2| \ll |\tau|$, with the object function defined as $\tau(\mathbf{r}) = k_0^2(1 - n(\mathbf{r}))$, where $k_0 = 2\pi/\lambda$. One can compute φ_s from τ using the Green's function $G(\mathbf{r}) = \exp(ikr)/(4\pi r)$ [17]

$$\varphi_s(\mathbf{r}) = -\frac{1}{\psi_0} \int d\mathbf{r}' \, G(\mathbf{r} - \mathbf{r}') \, \psi_0 \tau(\mathbf{r}'), \tag{2.57}$$

which in frequency space gives [17]

$$\tilde{\varphi}_s(k_x, k_y, z) = \frac{\exp\left[iz\left(\sqrt{k_0^2 - k_x^2 - k_y^2} - k_0\right)\right]}{2i\sqrt{k_0^2 - k_x^2 - k_y^2}}$$
$$\cdot \tilde{\tau}\left(k_x, k_y, \sqrt{k_0^2 - k_x^2 - k_y^2} - k_0\right). \tag{2.58}$$

For paraxial propagation with $\sqrt{k_0^2 - k_x^2 - k_y^2} \simeq k_0\left(1 - \frac{k_x^2}{2k_0^2} - \frac{k_y^2}{2k_0^2}\right)$, and for objects with thickness Δz sufficiently small it can be shown that [17]

$$\tilde{\varphi}_s(k_x, k_y, z) = \frac{1}{2ik_0} \exp\left[-\frac{i\lambda z}{2}(k_x^2 + k_y^2)\right] \tilde{\tau}(k_x, k_y, 0). \tag{2.59}$$

This last idea of considering the limit of small object thickness, is extremely useful and shall be approached in even simpler terms. To this end, let us first consider a plane wave $\exp(ikz)$ incident at $z = 0$ onto a homogeneous slab of thickness Δz, the wavefield behind the object (outgoing wave) can then be written as

$$\exp\left[ik_0\Delta z(1 - \delta + i\beta)\right] = \exp(ik_0\Delta z)$$
$$\cdot \exp(-ik_0\delta\Delta z) \exp(-k_0\beta\Delta z), \tag{2.60}$$

where $\exp(-ik_0\delta\Delta z)$ describes phase shift and $\exp(-k_0\beta\Delta z) = \exp(-\mu\Delta z/2)$ absorption of the wave in the medium (absorption coefficient μ). For an inhomogeneous medium and negligible diffraction inside the specimen, the complex-valued transmission function of a slab can then be expressed by the integral along the optical path

$$\tau(x, y) = \exp\left[-ik_0 \int_0^{\Delta z} (\delta(\mathbf{r}) - i\beta(\mathbf{r}))\, dz\right]. \tag{2.61}$$

The underlying approximation is known as the projection approximation. Its basic assumption is that the value of the wavefield is entirely determined by the phase and amplitude shifts accumulated along streamlines of the unscattered beam. In doing so, the spread of the wave by diffraction inside the sample is ignored. Thus, this is correct only down to a certain sample size, depending on the resolution element and photon energy E. The projection approximation is valid for sufficiently small spatial frequencies, fulfilling

$$\sqrt{k_0^2 - k_x^2 - k_y^2} - k_0 \ll \frac{\pi}{\Delta T}, \tag{2.62}$$

where ΔT denotes the thickness of the object. Propagation through arbitrary objects can be treated by sequences of small projection and propagation steps, as if matter came in form of thin slices lined up along the optical axis (multi-slice approach). For further details of this method see [1, 18]. A different approach to solve propagation through matter is propagation by finite difference equations which is presented below. Finite difference propagation and the multi-slice approach have been compared in [1].

2.1.6   Propagation by Finite Difference Equations

Next, we will present the basic scheme of finite difference equations (FD) for propagation in the framework of the paraxial wave equation. As the paraxial wave equation is a parabolic partial differential equation, initial and boundary data are required for its solution. But in contrast to elliptic equations such as the Helmholtz equation, there is no data required on a closed boundary. In other words, we can propagate from left to right along the optical axis, and we expect causality in the sense that the field is only determined by what has happened upstream, but not downstream. For two dimensional propagation problems (one dimension for the optical axis z plus a single lateral direction x), i.e. if the index of refraction is independent of y, following (2.19), the parabolic wave equation can be written as

$$\frac{\partial u}{\partial z} = A \frac{\partial^2 u}{\partial x^2} + C(x, z)\, u, \tag{2.63}$$

where

$$A := \frac{1}{2ik} \qquad \text{and} \qquad C(x, z) := \frac{k}{2i}\left(n^2(x, z) - 1\right)$$

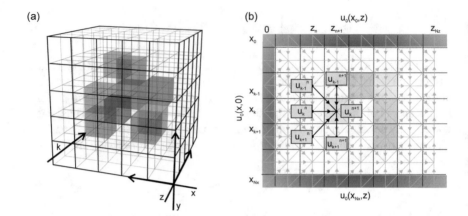

Fig. 2.4 **a** Three dimensional distribution of the refractive index in a voxelated space. The optical axis is parallel to the z axis and X-rays (wavevector **k**) are impinging on the sample. **b** Wave propagation using finite difference equations solved by the Crank-Nicolson algorithm. Boundary conditions/initial values are represented by thick gray lines. Colored regions denote regions of different refractive index. Note that because the parabolic wave equation is used, no boundary conditions for $u(x, z_{N_z} + 1)$ need to be set. In order to compute u_k^{n+1}, the values u_k^n, u_{k-1}^n, u_{k+1}^n, u_{k-1}^{n+1}, u_{k+1}^{n+1} are required

with boundary conditions

$$u(x, 0) = u_0(x, 0), \tag{2.64}$$

$$u(x_0, z) = u_0(x_0, z), \tag{2.65}$$

$$u(x_{N_z}, z) = u_0(x_{N_x}, z), \tag{2.66}$$

where $u_0(x, 0)$ is the field in the initial plane, while $u_0(x_0, z)$ and $u_0(x_{N_x}, z)$ are the bottom and top boundary conditions to be specified along lines parallel to the optical axis, at the sides of the computational field of view. The computational field of view is modeled such that it has $N_x + 1$ grid points in the lateral direction x, and $N_z + 1$ grid points along the optical axis. Importantly, and contrarily to elliptical partial differential equations, no values have to be set for $u_0(x, N_z + 1)$. The initial-boundary-value problem given in (2.63) and (2.66) can be solved numerically using finite-difference schemes [3, 19]. Here, the crucial point is the update scheme ('stencil'), see Fig. 2.4b. For the two dimensional case (one dimension for the optical axis, one lateral dimension), the stencil introduced by Crank and Nicolson [20] gives second order accuracy in steps Δz along the optical axis and Δx perpendicular to the optical axis [21]. Accordingly, (2.63) is approximated by the following finite-difference expressions [3]

$$\frac{\partial u}{\partial z} \rightarrow \frac{u_k^{n+1} - u_k^n}{\Delta z}, \tag{2.67}$$

$$A\frac{\partial^2 u}{\partial x^2} \rightarrow \frac{A}{2}\left(\frac{u_{k-1}^n - 2u_k^n + u_{k+1}^n}{\Delta x^2} + \frac{u_{k-1}^{n+1} - 2u_k^{n+1} + u_{k+1}^{n+1}}{\Delta x^2}\right), \tag{2.68}$$

$$C(x,z)u \rightarrow \frac{C_k^{n+\frac{1}{2}}}{2}\left(u_k^n + u_k^{n+1}\right) \tag{2.69}$$

with $u_k^n = u(x_k, z_n)$ and $C_k^{n+\frac{1}{2}} = C(x_k, z_{n+\frac{1}{2}})$, resulting in the finite-difference equation [3]

$$\frac{u_k^{n+1} - u_k^n}{\Delta z} = \frac{A}{2}\left(\frac{u_{k-1}^n - 2u_k^n + u_{k+1}^n}{\Delta x^2} + \frac{u_{k-1}^{n+1} - 2u_k^{n+1} + u_{k+1}^{n+1}}{\Delta x^2}\right)$$
$$+ \frac{C_k^{n+\frac{1}{2}}}{2}\left(u_k^n + u_k^{n+1}\right). \tag{2.70}$$

Furthermore, we abbreviate

$$a_x := A\frac{\Delta x}{\Delta x^2}, \qquad c_k^{n+\frac{1}{2}} := \frac{C_k^{n+\frac{1}{2}}\Delta z}{2}. \tag{2.71}$$

and (2.70) may now be written as a system of $N_x - 1$ linear equations [3]

$$M^n u^{n+1} = d^n \tag{2.72}$$

with

$$M^n = \begin{pmatrix} 1 + a_x - c_1^{n+\frac{1}{2}} & -\frac{a_x}{2} & 0 & \cdots & & \\ -\frac{a_x}{2} & 1 + a_x - c_2^{n+\frac{1}{2}} & -\frac{a_x}{2} & 0 & \cdots & \\ & & \ddots & & & \\ \cdots & 0 & -\frac{a_x}{2} & 1 + a_x - c_{N_x-2}^{n+\frac{1}{2}} & -\frac{a_x}{2} \\ & \cdots & 0 & -\frac{a_x}{2} & 1 + a_x - c_{N_x-1}^{n+\frac{1}{2}} \end{pmatrix},$$

$$\mathbf{u}^{n+1} = \begin{pmatrix} u_1^{n+1} \\ u_2^{n+1} \\ \vdots \\ u_{N_x-2}^{n+1} \\ u_{N_x-1}^{n+1} \end{pmatrix},$$

and

$$\mathbf{d}^n = \begin{pmatrix} \frac{a_x}{2}u_0^n + (1 - a_x + c_1^{n+\frac{1}{2}})u_1^n + \frac{a_x}{2}u_2^n + \frac{a_z}{2}u_0^{n+1} \\ \frac{a_x}{2}u_1^n + (1 - a_x + c_2^{n+\frac{1}{2}})u_2^n + \frac{a_z}{2}u_3^n \\ \vdots \\ \frac{a_x}{2}u_{N_x-3}^n + (1 - a_x + c_{N_x-2}^{n+\frac{1}{2}})u_{N_x-2}^n + \frac{a_x}{2}u_{M_x-1}^n \\ \frac{a_x}{2}u_{N_x-2}^n + (1 - a_x + c_{N_x-1}^{n+\frac{1}{2}})u_{N_x-1}^n + \frac{a_x}{2}u_{N_x}^n + \frac{a_z}{2}u_{N_x}^{n+1} \end{pmatrix}.$$

Since $\mathbf{d}^n$ has no field values with index $n + 1$ except for u_0^{n+1} and $u_{N_x}^{n+1}$, which are set by the boundary conditions (2.65) and (2.66), it is possible to compute the field in plane $n + 1$ from plane n, by solving a system of $N_z - 1$ linear equations. Furthermore, because M^n is tridiagonal, this can be carried out with $\mathcal{O}(N_x)$ operations [21]. Finally, the process is repeated sequentially for all N_z grid points, resulting in numerical complexity $\mathcal{O}(N_x \times N_z)$. The update scheme for propagation in three dimensions (3d) (i.e. 2d + 1d) is described in [1, 22, 23].

2.2   Coherent Image Formation

In early years, electron microscopy was severely limited by aberrations of electromagnetic objective lenses. To overcome these difficulties, D. Gabor proposed lens-less imaging by inline holography in 1948, inspired by L. Bragg's ideas of a coherent projection X-ray microscope, see [24] for a historical perspective. Gabor demonstrated his idea of coherent imaging without optics with visible light [25]. Instead of recording sharp images with a lens-based system, the (near field) interference pattern behind a (semi-transparent) object was recorded on a photographic plate. The pattern originates from interference of the direct beam passing through the object with the secondary waves diffracted by the object. Illuminating the (reversed) photographic plate, i.e. using the photographic plate as an object resulted in a clear image of the sample. In Gabor's time, this type of holographic reconstruction was by necessity of analogue optical nature, whereas nowadays a hologram recorded by a detection device can be reconstructed numerically.

While in electron microscopy the lens problems were eventually overcome, a similar challenge appeared in X-ray microscopy: X-ray lenses (diffractive or refractive) lack the efficiency and numerical aperture which we are used to from visible light lenses. At the same time, X-rays are attractive as a microscopy probe owing to short wavelength and hence potentially high resolution as well as high penetration power. Therefore, Gabor's idea of coherent lens-less imaging re-emerged in X-ray optics and microscopy, once that radiation of high brilliance and sufficient coherence became available by synchrotron sources, in particular after the invention of undulators. In contrast to coherent diffractive imaging (CDI), where the primary beam is blocked behind the sample and the diffracted radiation is recorded without interference with

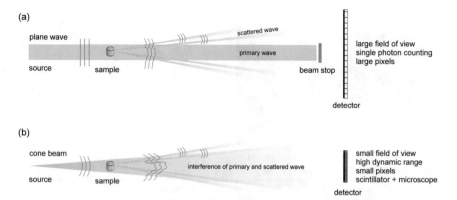

Fig. 2.5 Illustration of coherent diffractive imaging (optical far field) and holographic full field imaging (optical near field). **a** In conventional diffraction from crystalline or non-crystalline specimen, intensity scattered to high angles is measured and the direct beam is blocked. **b** Contrarily, holography uses the interference between diffracted waves and the direct beam, such that phase information is encoded in the intensity pattern. Note that the techniques require different detectors: single photon counting detectors with a large numerical aperture are preferred for (**a**), while (**b**) requires high resolution detection devices such as scintillator-based cameras

the primary or reference wave, and in direct analogy to Gabor's setup, X-ray holography is based on the interference between primary and scattered waves on the detector (see Fig. 2.5).

2.2.1 Holographic Imaging in Full Field Setting

Inline holography is a full field imaging technique, in which many resolution elements of object and detector are illuminated in parallel, typically employing a mega-pixel detector with sufficiently small pixel size to record the fine interference fringes between scattered and primary waves. While the field of view (FOV) can of course be further increased by lateral scanning, an image of large FOV can be acquired in a single exposure. This is in contrast to coherent techniques which are based on scanning the object in a focused beam, or which require a fully coherent illumination and are therefore limited to a correspondingly small field of view. For this reason, holographic imaging is of significant advantage in particular for tomography, where compared to the net counting time, motor overhead and detector readout for three degrees of freedom become a dominant time factor in recording data. Furthermore, time-resolved imaging is more easily implemented in parallel than serial acquisition.

Two geometries are commonly used for inline X-ray holography: (quasi-) parallel and (quasi-)spherical illumination. The first case is implemented at synchrotron beamlines using almost the full undulator beam without focusing. The divergence is then small enough so that in the object and detector plane the beam is almost of the

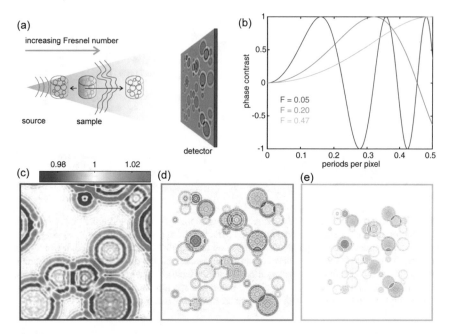

Fig. 2.6 a Cone-beam setting for holographic full field imaging. The object is illuminated by diverging wavefronts. **b** Placing the sample at different defocus positions results in oscillatory contrast of spatial frequencies in the recorded holograms. **c–e** Illustration of the effect of geometrical magnification: Placing the sample close to the focus (**c**) increases magnification and the number of contrast oscillations when plotting the CTF as a function of spatial frequency (dark blue curve in **b**). Placing the sample closer to the detector (**d–e**) decreases magnification and reduces the number of contrast oscillations (light blue and cyan curves in **b**)

same lateral size. The FOV is adjusted by slits upstream from the object. This geometry is used for holography/tomography of large objects with resolution elements at the size of the detector pixels.

The second case is cone beam illumination used for high resolution holographic imaging, since it offers variable geometric magnification and FOV by adjusting the distance between focus and object, the so-called *defocus* distance. Resolution elements are of the detector pixel size scaled by the inverse geometric magnification. Figure 2.6 illustrates magnification and contrast evolution for a phantom consisting of an assembly of spheres. The Fresnel scaling theorem was used to compute the contrast in an effective parallel beam setup which is numerically more convenient.

2.2.2   Contrast in X-ray Holograms

As discussed above, elastic interaction of light and matter can be accounted for by the complex-valued index of refraction, which we rewrite here as

$$n(\lambda) = 1 - \delta(\lambda) + i\beta(\lambda) \tag{2.73}$$

to stress its wavelength dependency. For forward scattered radiation and away from absorption edges, δ is directly proportional to the electron density ρ of the sample [13]

$$\delta(\lambda) = \frac{r_0}{2\pi}\lambda^2\rho, \tag{2.74}$$

with the classical electron radius r_0. In this (often warranted) approximation, δ varies quadratically with wavelength, independent of sample stoichiometry, in contrast to the imaginary part of the index of refraction $\beta(\lambda)$, which strongly depends on the elemental composition and also exhibits much larger jump discontinuities at absorption edges. For hard X-rays and low-Z materials such as in soft and biological matter, $\delta(\lambda)$ is up to three orders of magnitude larger than $\beta(\lambda)$. For this reason, interference contrast in holograms due to phase shift often dominates over absorption contrast. By enhancing radiographic imaging with phase contrast, fine details of weakly absorbing components become detectable, which would otherwise be invisible in classical absorption based X-ray radiography. The effect of the object on the phase of the so-called *exit wave* $\psi_\lambda(x, y)$ directly behind the object, becomes apparent by considering a monochromatic plane wave (wavelength λ) passing through an object which is homogeneous in z and has a thickness ΔT:

$$\psi_\lambda(x, y) = \exp\left[-ik\delta(\lambda, x, y)\Delta T\right] \cdot \exp\left(-\frac{1}{2}\mu(\lambda, x, y)\Delta T\right), \tag{2.75}$$

where $k = \frac{2\pi}{\lambda}$ and $\mu(\lambda, x, y)$ is the absorption coefficient with $\mu = 2k\beta$. Neglecting absorption and replacing δ by (2.74) results in

$$\psi_\lambda(x, y) \approx \exp\left[-ir_0\lambda\rho(x, y)\Delta T\right]. \tag{2.76}$$

Phase retrieval techniques aim at recovering the phase in the exit plane

$$\varphi(x, y, \lambda) = -r_0\lambda\rho(x, y)\Delta T \tag{2.77}$$

from measured intensity distributions. By (2.77), it becomes clear that the phase of the exit plane is directly proportional to the local electron density, if the object is homogeneous in z. More generally, for inhomogeneous (thin) objects, the phase is proportional to the projected electron density, as $\int \rho(x, y, z)dz$.

The lateral distribution of $\varphi(x, y)$ then results in diffraction behind the object, and as the wave propagates, the phase information is converted to an intensity pattern.

The relative contrast of features on different length scales (spatial frequencies) is found to change in a characteristic manner with the free-space propagation distance between object and detector, or more generally with Fresnel number. This evolution of contrast can be analyzed by inspecting the Fourier transform of the hologram. What one finds is an oscillatory contrast as a function of spatial frequency. For an optically weak object (i.e. weakly varying phase and small absorption) it can be shown that in Fourier space, intensity is given as a linear combination of the object's phase and absorption map [26–29]

$$\mathcal{F}[I(x, y, z)/I_0 - 1] = 2 \sin\left(\frac{z\lambda}{4\pi}\left(k_x^2 + k_y^2\right)\right) \mathcal{F}[\varphi(x, y, \lambda)]$$
$$+ \cos\left(\frac{z\lambda}{4\pi}\left(k_x^2 + k_y^2\right)\right) \mathcal{F}[\mu(x, y, \lambda)], \quad (2.78)$$

with z the distance between sample and detector and $k = 2\pi\nu$ the angular spatial frequency in the object plane. To obtain this linear relationship, the signal is normalized to the incident plane, subtracted by one. If the diffraction of the illuminating wave is neglected, the expression can be generalized to an inhomogeneous illumination field $I_0 \rightarrow I(x, y, 0)$, see [30, 31] for a discussion of empty beam division. The structure of (2.78) suggests to define linear filter functions describing the evolution of contrast. For the case of weak phase objects (i.e. with weakly varying phase and negligible absorption), the contrast transfer function (CTF) can be defined as

$$\mathcal{F}[I(x, y, z)/I(x, y, 0) - 1] = \text{CTF}_p(\nu_x, \nu_y, F) \cdot \mathcal{F}[\varphi(x, y, \lambda)], \quad (2.79)$$

where $\mathcal{F}[\varphi(x, y, \lambda)]$ is the Fourier transform of the object function and

$$\text{CTF}_p(\nu_x, \nu_y, F) = 2\left\{\sin\left[\frac{\pi}{F}\left(\nu_x^2 + \nu_y^2\right)\right]\right\}. \quad (2.80)$$

If phase and absorption are proportional in each pixel (single material object), this can be generalized to [26, 28, 29]

$$\text{CTF}(\nu_x, \nu_y, F) = 2\left\{\sin\left[\frac{\pi}{F}\left(\nu_x^2 + \nu_y^2\right)\right] + \frac{\beta}{\delta}\cos\left[\frac{\pi}{F}\left(\nu_x^2 + \nu_y^2\right)\right]\right\}. \quad (2.81)$$

Once absorption becomes important, the cosine term is non-negligible including a scaling with the ratio $\frac{\beta}{\delta}$, i.e. the ratio between energy dependent absorbing and phase shifting properties of the sample [28, 29]. Note also that spatial frequencies (ν_x, ν_y) are now expressed in natural dimensionless units, by $\nu = ka/2\pi$ with a the pixel size. In this way, the dependence on the Fresnel number $F = \frac{a^2}{\lambda z}$ is highlighted. The spacing of discrete sampling points in reciprocal space is $\Delta\nu = 1/N$, with N being the number of pixels in horizontal or vertical direction. An illustration is provided in Fig. 2.6, showing holograms simulated for different distances and magnification in cone-beam geometry (a), with corresponding phase CTF_p (only the sine part of the

CTF) for the three indicated positions or equivalently F (b). Holograms for the three object positions are shown in (c–e). When the Fresnel number is large, the image contrast is dominated by edges (edge enhancement). For example, the cyan curve in (b) is maximal at high spatial frequencies and hence edges in (e) are enhanced with respect to areas. Shifting the sample closer to the source (light blue and dark blue curves in (b) and holograms in (c–d) shows oscillating contrast for lower image frequencies than in (e).

2.3 Solving the Phase Problem in the Holographic Regime

Consider holographic intensity distributions as shown in Fig. 2.6. If phase and amplitude were directly measurable by a detector, the complex wavefield could be numerically back-propagated to obtain the wavefield in the exit plane of the object, where the phase represents a sharp image of the projected electron density. Measurement of the phase is of course completely impossible as the frequency of hard X-rays is on the order of 10^{18} Hz. However, the intensity pattern in the detection plane is directly related to the phase in the exit plane. Unfortunately, the corresponding set of equations is way too large to be solved directly, as it equals to the number of pixels. Furthermore, the interference between primary and scattered wave and the self-interference of the scattered wave render the equations non-linear. Finally, given a single detector image, the system of equations is under-determined, because the amplitude and phase maps in the exit plane contain twice as many unknowns as the available (real-valued) intensity map in the detection plane. The first concern of holographic imaging must therefore be to design the experiment such that the measured data is sufficient, i.e. to achieve uniqueness. Several detector images with sufficient diversity in the data can be generated by variation of the Fresnel number, but in practice a second strategy is more common: The number of unknowns is reduced when sufficient constraints can be formulated, restricting the solution. For example when the phase map can be set be zero outside a known support (support constraint), or when absorption in the object is negligible and the amplitude can be assumed to be one everywhere (pure phase contrast constraint), or finally when the object is approximated to have identical stoichiometry resulting in a coupling of phase and absorption (single material constraint), the phase problem becomes manageable.

Once this first concern of sufficient data and constraints is met, the second concern is to decode the holographic images and retrieve the phase map, i.e. the solution of the phase problem as an inverse problem. This phase retrieval process requires knowledge about image formation (i.e. the forward problem based on the wave equation), including all experimental parameters, as well as about the constraints which can be formulated based on properties of the object. In the following, two basic approaches of phase retrieval will be introduced: Deterministic single step and iterative algorithms using alternating projections onto constraint sets.

2.3.1 Single-Step Phase Retrieval

Single-step deterministic reconstruction techniques invert the process of holographic image formation under certain approximations. An example is phase retrieval based on the contrast transfer function (CTF) given in (2.81). In cases where the Fourier transform of a hologram can be expressed as a product of the object function and the oscillating CTF

$$\mathcal{F}[I(\cdot, \cdot, z)/I(\cdot, \cdot, 0) - 1](\nu_x, \nu_y, z) = \text{CTF}(\nu_x, \nu_y, F) \cdot \mathcal{F}[\varphi(\cdot, \cdot, \lambda)](\nu_x, \nu_y, \lambda), \tag{2.82}$$

a simple ansatz is to divide the Fourier transformed intensities by the analytically determined CTF [28]:

$$\varphi(x, y, \lambda) = \mathcal{F}^{-1}\left[\frac{\mathcal{F}[I(\cdot, \cdot, z)/I(\cdot, \cdot, 0) - 1]}{\text{CTF}(\cdot, \cdot, F)}\right](x, y, \lambda). \tag{2.83}$$

This separation firstly requires one of the following three constraints: pure phase contrast, pure amplitude contrast, or single material (coupled contrast). Secondly, an analytical expression for the CTF is only possible by linearization of the optical properties of the sample. Note that the original assumption of a weak phase shift in the derivation of the CTF was later lifted to a weakly varying phase [27]. Furthermore, even if the assumptions are justified, proper implementation of CTF-based phase retrieval requires regularization in order to compensate for the zero-crossings of the CTF [26, 28, 29]. A detailed description on the implementation of the CTF formalism including the extension to multi-distances can be found in [32].

2.3.2 Iterative Phase Retrieval

Analytic, single step reconstruction techniques are quick but come at a price: They require very restrictive assumptions or approximations, as detailed above. The applicability of phase retrieval can be significantly extended by iterative methods which are computationally expensive but compatible with a wider range of constraints and valid for more general X-ray optical properties of the object. Iterative algorithms cycle between object and detector plane, and are alternatively subjected to an object constraint and the constraint that the solution has to satisfy the measured data. In the following, some of the most common constraints will be introduced.

Compatibility with the measured data is assured by the so-called **magnitude constraint**: A solution to the phase problem must satisfy the measured intensity distribution of the hologram $I(x, y)$. To this end, the wavefield $\psi(x, y)$ is modified such that

$$[\mathcal{P}_M \psi](x, y) := \left[\mathcal{D}_{-z}\hat{\psi}\right](x, y), \qquad \hat{\psi}(\hat{x}, \hat{y}) = \frac{\left[\mathcal{D}_z \psi\right](\hat{x}, \hat{y})}{|\left[\mathcal{D}_z \psi\right](\hat{x}, \hat{y})|} \cdot \sqrt{I(\hat{x}, \hat{y})}.$$

(2.84)

Note that in order to formulate phase retrieval only in terms of projections and to treat all constraints on equal footing, the Fresnel propagation $\mathcal{D}_z$ of the wavefield between object and detector (and back) is incorporated into the projection operator.

Next, we need at least a second constraint, notably in the object plane (or more precisely the exit plane). A very general constraint is the **range constraint**, which sets the magnitude behind the object strictly equal to one (after normalization to incoming intensity). In other words, one assumes the object to be of pure phase contrast. In a more general setting one solely requires the amplitude to be smaller than one (i.e. thereby fixing the range of the amplitude). This is justified, since right behind the object, no interference effects have yet evolved, the wavefield amplitudes can only be smaller (absorbing components in the object) than or equal (transparent components) to unity

$$[\mathcal{P}_A \psi](x, y) := \begin{cases} 1 & \text{if } |\psi(x, y)| > 1 \\ \psi(x, y) & \text{else.} \end{cases}$$

(2.85)

However, some caution is advised. Application of the range constraint requires normalization of intensity by an independent measurement of the empty beam intensity and not just normalization by the mean intensity.

Another straightforward but not always suitable constraint is the **support constraint**. The reconstructed sample is only allowed to cover a limited part D of the field of view

$$[\mathcal{P}_S \psi](x, y) := \begin{cases} \psi(x, y) & \text{if } (x, y) \in D \\ 0/1 & \text{else (Fourier/Fresnel imaging).} \end{cases}$$

(2.86)

In phase retrieval, this constraint is very powerful, for example in view of recovering spatial frequencies corresponding to zeros in the CTF [33]. Unfortunately, it is not applicable to extended samples.

A compact support can be regarded as a special case of sparsity. Obviously, the pixels with non-zero density are sparse, if the support is small. More generally, the object may be sparse in very different ways, i.e. the object or its projection may be specified by a set of independent values much smaller than the number of pixels (voxels). **Sparsity constraints** enforce image properties to be sparse in some sense, without being too restrictive and specific. A suitable way to enforce sparsity for X-ray holography is the **shearlet constraint** [34]. Shearlets are deformed (scaled, translated and sheared) wavelet-type basis functions [35, 36]. They are particularly useful to represent so-called cartoon-like images (compactly supported and twice continuously differentiable functions) [37–39]. Hence, in case that amplitude and phase of the object's transmission function can be categorized as cartoon-like, a sparse representation by a linear combination of shearlets is possible, and this information can be used as an additional constraint for phase retrieval [34]. Yet the applicability of the shearlet constraint is not limited to cartoon-like objects, but the set of shearlets required is much smaller in this case, and the constraint can therefore be formulated more 'strictly', and will thus be more powerful for phase retrieval.

If no reasonable constraint can be formulated for the object, i.e. if the object is extended, exhibits uncoupled variations in phase and amplitude and is not sparse, additional data has to be acquired and used as input for phase retrieval. One way to do this is by translations of the object or detector (either longitudinal or lateral) and successive image acquisitions. Alternating projections on such multi-measurements by various update schemes are denoted as **multi-magnitude projection** (mmp) [30, 40, 41]. In mmp, the illumination (probe) has to be perfect or known beforehand, for example by a complete mmp series acquired at different detector positions [40]. This is often difficult to accomplish.

A suitable alternative is to generate multi-measurements by translation of the object and to use ptychographic algorithms for phasing by enforcing separability of object and probe afterwards. This so-called **separability constraint** enables simultaneous phase retrieval of object and probe, and can thereby account for the fact that aberrations inherent in the illumination interfere with the modulations imposed by the sample, which otherwise result in degradation of image quality and resolution [30, 31]. The separability constraint is key in all ptychographic algorithms [42–44], and was introduced in X-ray holography in several different ways, based on using either a wavefront diffuser and lateral translations [45], or longitudinal translations [46], or a combination of lateral and longitudinal translations [47]. Note that the last case is least restrictive in terms of probe properties, see also [48] for a detailed comparison and discussion.

Figure 2.7 presents a schematic of an iterative phase retrieval algorithm. It is composed of alternating projections and reflections onto sets of functions that fulfill two or more constrains. The final goal is to ultimately decode the holographic intensities and reveal a solution proportional to the electron density distribution of the sample. The precise update scheme of a specific reconstruction algorithm (step 6 in Fig. 2.7), has significant effect on the convergence. As an illustration, three common update techniques will be briefly introduced.

The oldest such schemes are known as **Gerchberg-Saxton-type (GS)** algorithms [49] and consist of alternating projections onto sets of functions fulfilling magnitude or range-constraints as given in (2.84). Holograms $M_1(x, y)$ and $M_2(x, y)$ recorded at two different defocus positions provide the required data, but a single measurement can be sufficient, for example in case of pure phase contrast. The complete algorithm can be written as

$$\psi^{(n+1)}(x, y) = \left[\mathcal{P}_{M_1}\mathcal{P}_{M_2}\psi^{(n)}\right](x, y). \tag{2.87}$$

Alternating projections are stopped, once a fixed point solution $\psi^{(n+1)}(x, y) = \psi^{(n)}(x, y)$ is found. In case of a pure phase object, the range constraint for the object domain is replaced by (2.85). It can be shown that the GS algorithm corresponds to a local optimization, which is therefore highly dependent on the initial guess. A very similar algorithm based on the iterative application of two projections is the **Error**

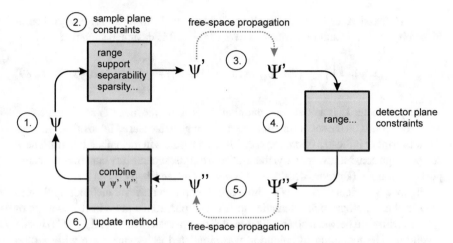

Fig. 2.7 Illustration of iterative phase retrieval: An initially guessed wavefield ψ exiting the object of interest (1.) is forced to fulfill constraints right behind the sample (2.), has to fulfill the wave equation (3. and 5.), has to match the measured intensities (4.) and is updated to form a new guessed field ψ (6.)

Reduction Algorithm which alternates the magnitude (measurement) constraint and a support constraint.

To avoid stagnation in side minima, a more general update scheme can be formulated, with non-local search properties. The first such example was provided by the **Hybrid-Input-Output** algorithm, which was originally formulated for far field coherent diffractive imaging and objects limited to a finite support [50]. The main idea is to formulate the update as a linear combination of the wavefield of the previous iterate (the input), and the wavefield resulting from projection onto the measurement and application of a support constraint (the output). This linear combination is governed by the parameter $\beta \in [0, 1]$, gradually pulling the wavefield outside D to zero

$$
\psi^{(n+1)}(x, y) = \begin{cases} \left[\mathcal{P}_M \psi^{(n)} \right](x, y) & \text{if } (x, y) \in D \\ \psi^{(n)}(x, y) - \beta \left[\mathcal{P}_M \psi^{(n)} \right](x, y) & \text{else.} \end{cases} \tag{2.88}
$$

As a consequence, the solution $\psi^{(n+1)}(x, y)$ does neither exactly fulfill the magnitude constraint, nor does it fulfill the support constraint. Yet, it can overcome stagnation and finally find a solution that is consistent with both, support and range. Whereas the classical HIO algorithm was designed for far field imaging (i.e. the propagated wavefields are calculated by Fourier transforms), a modified Hybrid-Input-Output algorithm was formulated for full field holography [33]. The differences to its original version are the replacement of the Fourier transforms by the Fresnel near field propagator and the fact that both phase and amplitude of the wavefield outside the support are subjected to a support constraint [33] pulling the amplitudes to one and the phases to zero.

The **Relaxed-Averaged-Alternating-Reflections (RAAR)** algorithm [51] combines projections $\mathcal{P}$ and reflections $\mathcal{R} = 2 \cdot \mathcal{P} - 1$. It is designed such that

$$\psi^{(n+1)}(x, y) = \left[\left(\frac{\lambda_n}{2}\left(\mathcal{R}_O\mathcal{R}_M + \text{Id}\right) + (1 - \lambda_n)\mathcal{P}_M\right)\psi^{(n)}\right](x, y). \qquad (2.89)$$

The parameter λ_n is a relaxation parameter. As above the index M refers to projections/reflections on the set of functions reproducing the measured intensities, whereas the subscript O indicates operations fulfilling constraints in the object domain. These can be range constraints, sparsity/shearlet constraints, separability constraints or support constraints. For more details on this algorithm see Chaps. 6 and 23.

Figure 2.8 depicts the results of three different phase retrieval methods applied to the simulated holographic intensities in (a). The phantom shown in (b) consists of spheres with radii between 50 and 200 nm of different materials (Al, Al_2O_3, Ca) inside a volume. The incoming illumination was simulated as a plane wave with photon energy of 7keV. The fact that the spheres are not made of a single material violates the assumptions required to derive single step CTF-based phase retrieval (2.79). Hence, the reconstruction shown in (d) depicts blurry regions (absorbing and phase shifting components β and δ were set to the mean values of the given materials). The iterative modified Hybrid-Input-Output (e) and the Relaxed-Averaged-Alternating-Reflections algorithm can reveal features of the projected phase more distinctly and with less blur. A detailed comparison between iterative and analytic phase retrieval for experimental data can be found in [52].

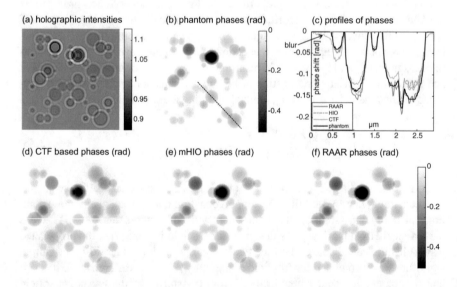

Fig. 2.8 **a** Simulated holographic intensities of a sample consisting of multi-material spheres. **b** True projected phases of the sample. **c** Line profiles through the phantom and the reconstructions shown in (**d**–**f**). **d** Reconstructed phase by the CTF-based single step technique. **e** Phase retrieval result by the modified Hybrid-Input-Output algorithm (mHIO). **f** Phase retrieval result by the Relaxed-Averaged-Alternating-Reflections (RAAR) algorithm

2.4 From Two to Three Dimensions: Tomography and Phase Retrieval

Inverting the holographic intensity distributions via phase retrieval techniques, one arrives at a two dimensional (2d) image proportional to the projected index variation, or equivalently—when considering the real part only—the projected electron density. But for a three dimensional (3d) object $f(x, y, z)$, we would also like to find the 3d structure as given by the refractive index $n(\mathbf{r})$. For this purpose, many projections of the sample have to be recorded under different angles, see Fig. 2.9a. Projected gray values onto a single line s under an angle θ shown in Fig. 2.9b correspond to

$$\mathcal{R}\,[f(\mathbf{r})]\,(s) := \int f(\mathbf{r})\,\delta_{\text{Dirac}}(\mathbf{r}\cdot\mathbf{n}_\theta - s)d^2x, \qquad (2.90)$$

where $\mathbf{r}$ is a vector in a 2d hyper-plane of the 3d volume, δ_{Dirac} is the Dirac delta distribution and $\mathbf{n}_\theta$ is the unit length vector pointing in direction of s. The operator $\mathcal{R}\,[f(\mathbf{r})]\,(s)$ is denoted as the Radon transform. It results in the 1d projection of the sample onto the line s. Figure 2.9b illustrates the discrete version of (2.90) (integral is replaced by sum). It shows the top view of a selected slice (xy-plane) through the sample f (gray shaded region). The operator $\mathcal{R}\,[f(\mathbf{r})]\,(s)$ computes line integrals through the object: Whenever the scalar product $\mathbf{r}\cdot\mathbf{n}_\theta$ equals a distinct value s (here: s_a and s_b) corresponding to the projection of $\mathbf{r}$ (here illustrated by $\mathbf{a_n}$, $\mathbf{a_{n+m}}$, $\mathbf{b_n}$, $\mathbf{b_{n+m}}$) onto $\mathbf{n}_\theta$, the value of the object $f(\mathbf{r})$ contributes to the projected gray value in a specific bin of the detector (here: the gray values at s_a and s_b).

There are different techniques to invert the Radon transform, i.e. to reconstruct $f(\mathbf{r})$ from a set of projections. The most popular and widely used method is the so-called filtered back-projection. Its basic ingredients are a Fourier-filtering step of the projection data by a ramp filter and the back-projection or smearing out of the filtered projection values along straight lines equal to the paths of the line integrals of the Radon transform. In-depth literature about tomography can be found in [14, 53, 54]. Finally, Fig. 2.9c summarizes the basic steps of three-dimensional holographic X-ray imaging (holographic tomography): A small sample is rotated in a (partially) coherent beam (see Fig. 2.6a), and for each angle a holographic intensity distribution is recorded. Each of the holograms needs to be processed by a suitable phase retrieval technique resulting in an image proportional to the object's projected electron density. Finally, the 3d object is computed using the information collected by all of the projections.

A different approach is to combine phase retrieval and tomographic reconstruction. Instead of performing phase retrieval of all projections first and in a second step the inverse Radon transform (e.g. filtered back-projection), these two steps can be intertwined iteratively: **Iterative Reprojection Phase retrieval** (IRP) [55]. In this way, an iterate of the full 3d object exists at all times during the process, which ensures tomographic consistency of all projections. This was found to facilitate phase retrieval, effectively acting as a constraint of its own.

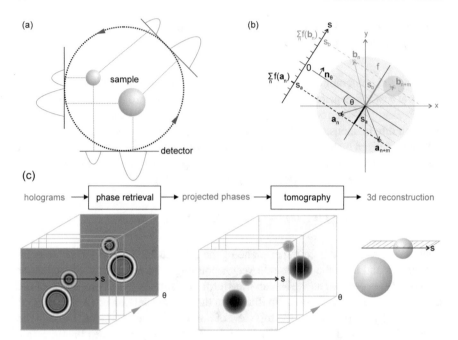

Fig. 2.9 From two to three dimensions: **a** Projections of a sample (consisting of spheres) are recorded under different angles. This can be realized by either rotating the sample or the detector. **b** Illustration of (2.90): The delta-distribution in combination with the scalar product $\mathbf{r} \cdot \mathbf{n}_\theta$ selects points with a component of length s in direction of $\mathbf{n}_\theta$. These point lie on lines pointing to the detector under the angle θ. The summed values of f (shaded region) along these lines are recorded by the detector. Note that (**b**) illustrates a single plane/slice through the three dimensional object perpendicular to the axis of rotation—see outermost right sketch in (**c**) for clarity. **c** Holographic X-ray microscopy is used for high resolution tomography. Holographic intensity distributions are collected under different angles. Each single projection is processed by phase retrieval algorithms. The reconstructed phases are used for inversion of the Radon transform, i.e. for three dimensional tomographic reconstruction

Since IRP is computationally very involved, another simultaneous operation of tomographic reconstruction and phase retrieval by CTF was proposed, which also couples the two previously sequential operations [56]. It relies on propagation of the transmission function of an entire 3d object—the **propagated object**—and requires linearization of the object's optical properties. If this approximation is justified, the concept of propagating the entire 3d object in parallel is extremely useful for holo-tomography, an shall be briefly introduced here, following [56].

To this end, let us consider a plane wave ψ_0 along the optical axis z impinging on a 3d object, parameterized by the spatial distribution of the index of refraction $n(\mathbf{r})$. If the projection approximation holds, the exit wave $\psi(x, y, z = 0)$ is determined by the projection of the index of refraction onto a plane perpendicular to the optical axis, which can be written in terms of the Radon operator $\mathcal{R}(n - 1) := \int (n - 1) \, \mathrm{d}z$ as

$$\psi(x, y, z = 0) = \psi_0(x, y) \exp\left[ik_0 \mathcal{R}(n-1)\right]$$
$$\approx \psi_0(x, y)\left[1 + ik_0 \mathcal{R}(n-1)\right], \tag{2.91}$$

where the approximation of an optically weak object with $|k_0 \mathcal{R}(n-1)| \ll 1$ was used to linearize the exponential function. Following the angular spectrum approach in Sect. 2.1.2, the free-space propagation to the detector plane at distance z (using the propagator P_z) can be formulated as

$$\psi(x, y, z) = \left[\mathcal{F}_{2d}^{-1} P_z \mathcal{F}_{2d} \psi(\cdot, \cdot, 0)\right](x, y, z)$$
$$\approx \psi_0(x, y) \exp\left[ik_0 z\right] + ik_0 \psi_0(x, y)\left[\mathcal{F}_{2d}^{-1} P_{2d}(\cdot, \cdot, z)\mathcal{F}_{2d}\mathcal{R}(n-1)\right](x, y), \tag{2.92}$$

with the radially symmetric free-space propagator function $P_{2d}(k_x, k_y, z) := \exp(iz[k_0^2 - k_x^2 - k_y^2]^{1/2})$ (Fourier space coordinates k_x, k_y and wavenumber $k_0 = 2\pi/\lambda$), followed by a 2d Fourier back-transformation $\mathcal{F}_{2d}^{-1}$. Next, we consider the 2d Fourier transform of a projection of the object, e.g. along z, which can be rewritten as

$$\mathcal{F}_{2d}\left[\mathcal{R}(n-1)\right] = \int\!\!\!\int\!\!\!\int_{-\infty}^{\infty} (n-1)\, \mathrm{d}z \cdot \exp\left[-i(k_x x + k_y y)\right]\, \mathrm{d}x\, \mathrm{d}y$$
$$= \left[\mathcal{F}_{3d}(n-1)\right]_{k_z=0}. \tag{2.93}$$

From (2.93), it can be inferred that the central slice of the 3d Fourier transform of the object (given by $n-1$), equals the 2d Fourier transform of a projection normal to the slice. This is an important geometric relation in tomography, known as the **Fourier slice theorem**. Here we use it to show that the order of projection and propagation $\mathcal{D}$ can be inverted [56]. We consider the 2d propagation $\mathcal{D}_{2d}$ of the projection $\mathcal{R}(n-1)$ [56]

$$\mathcal{D}_{2d}\mathcal{R}(n-1) = \mathcal{F}_{2d}^{-1}\left\{P_{2d} \cdot \mathcal{F}_{2d}\left[\mathcal{R}(n(x, y, z) - 1)\right]\right\}$$
$$= \mathcal{F}_{2d}^{-1}\left\{P_{2d} \cdot \left[\mathcal{F}_{3d}(n-1)\right]_{k_z=0}\right\}$$
$$= \mathcal{F}_{2d}^{-1}\left\{P_{3d} \cdot \mathcal{F}_{3d}[n-1] \cdot \delta_{\mathrm{Dirac}}(k_z)\right\}$$
$$= \mathcal{F}_{2d}^{-1}\left\{\left[P_{3d} \cdot \mathcal{F}_{3d}(n-1)\right]_{k_z=0}\right\}$$
$$= \mathcal{F}_{2d}^{-1}\left\{\mathcal{F}_{2d}\left[\mathcal{R}\left(\mathcal{F}_{3d}^{-1}(P_{3d} \cdot \mathcal{F}_{3d}(n-1))\right)\right]\right\}$$
$$= \mathcal{R}\left[\mathcal{D}_{3d}(n-1)\right]. \tag{2.94}$$

The term $P_{3d} = \exp\left[i\Delta\sqrt{k_0^2 - k_x^2 - k_y^2 - k_z^2}\right]$ is the 3d propagator function. As detailed in [56], this result is the starting point to perform CTF approximations and fast iterative phase retrieval directly in 3d, i.e. simultaneously with tomographic reconstruction. Within a few iterations, tomographic consistency can be enforced and serves as constraint for phase retrieval.

Acknowledgements We thank present and past group members for insights and discussions, in particular Lars Melchior, Aike Ruhlandt and Johannes Hagemann, for the progress they have provided in FD propagation (L.M.), numerical free-space propagators and 3d propagation (A.R.), and in iterative methods of holographic phase retrieval (J.H.). We also acknowledge the fruitful exchange with our colleagues from mathematics, in particular Russell Luke, Thorsten Hohage, Carolin Homann, Gerlind Plonka-Hoch, Stefan Look, and Simon Maretzke.

References

1. Melchior, L., Salditt, T.: Finite difference methods for stationary and time-dependent X-ray propagation. Opt. Express **25**, 32090–32109 (2017)
2. Bergemann, C., Keymeulen, H., van der Veen, J.F.: Focusing x-ray beams to nanometer dimensions. Phys. Rev. Lett. **91**(20), 204801 (2003)
3. Fuhse, C., Salditt, T.: Finite-difference field calculations for one-dimensionally confined X-ray waveguides. Phys. B **357**(1–2), 57–60 (2005)
4. Kopylov, Y.V., Popov, A.V., Vinogradov, A.V.: Application of the parabolic wave equation to X-ray diffraction optics. Opt. Commun. **118**(5–6), 619–636 (1995)
5. Husakou, A.: Nonlinear phenomena of ultrabroadband radiation in photonic crystal fibers and hollow waveguides. Ph.D. thesis, Freie Universität Berlin (2002)
6. Paganin, D.M.: Coherent X-ray Optics. Oxford University, New York (2006)
7. Ruhlandt, A.: Time-resolved x-ray phase-contrast tomography. Ph.D. thesis, Universität Göttingen (2018)
8. Bartels, M., Krenkel, M., Haber, J., Wilke, R.N., Salditt, T.: X-ray holographic imaging of hydrated biological cells in solution. Phys. Rev. Lett. **114**, 048103 (2015)
9. Döring, F., Robisch, A.L., Eberl, C., Osterhoff, M., Ruhlandt, A., Liese, T., Schlenkrich, F., Hoffmann, S., Bartels, M., Salditt, T., Krebs, H.U.: Sub-5 nm hard x-ray point focusing by a combined Kirkpatrick-Baez mirror and multilayer zone plate. Opt. Express **21**(16), 19311–19323 (2013)
10. Voelz, D.G.: Computational Fourier Optics: A MATLAB Tutorial (SPIE Tutorial Texts Vol. TT89). SPIE press (2011)
11. Voelz, D.G., Roggemann, M.C.: Digital simulation of scalar optical diffraction: revisiting chirp function sampling criteria and consequences. Appl. Opt. **48**(32), 6132–6142 (2009)
12. Matsushima, K., Shimobaba, T.: Band-limited angular spectrum method for numerical simulation of free-space propagation in far and near fields. Opt. Express **17**(22), 19662–19673 (2009)
13. Als-Nielsen, J., McMorrow, D.: Elements of Modern X-ray Physics, 2nd edn. Wiley (2011)
14. Salditt, T., Aspelmeier, T., Aeffner, S.: Biomedical Imaging: Principles of Radiography, Tomography and Medical Physics. Walter de Gruyter GmbH & Co KG (2017)
15. Davis, T.J.: Dynamical X-ray diffraction from imperfect crystals: a solution based on the Fokker-Planck equation. Acta Crystallogr. Sec. A **50**(2), 224–231 (1994)
16. Gureyev, T.E., Davis, T.J., Pogany, A., Mayo, S.C., Wilkins, S.W.: Optical phase retrieval by use of first Born- and Rytov-type approximations. Appl. Opt. **43**(12), 2418–2430 (2004)
17. Sung, Y., Barbastathis, G.: Rytov approximation for x-ray phase imaging. Opt. Express **21**(3), 2674–2682 (2013)
18. Li, K., Wojcik, M., Jacobsen, C.: Multislice does it all-calculating the performance of nanofocusing x-ray optics. Opt. Express **25**(3), 1831–1846 (2017)
19. Scarmozzino, R., Osgood, R.M.J.: Comparison of finite-difference and Fourier-transform solutions of the parabolic wave equation with emphasis on integrated-optics applications. J. Opt. Soc. Am. A **8**(5), 724–731 (1991)
20. Crank, J., Nicolson, P.: A practical method for numerical evaluation of solutions of partial differential equations of the heat-conduction type. Proc. Camb. Philos. Soc. **43**, 55–67 (1947)

21. Thomas, J.W.: Numerical Partial Differential Equations: Finite Difference Methods, vol. 22. Springer Science & Business Media (2013)
22. Fuhse, C.: X-ray waveguides and waveguide-based lensless imaging. Ph.D. thesis (2006)
23. Fuhse, C., Salditt, T.: Finite-difference field calculations for two-dimensionally confined x-ray waveguides. Appl. Opt. **45**(19), 4603–4608 (2006)
24. Spence, J.C.: Lawrence Bragg, microdiffraction and X-ray lasers. Acta Crystallogr. Sect. A Found. Crystallogr. **69**(1), 25–33 (2013)
25. Gabor, D.: A new microscopic principle. Nature **161**, 777–778 (1948)
26. Cloetens, P., Ludwig, W., Baruchel, J., Guigay, J.P., Pernot-Rejmankova, P., Salome-Pateyron, M., Schlenker, M., Buffiere, J.Y., Maire, E., Peix, G.: Hard x-ray phase imaging using simple propagation of a coherent synchrotron radiation beam. J. Phys. D **32**(10A), A145–A151 (1999)
27. Guigay, J.P.: Fourier transform analysis of Fresnel diffraction patterns and in-line holograms. Optik **49**(1), 121–125 (1977)
28. Turner, L.D., Dhal, B.B., Hayes, J.P., Mancuso, A.P., Nugent, K.A., Paterson, D., Scholten, R.E., Tran, C.Q., Peele, A.G.: X-ray phase imaging: demonstration of extended conditions for homogeneous objects. Opt. Express **12**(13), 2960–2965 (2004)
29. Zabler, S., Cloetens, P., Guigay, J.P., Baruchel, J., Schlenker, M.: Optimization of phase contrast imaging using hard x rays. Rev. Sci. Instrum. **76**(7), 073705 (2005)
30. Hagemann, J., Robisch, A.-L., Luke, D.R., Homann, C., Hohage, T., Cloetens, P., Suhonen, H., Salditt, T.: Reconstruction of wave front and object for inline holography from a set of detection planes. Opt. Express **22**(10), 11552–11569 (2014)
31. Homann, C., Hohage, T., Hagemann, J., Robisch, A.-L., Salditt, T.: Validity of the empty-beam correction in near-field imaging. Phys. Rev. A **91**, 013821 (2015)
32. Krenkel, M.: Cone-beam x-ray phase-contrast tomography for the observation of single cells in whole organs. Ph.D. thesis, Universität Göttingen (2015)
33. Giewekemeyer, K., Krüger, S.P., Kalbfleisch, S., Bartels, M., Beta, C., Salditt, T.: X-ray propagation microscopy of biological cells using waveguides as a quasipoint source. Phys. Rev. A **83**(2), 023804 (2011)
34. Loock, S., Plonka, G.: Phase retrieval for Fresnel measurements using a shearlet sparsity constraint. Inverse Probl. **30**(5), 055005 (2014)
35. Guo, K., Kutyniok, G., Labate, D.: Sparse multidimensional representations using anisotropic dilation and shear operators. In: Chen, G., Lai, M. (eds.) Wavelets and Splines. Nashboro Press, pp. 189–201 (2006)
36. Labate, D., Lim, W.Q., Kutyniok, G., Weiss, G.: Sparse multidimensional representation using shearlets. In: Papadakis, M., Laine, A.F., Unser M.A. (eds.) Wavelets XI, Proceedings of the SPIE, vol. 5914, pp. 254–262 (2005)
37. Donoho, D.L.: Sparse components of images and optimal atomic decomposition. Constr. Approx. **17**, 353–382 (2001)
38. Kutyniok, G., Labate, D. (eds.): Shearlets: Multiscale Analysis for Multivariate Data. Birkhäuser (2012)
39. Pein, A., Loock, S., Plonka, G., Salditt, T.: Using sparsity information for iterative phase retrieval in x-ray propagation imaging. Opt. Express **24**(8), 8332–8343 (2016)
40. Hagemann, J., Robisch, A.-L., Osterhoff, M., Salditt, T.: Probe reconstruction for holographic X-ray imaging. J. Synchrotron Rad. **24**(2), 498–505 (2017)
41. Hagemann, J., Salditt, T.: Divide and update: towards single-shot object and probe retrieval for near-field holography. Opt. Express **25**(18), 20953–20968 (2017)
42. Guizar-Sicairos, M., Fienup, J.R.: Phase retrieval with transverse translation diversity: a nonlinear optimization approach. Opt. Express **16**(10), 7264–7278 (2008)
43. Maiden, A.M., Rodenburg, J.M.: An improved ptychographical phase retrieval algorithm for diffractive imaging. Ultramicroscopy **109**(10), 1256–1262 (2009)
44. Thibault, P., Dierolf, M., Menzel, A., Bunk, O., David, C., Pfeiffer, F.: High-resolution scanning x-ray diffraction microscopy. Science **321**(5887), 379–382 (2008)
45. Stockmar, M., Cloetens, P., Zanette, I., Enders, B., Dierolf, M., Pfeiffer, F., Thibault, P.: Near-field ptychography: phase retrieval for inline holography using a structured illumination. Sci. Rep. **3**, 1927 (2013)

46. Robisch, A.-L., Salditt, T.: Phase retrieval for object and probe using a series of defocus near-field images. Opt. Express **21**(20), 23345–23357 (2013)
47. Robisch, A.-L., Kröger, K., Rack, A., Salditt, T.: Near-field ptychography using lateral and longitudinal shifts. New J. Phys. **17**(7), 073033 (2015)
48. Robisch, A.-L.: Phase retrieval for object and probe in the optical near-field. Ph.D. thesis, Universität Göttingen (2016)
49. Gerchberg, R.W., Saxton, W.O.: A practical algorithm for the determination of phase from image and diffraction plane pictures. Optik **35**(2), 237–246 (1972)
50. Fienup, J.R.: Reconstruction of an object from the modulus of its Fourier transform. Opt. Lett. **3**(1), 27–29 (1978)
51. Luke, D.R.: Relaxed averaged alternating reflections for diffraction imaging. Inverse Probl. **21**(1), 37 (2005)
52. Krenkel, M., Toepperwien, M., Alves, F., Salditt, T.: Three-dimensional single-cell imaging with X-ray waveguides in the holographic regime. Acta Crystallogr. Sec. A **73**(4), 282–292 (2017)
53. Buzug, T.: Computed Tomography: From Photon Statistics to Modern Cone-Beam CT. Springer (2008)
54. Natterer, F.: The Mathematics of Computerized Tomography. Classics in Applied Mathematics. Society for Industrial and Applied Mathematics (2001)
55. Ruhlandt, A., Krenkel, M., Bartels, M., Salditt, T.: Three-dimensional phase retrieval in propagation-based phase-contrast imaging. Phys. Rev. A **89**, 033847 (2014)
56. Ruhlandt, A., Salditt, T.: Three-dimensional propagation in near-field tomographic X-ray phase retrieval. Acta Crystallogr. Sec. A **72**(2), 215–221 (2016)

Chapter 3
X-ray Focusing and Optics

Tim Salditt and Markus Osterhoff

*Es ist aber leicht einzusehen, daß bei nahezu streifender
Inzidenz der Röntgenstrahlen im Falle n < 1 eine nachweisbare
Totalreflexion auftreten muß.*
— Albert Einstein, 21st March, 1918

3.1 General Aspects of X-ray Optics and Focusing

X-ray optics can be considered as optics in the "vacuum limit". In fact, the index of refraction $n = 1 - \delta + i\beta$ asymptotically approaches one for high photon energy E, as δ and β decrease algebraically for $E \geq E_r$, where E_r stands for an atomic resonance, i.e. an absorption edge given by the corresponding electronic binding energy. For all materials, the X-ray regime is hence characterized by extremely small differences in the indices of refraction. This can be a blessing in terms of penetration power, or the validity of various approximations, such as kinematic scattering (neglect of multiple scattering) or the projection approximation, as addressed in Chap. 2. At the same time it can also be a curse, as one readily realizes the challenge of focusing radiation when the index difference between a lens and air or vaccuum goes to zero. More generally, not only focusing but any type of optical element and function is heavily constrained by the small differences in the index of refraction. For this reason, it is not yet possible to focus down to X-ray wavelength λ. In Abbe's sense, the diffraction limit is not in λ but in the achievable numerical aperture. In other words, the *diffraction limit* for X-rays is a limit of the *diffraction structure*. This has raised the question of a fundamental resolution limit for X-rays existing above the wavelength. In 2003, Bergemann and van der Veen had conjectured a fundamental

T. Salditt (✉) · M. Osterhoff
Institute for X-ray Physics, Universität Göttingen, Friedrich -Hund -Platz 1, 37077 Göttingen, Germany
e-mail: tsaldit@gwdg.de

M. Osterhoff
e-mail: mosterh1@gwdg.de

© The Author(s) 2020
T. Salditt et al. (eds.), *Nanoscale Photonic Imaging*, Topics in Applied Physics 134,
https://doi.org/10.1007/978-3-030-34413-9_3

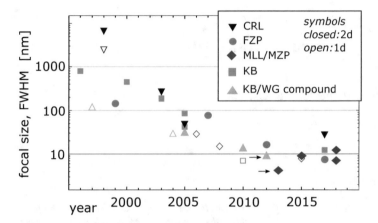

Fig. 3.1 Illustration of the rapid progress in X-ray nano-focusing, following the advent of 3rd generation synchrotron radiation. Major benchmarks are displayed as full symbols (2d focusing) or empty symbols (1d focusing), representing the following references (from left to right): FZP optics (brown circles) [6–9], CRL (black downwards-pointed triangles) [10–15], KB mirror optics [16–22], MLL/MZP (purple diamonds) [4, 23–28], and WG or KB/WG compound optics (orange triangles) [29–33], respectively. The arrows mark results obtained by the CRC 'Nanoscale Photonic Imaging' in Göttingen

length scale Λ given by the decay length of evanescent waves, which should present a lower limit for any X-ray focus size well above the wavelength λ. This critical length prototypically appears in waveguide optics in terms of the minimum width to which a mode can be confined, as explained further below, and roughly ranges in between 8 and 15 nm depending on the density of the material used for focusing [1]. This limit was later rejected and successively disproved [2, 3]. Yet, the idea is correct that the small contrast in the index of refraction for X-rays and the correspondingly large decay length of evanescent waves indeed significantly constrain our focusing capabilities. More than fifteen years after the postulation of the "Bergemann limit" we must realize that we still do not know the minimum focus size nor the maximum local field enhancement (gain) in focusing X-ray radiation. Experimentally, however, 10 nm focal size has become a reality also for hard X-rays [4, 5]. Figure 3.1 illustrates the rapid development of hard X-ray focusing over the last two decades. This progress became possible after the advent of high brilliance (3rd generation) synchrotron radiation, which provide the necessary coherence for diffraction-limited or near-diffraction limited focusing.

The primary challenge for X-ray microscopy is hence to narrow down the gap between the theoretical resolution limit associated with the wavelength λ and the actual resolution limited by the optical systems. Fresnel zone plates (FZP) have been developed as X-ray focusing and objective lenses by G. Schmahl and colleagues in Göttingen, initially for X-ray microscopy in the soft X-ray range (0.2–1.2 keV, $\lambda = 1$–7 nm). Spot sizes of soft X-ray microscopes of around 30 nm are common; "best values" are about two times smaller, in the range 10–15 nm [34]. Hard X-ray zone-plate optics was for a long time limited to above $\simeq$0.25 μm, but over the last fifteen years, significant progress as been achieved by several advanced concepts which realize FZP optics based on multilayer deposition on planar solids as well as on thin wires. Such structures are denoted as multilayer Laue lens (MLL) and multilayer zone plates (MZP), respectively, and will be discussed in detail in Sect. 3.5. Progress within the present collaborative research center (CRC *Nanoscale Photonic Imaging*), in particular, has resulted in 5 nm point focusing by a MZP optics. Compared to diffractive optics, refractive optics as used for visible and UV light seem at the first glance, unsuitable due to the small X-ray refractive index, with δ ranging in the order of 10^{-5} for hard X-rays. To realize refraction comparable to that of lenses for visible light, a multitude of lenses must be lined up; this concept of *compound refractive lenses* (CRL) was invented in the 1990s by A. Snigirev and B. Lengeler at the European Synchrotron Radiation Facility (ESRF) in Grenoble [15], and has been thriving since. Today, CRLs made out of Beryllium are found almost at every synchrotron beamline. For nano-focusing, CRLs fabricated by electron beam (e-beam) lithography in silicon have been developed by C. Schroer, and reach spot sizes down to 50 nm [14]. Next to diffractive and refractive optics, reflective optics can be implemented for hard X-rays, taking advantage of grazing-incidence total reflection or multilayer-constructive reflection. Since long, curved mirrors have been appreciated as high efficiency and non-dispersive focusing elements for synchrotron radiation. In the 1990s, with advent of 3rd generation synchrotron sources, mirror-based optics reached spot sizes in the range of 1–5 μm. With novel polishing tools for highly curved mirrors developed by the group of K. Yamauchi in Osaka [35], and alternatively of Kirkpatrick-Baez (KB) mirrors with adaptive bending as implemented by O. Hignette at ESRF [18], sub-100 nm focusing became available ten years ago. At the same time, first compound optics with two-stage focusing or collimation was implemented for hard X-rays. Using a combination of high gain KB mirrors and X-ray waveguide optics, a $25 \times 47 \, nm^2$ exit beam with clean background and high degree of coherence was demonstrated in [30]. In the course of subsequent research within CRC *Nanoscale Photonic Imaging*, waveguide optics has been significantly improved, and point focusing down to 10 nm (in the exit plane of the waveguide) is now possible. At the same time efficiency has also been significantly improved. As a result, X-ray micro- and nanofocussing can be implemented today by either diffractive (example: Fresnel zone plates), reflective (examples: Kirckpatrick-Baez mirror, waveguides) and refractive optical elements (example: compound refractive lenses), and/or combinations thereof. X-ray optics and in particular nanofocusing has been an enabling tool to extend X-ray microscopy over the recent years, in spectral range, in resolution and in contrast mechanism. This is true not only for the classical full-field scheme of transmission X-ray microscopy (TXM) which is based on objective zone plates, or scanning X-ray transmission microscopy (STXM), but also for coherent diffraction imaging (CDI) and holography, which also take advantage of X-ray focusing, even if the resolution limits are no longer limited by the focal size. Figure 3.2 illustrates the rapid development of hard X-ray focusing over the last two decades, following the advent of high brilliance (3rd generation) synchrotron radiation, which had provided the necessary coherence for diffraction-limited or near-diffraction limited focusing.

Fig. 3.2 Number of scientific articles over the last 20 years, as retrieved by a Google Scholar search using the filter X-ray AND nano-focus. The increase nicely illustrates the progress and interest in this field

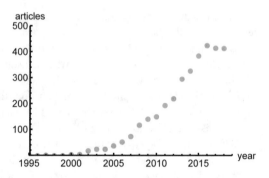

In this chapter, we give an introduction into reflective and diffractive X-ray optics, to provide basic knowledge for further chapters. For refractive optics, we refer to the excellent reviews in [36]. In Sect. 3.2, we first present the basics of X-ray reflectivity, followed by a section on mirrors (Sect. 3.3) and X-ray waveguides (Sect. 3.4). Section 3.5 then presents Fresnel zone plate (FZP) optics, and Sect. 3.6 an introduction to coherence. We close by briefly addressing compound optics and different variants of X-ray microscopes (Sect. 3.7).

3.2 X-ray Reflectivity and Reflective X-ray Optics

In Chap. 2 we have justified the use of scalar wave theory in the hard X-ray spectral range. Therefore, also Fresnel reflectivity can be accounted for simply by considering the boundary conditions of a scalar wave ψ at interfaces of layered materials. More generally, one has to differentiate between the different polarisation states. The scalar approximation holds, since the decrements of the index of refraction δ and β are much smaller than unity, and only small angles (much smaller than the Brewster angle) are relevant in X-ray reflectivity. In fact, small-angle approximation is also warranted in most cases. There are excellent treatments of X-ray reflectivity [37–39]. In this section, we follow the derivation presented in the textbook 'Elements of modern X-ray physics' by Als-Nielsen and McMorrow [37].

3.2.1 X-ray Reflectivity of an Ideal Single Interface

Consider a scalar wave with wave vector $\mathbf{k}_I$ and an amplitude a_I, impinging from vacuum onto a semi-infinite medium with a sharp interface. The reflected wave is denoted by $\mathbf{k}_R$ and a_R, and the transmitted wave by $\mathbf{k}_T$ and a_T. As boundary conditions we require the wave ψ and its derivative $\nabla\psi$ to be continuous at the interface between the two media (Fig. 3.3)

Fig. 3.3 An incoming wave $\psi_I = a_I e^{i\mathbf{k}_I \cdot \mathbf{r}}$ at an incident angle α is partly reflected under the same angle α forming the wave $\psi_R = a_R e^{i\mathbf{k}_R \cdot \mathbf{r}}$ and partly transmitted under an angle α' forming a transmitted wave $\psi_T = a_T e^{i\mathbf{k}_T \cdot \mathbf{r}}$ following [37]

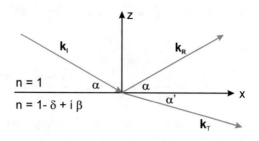

$$a_I + a_R = a_T \tag{3.1}$$

and

$$a_I \mathbf{k}_I + a_R \mathbf{k}_R = a_T \mathbf{k}_T. \tag{3.2}$$

The wave number is $k = |\mathbf{k}_{I,R}|$ in vacuum, and $nk = |\mathbf{k}_T|$ in the medium. Considering the components of the wave vector parallel and perpendicular to the surface yields

$$a_I k \cos \alpha + a_R k \cos \alpha = a_T \, nk \cos \alpha' \tag{3.3}$$

and

$$-(a_I - a_R) k \sin \alpha = -a_T \, nk \sin \alpha'. \tag{3.4}$$

From the above equations, Snell's law is obtained

$$\cos \alpha = n \cos \alpha'. \tag{3.5}$$

Approximating the cosines for small angles using $\cos \alpha = 1 - \alpha^2/2$, and $n = 1 - \delta + i\beta$, one finds

$$\alpha^2 = \alpha'^2 + 2\delta - 2i\beta$$
$$= \alpha'^2 + \alpha_c^2 - 2i\beta. \tag{3.6}$$

Here, $\alpha_c^2 = \sqrt{2\delta}$ denotes the critical angle of total external reflection from the optically thicker (here: vacuum) to the optically thinner medium. Using (3.1) and (3.4), we have

$$\frac{a_i - a_R}{a_I + a_R} = n\frac{\sin \alpha'}{\sin \alpha} \simeq \frac{\alpha'}{\alpha}. \tag{3.7}$$

With

$$\frac{1 - r}{1 + r} = \frac{\alpha'}{\alpha} \tag{3.8}$$

and $1 + r = t$, this leads to the Fresnel equations

$$r := \frac{a_R}{a_I} = \frac{\alpha - \alpha'}{\alpha + \alpha'}, \tag{3.9}$$

$$t := \frac{a_T}{a_I} = 1 + r = 1 + \frac{\alpha - \alpha'}{\alpha + \alpha'} = \frac{\alpha + \alpha' + \alpha - \alpha'}{\alpha + \alpha'} = \frac{2\alpha}{\alpha + \alpha'}, \tag{3.10}$$

where r denotes the amplitude reflectivity and t the amplitude transmission function. The intensity reflectivity is expressed by

$$R_F = \left| \frac{\alpha - \alpha'}{\alpha + \alpha'} \right|^2 = \left| \frac{q - q'}{q + q'} \right|^2, \tag{3.11}$$

where $q = 2k \sin \alpha$ and $q' = 2k \sin \alpha'$ denote the momentum transfer, which is always vertical to the interface. The reflectivity as a function of q is unity up to the critical wave vector $q_c = 2k/\sqrt{2\delta}$ (discarding absorption) and then decreases algebraically with q^{-4}. This characteristic makes X-ray reflectometry a powerful tool for the study of surfaces and interfaces of materials, since weak signals of interface disturbances can interfere with this "carrier wave", such that the signal of a single atomic layer becomes observable. The transmission function $T = |t|^2$ increases from zero to q_c, where it reaches a maximum of four (discarding absorption), and then decreases again to unity for $q \gg q_c$. The propagation angle in the medium α' is a complex number, which can hence be decomposed into

$$\alpha' = \text{Re} \left(\alpha' \right) + i \, \text{Im} \left(\alpha' \right). \tag{3.12}$$

Accordingly, the transmitted wave can be expressed by

$$a_T e^{ik\alpha' z} = a_T \, e^{ik\text{Re}(\alpha')z} \, e^{-k\text{Im}(\alpha')z}. \tag{3.13}$$

Hence, the intensity falls off with a $1/e$ penetration depth Λ given by

$$\Lambda = \frac{1}{2k \, \text{Im}(\alpha')} = \frac{1}{q_c \, \text{Im}(q')}. \tag{3.14}$$

Below the critical angle, the real term is zero and the wave is purely evanescent with a decay length which goes to $1/q_c$ for $\alpha \ll \alpha_c$. This localisation of intensity to the immediate sub-surface region is exploited in grazing incidence diffraction (GID) [40], and grazing incidence small-angle scattering (GISAXS) [41] (Fig. 3.4).

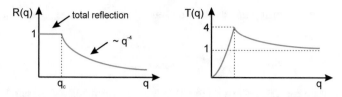

Fig. 3.4 Fresnel reflectivity R and transmission T as a function of momentum transfer q, expressed in natural (dimensionless) unit q/q_c

3.2.2 Multiple Interfaces and Multilayers

Let us first consider reflectivity in case of a sample with one layer above the substrate, still following [37]. In the following, n_0 is the index of refraction of vacuum, n_1 the index of refraction of the layer and n_2 the index of refraction of the substrate. In contrast to the case of reflection from a pure substrate, there is now a series of possible reflections:

(i) Firstly, reflection at the interface 0 to 1 (interface vacuum/layer), amplitude reflectivity is r_{01}.

(ii) Secondly, transmission at the interface 0 to 1, t_{01}, then reflection at the interface 1 to 2, r_{12}, followed by transmission at the interface 1 to 0, t_{10}. By adding this wave to the above, it is necessary to include the phase factor $p^2 = e^{iq\Delta}$, where Δ is the thickness of the layer.

(iii) Thirdly, transmission at the interface 0 to 1, t_{01}, then reflection at the interface 1 to 2, r_{12}, followed by reflection at the interface 1 to 0, r_{10}, then another reflection at the interface 1 to 2, r_{12}, finally followed by transmission 1 to 0, t_{10}. The total phase factor for this wave is p^4.

Hence, the total amplitude reflectivity is

$$
\begin{aligned}
r_{\text{layer}} &= r_{01} + t_{01}t_{10}r_{12}p^2 + t_{01}t_{10}r_{10}r_{12}^2 p^4 + t_{01}t_{01}r_{10}^2 r_{12}^3 p^6 + \dots \\
&= r_{01} + t_{01}t_{10}r_{12}p^2 \left[1 + r_{10}r_{12}p^2 + r_{10}^2 r_{12}^2 p^4 + \dots \right] \\
&= r_{01} + t_{01}t_{10}r_{12}p^2 \sum_{m=0}^{\infty} (r_{10}r_{12}p^2)^m \\
&= r_{01} + t_{01}t_{10}r_{12}p^2 \frac{1}{1 - r_{10}r_{12}p^2},
\end{aligned}
\tag{3.15}
$$

where the geometrical series has been used in the last line. Using the definitions of r and t, as presented in the previous subsection, we obtain

$$
r_{01} + t_{01}t_{10} = \frac{(q_0 - q_1)^2}{(q_0 + q_1)^2} + \frac{2q_0^2 2q_1^2}{(q_0 + q_1)^2} = \frac{(q_0 + q_1)^2}{(q_0 + q_1)^2} = 1,
\tag{3.16}
$$

with $r_{01} = -r_{10}$. Inserting this expression into the equation of r_{layer} leads to

$$r_{\text{layer}} = \frac{r_{01} + r_{12}p^2}{1 + r_{01}r_{12}p^2}. \tag{3.17}$$

The equation for the reflectivity of a thin layer (layer thickness Δ) can be further simplified for the case of identical materials on either side of the layer. In this case, $r_{01} = -r_{12}$ holds and (3.17) is simplified to

$$r_{\text{layer}} = \frac{r_{01}(1 - p^2)}{1 - r_{01}^2 p^2}. \tag{3.18}$$

While the above equation is exact, further approximation can be performed when considering an angular range where refraction can be neglected (angle sufficiently large compared to critical angle). In this case $|r_{01}| \ll 1$ ($q \gg 1$), and the amplitude reflectivity $r(q)$ can be written as

$$r(q) \approx \left(\frac{q_c}{2q}\right)^2 \tag{3.19}$$

Using these assumptions, the amplitude reflectivity of a thin layer becomes

$$r_{\text{layer}} = \frac{r_{01}(1 - p^2)}{1 - r_{01}^2 p^2} \approx r_{01}(1 - p^2) \approx \left(\frac{q_c}{2q}\right)^2 (1 - e^{iq\Delta}) \tag{3.20}$$

This can be rewritten as

$$r_{\text{layer}} = -\frac{16\pi \rho r_0}{4q^2} e^{iq\Delta/2}(e^{iq\Delta/2} - e^{-iq\Delta/2}) \tag{3.21}$$

$$= \left(\frac{16\pi \rho r_0 \Delta}{2q}\right) \frac{e^{iq\Delta/2}}{2(q\Delta/2)} - i\frac{e^{iq\Delta/2} - e^{-iq\Delta/2}}{2i} \tag{3.22}$$

$$= -i\left(\frac{4\pi \rho r_0 \Delta}{q}\right)\left(\frac{\sin(q\Delta/2)}{q\Delta/2}\right) e^{iq\Delta/2}. \tag{3.23}$$

As the equation is supposed to describe the properties of a *thin layer* (layer thickness Δ), we assume $q\Delta \ll 1$, which results in

$$r_{\text{thin layer}} \approx -i\frac{4\pi \rho r_0 \Delta}{q} = -i\frac{\lambda \rho r_0 \Delta}{\sin(\alpha)}, \tag{3.24}$$

using

$$q = \frac{4\pi \lambda}{\sin(\alpha)}. \tag{3.25}$$

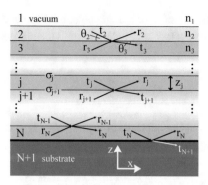

Fig. 3.5 Reflectivity of a stratified medium (e.g. multi-layerd thin film), with reflected and transmitted beams, as required for the Parratt algorithm. The reflectivity is calculated recursively from bottom to top. Each layer is parameterized with thickness t_j, roughness σ_j, and complex-valued index of refraction n_j. Profiles which are not step-wise constant can be approximated by a sequence of sufficiently thin slabs. Illustration following [39]

The expression for the reflectivity of a thin layer in (3.24) is known as the *kinematical reflectivity*. Note that this equation only holds for angles sufficiently above the critical angle.

Next, we consider multiple layers (N layers) on top of an infinitely thick substrate, still following [37]. The reflectivity can be calculated using the Parratt algorithm [42], which is based on recursion. By definition, the Nth layer is on top of the substrate (see Fig. 3.5). The z-component of the wave vector, $k_{z,j}$ in the layer denoted j is determined by the wave vector k_j and its x-component $k_{x,j}$, which is conserved through all layers $k_{x,j} = k_x$:

$$k_{z,j} = (n_j k)^2 - k_x^2 = (1 - \delta_j + i\beta_j)^2 k^2 \simeq k_z^2 - 2\delta_j k^2 + 2i\beta_j k^2. \quad (3.26)$$

The wave vector in the jth layer yields

$$q_j = \sqrt{q^2 - 8k^2\delta_j + i8k^2\beta_j}. \quad (3.27)$$

In a first step, the reflectivity is calculated for the interface of the Nth layer/substrate yielding

$$r'_{N,s} = \frac{q_N - q_s}{q_N + q_s}. \quad (3.28)$$

Note that no multiple reflections have to be taken into account, since the substrate is assumed to be infinitely thick. Then, the reflectivity at the interface Nth layer/$N-1$th is considered, which can be written as

$$r_{N-1,N} = \frac{r'_{N-1,N} + r'_{N,s}p_N^2}{1 + r'_{N-1,N}r'_{N,s}p_N^2}, \quad (3.29)$$

where the reflectivity expression for a single layer has been used. Here $r'_{j,j+1}$ denotes the reflectivity at the respective interface without considering multiple reflections, given by

$$r'_{j,j+1} = \frac{q_j - q_{j+1}}{q_j + q_{j+1}}. \tag{3.30}$$

Next, the reflectivity at the interface of layers $(N-1)$ and $(N-2)$ is calculated using

$$r_{N-2,N} = \frac{r'_{N-2,N-1} + r'_{N,N-1} p^2_{N-1}}{1 + r'_{N-2,N-1} r'_{N-1,N} p^2_{N-1}}. \tag{3.31}$$

This procedure of determining the respective reflectivities can be repeated until the total reflectivity amplitude r_{01} at the top interface 1st layer/vacuum is obtained. This iterative solution is the basis of reflectivity codes such as IMD by Windt [43]. Note that not only the intensity reflectivity as shown here, but the full fields inside the structure can be computed, by this or equivalent matrix methods with field vectors in each layer and boundary conditions at the interfaces taken into account. Typical reflectivity curves of periodic multilayers exhibit strong multilayer Bragg peaks, as well as total thickness oscillations known as *Kiessig fringes*. They reflect the interference of the reflected waves from the vacuum/layer and layer/substrate interfaces. From the period of these oscillations, the thickness of the layer can be determined.

3.2.3 Interfacial Roughness

Generalizing the results obtained for sharp or flat interface, where the density profile along z can be described by a step function, we now consider interfacial roughness, still following [37]. For real materials, we need to model a graded or rough interface. In this case, the density profile at the interface has to be modified. The density profile as a function of depth z can now be better described by an error function. Accordingly, the reflectivity of an ideal flat interface, which is given by (3.11), is modified in case of a rough interface by

$$R = R_F e^{-\sigma^2 q_z^2}. \tag{3.32}$$

We can derive this expression the following way (see [37]): First, we model the density profile of the interface by a function $\rho(z)$ which fulfills $\rho(z) \to 1$ for $z \to \infty$ and $\rho(z) \to 0$ for $z \to -\infty$ (see Fig. 3.6). Most commonly, $\rho(z)$ will be the error function (see below). Now we consider the contribution $\delta r(q_z)$ to the amplitude reflectivity $r(q)$ from an infinitesimal thin slab at depth z

$$\delta r(q_z) = -i\frac{q_c^2}{4q}\rho(z)\,dz \tag{3.33}$$

and integrate over all infinitesimal thin layers to obtain the amplitude reflectivity $r(q)$ as a superposition

$$r(q) = -i\frac{q_c^2}{4q}\int_{-\infty}^{\infty}\rho(z)\,e^{iqz}\,dz\ . \tag{3.34}$$

Using partial integration, we find

$$r(q) = -i\frac{1}{iq}\frac{q_c^2}{4q}\int_{-\infty}^{\infty}\frac{d\rho}{dz}\,e^{iqz}\,dz\ . \tag{3.35}$$

With 3.19 (limit of a perfect interface, $q \gg 1$), this yields

$$r(q) = r_F(q)\,\Phi(q), \tag{3.36}$$

using the definition

$$\Phi(q) \equiv \int_{-\infty}^{\infty}\frac{d\rho}{dz}\,e^{iqz}\,dz. \tag{3.37}$$

The function $\Phi(q)$ describes the structure of the interface in reciprocal space (when modeling $\Phi(r)$ with an error function in real space, as described below, its derivative $d\rho/dz$ will have the form of a Gaussian). The reflectivity (intensity) $R(q)$, as measured in an experiment, is described by the so-called *master formula* [37]

$$\frac{R(q)}{R_F(q)} = \left|\frac{r(q)}{r_F(q)}\right|^2 = \left|\int_{-\infty}^{\infty}\left(\frac{d\rho}{dz}\right)e^{iqz}\,dz\right|^2, \tag{3.38}$$

which not only holds for a profile broadened by roughness, but more generally for any structured interface profile, within kinematic approximation. A common choice for the density profile of the interface $\rho(z)$ is the error function erf(z) (see Fig. 3.6):

Fig. 3.6 A vertically smeared out density profile $\rho(z)$ is used to model a rough interface. Illustration following [39]

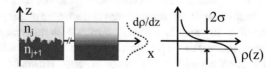

$$\rho(z) = \mathrm{erf}\left(\frac{z}{\sqrt{2}\sigma}\right). \tag{3.39}$$

The parameter σ gives a measure for the width of the graded region of the interface. This smeared out density profile can be regarded as an averaging of the rough surface. The derivative of the error function, $d\rho/dz$, is a Gaussian:

$$\frac{d\rho}{dz} = \frac{1}{\sqrt{2\pi\sigma^2}}e^{-\frac{1}{2}\left(\frac{z}{\sigma}\right)^2}. \tag{3.40}$$

Hence we obtain

$$\frac{r(q)}{r_F(q)} = \int_{-\infty}^{\infty}\frac{d\rho}{dz}(z)\,e^{iqz}\,dz = \frac{1}{\sqrt{2\pi\sigma^2}}\int_{-\infty}^{\infty}e^{-\frac{1}{2}\left(\frac{z}{\sigma}\right)^2}e^{iqz}dz. \tag{3.41}$$

By definition, the right hand side of (3.41) is the Fourier transform of $d\rho(z)/dz$. By computing the integral for the Gaussian case, we obtain

$$\frac{R(q)}{R_F(q)} = \left|\frac{r(q)}{r_F(q)}\right|^2 = e^{-q^2\sigma^2}. \tag{3.42}$$

We can now discern two cases

(i) $q_z\sigma \gg 1$, the surface is optically rough
(ii) $q_z\sigma \ll 1$, the surface is optically flat.

Therefore, X-ray reflectivity can be used to quantify the roughness of a surface or interface. More importantly in the present context, mirror roughness severely affects the focusing intensity and field distribution.

3.3 X-ray Mirrors

Reflective optics in form of planar and curved mirrors are indispensable tools for synchrotron radiation science. Mirrors are encountered in almost every beamline for rejection of harmonics, which would also fullfill the Bragg condition of the monochromator. At fixed grazing angles of incidence α_i, higher harmonics impinge above their critical angle α_c, and are hence only very weakly reflected, while the fundamental is still below its critical angle and hence has a reflectivity r close to one. Mirror optics are also often preferred as the first optical element to take the white synchrotron beam, since a large surface area under grazing incidence can be used for cooling. In many beamlines, mirrors with moderate curvature are used to focus the beam to the desired position in the experimental hutch, in particular in the horizontal direction where the divergence is large. However, this type of focusing with large mirrors and large focal distances are designed for focal beam sizes of a few mm.

Contrarily, micro- or nano-focusing for X-ray microscopy requires much shorter focal distances and much smaller radii of curvature. The most common arrangement of focusing mirrors for this purpose at synchrotron and FEL facilities is known as the Kirkpatrick-Baez (KB) mirror system, which we discuss in this section. Two major properties apply to KB focusing as to mirror optics in general: Firstly, it is non-dispersive, and hence well suited for broad bandpass or photon energy variation. Secondly, the efficiency is high since $r \simeq 1$ for $\alpha \leq \alpha_c$.

3.3.1 Kirkpatrick-Baez Geometry

A KB system consists of two crossed elliptically shaped mirrors [44], as sketched in Fig. 3.7. The mirror length is typically a factor of ten shorter than the large beamline mirrors, often around 10 cm. The mirror surface is polished to an elliptical shape. The ellipse is designed to have the first focal point at the radiation source, for example at the undulator exit, and the second at the focal plane of the experiment (sample position). Since ellipsoidal surfaces with two principle planes of curvature are difficult to fabricate, the two mirrors are elliptically curved only in one plane and are assembled perpendicular to each other. Rays are sequentially reflected off this orthogonal mirror pair, emulating a 3d ellipsoidal mirror surface.

In the design of a KB system, the following requirements must be considered. The mirrors must have

- a suitable reflectivity—so the grazing angle of incidence α is bounded by the critical angle $\alpha_c \sim$ mrad;
- a homogeneous phase of the reflected beam—so a well-shaped mirror with negligible figure errors to minimise aberrations;

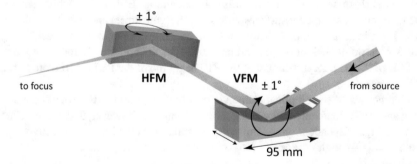

Fig. 3.7 Geometry of KB focusing. A vertically (VFM) and horizontally focusing mirror (HFM), each with elliptical shape function, are aligned behind each other in orthogonal planes. Orthogonality and Bragg angles must be carefully aligned. Fixed curvature by polishing of the substrate and/or adaptively curved mirrors are both common

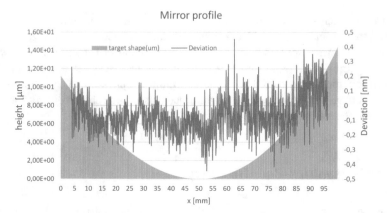

Fig. 3.8 Example of an elliptical KB mirror profile, with the surface (blue) polished to elliptical shape as a function of the position on the mirror (x-axis). The mirror is positioned at about 87 m behind the undulator at the P10 beamline of the PETRAIII storage ring [45], and has a focal length of about 300 mm. The deviations from the perfect height profile are also shown (red curve) and have a rms roughness of only $\sigma \simeq 0.1$ nm. The mirror is made of SiO_2 and coated with Rh

- a well-polished surface—to reduce scattering which leads to artefacts for example in holographic imaging.

The first point limits the numerical aperture (NA) of reflective optics; since the critical angle scales linearly with X-ray wavelength, the achievable resolution $\lambda/\vartheta_c \sim 10$ nm is approximately constant with photon energy, and only depends on the material. This length scale would then be just one example of the more general limit postulated by Bergemann et al. for *all* kinds of X-ray focussing [1], as discussed in the introduction. The second and third points have been solved by technological progress. An important break-through has been achieved by the group of Yamauchi, by the development of the elastic emission machining (EEM) [35], which enabled the fabrication of elliptical surfaces with sub-nm figure errors and few Å roughness, even for mirror lengths of 100 mm and longer. As an example of this technology, Fig. 3.8 shows the height profile and deviations for the horizonally focusing mirror (HFM) of the GINIX instrument at the P10 beamline of the PETRAIII storage ring [45].

Geometrically, the elliptical shape yields a perfect point focus, providing a constant and real-valued reflectivity along the active surface. However, under total reflection, an angle-dependent phase-shift $\varphi(\alpha)$ occurs. From the Fresnel reflectivity formula $r = \frac{\alpha - \alpha'}{\alpha + \alpha'}$ with $\alpha' = \cos^{-1}(\cos \alpha/n) \in i\,\mathbb{R}$ for $\alpha < \alpha_c$ and $n < 1$, we obtain as phase shift $\varphi(\alpha)$

$$\varphi(\alpha) = 2\tan^{-1}\left(\frac{-\sqrt{2\delta - \sin^2 \alpha}}{\sin \alpha}\right), \tag{3.43}$$

where α varies along the mirror's surface [46]. This phase gradient, $\nabla \varphi(\alpha)$, leads to a small shift of the beam. It is connected to the Goos-von Hänchen effect. Although totally reflected, an evanescent wave enters the medium to experience a small phase-lag. Numerically, this lateral shift of the focal spot is only on the order of a few nm. In addition, the index of refraction $n = 1 - \delta + i\beta$ has an imaginary part due to absorption. By ways of this imaginary component, the angle α' changes slightly, yielding a second phase contribution to $\varphi(\alpha)$. Again, the effect on the lateral position of the focal spot is in the nm range. The spot size, however, is unaffected. Hence, albeit the evanescent wave and absorption of the reflecting material, an elliptically shaped mirror operating under total external reflection provides efficient point-to-point focusing. However, since $\alpha_i \leq \alpha_c$ must be fulfilled for all points on the reflecting surface, the numerical aperture is quite limited. To overcome this limitation without severe reduction in r, multilayer (ML) coatings are used.

3.3.2 Multilayer Mirrors

For "simple" mirrors based on total external reflection, the numerical aperture is limited by the critical angle $\vartheta_c \sim 4$ mrad for hard X-rays and typical coating materials, e.g. at 14 keV and Rh coating. Hence, also the focal spot size has a lower limit of about 50 nm, if we pose reasonable bounds on all other geometrical properties. To enhance the reflectivity at higher angles of incidence, multilayer coatings with alternating high and low density layers are applied. As known from planar multilayers, the first Bragg peak assures high reflectivity at angles of incidence which can be easily a factor of ten higher than α_c, depending on the multilayer period Λ. Common materials for hard X-rays are e.g. W, Mo, Ta for the high density layers, and B_4C, C or Si for the low density layers. For a KB system, one expects that these layers and the substrate must follow the shape functions of conformal ellipses, with the X-ray source (undulator) and focal spot as the two focal points. However, due to refraction inside the multilayer structure, the layer shapes need to be slightly modified and varied across the mirror surface [47, 48]. Using such multilayer mirrors with a laterally graded layer period, it was for the first time possible to "Break the 10 nm barrier in hard X-ray focusing" [5].

In order to design optimal multilayer mirrors, e.g. for the upgraded beamline ID16a at the ESRF, an analytical treatment of dynamic X-ray diffraction inside such a graded multilayer structure in elliptical geometry has been developed in [49, 50]. Here we briefly describe this wave-optical theory of nano-focusing X-ray multilayer mirrors based on the Takagi-Taupin theory of strained crystals. The geometry and system of coordinates is shown in Fig. 3.9. As a natural choice, we use elliptical coordinates (t, s) given by

$$t := \frac{r_0 + r_1}{2}, \qquad s := \frac{r_0 - r_1}{2}, \tag{3.44}$$

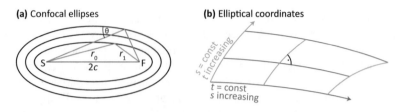

Fig. 3.9 a Multilayer coatings of KB mirrors approximately follow the shape functions of confocal ellipses. **b** Definition of the elliptical coordinates used for the Tagaki-Taupin theory

where r_0 is the distance of a point from the source S, and r_1 is the distance of this point to the focus F. We derive the Takagi-Taupin (TT) equations from the Helmholtz equation of a scalar field $\psi(s, t)$, here written not with the index of refraction n, but with the susceptibility $\chi = n^2 - 1$:

$$\nabla^2 \psi(s, t) + k^2 [1 + \chi(s, t)] \psi(s, t) = 0 . \tag{3.45}$$

For a (quasi-)periodic structure, we write the susceptibility as a truncated Fourier series to first order

$$\chi(s, t) = \chi_0 + \chi_{\bar{1}} \exp(-2ikt) + \chi_1 \exp(2ikt) . \tag{3.46}$$

Then, we decompose the field ψ into two components: the incoming wave $\psi_0 \exp(ikr_0)$ diverging from the source, and the reflected wave $\psi_1 \exp(-ikr_1)$ converging to the focus. To re-write the Helmholtz equation, we need the folowing expressions:

$$\psi(s, t) = \psi_0 \exp[ik(s + t)] + \psi_1 \exp[ik(s - t)] \tag{3.47}$$

$$\chi\psi \approx (\chi_0\psi_0 + \chi_1\psi_1) \exp[ik(s + t)] \tag{3.48}$$

$$+ (\chi_0\psi_1 + \chi_{\bar{1}}\psi_0) \exp[ik(s - t)] \tag{3.49}$$

$$\nabla^2\psi(s, t) = \alpha^2(\partial_s^2\psi) + \beta^2(\partial_t^2\psi) + \frac{t(\partial_t\psi) - s(\partial_s\psi)}{t^2 - s^2} \tag{3.50}$$

$$\alpha^2(s, t) = \frac{c^2 - s^2}{t^2 - s^2} = \cos^2\vartheta, \quad \beta^2(s, t) = \sin^2\vartheta \tag{3.51}$$

Here, ϑ is the local angle of incidence, and $2c$ the distance between source and focus.

Assuming slowly varying envelopes, $\nabla^2\psi_{0,1}(s, t) \approx 0$, and defining $u_h := k\chi_h/2$, we obtain

$$(\alpha^2\partial_s + \beta^2\partial_t)\psi_0 = i(u_0\psi_0 + u_1\psi_1) - \psi_0/2(t + s), \tag{3.52}$$

$$(\alpha^2\partial_s - \beta^2\partial_t)\psi_1 = i(u_0\psi_1 + u_{\bar{1}}\psi_1) - \psi_1/2(t - s) . \tag{3.53}$$

With α, $\beta = \text{const}$, these Takagi-Taupin equations are valid in the flat case; here, these coefficients are dependent on coordinates as given above. When applied to curved multilayer mirrors, the bilayer period Λ^B of the stacked system following Bragg's law is given by

$$\Lambda^B = \frac{\lambda}{2\sin\vartheta(s,t)} = \frac{\lambda}{2\beta}. \tag{3.54}$$

Now we take refraction of the X-ray beam due to the average index of refraction inside the ML structure into account. The modified Bragg condition then reads

$$\Lambda^{\text{mB}} = \frac{\lambda}{2\sqrt{n^2 - \cos^2\vartheta}} \approx \frac{\lambda}{2\sqrt{\beta^2 - 2\delta}}, \tag{3.55}$$

where $\delta = (\delta_1 + \delta_2)/2$ is the average decrement, assuming equal thicknesses of the bilayers. For $\vartheta_B \geq 3\vartheta_c \approx 3\sqrt{2\delta}$, a good approximation is given by

$$\Lambda^{\text{mB}} \approx (1 + \delta/\beta^2)\,\Lambda^B. \tag{3.56}$$

The increased layer thickness is accounted for by using a pseudo-Fourier series of $\chi(s,t)$, in which the exponentials are modified according to $\exp(\pm 2ikt) \mapsto \exp(\pm 2ikt(1 - \delta/\beta^2))$. Replacing further ψ_1 by $\psi_1' := \psi_1 \exp(-2ikt\delta/\beta^2)$, the modified propagation constant u_0 in the *second* TT-equation is replaced by

$$u_0' \approx u_0 - 2(\alpha^2\partial_s\varphi - \beta^2\partial_t\varphi)k, \qquad \varphi = (t - t_0)\delta/\beta^2. \tag{3.57}$$

Assuming a constant ϑ, it can be shown that $u_0' = -u_0^*$; in other words, while the *first* TT equation gives rise to a phase-lag due to refraction; the *modified second* equation now yields an anti-phase lag of the reflected wave, in fact correcting for refraction. In the curved case, the next-to-leading order term reads

$$u_0' \approx u_0 + 2\delta k\left[1 - 2\frac{\alpha^2\Delta t}{\beta^2(t+s)}\right], \qquad \Delta t := t - t_0, \tag{3.58}$$

with $t = t_0$ along the entrance surface. For realistic parameters, this curvature term leads to a small numerical correction on the required bi-layer spacing.

Reflectivity curves in dynamical diffraction are not symmetric; in particular, the peak intensity does not occur at the nominal Bragg angle. For further numerical optimisation, a scaling factor f interpolating between Bragg and modified Bragg layer spacing is now introduced; we define

$$\Lambda(f) := \Lambda^B + f \times (\Lambda^{\text{mB}} - \Lambda^B), \qquad f \in \mathbb{R}. \tag{3.59}$$

The TT system of coupled differential equations is solved numerically and for different parameters f. Figure 3.10 summarises a simulation of a ML mirror for the

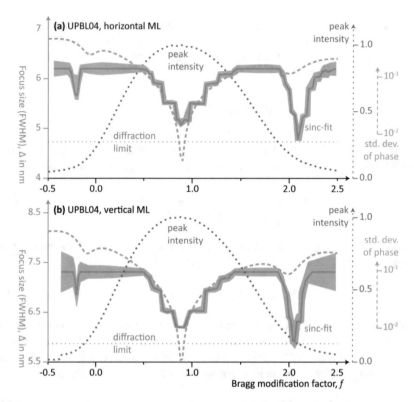

Fig. 3.10 Simulated focus sizes for the geometry of the ESRF KB nano-focus upgrade beamline: **a** horizontal ML and **b** vertical ML, as a function of Bragg modification factor f. The focus sizes (red curve with reddish error bands) have been obtained from $sinc^2$-fits to the intensity in the focus. Standard deviation of reflected phase (along the ML surface) is shown in green on a logarithmic scale; peak intensity in the focus is shown in blue. Dotted line shows the diffraction limit. Simulation has been carried out for a point-source. From [51]

ESRF beamline ID-16a. Both the focal spot size Δ (red line), peak intensity in the focus (blue points), and the standard deviation of the reflected phase (green dashes) is shown as a function of optimisation parameter f. Based on the simulations, a value of $f = 0.9$ yields the best results, and a theoretical focal spot size of about 5 nm (FWHM).

3.4 X-ray Waveguides

Compared to other spectral domains, notably that of visible and infrared light, waveguide optics is much less developed in the X-ray range. Total reflection in a thin film of low electron density surrounded by high electron density is the basis for guid-

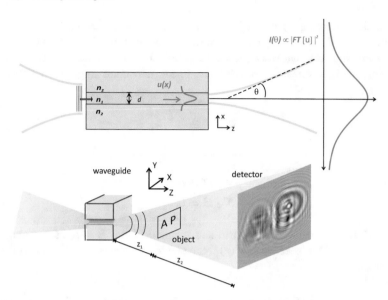

Fig. 3.11 Basic schematic of using a waveguide exit beam for holographic X-ray imaging. By selecting the waveguide to sample distance, objects can be imaged in a full field configuration and geometrically magnified holograms of nanoscale objects can be recorded. The exit field behind the sample is reconstructed by phase retrieval algorithms, which are often more robust in this regime, compared to far-field coherent diffractive imaging

ing X-ray radiation. The first waveguiding effects for X-rays used propagation in planar (straight) thin film structures [33, 52–56], followed by the development of two-dimensional channel waveguides [30, 57], which have posed significant fabrication challenges up to recently [58–60]. Progress in waveguide fabrication has led to a usable exit flux outpassing in some case 10^9 photons per second [58]. If optimized for small beam size, X-ray waveguides with beam confinement of sub-10 nm (FWHM) have also been demonstrated [32]. In the context of this volume, the use of X-ray waveguides to create a monochromatic and fully coherent secondary quasi-point source is of particular importance. This coherent point source is ideally suited for X-ray holography and coherent imaging techniques. This has resulted in (holographic) propagation imaging at unprecedented resolution and image quality [61]. Figure 3.11 illustrates the basic geometry of using X-ray waveguides to record in-line X-ray holograms.

3.4.1   Waveguide Modes: The Basics

X-ray waveguides can be treated as a special case of the general theory of electro-magnetic waveguides, as presented in the classic textbook by Marcuse [62], which we follow here. The only particularities associated with waveguiding in the X-ray regime derive from the nature of the X-ray index of refraction and the short wavelength. Adapted to the X-ray case, notation and parameterisations used here, closely follow previous original work presented in [54, 57, 63]. We start from Maxwell's equations, written down for an isotropic, linear, nonconducting, and nonmagnetic medium as

$$\nabla \times \mathcal{E} = -\frac{\partial \mathcal{B}}{\partial t}, \tag{3.60}$$

$$\nabla \times \mathcal{H} = \frac{\partial \mathcal{D}}{\partial t}, \tag{3.61}$$

$$\nabla \cdot \mathcal{D} = 0, \tag{3.62}$$

$$\nabla \cdot \mathcal{B} = 0, \tag{3.63}$$

where $\mathcal{E}$ and $\mathcal{H}$ denote the electric and magnetic field, $\mathcal{B} = \mu_0 \mathcal{H}$ the magnetic induction, and $\mathcal{D} = \varepsilon_0 \varepsilon \mathcal{E} = \varepsilon_0 n^2(r) \mathcal{E}$ the electric displacement. We then take the curl of (3.60)

$$\nabla \times (\nabla \times \mathcal{E}) = -\mu_0 \partial_t (\nabla \times \mathcal{H}) \overset{(3.61)}{=} -\mu_0 \varepsilon_0 n^2 \partial_t^2 \mathcal{E} \tag{3.64}$$

and use $\nabla \times (\nabla \times \mathcal{E}) = \nabla(\nabla \cdot \mathcal{E}) - \nabla^2 \mathcal{E}$ to obtain the source-free (homogeneous) wave equation

$$\nabla^2 \mathcal{E} - \mu_0 \varepsilon_0 n^2 \partial_t^2 \mathcal{E} = 0, \tag{3.65}$$

which holds for $\nabla \cdot \mathcal{E} = 0$. This is warranted for section-wise constant index n, or in the approximation of a weakly varying index, since in this case we can neglect $\nabla \cdot \mathcal{E} \simeq 0$, as can be verified from

$$0 = \nabla \cdot \mathcal{D} = \varepsilon_0 (\nabla n^2) \cdot \mathcal{E} + \varepsilon_0 n^2 \nabla \cdot \mathcal{E} \simeq \varepsilon_0 n^2 \nabla \cdot \mathcal{E} . \tag{3.66}$$

Analogous to (3.65), we can also obtain the wave equation for the magnetic field

$$\nabla^2 \mathcal{H} = -\mu_0 \varepsilon_0 n^2 \partial_t^2 \mathcal{H} . \tag{3.67}$$

Next we consider a solution of Maxwell's equation which has the particular form of a guided wave, i.e.

$$\mathcal{E} = \mathbf{E}(\mathbf{r}_\perp) \, e^{i(\omega t - \mathbf{k} \cdot \mathbf{r})}, \tag{3.68}$$

$$\mathcal{H} = \mathbf{H}(\mathbf{r}_\perp) \, e^{i(\omega t - \mathbf{k} \cdot \mathbf{r})}, \tag{3.69}$$

where ω is the angular frequency, $\mathbf{k} = k \mathbf{e_k}$ the wave vector with magnitude (wave number) $2\pi/\lambda$, and $\mathbf{r}_\perp = \mathbf{r} - (\mathbf{e_k} \cdot \mathbf{r})\mathbf{e_k}$ is a position vector perpendicular to $\mathbf{k}$. In other words, in a guided mode we require the field to be stationary with respect to the propagation axis, i.e. $\mathcal{E}$ and $\mathcal{H}$ are only functions of the coordinates perpendicular to

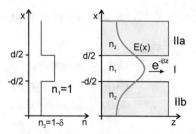

Fig. 3.12 Sketch of a simple slab waveguide geometry. The guiding layer material with high index of refraction is sandwiched by a low index surrounding material ("cladding"). For X-rays, air/vacuum with $n_1 \approx 1$ is an ideal guiding medium, and the index of refraction of the metal or semiconductor cladding is given by $n_2 = 1 - \delta$

the propagation axis. Unless they are constant (plane wave), this requires the presence of matter. More precisely, a distribution of the index of refraction $n(\mathbf{r}_\perp)$ with the translational symmetry along $\mathbf{k}$, which is chosen such that the field is guided along the propagation axis while being confined in the orthogonal direction(s). A simple example is sketched in Fig. 3.12, with z as the propagation axis, and a stepwise constant index of refraction profile $n(x)$

$$n(x) = \begin{cases} n_1 & \text{if } -d/2 < x < d/2, \\ n_2 & \text{else}, \end{cases} \qquad (3.70)$$

describing a simple planar waveguide geometry with guiding layer of refractive index n_1 and thickness d (guiding core) sandwiched between two semi-infinite cladding regions of refractive index $n_2 < n_1$. The profile function $n(x)$ parameterizes a planar waveguide with one-dimensional beam confinement (1DWG), while two-dimensional confinement would require a corresponding two-dimensional profile function $n(x, y)$, describing for example a channel waveguide (2DWG) of cylindrical or rectangular cross section. For the given geometry of a planar waveguide, we hence have

$$\mathcal{E}_j = E_j(x)e^{i(\omega t - \beta z)}, \qquad (3.71)$$
$$\mathcal{H}_j = H_j(x)e^{i(\omega t - \beta z)}. \qquad (3.72)$$

Inserting this ansatz in (3.60) and (3.61) yields six differential equations for the field components, out of which two sets are uncoupled, describing the transverse electric (TE) modes (modes without electric component in propagation direction)

$$i\beta E_y = -i\omega\mu_0 H_x, \qquad (3.73)$$
$$\partial_x E_y = -i\omega\mu_0 H_z, \qquad (3.74)$$
$$-i\beta H_x - \partial_x H_z = i\omega\varepsilon_0 n^2(x) E_y, \qquad (3.75)$$

and the transverse magnetic (TM) modes (modes without magnetic component in propagation direction)

$$-i\beta E_x - \partial_x E_z = -i\omega\mu_0 H_y, \tag{3.76}$$

$$i\beta H_y = i\omega\varepsilon_0 n^2(x) E_x, \tag{3.77}$$

$$\partial_x H_y = i\omega\varepsilon_0 n^2(x) E_z . \tag{3.78}$$

For TE modes E_x, E_z, $H_y = 0$ (for TM modes E_y, H_x, $H_z = 0$). Expressing the field components along x and z by those in y, we obtain for the two sets of modes

$$\partial_x^2 E_y + \left(k^2 n^2(x) - \beta^2\right) E_y = 0, \tag{3.79}$$

$$\partial_x^2 H_y + \left(k^2 n^2(x) - \beta^2\right) H_y = 0 . \tag{3.80}$$

These equations are sometimes denoted as reduced wave equations. For X-rays, the propagation is extremely forward directed, i.e. the internal reflections angles are on the order of a few mrad, much smaller than the Brewster angle. For this reason, the TM and TE solutions degenerate, and scalar diffraction theory holds. Correspondingly, it is sufficient to consider a single scalar field ψ. In fact, we could also start directly from the scalar wave equation

$$\nabla^2\psi - \frac{n^2}{c^2}\partial_t^2\psi = 0 \quad \forall \psi \in \{E_x, E_y, E_z, H_x, H_y, H_z\}, \tag{3.81}$$

with $c = \frac{1}{\sqrt{\varepsilon_0\mu_0}}$ the speed of light, and $n = \sqrt{\varepsilon_r\mu_r}$ the refractive index, keeping in mind that the components are in general not independent. For forward directed propagation of X-rays, the scalar wave equation (3.81) is, an excellent approximation. The field ψ can be written as a superposition of monochromatic fields Ψ_ω (spectral decomposition)

$$\psi = \psi(\mathbf{r}, t) = \frac{1}{2\pi} \int_0^\infty \psi_\omega(\mathbf{r}) \, e^{-i\omega t} \, d\omega, \tag{3.82}$$

if stationary quasi-monochromatic waves are considered, i.e. $\psi \to \psi_\omega$ and $n \to n_\omega$. In this case time dependence is harmonic

$$\psi(\mathbf{r}, t) = U(\mathbf{r})e^{-i\omega t}, \quad u(\mathbf{r}) \in \mathbb{C}, \tag{3.83}$$

and we can write down a differential equation only for the complex amplitude $U(\mathbf{r})$ by inserting (3.83) in (3.82)

$$\left(\nabla^2 U(\mathbf{r})\right) e^{-i\omega t} + \frac{n^2}{c^2}\omega^2 U(\mathbf{r}) \, e^{-i\omega t} = 0 \tag{3.84}$$

$$\Rightarrow \nabla^2 U(\mathbf{r}) + n^2(\mathbf{r})k_0^2 \, U(\mathbf{r}) = 0 . \tag{3.85}$$

Equation (3.85) is the scalar Helmholtz equation (HE). Here, the wave number in vacuum is given by $k_0 = \frac{\omega}{c} = \frac{2\pi}{\lambda}$. Another notation is $k(\mathbf{r}) = n(\mathbf{r})k_0$ where $k(\mathbf{r})$ is the wave number in a medium. To simplify the notation, we will write k for k_0 in the following, and refer k to the absolute value of the wave number in vacuum. The component k_z will be denoted as $k_z = \beta$. Again, we use a separation ansatz for guided modes

$$U(x, z) = u(x) e^{-i\beta z}, \tag{3.86}$$

where β is called the propagation constant. Insertion in (3.85) then yields the one-dimensional reduced wave equation for $u(x)$,

$$\frac{d^2}{dx^2} u(x) + \left(n^2(x)k^2 - \beta^2\right)u(x) = 0, \tag{3.87}$$

which has the same form as (3.80). Hence we can either first work with Maxwell's equation and then use scalar approximation later in the reduced Helmholz equation, or start from the scalar wave equation, and arrive at the same result. Note that in order to have β real, we have to assume the refractive index to be real and thus at least initially ignore absorption. More generally, the modes will also be affected by the imaginary part of the index, but in practice it is sufficient to treat absorption *a posteriori* by an effective (weighted) absorption coefficient for each mode. Even though this does not matter in scalar approximation, $u(x)$ could be taken to represent the horizontal component of the electric field, considering the so-called transverse electric (TE) modes of the waveguide. This requires $u(x)$ to be continuous at the interfaces. Furthermore, for guided modes we require the field to vanish far inside the cladding, i.e., $u(x)$ must approach zero in the limit $|x| \rightarrow \infty$. For symmetric potential functions (here: index of refraction profiles), the eigenfunctions (modes) have defined parity, i.e. are symmetric or anti-symmetric. A general form of a symmetric function which solves 3.87 and which does not diverge, is given by

$$u(x) = \begin{cases} A \cos(\kappa x) & \text{if } 0 < |x| \le d/2 \\ C\, e^{\gamma|x|} & \text{if } |x| > d/2 \end{cases} \tag{3.88}$$

where A and C are constants. Requiring $u(x)$ and its derivative $u'(x)$ to be continuous functions at $x = \pm d/2$, we get

$$A \cos(\kappa d/2) = C\, e^{-\gamma d/2}, \tag{3.89}$$

$$-A\kappa \sin(\kappa d/2) = -\gamma C\, e^{-\gamma d/2}. \tag{3.90}$$

Dividing (3.90) by (3.89), we obtain a transcendental equation

$$\kappa \tan\left(\kappa \frac{d}{2}\right) = \gamma. \tag{3.91}$$

Symmetric modes are found by solving this transcendental equation. For anti-symmetric modes

$$u(x) = \begin{cases} B \sin(\kappa x) & \text{if } 0 < |x| \le d/2, \\ C\,e^{\gamma|x|} & \text{if } |x| > d/2, \end{cases} \tag{3.92}$$

we have correspondingly

$$-\kappa \cot\left(\kappa\frac{d}{2}\right) = \gamma. \tag{3.93}$$

Using the definitions

$$\xi := \frac{\kappa d}{2} = \sqrt{k^2 n_1^2 - \beta^2}\,\frac{d}{2} \tag{3.94}$$

and

$$V := \sqrt{n_1^2 - n_2^2}\,\kappa d \approx \sqrt{2\delta_2 - 2\delta_1}\,\kappa d \tag{3.95}$$

the transcendental equation can be rewritten as

$$\xi \tan(\xi) = \sqrt{\frac{V^2}{4} - \xi^2} \tag{3.96}$$

for symmetric modes and

$$-\xi \cot(\xi) = \sqrt{\frac{V^2}{4} - \xi^2} \tag{3.97}$$

for antisymmetric modes, respectively. The transcendental equation determines a discrete set of modes ξ_m, with $0 \le m \le N-1$. The total number of guided modes N is given by

$$N = \left\lceil \frac{V}{\pi} \right\rceil, \tag{3.98}$$

where $\lceil \rceil$ denotes the Gauss bracket (rounding to the next integer). The recipe to compute a mode, is then to solve the transcendental equation, and to compute the parameters in sequence $\xi_m \to \kappa_m \to \beta_m \to \gamma_m$, to obtain $u(x)$. The smallest ξ_0 which solves (3.96) determines the fundamental mode

$$u_0(x) = \begin{cases} \cos(\kappa_0 x) & \text{if } |x| < d/2 \\ \cos(\kappa_0 d/2)e^{-\gamma_0(|x|-d/2)} & \text{else.} \end{cases}$$

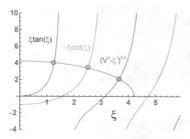

Fig. 3.13 Graphical representation of the transcendental mode equation with the waveguide parameter V. Each intersection corresponds to a mode. Symmetric modes are indicated in blue, antisymmetric modes by in orange. For given $V = 8.44$, the waveguide supports $N = \lceil \frac{V}{\pi} \rceil = 3$ modes

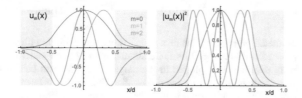

Fig. 3.14 Mode amplitudes (left) and intensities (right), corresponding to the solution of the transcendental equation shown in Fig. 3.13. The cladding is shaded in gray

In order to interpret a mode in a geometric optical picture it is helpful to consider the complete field in the guiding layer, e.g. of a symmetric mode, with mode envelope $u(x) = \cos(\kappa x)$ (Fig. 3.14)

$$U(x, z) = u(x)\, e^{i(\omega t - \beta_m z)} = \frac{A}{2} \left(e^{i(\omega t - \beta_m z - \kappa x)} + e^{i(\omega t - \beta_m z + \kappa x)} \right) . \qquad (3.99)$$

The right hand side corresponds to two internally reflected plane waves (guided by total reflection), or beams in the geometric optical model, with wave vectors $\mathbf{k}$

$$\mathbf{k}_\pm = \begin{pmatrix} \beta \\ 0 \\ \pm \kappa \end{pmatrix}, \qquad (3.100)$$

as sketched in Fig. 3.17.

3.4.2 Coupling and Propagation

The modes $u_m(z)$ are eigenfunctions of the waveguide potential. For a rectangular profile they consist of a sine or cosine term with $m + 1$ antinodes in the guiding layer,

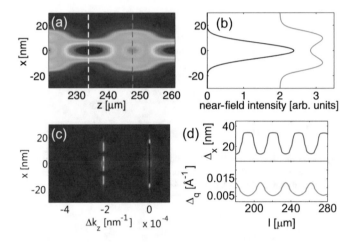

Fig. 3.15 **a** Multi-modal propagation inside a planar waveguide with thin film sequence Ge/Mo [$d_i = 30$ nm]/C [$d = 35$ nm]/Mo[$d_i = 30$ nm]/Ge, simulated for 17.5 keV and front coupling (plane wave). The intensity distribution is plotted in the range of 221–261 μm behind the waveguide entrance, showing a mode beating along the propagation direction z. **b** Intensity profiles corresponding to the dashed lines in **a** illustrating the interference due to multi-modal propagation. **c** A Fourier transformation with respect to z reveals both the shape of the guided modes (vertical profile), and the propagation constant (proportional to the horizontal offset). **d** FWHM of the simulated near-field distribution (top) and far-field distribution (bottom) as a function of z. Adapted from [32]

and an exponentially decaying evanescent wave in the cladding, as derived above. For more general potential shapes, the mode function can also be found numerically by integration via Numerov's method (shooting method) [64]. For given geometry and boundary conditions, propagation can be calculated by finite difference (FD) calculations as presented in Chap. 2, and the different modes can be dissected by means of Fourier transformation along z, see Fig. 3.15. Neglectingmodes and the corresponding interference effects is well described by linear combination of all N guided modes

$$\psi(x, z) = \sum_{m=0}^{m_{\max}} c_m u_m(x) \exp\left(-i\beta_m z\right). \tag{3.101}$$

In front-coupled waveguides, the coefficient c_m is given by an overlap integral of the incident field ψ_{in} and u_m [29, 52]

$$c_0 = \frac{1}{||u_m||^2} \int \psi_{\text{in}}(x) u_m(x)\, dx \ .$$

Absorption can be accounted for by a factor $\exp\left(-\mu_{\text{eff},m} x\right)$ in the right hand side of equation (3.101), with an "effective linear absorption coefficient" $\mu_{\text{eff},m}$ given by a mode-weighted average of the absorption coefficient profile $\mu(x)$ [65]

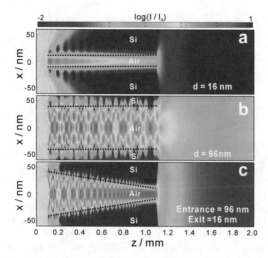

Fig. 3.16 Field intensity distribution (logarithmic color code) in **a** a nearly mono-modal, **b** a multi-modal air/silicon waveguide, simulated for plane wave incoming 8keV radiation with unit intensity. Due to higher absorption of the $m = 1$ mode, only the fundamental mode $m = 0$ persists in the $d = 16$ nm guide, while $N = 5$ modes propagate in the $d = 96$ nm guide. **c** By tapering the exit intensity can be increased, and single mode radiation is achieved at the exit. The intensity gain between **a** and **c** for same exit width is directly visible

$$\mu_{\text{eff},m} = \frac{1}{||u_m||^2} \int |u_m(x)|^2 \mu(x) \, dx.$$

For a vacuum guide, only the intensity fraction in the cladding contributes to the absorption of the mode. The transition from multi-modal to mono-modal regime as a function of guiding layer thickness d is illustrated by Fig. 3.16a, b. Note that d can also be tapered along the optical axis as in (c) to concentrate the field. Instead of coupling from the side, a beam can also be coupled in through the cladding, via the so-called resonant beam coupler (RBC) geometry, see Fig. 3.17a. In this case, modes can be excited selectively, even if the waveguide support multiple modes. Figure 3.17 also shows a simulation depicting the position of a waveguide in the focal plane of a KB-mirror. By computing the propagation for different incoming realisations of the (stochastic) field, the guiding and filtering of a waveguide can be studied [66].

3.4.3 Fabrication and Characterisation of X-ray Waveguides

To isolate a guided X-ray beam with a cross section down to about 10 nm close to the fundamental limit [1], long channels are needed with aspect ratios (length to width) in the range of $10^4 - 10^6$, depending on the photon energy E and cross section d. This is because the radiation entering at the sides of the over-illuminated channel

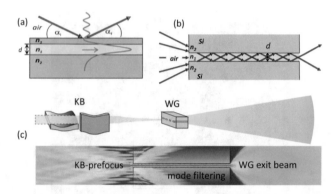

Fig. 3.17 Sketch of different coupling geometries. **a** Resonant beam coupling (RBC). The waveguide is illuminated by an incoming plane wave under grazing incidence with α_i tuned to a mode. Modes can be exited in an interval $\alpha_c^{core} \leq \alpha_i \leq \alpha_c^{cladd}$, where α_c^{core} and α_c^{cladd} denote the critical angles of total reflection for waveguide core and cladding, respectively. The mode is exited via an evanescent wave in the top cladding. **b** Front coupling scheme, illustrated in a ray-optical picture, with the mode formed by up and down reflected rays in the guiding core, according to (3.100). **c** Finite difference simulation in coupling a pre-focused beam (KB mirrors) into a silicon-air waveguide, propagation and out coupling, adapted from [66]

entrance (radiative modes) has to be absorbed in the cladding material. Not only the small cross section, but also the high aspect ratios impose a significant challenge in fabrication. Waveguide structures for one-dimensional beam confinement by planar waveguides (1DWG) are easily obtained by thin film deposition techniques, but most applications require two-dimensional waveguides (2DWG). Using guiding channels of polymer structured by e-beam lithography and coated with metal or semiconductor cladding, 2DWGs were first realized in [57] and later improved by [30]. An alternative fabrication scheme based on dry etching of channels into silicon wafers and subsequent capping by wafer bonding makes it possible to employ an empty guiding core (air or vacuum) and hence to minimize absorption in particular for lower photon energies [60]. This has enabled a waveguide exit flux on the order of 10^8 ph/s (P10 beamline of the PETRA III storage ring of DESY [67].

Figure 3.18 illustrates the fabrication of waveguide channels in silicon by e-beam lithography and subsequent wafer bonding, according to [60]. A spin-coated poly-methyl-methacrylate (PMMA) is used as positive e-beam resist. The desired pattern of an array of waveguide channels is written by moving an interferometric laser stage below a stationary electron beam, in order to achieve the required channel length (of a few mm's) without stitching errors. The developed resist then provides the etching mask for pattern transfer into the semiconductor substrate by reactive ion etching (RIE). Subsequently, the mask is removed and the channels a capped by a second wafer via hydrophilic wafer bonding [60]. An alternative fabrication scheme has been demonstrated in [31], where two planar waveguides (1DWG), which each confine the beam in an orthogonal direction, were combined in a crossed geometry to form an effective two-dimensional quasi-point source for holographic imaging. This crossed

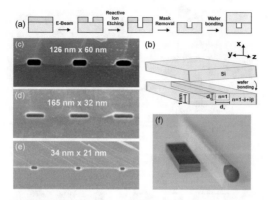

Fig. 3.18 Fabrication of lithographic waveguides. **a** Sketch of waveguide processing sequence: resist deposition, e-beam exposure, reactive ion etching, mask removal, and finally wafer bonding. **b** Schematic of air-filled channel capped by a top wafer bonded to the substrate. **c, d, e** SEM micrographs of waveguide channel entrances. **f** Photograph of waveguide chip as cut by the wafer dicing machine. From [60]

two-dimensional waveguide (c2DWG) scheme is compatible with fabrication by thin layer deposition. Hence, smaller guiding layers, a wider range of materials, and more complex layer sequences can be realized, including a two-component cladding optimized for high transmission [68]. Using for example an interlayer made of Mo, placed between the guiding core (C), and a high absorption cladding (Ge), this scheme provides excellent waveguides for the photon energy range between the Ge L-edges and the Mo K-edge, see Fig. 3.19. Figure 3.20 shows the measured far-field pattern of a Mo/C/Mo c2DWG system with guiding layer thickness $d = 35$ nm. The far-field exhibits a relatively uniform intensity distribution in the center along with a characteristic arrangement of fringes in the tails. The large divergence reflects the small focal width of the waveguide as quantified reconstruction of the near-field intensity distribution by the error reduction (ER) algorithm [31]. The calculation of the field's auto-correlation function by Fourier transformation of the far-field intensity can be used as a verification, since its width should give the value as the auto-correlation of the ER result.

3.4.4 Advanced Waveguide Configurations

Waveguide optics enables a variety of optical functions, such as filtering, confining, guiding, coupling or splitting of beams. Advanced X-ray waveguides now begin to exploit such advanced functionalities, beyond simple filtering the mode structure of a synchrotron beam, which is already well established. Based on an array of waveguide channels, *X-ray optics on a chip* has been proposed in [69]. Beam concentration by tapering [58], guiding beams around a bent [69], and beam splitting

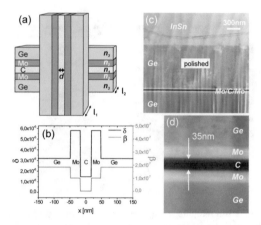

Fig. 3.19 Crossed planar waveguides. **a** Schematic. **b** Profiles Re(n) and Im(n) of the index of refraction $n = 1 - \delta + i\beta$, for photon energy $E = 17.5$ keV. Transmission of the guided modes is increased by the high δ but relatively low β of Mo. **c** Scanning electron microscopy (SEM) image (magnification 52.85 kx) showing the Mo/C/Mo layers in between Ge. The In52Sn48 alloy serves as bond material to an additional Ge cap wafer. **d** SEM micrograph 200 kx magnification. From [31]

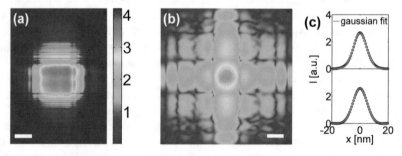

Fig. 3.20 X-ray waveguide beam with cross section at around 10 nm. **a** Fraunhofer far-field diffraction pattern of a beam exiting a crossed two-dimensional waveguide system (c2DWG) consisting of Mo/C/Mo layers, recorded with a pixel detector (Pilatus, Dectris Inc.) at a distance of about 5 m behind the waveguide exit ($E = 13.8$ keV, logarithmic scale, scale bar 0.02 Å^{-1}, 100 s dwell time). The two orthogonal slices had a guiding layer thickness of 35 nm, and a thickness of $l = 490$ μm (vertical slice: $l_1 = 270$ μm, horizontal slice: $l_2 = 220$ μm). A maximum (output) photon flux of 1.0×10^8 ph/s in the c2DWG beam was achieved by focusing a KB beam onto the waveguide entrance. **b** The near-field intensity distribution in the effective focal plane, obtained by inverting the diffraction pattern based on phase reconstruction by the error reduction (ER) algorithm (logarithmic scale, scale bar 20 nm). A high beam confinement in the effective confocal plane is achieved by multi-modal interference. **c** Line scans with corresponding Gaussian fits yield a FWHM of 10.7 and 11.4 nm in horizontal and vertical direction, respectively. Adapted from [32]

for nano-interferometry [59], have also been demonstrated. In contrast to refractive or diffractive optics, X-ray waveguides are non-dispersive and can thus support broader bandpass. An advanced fabrication scheme with improved lithography, etching and wafer bonding steps has now paved the way to develop this field further [59].

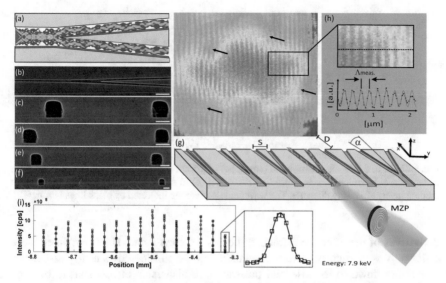

Fig. 3.21 Minitaturized beam splitter based on X-ray waveguides. **a** Finite differences simulation of a beam splitter. **b** Top view SEM image of a splitting structure before wafer bonding. Scale bar denotes $1\,\mu m$. **c–f** SEM images of the exit side of beam splitters with different spacings S. Scale bars denote $100\,nm$. **g** Schematic of the experimental geometry showing the coupling of the focused X-ray beam into the entrance of the beam splitter, the subsequent guiding in the two channels, the free space propagation behind the chip, and finally the far-field detector at a distance D. The far-field pattern shows the characteristic double slit interference pattern, modulated with features of the waveguide modal structure. Arrows mark bifurcations in the interference fringes (fork-shaped structures). Length and angles are not to scale. **h** Enlarged view of the interference pattern with a sinusodial fit to the intensity oscillations. **i** Scan in y-direction indicating the position of different beam splitters which have all been defined on the same chip with different geometric parameters, and which can be selected by translating the chip in the FZP focus. Detailed scan profile of a single channel with a width (FWHM) of $282.6\,nm$ giving an upper limit for the beam size in the horizontal direction. From [59]

Multiplexed beamlets can be particularly useful for coherent imaging [70], and possibly also X-ray quantum optical experiments [71]. As an example, the function of a waveguide beam splitter is illustrated in Fig. 3.21.

X-ray waveguides are also promising optical devices for the emerging field of ultra-fast X-ray optics at free electron laser (FEL) and higher harmonic generation (HHG) sources, since they support nearly dispersion-free pulse propagation down to ultra-short pulse width in the range of $0.1\,fs$ [72]. FEL or HHG beam splitters with attosecond delay would be orders of magnitude smaller than macroscopic pulse delay stages. Spatial and temporal splitting of a pulse into two reflected beams, one displaced along the surface with respect to the other, can be also achieved by X-ray waveguides in resonant beam coupling geometry, based on a giant Goos-Hänchen effect [73]. As shown above for the stationary case, propagation is described by a finite number of guided modes, each with its own propagation constant and effective absorption index. The propagation of a short pulse is therefore governed by the effective dispersion and group velocity of the excited modes, which depend on the

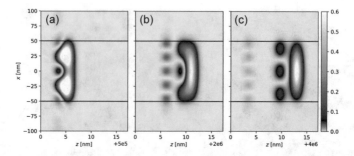

Fig. 3.22 a, b, c Intensity distribution (envelope) of a 5 attosecond beam propagating at different distances $z \approx 0.5$ mm (**a**), $z \approx 2$ mm (**b**), $z \approx 4$ mm (**c**) in a silicon slab waveguide (air/vacuum guiding layer) of 100 nm diameter. The waveguide's edges are indicated by black lines. From [72]

derivatives of the effective refractive indices for each mode. However, since these differ only very slightly, X-ray waveguides can be considered as nearly dispersion free optics down to femtosecond pulses, while dispersion effects start to become visible in form of mode separation only for attosecond pulses [72]. An example of pulse propagation in X-ray waveguides is shown in Fig. 3.22. A 12 keV pulse width of 5 attoseconds is simulated in a planar silicon (slab) waveguide with vacuum guiding layer of $d = 100$ nm. The modes separate spatially by a few nm after several mm of propagation distances. Even if the pulse spectrum covers an absorption edge of the cladding material, modal dispersion could would manifest itself only for a pulse width of 0.3 fs, according to simulations by time-dependent finite difference propagation in [72].

3.5 Diffractive Optics and Zone Plates

In this section, we first recall the basic theory of Fresnel zone plate (FZP) optics, and then present different approaches of FZP fabrication. With the advent of improved fabrication techniques, smaller zones can now be achieved. However, this also required advanced optical design concepts and numerical methods for simulation, as presented in the last part of this section. Here we limit the discussion to the experimentally relevant case of *binary* zone plates, which are fabricated from two different materials; one of low and one of high density. The low density material can also be air or vacuum.

3.5.1 Basic Theory of Fresnel Zone Plates

We assume a plane wave of wavelength λ propagating along the optical axis and impinging on a circular aperture. The wave shall be focused to a point a distance

f downstream the aperture. This focused wave is given as a sector of a spherical wave. The focus is formed by constructive interference of waves transmitted through rings around the optical axis, with radius $R = f + n\lambda/2, n \in 2\mathbb{N}_0$. Rings with $n \in 2\mathbb{N}_0 + 1$, on the other hand, would interfere destructively. The rings (or annuli) of different n form the so-called Fresnel Zones. These zones form concentric circles with radii

$$r_n = \sqrt{n\lambda f + \left(\frac{n\lambda}{2}\right)^2}. \tag{3.102}$$

For $n \ll f/\lambda = \mathcal{O}(10^7)$ for typical X-ray zone plates, the second term can be neglected. If now the "odd zones" with $n \in 2\mathbb{N}_0 + 1$ are blocked out in the aperture, the remaining waves interfere constructively in the focal spot. By Babinet's principle, blocking the "even zones" will lead to the same intensity. An optical device which focuses light by absorbing light from the opaque rings is called an *absorbing Fresnel Zone Plate*. By blocking light in some areas, a bright spot appears on the optical axis. Jean-Auguste Fresnel was the first to obtain this result from calculation, as an extension of the optical phenomenon of Arago's spot. As straightforward calculation shows, however, the focusing efficiency of such an absorbing FZP is limited to $1/\pi^2 \approx 10\%$ only (Fig. 3.23).

Proposed by Lord Rayleigh in 1888, and first demonstrated by Wood ten years later, phase-reversing zones increase the efficiency to 40%. Instead of absorbing every other zone by a thick material, a relative phase-shift of π is introduced. At hard X-ray energies of e.g. $E = 12.4\,\text{keV}$, it is challenging to achieve a full phase-shift of π. For example, for iridium with $n = 1-2.19 \times 10^{-5}$, an optical thickness of $2.28\,\mu\text{m}$ would be required. We discuss fabrication techniques and their advantages in the next subsection. The efficiency in the general case of a mixed absorbing/phase shifting zone plate follows further below.

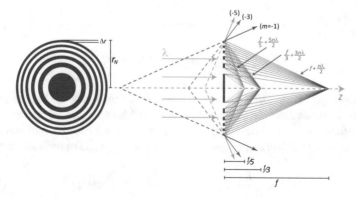

Fig. 3.23 Schematic of a Fresnel zone plate (left) in the aperture plane, and (right) in a plane containing the optical axis. Different positive and negative diffraction orders m are obtained by positive interference. From [74]

Higher diffraction orders: Apart from the nominal focus at a distance f from the zone plate, higher orders at distances f/m, $m \in \mathbb{N}$, occur. Generalizing the zone plate law, the zone radii can be written as

$$r_n^2 \approx \frac{n}{m}\lambda f, \qquad n \in m\mathbb{N}_0. \tag{3.103}$$

If now $m = 2$, then $n/m \in \mathbb{N}_0$, and the neighbouring zones interfere destructively. So there is no focal spot at $f/2$ (in the thin zone plate approximation; the even spots can appear by volume effects, see later). This argument also holds for higher even numbers. For odd numbers, e.g. $m = 3$, the condition for constructive interference is partly fulfilled for most of the zones. Hence, higher-order focal spots at $f/3$, $f/5$ etc. appear.

Negative diffraction orders: In the binary zone plates constructed as above, spherical waves converge onto the focal spot (and its higher order siblings). Applying the symmetry operations of time-reversal and inversion, however, also diverging waves are supported by the condition of constructive interference. Apart from the positive focal spots at f/m, also "negative orders" virtually emerging from spots located along the optical axis at $-f/m$ appear. These yield purely diverging waves. Usually, the negative and higher orders are blocked by a pinhole, the order sorting aperture (OSA).

Efficiency: In 1974, J. Kirz has presented a thorough treatment of Fresnel zone plates for soft X-rays, including the case of imaging at finite distances. Also, instead of purely absorbing or phase-shifting zones, the general case for a material with $n = 1 - \delta + i\beta$ was studied. Introducing the ratio $\eta := \beta/\delta$, and using Fresnel integrals, the intensity of the first pair of zones can be written as

$$I = |A_0 + A_1|^2 = \frac{1}{2\pi}\int_0^\pi e^{is}\, ds + \frac{1}{2\pi}e^{-2\pi\beta t/\lambda}\int_\pi^{2\pi} e^{i(s-2\pi t\delta/\lambda)}\, ds \tag{3.104}$$

$$= \frac{1}{\pi^2}\left(1 + e^{-4\pi\beta t/\lambda} - 2e^{-2\pi\beta t/\lambda}\cos(2\pi t\delta/\lambda)\right) \tag{3.105}$$

$$= \frac{1}{\pi^2}\left(1 + e^{-2\eta\varphi} - 2e^{-\eta\varphi}\cos\varphi\right), \tag{3.106}$$

with wavelength λ, optical thickness t and $\varphi := 2\pi t\delta/\lambda$. For all higher pairs of zones, the integrals evaluate to the same result, which can hence be regarded as the overall efficiency. We can deduce that even orders are not excited, and that higher (and negative) odd orders m are suppressed by a factor $1/m^2$. For optimal efficiency,

$$0 = \frac{\partial I_1}{\partial \varphi^*} \propto 1 + e^{-2\eta\varphi^*} - 2e^{-\eta\varphi^*}\cos\varphi^*; \tag{3.107}$$

for $\eta \to 0$, φ^* approaches π. The optimal optical thickness t^* can be calculated as $t^* = \varphi^*\lambda/(2\pi\delta)$.

3.5.2 Fabrication Techniques

X-ray microscopy has for long been limited by the difficulties of fabricating high resolution and high quality X-ray lenses, notably Fresnel zone plates. In fact, soft X-ray microscopy started with FZP fabrication by holographic laser illumination, pioneered at Institut für Röntgenphysik by G. Schmahl and colleagues in the 1960s and 1970s [76–78]. Subsequently, this fabrication technique was replaced by e-beam lithography in the 1980ies, achieving a lateral resolution which was no longer limited by visible light. The different steps of FZP fabrication by e-beam lithography are illustrated in Fig. 3.24. Major challenges were both in the writing process, e.g. a suitable pattern generator, write-field limitations, and interferometric positioning minimizing stitching errors, as well as in the structure transfer by reactive ion etching (RIE). Continuous efforts have pushed the limits towards the 10 nm range for soft X-rays [34]. For hard X-rays, however, fabrication with larger aspect ratio (zone height to depth ratio) required to achieve the necessary phase shifts becomes much more demanding. Nevertheless, by seminal work of C. David and his group at the Paul-Scherrer-Institut diffractive optics is today also established in the hard X-ray regime. Special fabrication techniques such as zone-doubling have helped to increase the aspect ratio [79], and progress has cumulated in record focal spot sizes down to 17 nm (point focus) [80]. To push beyond these values, diffractive optics must be fabricated by thin film deposition and subsequent dicing. With magnetron sputtering (MS), for example, large thin films can be grown on a flat substrate. Two materials, one optically "thin" and one "thick", can be deposited alternatively; this yields so-called multilayer Laue lenses (MLLs) of virtually unlimited size [24, 25, 81]. Tens

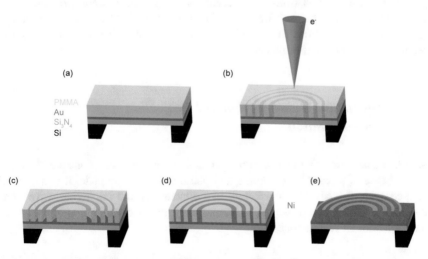

Fig. 3.24 Different steps in FZP fabrication by e-beam lithography. **a** A resist film is deposited by spin-coating. **b** The FZP pattern is written by the e-beam. **c** The illuminated resist is developed, leaving behind a pattern of circular trenches. **d** Metal (e.g. Ni) is grown by electrochemical methods. Electric conductivity is assured by the thin Au layer below. **e** The remaining resist is removed by solvent. From [75]

of thousands of bi-layer can be deposited with high accuracy. The final lens is then prepared by cutting out a slice of desired optical thickness using a focused ion beam (FIB) facility. Two such lamellae can then be used in series to form a two-dimensional focus. For a benchmark study with sub-10 nm point focus, see [23].

Contrarily, thin film deposition on a wire is called multilayer zone plate (MZPs). This goes back to an old idea [82, 83], which was also first implemented by magnetron sputtering (MS) and subsequent dicing [84]. This sputter-slice technique, however, was in most cases hampered by cumulative roughness, and the dicing also introduced severe artifacts. Only in recent years, these difficulties could be overcome by use of pulsed laser deposition (PLD). By this approach, the group of U. Krebs in Göttingen demonstrated cumulative smoothening of roughness [85] and was able to grow smooth multilayers with ultrathin layers. Using a FIB, the final lamella can be precisely cut to the desired optical thickness. Aspect ratios of one to several thousands can be achieved [86], and MZP optics has been implemented for hard X-ray energies in the broad range from 8 keV up to above 100 keV [87]. Figure 3.25a shows a sketch of MZP fabrication by PLD. An intense laser pulse is focused onto the target material (not shown), which then evaporates. A plasma plume forms, from which gas atoms are deposited on the substrate. Smoothing is favored by highly energetic particles with kinetic energies of up to 100 eV, resulting in high mobility and enhanced diffusion on the substrate surface. Advanced focused ion beam (FIB) cutting and manipulation protocols yield well positioned and mounted MZPs [86, 88]. For the MZP shown in Fig. 3.25b, a computer controlled KrF excimer laser (wavelength of 248 nm) was used with pulse duration of 30 ns and repetition rate of 10 Hz. The laser beam was focused onto the different targets in ultrahigh vacuum of about 10^{-8} mbar. The targets were moved constantly following an algorithm that allows uniform ablation from different directions. The films were grown at room temperature at a target-to-substrate distance of 6.5 cm [88]. The latest generation of lenses are fabricated from Ta_2O_5 and ZrO_2. For more information, see the progress report in the second part of this book.

3.5.3 Diffractive Optics Beyond the Projection Approximation

Above, we have described the working principle of optically thin zone plates. In the general case of a partially absorbing and phase-shifting zone plate, it is modelled as a complex-valued phase mask τ in two dimensions; the impinging wave-front ψ is modulated by this phase mask. Numerically, this is calculated as a pixel-wise multiplication of two matrices:

$$\psi' := \tau\psi, \qquad \psi'_{i,j} := \tau_{i,j} \cdot \psi_{i,j}. \tag{3.108}$$

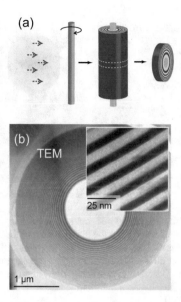

Fig. 3.25 **a** Schematic of thin film deposition on a rotating wire, e.g. by pulsed laser deposition. The zone plate is subsequently diced by a focused ion beam to the desired optical length. Adapted from [75]. **b** Transmission electron micrograph of a MZP lens, consisting of alternating thin layers of ZrO_2 and Ta_2O_5 fabricated by PLD and FIB on a rotating pulled glass wire with diameter $2r_0 = 1.2\,\mu m$. The diameter is $D = 3.2\,\mu m$, the outer-most zone width is $dr_{81} = 10.0\,nm$, the focal length for the photon energy $E = 18\,keV$ was $f = 470\,\mu m$ [27]

In the soft X-ray regime, where the optical thickness of FZPs is usually on the order of a few hundred nanometres, this model can usually be justified. We define the zone plate Fresnel number F_{ZP} as

$$F_{ZP} := \frac{(\Delta r_N)^2}{\lambda t} \qquad (3.109)$$

with outermost zone width $\Delta r_N = r_N - r_{N-1}$, wavelength λ, and thickness t. For $\Delta r_N \geq 30\,nm$, $\lambda \approx 3\,nm$ and $t \leq 300\,nm$ as an example of a soft X-ray FZP, $F_{ZP} \geq 1$; hence the treatment of a thin zone plate based on the projection approximation is completely adequate. For hard X-rays, however, we easily achieve $\Delta r_N = 5\,nm$, $\lambda = 0.1\,nm$ and $t = 5\,\mu m$, resulting in $F_{ZP} = 0.05$. This gives a clear indication that diffraction effects within the FZP itself have to be accounted for. More specifically, the kinematic or Born approximation of single diffraction at the phase mask τ has to be replaced by dynamical diffraction theory. For such optically thick optics, volume effects have to be taken into account.

A Takagi-Taupin based theory for MLL optics has been derived by Yan et al. [89], extending previous dynamical treatments denoted as coupled wave theory [2, 90, 91]. Here we briefly summarise their model and findings of [89]. The derivation is similar to that presented above for multilayer mirrors (MLMs) and starts with the Helmholtz equation of a scalar or vector field amplitude that interacts with the

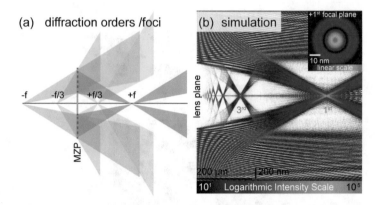

Fig. 3.26 a Schematic of diffraction orders and multi-order focusing. Positive orders are shown in red, negative in green. The orders result from the binary nature of the zone plate. For FZP imaging, order sorting apertures can be used to select only the $m = +1$ diffraction order. Alternatively, algorithmic schemes to dissect different orders can be used, which is, however, still challenging [27]. The reconstruction of the focal field can be achieved by iterative projection algorithms [4, 27]. **b** Simulation of the focused intensity, on a logarithmic false-color representation along the optical axis; compare to (**a**). From [27]

pseudo-periodic susceptibility $\chi(\mathbf{r})$. For MLMs, the Fourier series can be truncated after one term, and only two beams (incoming and reflected) are considered. MZPs, on the other hand, show multiple diffractive orders, and hence multiple beams and more Fourier orders need to be taken into account. A further complication arises since $\chi(\mathbf{r})$ is not a simple periodic function, but changes according to the zone plate law. Nevertheless, Yan and co-workers argue that the zone plate can be considered as a "strained crystal" with a varying d-spacing of $d = 2\Delta r_n$. They use the coordinate transformation (Fig. 3.26)

$$x \mapsto \frac{T}{\lambda}\left(\sqrt{x^2 + f^2} - f\right), \tag{3.110}$$

where T is the period of the new, unstrained and fully periodic lattice. The transformation yields a phase-factor

$$\exp(i\,\phi_h) := \exp\left(ikh(\sqrt{x^2 + f^2} - f)\right) \tag{3.111}$$

where h is the index of the diffractive order under consideration. Decomposing the field $\mathbf{E}$ into components $\mathbf{E}_h$, and using the a truncated series expansion for χ, yields a set of coupled partial differential equations describing the system. Within the Takagi-Taupin formalism, the gradient of the phase ϕ_h, is equal to the local reciprocal lattice vector:

$$\rho_h := \frac{2\pi h}{d}\hat{e}_x = \nabla\phi_h. \tag{3.112}$$

This shows that apart from the geometrical considerations for constructive interference discussed above, zone plates can also be considered as crystal optics where the local d-spacing is chosen such that all diffracted beams of a certain order h point to the same focal spot. When volume diffraction occurs, the diffracted X-ray beams are disturbed significantly within the structure. Geometrically speaking, a ray diffracted at the entrance of the zone plate at a specific zone would enter another zone while traversing the zone plate. Then, multiple diffraction will not follow Bragg's law. To match the diffraction angles, the originally parallel zones have to be varied along the optical axis. To this end, Yan et al. discuss tilted, wedged, and curved zones. Based on their computations, a focusing efficiency of 67% at sub-1 nm spot sizes at a photon energy of 19.5 keV is possible using MLLs fabricated from Si and WSi_2 [89].

3.6 Basic Coherence Theory and Simulations for X-ray Optics

Coherence of light beams refers to their ability to exhibit interference effects. Already in the first interference experiment of light, the famous double slit-experiment of 1801, Thomas Young discussed the "visibility of fringes", which today is referred to as the degree of coherence $\gamma(\mathbf{r}_1, t_1, \mathbf{r}_2, t_2)$ between two time-space-points $(\mathbf{r}_1, t_1)$ and $(\mathbf{r}_2, t_2)$. Whenever light waves emerging from these two points superimpose at a third point, the total intensity $I_{1,2}$ in general differs from the sum of the single intensities, $I_{1,2} \neq I_1 + I_2$. This is immediately clear since Maxwells' equations are linear in the *amplitudes* u, but not the intensity $|u|^2$

$$u_{1,2} = u_1 + u_2$$
$$I_{1,2} = |u_1 + u_2|^2 = |u_1|^2 + |u_2|^2 + 2\gamma_{1,2}|u_1||u_2| \neq |u_1|^2 + |u_2|^2.$$

In the following, we first give the basic definitions of coherence functions from literature; afterwards, we will shortly discuss an analytical treatment for synchrotron radiation. Using a stochastic model suitable for the numerical treatment of partially coherent propagation of light through various optical elements and samples, we will show that X-ray waveguides can indeed be used as coherence filtering devices.

3.6.1 Basic Definitions

Consider a scalar wave-field with complex amplitude $u(\mathbf{r}, t)$. Then the *intensity* $I(\mathbf{r}, t)$ at the space-time-point $(\mathbf{r}, t)$ can be defined as

$$I(\mathbf{r}, t) := u^*(\mathbf{r}, t)\, u(\mathbf{r}, t), \tag{3.113}$$

and the *mutual intensity* $\Gamma(\mathbf{r}_1, t_1, \mathbf{r}_2, t_2)$ between the given two space-time-points as

$$\Gamma(\mathbf{r}_1, t_1, \mathbf{r}_2, t_2) := u^*(\mathbf{r}_1, t_1)\, u(\mathbf{r}_2, t_2). \tag{3.114}$$

Note that higher-order correlations involving more than two space-time points can be defined, but are rather uncommon in practice. The temporal Fourier transform $W_{1,2} = \int d\tau\, \Gamma \exp(i\omega\tau)$ of the mutual intensity is known as the *cross-spectral density*. Note that the indices $(1, 2)$ are often used for notational simplification to denote the two space-time-points. The *normalized mutual intensity*

$$\gamma(\mathbf{r}_1, t_1, \mathbf{r}_2, t_2) := \frac{\Gamma(\mathbf{r}_1, t_1, \mathbf{r}_2, t_2)}{\sqrt{I(\mathbf{r}_1, t_1)\, I(\mathbf{r}_2, t_2)}} \tag{3.115}$$

is a measure of the cross-correlation of the fields at two different points in space and time. For stationary signals, γ depends only on the time difference $\tau = t_2 - t_1$ and is also denoted as the complex degree of coherence. Further, for quasi-monochromatic waves, it is sufficient to consider the mutual intensity at the same-time $t_1 = t_2$, since the time-harmonic variation of the fields is trivial. This *same-time mutual intensity* depends only on the spatial coordinates of two points

$$J(\mathbf{r}_1, \mathbf{r}_2) := \langle u^*(\mathbf{r}_1)\, u(\mathbf{r}_2)\rangle_T, \tag{3.116}$$

where $\langle \ldots \rangle_T$ denotes the time-average over at least a period T, or for practical purposes the illumination time of the experiment. The mutual intensity J contains all information about measurable intensities. Finally, the *normalized same-time mutual intensity* is defined as

$$j(\mathbf{r}_1, \mathbf{r}_2) := \frac{J(\mathbf{r}_1, \mathbf{r}_2)}{\sqrt{J(\mathbf{r}_1, \mathbf{r}_1)\, J(\mathbf{r}_2, \mathbf{r}_2)}}. \tag{3.117}$$

We use the normalized quantity j, if we are interested in the visibility of interference fringes, for example, not the absolute intensity values. If one considers a Young's double slit experiment with quasi-monochromatic light and two slits at points $\mathbf{r}_1$ and $\mathbf{r}_2$, the emitted spherical waves yield an interference pattern, with the fringe visibility (i.e. the Michelson contrast of the fringes) given by $|j|$. For $|j| < 1$ we call the light field *partially coherent*, whereas $|j| = 1$ and $|j| = 0$ denote fully coherent and incoherent light, respectively. These limiting cases can in fact never be completely realized.

In many practical problems, we are furthermore primarily interested in evaluating j in a plane orthogonal to the optical axis, e.g. to study coherence in the plane of an optic, sample or detector. The coherence properties in any one of such planes, however, evolve as the beam propagates. For matter of concreteness, let z be the optical axis, and let y denote the lateral direction of interest. For simplicity, we drop the dependence on x, and consider z as a parameter. In view of interference, we are often interested in the field correlation between point y_1 and another point y_2 at a lateral distance $y_2 = y_1 + d$. For linear optical systems which are characterized by lateral shift invariance, the degree of coherence is homogeneous in planes perpendicular to the optical axis, and hence

$$j_z(d) := j((y, z), (y + d, z)).$$ (3.118)

For experiments it is then sufficient, for example, to fix one point to the optical axis and to measure $j(d) := j((0, z), (d, z))$. What can we say about the functional dependence of $j(d)$? We expect the correlation to decrease with distance d (must not always be the case!), and would like to associate a characteristic length ξ to the decay of $j(d)$. For the simplest case of an incoherent field in plane $z = 0$ and paraxial propagation, it can be shown that the degree of coherence is given by a Fourier transform of the intensity in the source plane (*van Cittert-Zernike theorem*)

$$j_z(d) = \int_{-\infty}^{\infty} dy' \, I(y') \, \exp\left(i \frac{2\pi}{\lambda z} y' d \right),$$ (3.119)

i.e. we can easily predict how the coherence evolves by propagation. By ways of this Fourier relationship, we see that the *spatial coherence length* scales as

$$\xi_\perp = \frac{\lambda z}{2s},$$ (3.120)

where s is the source size. Correspondingly, ξ/z defines a "coherence angle". Note that a precise definition of ξ would require us to be more precise about the cut-off value, to which j would be allowed to decrease, as well as more information or assumption on the source distribution $I(y')$. However, this may all be incorporated into a prefactor. As an example, let us consider the 3rd generation synchrotron source PETRA III with a horizontal source size of $\sigma = 36$ nm (1σ) in the low β sections. For $\lambda = 0.1$ nm the coherence angle is $\vartheta_{coh} \sim 2\,\mu$rad. Typical optics, however, accept a beam angle of about $5\,\mu$rad, and the full beam has an opening angle of $100\,\mu$rad. One is thus facing the situation of rather reduced partial coherence, and small entrance slits or pinholes are required for coherent imaging or photon correlation spectroscopy experiments. For more information and the useful Gaussian shell model (GSM) to describe coherence properties of synchrotron beams, we refer to [92, 93]. Note that synchrotron sources are actually not fully incoherent; the radiation by the ultra-relativistic electron beam with $\gamma = \left(1 - (v/c)^2\right)^{-1/2} \sim 10^4$ is confined to a small cone with opening $\vartheta \sim 10^{-4}$ rad. This already yields a considerable correlation "already in the source plane", which can be incorporated in the GSM by an additional parameter. Nevertheless, at the position of experiments, ξ is dominated by propagation. While $\xi_\perp$ is the length scale characteristic for the correlations across a wavefront, coherence along the direction parallel to the optical axis is characterized by the *longitudinal or temporal coherence length*

$$\xi_z = \frac{\lambda^2}{2\Delta\lambda},$$ (3.121)

where $\Delta\lambda$ is the spectral width. This relationship follows from the Wiener-Khinchin theorem.

3.6.2 Stochastic Model

How can we model the emission of "chaotic sources"? Such sources do not emit plane waves, but rather wave-trains of finite duration, and hence of finite bandwidth. In addition, each emitter acts on its own, independent of its neighbours. Say the temporal duration of a wave-train is τ_{train}. But during the acquisition time $\tau_{\mathrm{acq}} \gg \tau_{\mathrm{train}}$, a detector registers light from many subsequent wave-trains, each of which has been emitted with a random phase. While a single wave-train may give rise to a fully coherent interference pattern, the patterns of different wave-trains are out of phase and hence shifted spatially. Over sufficiently long times, the patterns wash out, the interference fringes vanish. Usually, for monochromatized synchrotron radiation the relative bandwidth is on the order of 10^{-4} (common Si-111 monochromator). At X-ray frequencies of $\omega \sim 10^{18}\mathrm{s}^{-1}$, this yields correlation times of $\tau_c \sim 10^{-14}\mathrm{s}$ or shorter. This is of course well below the response time of typical detectors and shorter than the pulse duration. At X-ray free electron laser sources, however, extremely short pulses can be produced, and we can expect full temporal coherence. The finite temporal or longitudinal coherence time directly translates to the corresponding length $\xi_z = c\tau_c \sim 1\,\mu\mathrm{m}$, and hence also the largest possible path length difference between two interfering waves is still much larger than molecular length scales.

Based on this simple picture of finite wave trains with stochastic phases varying in time and space, we can introduce a simple stochastic model to treat partial coherence numerically. We replace the continuous extended source of size D by a set of N independent point-sources. The field by each point-source can be propagated numerically through an optical system, e.g. by solving the Fresnel-Kirchhoff integral for focussing mirrors, or the paraxial wave-equation for waveguides. For each point-source, a complex-valued field in some region of interest is thus obtained, denoted by $u_n(x, y)$ for the nth point-source and given in a two-dimensional area (the generalisation to three dimensions is obvious). We define a single stochastic *realisation* as the random superposition

$$u_{\mathrm{speckle}} := \sum_n w_n\, c_n\, u_n \qquad (3.122)$$

with random phase factors $c_n = \exp(i\varphi_n)$, $\varphi_n \in (0, 2\pi]$, and with weighting factors w_n corresponding to the intensity envelope of the source. A single realisation corresponds to the interference pattern of a short wave-train with coherence time τ_c; the pattern of a long exposure time $\tau \gg \tau_c$ is modelled for the time-average $\langle \ldots \rangle_\tau$ over superpositions with random phase coefficients c_n. Numerically, a few thousand realisations should be taken into account. From such an ensemble, we calculate the degree of coherence j as

$$j_{1,2} := \frac{\langle u_1^* u_2 \rangle}{\sqrt{\langle u_1^* u_1 \rangle \langle u_2^* u_2 \rangle}} = \frac{\langle u_1^* u_2 \rangle}{\sqrt{I_1 I_2}}. \tag{3.123}$$

3.6.3 Coherence Propagation and Filtering

Using the stochastic model, the partially coherent X-ray intensity and the degree of coherence can be propagated through an optical system. Here we briefly illustrate this with an example of mode propagation in X-ray waveguides, which can be used as coherence filters [66]. We also address nano-focusing of partially coherent radiation, using the example of the KB focus of the GINIX endstation at the P10 beamline of the PETRAIII storage ring at DESY [45, 51, 94]. Figure 3.27a shows the simulated intensity distribution $I(x, y) = J((x, y), (x, y))$ in the focal region (colour-coded) for X-ray photon energy $E = 7.9$ keV, and the geometric parameters of the horizontally focusing mirror (HFM). The green curves show iso-lines of the degree of coherence $|j(d, z)| = |j(x = 0, z, x = d, z)|$ along the optical axis. As can be seen, only the central part of the intensity distribution is coherent. In fact, the spatial coherence drops to 0.5 for separations largerthan the predicted coherence length ξ, which in the focal plane is 74 nm, while the FWHM beam size is 220 nm. Using Talbot interferometry, a beam size of 203 nm (FWHM of a Gaussian fit) and a coherence length of 0.37×203 nm $= 75$ nm have been measured [94]. To filter the coherent part of the illumination, an X-ray waveguide can be placed into the focal spot. In the simulation, a $D = 50$ nm guiding layer (vacuum) in Si is illuminated by independently propagated point-sources, and the complex valued amplitudes are propagated using the paraxial methods described above. On this set of basic fields u_n, a stochastic ensemble is performed. The result is shown in Fig. 3.27b. Again, the colour-code shows the partially coherent intensity $I(x, z) = J(x, z, x, z)$; as can be seen, only the central part of the focused beam is coupled into the guiding layer. As the iso-lines of coherence along the optical axis show, the beam is now fully coherent. This effect has also been demonstrated experimentally [94].

Using X-ray waveguides, it is also possible to directly measure the intensity distribution, by scanning them through the focal plane. This is of interest for characterisation of X-ray nano-focus instruments. Figure 3.28 shows results obtained in a characterisation of the KB optics at the NanoMAX beamline at MAX IV synchrotron in Lund. With a high dynamical range of larger than $1 : 10^4$, it is even possible to resolve interference patterns due to slit scattering of the KB entrance slit. From the visibility of these fringes, coherence properties can be quantified. This finite, or partial, coherence reduces the visibility of interference fringes. The resulting intensity pattern in the focal plane has been calculated based on an analytical model. Together with the measured profiles, the degree of coherence was measured as a function of (secondary) source size.

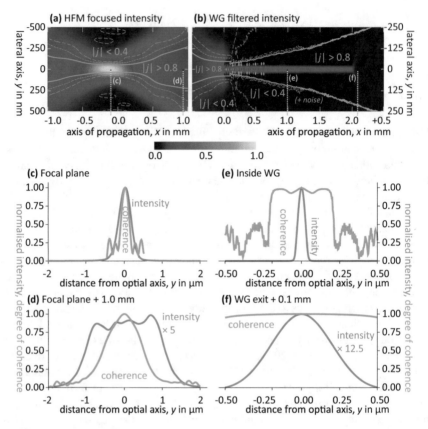

Fig. 3.27 Focus and coherence properties of KB focused (left) and WG filtered (right) X-ray beams: **a** shows the color-coded intensity distribution in the focal spot of the KB mirror, with isolines of the degree of coherence ($j = \{0.4, 0.6, 0.8\}$); in **b**, the focused field is coupled into a WG. Plots **c–f** are cuts of intensity (red) and coherence (green) at the indicated positions (focal spot, after 1 μm; at the center of the WG, after 0.1 μm). From [66]

3.7 Putting It All Together: Optics and X-ray Instrumentation

As a last note, we want to stress the importance of integrating X-ray optics into an instrument, be it an X-ray microscope, diffractometer or spectrometer. Notwithstanding the importance of its individual optical components, such as mirrors, waveguides, CRLs or Fresnel zone plates, the larger challenge is to put it all together into a fully working synchrotron instrument and beamline. The tasks are many: optical design and simulation, instrument control, radiation safety, precision, as well as data acquisition and management. Unlike other analytical techniques, instrumentation development is still largely in the hands of research groups, rather than commercial providers. The diversity of SR instruments is impressive. For a comprehensive view, we refer to the online documentation provided by almost all SR facilities, the beam-

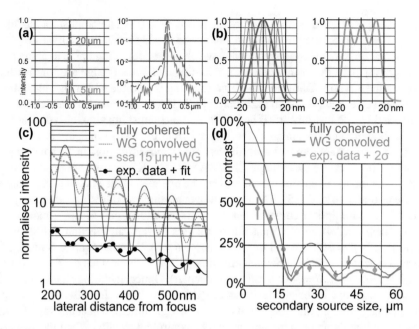

Fig. 3.28 Characterisation of focus and lateral coherence of a hard X-ray nanoprobe (NANOMAX beamline, MAXIV synchrotron). The focal intensity distribution is measured by scanning a waveguide through the focus. The coherence properties are quantified by analysis of the fringe visibility of the KB tails. **a** Two measured focus profiles of the KB mirror for partially coherent setting (secondary source size 20 μm, blue dashed curve) and quasi-coherent setting (secondary source size 5 μm, red line), shown on linear (left) and logarithmic (right) scale. **b** Intensity profiles of the three guided waveguide modes (left), and superposition of the three modes (right). **c** Simulated and experimental focus tails, for different coherence settings. The blue line shows the simulated intensity in a fully coherent setting; the red dashed line is convolved with the waveguide mode structure; the green pointed line is calculated for a secondary source size of 15 μm (and convolved with the waveguide guiding channel). The black circles represent experimental data for a secondary source size of 15 μm, with a sinusoidal fit shown as a black line. Experimental data has been shifted by a factor of 1 : 3.5 vertically for clarity. **d** Fringe contrast, quantifying the degree of coherence, as a function of the secondary source size. The blue thin line is simulated for ideal sampling, while the red thick line accounts for the convolution with the WG modes. The orange circles show contrast fits for experimental data; orange lines correspond to 2σ error bars of the fit. From [95]

line articles of IUCr's *Journal of Synchrotron Radiation*, and the proceedings of the recent international conference on *SR instrumentation*. Instead of even trying to give an overview, we present an example: the Göttingen Instrument for Nano-Imaging with X-rays (GINIX) at the P10 coherence beamline of the PETRA III storage ring at DESY in Hamburg [45]. The instrument has been designed based for nanoscale focusing using compound optical systems. This can be either the combination of a focusing and a filtering step [30, 31, 67], or by two sequential focusing steps [4, 27, 81].

Figure 3.29 illustrates (a) the optical path of the P10 beamline, with its components and respective distances from the undulator source, as well as (b, c) the nano-focus optics and sample stage of the instrument. GINIX comprises a modular compound nano-focus optical system, composed of a high gain fixed curvature (KB) mirror and an X-ray waveguide module, which is used for holographic imaging and tomography [61]. For scanning SAXS recordings [98], a soft-edge aperture is used to clean the KB tails. Ultimate sub-10 nm focusing is possible with a high resolution scanning stage, equipped with MZP optics [27]. Three major imaging modalities are supported by the instrument: (i) near-field phase contrast imaging, also denoted as in-line holographic imaging, (ii) far-field coherent diffractive imaging (CDI) with ptychographic phase retrieval, and (iii) scanning nano-diffraction, in the small angle or wide-angle regime (scanning SAXS/WAXS). The KB mirrors are positioned at ~85 m behind the undulator source, and can be operated in the photon energy range between 6 and 14 keV [45]. The two orthogonal mirrors with Rh coating are polished to fixed elliptical curvature are each 100 mm long, and accept a maximum beam size of $\simeq$0.4 mm. The first mirror focuses in vertical direction (VFM) with focal length $f = 302$ mm and incidence angle $\theta = 3.954$ mrad (mirror center), the second mirror focuses in horizontal direction (HFM) with focal length $f = 200$ mm and incidence angle $\theta = 4.05$ mrad (mirror center). Optical metrology has confirmed a surface quality with height deviation of $\leq$0.5 nm peak-to-valley. To reduce beam-induced degradation, the KB system is operated in a ultra high vacuum vessel. The beam size impinging onto the mirrors can be controlled by slits. When the slits are opened, the mirrors are operated under conditions of partial coherence, since the

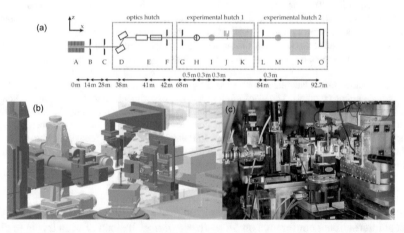

Fig. 3.29 The GINIX endstation, installed in the second experimental hutch (EH2) of the PETRAIII/P10 beamline at DESY. **a** Schematic of the beamline's optical path, with (A) undulator, (B) primary slits, (C) secondary slits, (D) double crystal monochromator, (E) horizontal mirrors, (F) girder system with slits, (G) slits, (H) fast shutter, (I) monitor, (J) attenuators, (K) experimental setup in EH1, (L) slits, (M) monitor, (N) experimental setups in EH2 (GINIX or diffractometer), (O) rear detector bench. From [96]. **b** CAD drawing, showing (from right to left) the KB mirror tank (light grey), the hexpod with the waveguide optics (dark grey), the mounting for the cleaning apertures (dark grey top), online optical microscopes for alignment and inspection (yellow, brown), and the tomography sample stage (blue). **c** Photograph of the instrument

geometric acceptance exceeds the spatial coherence length. However, at the expense of flux density, one can select the coherent fraction by closing the slits in front of the KB [99]. The focus then becomes diffraction limited, i.e. fully coherent, which is important for coherent diffractive imaging (CDI) and ptychography. Depending on orbit parameters, slit settings and alignment status, focal spot sizes down to about ≃200 nm × 200 nm (FWHM, as measured by waveguide scans) can be achieved with a flux larger than 10^{11} ph/s [94]. Coherent illumination of the mirrors by closing the slit in front of the KB results in focal spot sizes focus sizes in the range of

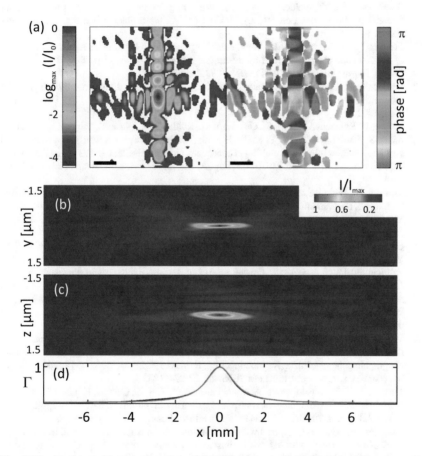

Fig. 3.30 Ptychographic reconstruction of the KB probe at 13.8 keV with KB entrance slits 100 μm × 100 μm, reconstructed from data recorded by the Lambda detector and without pinhole (dataset 3 in [97]). **a** Amplitude and phase are drawn according to the colorbar next to the image. Vertical and horizontal line-cuts through the intensity in the focal plane yield FWHM = 217 nm and FWHM = 136 nm, respectively. **b, c** Intensity distribution as calculated from numerical propagation of the reconstructed field, in the **b** xy and **c** the xz planes. **d** Normalized sharpness as obtained from an area integral of the squared intensity, along with a Lorentzian fit, yielding a depth of focus (DOF) of 1.53 mm (FWHM). Scale bar in **a** 500 nm. From [45]

150–500 nm [97, 99–101]. A ptychographic probe reconstruction is shown in Fig. 3.30. Different ptychographic and multi-plane probe reconstruction methods have been compared in [102]. As one can see, quite a bit of effort is required to realize optics in the "vacuum limit"!

References

1. Bergemann, C., Keymeulen, H., van der Veen, J.F.: Focusing X-ray beams to nanometer dimensions. Phys. Rev. Lett. **91**(20), 204,801 (2003). http://link.aps.org/abstract/PRL/v91/e204801
2. Pfeiffer, F., David, C., van der Veen, J.F., Bergemann, C.: Nanometer focusing properties of Fresnel zone plates described by dynamical diffraction theory. Phys. Rev. B **73**, 245,331 (2006). https://doi.org/10.1103/PhysRevB.73.245331
3. Schroer, C.G., Lengeler, B.: Focusing hard X-rays to nanometer dimensions by adiabatically focusing lenses. Phys. Rev. Lett. **94**(5), 054,802 (2005). https://doi.org/10.1103/PhysRevLett.94.054802
4. Döring, F., Robisch, A., Eberl, C., Osterhoff, M., Ruhlandt, A., Liese, T., Schlenkrich, F., Hoffmann, S., Bartels, M., Salditt, T., Krebs, H.: Sub-5 nm hard x-ray point focusing by a combined Kirkpatrick-Baez mirror and multilayer zone plate. Opt. Express **21**(16), 19311–19323 (2013). https://doi.org/10.1364/OE.21.019311, http://www.opticsexpress.org/abstract.cfm?URI=oe-21-16-19311
5. Mimura, H., Handa, S., Kimura, T., Yumoto, H., Yamakawa, D., Yokoyama, H., Matsuyama, S., Inagaki, K., Yamamura, K., Sano, Y., Tamasaku, K., Nishino, Y., Yabashi, M., Ishikawa, T., Yamauchi, K.: Breaking the 10 nm barrier in hard-X-ray focusing. Nat. Phys. **6**(2), 122–125 (2010). https://doi.org/10.1038/nphys1457
6. Chen, Y.T., Lo, T.N., Chiu, C.W., Wang, J.Y., Wang, C.L., Liu, C.J., Wu, S.R., Jeng, S.T., Yang, C.C., Shiue, J., Chen, C.H., Hwu, Y., Yin, G.C., Lin, H.M., Je, J.H., Margaritondo, G.: Fabrication of high-aspect-ratio Fresnel zone plates by e-beam lithography and electroplating. J. Synchrotron Radiat. **15**(2), 170–175 (2008). https://doi.org/10.1107/S0909049507063510
7. Mohacsi, I., Vartiainen, I., Rösner, B., Guizar-Sicairos, M., Guzenko, V.A., McNulty, I., Winarski, R., Holt, M.V., David, C.: Interlaced zone plate optics for hard X-ray imaging in the 10 nm range. Sci. Rep. **7**, 43,624 (2017). https://doi.org/10.1038/srep43624
8. Vila-Comamala, J., Diaz, A., Guizar-Sicairos, M., Mantion, A., Kewish, C.M., Menzel, A., Bunk, O., David, C.: Characterization of high-resolution diffractive X-ray optics by ptychographic coherent diffractive imaging. Opt. Express **19**(22), 21,333–21,344 (2011). http://www.opticsexpress.org/abstract.cfm?URI=oe-19-22-21333
9. Yun, W., Lai, B., Krasnoperova, A.A., Di Fabrizio, E., Cai, Z., Cerrina, F., Chen, Z., Gentili, M., Gluskin, E.: Development of zone plates with a blazed profile for hard x-ray applications. Rev. Sci. Instrum. **70**(9), 3537–3541 (1999). https://doi.org/10.1063/1.1149956
10. Lengeler, B., Schroer, C.G., Richwin, M., Tümmler, J., Drakopoulos, M., Snigirev, A., Snigireva, I.: A microscope for hard x-rays based on parabolic compound refractive lenses. Appl. Phys. Lett. **74**(26), 3924–3926 (1999). https://doi.org/10.1063/1.124225
11. Lengeler, B., Tümmler, J., Snigirev, A., Snigireva, I., Raven, C.: Transmission and gain of singly and doubly focusing refractive x-ray lenses. J. Appl. Phys. **84**(11), 5855–5861 (1998). https://doi.org/10.1063/1.368899
12. Patommel, J., Klare, S., Hoppe, R., Ritter, S., Samberg, D., Wittwer, F., Jahn, A., Richter, K., Wenzel, C., Bartha, J.W., Scholz, M., Seiboth, F., Boesenberg, U., Falkenberg, G., Schroer, C.G.: Focusing hard x rays beyond the critical angle of total reflection by adiabatically focusing lenses. Appl. Phys. Lett. **110**(10), 101,103 (2017). https://doi.org/10.1063/1.4977882

13. Schroer, C.G., Kuhlmann, M., Hunger, U.T., Gönzler, T.F., Kurapova, O., Feste, S., Frehse, F., Lengeler, B., Drakopoulos, M., Somogyi, A., Simionovici, A.S., Snigirev, A., Snigireva, I., Schug, C., Schröder, W.H.: Nanofocusing parabolic refractive x-ray lenses. Appl. Phys. Lett. **82**(9), 1485–1487 (2003). https://doi.org/10.1063/1.1556960
14. Schroer, C.G., Kurapova, O., Patommel, J., Boye, P., Feldkamp, J., Lengeler, B., Burghammer, M., Riekel, C., Vincze, L., van der Hart, A., et al.: Hard x-ray nanoprobe based on refractive x-ray lenses. Appl. Phys. Lett. **87**(12), 124,103 (2005). https://doi.org/10.1063/1.2053350
15. Snigirev, A., Kohn, V., Snigireva, I., Lengeler, B.: A compound refractive lens for focusing high-energy X-rays. Nature **384**(6604), 49–51 (1996). https://doi.org/10.1038/384049a0
16. Engström, P., Fiedler, S., Riekel, C.: Microdiffraction instrumentation and experiments on the microfocus beamline at the esrf. Rev. Sci. Instrum. **66**(2), 1348–1350 (1995). https://doi.org/10.1063/1.1145971
17. Hignette, O., Cloetens, P., Lee, W.K., Ludwig, W., Rostaing, G.: Hard X-ray microscopy with reflecting mirrors status and perspectives of the ESRF technology. J. Phys. IV France **104**, 231–234 (2003). https://doi.org/10.1051/jp4:200300068
18. Hignette, O., Cloetens, P., Rostaing, G., Bernard, P., Morawe, C.: Efficient sub 100 nm focusing of hard x-rays. Rev. Sci. Instrum. **76**(6), 063709 (2005). https://doi.org/10.1063/1.1928191, http://link.aip.org/link/?RSI/76/063709/1
19. Ice, G.E., Chung, J.S., Tischler, J.Z., Lunt, A., Assoufid, L.: Elliptical x-ray microprobe mirrors by differential deposition. Rev. Sci. Instrum. **71**(7), 2635–2639 (2000). https://doi.org/10.1063/1.1150668
20. Iida, A., Hirano, K.: Kirkpatrick-Baez optics for a sub-μm synchrotron X-ray microbeam and its applications to X-ray analysis. Nucl. Instrum. Methods Phys. Res. Sect. B Beam Interact. Mater. At. **114**(1–2), 149–153 (1996). https://doi.org/10.1016/0168-583X(96)00138-3
21. Mimura, H., Morita, S., Kimura, T., Yamakawa, D., Lin, W., Uehara, Y., Matsuyama, S., Yumoto, H., Ohashi, H., Tamasaku, K., Nishino, Y., Yabashi, M., Ishikawa, T., Ohmori, H., Yamauchi, K.: Focusing mirror for x-ray free-electron lasers. Rev. Sci. Instrum. **79**(8), 083,104 (2008). https://doi.org/10.1063/1.2964928
22. da Silva, J.C., Pacureanu, A., Yang, Y., Bohic, S., Morawe, C., Barrett, R., Cloetens, P.: Efficient concentration of high-energy x-rays for diffraction-limited imaging resolution. Optica **4**(5), 492–495 (2017). https://doi.org/10.1364/OPTICA.4.000492, http://www.osapublishing.org/optica/abstract.cfm?URI=optica-4-5-492
23. Bajt, S., Prasciolu, M., Fleckenstein, H., Domaracký, M., Chapman, H.N., Morgan, A.J., Yefanov, O., Messerschmidt, M., Du, Y., Murray, K.T., et al.: X-ray focusing with efficient high-NA multilayer Laue lenses. Light Sci. Appl. **7**(3), 17,162 (2018). https://doi.org/10.1038/lsa.2017.162
24. Kang, H.C., Maser, J., Stephenson, G.B., Liu, C., Conley, R., Macrander, A.T., Vogt, S.: Nanometer linear focusing of hard x-rays by a multilayer Laue lens. Phys. Rev. Lett. **96**, 127,401 (2006). https://doi.org/10.1103/PhysRevLett.96.127401
25. Kang, H.C., Yan, H., Winarski, R.P., Holt, M.V., Maser, J., Liu, C., Conley, R., Vogt, S., Macrander, A.T., Stephenson, G.B.: Focusing of hard x-rays to 16 nm with a multilayer Laue lens. APL **92**(22), 221114 (2008). https://doi.org/10.1063/1.2912503, http://link.aip.org/link/?APL/92/221114/1
26. Morgan, A.J., Prasciolu, M., Andrejczuk, A., Krzywinski, J., Meents, A., Pennicard, D., Graafsma, H., Barty, A., Bean, R.J., Barthelmess, M., et al.: High numerical aperture multilayer Laue lenses. Sci. Rep. **5**, 9892 (2015). https://doi.org/10.1038/srep09892
27. Osterhoff, M., Eberl, C., Döring, F., Wilke, R.N., Wallentin, J., Krebs, H.U., Sprung, M., Salditt, T.: Towards multi-order hard X-ray imaging with multilayer zone plates. J. Appl. Crystallogr. **48**(1) (2015). https://doi.org/10.1107/S1600576714026016
28. Yan, H., Bouet, N., Zhou, J., Huang, X., Nazaretski, E., Xu, W., Cocco, A.P., Chiu, W.K.S., Brinkman, K.S., Chu, Y.S.: Multimodal hard x-ray imaging with resolution approaching 10 nm for studies in material science. Nano Futur. **2**(1), 011,001 (2018). https://doi.org/10.1088/2399-1984/aab25d, http://stacks.iop.org/2399-1984/2/i=1/a=011001

29. Fuhse, C., Jarre, A., Ollinger, C., Seeger, J., Salditt, T., Tucoulou, R.: Front-coupling of a prefocused x-ray beam into a monomodal planar waveguide. Appl. Phys. Lett. **85**(11), 1907–1909 (2004). https://doi.org/10.1063/1.1791736, http://link.aip.org/link/?APL/85/1907/1

30. Jarre, A., Fuhse, C., Ollinger, C., Seeger, J., Tucoulou, R., Salditt, T.: Two-dimensional hard X-ray beam compression by combined focusing and waveguide optics. Phys. Rev. Lett. **94**(7), 074801 (2005). https://doi.org/10.1103/PhysRevLett.94.074801, http://link.aps.org/abstract/PRL/v94/e074801

31. Krüger, S.P., Giewekemeyer, K., Kalbfleisch, S., Bartels, M., Neubauer, H., Salditt, T.: Sub-15 nm beam confinement by twocrossed x-ray waveguides. Opt. Express **18**(13), 13492–13501 (2010). https://doi.org/10.1364/OE.18.013492, http://www.opticsexpress.org/abstract.cfm?URI=oe-18-13-13492

32. Krüger, S.P., Neubauer, H., Bartels, M., Kalbfleisch, S., Giewekemeyer, K., Wilbrandt, P.J., Sprung, M., Salditt, T.: Sub-10 nm beam confinement by X-ray waveguides: design, fabrication and characterization of optical properties. J. Synchrotron Rad. **19**(2), 227–236 (2012). https://doi.org/10.1107/S0909049511051983

33. Lagomarsino, S., Cedola, A., Cloetens, P., Di Fonzo, S., Jark, W., Soullie, G., Riekel, C.: Phase contrast hard x-ray microscopy with submicron resolution. Appl. Phys. Lett. **71**(18), 2557–2559 (1997). https://doi.org/10.1063/1.119324, http://link.aip.org/link/?APL/71/2557/1

34. Chao, W., Harteneck, B.D., Liddle, J.A., Anderson, E.H., Attwood, D.T.: Soft X-ray microscopy at a spatial resolution better than 15 nm. Nature **435**(7046), 1210–1213 (2005). https://doi.org/10.1038/nature03719

35. Yamauchi, K., Mimura, H., Inagaki, K., Mori, Y.: Figuring with subnanometer-level accuracy by numerically controlled elastic emission machining. Rev. Sci. Instrum. **73**(11), 4028–4033 (2002). https://doi.org/10.1063/1.1510573

36. Lengeler, B., Schroer, C.G., Kuhlmann, M., Benner, B., Günzler, T.F., Kurapova, O., Zontone, F., Snigirev, A., Snigireva, I.: Refractive x-ray lenses. J. Phys. D Appl. Phys. **38**(10A), A218 (2005). http://stacks.iop.org/0022-3727/38/i=10A/a=042

37. Als-Nielsen, J., McMorrow, D.: Elements of Modern X-Ray Physics, 2nd edn. Wiley (2011)

38. Stangl, J., Mocuta, C., Chamard, V., Carbone, D.: Nanobeam X-Ray Scattering: Probing Matter at the Nanoscale. Wiley (2013)

39. Tolan, M.: X-Ray Scattering from Soft-Matter Thin Films: Materials Science and Basic Research. Springer Tracts in Modern Physics. Springer, Berlin (1999). https://cds.cern.ch/record/445084

40. Dosch, H.: Critical Phenomena at Surfaces and Interfaces: Evanescent X-Ray and Neutron Scattering. Springer Tracts in Modern Physics. Springer, Berlin (1992). http://cds.cern.ch/record/445932

41. Rauscher, M., Salditt, T., Spohn, H.: Small-angle x-ray scattering under grazing incidence: the cross section in the distorted-wave Born approximation. Phys. Rev. B **52**(23), 16855–16863 (1995). https://doi.org/10.1103/PhysRevB.52.16855

42. Parratt, L.G.: Surface studies of solids by total reflection of x-rays. Phys. Rev. **95**(2), 359–369 (1954). https://doi.org/10.1103/PhysRev.95.359

43. Windt, D.L.: IMD-Software for modeling the optical properties of multilayer films. Comput. Phys. **12**(4), 360–370 (1998). https://doi.org/10.1063/1.168689, http://link.aip.org/link/?CIP/12/360/1

44. Kirkpatrick, P., Baez, A.V.: Formation of optical images by X-rays. J. Opt. Soc. Am. **38**(9), 766–773 (1948). http://www.opticsinfobase.org/abstract.cfm?URI=josa-38-9-766

45. Salditt, T., Osterhoff, M., Krenkel, M., Wilke, R.N., Priebe, M., Bartels, M., Kalbfleisch, S., Sprung, M.: Compound focusing mirror and X-ray waveguide optics for coherent imaging and nano-diffraction. J. Synchrotron Rad. **22**(4), 867–878 (2015). https://doi.org/10.1107/S1600577515007742

46. Kewish, C.M., Guizar-Sicairos, M., Liu, C., Qian, J., Shi, B., Benson, C., Khounsary, A.M., Vila-Comamala, J., Bunk, O., Fienup, J.R., Macrander, A.T., Assoufid, L.: Reconstruction of an astigmatic hard X-ray beam and alignment of K-B mirrors from ptychographic coherent diffraction data. Opt. Express **18**(22), 23,420–23,427 (2010). http://www.opticsexpress.org/abstract.cfm?URI=oe-18-22-23420

47. Morawe, C., Osterhoff, M.: Curved graded multilayers for X-ray nano-focusing optics. Nucl. Instrum. Methods Phys. Res. Sect. A Accel. Spectrometers Detect. Assoc. Equip. (2010, In Press, Corrected Proof). http://www.sciencedirect.com/science/article/B6TJM-4XSVR2S-8/2/ed428190e29ea617d3993cc7b85d3e73

48. Morawe, C., Osterhoff, M.: Hard x-ray focusing with curved reflective multilayers. X-Ray Opt. Instrum. **2010** (2010)

49. Osterhoff, M., Morawe, C., Ferrero, C., Guigay, J.P.: Wave-optical theory of nanofocusing x-ray multilayer mirrors. Opt. Lett. **37**(17), 3705–3707 (2012). https://doi.org/10.1364/OL.37.003705, http://ol.osa.org/abstract.cfm?URI=ol-37-17-3705

50. Osterhoff, M., Morawe, C., Ferrero, C., Guigay, J.P.: Optimized x-ray multilayer mirrors for single nanometer focusing. Opt. Lett. **38**(23), 5126–5129 (2013)

51. Osterhoff, M.: Wave optical simulations of x-ray nano-focusing optics. Ph.D. thesis, Universität Göttingen (2012)

52. Bongaerts, J.H.H., David, C., Drakopoulos, M., Zwanenburg, M.J., Wegdam, G.H., Lackner, T., Keymeulen, H., van der Veen, J.F.: Propagation of a partially coherent focused X-ray beam within a planar X-ray waveguide. J. Synchrotron Rad. **9**(6), 383–393 (2002). https://doi.org/10.1107/S0909049502016308

53. Feng, Y.P., Sinha, S.K., Deckman, H.W., Hastings, J.B., Siddons, D.P.: X-ray flux enhancement in thin-film waveguides using resonant beam couplers. Phys. Rev. Lett. **71**(4), 537–540 (1993). https://doi.org/10.1103/PhysRevLett.71.537

54. Pfeiffer, F., Salditt, T., Høghøj, P., Anderson, I., Schell, N.: X-ray waveguides with multiple guiding layers. Phys. Rev. B **62**(24), 16,939 (2000)

55. Spiller, E., Segmuller, A.: Propagation of x-rays in waveguides. Appl. Phys. Lett. **24**(2), 60–61 (1974). https://doi.org/10.1063/1.1655093, http://link.aip.org/link/?APL/24/60/1

56. Zwanenburg, M.J., Peters, J.F., Bongaerts, J.H.H., de Vries, S.A., Abernathy, D.L., van der Veen, J.F.: Coherent propagation of x-rays in a planar waveguide with a tunable air gap. Phys. Rev. Lett. **82**(8), 1696–1699 (1999). https://doi.org/10.1103/PhysRevLett.82.1696

57. Pfeiffer, F., David, C., Burghammer, M., Riekel, C., Salditt, T.: Two-dimensional X-ray waveguides and point sources. Science **297**(6), 063709 (2002). https://doi.org/10.1126/science.1071994. http://www.sciencemag.org/cgi/content/abstract/297/5579/230

58. Chen, H.Y., Hoffmann, S., Salditt, T.: X-ray beam compression by tapered waveguides. Appl. Phys. Lett. **106**(19), 194105 (2015). https://doi.org/10.1063/1.4921095

59. Hoffmann-Urlaub, S., Salditt, T.: Miniaturized beamsplitters realized by X-ray waveguides. Acta Crystallogr. A **72**(5), 515–522 (2016). https://doi.org/10.1107/S205327331601144X

60. Neubauer, H., Hoffmann, S., Kanbach, M., Haber, J., Kalbfleisch, S., Krüger, S.P., Salditt, T.: High aspect ratio x-ray waveguide channels fabricated by e-beam lithography and wafer bonding. J. Appl. Phys. **115**(21), 214,305 (2014). https://doi.org/10.1063/1.4881495

61. Bartels, M., Krenkel, M., Haber, J., Wilke, R.N., Salditt, T.: X-ray holographic imaging of hydrated biological cells in solution. Phys. Rev. Lett. **114**, 048,103 (2015). https://doi.org/10.1103/PhysRevLett.114.048103

62. Marcuse, D.: Theory of Dielectric Optical Waveguides. Academic Press, New York (1974)

63. Fuhse, C.: X-ray waveguides and waveguide-based lensless imaging (2006). http://webdoc.sub.gwdg.de/diss/2006/fuhse/

64. Osterhoff, M., Salditt, T.: Real structure effects in X-ray waveguide optics: the influence of interfacial roughness and refractive index profile on the near-field and far-field distribution. Opt. Commun. **282**(16), 3250–3256 (2009). https://doi.org/10.1016/j.optcom.2009.05.008, http://www.sciencedirect.com/science/article/B6TVF-4WC500H-1/2/76173eb0a2d86851e482a82f8ae7b644

65. Fuhse, C., Salditt, T.: Propagation of X-rays in ultra-narrow slits. Opt. Commun. **265**(1), 140–146 (2006). http://www.sciencedirect.com/science/article/B6TVF-4JKRVYM-4/2/3ae9173e37c896d867819c214a57b995

66. Osterhoff, M., Salditt, T.: Coherence filtering of x-ray waveguides: analytical and numerical approach. New J. Phys. **13**(10), 103,026 (2011). http://stacks.iop.org/1367-2630/13/i=10/a=103026

67. Giewekemeyer, K., Neubauer, H., Kalbfleisch, S., Krüger, S.P., Salditt, T.: Holographic and diffractive x-ray imaging using waveguides as quasi-point sources. New J. Phys. **12**(3), 035,008 (2010). http://stacks.iop.org/1367-2630/12/i=3/a=035008

68. Salditt, T., Kruger, S.P., Fuhse, C., Bahtz, C.: High-transmission planar X-ray waveguides. Phys. Rev. Lett. **100**(18), 184,801–184,804 (2008). http://link.aps.org/abstract/PRL/v100/e184801

69. Salditt, T., Hoffmann, S., Vassholz, M., Haber, J., Osterhoff, M., Hilhorst, J.: X-ray optics on a chip: guiding x-rays in curved channels. Phys. Rev. Lett. **115**, 203,902 (2015). https://doi.org/10.1103/PhysRevLett.115.203902

70. Fuhse, C., Ollinger, C., Salditt, T.: Waveguide-based off-axis holography with hard X-rays. Phys. Rev. Lett. **97**(25), 254801 (2006). https://doi.org/10.1103/PhysRevLett.97.254801, http://link.aps.org/abstract/PRL/v97/e254801

71. Röhlsberger, R., Schlage, K., Klein, T., Leupold, O.: Accelerating the spontaneous emission of x-rays from atoms in a cavity. Phys. Rev. Lett. **95**, 097,601 (2005). https://doi.org/10.1103/PhysRevLett.95.097601

72. Melchior, L., Salditt, T.: Finite difference methods for stationary and time-dependent x-ray propagation. Opt. Express **25**, 32,090 (2017). https://doi.org/10.1364/OE.25.032090

73. Zhong, Q., Melchior, L., Peng, J., Huang, Q., Wang, Z., Salditt, T.: Goos-hänchen effect observed for focused x-ray beams under resonant mode excitation. Opt. Express **25**(15), 17431–17445 (2017). https://doi.org/10.1364/OE.25.017431, http://www.opticsexpress.org/abstract.cfm?URI=oe-25-15-17431

74. Liese, T.: Multilayer based transmission optics for x-ray microscopy. Ph.D. thesis (2012)

75. Eberl, C.: Multilayer zone plates for hard x-ray microscopy. Ph.D. thesis (2016)

76. Niemann, B., Rudolph, D., Schmahl, G.: Soft x-ray imaging zone plates with large zone numbers for microscopic and spectroscopic applications. Opt. Commun. **12**(2), 160–163 (1974). https://doi.org/10.1016/0030-4018(74)90381-2, http://www.sciencedirect.com/science/article/pii/0030401874903812

77. Niemann, B., Rudolph, D., Schmahl, G.: X-ray microscopy with synchrotron radiation. Appl. Opt. **15**(8), 1883–1884 (1976)

78. Schmahl, G., Rudolph, D., Niemann, B., Christ, O.: Zone-plate X-ray microscopy. Q. Rev. Biophys. **13**(3), 297–315 (1980)

79. Jefimovs, K., Vila-Comamala, J., Pilvi, T., Raabe, J., Ritala, M., David, C.: Zone-doubling technique to produce ultrahigh-resolution x-ray optics. Phys. Rev. Lett. **99**(26), 264,801 (2007). https://doi.org/10.1103/PhysRevLett.99.264801

80. Vila-Comamala, J., Pan, Y., Lombardo, J.J., Harris, W.M., Chiu, W.K.S., David, C., Wang, Y.: Zone-doubled fresnel zone plates for high-resolution hard x-ray full-field transmission microscopy. J. Synchrotron Rad. **19**(5), 705–709 (2012). https://doi.org/10.1107/S0909049512029640

81. Ruhlandt, A., Liese, T., Radisch, V., Krüger, S.P., Osterhoff, M., Giewekemeyer, K., Krebs, H.U., Salditt, T.: A combined Kirkpatrick-Baez mirror and multilayer lens for sub-10 nm x-ray focusing. AIP Adv. **2**(1), 012,175–7 (2012). https://doi.org/10.1063/1.3698119

82. D.Rudolph B.Niemann, G.: Status of the sputtered sliced zone plates for x-ray microscopy (1982). https://doi.org/10.1117/12.933141

83. Yun, W., Lai, B., Cai, Z., Maser, J., Legnini, D., Gluskin, E., Chen, Z., Krasnoperova, A.A., Vladimirsky, Y., Cerrina, F., Di Fabrizio, E., Gentili, M.: Nanometer focusing of hard x rays by phase zone plates. Rev. Sci. Instrum. **70**(5), 2238–2241 (1999). https://doi.org/10.1063/1.1149744

84. Koyama, T., Takano, H., Konishi, S., Tsuji, T., Takenaka, H., Ichimaru, S., Ohchi, T., Kagoshima, Y.: Circular multilayer zone plate for high-energy x-ray nano-imaging. Rev. Sci. Instrum. **83**(1), 013,705–013,705–4 (2012)

85. Röder, J., Liese, T., Krebs, H.U.: Material-dependent smoothing of periodic rippled structures by pulsed laser deposition. J. Appl. Phys. **107**(10), 103,515–103,515-5 (2010). https://doi.org/10.1063/1.3388591

86. Liese, T., Radisch, V., Krebs, H.U.: Fabrication of multilayer Laue lenses by a combination of pulsed laser deposition and focused ion beam (2010). https://doi.org/10.1063/1.3462985
87. Osterhoff, M., Soltau, J., Eberl, C., Krebs, H.U.: Ultra-high-aspect multilayer zone plates for even higher x-ray energies. In: Proceedings of the SPIE, vol. 10386, p. 1038608 (2017). https://doi.org/10.1117/12.2271139
88. Eberl, C., Döring, F., Liese, T., Schlenkrich, F., Roos, B., Hahn, M., Hoinkes, T., Rauschenbeutel, A., Osterhoff, M., Salditt, T., Krebs, H.U.: Fabrication of laser deposited high-quality multilayer zone plates for hard x-ray nanofocusing. Appl. Surf. Sci. **307**, 638–644 (2014). http://www.sciencedirect.com/science/article/pii/S016943321400854X
89. Yan, H., Maser, J., Macrander, A., Shen, Q., Vogt, S., Stephenson, G.B., Kang, H.C.: Takagitaupin description of x-ray dynamical diffraction from diffractive optics with large numerical aperture. Phys. Rev. B **76**, 115,438 (2007). https://doi.org/10.1103/PhysRevB.76.115438
90. Maser, J., Schmahl, G.: Coupled wave description of the diffraction by zone plates with high aspect ratios. Opt. Commun. **89**(2), 355–362 (1992). https://doi.org/10.1016/0030-4018(92)90182-Q, http://www.sciencedirect.com/science/article/pii/003040189290182Q
91. Schneider, G.: Zone plates with high efficiency in high orders of diffraction described by dynamical theory. Appl. Phys. Lett. **71**(16), 2242–2244 (1997). https://doi.org/10.1063/1.120069
92. Singer, A., Vartanyants, I.A., Kuhlmann, M., Duesterer, S., Treusch, R., Feldhaus, J.: Transverse-coherence properties of the free-electron-laser FLASH at DESY. Phys. Rev. Lett. **101**, 254,801 (2008)
93. Vartanyants, I.A., Singer, A.: Coherence properties of hard x-ray synchrotron sources and x-ray free-electron lasers. New J. Phys. **12**(3), 035,004 (2010). https://doi.org/10.1088/1367-2630/12/3/035004
94. Salditt, T., Kalbfleisch, S., Osterhoff, M., Krüger, S.P., Bartels, M., Giewekemeyer, K., Neubauer, H., Sprung, M.: Partially coherent nano-focused x-ray radiation characterized by Talbot interferometry. Opt. Express **19**(10), 9656–9675 (2011). https://doi.org/10.1364/OE.19.009656, http://www.opticsexpress.org/abstract.cfm?URI=oe-19-10-9656
95. Osterhoff, M., Robisch, A.L., Soltau, J., Eckermann, M., Kalbfleisch, S., Carbone, D., Johansson, U., Salditt, T.: Focus characterization of the NanoMAX Kirkpatrick-Baez mirror system. J. Synchrotron Rad. **26**(4), (2019)
96. Kalbfleisch, S.: A dedicated endstation for waveguide-based x-ray imaging. Ph.D. thesis, Universität Göttingen (2012)
97. Wilke, R.N., Wallentin, J., Osterhoff, M., Pennicard, D., Zozulya, A., Sprung, M., Salditt, T.: High flux ptychographic imaging using the new 55 μm-pixel detector 'lambda' based on the medipix3 readout chip. Acta Crystallogr. A **70**, 552–562 (2014). https://doi.org/10.1107/S2053273314014545
98. Nicolas, J.D., Bernhardt, M., Krenkel, M., Richter, C., Luther, S., Salditt, T.: Combined scanning X-ray diffraction and holographic imaging of cardiomyocytes. J. Appl. Crystallogr. **50**(2), 612–620 (2017). https://doi.org/10.1107/S1600576717003351
99. Giewekemeyer, K., Wilke, R.N., Osterhoff, M., Bartels, M., Kalbfleisch, S., Salditt, T.: Versatility of a hard x-ray Kirkpatrick-Baez focus characterized by ptychography. J. Synchrotron Rad. **20**(3), 490–497 (2013). https://doi.org/10.1107/S0909049513005372
100. Giewekemeyer, K., Philipp, H.T., Wilke, R.N., Aquila, A., Osterhoff, M., Tate, M.W., Shanks, K.S., Zozulya, A.V., Salditt, T., Gruner, S.M., Mancuso, A.P.: High-dynamic-range coherent diffractive imaging: ptychography using the mixed-mode pixel array detector. J. Synchrotron Rad. **21**(5), 1167–1174 (2014). https://doi.org/10.1107/S1600577514013411
101. Wilke, R.N., Vassholz, M., Salditt, T.: Semi-transparent central stop in high-resolution x-ray ptychography using Kirkpatrick-Baez focusing. Acta Crystallogr. A **69**(5), 490–497 (2013). https://doi.org/10.1107/S0108767313019612
102. Hagemann, J., Robisch, A.L., Osterhoff, M., Salditt, T.: Probe reconstruction for holographic X-ray imaging. J. Synchrotron Rad. **24**(2), 498–505 (2017). https://doi.org/10.1107/S160057751700128X

Chapter 4
Statistical Foundations of Nanoscale Photonic Imaging

Axel Munk, Thomas Staudt and Frank Werner

Essentially, all models are wrong, but some are useful.
— George Box

4.1 Introduction

4.1.1 Background and Examples

The term 'photonic imaging' describes an optical imaging setup where the available measurement data Y are counts of detected photons. The origin of these photons can be diverse in its nature. In coherent X-ray imaging (see e.g. Chap. 2), photons emitted by an X-ray source (like a free electron laser) are scattered (and/or absorbed) by a specimen. In fluorescence microscopy (see e.g. Chap. 1 or Chap. 7), marker molecules are excited by an excitation pulse and emit photons with a certain probability. These two examples are characteristic for the wide range of scenarios arising in photonic imaging: in coherent X-ray imaging we have on the one hand single-molecule diffraction data composed of only few photons [1], and on the other hand

A. Munk (✉) · T. Staudt · F. Werner
Institute for Mathematical Stochastics, Universität Göttingen, Goldschmidtstr. 7,
37077 Göttingen, Germany
e-mail: munk@math.uni-goettingen.de

T. Staudt
e-mail: thomas.staudt@uni-goettingen.de

F. Werner
e-mail: f.werner@math.uni-goettingen.de

A. Munk · F. Werner
Max Planck Institute for Biophysical Chemistry, Am Faßberg 11,
37077 Göttingen, Germany

© The Author(s) 2020
T. Saalditt et al. (eds.), *Nanoscale Photonic Imaging*, Topics in Applied Physics 134,
https://doi.org/10.1007/978-3-030-34413-9_4

holographic experiments where millions of photons can be collected from one sample [2]. In fluorescence microscopy, the number of photons is intrinsically limited to a few hundred or thousand per marker due to bleaching effects, and in case of temporally resolved measurements, only a handful of photons is available per time step [3]. Similar restrictions arise in related imaging modalities, including those based on Förster resonance energy transfer (FRET) or metal induced energy transfer (MIET), see e.g. Chap. 8 or [4, 5] for a discussion. Although not within the context of nanoscale imaging, statistically related is astrophysical imaging. Here, there is no a priori limit for the observation time and hence for the number of photons. However, the former is practically limited to several minutes to avoid severe motion blur, see e.g. [6, 7] for examples. We also mention positron emission tomography (PET), where the total number of emitted photons should be as small as possible to minimize the radiation dose for the patient [8]. In all of these applications, detected photons can also originate from undesired background contributions, whose nature strongly depends on the experimental setup, adding additional noise to the observations.

4.1.2 Purpose of the Chapter

The aim of this chapter is to give an overview over prototypical approaches to model the data emerging in photonic imaging from a statistical point of view, based on the physical modeling of photon observation. A sketch of the typical imaging setup we consider is presented in Fig. 4.1.

We assume that the imaging process is described by an underlying photon intensity $\lambda : \Omega \times [0, T] \to [0, \infty)$ at the detector interface, where Ω is the spa-

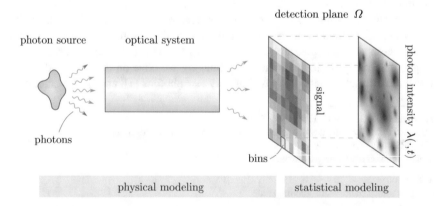

Fig. 4.1 Sketch of the imaging process. A source emits photons that are mapped on a (binned) detection interface Ω through the optical system. The underlying photon intensity $\lambda(\cdot, t)$ at time t, which is determined by the physics of the specific imaging setup, is used for statistical modeling of the detected signal

tial domain of observation (which can be two- or three-dimensional) and T is the total observation time. Let us enumerate the emitted photons by $1, \ldots, N$ and denote their specific detection position and time by $(\mathbf{x}_i, t_i) \in \Omega \times [0, T]$. For a given (measurable) subset $A \subset \Omega$ and time interval $I \subset [0, T]$ we write $Y(A \times I) := \# \{ 1 \le i \le N \mid \mathbf{x}_i \in A, t_i \in I \}$ to denote the number of photons observed in A during I. The expected number of photons detected in $A \times I$ is by definition of λ given by

$$\mathbb{E}\left[Y(A \times I)\right] = \int_I \int_A \lambda(\mathbf{y}, t) \, d\mathbf{y} \, dt . \tag{4.1}$$

Note that this includes all detected photons, including all background contributions. We will always assume $\lambda \ge 0$, which ensures that the integral in (4.1) is well-defined (however it might be ∞).

Throughout this manuscript, we will discuss statistical models for the distribution of the observations Y, depending on the physical measurement setup. We assume λ to be given, as deriving or estimating λ and/or other model parameters described (implicitly) by λ is the topic of other expositions (see e.g. Chap. 5 or Chap. 11).

4.1.3 Measurement Devices

Depending on the type of sensor used for photon detection, different models for photonic imaging settings have been proposed. One commonality of all measurement setups is that the spatial domain of observation Ω is discretized into detector regions, so-called bins. We will assume that the detectors on all bins have identical physical properties, and we denote the centers of such bins by $\mathbf{x} \in \Xi$ with Ξ being the set of all bin centers. If a charge-coupled device (CCD) camera is used for detection, all bins (the pixels of the sensor) can be observed simultaneously. This is e.g. the case in most coherent X-ray experiments or astrophysical imaging. PET requires a tomographic setup consisting of several photomultiplier tubes (PMT) surrounding the patient (see e.g. [9]). In confocal fluorescence microscopy the most widely applied detectors are based on avalanche photodiodes, which can measure photons in one bin at a time only. Hence, the domain of observation Ω is typically scanned by physically moving the specimen (or detector) at a fast pace. Temporal simultaneous photons can be measured as well, requiring a different experimental setup (see e.g. [10]).

Most photon detectors rely on the photoelectric effect. With a certain probability (the quantum efficiency), incident photons will release photo electrons on the detector surface. Since single electrons cannot be detected reliably, the signal is typically amplified by a cascade of electron multiplying systems. This introduces additional noise due to the stochastic nature of the multiplying steps. Another complication is the existence of dead times. The dead time of a detection device refers to the time interval (after activation) during which it is unable to record another event. Dead times can, for example, arise due to the necessity to recharge conductors in-between measurements, or due to time delays caused by analog-to-digital conversion and data storage. Details on the statistics of different detectors can be found in [11, Cpt. 12].

4.1.4 Structure and Notation

For the remainder of this chapter we will develop and discuss models for the right part in Fig. 4.1 with different degrees of accuracy. The model choice mainly depends on the total number of detected photons and on the spatial and temporal dependency structure of the randomly generated photons. We will start with the Poisson model, which is well-known and most common for many applications. It can be derived immediately from (4.1) under the assumption of independence, which explains its wide use in photonic imaging (see e.g. the reviews [7, 12] and the references therein). However, if it is necessary to count photons on small time scales, or if independence is not given, a more refined modeling is on demand. In these situations, we turn towards Bernoulli and Binomial models subsequently, and discuss to what extend they are compatible with the aforementioned Poisson model. Finally we turn to the case of large counting rates, which lead to Gaussian models based on asymptotic normality. We discuss differences and commonalities arising from the different base models and indicate in which situation which model should be used. This will be linked to different examples from this book, where we argue if our assumptions are met or not.

Let us introduce the basic notation used in this chapter. We will always assume that any observation y is the realization of a random variable Y, and we will denote by $\mathbb{P}$ probabilities w.r.t. this random object. By $\mathbb{E}$ and $\mathbb{V}$ we will denote the expectation and variance w.r.t. $\mathbb{P}$, respectively. The letters $\mathcal{P}$, $\mathcal{B}$ and $\mathcal{N}$ will denote the Poisson, Binomial and normal distribution introduced below. Random variables will always be denoted by capital letters X, X_i, Z etc., and if we write i.i.d. for a sequence $X_1, X_2, \ldots$ of random variables, this stands for independent identically distributed.

4.2 Poisson Modeling

Suppose we have a perfect photon detector that registers the individual arrival times of all emitted photons reaching a bin without missing any. We will focus on describing a single bin for the moment to avoid notational difficulties. In this situation, the total number of collected photons often can be modeled as Poissonian. A random variable X follows a Poisson law with parameter (intensity) $\mu \geq 0$, if

$$\mathbb{P}[X = j] = \frac{\mu^j}{j!} \exp(-\mu), \quad j \in \mathbb{N}_0.$$

We write $X \sim \mathcal{P}(\mu)$. The following fundamental theorem about point processes explains why the Poisson distribution often comes into play when modeling photon counts:

Theorem 4.1 *Suppose we observe a random number N of photons at random arrival times $0 \le t_1 < \cdots < t_N \le T$ such that*

(a) *for each choice of disjoint intervals $I_1, \ldots, I_n \subset [0, T]$, the random variables $\#\{1 \le k \le N \mid t_k \in I_i\}$, $1 \le i \le n$, corresponding to the number of observed photons during I_i are independent, and*
(b) *there exists some integrable function μ on $[0, T]$ such that for any choice $0 \le a < b \le T$ it holds*

$$\mathbb{E}\left[\#\{1 \le k \le N \mid a \le t_k \le b\}\right] = \int_a^b \mu(t) \, dt.$$

Then, for all $0 \le a < b \le T$, the number of photons observed between time a and time b is Poisson distributed with parameter $\int_a^b \mu(t) \, dt$, i.e.

$$\#\{1 \le k \le N \mid a \le t_k \le b\} \sim \mathcal{P}\left(\int_a^b \mu(t) \, dt\right).$$

For the proof we refer to [13, Theorem 1.11.8]. In terms of probability theory, this theorem implies that the point process $X := \sum_{i=1}^N \delta_{t_i}$, with δ_t denoting the Dirac measure at t, is a Poisson point process with intensity μ if the stated assumptions are satisfied.

Let us discuss these assumptions. Condition (b) underlies our whole modeling procedure as described in (4.1) and seems universally evident. Temporal independence of the arrival times in (a) is more critical but seems (at least approximately) reasonable in many imaging modalities where photons arise from a high-intensity source, including coherent X-ray imaging. However, if the photons arise from fluorescent markers, temporal independence can be violated due to hidden internal states of the fluorophores, energy transfer between different fluorophores on small time and spatial scales (e.g. FRET), or dead times of the detectors.

If temporal independence is given, then Theorem 4.1 states that the number $Y_{x,t}$ of collected photons within a bin B_x until time $t \in [0, T]$ can naturally be modeled by a Poissonian random variable with intensity $\int_0^t \int_{B_x} \lambda(y, \tau) \, dy \, d\tau$. This gives rise to the following model:

Poisson model

Let the spatial domain of observation Ω be discretized into bins $B_{\mathbf{x}}$ with centers $\mathbf{x} \in \Xi$. We assume that our observations are given by a field $Y_t := \left(Y_{\mathbf{x},t}\right)_{\mathbf{x} \in \Xi}$ of random variables such that

$$Y_{\mathbf{x},t} \sim \mathcal{P}\left(\int_0^t \int_{B_{\mathbf{x}}} \lambda(\mathbf{y}, \tau) \, d\mathbf{y} \, d\tau\right), \qquad \mathbf{x} \in \Xi, t \in [0, T] \qquad (4.2)$$

for some intensity function $\lambda \geq 0$.

This is the basis of many popular models covering a variety of distinct applications. Examples include PET (see Vardi et al. [9]), astronomy and fluorescence microscopy (see Bertero et al. [7] or Hohage and Werner [12]), or a more subtle model for CCD cameras due to Snyder et al. [14, 15].

Note that so far we have assumed that all arriving photons are collected by the detector. This will however be never the case due to several physical limitations, see Fig. 4.2.

The specific efficiency depends strongly on the setup and can vary considerably. Additionally to different quantum efficiencies of different detectors, it might also happen that the detector does not cover all of Ω or has some dead subregions (like interfaces between individual elements). This causes a loss of measured photons and hence a statistical thinning of the random variable $Y_{\mathbf{x},t}$. In this case, the actually

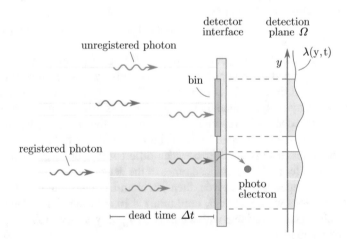

Fig. 4.2 Statistical photon thinning at the detector interface. Photons that reach the detection plane can stay undetected due to various reasons. For example, they can fail to free a photo electron, miss the sensitive regions of the detector bins, or arrive during the dead time caused by a previously recorded photon

observed random variable $\widetilde{Y}_{\mathbf{x},t}$ can be written as

$$\widetilde{Y}_{\mathbf{x},t} = \sum_{i=1}^{Y_{\mathbf{x},t}} X_i \tag{4.3}$$

with Bernoulli random variables X_i having success probability $\eta_i \in [0, 1]$ where each X_i indicates if the ith photon has been detected. If, in addition, the thinning happens identically and independently for each photon, i.e. $X_i \overset{\text{i.i.d.}}{\sim} \mathcal{B}(1, \eta)$, only the parameter in the Poisson law (4.2) changes, but not its distributional structure. More precisely, in this case it follows (see the Appendix) that

$$\widetilde{Y}_{\mathbf{x},t} \sim \mathcal{P}\left(\eta \int_0^t \int_{B_{\mathbf{x}}} \lambda(\mathbf{y}, \tau) \, d\mathbf{y} \, d\tau\right).$$

Consequently, the imperfectness of a detector (as long as the induced thinning happens independent for each photon) can be seen as a scaling of the underlying photon intensity λ by an efficiency factor $\eta \in (0, 1]$. In agreement with Fig. 4.1 we can hence assume that all physical processes causing a thinning have already been treated when modeling λ in the following.

Besides this kind of independent thinning, a further important issue in many imaging modalities is the dead time Δt of the employed detector. Dead times can vary significantly depending on the type of detector, but usually are in the range of nanoseconds. If a photon arrives at time $t \in [0, T)$, the detector will only be able to record the next photon arriving after $t + \Delta t$. Note that whenever $\Delta t > 0$, at most $T/\Delta t$ photons can be detected during the whole measurement, which contradicts (4.2) in the sense that $\mathbb{P}[Y_{\mathbf{x},T} > T/\Delta t] = 0$ in this case. Such an upper limit on the total number of detected photons can crucially change the distribution, which can, e.g., be seen from the following fact proven in the appendix:

Theorem 4.2 *Fix* $\mathbf{x} \in \Xi$ *and let* $I_1, \ldots, I_m$ *be a decomposition of* $[0, T]$ *into disjoint intervals. Denote by* X_i *the number of photons observed during* I_i *in bin* $B_{\mathbf{x}}$. *Assume model* (4.2), *and suppose that* $X_1, \ldots, X_m$ *are independent. Then the conditional distribution given* $Y_{\mathbf{x},T} = N$ *of* $(X_1, \ldots, X_m)$ *is multinomial with parameter* N *and probability vector* $(p_1, \ldots, p_m)$ *where*

$$p_i = \frac{\int_{I_i} \int_{B_{\mathbf{x}}} \lambda(\mathbf{y}, \tau) \, d\mathbf{y} \, d\tau}{\int_0^T \int_{B_{\mathbf{x}}} \lambda(\mathbf{y}, \tau) \, d\mathbf{y} \, d\tau}.$$

In other words, Theorem 4.2 states that, conditioning on the total number of photons, the arrival times of individual photons behave like a Bernoulli process with intensity $\tau \mapsto \int_{B_{\mathbf{x}}} \lambda(\mathbf{y}, \tau) \, d\mathbf{y}$. This implies that conditioning on the total number of photons introduces a dependency structure between the number of counts during different time intervals. Consequently, if Δt cannot be neglected, temporal independence is not given anymore, hence corrupting the Poisson law, and different modeling approaches are needed.

4.3 Bernoulli Modeling

To measure the temporal structure of the incoming photons, counting as described above is not sufficient. In such cases, photons are consecutively counted during (short) time frames. We suppose that the discretization of the temporal measurement process is refined such that temporal aggregation underlying the Poisson model is not appropriate anymore. This is described by (equidistant) time frames, which are consecutive intervals $I_1, I_2, \ldots, I_n \subset [0, T]$ of equal length $\delta > 0$, chosen such that the probability to observe more than one photon in each bin B_x during any interval is sufficiently close to 0, and separated by a waiting time $\epsilon > 0$, which allows to ignore the dead time. In this situation, the following model is a reasonable approximation:

Bernoulli model

For $\mathbf{x} \in \varXi$ and $1 \leq i \leq n$ the random variable $Y_{\mathbf{x},i}$ indicating if a photon arrives in bin $B_{\mathbf{x}}$ during the time interval I_i follows a Bernoulli distribution,

$$Y_{\mathbf{x},i} \sim \mathcal{B}\left(1, p_{\mathbf{x},i}\right), \qquad (4.4)$$

with success probability

$$p_{\mathbf{x},i} \approx \int\limits_{I_i} \int\limits_{B_{\mathbf{x}}} \lambda\left(\mathbf{y}, \tau\right) \, \mathrm{d}\mathbf{y} \, \mathrm{d}\tau. \qquad (4.5)$$

As mentioned before, the detector will hardly count all arriving photons, which causes a statistical thinning as in (4.3). If the thinning happens independently of the photon arrivals, we obtain $\widetilde{Y}_{\mathbf{x},i} \sim \mathcal{B}\left(1, \eta \cdot p_{\mathbf{x},i}\right)$ with the probability η that an incident photon is detected, which immediately follows from $X \cdot Z \sim \mathcal{B}\left(1, pp'\right)$ if $X \sim \mathcal{B}\left(1, p\right)$ is independent of $Z \sim \mathcal{B}\left(1, p'\right)$.

In many imaging setups, it would be difficult to store the whole time series $Y_{\mathbf{x},i}$, for instance due to memory limitations. Examples include fluorescence microscopy setups like confocal, STED or 4Pi microscopy, or coherent X-ray imaging, where millions of photons are observed in short times, which would require an unreasonably fine time discretization. For other examples like SMS microscopy, however, the temporal structure can be important (e.g. for adjusting temporal drifts, see e.g. [16, 17]) and hence most of the data of the above model has to be used. If temporal dependencies are less important, it is sufficient to count photon arrivals in some interval $I \subset [0, T]$ larger than δ, i.e. to consider $Y_{\mathbf{x},I} := \sum_{I_i \subset I} Y_{\mathbf{x},i}$. The distribution of $Y_{\mathbf{x},I}$ depends strongly on the temporal dependency structure of the $Y_{\mathbf{x},i}$. In case that they are independent and $p_{\mathbf{x},i} \equiv p_{\mathbf{x}}$ for all $1 \leq i \leq n$, we obtain a Binomial model:

Binomial model

For $\mathbf{x} \in \Xi$ and $I \subset [0, T]$, the number of photons observed in the bin centered at $\mathbf{x}$ during the time interval I is

$$Y_{\mathbf{x},I} \sim \mathcal{B}\left(\#\{I_i \subset I\}, p_{\mathbf{x}}\right) \tag{4.6}$$

with $p_{\mathbf{x},i} \equiv p_{\mathbf{x}}$ for all $1 \leq i \leq n$ and $p_{\mathbf{x},i}$ as in (4.5).

Note that if we proceed similarly with the thinned observations $\widetilde{Y}_{\mathbf{x},i}$, we obtain $\widetilde{Y}_{\mathbf{x},I} \sim \mathcal{B}\left(\#\{I_i \subset I\}, \eta p_{\mathbf{x}}\right)$, which is the canonical thinning of (4.6), see e.g. [18].

Independence of the $Y_{\mathbf{x},i}$ is strongly connected to the photon source, as discussed above. If $\epsilon \geq \Delta t$, the dead times of the detectors have no influence on the temporal dependency structure anymore. The second assumption, $p_{\mathbf{x},i} \equiv p_{\mathbf{x}}$ for all $1 \leq i \leq n$, is equivalent to stationarity of the underlying photon source, which again depends on the imaging modality. If, e.g., a freeze-dried sample is imaged sufficiently fast, then this assumption is reasonable.

Besides temporal dependencies, the field of random variables can also have a spatial dependency structure. In many modalities the random variables are independent for different pixels or voxels $\mathbf{x}$, but on sufficiently small scales some dependency can occur, e.g., due to energy transfer between molecules.

4.3.1 Law of Small Numbers

It is a fundamental and well-known fact that a Binomial distribution can in certain situations be approximated by a Poissonian distribution. In this section, we will discuss how this provides a link between the initial Poisson modeling (4.2) and the preceding Bernoulli modeling (4.6). To this end, we recall the so-called *law of small numbers*, which will be stated in terms of Le Cam's theorem [19]. For the moment we suppress dependencies on $\mathbf{x}$ and consider only a single Binomial random variable, corresponding to a fixed bin.

Theorem 4.3 (Law of small numbers) *Let $X_1, \ldots, X_m$ be independent and Bernoulli distributed with success probabilities $q_1, \ldots, q_m$. Then the distribution of $X :=$ $X_1 + \cdots + X_m$ can be approximated by $\mathcal{P}(\lambda_m)$ with $\lambda_m = -\sum_{i=1}^{m} \log(1 - q_i)$. More precisely it holds that*

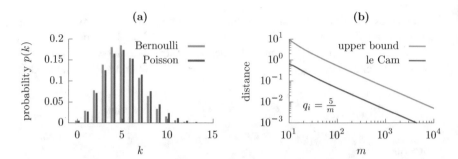

Fig. 4.3 Law of small numbers. Figure **(a)** shows the probabilities for a sum of 50 Bernoulli variables with $q_i = 0.1$ and the respective Poisson approximation with $\lambda = -50 \log(0.9) \approx 5$. Figure **(b)** depicts the left hand side of (4.7) and the corresponding upper bound (right hand side of (4.7)) of Theorem 4.3 for increasing m and $q_i := 5/m$

$$\sum_{k=0}^{\infty} \left| \mathbb{P}[X = k] - \frac{\lambda_m^k}{k!} \exp(-\lambda_m) \right| \leq 2 \sum_{i=1}^{m} (\log(1 - q_i))^2. \tag{4.7}$$

For a textbook proof we refer to [20, Theorem 5.1]. Figure 4.3 visualizes the law of small numbers. Note that the bound on the right-hand side of (4.7) can be simplified by using $\log(1 - x) \leq -x$, resulting in $\sum_{i=1}^{m} q_i^2$. We furthermore refer to [21, Proposition 4.3 and 4.4], where bounds of the supremum instead of the sum over k on the left-hand side are given. Note that Theorem 4.3 can be generalized to dependent Bernoulli random variables at the price of a worse upper bound, see e.g. [20, Theorem 5.5].

A classical example for this law is the situation when $q_i \equiv q_m$ for all $1 \leq i \leq m$ and $q_m \cdot m$ converges to some $\lambda > 0$, i.e., $q_m \sim 1/m$. In this case we may use $\log(1 - x) \approx -x$ for small x to obtain $\lambda_m \approx m q_m \to \lambda$ and $2 \sum_{i=1}^{m} (\log(1 - q_i))^2 \approx 2 \sum_{i=1}^{m} q_i^2 = 2m q_m^2 \sim 1/m \to 0$ as $m \to \infty$, i.e., the Binomial distribution of X converges rapidly to the Poisson distribution with parameter λ.

On the other hand, if the success probabilities $q_i \equiv q \in (0, 1)$ are fixed, the right-hand side of (4.7) diverges. This seems intuitive, as in this situation convergence towards a normal distribution has to be expected (cf. Sect. 4.4.1 below). This is in line with the observation that a Poisson distribution with growing parameter $\lambda_m = -m \log(1 - q)$ converges towards a normal distribution (cf. Sect. 4.4.2 below).

Let us now compare the two Poisson laws arising from Theorem 4.3 and (4.2). According to (4.4), our observations are Binomial random variables with success probability

$$p_{\mathbf{x},i} \approx \int_{I_i} \int_{B_{\mathbf{x}}} \lambda(\mathbf{y}, \tau) \, d\mathbf{y} \, d\tau,$$

where we used that the probability to observe more than one photon is close to 0. Hence, if we denote the largest time in I_m by t_m, and use again $\log(1 - x) \approx -x$, then the number of photons observed until time t_m is approximately Poisson distributed with parameter $\lambda_{x,m} = p_{x,1} + \cdots + p_{x,m} \approx \int_0^{t_m} \int_{B_x} \lambda(y, \tau) \, dy \, d\tau$ ignoring the waiting times. This is in good agreement with (4.2). According to (4.7), the error in this approximation is bounded by

$$2 \sum_{i=1}^{m} \left(\log \left(1 - p_{x,i} \right) \right)^2 \leq 2\lambda_{x,m} \max_{1 \leq i \leq m} p_{x,i} \leq C \max_{1 \leq i \leq m} |I_i| \,,$$

revealing it valid whenever the temporal discretization is sufficiently fine.

4.4 Gaussian Modeling

4.4.1 As Approximation of the Binomial Model

Besides the approximation by a Poisson distribution, it is well-known that a Binomial model can also be approximated by a Gaussian one under suitable circumstances. Let us start with the Bernoulli model (4.4) and suppose that all $Y_{x,i}$ are independent with $p_{x,i} \equiv p_x$. If we are interested in the total number of counts $Y_x := \sum_{i=1}^{n} Y_{x,i}$ in bin x, the de Moivre-Laplace theorem states that

$$\frac{Y_x - np_x}{\sqrt{np_x (1 - p_x)}} \longrightarrow Z \quad \text{as} \quad n \to \infty \tag{4.8}$$

in distribution, where $Z \sim \mathcal{N}(0, 1)$ follows a standard normal distribution. Note that $\frac{Y_x - np_x}{\sqrt{np_x(1-p_x)}}$ is just the centered and standardized version of the total number of counts Y_x. This implies that the distribution of Y_x can be approximated by a Gaussian distribution with mean np_x and variance $np_x (1 - p_x)$ if n is sufficiently large. This gives rise to a first Gaussian model:

Gaussian model I

For each $x \in \Xi$, the number of photons observed in the bin centered at x up to time T is

$$Y_x \sim \mathcal{N}(np_x, np_x (1 - p_x)) \tag{4.9}$$

where $n = n(T) \sim T/\delta$ with the length δ of the individual time frames.

The rate of convergence in (4.8) can be made more precise. For instance a special case of the Berry-Esseen theorem states

$$\sup_{y \in \mathbb{R}} \left| \mathbb{P} \left[\frac{Y_{\mathbf{x}} - np_{\mathbf{x}}}{\sqrt{np_{\mathbf{x}} (1 - p_{\mathbf{x}})}} \le y \right] - \Phi(y) \right| < \frac{\sqrt{10} + 3}{6\sqrt{2\pi}} \frac{p_{\mathbf{x}}^2 + (1 - p_{\mathbf{x}})^2}{\sqrt{np_{\mathbf{x}} (1 - p_{\mathbf{x}})}} \qquad (4.10)$$

where Φ denotes the distribution function of $\mathcal{N}(0, 1)$, i.e.,

$$\Phi(y) = \frac{1}{\sqrt{2\pi}} \int_{-\infty}^{y} \exp\left(-\frac{x^2}{2}\right) dx.$$

In fact, the constant on the right-hand side of (4.10) cannot be improved [22]. An interpretation of this theorem is that the approximation leading to the model (4.9) is reasonable as soon as $np_{\mathbf{x}} (1 - p_{\mathbf{x}}) > 9$, which implies the right-hand side of (4.10) to be bounded by $\frac{\sqrt{10}+3}{18\sqrt{2\pi}} \approx 0.137$.

If the success probabilities $p_{\mathbf{x},i}$ do vary in i, the de Moivre-Laplace theorem (4.8) cannot be applied immediately. However, it is still possible, under certain conditions, to derive an approximate Gaussian model of the form (4.9) by applying the Lindeberg central limit theorem (see e.g. [23]). It states that the sum $Y_{\mathbf{x}}$, after centralization and standardization, still converges to $\mathcal{N}(0, 1)$ in distribution even for non identically distributed $Y_{\mathbf{x},i}$. This motivates a second Gaussian model:

Gaussian model II

For each $\mathbf{x} \in \Xi$, the number of photons observed in the bin centered at $\mathbf{x}$ up to time T is

$$Y_{\mathbf{x}} \sim \mathcal{N} \left(\sum_{i=1}^{n} p_{\mathbf{x},i}, \sum_{i=1}^{n} p_{\mathbf{x},i} \left(1 - p_{\mathbf{x},i}\right) \right). \qquad (4.11)$$

Note that, if the random variables $Y_{\mathbf{x},i}$ are dependent, the type of dependency very much determines whether a central limit theorem is still valid (with different limiting variance), see e.g. [24] or [25–27] for mixing sequences, and [28] for martingale difference sequences, to mention two large classes of examples.

4.4.2 As Approximation of the Poisson Model

The Poisson model in (4.2) can also be approximated by a Gaussian one. This relies on the fact that the Poisson distribution is infinitely divisible, which means that whenever $X \sim \text{Poi}(\mu)$, then X can be represented as $X = X_1 + \cdots + X_n$ for any $n \in \mathbb{N}$ with i.i.d. random variables $X_1, \ldots, X_n \sim \text{Poi}(\mu/n)$. Consequently, the central limit theorem states that

$$\frac{X - \mu}{\sqrt{\mu}} \longrightarrow Z, \quad \text{as} \quad \mu \to \infty$$

with $Z \sim \mathcal{N}(0, 1)$. The general Berry-Esseen theorem can also be used to bound the error of an approximation of $\frac{X-\mu}{\sqrt{\mu}}$ by Z, namely one obtains (see also [29])

$$\sup_{y \in \mathbb{R}} \left| \mathbb{P}\left[\frac{X - \mu}{\sqrt{\mu}} \leq y \right] - \Phi(y) \right| < \frac{5}{2} \frac{1}{\sqrt{\mu}}. \tag{4.12}$$

Hence, if μ is sufficiently large, the distribution of X can be approximated by a Gaussian distribution with mean and variance μ. If we suppose that $Y_{\mathbf{x},t}$ satisfies (4.2) and that $\int_0^t \int_{B_{\mathbf{x}}} \lambda(\mathbf{y}, \tau) \, d\mathbf{y} \, d\tau \to \infty$ as $t \to \infty$, then the above reasoning gives rise to another Gaussian model:

Gaussian model III

For each $\mathbf{x} \in \varXi$, the number of photons observed in the bin centered at $\mathbf{x}$ up to time t is

$$Y_{\mathbf{x},t} \sim \mathcal{N}\left(\int_0^t \int_{B_{\mathbf{x}}} \lambda(\mathbf{y}, \tau) \, d\mathbf{y} \, d\tau, \int_0^t \int_{B_{\mathbf{x}}} \lambda(\mathbf{y}, \tau) \, d\mathbf{y} \, d\tau \right). \tag{4.13}$$

4.4.3 Comparison

Let us briefly compare the Gaussian models I-III in (4.9), (4.11) and (4.13) respectively. It is clear that (4.11) is a generalization of (4.9) to the case of non-identical success probabilities $p_{\mathbf{x},i}$, and both coincide if $p_{\mathbf{x},i}$ is independent of i. To compare (4.11) with (4.13), we recall our previous computation that $p_{\mathbf{x},1} + \cdots + p_{\mathbf{x},n} = \int_0^{t_n} \int_{B_{\mathbf{x}}} \lambda(\mathbf{y}, \tau) \, d\mathbf{y} \, d\tau \to \infty$ where t_n is the largest time in the sub-interval I_n. Consequently, (4.11) and (4.13) differ only in the variance by $1 - p_{\mathbf{x},i}$, which is usually small. Hence, all three Gaussian models are in good agreement, and (4.13) can be considered the most simple one which should be used.

4.4.4 Thinning

Taking into account the detection efficiency $\eta \in [0, 1]$ as discussed before, we will arrive at models similar to (4.9), (4.11) and (4.13) with the only difference being that $p_{\mathbf{x}}$, $p_{\mathbf{x},i}$ or λ are multiplied by η. In this sense, the canonical thinning of the Poisson or Binomial models carries over to the Gaussian one.

4.4.5 Variance Stabilization

Note that the variance in the Gaussian models I-III is always inhomogeneous, which hinders data analysis with standard methods and causes further difficulties. This can be overcome by variance stabilization. The most popular choice is the celebrated Anscombe transform, which is applied to the Poisson model (4.2) to obtain asymptotically a normal distribution with variance 1. It is based on the following result (see e.g. [30, Lemma 1]):

Lemma 4.1 (Anscombe's transform) *Let $\mu > 0$ and $Y \sim \mathcal{P}(\mu)$ be a Poisson distributed random variable. Then it holds for all $c \geq 0$ that*

$$\mathbb{E}\left[2\sqrt{Y+c}\right] = 2\sqrt{\mu} + \frac{4c-1}{4\sqrt{\mu}} + \mathcal{O}\left(\frac{1}{\mu^{\frac{3}{2}}}\right),$$

$$\mathbb{V}\left[2\sqrt{Y+c}\right] = 1 + \frac{3-8c}{8\mu} + \mathcal{O}\left(\frac{1}{\mu^2}\right).$$

From this we can conclude that the choice $c = 3/8$ ensures that the variance of $2\sqrt{Y+c}$ does no longer depend on the parameter μ up to second order. To reduce the bias, $c = 1/4$ is the best choice. Furthermore, applying this result to the Poisson model in (4.2) gives rise to a fourth Gaussian model:

Gaussian model IV

For each $\mathbf{x} \in \varXi$, denote the number of photons observed in the bin centered at $\mathbf{x}$ up to time t by $Y_{\mathbf{x},t}$. Then we assume

$$2\sqrt{Y_{\mathbf{x},t} + \frac{3}{8}} \sim \mathcal{N}\left(2\left(\int_0^t \int_{B_{\mathbf{x}}} \lambda(\mathbf{y}, \tau) \, \mathrm{d}\mathbf{y} \, \mathrm{d}\tau\right)^{1/2}, 1\right) \qquad (4.14)$$

for each $\mathbf{x} \in \varXi$.

We emphasize the importance of the model (4.14) in statistics, as it turns out to be equivalent in a strict sense to the previously discussed Poisson model (4.2) as the total number of photons (and hence the parameter t) tends to ∞ (see e.g. [31–33]).

4.5 Conclusion

In this chapter we introduced models for photonic imaging setups with different degrees of accuracy. The most common and basic Poisson model (4.2) is accurate as soon as the temporal dependency can be neglected and the detector has no significant dead time. If furthermore the number of observed photons is sufficiently large on each bin, then the Gaussian model (4.13) can be used. In case of significant temporal dependency, the Bernoulli model (4.4) with time resolved individual photon arrivals or the resulting Binomial model (4.6) should be considered instead.

An overview about appropriate model choices for the various imaging techniques discussed previously is provided in Fig. 4.4.

In fluorescence microscopy, STED based methods, which scan the sample pixelwise, record about 10–100 photons per fluorescent marker. Due to low temporal dependencies, we are thus in the scope of the binomial or Poisson models [3]. Even though a Gaussian approximation seems questionable as in regions of low intensities only a few photons per bin can be collected, it has been successfully applied employing variance stabilizing techniques [34]. In order to analyze STORM/PALM data, the full range of modeling approaches is applied. Individual frames contain spots with single or several photons and weak temporal dependency, calling for Bernoulli, binomial, or Poisson models, while Gaussian approximations are used successfully for drift and rotational corrections [17]. FRET/MIET based imaging heavily relies on the interactions of fluorescent markers, so that the assumption of temporal independence is violated. This makes the Bernoulli model the model of choice, or if more photons are counted, also the Binomial model can be applied [4, 5].

Another example in the scope of the Bernoulli model is the 3-photon correlation technique (see e.g. Chap. 16), where molecular structures are probed by femtosecond X-ray pulses. This leads to a high number of images consisting of a few photons

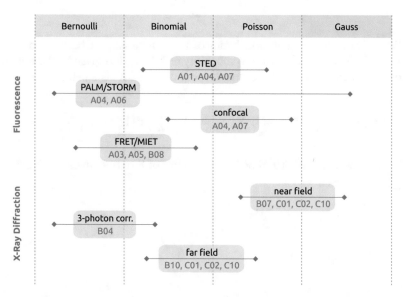

Fig. 4.4 Overview over viable model choices for different imaging methods. Projects of the SFB 755 associated to the respective methods are marked in gray

only, out of which only triples are used. Inference based on this sequence of images is additionally complicated by rotations of the single target molecules [1].

X-ray diffraction imaging also allows for a whole range of models. On first glance it seems that a Gaussian model is sufficient, as in total millions of photons are collected. However, depending on the specific setup, the photon intensity λ may vary strongly over the detection region. If imaging is performed in a near-field regime, as e.g. in many X-ray microscopy setups, the number of photons in the lower intensity regions is about one order of magnitude lower than in the high intensity regions, allowing for a Gaussian model. In contrast to this are far field methods where on high intensity bins 10^4 photons can be collected, but in low intensity regions only a handful of photons arrives, revealing a Binomial and/or Poisson model more suitable [12].

Acknowledgements We are grateful to Simon Maretzke, Tim Salditt and Britta Vinçon for several helpful comments.

Appendix: Poisson Thinning

Let $\mu > 0$, $\eta \in (0, 1)$ and suppose $Y \sim \mathcal{P}(\mu)$, $X_1, X_2, \ldots \sim \mathcal{B}(1, \eta)$ independent. The η-thinning of Y is defined as

$$\tilde{Y} := \sum_{i=1}^{Y} X_i.$$

We will now show that the distribution of $\tilde{Y}$ is still Poissonian, but with parameter $\eta \cdot \mu$. To this end, observe that the probability of $\tilde{Y}$ being k is given by the sum over all probabilities of Y being l and exactly k out of the first l X_i's being 1, i.e.

$$\mathbb{P}\left[\tilde{Y} = k\right] = \sum_{l=k}^{\infty} \mathbb{P}\left[Y = l, \sum_{i=1}^{l} X_i = k\right] = \sum_{l=k}^{\infty} \mathbb{P}[Y = l]\, \mathbb{P}\left[\sum_{i=1}^{l} X_i = k\right]$$

by independence. Inserting the Poisson distribution of Y and the Binomial distribution of $\sum_{i=1}^{l} X_i$ gives

$$\mathbb{P}\left[\tilde{Y} = k\right] = \sum_{l=k}^{\infty} \frac{\mu^l}{l!} \exp(-\mu) \binom{l}{k} \eta^k (1-\eta)^{l-k}$$

$$= \frac{\exp(-\mu)}{k!} (\eta\mu)^k \sum_{l=k}^{\infty} \frac{(\mu(1-\eta))^{l-k}}{(l-k)!}$$

$$= \frac{\exp(-\mu)}{k!} (\eta\mu)^k \exp(\mu(1-\eta))$$

$$= \frac{(\eta\mu)^k}{k!} \exp(-\mu\eta),$$

which proves $\tilde{Y} \sim \mathcal{P}(\eta\mu)$.

Appendix: Conditioned Poisson Processes

Suppose we observe a random number N of photons at random arrival times $0 \leq t_1 < \cdots < t_N \leq T$ such that the number of photons between time a and time b is Poisson distributed with parameter $\int_a^b \mu(t)\,dt$ for a fixed function $\mu \geq 0$. Given a decomposition of $[0, T]$ into disjoint intervals $I_1, \ldots, I_m$, denote by

$$Y_i := \#\left\{1 \leq j \leq N \mid t_j \in I_i\right\}, \qquad 1 \leq i \leq m$$

the number of photons observed during I_i. Assume furthermore that $Y_1, \ldots, Y_m$ are independent. We will now show that the conditional distribution given $N = n$ of $(Y_1, \ldots, Y_m)$ is multinomial with parameter n and probability vector $(p_1, \ldots, p_m)$ where

$$p_i = \frac{\int_{I_i} \mu(t)\,dt}{\int_0^T \mu(t)\,dt}.$$

Therefore let $n_1, \ldots, n_m \in \mathbb{N}_0$ such that $\sum_{i=1}^m n_i = n$. Then we have

$$\mathbb{P}\left[Y_1 = n_1, \ldots, Y_m = n_m \mid N = n\right] = \frac{\mathbb{P}\left[Y_1 = n_1, \ldots, Y_m = n_m, N = n\right]}{\mathbb{P}[N = n]}$$

$$= \frac{\prod_{i=1}^m \exp\left(-\int_{I_i} \mu(t)\,dt\right) \frac{1}{n_i!} \left(\int_{I_i} \mu(t)\,dt\right)^{n_i}}{\exp\left(-\int_0^T \mu(t)\,dt\right) \frac{1}{n!} \left(\int_0^T \mu(t)\,dt\right)^n}$$

$$= n! \prod_{i=1}^m \frac{p_i^{n_i}}{n_i!}$$

which proves the claim.

References

1. von Ardenne, B., Mechelke, M., Grubmüller, H.: Structure determination from single molecule x-ray scattering with three photons per image. Nat. Commun. **9**, 2375 (2018)
2. Bartels, M., Krenkel, M., Haber, J., Wilke, R.N., Salditt, T.: X-ray holographic imaging of hydrated biological cells in solution. Phys. Rev. Lett. **114**, 048,103 (2015). https://doi.org/10.1103/PhysRevLett.114.048103
3. Aspelmeier, T., Egner, A., Munk, A.: Modern statistical challenges in high-resolution fluorescence microscopy. Annu. Rev. Stat. Appl. **2**, 163–202 (2015)
4. Graen, T., Hoefling, M., Grubmüller, H.: Amber-dyes: Characterization of charge fluctuations and force field parameterization of fluorescent dyes for molecular dynamics simulations. J. Chem. Theory Comput. **10**(12), 5505–5512 (2014). https://doi.org/10.1021/ct500869p. PMID: 26583233
5. Michalet, X., Weiss, S., Jäger, M.: Single-molecule fluorescence studies of protein folding and conformational dynamics. Chem. Rev. **106**(5), 1785–1813 (2006). https://doi.org/10.1021/cr0404343
6. Adorf, H.M.: Hubble space telescope image restoration in its fourth year. Inverse Probl. **11**(4), 639 (1995). http://stacks.iop.org/0266-5611/11/i=4/a=003
7. Bertero, M., Boccacci, P., Desiderà, G., Vicidomini, G.: Image deblurring with Poisson data: from cells to galaxies. Inverse Probl. **25**(12), 025,004, 18 (2009). https://doi.org/10.1088/0266-5611/25/12/123006
8. Sawatzky, A., Brune, C., Wubbeling, F., Kosters, T., Schafers, K., Burger, M.: Accurate em-tv algorithm in pet with low snr. In: 2008 IEEE Nuclear Science Symposium Conference Record, pp. 5133–5137 (2008). https://doi.org/10.1109/NSSMIC.2008.4774392
9. Vardi, Y., Shepp, L.A., Kaufman, L.: A statistical model for positron emission tomography. J. Am. Stat. Assoc. **80**(389), 8–37 (1985). With discussion
10. Ta, H., Keller, J., Haltmeier, M., Saka, S.K., Schmied, J., Opazo, F., Tinnefeld, P., Munk, A., Hell, S.W.: Mapping molecules in scanning far-field fluorescence nanoscopy. Nat. Commun. **6**, 7977 (2015)
11. Pawley, J. (ed.): Handbook of Biological Confocal Microscopy. Springer (2006)
12. Hohage, T., Werner, F.: Inverse problems with poisson data: statistical regularization theory, applications and algorithms. Inverse Probl. **32**, 093,001, 56 (2016)
13. Kerstan, J., Matthes, K., Mecke, J.: Infinitely divisible point processes. Wiley Series in Probability and Mathematical Statistics. Wiley (1978)
14. Snyder, D.L., Helstrom, C.W., Lanterman, A.D., White, R.L., Faisal, M.: Compensation for readout noise in CCD images. J. Opt. Soc. Am. **12**(2), 272–283 (1995)
15. Snyder, D.L., White, R.L., Hammoud, A.M.: Image recovery from data acquired with a charge-coupled-device camera. J. Opt. Soc. Am. **10**(5), 1014–1023 (1993)
16. Geisler, C., Hotz, T., Schönle, A., Hell, S.W., Munk, A., Egner, A.: Drift estimation for single marker switching based imaging schemes. Opt. Express **20**(7), 7274–7289 (2012). https://doi.org/10.1364/OE.20.007274. http://www.opticsexpress.org/abstract.cfm?URI=oe-20-7-7274
17. Hartmann, A., Huckemann, S., Dannemann, J., Laitenberger, O., Geisler, C., Egner, A., Munk, A.: Drift estimation in sparse sequential dynamic imaging, with application to nanoscale fluoresence microscopy. J. Roy. Stat. Soc. Ser. B **78**(3), 563–587 (2016). https://doi.org/10.1111/rssb.12128
18. Harremoës, P., Johnson, O., Kontoyiannis, I.: Thinning and the law of small numbers. In: IEEE International Symposium on Information Theory, 2007. ISIT 2007, pp. 1491–1495. IEEE (2007)
19. Le Cam, L.: An approximation theorem for the poisson binomial distribution. Pac. J. Math. **10**(4), 1181–1197 (1960)
20. den Hollander, F.: Probability Theory: The Coupling Method (2012)
21. Novak, S.Y.: Extreme value methods with applications to finance. Monographs on Statistics and Applied Probability, vol. 122. CRC Press, Boca Raton, FL (2012)

22. Schulz, J.: The optimal berry-esseen constant in the binomial case. Ph.D. thesis, Univeristy of Trier (2016)
23. Billingsley, P.: Probability and Measure. Wiley (2008)
24. Peligrad, M.: On the central limit theorem for triangular arrays of ϕ-mixing sequences. In: Asymptotic Methods in Probability and Statistics (Ottawa, ON, 1997), pp. 49–55. North-Holland, Amsterdam (1998). https://doi.org/10.1016/B978-044450083-0/50005-8
25. Bradley, R.C.: Introduction to Strong Mixing Conditions, vol. 1. Kendrick Press, Heber City, UT (2007)
26. Bradley, R.C.: Introduction to Strong Mixing Conditions, vol. 2. Kendrick Press, Heber City, UT (2007)
27. Bradley, R.C.: Introduction to Strong Mixing Conditions, vol. 3. Kendrick Press, Heber City, UT (2007)
28. Shorack, G.R.: Probability for Statisticians. Springer Texts in Statistics. Springer (2000)
29. Lane, J.A.: The Berry-Esseen bound for the Poisson shot-noise. Adv. Appl. Probab. **19**(2), 512–514 (1987). https://doi.org/10.2307/1427432
30. Brown, L., Cai, T.T., Zhang, R., Zhao, L., Zhou, H.: The root-unroot algorithm for density estimation as implemented via wavelet block thresholding. Probab. Theory Relat. Fields **146**(3–4), 401–433 (2010)
31. Grama, I.: Gaussian approximation for nonparametric models. Tatra Mt. Math. Publ. **17**, 219–226 (1999)
32. Grama, I., Nussbaum, M.: Asymptotic equivalence for nonparametric regression. Math. Methods Stat. **11**(1), 1–36 (2002)
33. Ray, K., Schmidt-Hieber, J.: The le cam distance between density estimation, poisson processes and gaussian white noise. Math. Stat. Learn. **1**, 101–170 (2018)
34. Frick, K., Marnitz, P., Munk, A.: Statistical multiresolution estimation for variational imaging: with an application in Poisson-biophotonics. J. Math. Imaging Vis. **46**(3), 370–387 (2013). https://doi.org/10.1007/s10851-012-0368-5

Chapter 5
Inverse Problems

Thorsten Hohage, Benjamin Sprung and Frederic Weidling

> *The shortest path between two truths in the real domain passes through the complex domain.*
> —Jacques Hadamard

5.1 Introduction

5.1.1 What Is an Inverse Problem?

Generally speaking, inverse problems typically consist in the reconstruction of causes for observed effects. In imaging applications the cause is usually a probe and the effect are observed data. The corresponding forward problems then consists in predicting experimental data given perfect knowledge of the probe. In some sense solving an inverse problems means "computing backwards", which is usually more difficult then solving the forward problem.

To model these kind of problems mathematically we describe the imaging system or experimental setup by a forward operator $F\colon \mathbb{X} \to \mathbb{Y}$ between Banach spaces $\mathbb{X}$, $\mathbb{Y}$, which maps a probe $f \in \mathbb{X}$ to the corresponding effect $g \in \mathbb{Y}$. Then the inverse problem is, given data $g \in \mathbb{Y}$, to find a solution $f \in \mathbb{X}$ to the equation

$$F(f) = g. \tag{5.1}$$

T. Hohage (✉) · B. Sprung · F. Weidling
Institute for Numerical and Applied Mathematics, Universität Göttingen, Lotzestr. 16-18, 37083 Göttingen, Germany
e-mail: hohage@math.uni-goettingen.de

B. Sprung
e-mail: b.sprung@math.uni-goettingen.de

F. Weidling
e-mail: f.weidling@math.uni-goettingen.de

© The Author(s) 2020
T. Salditt et al. (eds.), *Nanoscale Photonic Imaging*, Topics in Applied Physics 134,
https://doi.org/10.1007/978-3-030-34413-9_5

5.1.2 Ill-Posedness and Regularization

A first obvious question to ask is whether or not a probe f is uniquely determined by the data g, i.e. if the operator F is injective. Such questions are addressed e.g. in Sect. 13.6.6. Given uniqueness, one might try to find the solution f to (5.1) by just applying the inverse operator F^{-1}, which gives another reason to call the problem anss inverse problem. However in practice this can cause several problems, due to the forward operator being *ill-posed*.

According to J. Hadamard a problem is called *well-posed* if the following conditions are satisfied:

1. There exists a solution.
2. The solution is unique.
3. The solution depends continuously on the data (stability).

Otherwise the problem is called *ill-posed*.

An inverse problem in the form of an operator equation (5.1) is well-posed if F is surjective (such that for all g there exists a solution), injective (such that the solution is unique) and if F^{-1} is continuous (guaranteeing stability). For many inverse problems in practice only the third condition is violated, and ill-posedness in the narrower sense often refers to this situation: The reconstruction of causes from observed effects is unstable since very different causes may have similar effects.

The remedy against ill-posedness is *regularization:* To obtain stable solutions of inverse problems one constructs a family of continuous operators $R_\alpha : \mathbb{Y} \to \mathbb{X}$ parameterized by a parameter $\alpha > 0$ converging pointwise to the discontinuous inverse F^{-1}:

$$R_\alpha \approx F^{-1}, \qquad \lim_{\alpha \to 0} R_\alpha(F(f)) = f \qquad \text{for all } f \in \mathbb{X}. \tag{5.2}$$

We will discuss several generic constructions of such families of stable approximate inverses R_α in §5.2.

5.1.3 Examples

5.1.3.1 Numerical Differentiation

In our first example we consider the forward operator given by integration. We fix the free integration constant by working in spaces of functions with mean zero, $L^2_\diamond([0, 1]) := \{f \in L^2([0, 1]): \int_0^1 f(x)\, dx = 0\}$. Let $F: L^2_\diamond([0, 1]) \to L^2_\diamond([0, 1])$ be given by

$$F(f)(y) = \int_0^y f(x)\, dx + c(f), \qquad y \in [0, 1]. \tag{5.3}$$

where $c(f)$ is such that $F(f) \in L_\diamond^2$. The corresponding inverse problem described by the operator equation (5.1) is to compute the derivative g'. If $g = F(f)$, then the existence of a solution to the inverse problem is guaranteed and this solution is unique. Now assume that instead of the exact data g we are given noisy data g^{obs} that fulfill

$$\|g^{obs} - g\|_{L^2} \le \delta,$$

for some small noise level $\delta > 0$. For example this could be one of the functions

$$g_n^{obs}(x) = g(x) + \delta\sqrt{2}\sin(\pi n x).$$

As the derivatives are given by $(g_n^{obs})'(x) = g'(x) + \pi n \delta \sqrt{2}\cos(\pi n x)$, we have

$$\|(g_n^{obs})' - g'\|_{L^2} = \pi n \delta.$$

This illustrates the typical ill-posedness of inverse problems: amplification of noise may be arbitrarily large for naive application of the inverse F^{-1}. This is the main difficulty one has to cope with. Our toy example illustrates another typical feature of inverse problems: Each function in the image of the operator F defined above is at least once differentiable. Also for many other inverse problems the forward operator is smoothing in the sense that the output function has higher smoothness than the input function, and this property causes the instability of the inverse problem.

5.1.3.2 Fluorescence Microscopy

In fluorescence microscopy one is interested in recovering the density f of fluorescent markers in some specimen in $\mathbb{R}^3$. The probe is sampled by a laser beam, and one detects fluorescent photons. In confocal microscopy, spatial resolution is achieved by focusing the laser beam by some lense and collecting fluorescent electrons by same lense such that out-of-focus fluorescent photons can be blocked by a pinhole.

Let $y \in \mathbb{R}^3$ be the focal point and assume that the probability (density) that for the focus point $y \in \mathbb{R}^3$ a fluorescent photon emitted by a marker at point $x \in \mathbb{R}^3$ is detected is $k(x - y)$. k is called the *point-spread function*, and we assume here that it is spatially invariant. Then our problem is described by the operator equation

$$g(y) = F(f)(y) = \int k(y - x) f(x) \, dx,$$

i.e. the observation g is given by a convolution of the marker density f with point spread function k. As convolution will usually blur an image, the forward operator is smoothing. Smoother kernels will lead to stronger smoothing.

5.1.4 Choice of Regularization Parameters and Convergence Concepts

Due to the ill-posedness discussed above it is essential to take into account the effects of noise in the observed data. Let $f^\dagger \in \mathbb{X}$ denote the unknown exact solution, and first assume data g^δ with deterministic errors such that

$$\|g^\delta - F(f^\dagger)\|_{\mathbb{Y}} \le \delta. \tag{5.4}$$

As mentioned above, regularization of an ill-posed operator equation (5.1) with an injective operator F consists in approximating the discontinuous inverse operator F^{-1} by a pointwise convergent family of continuous operators $R_\alpha : \mathbb{Y} \to \mathbb{X}, \alpha > 0$. This immediately gives rise to the question which operator in the family should be chosen for the reconstruction, i.e how to choose the parameter α. Usually the starting point of deterministic error analysis in regularization theory is the following splitting of the reconstruction error:

$$\|R_\alpha(g^\delta) - f^\dagger\| \le \|R_\alpha(g^\delta) - R_\alpha(F(f^\dagger))\| + \|R_\alpha(F(f^\dagger)) - f^\dagger\|. \tag{5.5}$$

The first term on the right hand side is called *propagated data noise error*, and the second term is referred to as *approximation error* or *bias*. Due to pointwise convergence (see (5.2)), the bias tends to 0 as $\alpha \to 0$. Hence, to control this error term, we should choose α as small as possible. However, as R_α converges pointwise to the discontinuous operator F^{-1}, the Lipschitz constant (or operator norm in the linear case) of R_α will explode as $\alpha \to 0$, and hence also the propagated data noise error. Therefore, α must not be chosen too small. This indicates that the choice of the regularization parameter must be a crucial ingredient of a reliable regularization method. Probably the most well-known parameter choice rule in the deterministic setting is *Morozov's discrepancy principle*:

$$\overline{\alpha}_{\mathrm{DP}}(\delta, g^\delta) := \sup\{\alpha > 0 : \|F(R_\alpha(g^\delta)) - g^\delta\| \le \tau\delta\} \tag{5.6}$$

with some parameter $\tau \ge 1$. In other words, among all estimators $R_\alpha(g^\delta)$ which can explain the data within the noise level (times τ), we choose the most stable one.

Definition 5.1 A family of operators $R_\alpha : \mathbb{Y} \to \mathbb{X}$ parameterized by a parameter $\alpha > 0$ together with some rule $\overline{\alpha} : [0, \infty) \times \mathbb{Y} \to (0, \infty)$ how to choose this parameter depending on the noise level δ and the data g^δ is called a *regularization method* if the worst case error tends to 0 with the noise level in the sense that

$$\lim_{\delta \to 0} \sup \left\{ \|R_{\overline{\alpha}(\delta, g^\delta)}(g^\delta) - f^\dagger\|_{\mathbb{X}} : g^\delta \in \mathbb{Y}, \|g^\delta - F(f^\dagger)\|_{\mathbb{Y}} \le \delta \right\} = 0 \tag{5.7}$$

for all $f^\dagger \in \mathbb{X}$.

The convergence (5.7) is a minimal requirement that one expects from a regularization method. However, it can be shown that for ill-posed problems this convergence may be arbitrarily slow depending on $f^\dagger$. This is of course not satisfactory. Fortunately, a-priori information on the solution $f^\dagger$, which is often available, may help. If we know a-priori that $f^\dagger$ belongs to some set $\mathcal{K} \subset \mathbb{X}$, then it is often possible to derive explicit error bounds

$$\sup \left\{ \|R_{\overline{\alpha}(\delta, g^\delta)}(g^\delta) - f^\dagger\|_{\mathbb{X}} : g^\delta \in \mathbb{Y}, \|g^\delta - F(f^\dagger)\|_{\mathbb{Y}} \le \delta \right\} \le \psi(\delta)$$

for all $f^\dagger \in \mathcal{K}$ with a function $\psi \in C([0, \infty))$ satisfying $\psi(0) = 0$.

Let us now consider statistical noise models instead of the deterministic noise model (5.4). Often statistical data G_t belong to a different space $\mathbb{Y}'$, e.g. a space of distributions. The distribution depends on some parameter t, and we assume that $G_t \to F(f^\dagger)$ in some sense as $t \to \infty$. Let t e.g. denote the number of observations or in photonic imaging the expected number of photons. As the estimator $R_\alpha(g^{\text{obs}})$ (where now R_α is a mapping from $\mathbb{Y}'$ to $\mathbb{X}$) will be a random variable, we have to use stochastic concepts of convergence, e.g. convergence in expectation. Other convergence concepts, in particular convergence in probability are also used frequently.

Definition 5.2 In the setting above, a family of operators $R_\alpha : \mathbb{Y}' \to \mathbb{X}$ parameterized by a parameter $\alpha > 0$ together with some parameter choice rule $\overline{\alpha} : [0, \infty) \times \mathbb{Y}' \to (0, \infty)$ is called a *consistent estimator* if

$$\lim_{t \to \infty} \mathbb{E}\left[\|R_{\overline{\alpha}(t, G_t)}(G_t) - f^\dagger\|_{\mathbb{X}}^2 \right] = 0$$

for all $f^\dagger \in \mathbb{X}$.

Again one may further ask not only for convergence, but even rates of convergence as $t \to \infty$ on certain subsets $\mathcal{K} \subset \mathbb{X}$.

5.2 Regularization Methods

In this section we will discuss generalized Tikhonov regularization, which is given by finding the minimum of

$$\hat{f}_\alpha \in \text{argmin}_{f \in \mathbb{X}} \left[\mathcal{S}_{g^{\text{obs}}}(F(f)) + \alpha \mathcal{R}(f) \right], \tag{5.8}$$

where $\mathcal{S}_{g^{\text{obs}}}$ is the *data fidelity functional*, which measures some kind of distance of $F(f)$ and the data g^{obs} and causes the minimizer $\hat{f}_\alpha$ to still explain the data well, whereas $\mathcal{R}$ is the *penalty functional* which penalizes certain properties of the minimizer. This approach is called variational regularization, as our regularized solution is found by minimization of a functional (usually an integral functional).

5.2.1 Variational Regularization

We start with a probabilistic motivation of generalized Tikhonov regularization. From here until the end of this section we will consider the finite-dimensional setting $\mathbb{X} = \mathbb{R}^n$, $\mathbb{Y} = \mathbb{R}^m$ and use boldface symbols to denote finite-dimensional vectors/mappings. We start from (5.1), where $\mathbf{f} \in \mathbb{R}^n$, $\mathbf{g} \in \mathbb{R}^m$ and $F : \mathbb{R}^n \to \mathbb{R}^m$ is some injective function. Given the data $\mathbf{g}$ we want to find the solution $\mathbf{f}$ of $F(\mathbf{f}) = \mathbf{g}$, but recall that we cannot just apply the inverse F^{-1} as discussed in Sect. 5.1. Instead we might estimate $\mathbf{f}$ by maximizing the likelihood function $\mathcal{L}(\mathbf{f}) = P(\mathbf{g}|\mathbf{f})$, i.e. the probability that for a certain preimage $\mathbf{f}$ the data $\mathbf{g}$ will occur. If we assume that our data is normally distributed with covariance matrix $\sigma^2 I$, then we can rearrange the problem by using the monotonicity of the logarithm, as well as the fact, that neither additive nor multiplicative constants change the extremal point:

$$\hat{\mathbf{f}}_{\text{ML}} \in \text{argmax}_{\mathbf{f} \in \mathbb{R}^n} P(\mathbf{g}|\mathbf{f}) = \text{argmax}_{\mathbf{f} \in \mathbb{R}^n} \log\left(P(\mathbf{g}|\mathbf{f})\right)$$

$$= \text{argmax}_{\mathbf{f} \in \mathbb{R}^n} \log\left(\frac{1}{\sqrt{2\pi}\sigma} \prod_{i=1}^{m} \exp\left(\frac{-(\mathbf{g}_i - F(\mathbf{f})_i)^2}{2\sigma^2}\right)\right)$$

$$= \text{argmin}_{\mathbf{f} \in \mathbb{R}^n} \frac{1}{2\sigma^2} \sum_{i=1}^{m} (\mathbf{g}_i - F(\mathbf{f})_i)^2 = \text{argmin}_{\mathbf{f} \in \mathbb{R}^n} \frac{1}{2\sigma^2} \|\mathbf{g} - F(\mathbf{f})\|_2^2.$$

This demonstrates the well-known fact that the maximum likelihood approach for Gaussian noise yields the least squares methods as first used by young Gauss to predict the path of the asteroid Ceres in 1801. However, as $\hat{\mathbf{f}}_{\text{ML}} = F^{-1}(\mathbf{g})$ for $\mathbf{g}$ in the range of F, this approach has no regularizing effect. In fact it is more reasonable to maximize $P(\mathbf{f}|\mathbf{g})$ instead of $P(\mathbf{g}|\mathbf{f})$, as our goal should be to find the solution $\mathbf{f}$ which is most likely to have caused the observation $\mathbf{g}$, instead of just finding any $\mathbf{f}$ which causes the observation $\mathbf{g}$ with maximal probability. This leads to the Bayesian perspective on inverse problem with the characteristic feature that prior to the measurements a probability distribution (the so-called *prior distribution*) is assigned to the solution space $\mathbb{X}$ modeling our prior knowledge on $\mathbf{f}$. By Bayes' theorem we have

$$P(\mathbf{f}|\mathbf{g}) = \frac{P(\mathbf{g}|\mathbf{f})P(\mathbf{f})}{P(\mathbf{g})} \quad \Leftrightarrow \quad \text{posterior} = \frac{\text{likelihood} \cdot \text{prior}}{\text{evidence}}.$$

Estimating $\mathbf{f}$ by maximizing the posterior $P(\mathbf{f}|\mathbf{g})$ is called maximum a posteriori probability (MAP) estimate. To use this approach we have to model the prior $P(\mathbf{f})$. If we assume that $\mathbf{f}$ is normally distributed with mean $\mathbf{f} \in \mathbb{R}^n$ and covariance matrix $\tau^2 I$, then we find

$$\hat{\mathbf{f}}_{\text{MAP}} \in \text{argmax}_{\mathbf{f} \in \mathbb{R}^n} P(\mathbf{f}|\mathbf{g}) = \text{argmax}_{\mathbf{f} \in \mathbb{R}^n} \left[\log\left(P(\mathbf{g}|\mathbf{f})\right) + \log\left(P(\mathbf{f})\right) \right]$$

$$= \text{argmin}_{\mathbf{f} \in \mathbb{R}^n} \frac{1}{2\sigma^2} \left[\sum_{i=1}^{m} (\mathbf{g}_i - F(\mathbf{f})_i)^2 + \frac{\sigma^2}{\tau^2} \sum_{j=1}^{m} (\mathbf{f}_j - (\mathbf{f}_0)_j)^2 \right]$$

$$= \text{argmin}_{\mathbf{f} \in \mathbb{R}^n} \left[\underbrace{\tfrac{1}{2}\|\mathbf{g} - F(\mathbf{f})\|_2^2 + \tfrac{\alpha}{2}\|\mathbf{f} - \mathbf{f}_0\|_2^2}_{J_\alpha(\mathbf{f})} \right]$$

where $\alpha = \frac{\sigma^2}{\tau^2}$. The functional $J_\alpha(\mathbf{f})$ is the standard (quadratic) Tikhonov functional, and therefore MAP and Tikhonov regularization coincide in this setting.

In photonic imaging the data is often given by photon counts, and these are typically Poisson distributed in the absence of read-out error (compare Chap. 4). Recall that a random variable $Z \in \mathbb{N}_0$ is called Poisson distributed with mean $\lambda > 0$, short $Z \sim \text{Pois}(\lambda)$, if $P(Z = g) = e^{-\lambda}\lambda^g/(g!)$ for all $g \in \mathbb{N}_0$. Hence the negative log-likelihood function is given by

$$- \log P(Z = g|\lambda) = \lambda - g\log(\lambda) + \log g! = \lambda - g + g \log\left(\frac{g}{\lambda}\right) + C_g$$

where C_g is a constant independent of λ. Now assume that $\mathbf{g}$ is a vector of independent Poisson distributed random variables such that $\mathbf{g}_i \sim \text{Pois}(F(\mathbf{f})_i)$. It follows that the negative log-likelihood $-\log(P(\mathbf{g}|\mathbf{f})) = -\sum_i \log P(\mathbf{g}_i|F(\mathbf{f})_i)$ is given (up to an additive constant independent of $\mathbf{f}$) by the Kullback-Leibler divergence

$$\text{KL}(\mathbf{g}, F(\mathbf{f})) := \sum_{i=1}^{m} \left[F(\mathbf{f})_i - \mathbf{g}_i + \mathbf{g}_i \log\left(\frac{\mathbf{g}_i}{F(\mathbf{f})_i}\right) \right].$$

If $\mathbf{f}$ has a probability distribution $P(f) = c\exp(-\mathcal{R}(\mathbf{f})/\tau^2)$ this leads to generalized Tikhonov regularization

$$\hat{\mathbf{f}}_\alpha \in \text{argmin}_{\mathbf{f} \in \mathbb{R}^n} \left[\mathcal{S}_{\mathbf{g}}(F(\mathbf{f})) + \alpha\mathcal{R}(\mathbf{f}) \right] \tag{5.9}$$

with fidelity term $\mathcal{S}_{\mathbf{g}} = \|\mathbf{g} - \cdot\|_2^2$ for normally distributed data, $\mathcal{S}_{\mathbf{g}} = \text{KL}(\mathbf{g}, \cdot)$ for Poisson distributed data and the penalty term $\alpha\mathcal{R}$ with regularization parameter $\alpha = \tau^{-2}$ for Poisson data and $\alpha = \sigma^2/\tau^2$ for Gaussian white noise.

Note that in the Bayesian setting above the regularization parameter α is uniquely determined by the prior distribution and the likelihood functional. However, often only qualitative prior knowledge on the solution is available, but the parameter τ is unknown. Then α has to be determined by a-posteriori parameter choice rules analogous to the discrepancy principle (5.6) for deterministic errors.

Let us discuss a few popular choices of the penalty functional $\mathcal{R}$, which allows the incorporation of prior information on the solution or is simply the negative logarithm of the density of the prior in the above Bayesian setting. For a-priori known sparsity of the solution one should choose the sparsity enforcing penalty $\mathcal{R}(\mathbf{f}) = \|\mathbf{f}\|_1$. If $\mathbf{f}$ is an image with sharp edges, then the total variation seminorm is a good choice of $\mathcal{R}$. However, we point out that for the total variation seminorm in a Bayesian setting there exists no straightforward useful infinite dimensional limit (see [16]). Bayesian prior modelling in infinite dimensional settings is often considerably more involved.

If $\mathbf{f}$ is a probability density, a frequent choice of the penalty functional is $\mathcal{R}(\mathbf{f}) = \mathrm{KL}(\mathbf{f}, \mathbf{f}_0)$, which naturally enforces nonnegativity of the solution because of the logarithm. Alternatively, more general inequality constraints $N(\mathbf{f}) \leq 0$ for some function N can be be incorporated into the penalty function by replacing $\mathcal{R}$ by

$$\widetilde{\mathcal{R}}(\mathbf{f}) = \begin{cases} \mathcal{R}(\mathbf{f}), & \text{if } N(\mathbf{f}) \leq 0 \\ \infty, & \text{else.} \end{cases}$$

5.2.1.1   Implementation

In this paragraph we will discuss several possibilities to compute the minimizer $\hat{\mathbf{f}}_\alpha$ of (5.9) for a linear forward operator denoted by $F = T$. In the case of quadratic Tikhonov regularization it follows from the first order optimality conditions that the Tikhonov functional J_α has the unique minimizer

$$\hat{\mathbf{f}}_\alpha = (T^*T + \alpha I)^{-1} (T^*\mathbf{g} + \alpha \mathbf{f}_0) \tag{5.10}$$

for all $\alpha > 0$. So in order to compute the regularized solution we have to solve the linear system of equations

$$A\mathbf{f} = \mathbf{b} \quad \text{with} \quad A := T^*T + \alpha I \text{ and } \mathbf{b} = T^*\mathbf{g} + \alpha \mathbf{f}_0.$$

Solving this directly by for example Gauss-Jordan elimination requires $\mathcal{O}(n^3)$ operations and we have to store the full matrix A. In imaging applications n is typically the number of pixels or voxels, which can be so large that storing A is impossible. Therefore, we have to resort to iterative methods which access A only via matrix-vector products. Such matrix-vector products can often be implemented efficiently without setting up the matrix A, e.g. by the fast Fourier transform (FFT) or by solving a partial differential equation. As A is positive definite, the most common method for solving $A\mathbf{f} = \mathbf{b}$ is the conjugate gradient (CG) method:

Algorithm 5.1 *(CG iteration for solving $A\mathbf{f} = \mathbf{b}$, $A > 0$)*
Initialization. Choose initial guess $\mathbf{f}_0 \in \mathbb{R}^n$. Set $\mathbf{s}_0 = \mathbf{b} - A\mathbf{f}_0$; $\mathbf{d}_0 = \mathbf{s}_0$.
General Step $(l = 0, 1, \ldots)$

$$\gamma_l = \frac{\|\mathbf{s}_l\|_2^2}{\langle \mathbf{d}_l, A\mathbf{d}_l \rangle}$$

$$\mathbf{f}_{l+1} = \mathbf{f}_l + \gamma_l \mathbf{d}_l$$

$$\mathbf{s}_{l+1} = \mathbf{s}_l + \gamma_l A\mathbf{d}_l$$

$$\beta_l = \frac{\|\mathbf{s}_{l+1}\|_2^2}{\|\mathbf{s}_l\|_2^2}$$

$$\mathbf{d}_{l+1} = \mathbf{s}_{l+1} + \beta_l \mathbf{d}_l.$$

If $A = T^*T + \alpha I$ as in Tikhonov regularization, the stopping criterion $\mathbf{s} \neq 0$ may be replaced by

$$\|\mathbf{s}_l\| > \frac{\text{TOL}}{\alpha}$$

for some tolerance parameter $\text{TOL} > 0$. It can be shown that $\mathbf{s}_l = \mathbf{b} - A\mathbf{f}_l$ for all l. As $\|A^{-1}\| \leq 1/\alpha$, this guarantees the error bound $\|\bar{\mathbf{f}} - \mathbf{f}_L\| \leq \text{TOL}$ to the exact minimum $\bar{\mathbf{f}} = A^{-1}b$ of the Tikhonov functional.

In the case of more general data fidelity terms $\mathcal{S}_\mathbf{g}$ and penalty terms $\mathcal{R}$ one can use a primal-dual algorithm suggested by Chambolle and Pock [3]. To formulate this algorithm, recall that for a functional $\mathcal{F}: \mathbb{R}^n \to \mathbb{R} \cup \{\infty\}$ the conjugate functional is given by

$$\mathcal{F}^*(\mathbf{s}) := \sup_{\mathbf{x} \in \mathbb{R}^n} [\langle \mathbf{s}, \mathbf{x} \rangle - \mathcal{F}(\mathbf{x})]. \tag{5.11}$$

If $\mathcal{F}$ is convex and continuous, then $\mathcal{F}^{**} = \mathcal{F}$. For more information on this and other basic notions of convex analysis used in the following we refer, e.g. to [17]. The algorithm is based on the saddle point formulation

$$\min_{\mathbf{f} \in \mathbb{R}^n} \max_{\mathbf{z} \in \mathbb{R}^m} [\langle T\mathbf{f}, \mathbf{z} \rangle + \alpha \mathcal{R}(\mathbf{f}) - \mathcal{S}_\mathbf{g}^*(\mathbf{z})] = \max_{\mathbf{z} \in \mathbb{R}^m} \min_{\mathbf{f} \in \mathbb{R}^n} [\langle T\mathbf{f}, \mathbf{z} \rangle + \alpha \mathcal{R}(\mathbf{f}) - \mathcal{S}_\mathbf{g}^*(\mathbf{z})].$$

Note that an analytic computation of the maximum leads to the original problem (5.9) whereas a computation of the minimum leads to the dual problem

$$\max_{\mathbf{z} \in \mathbb{R}^m} - \left[\mathcal{S}_\mathbf{g}^*(\mathbf{z}) + \alpha \mathcal{R}^* \left(-\frac{1}{\alpha} T^* \mathbf{z} \right) \right] \tag{5.12}$$

The algorithm requires the computation of so called *proximity operators* (see Chap. 6). For a functional $G: \mathbb{R}^n \to \mathbb{R}$ and a scalar $\lambda > 0$ the proximity operator $\text{prox}_{G,\lambda}: \mathbb{R}^n \to \mathbb{R}^n$ is defined by

$$\text{prox}_{G,\lambda} := (\mathbf{z}) = \text{argmin}_{\mathbf{x} \in \mathbb{R}^n} \left[\tfrac{1}{2} \|\mathbf{z} - \mathbf{x}\|_2^2 + \lambda G(\mathbf{x}) \right].$$

For many popular choices of $\mathcal{S}_\mathbf{g}$ and $\mathcal{R}$ the proximity operator can either be calculated directly via a closed expression or there are efficient algorithms for their computation.

To give a simple example, for $G(\mathbf{x}) = \frac{1}{2}\|x\|_2^2$ one can calculate the proximity operator just by (5.10) to be $(1 + \lambda)^{-1}\mathbf{z}$.

One also needs to evaluate the proximity operator of the *Fenchel convex conjugate* G^*, which can be done by Moreau's identity

$$\mathrm{prox}_{G^*,\lambda}(\mathbf{z}) = \mathbf{z} - \lambda\,\mathrm{prox}_{G,1/\lambda}\left(\tfrac{1}{\lambda}\mathbf{z}\right).$$

Algorithm 5.2 *(Chambolle-Pock primal dual algorithm)*
Initialization. Choose $(\mathbf{f}_0, \mathbf{p}_0) \in \mathbb{R}^n \times \mathbb{R}^m$ and set $\tilde{\mathbf{f}}_0 = \mathbf{f}_0$.
General Step $(k = 0, 1, \ldots)$ Choose parameters $\tau_X^k, \tau_Y^k > 0$, $\theta_k \in [0, 1]$ and let

$$\mathbf{p}_{k+1} = \mathrm{prox}_{\mathcal{S}_{\mathbf{g}}, \tau_Y^k}\left(\mathbf{p}_k + \tau_Y^k T \tilde{\mathbf{f}}_k\right)$$

$$\mathbf{f}_{k+1} = \mathrm{prox}_{\alpha\mathcal{R}, \tau_X^k}\left(\mathbf{f}_k - \tau_X^k T^* \mathbf{p}_k\right)$$

$$\tilde{\mathbf{f}}_{k+1} = \theta_k\left(\mathbf{f}_{k+1} - \mathbf{f}_k\right).$$

For constant parameters $\tau_X^k, \tau_Y^k > 0$ and $\theta_k = 1$ it has been shown [3] that $\mathbf{f}_k$ converges to a solution of (5.9) and $\mathbf{p}_k$ converges to a solution of the corresponding dual problem. Under certain assumptions on $\mathcal{S}_{\mathbf{g}}$ and $\mathcal{R}$ special choices for the parameters $\tau_X^k, \tau_Y^k > 0$ and θ_k will speed up the convergence.

To compute the minimizer of (5.9) with additional inequality constraints one can apply semismooth Newton methods for which we refer to [19].

5.2.2 Iterative Regularization

Whereas in the previous subsection we have discussed iterative methods to compute the minimum of a generalized Tikhonov functional, in the following we will discuss iterative methods for the solution of $F(\mathbf{f}) = \mathbf{g}$ without prior regularization. Typically the choice of the stopping index plays the role of the choice of the regularization parameter α. A motivation for the use of iterative regularization method is the fact that the Tikhonov functional is not convex in general for nonlinear operators. Therefore, it cannot be guaranteed that an approximation to the global minimum of the Tikhonov functional can be computed. For further information on iterative regularization methods we refer to the monograph [15].

5.2.2.1 Landweber Iteration

Landweber iteration can be derived as a method of steepest descent for the cost functional $J(\mathbf{f}) = \frac{1}{2}\|F(\mathbf{f}) - \mathbf{g}\|_2^2$. As the direction of steepest descent is given by the negative gradient $-J'(\mathbf{f}) = -F'(\mathbf{f})^*(F(\mathbf{f}) - \mathbf{g})$, this leads to the iteration formula

$$\mathbf{f}_{k+1} = \mathbf{f}_k - \mu F'(\mathbf{f}_k)^*\left(F(\mathbf{f}_k) - \mathbf{g}\right) \tag{5.13}$$

with a step size parameter μ. This parameter should be chosen such that $\mu \| F'(\mathbf{f})^* F'(\mathbf{f}) \| < 1$ for all $\mathbf{f}$. Since a too small choice of μ slows down convergence considerably, it is advisable to compute the operator norm of $F'(\mathbf{f})^* F'(\mathbf{f})$ by a few iterations of the power method.

Under certain conditions on the operator F it has been shown in [7] in a Hilbert space setting that Landweber iteration with the discrepancy principle as stopping rule is a regularization method in a sense analogous to Definition 5.1, i.e. the worst case error tends to 0 with the noise level for a sufficiently good initial guess.

5.2.2.2 Regularized Newton Methods

Although Landweber iteration often makes good progress in the first few iterations, asymptotic convergence is very slow. Faster convergence may be expected from Newton-type methods, which solve a linear system of equations or some minimization problem using the first order Taylor approximation

$$F(\mathbf{f}) \approx F(\mathbf{f}_k) + F'(\mathbf{f}_k)(\mathbf{f} - \mathbf{f}_k). \tag{5.14}$$

around a current iterate $\mathbf{f}_k$. Plugging this approximation into the quadratic Tikhonov functional and using the last iterate as initial guess leads to the *Levenberg-Marquardt-algorithm*

$$\mathbf{f}_{k+1} = \operatorname{argmin}_{\mathbf{f} \in \mathbb{R}^n} \left[\tfrac{1}{2} \| F'(\mathbf{f}_k)(\mathbf{f} - \mathbf{f}_k) + F(\mathbf{f}_k) - \mathbf{g} \|_2^2 + \tfrac{\alpha_k}{2} \| \mathbf{f} - \mathbf{f}_k \|_2^2 \right].$$

For a convergence analysis we refer to [6]. The minimization problems can be solved efficiently by Algorithm 5.1 without the need to set up the full Jacobi matrices $F'(\mathbf{f}_k)$ in each step. Newton-type methods converge considerably faster than Landweber iteration. In fact the number of Landweber steps which is necessary to achieve an accuracy comparable to k Newton steps increases exponentially with k (cf. [4]). On the other hand, each iteration step is more expensive. Which of the two methods is more efficient may depend on the size of the noise level. Newton-type methods are typically favorable for small noise levels. Plugging the first order Taylor approximation (5.14) into the generalized Tikhonov functional (5.8) leads to the iteration formula

$$\mathbf{f}_{k+1} \in \operatorname{argmin}_{\mathbf{f}} \left[\mathcal{S}_{g^{\mathrm{obs}}}(F(\mathbf{f}_k) + F'(\mathbf{f}_k)(\mathbf{f} - \mathbf{f}_k)) + \alpha \mathcal{R}(f) \right]. \tag{5.15}$$

If $\mathcal{S}_{g^{\mathrm{obs}}}(\mathbf{g}) = \tfrac{1}{2} \| \mathbf{g} - g^{\mathrm{obs}} \|^2$ and $\mathcal{R}(\mathbf{f}) = \tfrac{1}{2} \| \mathbf{f} - \mathbf{f}_0 \|$, this leads the commonly used *iteratively regularized Gauss-Newton method* where compared to the Levenberg-Marquardt method the penalty term is replaced by $\tfrac{\alpha_k}{2} \| \mathbf{f} - \mathbf{f}_0 \|_2^2$. Note that if $\mathcal{S}_{g^{\mathrm{obs}}}$ and $\mathcal{R}$ are convex, a convex optimization problems has to be solved in each Newton step, which can be done, e.g., by Algorithm 5.2. For a convergence analysis of (5.15) including the case of Poisson data we refer to [12].

5.3 Error Estimates

5.3.1 General Error Bounds for Variational Regularization

Under the deterministic noise model (5.4) we are now looking for error bounds of
the form

$$\|\hat{f}_\alpha - f^\dagger\|^2 \le \phi(\delta, \alpha) \tag{5.16}$$

for the Tikhonov estimator $\hat{f}_\alpha$. In a second step the right hand side may be minimized over α. As discussed in Sect. 5.1.4 such estimates can only be obtained under
additional conditions on $f^\dagger$, which are called source conditions. There are several
forms of such conditions. Nowadays, starting with [8] such conditions are often formulated in the form of variational inequalities. For the sake of simplicity we confine
ourselves to the case of Hilbert spaces with quadratic functionals $\mathcal{R}$ and $\mathcal{S}_{g^{obs}}$ here,
although the concepts can be generalized with little additional effort to general con-
vex $\mathcal{R}$ and $\mathcal{S}_{g^{obs}}$. For a concave, monotonically increasing and continuous function
$\psi : [0, \infty) \to [0, \infty)$ with $\psi(0) = 0$ we require that

$$\forall f \in \mathbb{X} \quad \frac{1}{4}\|f - f^\dagger\|^2 \le \frac{1}{2}\|f\|^2 - \frac{1}{2}\|f^\dagger\|^2 + \psi\left(\|F(f) - F(f^\dagger)\|^2\right). \tag{5.17}$$

Such conditions are not easy to interpret at first sight, and we will come back to this in
Sect. 5.3.2. However, they have been shown to be necessary for certain convergence
rates of Tikhonov regularization and other regularization methods (see [11]), and
sufficiency can be shown quite easily:

Theorem 5.3 If $f^\dagger$ fulfill (5.17), then the Tikhonov estimator $\hat{f}_\alpha$ in (5.8) satisfies
the error estimate

$$\frac{1}{4}\|\hat{f}_\alpha - f^\dagger\|^2 \le \frac{\delta^2}{\alpha} + (-\psi)^*\left(-\frac{1}{4\alpha}\right) \tag{5.18}$$

where $\psi^*(t) := \sup_{s\ge 0}\left(st - \psi(t)\right)$ denotes the conjugate function (see (5.11)).

Proof By definition of $\hat{f}_\alpha$ and our noise model we have

$$\frac{1}{2}\|F(\hat{f}_\alpha) - g^{obs}\|^2 + \frac{\alpha}{2}\|\hat{f}_\alpha\|^2 \le \frac{1}{2}\|F(f^\dagger) - g^{obs}\|^2 + \frac{\alpha}{2}\|f^\dagger\|^2 \le \frac{\delta^2}{2} + \frac{\alpha}{2}\|f^\dagger\|^2,$$

so together with our assumption (5.17) we find

$$\frac{1}{4}\|\hat{f}_\alpha - f^\dagger\|^2 \le \frac{1}{2}\|\hat{f}_\alpha\|^2 - \frac{1}{2}\|f^\dagger\|^2 + \psi\left(\|F(\hat{f}_\alpha) - F(f^\dagger)\|^2\right)$$

$$\le \frac{\delta^2}{2\alpha} - \frac{1}{2\alpha}\|F(\hat{f}_\alpha) - g^{obs}\|^2 + \psi\left(\|F(\hat{f}_\alpha) - F(f^\dagger)\|^2\right).$$

By the parallelogram law we have for all $x, y, z \in \mathbb{X}$ that

$$\|x - y\|^2 \le 2\|x - z\|^2 + 2\|z - y\|^2 - \|x + y - 2z\|^2 \le 2\|x - z\|^2 + 2\|z - y\|^2.$$

Apply this with $x = F(\hat{f}_\alpha)$, $y = F(f^\dagger)$ and $z = g^{\mathrm{obs}}$ to find

$$\frac{1}{4\alpha}\|F(\hat{f}_\alpha) - F(f^\dagger)\|^2 \le \frac{\delta^2}{2\alpha} + \frac{1}{2\alpha}\|F(\hat{f}_\alpha) - g^{\mathrm{obs}}\|^2,$$

so that finally we have

$$\frac{1}{4}\|\hat{f}_\alpha - f^\dagger\|^2 \le \frac{\delta^2}{\alpha} - \frac{1}{4\alpha}\|F(\hat{f}_\alpha) - F(f^\dagger)\|^2 + \psi\left(\|F(\hat{f}_\alpha) - F(f^\dagger)\|^2\right)$$

$$\le \frac{\delta^2}{\alpha} + \sup_{s \ge 0}\left[\frac{-s}{4\alpha} + \psi(s)\right] = \frac{\delta^2}{\alpha} + (-\psi)^*\left(-\frac{1}{4\alpha}\right). \qquad \square$$

The two most commonly used types of functions ψ in the literature,

$$\psi_\nu(t) = t^{\nu/2} \quad \text{and} \quad \psi_p^{\log}(t) = (-\log t)^p(1 + o(1)), \quad \text{as } t \to 0$$

are referred to as *Hölder* and *logarithmic* source function, respectively. For these functions we obtain

$$(-\psi_\nu)^*\left(-\frac{1}{t}\right) = ct^{\frac{\nu}{2-\nu}}$$

$$(-\psi_p^{\log})^*\left(-\frac{1}{t}\right) = (-\log t)^p(1 + o(1)), \quad \text{as } t \to 0.$$

Note that the two terms on the right hand side of (5.18) correspond to the error splitting (5.5). The following theorem gives an optimal choice of α balancing these two terms:

Theorem 5.4 *If under the assumptions of Theorem 5.3 ψ is differentiable, the infimum of the right hand side of (5.18) is attained at $\bar{\alpha}(\delta)$ if and only if*

$$\frac{1}{4\bar{\alpha}(\delta)} = \psi'(4\delta^2),$$

and

$$\frac{1}{4}\|\hat{f}_{\bar{\alpha}} - f^\dagger\|^2 \le \psi(4\delta^2).$$

Proof Note from the definition of $(-\psi)^*$ that $(-\psi)^*(t^*) \ge tt^* + \psi(t)$ for all $t \ge 0$, $t^* \in \mathbb{R}$. Further equality holds true if and only if $t^* = -\psi'(t)$, as for this choice of t the concave function $\tilde{t} \mapsto \tilde{t}t^* + \psi(\tilde{t})$ attains its unique maximum at t and thus in particular $tt^* + \psi(t) \ge (-\psi)^*(t^*)$. Therefore we have with $t = 4\delta^2$ and $t^* = -\frac{1}{4\alpha}$ that

$$\frac{\delta^2}{\alpha} + (-\psi)^* \left(-\frac{1}{4\alpha}\right) \geq \psi(4\delta^2) \qquad \text{for all } \alpha > 0$$

and $\frac{\delta^2}{\alpha} + (-\psi)^* \left(-\frac{1}{4\alpha}\right) = \psi(4\delta^2)$ if and only if $-\frac{1}{4\alpha} = -\psi'\left(4\delta^2\right)$. □

It can be shown that the same type of error bound can be obtained by a version of the discrepancy principle (5.6) which does not require knowledge of the function ψ describing abstract smoothness of the unknown solution $f^\dagger$ [2]. This is an advantage in practice, because such knowledge is often unrealistic.

5.3.2 Interpretation of Variational Source Conditions

5.3.2.1 Connection to Stability Estimates

Variational source conditions (5.17) are closely related to so called *stability estimates*. In fact if (5.17) holds true for all $f^\dagger \in \mathcal{K} \subset \mathbb{X}$, then all $f_1, f_2 \in \mathcal{K}$ satisfy the stability estimate

$$\frac{1}{4}\|f_1 - f_2\|^2 \leq \psi\left(\|F(f_1) - F(f_2)\|^2\right),$$

with the same function ψ, since one of the terms $\pm\left(\|f_1\|^2 - \|f_2\|^2\right)$ will be non-positive. There exists a considerable literature on such stability estimates (see e.g. [1, 14]). However it is unclear if stability estimates also imply variational source conditions as two difficulties have to be overcome. Firstly the term $\|f\|^2 - \|f^\dagger\|^2$ might be negative and secondly one would have to extend the estimate from the set $\mathcal{K}$ to the whole space $\mathbb{X}$.

5.3.2.2 General Strategy for the Verification of Variational Source Conditions

In general the rate at which the error of reconstruction methods converges to 0 as the noise level tends to 0 in inverse problems depends on two factors: The smoothness of the solution $f^\dagger$ and the degree of ill-posedness of F. We will describe both in terms of a family of finite dimensional subspaces $V_n \subset \mathbb{X}$ or the corresponding orthogonal projections $P_n : \mathbb{X} \to V_n$. The smoothness of $f^\dagger$ will be measured by how fast the best approximations $P_n f^\dagger$ in V_n converge to $f^\dagger$:

$$\|(I - P_n)f^\dagger\|_{\mathbb{X}} \leq \kappa_n \tag{5.19a}$$

$$\lim_{n \to \infty} \kappa_n = 0 \tag{5.19b}$$

Inequalities of this type are called Bernstein inequalities, and they are well studied for many types of subspaces V_n such as spaces of polynomials, trigonometric polynomials, splines, or finite elements. We will illustrate this for the case of trigonometric polynomials below.

Concerning the degree of ill-posedness, recall that any linear mapping on a finite dimensional space is continuous. Therefore, a linear, injective operator T restricted to a finite dimensional space V_n has a continuous inverse $(T|_{V_n})^{-1}$ defined on $T(V_n)$. However, the norm of these operators will grow with n, and the rate of growth may be used to measure the degree of ill-posedness. In the nonlinear case we may look at Lipschitz constants σ_n such that $\|P_n f^\dagger - P_n f\| \leq \sigma_n \|F(P_n f^\dagger) - F(P_n f)\|$. However, to obtain optimal results it turns out that we need estimates of inner products of $P_n f^\dagger - P_n f$ with $f^\dagger$. Moreover, on the right hand side we have to deal with $F(f^\dagger) - F(f)$ rather than $F(P_n f^\dagger) - F(P_n f)$:

$$\langle P_n f^\dagger, f^\dagger - f\rangle \leq \sigma_n \|F(f^\dagger) - F(f)\|_{\mathbb{Y}} \tag{5.19c}$$

The growth rate of σ_n describes what we will call *local degree of ill-posedness* of F at $f^\dagger$.

Theorem 5.5 *Let $\mathbb{X}$ and $\mathbb{Y}$ be Hilbert spaces and suppose that there exists a sequence of projection operators $P_n\colon \mathbb{X} \to \mathbb{X}$ and sequences $(\kappa_n)_{n\in\mathbb{N}}$, $(\sigma_n)_{n\in\mathbb{N}}$ of positive numbers such that (5.19) holds true for all $n \in \mathbb{N}$.*

Then $f^\dagger$ fulfills a variational source condition (5.17) with the concave, continuous, increasing function

$$\psi(\tau) := \inf_{n\in\mathbb{N}} \left[\sigma_n \sqrt{\tau} + \kappa_n^2\right], \tag{5.20}$$

which satisfies $\psi(0) = 0$.

Proof By straightforward computations we see that the variational source condition (5.17) has the equivalent form

$$\forall f \in \mathbb{X}: \quad \langle f^\dagger, f - f^\dagger\rangle \leq \frac{1}{4}\|f - f^\dagger\|^2 + \psi\left(\|F(f) - F(f^\dagger)\|^2\right).$$

Using (5.19a), (5.19c), and the Cauchy-Schwarz inequality we get for each $n \in \mathbb{N}$ that

$$\begin{aligned}
\langle f^\dagger, f^\dagger - f\rangle &= \langle P_n f^\dagger, f^\dagger - f\rangle + \langle (I - P_n)f^\dagger, f^\dagger - f\rangle \\
&\leq \sigma_n \|F(f^\dagger) - F(f)\|_{\mathbb{Y}} + \kappa_n \|f^\dagger - f\|_{\mathbb{X}} \\
&\leq \sigma_n \|F(f^\dagger) - F(f)\|_{\mathbb{Y}} + \kappa_n^2 + \frac{1}{4}\|f^\dagger - f\|_{\mathbb{X}}^2 \\
&\leq \sigma_n \|F(f^\dagger) - F(f)\|_{\mathbb{Y}} + \kappa_n^2 + \frac{1}{4}\|f - f^\dagger\|^2.
\end{aligned}$$

Taking the infimum over the right hand side with respect to $n \in \mathbb{N}$ yields (5.20) with $\tau = \|F(f^\dagger) - F(f)\|_{\mathbb{Y}}^2$. As ψ is defined by an infimum over concave and increasing functions, it is also increasing and concave. Moreover, (5.19b) implies $\psi(0) = 0$. $\square$

5.3.2.3 Example: Numerical Differentiation

Recall that the trigonometric monomials $e_n(x) := e^{2\pi n i x}$ form an orthonormal basis $\{e_n : n \in \mathbb{Z} \setminus \{0\}\}$ of $L^2_\diamond([0, 1])$, i.e. every function $f \in L^2_\diamond([0, 1])$ can be expressed as $f(x) = \sum_{n \in \mathbb{Z} \setminus \{0\}} \hat{f}(n) e_n(x)$ with $\hat{f}(n) = \int_0^1 f(x)\overline{e_n(x)}\, dx$. Further note that from the definition (5.3) of the forward operator we get

$$F(e_n) = \frac{1}{2\pi i n} e_n, \quad n \in \mathbb{Z} \setminus \{0\}.$$

and that the kth derivative of f has Fourier coefficients $\widehat{f^{(k)}}(n) = (2\pi i n)^k \hat{f}(n)$. Therefore, the norm

$$\|f\|_{H^s} := \left(\sum_{n \in \mathbb{Z} \setminus \{0\}} (2\pi n)^{2s} \left| \hat{f}(n) \right|^2 \right)^{1/2}$$

(called *Sobolev norm* of order $s \geq 0$) fulfills $\|f\|_{H^k} = \|f^{(k)}\|_{L^2}$ for $k \in \mathbb{N}_0$, but it also allows to measure non-integer degrees of smoothness of f.

We choose P_n as the orthogonal projection

$$P_n f := \sum_{0 < |m| \leq n} \hat{f}(m) e_m.$$

Suppose that the kth distributional derivative belongs to $L^2_\diamond([0, 1])$. Then

$$\|(I - P_n)f^\dagger\|_{L^2}^2 = \sum_{|m|>n} \left| \widehat{f^\dagger}(m) \right|^2 = \sum_{|m|>n} (2\pi m)^{-2s} \left| \widehat{(f^\dagger)^{(s)}}(m) \right|^2$$
$$\leq (2\pi n)^{-2s} \|f^\dagger\|_{H^s}^2$$

which shows that (5.19a) and (5.19b) are satisfied with $\kappa_n = (2\pi n)^{-k}$. Moreover, we have that

$$\langle P_n f^\dagger, f^\dagger - f \rangle = \sum_{0 < |n| \leq m} -2\pi i n \widehat{f^\dagger}(n) \overline{\left(\frac{1}{2\pi i n} \left(\widehat{f^\dagger}(n) - \hat{f}(n) \right) \right)}$$
$$\leq \left(\sum_{0 < |n| \leq m} (2\pi n)^2 |\widehat{f^\dagger}(n)|^2 \right)^{1/2} \left\| F(f^\dagger) - F(f) \right\|$$
$$\leq (2\pi m)^{1-s} \|f^\dagger\|_{H^s} \left\| F(f^\dagger) - F(f) \right\|.$$

for $s \in (0, 1]$, so (5.19c) is satisfied with $\sigma_n := (2\pi m)^{1-s}$. Choosing the parameter $m \approx \|f^\dagger\|_{H^s}^{1/(1+s)} \tau^{-1/2(1+s)}$ at which the infimum in (5.20) is attained approximately we see that there exists a constant $C > 0$ such that the variational source condition (5.17) is satisfied with $\psi(t) = C\|f^\dagger\|_{H^s}^{2/(1+s)} t^{s/(1+s)}$, and from Theorem 5.4 we obtain the error bound

$$\|\hat{f}_\alpha - f^\dagger\|_{L^2} = \mathcal{O}\left(\delta^{\frac{s}{s+1}}\right). \tag{5.21}$$

It can be shown that this rate is optimal in the sense that there exists no reconstruction method R for which

$$\inf_R \sup\left\{\left\|R(g^{\text{obs}}) - f^\dagger\right\| : f^\dagger \in H^s, \left\|F(f^\dagger) - g^{\text{obs}}\right\| \leq \delta\right\} = o\left(\delta^{\frac{s}{s+1}}\right).$$

5.3.2.4   Example: Fluorescence Microscopy

Similarly one can proceed for the example of fluorescence microscopy. As one has to work here with L^2 spaces rather than $L_\diamond^2$ spaces, the Sobolev norm is defined by $\|f\|_{H^s}^2 = \sum_{n\in\mathbb{Z}}(1 + n^2)^s \left|\widehat{f}(n)\right|^2$. Assuming that the convolution kernel is a-times smoothing ($a > 0$) in the sense that $\|F(f)\|_{H^a} \sim \|f\|_{H^0}$, which is equivalent to the existence of two constant $0 < c < C$ such that the Fourier transform of the convolution kernel k fulfills

$$c(1 + |\xi|^2)^{-a/2} \leq |\widehat{k}(\xi)| \leq C(1 + |\xi|^2)^{-a/2}$$

one can show that $\|f^\dagger\|_{H^s} < \infty$ for $s \in (0, a]$ implies a variational source condition with $\psi(t) \sim t^{\frac{s}{s+a}}$ and an error bound

$$\|\hat{f}_\alpha - f^\dagger\|_{L^2} = \mathcal{O}\left(\delta^{\frac{s}{s+a}}\right). \tag{5.22}$$

Again this estimate is optimal in the sense explained above.

5.3.2.5   Extensions

Variational source conditions with a given Hölder source function actually hold true on a slightly larger set. In the typical situation where the marker density of the investigated specimen is constant (or smooth) up to jumps, then it fulfills the same variational source condition with $s = 1/2$, although $f^\dagger \in H^s$ if and only if $s < 1/2$. The sets on which a variational souce condition is satisfied can be characterized in terms of Besov spaces $B_{2,\infty}^s$, and bounded subsets of such spaces are also the largest sets on which a Hölder-type error bound like (5.21) and (5.22) are satisfied with uniform constants (see [11]).

In the case where the convolution kernel is infinitely smoothing, e.g. if the kernel is a Gaussian, then we cannot expect to get a variational source condition with a Hölder source function under Sobolev smoothness assumptions. Instead one obtains logarithmic source functions $\psi_p^{\log}$ as introduced above, which will again be optimal and lead to very slowly decaying error estimates as $\delta \searrow 0$ [11].

Note that the rates (5.21) and (5.22) are restricted to smoothness indices $s \in (0, 1]$ and $s \in (0, a]$, respectively. This restriction to low order rates is a well-known shortcoming of variational source conditions. Higher order rates can be obtained by imposing a variational source condition on the dual problem (5.12), which can again be again by verified by Theorem 5.5 (see [5, 18]).

With some small modifications the strategy in Theorem 5.5 can be extended to Banach space settings [9, 21] and nonlinear forward operators F, in particular those arising in inverse scattering problems [10, 20].

5.3.3 Error Bounds for Poisson Data

We already briefly discussed discrete versions of inverse problems with Poisson in Sect. 5.2.1. Such problems arise in many photonic imaging modalities such as fluorescence microscopy, coherent x-ray imaging, positron emission tomography, but also electron microscopy. In the following we briefly discuss a continuous setting for such problems.

We consider a forward operator $F \colon \mathbb{X} \to \mathbb{Y}$ that maps an unknown sample $f^\dagger \in \mathbb{X}$ a photon density $g^\dagger \in \mathbb{Y} = L^1(\mathbb{M})$ generated by $f^\dagger$ on some measurement manifold $\mathbb{M} \subset \mathbb{R}^d$. The given data are modeled by a Poisson process

$$G_t = \sum_{k=1}^{N} \delta_{x_k},$$

with density $tg^\dagger$. Here $\{x_1, \dots, x_N\} \subset \mathbb{M}$ denote the positions of the detected photons and $t > 0$ can be interpreted as exposure time. Note that $t \sim \mathbf{E}(N)$, i.e. the exposure time is proportional to the number of detected photons.

Now to discuss error bounds we first need some notion of noise level. But what is the "noise" in our setting? Our data G_t do not belong to $\mathbb{Y} = L^1(\mathbb{M})$ and the "noise" is not additive. However, it follows from the properties of Poisson processes that

$$\mathbf{E}\left[\langle \tfrac{1}{t} G_t, h \rangle\right] = \int_{\mathbb{M}} hg^\dagger \, dx,$$

and the variance of $\langle \tfrac{1}{t} G_t, h \rangle$ is proportional to $\tfrac{1}{t}$. This suggests that $\tfrac{1}{\sqrt{t}}$ plays the role of the noise level. More precisely, it is possible to derive concentration inequalities

$$\mathbf{P}\left(d(\tfrac{1}{t} G_t, g^\dagger) > \frac{r + C}{\sqrt{t}}\right) \leq \exp\left(-cr\right)$$

where the distance function d is defined by a negative Besov norm (see [13, 22] for similar results).

As a reconstruction method we consider generalized Tikhonov regularization as in (5.8) with S_{G_t} given by a negative (quasi-)log-likelihood corresponding to the Poisson data. As discussed in Sect. 5.2 this amounts to taking the Kullback-Leibler distance as data fidelity term in the finite dimensional case, and in particular in the implementations of this method. (Sometimes a small shift is introduced in the Kullback-Leibler divergence to "'regularize" this term.) By assuming the VSC (5.17) with $\psi(\|F(\mathbf{f}^\dagger) - F(\mathbf{f})\|^2)$ replaced by $\psi\left(\mathrm{KL}(F(\mathbf{f}^\dagger), F(\mathbf{f}))\right)$ and an optimal parameter choice $\bar{\alpha}$ one can than show the following error estimate in expectation

$$\mathbf{E}\left(\|\hat{f}_{\bar{\alpha}} - f^\dagger\|_{\mathbb{X}}\right) = \mathcal{O}\left(\psi\left(\frac{1}{\sqrt{t}}\right)\right), \quad t \to \infty.$$

References

1. Alessandrini, G.: Stable determination of conductivities by boundary measurements. Appl. Anal. **27**, 153–172 (1988)
2. Anzengruber, S.W., Hofmann, B., Mathé, P.: Regularization properties of the sequential discrepancy principle for Tikhonov regularization in Banach spaces. Appl. Anal **93**(7), 1382–1400 (2014). https://doi.org/10.1080/00036811.2013.833326
3. Chambolle, A., Pock, T.: A first-order primal-dual algorithm for convex problems with applications to imaging. J. Math. Imaging Vis. **40**(1), 120–145 (2011). https://doi.org/10.1007/s10851-010-0251-1
4. Deuflhard, P., Engl, H.W., Scherzer, O.: A convergence analysis of iterative methods for the solution of nonlinear ill-posed problems under affinely invariant conditions. Inverse Probl. **14**(5), 1081–1106 (1998). https://doi.org/10.1088/0266-5611/14/5/002
5. Grasmair, M.: Variational inequalities and higher order convergence rates for Tikhonov regularisation on Banach spaces. J. Inverse Ill-Posed Probl. **21**(3), 379–394 (2013). https://doi.org/10.1515/jip-2013-0002
6. Hanke, M.: A regularizing Levenberg-Marquardt scheme, with applications to inverse groundwater filtration problems. Inverse Probl. **13**, 79–95 (1997)
7. Hanke, M., Neubauer, A., Scherzer, O.: A convergence analysis of the Landweber iteration for nonlinear ill-posed problems. Numer. Math. **72**, 21–37 (1995)
8. Hofmann, B., Kaltenbacher, B., Pöschl, C., Scherzer, O.: A convergence rates result for Tikhonov regularization in Banach spaces with non-smooth operators. Inverse Probl. **23**(3), 987–1010 (2007). https://doi.org/10.1088/0266-5611/23/3/009
9. Hohage, T., Miller, P.: Optimal convergence rates for sparsity promoting wavelet-regularization in Besov spaces. Technical report. https://arxiv.org/abs/1810.06316 (2018)
10. Hohage, T., Weidling, F.: Verification of a variational source condition for acoustic inverse medium scattering problems. Inverse Probl. **31**(7), 075,006 (2015). https://doi.org/10.1088/0266-5611/31/7/075006
11. Hohage, T., Weidling, F.: Characterizations of variational source conditions, converse results, and maxisets of spectral regularization methods. SIAM J. Numer. Anal. **55**(2), 598–620 (2017). https://doi.org/10.1137/16M1067445
12. Hohage, T., Werner, F.: Iteratively regularized Newton-type methods for general data misfit functionals and applications to Poisson data. Numer. Math. **123**, 745–779 (2013). https://doi.org/10.1007/s00211-012-0499-z
13. Hohage, T., Werner, F.: Inverse problems with poisson data: statistical regularization theory, applications and algorithms. Inverse Probl. **32**(9), 093,001 (2016). https://doi.org/10.1088/0266-5611/32/9/093001

14. Isaev, M.I., Novikov, R.G.: Effectivized Hölder-logarithmic stability estimates for the Gel'fand inverse problem. Inverse Probl. **30**(9), 095,006 (2014). https://doi.org/10.1088/0266-5611/30/9/095006
15. Kaltenbacher, B., Neubauer, A., Scherzer, O.: Iterative Regularization Methods for Nonlinear ill-posed Problems. Radon Series on Computational and Applied Mathematics. de Gruyter, Berlin (2008)
16. Lassas, M., Siltanen, S.: Can one use total variation prior for edge-preserving bayesian inversion? Inverse Probl **20**(5), 1537–1563 (2004). https://doi.org/10.1088/0266-5611/20/5/013
17. Rockafellar, R.T.: Convex Analysis. Princeton University Press (2015)
18. Sprung, B., Hohage, T.: Higher order convergence rates for Bregman iterated variational regularization of inverse problems. Numer. Math. **141**, 215–252 (2019). https://doi.org/10.1007/s00211-018-0987-x
19. Ulbrich, M.: Semismooth Newton Methods For Variational Inequalities And Constrained Optimization Problems In Function Spaces, MOS-SIAM Series on Optimization, vol. 11. Society for Industrial and Applied Mathematics (SIAM), Philadelphia, PA (2011). https://doi.org/10.1137/1.9781611970692
20. Weidling, F., Hohage, T.: Variational source conditions and stability estimates for inverse electromagnetic medium scattering problems. Inverse Probl. Imaging **11**(1), 203–220 (2017). https://doi.org/10.3934/ipi.2017010
21. Weidling, F., Sprung, B., Hohage, T.: Optimal convergence rates for Tikhonov regularization in Besov spaces. Technical report (2018). https://arxiv.org/abs/1803.11019
22. Werner, F., Hohage, T.: Convergence rates in expectation for Tikhonov-type regularization of inverse problems with Poisson data. Inverse Probl. **28**(10), 104,004 (2012). https://doi.org/10.1088/0266-5611/28/10/104004

Chapter 6
Proximal Methods for Image Processing

An Introduction to ProxToolbox for Testing Algorithms on the Göttingen Datasets

D. Russell Luke

Was ist dann eigentlich ein Hilbertischer Raum?
– David Hilbert to John von Neumann [1]

6.1 All Together Now

A major challenge in building and maintaining collaborations across disciplines is to establish a common language. Sounds simple enough, but even a common language is not helpful without a common understanding of the basic elements. When it comes to the day-to-day exchange of data and software, this means building a common data management and processing environment. Try to do this, however, and you learn very quickly that even for something as concrete as building software that everyone can use, there are different ways of interpreting and understanding what the software does. In the context of X-ray diffraction, for instance, what a physicist might understand as a software routine that simulates the propagation of an X-ray through an optical device, a mathematician would understand as an operator with certain mathematical properties.

The first successful algorithms for phase retrieval were developed and understood by physicists as iterative procedures that simulate the forward and backward propagation of a wave through an optical device, where in each iteration the computed wave is adjusted to fit either measurement data or some experimental constraint, like the shape of the aperture or the illuminating beam. Later, mathematicians reinterpreted these operations in terms of the application of *projectors* to iterates of a *fixed point mapping*. Of most recent vintage is an effort by a new generation of applied mathematicians to sidestep the more interesting aspects of the physicists'

D. R. Luke (✉)
Institute for Numerical and Applied Mathematics, Universität Göttingen,
Lotzestrasse 16-18, 37083 Göttingen, Germany
e-mail: r.luke@math.uni-goettingen.de

© The Author(s) 2020
T. Salditt et al. (eds.), *Nanoscale Photonic Imaging*, Topics in Applied Physics 134,
https://doi.org/10.1007/978-3-030-34413-9_6

algorithms—namely that they occasionally don't work—by lifting the problem to a space that is too high-dimension for any practical purposes, and then *relaxing* the underlying problem to something with theoretically nicer properties, but whose solution bears little meaningful relationship to the problem at hand. At this point, one is reminded of Richard Courant's lamentation, "the broad stream of scientific development may split into smaller and smaller rivulets and dry out." [2] Here's to swimming against the current.

6.1.1   What Seems to Be the Problem Here?

The story of computational phase retrieval in X-ray imaging is a perfect example of how different communities can come to understand the same things in very different ways. This serves as a sort of origin story for the ProxToolbox [3]

```
http://num.math.uni-goettingen.de/proxtoolbox/
```

which is the subject of this chapter. The setting for phase retrieval has been presented in Chap. 2 and this will be the stuff cooked in the mathematical crucible that the ProxToolbox represents.

The measurements are intensity readings which are denoted simply as a nonnegative-valued vector I, with n elements corresponding to the pixels in the CCD array (see 2.12). The model for these measurements is developed in Sect. 2.1. The various modalities (near field/far field) have the general form

$$\left\|\left(\mathcal{D}_F(\psi)\right)(x_k, y_j)\right\| = \sqrt{I(x_k, y_j)}, \quad (k = 1, 2, \ldots, n_x)\, (j = 1, 2, \ldots, n_y). \quad (6.1)$$

In (2.12) the model is given in the continuum with the intensity $I(x, y)$ at the position (x, y) in the measurement plane. The actual measurements consist of pixels indexed by $k = 1, 2, \ldots, n_x$ and $j = 1, 2, \ldots, n_y$ corresponding to positions (x_k, y_j) in the measurement plane. The model then represents a system of $n \equiv n_x n_y$ equations in $2n$ unknowns, the real and imaginary part of $\mathcal{D}_F(\psi)$ at each of the n pixels. With this discretization, the mapping $\mathcal{D}_F$ is understood as a discrete *Fresnel propagator* with Fresnel number F (see 2.84). To make things simpler, the indexes k and j are combined uniquely into a single index $i = 1, 2, \ldots, n$ so that the data model is just

$$\left\|\left(\mathcal{D}_F(\psi)\right)_i\right\| = \sqrt{I_i}, \quad \forall\, i = 1, 2, \ldots, n. \quad (6.2)$$

The solution to the phase retrieval problem as presented here is the complex-valued vector $\psi \in \mathbb{C}^n$ that satisfies (6.2) or, more generally (6.3). With only a single measurement the problem is *underdetermined*, that is, too many unknowns and not enough equations. To further constrain the problem, there are a number of possibilities. First, one could (and should) include a priori information implicit in the experiment (support, nonnegativity, sparsity, etc). The next thing one could try is to adjust the instrument in some known fashion and take several measurements of the same object. The mapping $\mathcal{D}_F$ is the model for the mapping of the electromagnetic field at the object to the field where the intensity is measured, either near field or far field. For the purpose of this tutorial, these are mappings from vectors in $\mathbb{C}^n$ to vectors in $\mathbb{C}^n$. This can be modified in a number of different ways. Representing n-dimensional complex-valued vectors instead as two-dimensional vectors on an n-dimensional product space, the phase retrieval problem is to find $\psi = (\psi_1, \psi_2, \ldots, \psi_n) \in \left(\mathbb{R}^2\right)^n$ ($\psi_i \in \mathbb{R}^2$) satisfying (6.2) for all i in addition to qualitative constraints and/or additional measurements.[1]

Ptychography was briefly mentioned in Chap. 2. Ptychography is harder to say than to describe. A quilt is a ptychogram of sorts. Or put in more technical terms, ptychography is a combination of blind deconvolution and computed tomography for phase retrieval. The original idea proposed by Hegerl and Hoppe [4], was just the computed tomography part: to stitch together the original object ψ from many measurements at different settings of the instrument, modeled by $\mathcal{D}_j$, the j indexing the setting. One of the implicit complications of conventional ptychography, which differs from the original is that the illuminating beam is also unknown—this is the blind deconvolution part.

In the above, different Fresnel numbers correspond to collecting the intensities I_i at different planes orthogonal to the direction of propagation. This further constrains the problem. The model is only a little more involved than simple phase retrieval (6.2). For m different intensity measurements, each consisting of n pixels, the generalized ptychography/phase diversity model takes the form

$$\| \left(\mathcal{D}_j(\psi)\right)_i \| = \sqrt{I_{i,j}} \quad \forall i = 1, 2, \ldots, n, \forall j = 1, 2, \ldots m. \tag{6.3}$$

This fits the first ptychographic reconstruction procedure [4] which assumed that the illuminating beam, characterized by $\mathcal{D}_j$, was known. In the *blind* ptychography problem – analogous to the blind deconvolution—the beam is not completely known. This corresponds to what is commonly understood by ptychography in modern applications [5–8]. To account for the unknown beam characteristics, the mapping $\mathcal{D}_j$ is further decomposed:

$$\mathcal{D}_j(z, \psi) \equiv \mathcal{F}\left(S_j(z) \odot \psi\right) \tag{6.4}$$

Here $\mathcal{F}$ is a parameter-free propagator and $S_j : \mathbb{C}^n \to \mathbb{C}^n$ denotes the j-th linear operator representing some known adjustment to the beam—a lateral shift, or translation in the direction of propagation—$z \in \mathbb{C}^n$ is the unknown vector characterizing the probe, and $\odot$ is the elementwise Hadamard product.

The problem in blind ptychography is to reconstruct simultaneously the object ψ and illuminating beam u from a given ptychgraphic dataset. Near-field, or *in-line*

---

[1]The issue of whether to represent the vectors as points in $\mathbb{R}^{2n}$ or points in $\mathbb{C}^n$ is notational. The representation as points in $\mathbb{C}^n$ is more convenient for the purpose of explanation, but on a computer you will need to work with $\mathbb{R}^{2n}$.

ptychography [9] involves moving the imaging plane along the axis of propagation of the beam (i.e., away from the plane where the object lies). This is very similar to *phase diversity* in astronomy [10], but there one changes the *defocus* in the far-field instead of the imaging plane in the near field (also, one does not have to recover the beam in astronomy applications). Mathematically, however, the two instances have the same structure. In conventional far-field ptychographic experiments the beam is much smaller than the specimen. The different measurements consist of scans of ψ in the lateral direction with sufficient overlap between successive images. Lateral translations and translations along the axis of propagation were combined in [11]. Note that the last case is least restrictive in terms of probe properties, see also [12] for a detailed comparison and discussion.

The issue of *existence* of a ψ that satisfies all the equations is discussed in Chap. 23. For this chapter, existence is recast as *consistency* of the measurements and the physical model. The data is exact.[2] What is not entirely accurate is the model for the data and the *computational bandwidth*, e.g. finite precision arithmetic.

Though it might seem unintuitive, for algorithmic reasons, it is better to separate the aspects of the imaging model having to do with the field at the object plane from those having to do with the field at the image plane. Denote $\mathbf{u} \equiv (u_1, u_2, \ldots, u_m)$ with $u_j \in \mathbb{C}^n$ ($j = 1, 2, \ldots, m$) and define the measurement sets

$$M_j \equiv \left\{ u \in \mathbb{C}^n \mid \|(\mathcal{F}u)_i\| = \sqrt{I}_{j,i}, \ (i = 1, 2, \ldots, n) \right\} \quad (j = 1, 2, \ldots, m) \quad (6.5)$$

where $\mathcal{F}$ is a parameter-free propagator in (6.4). The sets M_j are nothing more than the *phase sets*, or the set of all vectors that *could* explain the data. Solutions to the most general physical model represented by (6.3) consists of any triple of vectors $(z, \psi, \mathbf{u}) \in \mathbb{C}^n \times \mathbb{C}^n \times (\mathbb{C}^n)^m$ in the set

$$\mathcal{M} \equiv \left\{ (z, \psi, \mathbf{u}) \in \mathbb{C}^n \times \mathbb{C}^n \times \mathbb{C}^{n \times m} \mid S_j(z) \odot \psi = u_j, \ j = 1, 2, \ldots, m \right\} \quad (6.6)$$

such that $u_j \in M_j$ and the beam z and object ψ satisfy any other reasonable qualitative constraints. The constraints on the unknowns z, ψ and $\mathbf{u}$ are separable and given by

$$X \equiv \{\text{qualitative constraints on the probe}\}, \quad (6.7a)$$

$$O \equiv \{\text{qualitative constraints on the specimen}\}, \quad (6.7b)$$

$$M \equiv M_1 \times M_2 \times \cdots \times M_m, \ (\text{measurement constraints}). \quad (6.7c)$$

As before, the qualitative constraints characterized by X and O are support, support-nonnegativity or magnitude constraints corresponding respectively to whether the illumination and specimen are supported on a bounded set, whether these (most likely only the specimen) are "real objects" that somehow absorb or attenuate the

---

[2]This is a minority opinion, but it seems to be the height of hubris to think that an empirical observation is an approximation to a theoretical model rather than the other way around. The only thing that is indisputable is that the instrument behavior and the predicted behavior don't match up as well as desired.

probe energy, or whether these are "phase objects" with a prescribed intensity but varying phase. A support constraint for the set X, for instance, would be represented by

$$X \equiv \{z = (z_1, z_2, \ldots, z_n) \in \mathbb{C}^n \mid |z_i| \leq R \text{ and for } i \notin \mathbb{I}_X \ z_i = 0\}, \qquad (6.8)$$

where $\mathbb{I}_X$ is the index set corresponding to which pixels in the field of view the probe beam illuminates and R is some given amplitude.

6.1.2 What Is an Algorithm?

One of the best known algorithms for phase retrieval is Fienup's Hybrid Input-Output algorithm (HIO) [13] discussed in 2.88. This algorithm illustrates just about every difficulty one encounters in collaborating across disciplines, and sets the stage for the rest of the tutorial below. In the present setting, HIO with a support constraint is given as

$$(\forall k \in \mathbb{N})(\forall i = 1, 2, \ldots, n) \ \psi_i^{k+1} = \begin{cases} (\mathcal{P}_M(\psi^k))_i, & \text{if } i \in D; \\ \psi_i^k - \beta_k(\mathcal{P}_M(\psi^k))_i, & \text{otherwise.} \end{cases} \qquad (6.9)$$

Here $\mathbb{N}$ is the set of counting numbers, D indicates an index set where it is imagined that some constraints are satisfied; the mapping $\mathcal{P}_M$ fits the current iterate ψ^k to the data by propagating this guess through a model optical system, fixing the amplitude at the measurement surface to match the observed intensity and then propagating the resulting field back to the plane of the object. Putting (2.84) in into the present context, with the set M consisting of just a single intensity image, $M \equiv M_1$, yields

$$\psi_+ \simeq \mathcal{P}_M(\psi) \simeq \mathcal{D}_{-F}(\widehat{\psi}), \quad \widehat{\psi}_i \simeq \sqrt{I_i} \frac{(\mathcal{D}_F(\psi))_i}{\|(\mathcal{D}_F(\psi))_i\|} \qquad (6.10)$$

The symbol $\simeq$ indicates that this is not really and equivalence relation: for the moment think of it as equivalence with exceptions.

One obvious exception is when $(\mathcal{D}_F(\psi))_i = 0$. If you are a physicist you might argue that $(\mathcal{D}_F(\psi))_i = 0$ on a set of measure zero—a fancy way for saying never, with infinite precision arithmetic. Except that electronic processors operate with finite precision arithmetic and zero is therefore enormous on a computer. In fact, with double precision, zero is not smaller than $1e - 16$. To see what kind of error this can lead to, suppose that $(\mathcal{D}_F(\psi))_i = -1e - 15$ with infinite precision arithmetic, but because of roundoff, the computer returns $1e - 15$. A very small difference locally. But suppose that, at this pixel, the measured intensity is $I_i = 10$. In computing the projection, the computer returns a point with $\widehat{y}_i = 10$ instead of $\widehat{y}_i = -10$. This makes an enormous difference. The typical user won't see this kind of error often, but in numerical studies in [14] it happened about 12% of the time, which is not insignificant. Anyone with experience in programming knows to be careful about dividing. Without thinking much more about it, one usually codes some exception to avoid problems when division gets too dicey.

But there is another reason to pay attention to this exceptional point: *convexity*. The mathematical understanding of this and other phase retrieval algorithms originating from the optics community begins with viewing the entire collection of points satisfying the data and the qualitative constraints as *sets* and then viewing the operations in the above iterative procedures as *metric projections* onto these sets. A metric projector of a point ψ onto a set C is simply an operator that maps ψ to all points in C that are nearest to ψ. The operation $\mathcal{P}_M$ in (6.10) was long called a projection in the optics literature, but it was not shown to be a *metric projector* until [14, Corollary 4.3] where it was pointed out that the projector is *set-valued* in general. Being more careful one *should* write

$$\psi_+ \in \mathcal{P}_M(\psi) = \left\{ \mathcal{D}_{-F}(\widehat{\psi}) \ \middle| \ \widehat{\psi}_i \in \begin{cases} \sqrt{I_i} \frac{(\mathcal{D}_F(\psi))_i}{\|(\mathcal{D}_F(\psi))_i\|}, & \text{if } (\mathcal{D}_F(\psi))_i \neq 0, \\ \sqrt{I_i}\mathbb{S}, & \text{if } (\mathcal{D}_F(\psi))_i = 0. \end{cases} \right\} \tag{6.11}$$

The symbol $\mathbb{S}$ denotes the unit sphere in the complex plane ($\mathbb{R}^2$) and $\sqrt{I_i}\mathbb{S}$ is the sphere of radius $\sqrt{I_i}$. The symbol $\in$ is a reminder that the right hand side is a set of elements, and the left hand side is just a selection—any selection—from this set. The change in notation from "$=$" to "$\in$" is not just pedantic nit-picking, but underscores the fact that the problem is fundamentally *nonconvex*, meaning that you can find two points in the set where the line segment joining the two points leaves the set. In (6.11) take two points on opposite ends of the sphere $\sqrt{I_i}\mathbb{S}$: the line joining them is not in $\sqrt{I_i}\mathbb{S}$. If the sets M were convex (line segments joining any two points in the set are contained in said set), then the projector would be single-valued and one could forget the whole technicality, not to mention any worries about numerical instability.

Returning to iterative algorithms just in the context of phase retrieval (the probe z is known and there is only a single measurement $M = M_1$ so that the variables $\mathbf{u}$ are not needed), the procedure (6.9) is a natural way to think of a numerical procedure when approaching things physically: one makes a guess for the object ψ^0, propagates it through the optical system according the model given by (6.10) and updates this guess to ψ^1 depending on whether the elements satisfy the data and some a priori constraint. And repeat. The user would stop the iteration either when he needed to go for coffee or when the iterates stop making progress, in some loosely defined way. The subtle point here is that (6.9) is not a *fixed point iteration*, but rather a simulation of a physical process.

Bauschke, Combettes and Luke [15] showed that the HIO algorithm (6.9) is equivalent to

$$\psi^{k+1} \in \tfrac{1}{2}\big(\mathcal{R}_{\ominus}(\mathcal{R}_M + (\beta_k - 1)\mathcal{P}_M) + \mathrm{Id} + (1 - \beta_k)\mathcal{P}_M\big)(\psi^k). \tag{6.12}$$

where Id is the identity mapping (does nothing to the point), $\mathfrak{S}$ is the set of vectors satisfying a bounded support constraint, and

$$\mathcal{R}_M \equiv 2\mathcal{P}_M - \text{Id} \quad \text{and} \quad \mathcal{R}_\mathfrak{S} \equiv 2\mathcal{P}_\mathfrak{S} - \text{Id} \tag{6.13}$$

with

$$(\mathcal{P}_\mathfrak{S}\psi)_j = \begin{cases} \psi_j & \text{if } j \in D \\ 0 & \text{else.} \end{cases} \tag{6.14}$$

The mappings $\mathcal{R}_\mathfrak{S}$ and $\mathcal{R}_M$ are called *reflectors* because they send points to their reflection points on the opposite side of the set onto which one projects. The iteration (6.12) is the Hybrid Projection Reflection (HPR) algorithm proposed in [15]. When $\beta_n = 1$ for all n, then the iteration takes the form [16]

$$\psi^{k+1} \in \tfrac{1}{2}(\mathcal{R}_\mathfrak{S}(\mathcal{R}_M + \text{Id}))(\psi^k). \tag{6.15}$$

This is the popular *Douglas-Rachford Algorithm* [17] following the formulation of Lions and Mercier in the context of *monotone operator equations* [18]. In both cases, the desired point is a *fixed point* of the mapping T, that is a point $\overline{\psi}$ such that $\overline{\psi} = T\overline{\psi}$ where either $T = \tfrac{1}{2}(\mathcal{R}_\mathfrak{S}(\mathcal{R}_M + (\beta_k - 1)\mathcal{P}_M) + \text{Id} + (1 - \beta_k)\mathcal{P}_M)$ or $T = \tfrac{1}{2}(\mathcal{R}_\mathfrak{S}(\mathcal{R}_M + \text{Id}))$.

Whether or not the iterations above converge to a fixed point, and what this point has to do with the problem at hand is the subject of Chap. 23. For the purposes of this tutorial, the algorithm will simply be run with the given mappings and the user will be left to interpret the result. A few tools are provided within the ProxToolbox to monitor the iterates according to mathematical, as opposed to physical, criteria. For the beginning reader all that is important to keep in mind is that, first of all, the algorithms don't always converge, and second of all, when they converge, the limit point is *not* a solution to the problem you thought you were solving, but you can usually get there easily from the limiting fixed point. Another issue to keep in mind is that the physical criteria that scientists apply to judge the quality of a computed solution usually does not correspond to the mathematical criteria used to characterize and quantify convergence of an algorithm. It is not uncommon to see pictures of "solutions" returned by various algorithms at iteration k, or to see a comparison of a root mean-squared error estimate of an iterate of various algorithms. This is, from a mathematical perspective, not really meaningful for several reasons. The first reason is that, unless the underlying optimization problem is to minimize the root mean-squared error of something, there is no reason to expect that an algorithm should do this. The second reason is that, as already mentioned, the iterates of some algorithms, like Douglas-Rachford, are not the points that approximate solutions to the desired optimization problem, but their *shadows*, defined as the projection of these points onto a relevant set, are. Comparison of the quality of solutions returned by algorithms is common, but it should be recognized that such comparisons are not mathematical, but rather *phenomenological*, if of any scientific significance at all.

The identification of the HIO algorithms with Douglas-Rachford, at least for one parameter value, made a lot of sense when it was first discovered. The HIO algorithm is famously unstable. The way most people use it today is to get themselves in the neighborhood of a good solution by running 10–40 iterations of HIO, at which point they switch to a more stable algorithm to clean up their images. The value of HIO or Douglas-Rachford is that they rarely get stuck in local minimums. The identification with Douglas-Rachford makes this phenomenon clear since it can be proved that, if the sets $\mathfrak{S}$ and M do not intersect, then Douglas-Rachford *does not possess fixed points*. If M were convex, then you could even prove that the iterates must *diverge to infinity* in the direction of the gap vector between best approximation pairs between the sets [19]. For nonconvex problems like noncrystallographic phase retrieval, the iterates need not diverge, but they *cannot* converge. In Chap. 23 the convergence theory is discussed in some detail.

As should be clear by now, there is no equation being solved here, but rather some point is sought, any point, that satisfies an equation and any other kind of requirement one might like to add. It is high time to bring the main character of this story to the stage. In the most general format (ptychography) this has the form

General Problem

$$\text{Find } (z, \psi, \mathbf{u}) \in \mathcal{M} \cap (X \times O \times M).$$

This is a *feasibility problem* and what algorithms like (6.12) and (6.15) aim to solve, if possible. For the moment, it is easiest to examine the more elementary phase retrieval problem (ptychography with one measurement):

$$\text{Find } \psi \in \mathfrak{S} \cap M. \tag{6.16}$$

The sets $\mathfrak{S}$ and M have the explicit characterizations

$$M \equiv \left\{ \psi \in \mathbb{C}^n \; \middle| \; \left|(\mathcal{D}_F\psi)_i\right| = \sqrt{I_i}, \; i = 1, 2, \ldots, n \right\} \tag{6.17}$$

and

$$\mathfrak{S} \equiv \left\{ \psi \in \mathbb{C}^n \mid \psi_i = 0 \text{ if } i \notin D \right\}. \tag{6.18}$$

The projectors onto these sets are given by (6.11) and (6.14). When the requirements become so narrow that no point can satisfy all of them, the problem is said to be *inconsistent*. The reason for the instability of HIO and Douglas-Rachford lies with the failure of the sets $\mathfrak{S}$ and M to have points in common. This indicates a fundamental *inconsistency* of the physical model.

The behavior of these algorithms in the presence of model inconsistency tends to be mistaken for another bugbear of inverse problems, namely *nonuniqueness*. For example, around the turn of the millennium, oversampling in the image domain was proposed to overcome the nonuniqueness of phase reconstructions for noncrystallo-graphic observations [20]. A few moments reflection on elementary Fourier analysis, and careful reading of Chap. 2 is all you need to convince yourself, however, that oversampling has less to do with uniqueness than with inconsistency. Increased, but still finite, sampling in the image domain just pushes the inconsistency, or gap, between the sets $\mathfrak{S}$ and M to some level below either your numerical or experimental precision. It might look like the iteration has converged, but what has really happened is that the movement of the iterates has become so small that it is no longer detectable with a fixed arithmetic precision. This is just a physical manifestation of the fact from Fourier analysis that objects with compact support do not have compactly supported Fourier transforms, and vice versa. Since the measurements are finite, the object that is recovered cannot be finite. This means that the only time phase retrieval can be *consistent* is when imaging periodic crystals.

To be sure, uniqueness is nice when you have it, but you first need to clear up the issue of uniqueness *of what*. *No wave ψ with compact support can generate the given intensity I.* When the problem is inconsistent, it suffices to find some point, any point, that comes as close to both sets as possible. This is a *best approximation problem* and takes the form of the usual optimization problem:

$$\begin{array}{ll}
\underset{\psi\in\mathbb{C}^n}{\text{minimize}} & \frac{\lambda}{2(1-\lambda)}\,\mathrm{dist}^2(\psi,\mathfrak{S})\\
\text{subject to} & \psi\in M
\end{array} \qquad\qquad (6.19)$$

The reason for minimizing the distance squared instead of the distance is to have a nice smooth objective function—it doesn't really change the problem. Nor does the factor $\frac{\lambda}{2(1-\lambda)}$ out front, but it has a huge impact on the next algorithm, which solves (6.19). So the question of uniqueness amounts to whether problem (6.19) has a unique solution. Experts in optimization don't often worry about uniqueness, but rather the existence of *local minima* to (6.19). This is one of the few remaining unresolved mathematical issues in phase retrieval.

While, in most applications, the projections onto the sets X, O and M in (6.7) have a closed form and can be computed very accurately and efficiently, there does not seem to be any method, analytic or otherwise, for computing the projection onto the set $\mathcal{M}$ defined by (6.6). This might be another good reason for avoiding a feasibility model. Indeed, if the projections are too difficult, or impossible to compute analytically, then the large part of the advantage of projection methods evaporates. Nevertheless, this essentially two-set feasibility model suggests a wide range of techniques within the family of projection methods, alternating projections, averaged projections and Douglas–Rachford being representative members. In contrast to these, methods based on optimization models can avoid the difficulty of computing a projection onto the set $\mathcal{M}$ by instead minimizing a nonnegative coupling function that takes the value 0 (only) on $\mathcal{M}$. The model for this is a *constrained least squares minimization* model

$$\text{Find } \left(\overline{z}, \overline{\psi}, \overline{\mathbf{u}} \right) \in \operatorname{argmin} \{ \mathbf{F} (z, \psi, \mathbf{u}) \mid (z, \psi, \mathbf{u}) \in X \times Y \times M \} \qquad (6.20)$$

where

$$\mathbf{F} (z, \psi, \mathbf{u}) \equiv \sum_{j=1}^{m} \| S_j (z) \odot \psi - u_j \|^2. \qquad (6.21)$$

What has happened here is that the set $\mathcal{M}$ defined by (6.6) has been replaced by the least squares objective to avoid the complication of computing the projection onto $\mathcal{M}$.

The relaxed averaged alternating reflections algorithm (RAAR[3]) first proposed in [21] addresses the instability of the Douglas-Rachford algorithm by anchoring the usual Douglas-Rachford iterates to one of the sets:

$$\psi^{k+1} \in \left(\tfrac{\lambda}{2} \left(\mathcal{R}_{\mathfrak{S}} (\mathcal{R}_M + \operatorname{Id}) + (1 - \lambda) \mathcal{P}_M \right) \right) (\psi^k). \qquad (6.22)$$

When $\lambda \in [0, 1)$ it was shown in [22] that this algorithm is equivalent to the Douglas-Rachford algorithm applied to (6.19). For the moment, just recognise that this is a convex combination of (6.15) with the projection onto the set M. If one wanted to play around further, the constraints $\mathfrak{S}$ and M can be changed without changing the form of the fixed point iterations. For instance, if the thing one is trying to recover, ψ, is actually an *electron density*, it should be a real-valued, positive vector; so instead of the set $\mathfrak{S}$, one should restrict the possible points to

$$\mathfrak{S}_+ \equiv \left\{ \psi \in \mathbb{C}^n \;\middle|\; \psi_i = \begin{cases} \max\{0, \Re(\psi_i)\} & \text{if } i \in D \\ 0 & \text{else} \end{cases} \right\}. \qquad (6.23)$$

The RAAR algorithm then takes the form

$$\psi^{k+1} \in \left(\tfrac{\lambda}{2} \left(\mathcal{R}_{\mathfrak{S}_+} (\mathcal{R}_M + \operatorname{Id}) + (1 - \lambda) \mathcal{P}_M \right) \right) (\psi^k),$$

which is hardly a change from (6.22). In fact, mathematically there is no qualitative difference between the two. When translated back to the format of the original HIO algorithm, this takes the form [21, Prop.2.1][4]: $(\forall k \in \mathbb{N})(\text{for } i = 1, 2, \ldots, n)$,

---

[3]The names for these algorithms have evolved since their first introduction. In [19] the procedure that is today known as the Douglas-Rachford algorithm was called the Averaged Alternating Reflections algorithm, which then explains the genesis of the name RAAR for (6.22). Since Douglas-Rachford is more or less the accepted name for (6.15), (6.22) is called DRλ in more recent matheamtical articles. Nevertheless, RAAR is more common in the physics literature, so that is the nomenclature used here. In the ProxToolbox, however, the DRλ nomenclature is used.

[4](6.24) corrects a sign error in the lower half of [21, Eq (14)]

$$\psi_i^{k+1} = \begin{cases} \left(\mathcal{P}_M(\psi^k)\right)_i, & \text{if } i \in D \text{ and } \left(\mathcal{R}_M(\psi^k)\right)_i \geq 0; \\ \lambda\psi_i^k + (1-2\lambda)\left(\mathcal{P}_M(\psi^k)\right)_i, & \text{otherwise.} \end{cases} \tag{6.24}$$

Comparing (6.24) with (6.9), it is clear that these are very different algorithms. If the object domain constraint were to change, (for instance, to a magnitude constraint in the object plane for a complex-valued object) the physical description analogous to (6.24) would change dramatically yet again, but the description as a fixed point mapping would always have the form

$$\psi^{k+1} \in T_{RAAR}\psi^k \equiv \left(\tfrac{\lambda}{2}\left(\mathcal{R}_O(\mathcal{R}_M + \mathrm{Id}) + (1-\lambda)\mathcal{P}_M\right)\right)(\psi^k) \tag{6.25}$$

where O is a placeholder for the constraint in the physical domain (see also (2.89)). The main point here is that the mathematical properties of the fixed point mapping T_{RAAR} depend on the properties of the sets M and O, but the algorithm is always the same.

From this point hence, the word *algorithm* will be used more or less synonymously with the phrase *fixed point iteration*. This will be a convenient way to pack several (hundreds of) lines of code into a single symbol T, for the *fixed point mapping*. This T takes a guess x^0 and replaces it with an update x^1. In mathematical terms, T *maps* $x^0 \in X$ to $x^1 \in X$ where X is the *domain* and *image* spaces of T, the shorthand for which is $T : X \to X$. The domain and image spaces need not be the same, but for fixed point iterations they are. One important feature of this way of thinking about things is that the guess and the update are the same kinds of objects with the same physical interpretation. This is different than a *function* or more generally a *relation* which can map a point to anything, for instance a number, a color, or a set. A *fixed point iteration* is the process of repeatedly applying the fixed point mapping T: given x^0, generate a sequence of points x^k via

$$(\forall k \in \mathbb{N}) \qquad x^{k+1} = Tx^k. \tag{6.26}$$

There are a number of accessories one can add to (6.26). These take the form: given x^0, and an update rule $\mathcal{A}_k$ ($k = 1, 2, \dots$), generate a sequence of points x^k via

$$(\forall k \in \mathbb{N}) \quad y^{k+1} = Tx^k; \\ x^{k+1} = \mathcal{A}_k\left(x^k, y^{k+1}\right). \tag{6.27}$$

The main difference between (6.26) and (6.27) is that in the latter the operations from one iteration to the next can evolve and adjust along with the iterations. These are invariably called *accelerations* because that is the name of the game.

6.1.3   What Is a Proximal Method?

A *proximal method* is a fixed point iteration (6.26) or its acceleration (6.27) where
the fixed point mapping T consists of *proximal mappings*. A proximal mapping has
a specific mathematical definition, but before giving this an intuitive description will
probably be more helpful. A proximal mapping sends a point to one or more points
that strike a balance between solving a minimization problem and staying close to
the original point. The projectors of the previous section are proximal mappings. To
see this, consider the function

$$\iota_\Omega(u) \equiv \begin{cases} 0, & \text{if } u \in \Omega, \\ +\infty, & \text{else.} \end{cases} \tag{6.28}$$

where Ω is some set. Allowing this function to take the value $+\infty$ is very convenient.
The optimization problem corresponding to minimizing ι_Ω while staying as close as
possible to $\bar{u}$ is

$$\underset{u}{\text{minimize}} \ \iota_\Omega(u) + \tfrac{1}{2\lambda}\|u - \bar{u}\|^2,$$

and the *solution* to this problem is written

$$\operatorname{argmin}_u \left\{ \iota_\Omega(u) + \tfrac{1}{2\lambda}\|u - \bar{u}\|^2 \right\}.$$

The parameter $\lambda > 0$ will become important in a moment, but it has no significance
in this context since the solution to the optimization problem above is the same for
all positive values of λ. The solution to this problem is the *set* of points in Ω that
are nearest to $\bar{u}$, or the set $\mathcal{P}_\Omega \bar{u}$. This should not be confused with the *optimal value*
of this problem, which in this context is just the distance of the point $\bar{u}$ to the set
Ω. For practical purposes one simply takes a selection from the set; this is denoted
$u^+ \in \mathcal{P}_\Omega \bar{u}$ and u^+ is called a *projection*.

The function ι_Ω is not the only function one could use, hence the more general
use of the terminology *proximal mapping* for the general function f [23]

$$\operatorname{prox}_{f,\lambda}(\bar{u}) \equiv \operatorname{argmin}_u \left\{ f(u) + \tfrac{1}{2\lambda}\|u - \bar{u}\|^2 \right\}. \tag{6.29}$$

Here the value of λ plays the role of dialing up or down the requirement of staying
close to the point $\bar{u}$. This is often understood as a *step-length parameter* in the context
of algorithms: the smaller λ is, the greater the penalty for moving away from $\bar{u}$.

The algorithm (6.15) written using the formalism of proximal mappings takes the
form

$$\psi^{k+1} = \tfrac{1}{2}\left(\mathcal{R}_{f_0,\lambda_0} \mathcal{R}_{f_1,\lambda_1} + \operatorname{Id} \right)(\psi^k). \tag{6.30}$$

Here $\mathcal{R}_{f_0,\lambda_0}$ and $\mathcal{R}_{f_1,\lambda_1}$ are called *proximal reflectors* defined by

$$\mathcal{R}_{f_0,\lambda_0} \equiv 2\operatorname{prox}_{f_0,\lambda_0} - \operatorname{Id} \quad \text{and} \quad \mathcal{R}_{f_1,\lambda_1} \equiv 2\operatorname{prox}_{f_1,\lambda_1} - \operatorname{Id}. \tag{6.31}$$

This is quite liberating because now, the same basic fixed point mapping can be applied without any changes in the mathematical theory to a much broader range of problem types.

It is worthwhile spending a few moments to marvel at prox. This is a mapping from points in a space X to points in the same space; using mathematical notation $\text{prox}_{f,\lambda} : X \to X$. But, look again at (6.29): this is the solution to another optimization problem. There are two important things to notice about this observation, first of which is that a *mapping* has been created out of an optimization problem. This is what a mathematician might call pretty. The second thing to notice is that the *value* of the optimization problem—"the answer to the ultimate question of life, the universe and everything" [24]—is beside the point.[5]

6.1.4   On Your Mark. Get Set...

There are three groups of readers envisioned for this tutorial. The first group is students, of either physics or mathematics, wishing to get hands-on numerical experience with classical algorithms for real-world problems in the physical sciences. The second group is optical scientists who already know what they want to do, but would like a repository of algorithms to see what works for their problem and what does not work. The third group is applied mathematicians who have new algorithmic ideas, but need to see how they perform in comparison to other known methods on real data. A stripped-down version of the ProxToolbox is used at the University of Göttingen to teach graduate and undergraduate courses in numerical optimization and mathematical imaging. What is omitted from the student version is the repository of algorithms and some of the prox mappings— the students are expected to write these themselves, with some guidance. Experienced researchers, it is expected, will extract the parts of the toolbox they need and incorporate these into their own software. To make it easy to identify the pieces, the toolbox has been organized in a highly modular structure. The modularity comes at the cost of an admittedly labyrinthine structure, which is the hardest thing to master and the main goal of the rest of this tutorial.

6.2   Algorithms

The two different models discussed above, feasibility and constrained optimization (6.19), lead to a natural classification of categories of algorithms. The development presented here follows [25]. To underscore the fact that the algorithms can be applied to problems other than X-ray imaging, the sets involved are denoted by Ω_j for $j = 0, 1, 2, \ldots, m$ and the points of interest are denoted with a u, instead of the

---

[5]In case you forgot, it's 42.

context-specific notation for a wavefield ψ. The sets Ω_j are subsets of the model space $\mathbb{C}^n$ (or $\mathbb{R}^{2n}$) and, since there can be more than just two sets, as in the case of *phase diversity* or *ptychography* the integer m is just a stand-in for the number of images and other qualitative constraints involved in an experiment.

6.2.1 Model Category I: Multi-set Feasibility

The multi-set feasibility problem is:

Feasibility

$$\text{Find } u \in \bigcap_{j=0}^{m} \Omega_j. \tag{6.32}$$

The numerical experience is that this model format leads to the most effective methods for solving phase-type problems. It is important to keep in mind, however, that for all practical purposes the intersection above is empty, so the algorithm is not really solving the problem since it has no solution.

Feasibility problems can be conveniently reformulated in an optimization format:

$$\min_{u \in \mathbb{C}^n} \sum_{j=0}^{m} \iota_{\Omega_j}(u), \tag{6.33}$$

where ι_{Ω_j} is the indicator function (6.28) of the set Ω_j. The fact that the intersection is empty is reflected in the fact that the optimal value to problem (6.33) is $+\infty$. For the purposes of this tutorial anything bigger than, say, 42 will be approximately infinity.

The easiest iterative procedure of all is the *Cyclic Projections* algorithm

Algorithm 6.2.1
Initialization. *Choose* $u^0 \in (\mathbb{C}^n)$.
General Step *(k = 0, 1, . . .)*

$$u^{k+1} \in T_{CP}(u^k) \quad \text{where} \quad T_{CP} \equiv P_{\Omega_0} P_{\Omega_1} \cdots P_{\Omega_m}.$$

In the context of phase retrieval with one observation and an object domain constraint this is called the *Gerchberg-Saxton* algorithm [26]. An early champion of projection methods for convex feasibility was Censor [27] who together with Cegielski has written a nice review of the extensive literature on these methods [28]. A more recent review of inconsistent feasibility can be found in [29]. The most complete analysis of this algorithm for consistent and inconsistent nonconvex problems has been established in [30] and is reviewed in Chap. 23. In the inconsistent case the fixed points generate cycles of smallest length locally over all other possible cycles generated by projecting onto the sets in the same order. *Rates* of convergence have been established generically for problems with this structure (see [30, Example 3.6]). Rates are important for estimating *how far* a particular iterate is from the solution. The most elementary convergence rate is *linear* convergence, also known as *geometric* or *exponential* convergence in various communities. A sequence $(u^k)_{k=0,1,2,\ldots}$ of points u^k is said to converge linearly (technically, Q-linearly) to a limit point u_* with a global rate $c < 1$ whenever

$$\|u^{k+1} - u_*\| < c\|u^k - u_*\| \qquad \forall k = 0, 1, 2, \ldots.$$

The Douglas-Rachford iteration given by (6.15) can only be applied directly to two-set feasibility problems,

$$\text{Find } x \in \Omega_0 \cap \Omega_1.$$

The fixed point iteration is given by

$$u^{k+1} \in T_{DR}u^k \quad \text{where} \quad T_{DR} \equiv \frac{1}{2}\left(\mathcal{R}_{\Omega_0}\mathcal{R}_{\Omega_1} + \text{Id}\right), \tag{6.34}$$

for $\mathcal{R}_{\Omega_0}$ and $\mathcal{R}_{\Omega_2}$ are generic set reflectors (see (6.13)). It is important not to forget that, even if the feasibility problem is consistent, the fixed points of the Douglas-Rachford Algorithm will not in general be points of intersection. Instead, the *shadows* of the iterates defined as $\mathcal{P}_{\Omega_1}(u^k)$, $k = 0, 1, 2, \ldots$, converge to intersection points, when these exist [19].

To extend this to more than two sets, Borwein and Tam [31, 32] proposed the following variant:

Algorithm 6.2.2 (Cyclic Douglas-Rachford—CDR)
Initialization. *Choose* $u^0 \in \mathbb{C}^n$.
General Step $(k = 0, 1, \ldots)$

$$u^{k+1} \quad \in \quad T_{CDR}u^k$$

where

$$T_{CDR} \equiv \left(\frac{1}{2}\left(\mathcal{R}_{\Omega_0}\mathcal{R}_{\Omega_1} + \mathrm{Id}\right)\right)\left(\frac{1}{2}\left(\mathcal{R}_{\Omega_1}\mathcal{R}_{\Omega_2} + \mathrm{Id}\right)\right)\cdots$$
$$\left(\frac{1}{2}\left(\mathcal{R}_{\Omega_m}\mathcal{R}_{\Omega_0} + \mathrm{Id}\right)\right).$$

Different sequencing strategies than the one presented above are possible. In [33] one of the pair of sets is held fixed. This has some theoretical advantages in a convex setting, though no advantage was observed for phase retrieval.

The relaxed Douglas-Rachford algorithm(6.25) takes the general form: for $\lambda \in (0, 1]$

$$(DR\lambda/RAAR) \qquad u^{k+1} \in \left(\frac{\lambda}{2}\left(\mathcal{R}_{\Omega_0}\mathcal{R}_{\Omega_1} + \mathrm{Id}\right) + (1-\lambda)\,\mathcal{P}_{\Omega_1}\right)(u^k). \quad (6.35)$$

Extending this to more than two sets yields the following algorithm, which was first proposed in [25], where it is called CDRλ.

Algorithm 6.2.3 (Cyclic Relaxed Douglas-Rachford CDRλ)
Initialization. *Choose $u^0 \in \mathbb{C}^n$ and $\lambda \in [0, 1]$.*
General Step $(k = 0, 1, \ldots)$

$$u^{k+1} \quad \in \quad T_{CDR\lambda}u^k$$
$$\textit{where}$$
$$T_{CDR\lambda} \equiv \left(\frac{\lambda}{2}\left(\mathcal{R}_{\Omega_0}\mathcal{R}_{\Omega_1} + \mathrm{Id}\right) + (1-\lambda)\,\mathcal{P}_{\Omega_1}\right)\cdot$$
$$\left(\frac{\lambda}{2}\left(\mathcal{R}_{\Omega_1}\mathcal{R}_{\Omega_2} + \mathrm{Id}\right) + (1-\lambda)\,\mathcal{P}_{\Omega_2}\right)$$
$$\cdots\left(\frac{\lambda}{2}\left(\mathcal{R}_{\Omega_m}\mathcal{R}_{\Omega_0} + \mathrm{Id}\right) + (1-\lambda)\,\mathcal{P}_{\Omega_0}\right).$$

The analysis for RAAR and its precursor, Douglas-Rachford is contained in [30, Sect. 3.2.2] and [34].

Another popular algorithm that can be derived from the Douglas-Rachford algorithm in the convex setting [35] is the Alternating Directions Method of Multipliers (ADMM, [36]). For nonconvex problems like phase retrieval the direct link between these methods is lost, though there have been some recent developments and studies [37–39]. The ADMM algorithm falls into the category of *augmented Lagrangian-based methods*. Here, problem (6.33) is reformulated as

$$\min_{x,u_j \in \mathbb{C}^n} \left\{ \iota_{\Omega_0}(x) + \sum_{j=1}^m \iota_{\Omega_j}(u_j) \mid u_j = x, \; j = 1, 2, \ldots, m \right\}, \quad (6.36)$$

so that one can apply ADMM to the augmented Lagrangian given by

$$\tilde{L}_\eta(x, u_j, v_j) \equiv \iota_{\Omega_0}(x) + \sum_{j=1}^m \left(\iota_{\Omega_j}(u_j) + \langle v_j, \; x - u_j \rangle + \tfrac{\eta}{2} \| x - u_j \|^2 \right), \quad (6.37)$$

where $\eta > 0$ is a penalization parameter and $v_j, \; j = 1, 2, \ldots, m$, are the multipliers which are associated with the linear constraints. The ADMM algorithm applied to finding the critical points of the corresponding augmented Lagrangian (see (6.37)) is given by

---

Algorithm 6.2.4 (Nonsmooth ADMM$_1$)
Initialization. *Choose $x^0, u_j^0, v_j^0 \in \mathbb{C}^n$ and fix $\eta > 0$.*
General Step $(k = 0, 1, \ldots)$

1. Update

$$x^{k+1} \in \operatorname{argmin}_{x \in \mathbb{C}^n} \left\{ \iota_{\Omega_0}(x) + \sum_{j=1}^m \left(\langle v_j^k, \; x - u_j^k \rangle + \frac{\eta}{2} \| x - u_j^k \|^2 \right) \right\}$$

$$= \mathcal{P}_{\Omega_0} \left(\frac{1}{m} \sum_{j=1}^m \left(u_j^k - \frac{1}{\eta} v_j^k \right) \right). \quad (6.38)$$

2. For all $j = 1, 2, \ldots, m$ update (in parallel)

$$u_j^{k+1} \in \operatorname{argmin}_{u_j \in \mathbb{C}^n} \left\{ \iota_{\Omega_j}(u_j) + \langle v_j^k, \; x^{k+1} - u_j \rangle + \frac{\eta}{2} \| x^{k+1} - u_j \|^2 \right\}$$

$$= \mathcal{P}_{\Omega_j} \left(x^{k+1} - \eta v_j^k \right). \quad (6.39)$$

3. For all $j = 1, 2, \ldots, m$ update (in parallel)

$$v_j^{k+1} = v_j^k + \eta \left(x^{k+1} - u_j^{k+1} \right). \quad (6.40)$$

This *can* be written as a fixed point iteration on triplets $(x^k, u^k, v^k) \in \mathbb{C}^n \times \mathbb{C}^{mn} \times \mathbb{C}^n$, but it is not very convenient to see things this way. Note that the projections in Step 2 of the algorithm can be computed in parallel, while the Cyclic Projections and Cyclic Douglas-Rachford Algorithms must be executed sequentially. Note also that the update of the block u^{k+1} incorporates the newest information from the block

x^{k+1} together with the old data v^k, while the update of the block v^{k+1} incorporates the newest information from both blocks x^{k+1} and u^{k+1}. This is in the same spirit as the Gauss-Seidel method for systems of linear equations. Obviously, there is an increase (by a factor of $3 + m$) of the number of variables, but this is a mild increase in complexity in comparison to some recent proposals for phase retrieval which involve *squaring* the number of variables! Indeed ADMM is starting point for just about all the most successful methods for large-scale optimization with linear constraints (see, for instance, [40] and references therein).

An ADMM scheme for phase retrieval has appeared in [41]. This is a terrible algorithm for phase retrieval. It is included here, however, as a point of reference to the Douglas-Rachford Algorithm.

6.2.2   Model Category II: Product Space Formulations

The second category of algorithms is a stepping stone to smooth optimization models, though this is not the most obvious way to motivate the strategy—the connection to smoothing only becomes apparent after some consideration. The idea is to lift the problem to the product space $(\mathbb{C}^n)^{m+1}$ which can be then formulated as a two-set feasibility problem

$$\text{Find } \mathbf{u}^* \in \Omega \cap \mathcal{I},$$

where $\mathbf{u}^* = \left(u_0^*, u_1^*, \ldots, u_m^*\right)$, $\Omega := \Omega_0 \times \Omega_1 \times \cdots \times \Omega_m$ and $\mathcal{I}$ is the diagonal set of $\mathbb{C}^{n(m+1)}$ which is defined by $\{\mathbf{u} = (u, u, \ldots, u) : u \in \mathbb{C}^n\}$. This also involves an increase in the number of unknowns, but only by a factor of m which, while not insignificant when m is large, can be managed through clever implementation. There are two important features of this formulation. First of these is that the projection onto the set Ω can be easily computed since

$$P_\Omega \mathbf{u} = \left(P_{\Omega_0} u_0, P_{\Omega_1} u_1, \ldots, P_{\Omega_m} u_m\right),$$

where P_{Ω_j}, $j = 1, 2, \ldots, m$, are given in (6.11). The second important feature is that $\mathcal{I}$ is a subspace which also has simple projection given by $P_\mathcal{I}(\mathbf{u}) = \bar{\mathbf{u}}$ where

$$\bar{u}_j = \frac{1}{m+1} \sum_{j=0}^{m} u_j.$$

This formulation immediately suggests the Cyclic Projections algorithm 6.2.1, which, in the case of just two sets, is often called *Alternating Projections*

Algorithm 6.2.5 (Alternating Projections—AP)
Initialization. *Choose* $\mathbf{u}^0 \in \mathbb{C}^{n(m+1)}$.
General Step ($k = 0, 1, \ldots$)

$$\mathbf{u}^{k+1} \in \mathcal{P}_\mathcal{I} \mathcal{P}_\Omega \mathbf{u}^k.$$

But Algorithm 6.2.5 is equivalent to

$$u_j^{k+1} \in \frac{1}{m+1} \sum_{j=0}^m \mathcal{P}_{\Omega_j} u_j^k, \quad j = 0, 1, \ldots, m;$$

in other words, the Alternating Projections algorithm on the product space is equivalent to the Averaged Projections algorithm 6.2.10 and the Alternating Minimization Algorithm 6.2.11. Also the popular *Projected Gradient* method reduces to averaged projections. To see this, consider the following minimization problem:

$$\underset{\mathbf{u}\in\Omega}{\text{minimize}} \; \frac{1}{2} \operatorname{dist}^2 (\mathbf{u}, \mathcal{I}) . \tag{6.41}$$

The objective above is convex and, because $\mathcal{I}$ is a closed and convex set, continuously differentiable with a Lipschitz continuous gradient given by $\nabla \operatorname{dist}^2 (\mathbf{u}, \mathcal{I}) = 2 (\mathbf{u} - \mathcal{P}_\mathcal{I} \mathbf{u})$ (see (6.46)). The Projected Gradient Algorithm applied to this problem follows easily.

Algorithm 6.2.6 (Projected Gradient—PG)
Initialization. *Choose* $\mathbf{u}^0 \in \mathbb{C}^{n(m+1)}$.
General Step ($k = 0, 1, \ldots$)

$$\mathbf{u}^{k+1} \in \mathcal{P}_\Omega \left(\mathbf{u}^k - \tfrac{1}{2} \nabla \operatorname{dist}_\mathcal{I}^2 (\mathbf{u}^k) \right) \iff \mathbf{u}^{k+1} \in \mathcal{P}_\Omega \mathcal{P}_\mathcal{I} \mathbf{u}^k$$

$$\iff y^{k+1} \in \frac{1}{m+1} \sum_{j=0}^m \mathcal{P}_{\Omega_j} y^k,$$

$$\textit{where } u_j^{k+1} = \mathcal{P}_{\Omega_j} y^{k+1}.$$

This is not surprising since the minimization problem (6.41) is equivalent to (6.48). To see this, note that by the definition of the distance function

$$\min_{\mathbf{u} \in \Omega} \frac{1}{2} \operatorname{dist}^2 (\mathbf{u}, \mathcal{I}) = \min_{\mathbf{u} \in \Omega} \min_{\mathbf{y} \in \mathcal{I}} \frac{1}{2} \|\mathbf{u} - \mathbf{y}\|^2 = \min_{\mathbf{u} \in \Omega} \min_{\mathbf{y} \in \mathbb{C}^n} \frac{1}{2} \sum_{j=0}^{m} \|y - u_j\|^2$$

$$= \min_{y, \mathbf{u}} \left\{ \sum_{j=0}^{m} \frac{1}{2} \|y - u_j\|^2 \, \Big| \, u_j \in \Omega_j, \quad j = 0, 1, \dots, m \right\}.$$

In the convex setting, the Projected Gradient Algorithm has the advantage that it can be accelerated [42, 43]. A Fast Projected Gradient Algorithm for problem (6.41) looks like:

Algorithm 6.2.7 (Fast Projected Gradient—FPG)
Initialization. *Choose* $\mathbf{u}^0, \mathbf{y}^1 \in \mathbb{C}^{n(m+1)}$ *and* $\alpha_k = \frac{k-1}{k+2}$ *for all* $k = 0, 1, 2, \dots$.
General Step $(k = 1, 2, \dots)$

$$\mathbf{u}^k \in \mathcal{P}_\Omega \left(\mathbf{y}^k - \frac{1}{2} \nabla \operatorname{dist}^2 \left(\mathbf{y}^k, \mathcal{I} \right) \right),$$
$$\mathbf{y}^{k+1} = \mathbf{u}^k + \alpha_k \left(\mathbf{u}^k - \mathbf{u}^{k-1} \right)$$

$$\Longleftrightarrow$$

$$\mathbf{u}^k \in \mathcal{P}_\Omega \mathcal{P}_\mathcal{I} \mathbf{y}^k,$$
$$\mathbf{y}^{k+1} = \mathbf{u}^k + \alpha_k \left(\mathbf{u}^k - \mathbf{u}^{k-1} \right).$$

There is no theory for the choice of acceleration parameter $\alpha_k, k = 0, 1, 2, \dots$, in Algorithm 6.2.7 for nonconvex problems, but numerical experience [25, 44] indicates that this works pretty well. All that is missing is an explanation.

In the product space setting the best approximation problem takes the form

$$\underset{\mathbf{u} \in \mathbb{C}^{n(m+1)}}{\text{minimize}} \left\{ \frac{\lambda}{2(1 - \lambda)} \operatorname{dist}^2 (\mathbf{u}, \mathcal{I}) + \iota_\Omega (\mathbf{u}) \right\}; \tag{6.42}$$

Since there are only two functions, the proximal Douglas-Rachford or the relaxed proximal Douglas-Rachford algorithms apply to (6.42) without any tricks:

Algorithm 6.2.8 (Relaxed Douglas-Rachford—DRλ/RAAR)
Initialization. *Choose* $\mathbf{u}^0 \in \mathbb{C}^{n(m+1)}$ *and* $\lambda \in [0, 1]$.
General Step $(k = 0, 1, \dots)$

$$\mathbf{u}^{k+1} \in \frac{\lambda}{2} \left(\mathcal{R}_\mathcal{I} \mathcal{R}_\Omega \mathbf{u}^k + \mathbf{u}^k \right) + (1 - \lambda) \mathcal{P}_\Omega \mathbf{u}^k. \tag{6.43}$$

The relaxed Douglas-Rachford Algorithm is exactly the proximal Douglas-Rachford algorithm applied to the problem (6.42) [22]; that is,

$$\frac{1}{2} \left(\mathcal{R}_1 \mathcal{R}_\Omega + \mathrm{Id}\right) = \frac{\lambda}{2} \left(\mathcal{R}_\mathcal{I} \mathcal{R}_\Omega + \mathrm{Id}\right) + (1 - \lambda)\, \mathcal{P}_\Omega,$$

where $\mathcal{R}_1 \equiv 2\operatorname{prox}_{1, f_\lambda}(\mathbf{u}) - \mathbf{u}$ is the *proximal reflector* of the function $f_\lambda(\mathbf{u}) \equiv \frac{\lambda}{2(1-\lambda)} \operatorname{dist}^2(\mathbf{u}, \mathcal{I})$.

A different kind of relaxation to the Douglas-Rachford algorithm was recently proposed and studied in [45]. This appears to be better than Algorithm 6.2.8. When the sets involved are affine, the algorithm is a convex combination of Douglas-Rachford and Alternating Projections, but generally it takes the form

---

Algorithm 6.2.9 (Douglas-Rachford-Alternating-Projections)
Initialization. *Choose* $\mathbf{u}^0 \in \mathbb{C}^{n(m+1)}$ *and* $\lambda \in [0, 1]$.
General Step $(k = 0, 1, \ldots)$

$$\mathbf{u}^{k+1} \in \mathcal{P}_\mathcal{I}\left((1 + \lambda)\,\mathcal{P}_\Omega \mathbf{u}^k - \lambda \mathbf{u}^k\right) - \lambda\left(\mathcal{P}_\Omega \mathbf{u}^k - \mathbf{u}^k\right). \tag{6.44}$$

This algorithm is denoted by DRAP in the demonstrations below.

6.2.3 Model Category III: Smooth Nonconvex Optimization

The next algorithm, *Averaged Projections*, could be motivated purely from the feasibility framework detailed above. But there is a more significant *smooth* interpretation of this model, which motivates the smooth model class.

Algorithm 6.2.10 (Averaged Projections—AvP)
Initialization. *Choose* $u^0 \in \mathbb{C}^n$.
General Step $(k = 0, 1, \ldots)$

$$u^{k+1} \in T_{AvP} u^k \quad \text{where} \quad T_{AvP} \equiv \frac{1}{m+1} \sum_{j=0}^{m} \mathcal{P}_{\Omega_j}.$$

The analysis of averaged projections for problems with this structure is covered by the analysis of nonlinear/nonconvex gradient descent. This is classical and can be found throughout the literature, but it is limited to guarantees of convergence to *critical points* [46, 47]. For phase retrieval it is not known how to guarantee that all critical points are global minimums, though this is a topic of intense interest at the moment.

Although, in general, averaged projections has a slower convergence rate than its sequential counterpart [48], there are two features that recommend this method. First, it can be run in parallel. Secondly, it appears to be more robust to problem incon-sistency. Indeed, Averaged Projections algorithm is equivalent to gradient-based schemes when applied to an adequate smooth and nonconvex objective function. This well-known fact goes back to [49] when the sets Ω_j, $j = 0, 1, \ldots, m$, are closed and convex. In particular, two very prevalent schemes are in fact equivalent to AvP.

To see this, consider the problem of minimizing the sum of squared distances to the sets Ω_j, $j = 0, 1, \ldots, m$, that is,

$$\underset{u \in \mathbb{C}^n}{\text{minimize}} \; f(u) \equiv \frac{1}{2(m+1)} \sum_{j=0}^{m} \text{dist}^2(u, \Omega_j). \tag{6.45}$$

Since the sets Ω_j, $j = 0, 1, \ldots, m$, are nonconvex, the functions $\text{dist}^2(u, \Omega_j)$ are clearly not differentiable, and hence, same for the objective function $f(u)$. However, in the context of phase retrieval, the sets Ω_j, $j = 0, 1, \ldots, m$, are *prox-regular* (i.e. the projector onto these sets is single-valued near the sets [50]). From elementary properties of prox-regular sets [51] it can be shown that the gradient of the squared distance is defined and differentiable with Lipschitz continuous derivative (that is, the corresponding Hessian) up to the boundary of Ω_j, $j = 0, 1, \ldots, m$, and points where the coordinate elements of the vector u vanish. Indeed, for f given by (6.45)

$$\nabla f(u) \equiv \frac{1}{m+1} \sum_{j=0}^{m} \left(\text{Id} - \mathcal{P}_{\Omega_j} \right)(u). \tag{6.46}$$

Thus, applying the gradient descent with *unit stepsize* to problem (6.45), one imme-diately recovers the avareged projection algorithm.

The objective in (6.45) is as nice as one could hope for: it has full domain, is smooth and nonegative and has the value zero at points of intersection. These kinds of models are for obvious reasons favored in applications; unfortunately, these reasons are a little old fashioned considering today's mathematical technology for dealing with nonsmooth objectives like (6.33).

Another way to approach problem (6.45) underscores connections with another fundamental algorithmic strategy. Consider the following problem:

$$\underset{u \in \mathbb{C}^n}{\min} \; f(u) \equiv \frac{1}{2} \sum_{j=0}^{m} \text{dist}^2(u, \Omega_j). \tag{6.47}$$

Using the definition of the function $\text{dist}^2(\cdot, \Omega_j)$, $j = 0, 1, \ldots, m$, problem (6.47) is equivalent to

$$\underset{x, u}{\min} \left\{ \sum_{j=0}^{m} \frac{1}{2} \| x - u_j \|^2 \; \middle| \; u_j \in \Omega_j, \quad j = 0, 1, \ldots, m \right\}, \tag{6.48}$$

where $\mathbf{u} = (u_0, u_1, \ldots, u_m) \in (\mathbb{C}^n)^{m+1}$. The number of variables has now increased $(m + 1)$-fold, which, for applications like ptychography starts to get worrying since m can be large. But this is more a conceptual issue than practical.

Problem (6.48) always has an optimal solution (the objective is continuous and the constraint is closed and bounded set, so by a theorem from Weierstrass the minimum is attained). The optimization problem (6.48) consists of constraint sets which are separable over the variables u_j, $j = 0, 1, \ldots, m$; this can be exploited to divide the optimization problem into a sequence of easier subproblems. Alternating Minimization (AM) does just this, and involves updating each variable sequentially:

Algorithm 6.2.11 (Alternating Minimization—AM)
Initialization. *Choose* $(y^0, u_0^0, u_1^0, \ldots, u_m^0) \in (\mathbb{C}^n)^{m+2}$.
General Step $(k = 0, 1, \ldots)$

1. Update

$$y^{k+1} = \operatorname{argmin}_{y \in \mathbb{C}^n} \sum_{j=0}^m \frac{1}{2} \| y - u_j^k \|^2 = \frac{1}{m+1} \sum_{j=0}^m u_j^k. \tag{6.49}$$

2. For all $j = 0, 1, \ldots, m$ update (in parallel)

$$u_j^{k+1} \in \operatorname{argmin}_{u_j \in \Omega_j} \frac{1}{2} \| u_j - y^{k+1} \|^2 = \mathcal{P}_{\Omega_j} y^{k+1}. \tag{6.50}$$

By combining (6.49) and (6.50), the algorithm is written compactly as

$$y^{k+1} \in \frac{1}{m+1} \sum_{j=0}^m \mathcal{P}_{\Omega_j} y^k, \tag{6.51}$$

which is just averaged projections, Algorithm 6.2.10! When $m = 1$, i.e., only one image is considered, the Alternating Minimization Algorithm above coincides with what is known as the *Error Reduction* Algorithm [52] in the optics community.

In [53] Marchesini studied an *augmented Lagrangian* approach to solving

$$\min_{\psi \in \mathbb{C}^n} \frac{1}{2n} \sum_{j=0}^m \sum_{i=1}^n \left(\| (\mathcal{D}_j(\psi))_i \| - \sqrt{I_{ij}} \right)^2. \tag{6.52}$$

This is a nonsmooth least-squares relaxation of problem (6.2). Generic linear least squares problems take the form

$$\min_{y \in \mathbb{C}^n} \frac{1}{2n} \sum_{j=0}^m \sum_{i=1}^n \left(\| (F_j(y))_i \| - b_{ij} \right)^2 \tag{6.53}$$

where F_j is a linear mapping from $\mathbb{C}^n$ to $\mathbb{C}^n$ and b_{ij} are positive scalars. When the sets Ω_j are defined by

$$\Omega_j \equiv \left\{ u \in \mathbb{C}^n \mid \left\| (F_j(u))_i \right\| = b_{ij} \right\} \tag{6.54}$$

and the variables $y \in \mathbb{C}^n$ and $\mathbf{u} = (u_1, u_2 \ldots, u_m) \in \mathbb{C}^{mn}$ with $u_j \in \mathbb{C}^n$ satisfy $y = u_j$ for each $j = 1, 2, \ldots, m$, the resulting primal-dual/ADMM Algorithm takes the form:

Algorithm 6.2.12 (AvP^2)
Initialization. *Choose any $x^0 \in \mathbb{C}^n$ and $\rho_j > 0$, $j = 0, 1, \ldots, m$. Compute $u_j^1 \in \mathcal{P}_{\Omega_j}(y^0)$ $(j = 0, 1, \ldots, m)$ and $y^1 \equiv (1/(m+1)) \sum_{j=0}^m u_j^1$.*
General Step. *For each $k = 1, 2, \ldots$ generate the sequence $\left\{ (y^k, \mathbf{u}^k) \right\}_{k=0,1,2,\ldots}$ as follows:*

- *Compute*

$$y^{k+1} = \frac{1}{m} \sum_{j=1}^m \left(u_j^k + \frac{1}{\rho_j} \left(y^k - y^{k-1} \right) \right). \tag{6.55}$$

- *For each $j = 1, 2, \ldots, m$, compute*

$$u_j^{k+1} = \mathcal{P}_{\Omega_j} \left(u_j^k + \frac{1}{\rho_j} \left(2y^k - y^{k-1} \right) \right). \tag{6.56}$$

This algorithm can be viewed as a smoothed/relaxed version of Algorithm 6.2.4, but, when you look at it for the first time, the most obvious thing that jumps out at you is that this is averaged projections with a tw-step recursion. This is why it has been called AvP^2 in [25].

The more general PHeBIE Algorithm 6.2.13 applied to the problem of blind ptychography [54] reduces to Averaged Projections Algorithm for phase retrieval when the illuminating field is known. To derive this method, note that for any fixed y and $\mathbf{u}$, the function $u \mapsto \mathbf{F}(z, y, \mathbf{u})$ given by (6.21) is continuously differentiable and its partial gradient, $\nabla_z \mathbf{F}(z, y, \mathbf{u})$, is Lipschitz continuous with moduli $L_z(y, \mathbf{u})$. The same assumption holds for the function $y \mapsto \mathbf{F}(z, y, \mathbf{u})$ when z and $\mathbf{u}$ are fixed. In this case, the Lipschitz moduli is denoted by $L_y(z, \mathbf{u})$. Define $L_z'(y, \mathbf{u}) \equiv \max \{ L_z(y, \mathbf{u}), \eta_z \}$ where η_z is an arbitrary positive number. Similarly define $L_y'(z, \mathbf{u}) \equiv \max \{ L_y(z, \mathbf{u}), \eta_y \}$ where η_y is an arbitrary positive number. The constant η_z and η_y are used to address the following issue: if the Lipschitz constants $L_z(y, \mathbf{u})$ and/or $L_y(z, \mathbf{u})$ are zero then one should replace them with positive numbers (for the sake of well-definedness of the algorithm). In practice, it is better to chose them to be small numbers but for the analysis it can be chosen arbitrarily.

Algorithm 6.2.13 (Proximal Heterogeneous Block Implicit-Explicit)
Initialization. *Choose $\alpha, \beta > 1$, $\gamma > 0$ and $\left(z^0, y^0, \mathbf{u}^0\right) \in X \times O \times M$.*
General Step $(k = 0, 1, \ldots)$

1. Set $\alpha^k = \alpha L_z'\left(y^k, \mathbf{u}^k\right)$ and select

$$z^{k+1} \in \operatorname{argmin}_{z \in X} \left\{ \langle z - z^k, \ \nabla_z \mathbf{F}\left(z^k, y^k, \mathbf{u}^k\right) \rangle + \frac{\alpha^k}{2} \|z - z^k\|^2 \right\},$$
$$(6.57)$$

2. Set $\beta^k = \beta L_y'\left(z^{k+1}, \mathbf{u}^k\right)$ and select

$$y^{k+1} \in \operatorname{argmin}_{y \in O} \left\{ \langle y - y^k, \ \nabla_y \mathbf{F}\left(z^{k+1}, y^k, \mathbf{u}^k\right) \rangle + \frac{\beta^k}{2} \|y - y^k\|^2 \right\},$$
$$(6.58)$$

3. Select

$$\mathbf{u}^{k+1} \in \operatorname{argmin}_{\mathbf{u} \in M} \left\{ \mathbf{F}\left(z^{k+1}, y^{k+1}, \mathbf{u}\right) + \frac{\gamma}{2} \|\mathbf{u} - \mathbf{u}^k\|^2 \right\}. \quad (6.59)$$

Algorithm 6.2.13, referred to as PHeBIE in Sect. 6.3.3, can be interpreted as a combination of the algorithm proposed in [55] and a slight generalization of the PALM Algorithm [47]. In the context of blind ptychography (6.4), the block of variables y is replaced with the object ψ and the function $\mathbf{F}$ is the least squares objective (6.21). A partially preconditioned version of PALM was studied in [56] for phase retrieval, with improved performance over PALM. The regularization parameters α^k and β^k, $k = 0, 1, 2, \ldots$, are discussed in [54]. These parameters are inversely proportional to the step size in Steps (6.57) and (6.58) of the algorithm. Noting that α_k and β_k, $k = 0, 1, 2, \ldots$, are directly proportional to the respective partial Lipschitz moduli, the larger the partial Lipschitz moduli the *smaller* the step size, and hence the slower the algorithm progresses.

This brings to light an advantage of blocking strategies that is discussed in Chap. 12: algorithms that exploit block structures inherent in the objective function achieve better numerical performance by taking heterogeneous step sizes optimized for the separate blocks. There is, however, a price to be paid in the blocking strategies that are explored here: namely, they result in procedures that pass *sequentially* between operations on the blocks, and as such are not immediately parallelizable. The ptychography application is very generous in that it permits parallel computations on highly segmented blocks.

Nonsmooth analysis can be applied to the objective in (6.53), but this is still not main stream enough to be the stuff of normal graduate training, so it remains rather exotic. A popular way around this, is to formulate (6.53) as a system of quadratic equations:

$$\left\| \left(F_j(u)\right)_i \right\|^2 = b_{ij}^2, \quad \forall \, j = 1, 2, \ldots, m, \ \forall \, i = 1, 2, \ldots, n. \quad (6.60)$$

The corresponding squared least squares residual of the quadratic model is plenty smooth:

$$\min_{u \in \mathbb{C}^n} G(u) \equiv \frac{1}{2} \sum_{j=0}^{m} \sum_{i=1}^{n} \left(\left\| (F_j(u))_i \right\|^2 - b_{ij}^2 \right)^2. \tag{6.61}$$

There is a trick from conic programming that allows you to recast a quadratic equation on $\mathbb{R}^n$ as a linear equation on the space of matrices, $\mathbb{R}^{n \times n}$. The idea in the context of phase retrieval is called *phase lift*. The problem here is that, though the quadratic equation has been replaced by a linear equation (albeit in a much larger space), the desired solution is rank 1. Even though the set of fixed rank matrices is a manifold, it is not convex, so there is conservation of difficulty. The way around this is to replace the rank constraint with a norm—otherwise known as *convex relaxation*. The reasoning here is that, for convex problems, all local solutions are also global solutions to the problem; so you solve your convex problem and—poof— you have the global solution to the nonconvex problem under certain (hard to verify) conditions that guarantee the correspondence of the two problems. This is attractive as an analytical strategy, but as an algorithmic strategy it is not practical. Blumensath and Davies [57, 58] were the first ones to ask the question whether the conditions that guarantee correspondence of the nonconvex problem and its convex relaxation are also sufficient to guarantee that the nonconvex problem doesn't have any critical points *other than* global minima. They answered this question in the affirmative for the projected gradient Algorithm 6.2.6 and Hesse, Luke and Neumann showed that this is also the case for alternating projections Algorithm 6.2.5 [59]. So, there is no need to resort to convex relaxations, which is good news indeed since the phase lift method is not implementable on standard consumer-grade architectures for any of the Göttingen data sets.

Other methods based on the quartic objective have gained popularity in the newer generation of phase retrieval studies in the applied mathematics community. Notable among these are methods called Wirtinger flow. Smoothness makes the analysis easier, but the quartic objective has almost no curvature around critical points, which makes convergence of first order methods much slower than first order methods applied to nonsmooth objectives. See [14, Sect. 5.2] for a discussion of this and [25] for numerical comparisons.

Accelerations. The formulation of problem (6.45) has a fixed weight between the various distances. An extension of this is a dynamically weighted average between the projections to the sets Ω_j, $j = 0, 1, \ldots, m$. This idea was proposed in [14] where it is called *extended least squares*. In the context of sensor localization a similar approach was also proposed in [60] where it is called Sequential Weighted Least Squares (SWLS). The underlying model in [14] is the negative log-likelihood measure of the sum of squared set distances:

$$\underset{u \in \mathbb{C}^n}{\text{minimize}} \sum_{j=0}^{m} \ln \left(\text{dist}^2 (u, \Omega_j) + c \right), \qquad (c > 0). \tag{6.62}$$

Gradient descent applied to this objective yields what was called the *Dynamically Reweighted Averaged Projections* Algorithm (DyRePr) in [25].

Algorithm 6.2.14 (Dynamically Reweighted Averaged Projections)
Initialization. *Choose $u^0 \in \mathbb{C}^n$ and $c > 0$.*
General Step $(k = 0, 1, \ldots)$

$$u^{k+1} \in u^k - \sum_{j=0}^{m} \frac{2}{\left(\text{dist}^2 \left(u^k, \Omega_j\right) + c\right)} \left(u^k - \mathcal{P}_{\Omega_j} u^k\right). \qquad (6.63)$$

The smoothness of the sum of squared distances (almost everywhere) opens the door to higher-order techniques from nonlinear optimization that accelerate the basic gradient descent method. Quasi-Newton methods, for instance, would do the trick, and as observed in [61], they work unexpectedly well even on nonsmooth problems.

Algorithm 6.2.15 (Limited Memory BFGS with Trust Region)

1. *(Initialization) Choose $\tilde{\eta} > 0$, $\zeta > 0$, $\bar{\ell} \in \{1, 2, \ldots, n\}$, $u^0 \in \mathbb{C}^n$, and set $\nu = \ell = 0$. Compute $\nabla f \left(u^0\right)$ and $\|\nabla f \left(u^0\right)\|$ for*

$$f(u) \equiv \frac{1}{2(m+1)} \sum_{j=0}^{m} \text{dist}^2 \left(u, \Omega_j\right), \quad \nabla f(u) \equiv \frac{1}{m+1} \sum_{j=0}^{m} \left(\text{Id} - \mathcal{P}_{\Omega_j}\right)(u).$$

2. *(L-BFGS step) For each $k = 0, 1, 2, \ldots$ if $\ell = 0$ compute u^{k+1} by some line search algorithm; otherwise compute*

$$s^k = -\left(M^k\right)^{-1} \nabla f \left(u^k\right),$$

   where M^k is the L-BFGS update [62], $u^{k+1} = u^k + s^k$, $f\left(u^{k+1}\right)$, and the predicted change (see, for instance [63]).
3. *(Trust Region) If $\rho\left(s^k\right) < \tilde{\eta}$, where*

$$\rho\left(s^k\right) = \frac{\textit{actual change at step } k}{\textit{predicted change at step } k},$$

   reduce the trust region Δ^k, solve the trust region subproblem for a new step s^k [64], and return to the beginning of Step 2. If $\rho\left(s^k\right) \geq \tilde{\eta}$ compute $u^{k+1} = u^k + s^k$ and $f\left(u^{k+1}\right)$.

---

4. *(Update) Compute* $\nabla f\left(u^{k+1}\right)$, $\left\|\nabla f\left(u^{k+1}\right)\right\|$,

$$y^k \equiv \nabla f\left(u^{k+1}\right) - \nabla f\left(u^k\right), \quad s^k \equiv u^{k+1} - u^k,$$

and $s^{k^T} y^k$. *If* $s^{k^T} y^k \leq \zeta$, *discard the vector pair* $\{s^{k-\ell}, y^{k-\ell}\}$ *from storage, set* $\ell = \max\{\ell - 1, 0\}$, $\Delta^{k+1} = \infty$, $\mu^{k+1} = \mu^k$ *and* $M^{k+1} = M^k$ *(i.e. shrink the memory and don't update); otherwise set* $\mu^{k+1} = \frac{y^{k^T} y^k}{s^{k^T} y^k}$ *and* $\Delta^{k+1} = \infty$, *add the vector pair* $\{s^k, y^k\}$ *to storage, if* $\ell = \bar{\ell}$, *discard the vector pair* $\{s^{k-\ell}, y^{k-\ell}\}$ *from storage. Update the Hessian approximation* M^{k+1} *[62]. Set* $\ell = \min\{\ell + 1, \bar{\ell}\}$, $\nu = \nu + 1$ *and return to Step 1.*

This looks complicated but is standard in nonlinear optimization. Convergence is still unexplained for the limited memory implementation.

6.3 ProxToolbox—A Platform for Creative Hacking

A platform for collecting and working with data should satisfy several objectives:

- data transfer
- sharing data processing algorithms
- comparing the performance of different algorithmic approaches
- teaching
- innovation.

The ProxToolbox has been used within the Collaborative Research Center Nanoscale Photonic Imaging (SFB 755) at the University of Göttingen for each of the points above. It is written to be able to incorporate new problems, data, and algorithms without abandoning the old knowledge. This type of built-in knowledge retention requires a structure that is burdensome for single-purpose users. Most colleagues and students prefer to cannibalize the ProxToolbox—hacking is positively encouraged. This tutorial and the demos in the toolbox are intended to put the user on a fast track to successfully disassembling and re-purposing the basic elements.

Our presentation of the toolbox here is without specific reference to commands and code to prevent this tutorial from being outdated within a few months. Certain aspects of the code will change as new applications and new features get added to the toolbox, but what will not change is the compartmentalization of various mathematically and computationally distinct tools.

To download the toolbox and the data go to

```
http://num.math.uni-goettingen.de/proxtoolbox/
```

Here you will find links to the Matlab and Python versions of the toolbox, along with documentation and literature. The toolbox has the following organizational structure:

- Nanoscale_Photonic_Imaging_demos
- Algorithms
- ProxOperators
- Drivers/Problems

  - ...
  - Phase
    · Demos
    · DataProcessors
    · ProxOperators
  - Ptychography
    · Demos
    · DataProcessors
    · ProxOperators

  - ...

- Utilities
- InputData
- Documentation

The *Nanoscale_Photonic_Imaging_demos folder*. This folder contains scripts to generate the figures shown in this tutorial. This is the rabbit you will follow down the hole.

The *Algorithms* folder. This folder contains a general *algorithm wrapper* that loops through the iterations calling the desired algorithm. This is exactly T in (6.26), and the T_* indicates which specific algorithm is run, from Algorithm 6.2.1 through Algorithm 6.2.9. After the specific fixed point operator is applied, a specialized *iterate monitor* is called. This will depend both on the problem and the algorithm being run. The default is a *generic iterate monitor* that merely checks the distance between successive iterates. By default, the stopping criterion for the fixed point iteration is when the step between successive iterates falls below a tolerance given by the user. But for some algorithms and some problems, this may not be the best or most informative data about the progress of the iteration. For instance, if the problem is a feasibility problem (6.32), then the *feasibility iterate monitor* not only computes the difference between successive iterates, but also the distance between sets (the gap) at a given iterate. In this context, a reasonable comparison between algorithms is not the step-size, but rather between the gap achieved by different algorithms. If one is running a Douglas-Rachford-type algorithm on a feasibility problem, then as explained above, the iterates themselves don't have to converge, but their *shadows*, defined as the projection of the iterates onto one of the sets, will give a good indication of convergence of some form. Still other algorithms, like ADMM 6.2.12, generate several sequences of iterates (three in the case of ADMM), only two of which converge nicely when everything goes well [65]. As much as possible, the iterate monitoring is automated so that the user does not have to bother with this. But users who are interested in algorithm development will want to pay close attention to this.

The *ProxOperators* folder. Some prox operators are generic, like a projector onto the diagonal of a product space $\mathcal{I}$ (otherwise known as averaging), or the prox of the ℓ_1-norm (soft thresholding), or the prox of the ℓ_0-function (hard thresholding). General prox operators are stored here. These always map an input to another point in the same space, but how they do this depends on the strucure of the input u (array, dimension, etc.). Some problems involve prox mappings that are specific to that problem, like Sudoku. These specific prox mappings are stored under the Problem/Drivers folder.

The *Drivers/Problems folder*. This is the folder where the specific problem instances are stored. The problems that are of interest for this chapter are phase and ptychography, though there are other problems, like computed tomography, sensor localization and Sudoku. The *Phase* subfolder contains a general problem family handler called "phase". Since all phase problems have similar features, this problem handler makes sure that all the inputs and outputs are processed in the same way. The toolbox works through input files stored in the *demos* subfolder. The input files contain names of data sets, data processors, algorithm names and parameters, and other user defined parameters like stopping tolerances, output choices and so forth. The input files might be augmented by a graphical user interface in the future. The link between the experimentalist and the mathematician is through the *data processor*. The data processor for the Göttingen data sets is easily identified and contains all the required parameter values for specific experiments conducted at the Institute for Physics in Göttingen. The data that the processor manipulates is not contained in the ProxToolbox release, but is stored separately and must be downloaded from the links provided on the Prox-Toolbox homepage. Prox operators specific to phase retrieval, such as the projection onto the intensity data (6.11), are also stored at this level.
The *InputData folder*. The data, which is intended to be stored or linked in the directory "InputData", is not included in the software toolbox in the interest of portability. This tutorial will only cover demonstrations with the *Phase* datasets and the *Ptychography* datasets. As these sets grow and develop, the links may change to reflect different hosts.

The *Utilities folder*. This is where generic image and data manipulation tools are stored.

6.3.1 Coffee Break

The first walk through the Toolbox is demonstrated on an image set produced by undergraduates in Tim Salditt's laboratory at the Institute for X-ray Physics at the University of Göttingen. The data is the *CDI* intensity datafile contained in the *Phase* dataset linked to the ProxToolbox homepage. There is a demonstration of the Cyclic Projections Algorithm 6.2.1 in the folder *Nanoscale_Photonic_Imaging_demos*. To run the demonstration, just type *Coffee_demo* at the Matlab prompt (assuming Matlab has the demo folder and all the data folders in its path) or *python Coffee_demo.py* at the shell prompt if you are working with Python. The data set presented here

(I_i, in problem (6.2) with $n = 128^2$) is a diffraction image of visible light shown in Fig. 6.1a (log intensity scale). The physical parameters of the image (magnification factor, Fresnel number, etc.) are not given, so the easiest thing to do is to assume a perfect imaging system and expert experimentalists. The imaging model is then just an unmodified Fourier transform, that is, far-field imaging. The object was a real, nonnegative obstacle, supported on some patch in the object plane, so that the qualitative constraint O is of the form (6.23). The only way to know that the algorithm is converging at least to a local best approximation point is to monitor the successive iterates and feasibility gap Fig. 6.1b–c. A small feasibility gap is not necessarily desireable, since this also means that the noise is being faithfully recovered. For the demonstration shown here, a low-pass filter is applied to the data since almost all of the recoverable information about the object is contained in the low-frequency elements. It might seem counterintuitive, but the larger the feasibility gap (i.e., the more *inconsistent the problem is*), the faster the algorithm converges. In Chap. 23 this is explained. The original object was a coffee cup which the generous reader can see if he tilts his head to the left and squints really hard (Fig. 6.1d right). When you run the demo, don't be surprised if your reconstruction is an upside down version of what is shown here—this is a symptom of nonuniqueness of solutions to the phase problem.

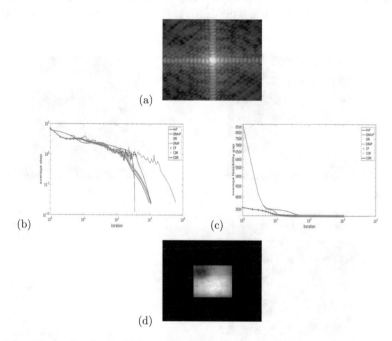

(a)

(b)   (c)

(d)

Fig. 6.1 **a** Observation (log scale) from optical diffraction experiment. **b** Step-size and **c** gap size between constraint sets versus iteraton for several algorithms [25]. **d** Typical recovery from the algorithms

6.3.2 Star Power

The next demonstration is of the reconstruction of a test object (the Siemens star) from near field X-ray data provided by Tim Salditt's laboratory at the Institute for X-ray Physics at the University of Göttingen. Here the structured illumination shown in Fig. 6.2a left is modeled by $\mathcal{D}_j$, $j = 1, 2, \ldots, m$, in problem (6.3) with $m = 1$. The image shown in Fig. 6.2a right is in the near field, so the mapping $\mathcal{D}_F$ in problem (6.2) is the near-field *Fresnel transform* [66]. In the model (6.3) this image is represented by I_{ij}, $j = 1, 2, \ldots, m$, with $m = 1$. The qualitative constraint is that the object is a *pure phase object*, that is, the field in the object domain has amplitude 1 at each pixel.

A reconstruction with this data that does not take noise into account is shown in Fig. 6.3. What is remarkable here is that if one only looks at the convergence of the algorithms and judges by the achieved gap before termination, it appears that the quasi-Newton accelerated average projections algorithm (QNAvP) is clearly the best Fig. 6.3b–c. But when you look at the reconstructions Fig. 6.3a the QNAvP reconstruction is the worst. The problem here is that the noise has also been faithfully recovered.

In [67] a regularization strategy is proposed that blows a ball around the data (either Euclidean or Kullback-Leibler, as appropriate) and takes any reasonable point within the ball. The justification is that, if the data is noisy anyway, you don't want to match

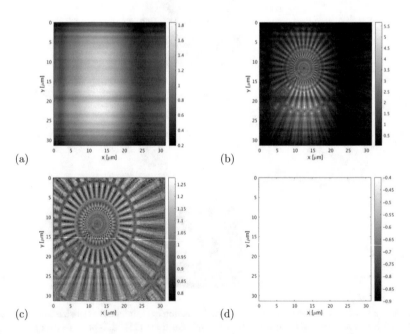

Fig. 6.2 Near field X-ray holography experiment with a Siemens test object [25]. **a** The empty beam. **b** Observed pattern. **c** Initial guess for object amplitude. **d** Initial guess for object phase

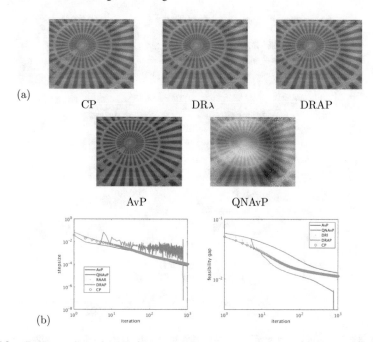

(a)

CP   DRλ   DRAP

AvP   QNAvP

(b)

Fig. 6.3 **a** Reconstruction of regularized near field holography experiment with empty beam correction for the same data shown in Fig. 6.2. **b** Step-size and **c** gap between the constraint sets versus iteration [25]

it exactly. For a more precise development of this intuition see Chap. 4. Though it is not obvious, projecting onto such fattened data sets is more expensive than projecting onto the original noisy data. The latter is viewed as an approximate and extrapolated projection onto the fattened set. The algorithm is terminated when the iterates are within a tolerable distance of the data. A demonstration of this is shown in Fig. 6.4. The theoretical justification for this strategy is quite technical, but effectively what one is doing is running the old algorithms with early termination.

6.3.3 E Pluribus Unum

The demonstration *Ptychography_demo* shown in Fig. 6.5 computes the probe and object from a far-field raster scan of 676 overlapping patches of the Siemens star, illuminated by a narrow X-ray beam. The mathematical problem is to minimize the objective function given in (6.3). The demonstration shows how the PHeBIE algorithm 6.2.13 does this. Since blind deconvolution has many local solutions, the process has two phases: the first conventional phase retrieval on the data with a probe

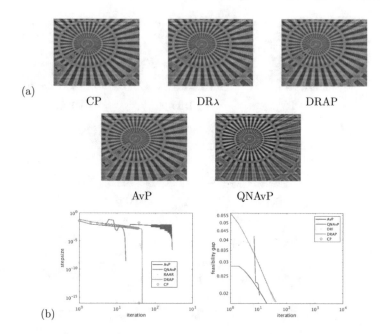

Fig. 6.4 **a** Reconstruction of *regularized* (i.e. early termination) near field holography experiment with empty beam correction for the same data shown in Fig. 6.2a. **b** Step-size and **c** gap between the constraint sets versus iteration

ansatz, the second phase simultaneous phase retrieval and probe determination in what is essentially a nonlinear blind deconvolution. In the *Ptychography_demo* the first phase is executed with the DRλ algorithm on the product space: 676 images of size (192^2) accounting for 26 equal translations of the beam side to side and 26 equal shifts of the beam top to bottom. The second phase is executed with the PHeBIE algorithm starting from the last iterate of the first phase.

6.4 Last Word

One of the most rewarding things about participating in the Nanoscale Photonic Imaging Collaborative Research Center at the University of Göttingen has been working with scientists from different disciplines with different sensibilities and intuition. Collaboration starts with mutual respect and an openness for new ways of thinking about things. This has resulted in better mathematics and better science, both grounded in real world experience but with an attention to abstract structures. This has forced the examination of aspects of abstract models that, at first glance, don't seem that important, but turn out to be decisive in practice.

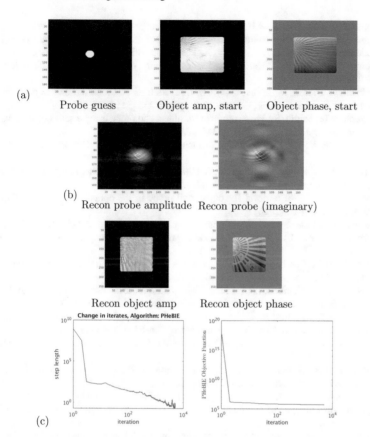

Fig. 6.5 **a** Initial probe and warm start object initialization of a far field ptychography experiment for the scan data. **b** Probe and object reconstructed by the PHeBIE algorithm 6.2.13. **c** Step size and objective function values versus iteration

Acknowledgements This work was supported by DFG grant SFB755.

References

1. Krantz, S.G.: Mathematical Apocrypha Redux. Mathematical Association of America, UK edition (2006)
2. Courant, R., Hilbert, D.: Methods of Mathematical Physics. Interscience Publishers, New York (1953)
3. Luke, D.R.: Proxtoolbox. http://num.math.uni-goettingen.de/proxtoolbox/ (2017)
4. Hegerl, R., Hoppe, W.: Dynamische Theorie der Kristallstrukturanalyse durch Elektronenbeugung im inhomogenen Primärstrahlwellenfeld. Ber. Bunsenges. Phys. Chem **74**(11), 1148–1154 (1970)
5. Maiden, A.M., Rodenburg, J.M.: An improved ptychographical phase retrieval algorithm for diffractiveimaging. Ultramicroscopy **109**, 1256–1262 (2009)

6. Rodenburg, J.M., Bates, R.H.T.: The theory of super-resolution electron microscopy via Wigner-distribution deconvolution. Philos. Trans. R. Soc. Lond. Ser. A **339**, 521–553 (1992)
7. Rodenburg, J.M., Hurst, A.C., Cullis, A.G., Dobson, B.R., Pfeiffer, F., Bunk, O., David, C., Jefimovs, K., Johnson, I.: Hard-X-ray lensless imaging of extended objects. Phys. Rev. Lett. **98**, 034801 (2007)
8. Thibault, P., Dierolf, M., Bunk, O., Menzel, A., Pfeiffer, F.: Probe retrieval in ptychographic coherent diffractive imaging. Ultramicroscopy **109**, 338–343 (2009)
9. Robisch, A.L., Salditt, T.: Phase retrieval for object and probe using a series of defocus near-field images. Opt. Express **21**(20), 23345–23357 (2013)
10. Gonsalves, R.A.: Phase retrieval and diversity in adaptive optics. Opt. Eng. **21**(5), 829–832 (1982)
11. Robisch, A.L., Kröger, K., Rack, A., Salditt, T.: Near-field ptychography using lateral and longitudinal shifts. New J. Phys. **17**(7), 073033 (2015)
12. Robisch, A.L.: Phase retrieval for object and probe in the optical near-field. Ph.D. thesis, Universität Göttingen (2016)
13. Fienup, J.R.: Phase retrieval algorithms: a comparison. Appl. Opt. **21**(15), 2758–2769 (1982)
14. Luke, D.R., Burke, J.V., Lyon, R.G.: Optical wavefront reconstruction: Theory and numerical methods. SIAM Rev. **44**, 169–224 (2002)
15. Bauschke, H.H., Combettes, P.L., Luke, D.R.: A hybrid projection reflection method for phase retrieval. J. Opt. Soc. Amer. A **20**(6), 1025–1034 (2003)
16. Bauschke, H.H., Combettes, P.L., Luke, D.R.: Phase retrieval, error reduction algorithm and Fienup variants: a view from convex feasibility. J. Opt. Soc. Amer. A **19**(7), 1334–1345 (2002)
17. Douglas Jr., J., Rachford Jr., H.H.: On the numerical solution of heat conduction problems in two or three space variables. Trans. Amer. Math. Soc. **82**, 421–439 (1956)
18. Lions, P.L., Mercier, B.: Splitting algorithms for the sum of two nonlinear operators. SIAM J. Numer. Anal. **16**, 964–979 (1979)
19. Bauschke, H.H., Combettes, P.L., Luke, D.R.: Finding best approximation pairs relative to two closed convex sets in Hilbert spaces. J. Approx. Theory **127**, 178–192 (2004)
20. Miao, J., Charalambous, P., Kirz, J., Sayre, D.: Extending the methodology of X-ray crystallography to allow imaging of micrometre-sized non-crystalline specimens. Nature **400**, 342–344 (1999)
21. Luke, D.R.: Relaxed averaged alternating reflections for diffraction imaging. Inverse Probl. **21**, 37–50 (2005)
22. Luke, D.R.: Finding best approximation pairs relative to a convex and a prox-regular set in Hilbert space. SIAM J. Optim. **19**(2), 714–739 (2008)
23. Moreau, J.J.: Fonctions convexes duales et points proximaux dans un espace Hilbertien. Comptes Rendus de l'Académie des Sciences de Paris **255**, 2897–2899 (1962)
24. Adams, D.: The Hitchhikers Guide to the Galaxy. Pan Books, New York (1980)
25. Luke, D.R., Sabach, S., Teboulle, M.: Optimization on spheres: models and proximal algorithms with computational performance comparisons. SIAM J. Math. Data Sci. **1**(3), 408–445 (2019)
26. Gerchberg, R.W., Saxton, W.O.: A practical algorithm for the determination of phase from image and diffraction plane pictures. Optik **35**, 237–246 (1972)
27. Censor, Y.: Row-action methods for huge and sparse systems and their applications. SIAM Rev. **23**, 444–466 (1981)
28. Censor, Y., Cegielski, A.: Projection methods: an annotated bibliography of books and reviews. Optimization **64**, 2343–2358 (2015). https://doi.org/10.1080/02331934.2014.957701
29. Censor, Y., Zaknoon, M.: Algorithms and convergence results of projection methods for inconsistent feasibility problems: a review. Pure Appl. Funct. Anal. (2019). https://arxiv.org/abs/1802.07529 (to appear)
30. Luke, D.R., Thao, N.H., Tam, M.K.: Quantitative convergence analysis of iterated expansive, set-valued mappings. Math. Oper. Res. **43**(4), 1143–1176 (2018). https://doi.org/10.1287/moor.2017.0898
31. Borwein, J.M., Tam, M.K.: A cyclic Douglas-Rachford iteration scheme. J. Optim. Theory Appl. **160**(1), 1–29 (2014)

32. Borwein, J.M., Tam, M.K.: The cyclic Douglas-Rachford method for inconsistent feasibility problems. J. Nonlinear Convex Anal. **16**(4), 537–584 (2015)
33. Bauschke, H.H., Noll, D., Phan, H.M.: Linear and strong convergence of algorithms involving averaged nonexpansive operators. J. Math. Anal. Appl. **421**(1), 1–20 (2015)
34. Luke, D.R., Martins, A.L., Tam, M.K.: Relaxed cyclic Douglas-Rachford algorithms for non-convex optimization. In: ICML Workshop: Modern Trends in Nonconvex Optimization for Machine Learning (2018). https://sites.google.com/view/icml2018nonconvex/papers
35. Gabay, D.: Augmented Lagrangian Methods: Applications to the Solution of Boundary- Value Problems. In: Applications of the Method of Multipliers to Variational Inequalities, pp. 299–331. North-Holland, Amsterdam (1983)
36. Glowinski, R., Marroco, A.: Sur l'approximation, par elements finis d'ordre un, et las resolution, par penalisation-dualitè, d'une classe de problemes de dirichlet non lineares. Revue Francais d'Automatique, Informatique et Recherche Opérationelle **9**(R-2), 41–76 (1975)
37. Bolte, J., Sabach, S., Teboulle, M.: Nonconvex Lagrangian-based optimization: monitoring schemes and global convergence. Math. Oper. Res. (2018). https://doi.org/10.1287/moor.2017.0900
38. Li, G., Pong, T.K.: Global convergence of splitting methods for nonconvex composite optimization. SIAM J. Optim. **25**(4), 2434–2460 (2015)
39. Themelis, A., Stella, L., Patrinos, P.: Forward-backward envelope for the sum of two nonconvex functions: Further properties and nonmonotone line-search algorithms. SIAM J Optim **28**, 2274–2303 (2018)
40. Shefi, R., Teboulle, M.: Rate of convergence analysis of decomposition methods based on the proximal method of multipliers for convex minimization. SIAM J. Optim. **24**(1), 269–297 (2014)
41. Liang, J., Stoica, P., Jing, Y., Li, J.: Phase retrieval via the alternating direction method of multipliers. IEEE Signal Process. Lett. **25**(1), 5–9 (2018)
42. Beck, A., Teboulle, M.: A fast iterative shrinkage-thresholding algorithm for linear inverse problems. SIAM J. Imaging Sci. **2**(1), 183–202 (2009). https://doi.org/10.1137/080716542
43. Nesterov, Y.E.: A method for solving the convex programming problem with convergence rate $O(1/k^2)$. Dokl. Akad. Nauk SSSR **269**, 543–547 (1983)
44. Pauwels, E.J.R., Beck, A., Eldar, Y.C., Sabach, S.: On Fienup methods for sparse phase retrieval. IEEE Trans. Signal Process. **66**(4) (2018)
45. Thao, N.H.: A convergent relaxation of the Douglas-Rachford algorithm. Comput. Optim., Appl (2018). https://doi.org/s10589-018-9989-y
46. Attouch, H., Bolte, J., Svaiter, B.F.: Convergence of descent methods for semi-algebraic and tame problems: proximal algorithms, forward-backward splitting, and regularized Gauss-Seidel methods. Math. Program. **137**, 91–129 (2013)
47. Bolte, J., Sabach, S., Teboulle, M.: Proximal alternating linearized minimization for nonconvex and nonsmooth problems. Math. Program. **146**(1), 459–494 (2014). https://doi.org/10.1007/s10107-013-0701-9
48. Lewis, A.S., Luke, D.R., Malick, J.: Local linear convergence of alternating and averaged projections. Found. Comput. Math. **9**(4), 485–513 (2009)
49. Zarantonello, E.H.: Projections on convex sets in Hilbert space and spectral theory. In: Zarantonello, E.H. (ed.) Contributions to Nonlinear Functional Analysis, pp. 237–424. Academic Press, New York (1971)
50. Luke, D.R., Martins, A.L.: Convergence Analysis of the Relaxed Douglas-Rachford Algorithm. SIAM J. Optim. https://arxiv.org/abs/1811.11590 (to appear)
51. Poliquin, R.A., Rockafellar, R.T., Thibault, L.: Local differentiability of distance functions. Trans. Amer. Math. Soc. **352**(11), 5231–5249 (2000)
52. Fienup, J.R.: Reconstruction of an object from the modulus of its Fourier transform. Opt. Lett. **3**(1), 27–29 (1978)
53. Marchesini, S.: Phase retrieval and saddle-point optimization. J. Opt. Soc. Am. A **24**(10) (2007)
54. Hesse, R., Luke, D.R., Sabach, S., Tam, M.: The proximal heterogeneous block implicit-explicit method and application to blind ptychographic imaging. SIAM J. Imaging Sci. **8**(1), 426–457 (2015)

55. Attouch, H., Bolte, J., Redont, P., Soubeyran, A.: Proximal alternating minimization and projection methods for nonconvex problems: an approach based on the Kurdyka-Lojasiewicz inequality. Math. Oper. Res. **35**(2), 438–457 (2010)
56. Chang, H., Marchesini, S., Lou, Y., Zeng, T.: Variational phase retrieval with globally convergent preconditioned proximal algorithm. SIAM J. Imaging Sci. **11**(1), 56–93 (2018)
57. Blumensath, T., Davies, M.: Iterative hard thresholding for compressed sensing. Appl. Comput. Harmon. Anal. **27**(3), 265–274 (2009)
58. Blumensath, T., Davies, M.: Normalised iterative hard thresholding; guaranteed stability and performance. IEEE J. Sel. Top. Signal Process. **4**(2), 298–309 (2010)
59. Hesse, R., Luke, D.R., Neumann, P.: Alternating projections and Douglas-Rachford for sparse affine feasibility. IEEE Trans. Signal. Process. **62**(18), 4868–4881 (2014). https://doi.org/10.1109/TSP.2014.2339801
60. Beck, A., Teboulle, M., Chikishev, Z.: Iterative minimization schemes for solving the single source localization problem. SIAM J. Optim. **19**(3), 1397–1416 (2008)
61. Lewis, A.S., Overton, M.L.: Nonsmooth optimization via quasi-Newton methods. Math. Program. **141**, 135–163 (2013)
62. Byrd, R.H., Nocedal, J., Schnabel, R.B.: Representations of quasi-Newton matrices and their use in limited memory methods. Math. Program. **63**, 129–156 (1994)
63. Nocedal, J., Wright, S.: Numerical Optimization. Springer, New York (1999)
64. Burke, J.V., Wiegmann, A.: Low-dimensional quasi-Newton updating strategies for large-scale unconstrained optimization. Department of Mathematics, University of Washington (1996)
65. Aspelmeier, T., Charitha, C., Luke, D.R.: Local linear convergence of the ADMM/Douglas-Rachford algorithms without strong convexity and application to statistical imaging. SIAM J. Imaging Sci. **9**(2), 842–868 (2016)
66. Hagemann, J., Robisch, A.L., Luke, D.R., Homann, C., Hohage, T., Cloetens, P., Suhonen, H., Salditt, T.: Reconstruction of wave front and object for inline holography from a set of detection planes. Opt. Express **22**, 11552–11569 (2014)
67. Luke, D.R.: Local linear convergence of approximate projections onto regularized sets. Nonlinear Anal. **75**, 1531–1546 (2012). https://doi.org/10.1016/j.na.2011.08.027

Part II
Progress and Perspectives

Chapter 7
Quantifying Molecule Numbers in STED/RESOLFT Fluorescence Nanoscopy

Jan Keller-Findeisen, Steffen J. Sahl and Stefan W. Hell

Abstract Quantification of the numbers of molecules of interest in the specimen has emerged as a powerful capability of several fluorescence nanoscopy approaches. Carefully relating the measured signals from STED or RESOLFT scanning nanoscopy data to the contribution of a single molecule, reliable estimates of fluorescent molecule numbers can be obtained. To achieve this, higher-order signatures in the obtained photon statistics are analyzed, as arise from the antibunched nature of single-fluorophore emissions or in the signal variance among multiple on/off-switching cycles. In this chapter, we discuss the concepts and approaches demonstrated to date for counting molecules in STED/RESOLFT nanoscopy.

7.1 Introduction

Ideally, a microscope discerns and maps all features in the specimen. Until the dawn of the 21st century, it was generally accepted that an optical microscope using lenses would not be able to discern features on the nanoscale [1]. By the mid 1990s, however, physical concepts for overcoming the longstanding diffraction barrier had been introduced, and the development of super-resolved far-field fluorescence microscopy (or 'nanoscopy') has since progressed tremendously. At a deep level, the reason for the vastly increased resolution capabilities of modern fluorescence nanoscopes—to resolve molecules with distances of even just a few nanometers—has been a major

J. Keller-Findeisen (✉) · S. J. Sahl · S. W. Hell
Department of NanoBiophotonics, Max Planck Institute for Biophysical Chemistry,
Am Faßberg 11, 37077 Göttingen, Germany
e-mail: jan.keller@mpibpc.mpg.de

S. J. Sahl
e-mail: steffen.sahl@mpibpc.mpg.de

S. W. Hell
e-mail: stefan.hell@mpibpc.mpg.de

S. W. Hell
Department of Optical Nanoscopy, Max Planck Institute for Medical Research,
Jahnstr. 29, 69120 Heidelberg, Germany

© The Author(s) 2020
T. Salditt et al. (eds.), *Nanoscale Photonic Imaging*, Topics in Applied Physics 134,
https://doi.org/10.1007/978-3-030-34413-9_7

paradigm shift: the discrimination of molecules is no longer realized by the focusing of the light in use. Rather, the molecules are transiently transferred to different states, usually fluorescence 'on' and fluorescence 'off' states, so that they are distinguishable when using a (diffraction-limited) illumination pattern to probe their signals [2].

An important aspect for mapping the molecules is that the transient state change can occur in a spatially controlled (i.e., coordinate-targeted) or in a spatially stochastic manner. The first kind is realized in methods called stimulated emission depletion (STED) [3, 4], saturated structured-illumination microscopy (SSIM) [5] and reversible saturable/switchable optical fluorescence transitions (RESOLFT) [6, 7]. In these approaches, a pattern of light with one or multiple intensity minima switches the molecules optically between an 'on' and an 'off' state, thus transferring all molecules to one of these states except those located at or near the intensity minima. Scanning the pattern of light across the specimen ensures that every molecule ends up in a subdiffraction-sized region at least once, and hence is for that time in a different state from its resolved neighbors.

The highest combined resolution along all three spatial dimensions has been achieved for STED or RESOLFT microscopes with two opposing lenses, where the illumination as well as the detected fluorescence light from both lenses can be combined in a coherent manner [8–10].

Spatially stochastic methods, among them photoactivated localization microscopy (PALM) [11] and stochastic optical reconstruction microscopy (STORM) [12] bring molecules in close proximity independently to an on-state in which the individual molecules can emit multiple and ideally a large number of fluorescence photons. The collected signal is then used to estimate the position of the isolated molecules with subdiffraction precision. Again, the use of two opposing lenses coherently combining the fluorescence light has delivered very high and close-to-isotropic 3D resolution [13].

A recent breakthrough development, MINFLUX, has attained the ultimate molecule-size, one-nanometer resolution scale at minimal fluxes of emitted fluorescence photons [14]. MINFLUX operates with spatially stochastic single-molecule switching, but makes use of one or more coordinate-giving intensity minima of excitation light to make the controlled, known position of a minimum coincide with the molecule position and determine it very efficiently in terms of registered fluorescence photons.

Although the spatially stochastic methods can provide molecular maps [15], counting molecules with stochastic methods is not as straightforward as it may appear. Molecules which do not emit sufficient numbers of photons while residing in the on-state to be detected, or which do not assume this state at all, are missed out completely. Other molecules might occupy the on-state repeatedly, and thus might be counted multiple times, thus requiring a careful and non-trivial calibration. For these reasons, fluorophores which assume the on-state only once would be favorable in principle, but such fluorophores would allow only a single super-resolution recording, meaning that the molecular counting would not be repeatable. As an additional aspect to consider, counting molecules one by one requires an extended recording time in which the molecules must remain stationary.

In contrast, STED and RESOLFT microscopy are not based on single-molecule detection and, by registering signals from all molecules from a given coordinate simultaneously, they provide a potential speed advantage. However, this very same fact has made the counting of molecules in ensemble-based nanoscopy imaging modes challenging. In most cases, the actual brightness of individual molecules in an experiment remained elusive, because the sample did not contain spatially sparse molecules found on their own. Furthermore, local environment heterogeneity can induce local variations of the molecular brightness. However, if the average contribution of a single molecule to the recorded image can be reliably estimated, the number of participating molecules can simply be deduced from the magnitude of the fluorescence signal.

A reliable method to extract the numbers of molecules in STED or RESOLFT microscopy is very desirable, and substantial progress has been achieved towards this goal. Indeed, a careful analysis of the photon arrival statistics in STED and RESOLFT imaging, especially the study of (1) occurrences of simultaneous arrivals of fluorescence photons in STED as well as the (2) fluctuations in signal of repeated recordings at the same scan position in RESOLFT, reveals higher-order dependencies of the recorded photon statistics on the number of molecules and their brightness. Such a careful analysis allows to disentangle number and brightness and thus map the number of molecules in an image. The effects and statistical signatures harnessed are fully compatible with the subdiffraction resolution of STED and RESOLFT and can therefore readily be applied also in a live-cell imaging regime.

In the following sections, these two relatively new quantitative nanoscopy methods will be presented, with an emphasis on the statistical modeling that goes beyond a purely 'classical' shot-noise description of photon statistics.

7.1.1 Molecular Contribution Function (MCF)

In analogy to the point spread function (PSF), which corresponds to the image of a theoretical point source, the molecular contribution function (MCF) can be defined as the quantitative spatial distribution of the average number of photons counted from a single fluorophore. With the knowledge of the MCF, all linear imaging systems where the recorded image is the sum of the contributions of all the individual molecules, counting molecules then becomes a conceptually straightforward and accessible task of normalizing the image by the MCF. For example, a space-invariant, linear imaging system expresses the measured image $Y(\mathbf{x})$ at each scan position $\mathbf{x}$ as the convolution of the number density $n(\mathbf{x})$ with the MCF$(\mathbf{x})$

$$Y(\mathbf{x}) = (n * \mathrm{MCF})(\mathbf{x}) + \varepsilon(\mathbf{x}), \tag{7.1}$$

where $*$ denotes the convolution operator and $\varepsilon(\mathbf{x})$ describes the measurement noise, i.e., the deviation of a particular recorded image from the average (noise-free) image.

Estimation of the unknown number density $n(\mathbf{x})$ from the data then amounts to simply deconvolving the data with a carefully calibrated MCF.

Counting the number of molecules in defined isolated regions, which in principle could be as small as the resolution scale, means a division of the summed image data in those regions by the average total signal of a single molecule, i.e., the integral over the MCF.

Even though a considerable part of the theory in this chapter is presented using continuous functions mostly for the sake of simplicity of notation, it is understood that all recorded microscopic data is pixelated with a pixel size smaller than the spatial resolution. The transition between continuous and discrete data grids is straightforward and may be realized implicitly wherever it is convenient.

Note that every fluorescence microscopy technique with linear imaging conditions and independent and identically behaving fluorophores features an MCF, which lends itself to mapping the number of fluorophores. However, the MCF and the total signal of a single molecule will depend on the used optics, measurement properties and chosen fluorophores. The main task of quantitative STED/RESOLFT nanoscopy therefore lies in determining the average signal per fluorophore intrinsically from the dataset itself.

7.2 STED Nanoscopy with Coincidence Photon Detection

Measurements of the statistics of simultaneous photon arrivals in fluorescence microscopy have been shown to identify individual fluorophores [16, 17], to improve the resolution of fluorescence microscopy [18, 19] and have been used to analyze individual clusters of molecules distributed in space [20, 21]. In the next section, a full imaging model of simultaneous photon arrivals in confocal and STED microscopy is derived, and afterwards a live-cell imaging experiment will be described.

In a scanning fluorescence microscope with a pulsed illumination light source (the preferred approach for photon coincidence measurements), the specimen at each recorded pixel position experiences a certain number of light pulses. If excitation of the molecules is accomplished with pulses with a duration much shorter than the fluorescence lifetime then each molecule can be excited at most once per pulse. The probability of a molecule to be excited and to yield a detected photon as the focal center of the scanning and the coordinate of the molecule coincide is named the molecular brightness λ. Reflecting changes in local environment, it may change with the location of a molecule, but here it is assumed that all molecules at a given location (within a resolved region) also share the same brightness and that $\lambda(\mathbf{x})$ remains constant over time. The optimal case of $\lambda = 1$ would be reached if the molecule contributed with a detected photon for every pulse. In practice, strong excitation leads to increased photo-bleaching and a widening of the spatial region in which fluorophores are in the saturated regime (broadening of the excitation spot), which must be avoided. The quantum yield of the fluorophores is limited and the detection optics cannot collect and detect every emitted photon, resulting in a molecular brightness considerably below one (typically on the order of 0.01). For molecules not at the center of the scanned focal spot, the effective molecular brightness has to be additionally scaled by the PSF (h). These aspects are contained in the MCF. The goal is to measure the statistics of the numbers of detected photons from the sample after each light pulse.

In a suitable experimental arrangement [22], the detected photons from all molecules at a scan coordinate are distributed by an array of beam splitters randomly onto four equally sensitive detectors in order to measure the numbers of detected photons for every pulse (see Fig. 7.3a). Detectors with at least one assigned photon are considered active and count as a detection event. The number of active detectors thus becomes the experimentally accessible value. A direct quantitative detection of the numbers of fluorescent photons with a time resolution of at least the duration between consecutive light pulses would simplify the statistical model as well as the experimental setup and would therefore be highly desirable, but at the time when these experiments were conducted an array of APD detectors capable of counting single photons with a dead time larger than the fluorescence lifetime was deemed an acceptable compromise of the ability to detect simultaneously arriving photons and experimental effort.

7.2.1 Statistical Model

The probability for a single molecule to emit more than one photon in the duration of the excitation pulse is negligible. Therefore the photon emission process can be well described by a multinomial random process with λ_i the molecular brightnesses of molecules $i = 1, .., N$ located at positions $\mathbf{u}_i$. The probability $\mathfrak{p}_i$ of a molecule to contribute with a photon to the detection at the current scanning position $\mathbf{x}$ is $\mathfrak{p}_i(\mathbf{x}) = \lambda_i h(\mathbf{x} - \mathbf{u}_i)$ with the PSF of the system h.

Due to superposition and independence of the molecular markers, for each scan position the number of contributed photons follows a discrete probability distribution of a sum of independent Bernoulli trials with parameters $\mathfrak{p}_i$. The probability that exactly k photons contribute during a single pulse is denoted by $Q_k(\mathbf{x})$. The expressions for $k = 1, 2$ become

$$Q_1(\mathbf{x}) = \sum_{i=1}^{N} \mathfrak{p}_i(\mathbf{x}) \prod_{j \neq i} \left(1 - \mathfrak{p}_j(\mathbf{x})\right)$$

$$Q_2(\mathbf{x}) = \sum_{i=1}^{N} \mathfrak{p}_i(\mathbf{x}) \sum_{j>i}^{N} \mathfrak{p}_j \prod_{k \neq i,j} (1 - \mathfrak{p}_k(\mathbf{x})) \tag{7.2}$$

expressing all possibilities of exactly one or two molecules contributing with a photon. The $p_i(\mathbf{x})$ are much less than unity, and one can simplify these expressions further by neglecting terms with higher orders in $p_i(\mathbf{x})$:

$$Q_1(\mathbf{x}) \simeq \sum_{i=1}^{N} p_i(\mathbf{x})$$

$$Q_2(\mathbf{x}) \simeq \frac{1}{2} \left[\left(\sum_{i=1}^{N} p_i(\mathbf{x}) \right)^2 - \sum_{i=1}^{N} p_i(\mathbf{x})^2 \right] \qquad (7.3)$$

Q_2 is approximately given by the probability to obtain two photons from any molecule minus the so-called 'antibunching term' $\sum_{i=1}^{N} p_i(\mathbf{x})^2$, which accounts for the unphysical case that two photons would originate from the very same molecule.

7.2.1.1 Distribution on Active Detectors

Because the utilized detectors are not able to quantitatively detect the number of incident photons, information about the numbers of contributing photons is partially lost. This loss can be taken into account by geometrical factors. For example, $\beta = (d-1)/d$ is the probability that two photons are registered on two different detectors for d available detectors. Let $D_i(\mathbf{x})$ be the mean number of active detectors at each scan position. Neglecting higher order terms of Q_i gives

$$D_1(\mathbf{x}) \simeq Q_1(\mathbf{x})$$
$$D_2(\mathbf{x}) \simeq \beta Q_2(\mathbf{x}) \qquad (7.4)$$

Using $d = 4$ detectors results in a loss of about 25% of the two-photon incidence events compared to the ideal case of detecting all contributing photons. $n(\mathbf{x})$ is denoted as the local fluorophore density and $\lambda(\mathbf{x})$ as the molecular brightness (defined only where $n(\mathbf{x}) > 0$). The transition to a continuous grid can be easily performed by

$$\sum_{i=1}^{N} p_i(\mathbf{x})^j \rightarrow (p^j n) * h^j$$

with $*$ the convolution operator. On a continuous grid, the mean number of active detectors per light pulse becomes

$$D_1^m(\mathbf{x}) = ((\lambda n) * h_m)(\mathbf{x})$$
$$D_2^m(\mathbf{x}) = \frac{\beta}{2} \left[((\lambda n) * h_m)^2 - (\lambda^2 n) * h_m^2 \right](\mathbf{x}) \qquad (7.5)$$

where m denotes the imaging mode ($m = \{c, s\}$ for confocal or STED recordings), which affects the width of the PSF. If $(\lambda n) * h$ is not much smaller than one, for example if the number of simultaneously recorded molecules is large, additional, higher-order terms must be included. Expressions including a possible Poissonian background contribution and higher-order terms are given in [22] (see Supplementary Note). D_2^m includes the interesting 'antibunching term' $(\lambda^2 n) * h_m^2$, which depends quadratically on $\lambda(\mathbf{x})$, thus making it a critical parameter.

7.2.1.2 Estimation of Molecule Density and Brightness Distribution

The average number of active detectors $D_{1,2}^m$ can be estimated empirically by the measured occurrences of active detector events in an experiment $Y_{1,2}^m(\mathbf{x})$ normalized by the number of repetitions t, i.e., the number of applied laser light pulses t per pixel. The desired molecule density and brightness distributions are ultimately extracted by fitting the data with the model in (7.5). This is accomplished by the fast proximal gradient algorithm FISTA [23], which minimizes the squared distance between the model and the experiment while also penalizing strong variations in $\lambda(\mathbf{x})$. The estimated molecule density $\hat{n}(\mathbf{x})$ and molecular brightness $\hat{\lambda}(\mathbf{x})$ are the solution of the constraint optimization problem:

$$\text{argmin}_{n,\lambda} \sum_{m=c,s} \sum_{i=1,2} \alpha_{im} \left\| D_i^m(n, \lambda) - Y_i^m/t \right\|^2 + \gamma\phi(\lambda)$$

$$n(\mathbf{x}) \geq 0, \lambda(\mathbf{x}) \geq 0 \tag{7.6}$$

with α_{ij} and γ positive weighting parameters and ϕ a typical penalization term, the Laplacian of the brightness ($\phi(\lambda) = \nabla^2\lambda$) in order to enforce smoothness in the brightness distribution.

In (7.6) data recorded in both the confocal and STED imaging mode of the same specimen is jointly optimized, which is advantageous in practical experiments. The confinement to only one of the imaging modes in the fit is straightforward. With the value of γ appropriately chosen, the penalization sufficiently stabilizes the solution of (7.6), preventing strong spatial oscillations in brightness on scales below the resolution. The weighting parameters α_{im} are chosen such that all least-square residuals are approximately on the same scale ($\alpha_{im} \simeq 1/\sum_{\mathbf{x}} Y_i^m(\mathbf{x})$). In order to incorporate the non-negativity constraints, $n(\mathbf{x})$ and $\lambda(\mathbf{x})$ are substituted by squared variables $m^2(\mathbf{x})$ and $q^2(\mathbf{x})$ and (7.6) was solved for the $m(\mathbf{x})$ and $q(\mathbf{x})$ instead. As starting point for the numerical optimization, a deconvolved single active detector image was chosen, which provided an initial molecular density for a given reasonable choice of the average molecular brightness.

The main cause of deviation between model and measurement is shot noise. Interestingly, the relative standard deviations (RSTD) of the estimated number of molecules $\hat{n}$ and brightness $\hat{\lambda}$ at a given position with molecular signals (such as from a single cluster) can be derived analytically (see [22])

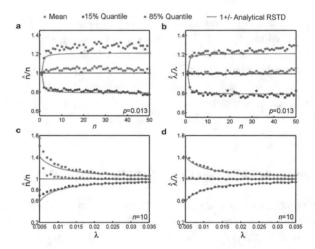

Fig. 7.1 Simulation of the error of the estimation of number and brightness for a single cluster of molecules in the confocal mode. **a, c** Relative estimated numbers of molecules $\hat{n}/n$ and **b, d** relative estimated brightness $\hat{\lambda}/\lambda$. The number of molecules n varies from 1–50 with a brightness of $\lambda = 0.013$ (**a, b**). The brightness varies (0.005–0.035) while keeping the number of molecules constant $n = 10$ (**b, d**). The number of illumination pulses is $2.5 \cdot 10^4$ (**a, b**) and $4 \cdot 10^4$ (**c, d**). The FWHM of the PSF is 240 nm. Mean, quantiles (15, 85%) and analytically derived rel. standard deviations are shown. Figure adapted from [22]

$$\mathrm{RSTD}(\hat{n}) = \mathrm{RSTD}(\hat{\lambda}) \propto \sqrt{(n-1)/(n\lambda^2 t)} \tag{7.7}$$

and is confirmed by simulations (see Fig. 7.1). Calculations and simulations both suggest that the RSTD is rather independent of the actual number of molecules and can be below ~20% for conditions provided by synthetic fluorophores ($\lambda \sim 0.01$, $t > 10^3$, i.e. more than 1000 pulses) [24].

Please note that in the imaging model (mean number of active detectors) in (7.5) and possible generalizations, higher-order terms represent convolutions of the molecular density with higher orders of the PSF h_m^n. In principle, these terms represent parts of the image with an increased resolution, however their contribution is rather low as they scale with higher orders of the molecular brightness. Measurements of these higher-order PSFs are displayed in Fig. 7.2.

7.2.2 Intrinsic Molecular Brightness Calibration

Double-stranded DNA (dsDNA) is easy to synthesize and label in a controlled manner. Labeling the dsDNA with up to four ATTO 647N fluorophores and immobilizing it sparsely on a glass surface allows controlled measurements as well as indepen-

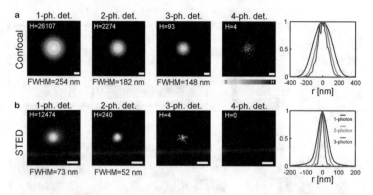

Fig. 7.2 The PSFs of 1–4 active detector events in confocal and STED mode. DNA origami with up to 24 ATTO 647N in two lines of 41 nm spacing was sparsely immobilized and measured in confocal (**a**, overlay of 64 single DNA images) and STED (**b**, overlay of 114 single DNA images) mode. On the right: normalized line profiles crossing the center of the PSFs. All overlays allowing a fit with a 2D Gaussian peak function display resulting FWHM values. H is the maximum value (events). Scale bars: 100 nm. Figure adapted from [22]

dent estimates of the number of fluorophores by observing individual bleaching steps. A schematic of a microscope with a fluorescence detection path split into four equivalent units is shown in Fig. 7.3a. For a measurement in confocal mode with $t = 3000$ excitation pulses per pixel, resulting in 6000–7500 collected photons per fluorophore in the resulting scanned image, the number of fluorophores can be estimated with ∼20% uncertainty and matches the number of observed bleaching steps (see Fig. 7.3). However, the advantages over observing abrupt changes in brightness by stepwise bleaching are that the molecules remain intact (except for the gradual bleaching during multiple scans) and that the observation of bleaching steps often suffers from uncertainty about the number of simultaneously bleached molecules.

7.2.3 Counting Transferrin Receptors in HEK293 Cells

The bleaching rate of ATTO 647N in the experiments shown in the previous section (Figs. 7.2 and 7.3) is as low as ∼3% per full scan [22]. This allows to perform a STED recording right afterwards, so that the photon statistics of both recordings can be combined and the optimization algorithm can use both datasets for the solution of (7.6). The combined data yields the molecular numbers, mostly derived from the statistical information contained in the confocal recordings, with a spatial 2D resolution of ∼50 nm mostly originating from the STED recording. Using the ability of fluorescence microscopes to look into the interior of cells, transferrin receptor (TfR) molecules in human embryonic kidney 293 (HEK293) cells were counted. The

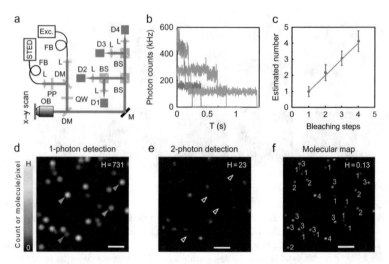

Fig. 7.3 Experimental setup and measurements on double-stranded DNA (dsDNA) labeled with up to four ATTO 647N. **a** Confocal/STED microscope equipped with four independent detection channels. BS 1:1 beam splitter, D_i i-th detector (i = 1–4). **b** Fluorescence bleaching steps of single dsDNA (corresponding spots are indicated in **d** by triangles of the same colors). **c** Comparison of the number of dye molecules (mean and s.d.) derived from photon incidence analysis with the detected number of bleaching steps of the same single dsDNA (red line: $y = x$). **d, e** Example image showing one- and two-photon detection events measured in the confocal mode. The dsDNA positions with only a single dye molecule are indicated by open triangles in (**e**). **f** Established map of the number of ATTO 647N molecules, derived from the data in (**d**) and (**e**). H is the maximum of the color scale, representing events in **d** and **e** and molecules in **f**. Scale bars: 1 μm. Figure reproduced from [22]

TfR molecules were labeled by an ATTO 647N-conjugated DNA aptamer because attaching a single fluorophore to each aptamer molecule can be performed with high precision [25]. With the aptamers designed to bind to their target in a one-to-one stoichiometry, quantification becomes straightforward. Images of the distribution of TfR clusters are shown in Fig. 7.4. Using information from the two active detector events, one can estimate that the total number of internalized TfRs in each cell is on the order of ∼100000. In addition, this counting method can determine molecular density variations of internalized TfRs in the cell, which is not possible by conventional measurements such as quantitative Western blots. With the spatial resolution delivered by the STED recording, most intracellular TfRs can be visualized as separate clusters. Interestingly, the number of estimated TfRs in each isolated cluster closely follows an exponential distribution with an expectation of ∼6.0 ± 1.9. It also indicates that a single cluster may have a capacity to accommodate more than 20 TfR molecules since the measurement fits an exponential distribution up to 20 fluorophores closely.

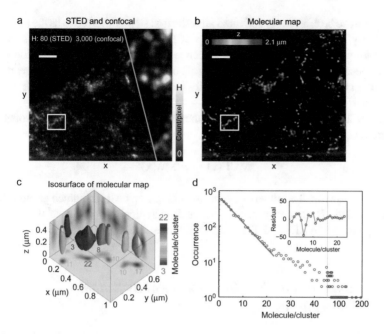

Fig. 7.4 Counting the number of transferrin receptors (TfR) in HEK293 cells. Living cells incubated with ATTO 647N-conjugated anti-TfR aptamer. After incubation, excess aptamer molecules were washed off and cells were chemically fixed. Stained receptors were imaged in confocal and STED mode, with 100 nm increments along z. **a** Summed axial projection of confocal and STED images (raw data) along 0.9 μm depth. **b** Estimated 3D molecular map resulting from photon statistics of both confocal and STED recordings. Colors present the axial position. **c** Isosurfaces of the molecular map in the boxed region in a and b. The isosurfaces include 70% of all molecules in this region. The number of molecules in each segment is displayed below. **d** The histogram of the estimated number of TfR receptors per recognized separated segment. The red line is an exponential fit of the number of occurrences up to 24 molecules per spot (inset shows the residual of the fit). Scale bars: 1 μm. Figure reproduced from [22]

7.3 Mean and Variance in RESOLFT Nanoscopy

RESOLFT experiments require switchable fluorophores with on- and off-states [2]. For simplicity, only the positive imaging mode is discussed here, i.e., where the MCF features a sharp, subdiffraction-sized positive signal peak. This is the commonly used imaging mode. Another common assumption is that the molecules independently switch between their states and independently emit fluorescence.

Two main steps can be differentiated. At first, fluorophores are transferred from the off-state, in which they reside initially, into the on-state only within a sharp, subdiffraction-sized region centered at the current scanning position **s**. In practice, this is achieved by first activating fluorophores with a diffraction-limited focus, and then deactivating fluorophores in the periphery using a doughnut-shaped focus with a central intensity minium [6]. The spatial distribution of the probability of a single fluorophore to be effectively activated after the first step is referred to as the activation probability $p(\mathbf{x})$. As a second main step of RESOLFT, activated fluorophores are then read out by recording their fluorescence signal. The spatial distribution of the average recorded fluorescence photon signal of a single activated fluorophore during the readout time is $\lambda(\mathbf{x})$, which is typically proportional to the confocal PSF but can also be performed in STED mode [26].

The MCF of RESOLFT nanoscopy is therefore given by the activation probability multiplied by the average readout signal,

$$\mathrm{MCF}(\mathbf{x}) = p(\mathbf{x})\lambda(\mathbf{x}). \tag{7.8}$$

In the following, the influence of the switching step on the obtained photon statistics is studied. To demonstrate the principal statistical properties of the two-step imaging process in RESOLFT, a simplified model is analyzed.

7.3.1 Cumulants of the Fluorescence of Switchable Fluorophores

Let us begin by assuming n molecules, all with the same activation probability p and the same brightness λ, which is the average signal that is obtained from a single activated fluorophore. Further, let us assume that there is no spatial dependency in the measurement process (e.g. a situation of an isolated cluster of molecules where the scanning position and the cluster position coincide). There are three unknown properties (number of molecules, activation probability and brightness) but only a single variable (fluorescence signal) that is measured. How can the unknown properties be retrieved with only a single measurement variable? The basic idea is to exploit higher-order characteristics (cumulants) of the measured signal that go beyond the mean in order to separate the unknown properties. Figure 7.5 illustrates this idea, showing that different choices of parameters resulting in the same mean signal can still differ in their variance (or other higher-order characteristics). For a purely Poisson-distributed signal, all cumulants revert to the same value, the mean of the signal. It turns out that the switching step in RESOLFT nanoscopy is the essential factor in separating the number and the brightness.

The goal is to calculate the cumulants of the fluorescence signal of a single molecule, given p and λ. The signal can be modeled as the result of a two-step stochastic process, where the first step is the activation and the second step is the photon emission by the activated fluorophore. The activation is modeled as a Bernoulli process with activation probability p. The activation state of the fluorophore is represented by a Bernoulli random variable A. If the fluorophore is in the on-state it will yield λ detected photons on average, which is modeled by a Poisson distribution. Therefore, the random variable describing the signal of a fluorophore B given the activation state A is

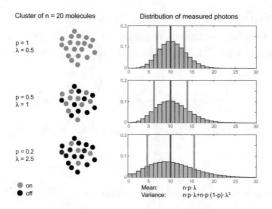

Fig. 7.5 Left: Depiction of a cluster of $n = 20$ molecules with different activation probabilities p and different brightnesses λ. Right: Photon counting histogram. The mean remains unchanged for all cases, while the variance differs. Calculating statistical parameters of the measured photon numbers allows estimating the number of molecules. Figure adapted from [27]

$$B|A = 1 \sim \text{Poisson}(\lambda)$$
$$B|A = 0 \sim 0 \tag{7.9}$$

The characteristic function $\phi(t)$ of a Bernoulli-distributed random variable is $1 - p + p \exp(it)$ and of a Poisson-distributed random variable $\exp(\lambda(e^{it} - 1))$. The characteristic function of B can be computed with a conditional expectation over A.

$$\phi(t) = \mathbb{E}_A \mathbb{E}_{B|A} \exp(it B) = 1 - p + p \exp\left(\lambda(e^{it} - 1)\right) \tag{7.10}$$

where $\mathbb{E}$ represents the expectation. The cumulant-generating function $H(t)$ is the logarithm of ϕ with a series expansion giving the cumulants

$$H(t) = \log \phi(t) = \sum_{n=10}^{\infty} \kappa_n \frac{(it)^n}{n!}. \tag{7.11}$$

The cumulants are polynomials with increasing order in p and λ.

$$\kappa_1 = p\lambda$$
$$\kappa_2 = p\lambda + p(1 - p)\lambda^2 \tag{7.12}$$
$$\kappa_3 = p\lambda - 3p(1 - p)\lambda^2 + p(1 - p)(1 - 2p)\lambda^3$$

The molecules are assumed to act independently from each other, so that the cumulant of the summed signal of all n molecules can conveniently be expressed as the sum of single-molecule cumulants. This means that the expressions in (7.12) are simply scaled by n.

If the first three cumulants would also be estimated empirically as $\hat{\kappa}_{1,2,3}$ one could solve the non-linear equation system in (7.12) to obtain estimates $\hat{n}$, $\hat{p}$, $\hat{\lambda}$ for the activation probability, the brightness and most importantly, for the number of molecules.

$$\hat{n} = \hat{\kappa}_1^{\,2}\left(\hat{\kappa}_1 - \hat{\kappa}_2\right) / \left(\hat{\kappa}_1^{\,2} - \hat{\kappa}_1\hat{\kappa}_2 - \hat{\kappa}_2^{\,2} + \hat{\kappa}_1\hat{\kappa}_3\right)$$

$$\hat{p} = \left(\hat{\kappa}_1^{\,2} - \hat{\kappa}_1\hat{\kappa}_2 - \hat{\kappa}_2^{\,2} + \hat{\kappa}_1\hat{\kappa}_3\right) / \left(\hat{\kappa}_1\hat{\kappa}_2 - 2\hat{\kappa}_2^{\,2} + \hat{\kappa}_1\hat{\kappa}_3\right) \qquad (7.13)$$

$$\hat{\lambda} = \left(\hat{\kappa}_1\hat{\kappa}_2 - 2\hat{\kappa}_2^{\,2} + \hat{\kappa}_1\hat{\kappa}_3\right) / \left(\hat{\kappa}_1\left(\hat{\kappa}_1 - \hat{\kappa}_2\right)\right)$$

This principle is exploited in the next section for a realistic RESOLFT imaging model.

7.3.2 Statistical Model

In the general case, the activation probability and signal in the active state is position-dependent. The following transformations hold

$$p \rightarrow p(\mathbf{x}_i - \mathbf{s})$$
$$\lambda \rightarrow \lambda(\mathbf{x}_i - \mathbf{s})$$

with $\mathbf{x}_i$ the position of the molecules and $\mathbf{s}$ the scanning position. Again, the signal is a result of a two-stage stochastic process. The activation strength and the average readout signal, however, both depend strongly on the position of the molecules relative to the focal center. Rewriting the mean $m(\mathbf{s})$ and variance $v(\mathbf{s})$ of the total signal using the results of the previous section (see (7.12)) with position-dependent parameters and adding a Poissonian background $d(\mathbf{s})$ gives

$$m(\mathbf{s}) = \sum_{i=1}^{N} p(\mathbf{x}_i - \mathbf{s})\lambda(\mathbf{x}_i - \mathbf{s}) + d(\mathbf{s})$$

$$v(\mathbf{s}) = m(\mathbf{s}) + \sum_{i=1}^{N} p(\mathbf{x}_i - \mathbf{s})\left(1 - p(\mathbf{x}_i - \mathbf{s})\right)\lambda^2(\mathbf{x}_i - \mathbf{s}) \qquad (7.14)$$

The transition to a continuous grid can be carried out analogously to the procedure in Sect. 7.2.1. With the transformation

$$\sum_{i=1}^{N} p(\mathbf{x}_i - \mathbf{s})\lambda(\mathbf{x}_i - \mathbf{s}) \rightarrow (n * p\lambda)(\mathbf{s})$$

the mean and variance of the total signal can conveniently be expressed as convolutions with the density of fluorophores $n(\mathbf{x})$. Additionally, the readout signal λ is written as a product of the focal brightness b (i.e. $\lambda(0)$) and the readout PSF h_{read}. The focal brightness is assumed constant for all the molecules.

$$m(\mathbf{s}) = b\,(n * h_{\text{read}} p)\,(\mathbf{s}) + d(\mathbf{s}),$$
$$v(\mathbf{s}) = m(\mathbf{s}) + b^2\left(n * h_{\text{read}}^2 p(1-p)\right)(\mathbf{s}). \tag{7.15}$$

The mean is the convolution of the number density with the MCF and background contributions while the variance exceeds the mean by an additional term with a convolution kernel $b^2 h_{\text{read}}^2 p\,(1-p)$ which is similar but not identical to the MCF. Note that for $p(\mathbf{x}) = 1$ the situation of non-switchable fluorophores is recovered, the excess variance term vanishes and a purely Poisson distributed signal is regained. In the case of photo-switchable fluorophores, the variance of the collected signal is augmented due to the stochastic nature of the activation in the preparation step. The excess variance or 'over-dispersion' term is proportional to the square of the focal brightness b, unlike the mean signal, which scales linearly with it. This relation can be used to estimate b directly from integrated mean and variance of the image data using a Method of Moments estimator (MME).

For a region X in the sample that comprises a conglomeration of molecules but is isolated from other such regions the position-dependent mean and variance of the signal can be summed ($M = \int_X m(\mathbf{s})\,d\mathbf{s}$, $V = \int_X v(\mathbf{s})\,d\mathbf{s}$) and the convolutions in (7.15) reduce to simple products.

$$M = bN_X H_1 + D,$$
$$V = M + b^2 N_X H_2, \tag{7.16}$$

with $H_1 = \int_{\mathbb{R}^2}(h_{\text{read}} p)\,d\mathbf{s}$ and $H_2 = \int_{\mathbb{R}^2}\left(h_{\text{read}}^2 p(1-p)\right)\,d\mathbf{s}$ being integrals of products of p and h_{read} and $D = \int_X d\,d\mathbf{s}$ the integrated background. If the integrated mean and variance can be estimated empirically as $\hat{M}$ and $\hat{V}$ then (7.16) can be solved for the focal brightness resulting in the MME $\hat{b}$:

$$\hat{b} = \frac{H_1}{H_2}\frac{\hat{V} - \hat{M}}{\hat{M} - D}. \tag{7.17}$$

The results of a simulation of the RESOLFT imaging process and the estimation of the focal brightness are shown in Fig. 7.6. Essentially, the relative errors of $\hat{b}$ and $\hat{n}$ depend mainly on the number of measurements, and are largely independent of the number of molecules in the image [27].

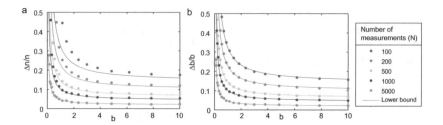

Fig. 7.6 Simulation of the error of the estimation of number and brightness for a single cluster of molecules. **a** Relative standard deviation of the estimated number of molecules (for 20 molecules in a single cluster) with an activation probability of 20% and a variable number of repetitions. **b** Relative standard deviation of the estimated brightness under the same conditions. Above a threshold on the molecular brightness, the counting error mainly depends on the number of repetitions. Figure adapted from [27]

7.3.3 Counting rsEGFP2 Fused α-tubulin Units in Drosophila Melanogaster

We illustrate the approach for RESOLFT nanoscopy imaging of *Drosophila melanogaster* larvae body wall muscles ubiquitously expressing rsEGFP2, a reversibly photoswitchable fluorescent protein (RSFP) [6], fused to α-tubulin. A commercial RESOLFT nanoscope was used for the recordings, and the preparation of the sample is described in [28]. Tubulin is known to form helices with a diameter of approximately 25 nm, each turn comprising 13 dimers of α- and β-tubulin that are spaced 8 nm apart [29]. The total density of α-tubulin along a single filament can thus be estimated to be ≈ 1625 per μm. To analyze the ratio of labeled to non-labeled α-tubulin subunits in body wall muscles of *Drosophila melanogaster*, L3-larvae were dissected to isolate body wall muscles, which were subsequently used as a sample for Western blot analysis. The ratio of rsEGFP2-α-tubulin to α-tubulin in the muscle tissue was estimated to be ~ 1:6. Therefore, the number density of rsEGFP2 molecules along a microtubule fiber is expected to be ~ 230 per μm. Can a RESOLFT imaging experiment confirm this independently measured molecule density using the framework laid out in Sect. 7.3.2?

7.3.3.1 Switching Kinetics of rsEGFP2

Measurements of the switching kinetics of rsEGFP2 were conducted on a confocal point scanning microscope with additional widefield illumination paths for excitation at 491 nm and activation at 375 nm (both continuous wave). The signal was detected with a point detector at 525 ± 25 nm. The advantage of widefield illumination and point-like detection is that averaging effects of the observed kinetics due to inhomogeneous intensity distributions within the detected volume can be excluded. rsEGFP2 is activated by UV light and switches off while being read out, which means that during the effective activation step in RESOLFT while molecules are switched off fluorescence can also be recorded.

In this measurement, molecules were activated by applying the UV light for a comparatively long time (several ms), thus strongly saturating the activation. Assuming that this prepares all the molecules in the on-state, the fluorescence signal at any given time during the off-switching indicates the remaining on-state population of rsEGFP2 at that time. An example of an off-switching curve is shown in Fig. 7.7. Systematic deviations from a single exponential decay are evident, especially for intermediate times, motivating the use of a Gamma distributed decay model with parameters α and β for fitting the off-switching curves.

$$s(t) = A \frac{\beta^{\alpha}}{(\beta + t)^{\alpha}} + b \qquad (7.18)$$

with the total switching rate k given by the inverse of the time $\beta(\exp(1/\alpha) - 1)$ where the signal drops to $1/e$ of the amplitude. The equilibrium signal after complete switching, normalized by the initial signal, estimates the equilibrium population ρ_{∞}. It is $\sim 2.5\%$, independent of the excitation strength. The switching rate seems to depend quite linearly on the excitation light intensity for intensities up to a few kW/cm^2. These are values typical for light intensities close to the center of the focal intensity minima, which determine the resolution enhancement in a RESOLFT microscope.

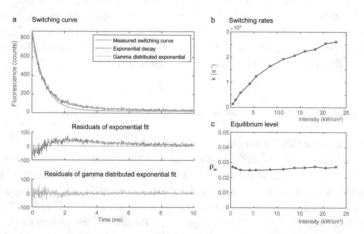

Fig. 7.7 Switching kinetics of rsEGFP2 in living *E. coli* cells overexpressing rsEGFP on a coverslip in an agarose gel. **a** Single switching curve with an off-switching light intensity of 500 W/cm^2. Fits with a single exponential decay (red) and a gamma distributed exponential decay (yellow) with residuals shown below. **b** Estimated switching rate k in dependence on the off-switching light intensity. **c** Equilibrium on-state level in dependence of the off-switching light intensity. Figure reproduced from [27]

7.3.3.2   Shape of the MCF

The MCF for rsEGFP2 molecules is modeled as a simple two-state switching process ('on'/'off'), with switching rates depending linearly on the applied light intensity and the frequently-made approximation that the doughnut of off-switching light features a parabolic intensity distribution close to the focal center [30]. The shape of the MCF is then a superposition of two Gaussian peaks that differ only in amplitude and width:

$$\text{MCF}(r) = b \left(\rho_\infty e^{-4 \ln(2) r^2 w_{\text{read}}^{-2}} + (p(0) - \rho_\infty) e^{-4 \ln(2) r^2 w_{\text{RESOLFT}}^{-2}} \right) \tag{7.19}$$

with the equilibrium activation level ρ_∞ and the focal activation level $p(0)$. The broader peak with diffraction-limited resolution w_{read} and low amplitude represents the signal from the nonzero equilibrium activation (see Fig. 7.7c) while the sharp peak with increased resolution w_{RESOLFT} represents the usable RESOLFT signal that originates from the subdiffraction-sized spot with an effective activation level above the equilibrium value. While the absolute scaling factor, the focal brightness b, is generally very difficult to predict, the relative amplitudes and sizes of the two Gaussian spots forming the MCF can typically be modeled reasonably well or retrieved from the data itself. Here, the structure consists of microtubule fibers with a width well below the resolution. For a sufficiently sparse distribution of filaments and high enough signal-to-noise ratio, the object can approximately be estimated from the data as a set of curved lines, even without quantitative knowledge of the MCF, simply by detecting lines in the image [31]. With the given image data and estimated object the shape of the MCF can be retrieved by a deconvolution and a fit of the shape model to the deconvolution result. Image data and retrieved object are shown in Fig. 7.8. The estimated FWHM of the diffraction-limited peak was 235 nm, and of the sharp RESOLFT peak 73 nm, and with the knowledge of the equilibrium activation level of $\sim$2.5% the focal activation level can be estimated to be $\sim$17%.

7.3.3.3   Counting the Number of Molecules Along Filaments

Applying the method of moments estimator for the focal brightness given in (7.17) on the whole image shown in Fig. 7.8a yields a value of $\hat{b}$ of $\sim$0.9 photon counts per activated rsEGFP2 molecule. Please note that the readout time in this experiment was set to a very short duration because rsEGFP2 switches off during readout, however, there are RSFPs that decouple the off-switching and the readout [32, 33]. The total average contribution of each fluorophore to the RESOLFT image can be estimated by integrating the MCF given by (7.19) using the estimated shape and estimated focal brightness. The average value of 3.58 photons per rsEGFP2 molecule can then be used to quantify the number of fluorophores in a region in the image. The results of applying the estimator to the regions marked in Fig. 7.8a are shown in Table 7.1. The average number of rsEGFP2 molecules per μm along a microtubule filament

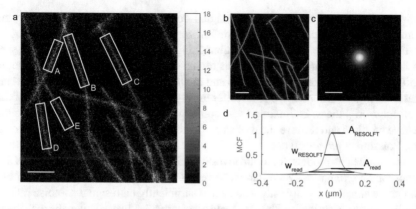

Fig. 7.8 Counting α-tubulin units in *Drosophila melanogaster*. **a** RESOLFT image of dissected body wall muscle cells expressing rsEGFP2-α-tubulin with regions (A–E) in which the number of rsEGFP2 molecules was counted. **b** Line detection of the tubulin structure. **c, d** PSF shape detection based on the image (**a**), the object detection (**b**) and the equilibrium on-state population (see Fig. 7.7). Scale bars, 1 μm (**a, b**), 200 nm (**c**). Figure reproduced from [27]

Table 7.1 Analysis of the regions A–E in the number density shown in Fig. 7.8a

Region	Number	Length in nm	Density per μm
A	268	1324	203
B	405	2065	196
C	379	2147	176
D	293	1713	171
E	237	1378	172
A–E	1582	8627	183

is ∼180, which is similar but below the expected number density. This might hint at a less efficient incorporation of rsEGFP2 fused to α-tubulin in microtubules. A detailed description of this experiment as well as a theory for the counting error is given in [27].

7.4   Summary

Ensemble-based, coordinate-targeted fluorescence microscopy methods like STED or RESOLFT deliver spatial resolution on the nanometer scale [34]. Registering all the molecules at a certain position at the same time in principle provides an advantage of recording speed. The repeatability and live-cell compatibility make quantitative applications of STED and RESOLFT highly desirable. Reaching this goal, however, has so far been challenging, in large measure because the observed

Poisson statistics of the fluorescence signal does not lend itself to separating the number of the molecules from their brightness.

Fortunately, effects like photon 'antibunching' or an additional on-switching step of fluorophores manifest themselves as higher-order terms in the observed photon statistics. A careful statistical analysis and modeling of the imaging process reveals that these higher-order components differ mostly by containing higher powers of the molecular brightness. Estimating these components empirically allows to solve the inverse problem and retrieve the molecular density and brightness from the observed non-Poissonian photon statistics.

The statistical models for both investigated effects for STED and RESOLFT nanoscopy are remarkably similar. However, as higher-order effects, their contribution to the signal is usually only weak and a sufficiently high signal-to-noise ratio is required for the described methods to yield precise results. In particular, the molecules should not photobleach during the experiment.

A central assumption is the independence and uniformity of behavior of the individual molecules, at least locally within a region of interest. A violation of this assumption or a compromised (false) estimation of the shape of the MCF will result in a biased counting procedure.

This chapter demonstrated that modeling and analysis of the obtained photon statistics are key elements to achieving quantitative subdiffraction-resolution information in STED and RESOLFT nanoscopy.

Acknowledgements The presented measurements and analyses were only possible thanks to the contributions from several colleagues, among them: Haisen Ta and Markus Haltmeier (STED nanoscopy with coincidence photon detection), as well as Lars Frahm, Sebastian Schnorrenberg and Miguel del Álamo Ruiz (RESOLFT nanoscopy with switchable fluorescent proteins).

References

1. Abbe, E.: Beiträge zur Theorie des Mikroskops und der mikroskopischen Wahrnehmung. Archiv für mikroskopische Anatomie **9**, 413–468 (1873)
2. Hell, S.W.: Far-field optical nanoscopy. Science **316**, 1153–1158 (2007)
3. Hell, S.W., Wichmann, J.: Breaking the diffraction limit by stimulated-emission; stimulated-emission-depletion fluorescence microscopy. Opt. Lett. **19**, 780–782 (1994)
4. Klar, T.A., Jakobs, S., Dyba, M., Egner, A., Hell, S.W.: Fluorescence microscopy with diffraction resolution barrier broken by stimulated emission. Proc. Natl. Acad. Sci. USA **97**, 8206–8210 (2000)
5. Gustafsson, M.G.L.: Nonlinear structured-illumination microscopy: wide-field fluorescence imaging with theoretically unlimited resolution. Proc. Natl. Acad. Sci. USA **102**, 13081–13086 (2005)
6. Grotjohann, T., Testa, I., Reuss, M., Brakemann, T., Eggeling, C., Hell, S.W., Jakobs, S.: rsEGFP2 enables fast RESOLFT nanoscopy of living cells. eLife **1**, e00248 (2012)
7. Hofmann, M., Eggeling, C., Jakobs, S., Hell, S.W.: Breaking the diffraction barrier in fluorescence microscopy at low light intensities by using reversibly photoswitchable proteins. Proc. Natl Acad. Sci. USA **102**, 17565–17569 (2005)
8. Böhm, U., Hell, S.W., Schmidt, R.: 4Pi-RESOLFT nanoscopy. Nat. Commun. **7**, 10504 (2016)

9. Dyba, M., Hell, S.W.: Focal spots of size $\lambda/23$ open up far-field fluorescence microscopy at 33 nm axial resolution. Phys. Rev. Lett. **88**, 163901 (2002)
10. Schmidt, R., Wurm, C.A., Jakobs, S., Engelhardt, J., Egner, A., Hell, S.W.: Spherical nanosized focal spot unravels the interior of cells. Nat. Methods **5**, 539–544 (2008)
11. Betzig, E., Patterson, G.H., Sougrat, R., Lindwasser, O.W., Olenych, S., Bonifacino, J.S., Davidson, M.W., Lippincott-Schwartz, J., Hess, H.F.: Imaging intracellular fluorescent proteins at nanometer resolution. Science **313**, 1642–1645 (2006)
12. Rust, M.J., Bates, M., Zhuang, X.W.: Sub-diffraction-limit imaging by stochastic optical reconstruction microscopy (STORM). Nat. Methods **3**, 793–795 (2006)
13. Aquino, D., Schönle, A., Geisler, C., von Middendorff, C., Wurm, C.A., Okamura, Y., Lang, T., Hell, S.W., Egner, A.: Two-color nanoscopy of three-dimensional volumes by 4Pi detection of stochastically switched fluorophores. Nat. Methods **8**, 353–359 (2011)
14. Balzarotti, F., Eilers, Y., Gwosch, K.C., Gynna, A.H., Westphal, V., Stefani, F.D., Elf, J., Hell, S.W.: Nanometer resolution imaging and tracking of fluorescent molecules with minimal photon fluxes. Science **355**, 606–612 (2017)
15. Annibale, P., Vanni, S., Scarselli, M., Rothlisberger, U., Radenovic, A.: Quantitative photo activated light localization microscopy: unraveling effects of photoblinking. PLoS One **6**, e22678 (2011)
16. Basche, T., Moerner, W.E., Orrit, M., Talon, H.: Photon antibunching in the fluorescence of a single molecule trapped in a solid. Phys. Rev. Lett. **69**, 1516–1519 (1992)
17. Kimble, H.J., Dagenais, M., Mandel, L.: Photon antibunching in resonance fluorescence. Phys. Rev. Lett. **39**, 691–695 (1977)
18. Hell, S.W., Soukka, J., Haenninen, P.E.: Two- and multiphoton detection as an imaging mode and means of increasing the resolution in far-field light microscopy. Bioimaging **3**, 65–69 (1995)
19. Schwartz, O., Levitt, J.M., Oron, D.: Fluorescence antibunching microscopy. Proc. SPIE **8228**, 822802–822804 (2012)
20. Ta, H., Kiel, A., Wahl, M., Herten, D.-P.: Experimental approach to extend the range for counting fluorescent molecules based on photon-antibunching. Phys. Chem. Chem. Phys. **12**, 10295–10300 (2010)
21. Tinnefeld, P., Müller, C., Sauer, M.: Time-varying photon probability distribution of individual molecules at room temperature. Chem. Phys. Lett. **345**, 252–258 (2001)
22. Ta, H., Keller, J., Haltmeier, M., Saka, S.K., Schmied, J., Opazo, F., Tinnefeld, P., Munk, A., Hell, S.W.: Mapping molecules in scanning far-field fluorescence nanoscopy. Nat. Commun. **6**, 7977 (2015)
23. Beck, A., Teboulle, M.: A fast iterative shrinkage-thresholding algorithm for linear inverse problems. SIAM J. Imaging Sci. **2**, 183–202 (2009)
24. Grussmayer, K.S., Herten, D.P.: Time-resolved molecule counting by photon statistics across the visible spectrum. Phys. Chem. Chem. Phys. **19**, 8962–8969 (2017)
25. Opazo, F., Levy, M., Byrom, M., Schäfer, C., Geisler, C., Groemer, T.W., Ellington, A.D., Rizzoli, S.O.: Aptamers as potential tools for super-resolution microscopy. Nat. Methods **9**, 938–939 (2012)
26. Danzl, J.G., Sidenstein, S.C., Gregor, C., Urban, N.T., Ilgen, P., Jakobs, S., Hell, S.W.: Coordinate-targeted fluorescence nanoscopy with multiple off states. Nat. Photonics **10**, 122–128 (2016)
27. Frahm, L., Keller-Findeisen, J., Alt, P., Schnorrenberg, S., Del Álamo Ruiz, M., Aspelmeier, T., Munk, A., Jakobs, S., Hell, S.W.: Molecular contribution function in RESOLFT nanoscopy. Opt. Express **27**, 21956–21987 (2019)
28. Schnorrenberg, S., Grotjohann, T., Vorbrüggen, G., Herzig, A., Hell, S.W., Jakobs, S.: In vivo super-resolution RESOLFT microscopy of Drosophila melanogaster. eLife **5**, e15567 (2016)
29. Amos, L.A., Klug, A.: Arrangement of subunits in flagellar microtubules. J. Cell Sci. **14**, 523–549 (1974)
30. Harke, B., Keller, J., Ullal, C.K., Westphal, V., Schönle, A., Hell, S.W.: Resolution scaling in STED microscopy. Opt. Express **16**, 4154–4162 (2008)

31. Steger, C.: An unbiased detector of curvilinear structures. IEEE Trans. Pattern Anal. Mach. Intell. **20**, 113–125 (1998)
32. Jensen, N.A., Danzl, J.G., Willig, K.I., Lavoie-Cardinal, F., Brakemann, T., Hell, S.W., Jakobs, S.: Coordinate-targeted and coordinate-stochastic super-resolution microscopy with the reversibly switchable fluorescent protein Dreiklang. ChemPhysChem **15**, 756–762 (2014)
33. Brakemann, T., Stiel, A.C., Weber, G., Andresen, M., Testa, I., Grotjohann, T., Leutenegger, M., Plessmann, U., Urlaub, H., Eggeling, C., Wahl, M.C., Hell, S.W., Jakobs, S.: A reversibly photoswitchable GFP-like protein with fluorescence excitation decoupled from switching. Nat. Biotechnol. **29**, 942–947 (2011)
34. Sahl, S.J., Hell, S.W., Jakobs, S.: Fluorescence nanoscopy in cell biology. Nat. Rev. Mol. Cell Biol. **18**, 685–701 (2017)

Chapter 8
Metal-Induced Energy Transfer Imaging

Alexey I. Chizhik and Jörg Enderlein

Abstract Super-resolution microscopy has seen a tremendous development over the last two decades. It has opened new perspectives for the application of fluorescence microscopy in the life sciences. Achieving a spatial resolution beyond the diffraction limit of light allowed one to observe many biological structures that are not resolvable in conventional fluorescence microscopy. However, despite recent development of super-resolution fluorescence microscopy techniques that allowed for squeezing the lateral resolution down to tens of nanometers, the much less axial resolution remains a key limiting factor for applications where z-sectioning of a sample is needed. In this chapter, we present the recently developed fluorescence imaging method that is called metal-induced energy transfer. It combines unprecedented nanometer resolution with technical simplicity that allows life science researchers to use it with standard microscopes. We discuss basic principle of the method, its theoretical background, and its applications for imaging of various sub-cellular structures.

PACS Subject Classification: 87.64.M- · 73.20.Mf

8.1 Introduction

Fluorescence imaging is one of the most commonly used techniques for investigation of biological systems. Among its key advantages are (i) possibility to observe live samples in real time, (ii) the technical simplicity that makes it accessible for a broad community of life-science researchers, and (iii) specific labeling allows one to directly visualize sub-cellular structures. However, the wave nature of light limits the spatial resolution of a conventional fluorescence microscope, i.e. it cannot resolve structures smaller than the diffraction-limit. In the visible spectral range, this

A. I. Chizhik (✉) · J. Enderlein
Third Institute of Physics - Biophysics, Universität Göttingen, Friedrich-Hund-Platz 1, 37077 Göttingen, Germany
e-mail: alexey.chizhik@phys.uni-goettingen.de

J. Enderlein
e-mail: jenderl@gwdg.de

© The Author(s) 2020
T. Salditt et al. (eds.), *Nanoscale Photonic Imaging*, Topics in Applied Physics 134,
https://doi.org/10.1007/978-3-030-34413-9_8

corresponds to a spatial resolution of roughly half a micrometer along the optical (z-) axis, and of about a quarter of a micrometer in the xy-plane. The field of super-resolution microscopy has seen a tremendous development over the last two decades and has opened up new advances for the application of fluorescence microscopy in bio-imaging. However, each of the existing methods are either technically challenging and require high light excitation intensities at the limit of what is tolerable for live-cell imaging, or are rather slow and require specialized labels and environmental conditions, which are not always compatible with live-cell microscopy. Moreover, the majority of these methods suffer from one common problem: Their axial resolution is by roughly one order of magnitude worse than their lateral resolution.

In this chapter, we present a new fluorescence-based method called metal-induced energy transfer (MIET), which is based on the energy transfer from an optically excited donor molecule to a thin metal film. It allows one to achieve an axial localization of a fluorophore with down to one nanometer accuracy. This goes far beyond the diffraction limit of light microscopy and surpasses in accuracy all known light-based techniques for enhancing the axial resolution. One of the key advantages of this method is that it does not require any hardware modification to a conventional fluorescence-lifetime imaging microscope (FLIM), thus preserving its full lateral resolution. The technical simplicity of MIET and its compatibility with live-cell imaging makes it applicable for broad range of studies.

This chapter partly overlaps with recent review papers that discuss various aspects of MIET imaging [1, 2]. However, in contrast to the previous publications that focus on specific points, such as single molecule imaging using MIET or its comparison with other methods for high resolution axial localization, this chapter provides readers with a general overview of basic principle of MIET and its potential for bio-imaging.

8.2   Basic Principle and Theory

It was predicted by Edward Purcell in 1946 [3] that placing a fluorescent molecule in the vicinity of a metal quenches its fluorescence emission and decreases its excited state lifetime. From a physics point of view, the mechanism behind this phenomenon is similar to that of FRET [4]: energy from the excited molecule is transferred, via electromagnetic coupling, into plasmons of the metal, where energy is either dissipated or re-radiated as light. This fluorophore-metal interaction was extensively studied in the 1970s and 1980s [5], and a quantitative theory developed on the basis of semi-classical quantum optics [6, 7]. The achieved quantitative agreement between experimental measurement and theoretical prediction was excellent (Fig. 8.1).

Owing to the fact that the energy transfer rate is dependent on the distance of a molecule from the metal layer, the fluorescence lifetime can be directly converted into a distance value (Fig. 8.2). The theoretical basis for the success of this conversion is the perfect quantitative understanding of MIET [8]. It is important to emphasize that the energy transfer from the molecule to the metal is dominated by the interaction of the molecule's near-field with the metal and is thus a thoroughly near-field effect,

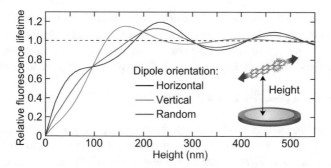

Fig. 8.1 Calculated dependence of fluorophore lifetime on its height over the metal film. Curves are calculated for an emission wavelength of 650 nm and a gold film thickness of 20 nm deposited on the glass substrate

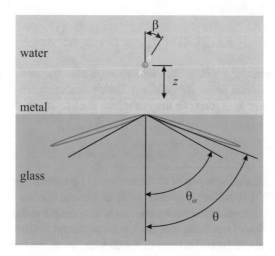

Fig. 8.2 Geometry of MIET sample: A fluorophore is placed above of a thin metal film deposited on glass. Fluorescence detection is done with a high numerical aperture objective lens from the glass side. Fluorescence excitation is performed by the same lens. In the figure, an electric dipole emitter is placed at a distance z from the metal film. Its orientation is described by the angle β between its dipole axis and the optical (vertical) axis. The angular distribution of radiation into the glass is depicted as a red curve and is a function of angle θ. The critical angle θ_{cr} of total internal reflection between glass and water is also shown

similar to FRET. However, due to the planar geometry of the metal film, which acts as the acceptor, the distance dependency of the energy transfer efficiency is much weaker than the sixth power of the distance, which leads to a monotonous relation between lifetime and distance over a size range between zero and ~250 nm above the surface.

Evaluation of MIET measurements can be done by modeling the emission properties of an emitter above a metal surface. The geometry of the modeled situation is shown in Fig. 8.2.

Let us consider the emission of a single molecule with orientation angles (α, β, where β denotes the inclination towards the vertical axis and α the angle around that axis. The molecule is assumed to be an electric dipole emitter. Then, the electric field amplitude of its emission into direction (θ, ϕ) is given by the general formula

$$\mathbf{E}_{em} = \hat{\mathbf{e}}_p \left[A_\perp \cos \beta + A_\parallel^c \sin \beta \cos (\phi - \alpha) \right] + \hat{\mathbf{e}}_s A_\parallel^s \sin \beta \sin (\phi - \alpha) \qquad (8.1)$$

where the are functions of emission angle θ but not on α, β or ϕ. Explicit expressions for can be found in a standard way by expanding the electric field of the dipole emission into a plane wave superposition and tracing each plane wave component through the planar structures using Fresnel's relations, for details see [9–12]. It is important to note that the functions depend also on wavelength. Knowing the electric field amplitude of the emission into a given direction (θ, ϕ), one can then derive the total power of emission as

$$S_{\text{total}}(\beta, \alpha) \propto B_\perp \cos^2 \beta + B_\parallel \sin^2 \beta \qquad (8.2)$$

with weight factors $B_{\perp,\parallel}$ which take into account also the absorption of emitted energy within the metal layer, for details of their calculation see [9]. Knowing the total emission power S_{total}, one can then calculate the lifetime of the molecule by

$$\frac{\tau}{\tau_0} = \frac{S_0}{\Phi S_{\text{total}} + 1 - \Phi} \qquad (8.3)$$

where S_0 is the total emission power of the emitter in free space (sample space), Φ is the quantum yield, τ_0 is the free space excited state lifetime lifetime of the emitter. For calculating the lifetime-distance curve, one has to average the result over all possible molecular orientations (assuming that there is no preferred molecular orientation in the sample) and the emission spectrum of the emitter (using the free-space emission spectrum as weight function).

Experimentally, one needs a standard scanning confocal microscope that allows one to do fluorescence lifetime imaging (Fig. 8.3), that is, equipped with a pulsed excitation laser and a single photon avalanche diode. The only addition that is required for MIET imaging is coating the substrate with a semitransparent metal film, typically 10–15 nm. Gold as a coating material combines such crucial properties as non-toxicity for living cells, absence of oxidation, and high transparency compared to other metals.

8.3 The MIET-GUI Software

We have developed a Matlab-based MIET-GUI for analysis of measured data. The MIET-GUI is a graphical user interface designed for various types of data evaluation, for instance the conversion of the raw FLIM data into a MIET image. The software

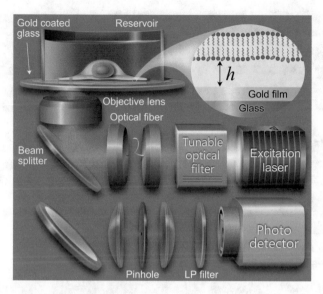

Fig. 8.3 Schematic of the experimental set-up for MIET imaging

can be downloaded via the link www.joerg-enderlein.de/MIET/MIETGUI.zip. The
MIET-GUI accepts .ht3 and .ptu files generated by the FLIM-hardware HydraHarp
of PicoQuant GmbH (Berlin), from which it calculates the lifetime and intensity for
every pixel of an image, elliptical regions of interest (ROI) or the patterns generated
by scanning the excitation light over single dipole emitters. These lifetimes are converted into height information via the MIET lifetime versus height calibration curve
(Fig. 8.4).

As a first step, the user has to choose the general type of evaluation, pixel-by-pixel
or one of the more elaborate ROI/pattern techniques. In the pixel-by-pixel mode, the
time-correlated single photon counting (TCSPC) histogram of each pixel with more
than 25 photons is assembled. The shape of these histograms can be described by a
steep rise followed by a peak and then an exponential decay. By setting a cutoff after
which the curve is purely exponential and calculating the mean arrival time of the
photons after this cutoff, one gets the lifetime value for this pixel. In the ROI mode,
the user specifies an elliptical region of interest believed to belong to molecules
with the same lifetime. The photons from all pixels within the ROI are collected
into a single histogram, which is less prone to noise problems than histograms for
single pixels. For this reason, the histogram can be fit with either mono- or multi-
exponential decay curves, thus finding the lifetime of the molecules in the ROI. The
most sophisticated mode is the pattern matching mode. Here, the user has to specify
the parameters of the excitation light such as the wavelength, the polarization mode of
the laser, the numerical aperture of the objective and the defocusing of the objective.
From these parameters, the patterns generated by scanning the excitation beam over
molecules with different angular orientations can be calculated. The intensity image
obtained by integrating the TCSPC data over time is now fitted with the simulated

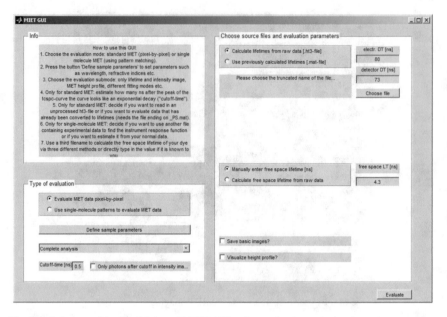

Fig. 8.4 Interface of the Matlab-based MIET-GUI software for the analysis of raw data

patterns to determine the position and orientation of each single dipole emitter. The photons from all the pixels assigned to a molecule's pattern are grouped into a single histogram and fitted as in the ROI mode.

In the second step, the lifetime information is converted into height information. To this end, the user has to specify the emission wavelength, the quantum yield and the excited state lifetime of the emitters as well as the thicknesses and complex refractive indices of all materials in the sample (e.g. metal-coated glass cover slides, buffer solutions etc.). As described above, this data can be used to calculate the observed lifetime as a function of the dipole's height above the interface and its angle with the optical axis. In the pixel-by-pixel evaluation mode, nothing is known about the particle's orientation, so a random orientation is assumed and the calibration curve calculated accordingly. In the pattern matching mode, the particle's orientation is known and the correct curve is used for the evaluation. If the emission spectrum of the fluorescent probe is known, the calibration curves obtained for all wavelengths that are able to pass the optical filters are calculated and averaged according to the spectrum. A complication arises from the fact that the lifetime versus height curve oscillates, meaning that some lifetime values cannot be matched unambiguously to a height value. The first possibility for solving this problem is to crop the calibration curve at the largest unique value and to mark all longer lifetimes as 'not a number' in the height image. If there is a prior knowledge about the sample states that no height values larger than the value corresponding to the first peak in the calibration curve can exist, it is possible to crop the calibration curve at this peak. The height information gained through this progress can then be visualized or used for further analysis.

8.4   Metal-Induced Energy Transfer for Biological Imaging

The applicability of MIET for live-cell imaging has been first shown by mapping the basal membrane of living cells with nanometer accuracy [13]. Knowledge of the precise cell-substrate distance as a function of time and location with unprecedented resolution provides a new means to quantify cellular adhesion and dynamics, as is required for a deeper understanding of fundamental biological processes such as cell differentiation, tumor metastasis and cell migration.

As a biological model system three adherent cell lines were chosen: MDA-MB-231 human mammary gland adenocarcinoma cells and A549 human lung carcinoma cells, which are able to form metastasis in vivo models, as well as MDCK-II from canine kidney tissue as a benign epithelial cell line. Interestingly, significant differences in the cell–interface distance between a normal epithelial cell and cancerous cell lines were observed.

Figure 8.5a and b show the measured intensity and lifetime images that were used to obtain the 3D reconstruction of the basal cell membrane. Because the variation of the fluorescence intensity is not only dependent on the metal-induced quenching, but also on the homogeneity of labelling, exclusively the lifetime information was used for reconstructing a three-dimensional map of the basal membrane. On the other hand, the intensity distribution was used to discriminate the membrane fluorescence against the background. Regions with no cells are difficult to identify from the lifetime images alone, as the lifetime values can become exceedingly scattered at low signal-to-noise ratios. Figure 8.5c shows the result of recalculation of the lifetime image into the 3d height profile.

A relatively fast scanning speed of a confocal microscope that is used for MIET imaging allows to monitor dynamic processes. Figure 8.6 shows the spreading behaviour of MDCK-II cells. Generally, the spreading process of adherent cells can be divided into three distinct temporal phases. The first phase is characterized by the formation of initial bonds between adhesion molecules and molecules of the extracellular matrix. This process of tethering is followed by the second phase, which

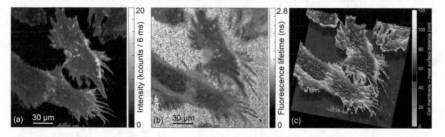

Fig. 8.5 Simultaneously acquired fluorescence intensity (**a**) and lifetime (**b**) images of the basal membrane of living MDA-MB-231 cells grown on a gold-covered glass substrate, acquired with a standard confocal microscope. **c** Three-dimensional reconstruction of the basal cell membrane. Three-dimensional profiles computed from the fluorescence lifetime image (**b**)

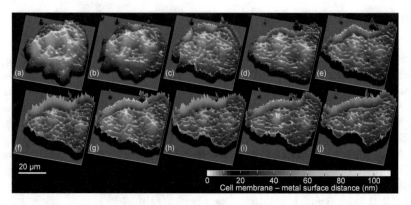

20 μm

0   20   40   60   80   100
Cell membrane – metal surface distance (nm)

Fig. 8.6 Time elapsed MIET images recorded in 5 min time intervals showing the late stages of cell (MDCK-II) spreading on gold. The cell forms tightly attached protrusions/lamellipodia away from the center of the cell. The cell occupies a larger area with time and presses down more closely. A darker color refers to lower cell-substrate distance. At later stages (k-n) first lamellipodia are formed that exhibit a low cell-substrate distance

comprises the initial cell spreading and that is driven by actin polymerization. The latter forces the cell surface area to increase by drawing membrane from a reservoir of folded regions. The third phase encompasses recruitment of additional plasma membrane from the internally stored membrane buffer and extension of lamellipodia to occupy a larger area.

The axial resolution of the recorded images can be determined by calculating the standard deviation of cell-substrate distance. The resolution depends on the photon rate and varies between 2 and 4 nm for typical fluorescence intensities measured and can be further enhanced to 1 nm by increasing the number of detected photons.

The unprecedented axial resolution of MIET allowed us to monitor the cell-substrate distance of epithelial NMuMG cells during the biological process of the epithelial-tomesenchymal transition (Fig. 8.7) [14]. EMT allows epithelial cells to enhance their migratory and invasive behavior and plays a key role in embryogenesis, fibrosis, wound healing, and metastasis. Among the multiple biochemical changes from an epithelial to a mesenchymal phenotype, the alteration of cellular dynamics in cell-cell as well as cell-substrate contacts is crucial. It was shown that, in the very first hours of the transition, the cell-substrate distance increases by several tens on nanometers, but later in the process after reaching the mesenchymal state, this distance is reduced again to the level of untreated cells.

Dual-color MIET allowed for reconstructing the 3D profile of the nuclear envelope over the whole basal area of HeLa cells [15]. The profilometry was done by measuring the axial distance between the proteins Lap2β and Nup358 as components of the nuclear envelope and the nuclear pore complex, with defined localizations at the inner nuclear membrane and the cytoplasmic side of the protein complex, respectively (Fig. 8.8). The obtained thickness of the nuclear envelope of 30–35 nm is in very good agreement with the values that were obtained using electron microscopy. This study

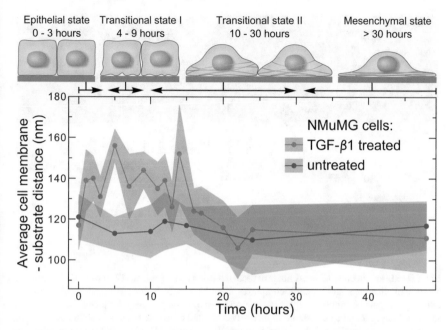

Fig. 8.7 Average cell membrane-substrate distance of untreated (blue) and TGF-β1 treated NMuMG cells (red) over time. NMuMG cells detach from the surface by more than 20 nm on average in response to TGF-β1 administration. After 20 h the initial cell-substrate distance is restored. The standard error of mean is illustrated as colored area around the data points

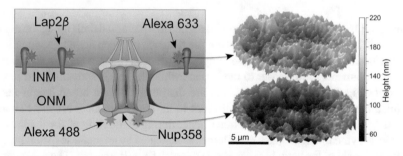

Fig. 8.8 Schematic of the positions of Lap2β and Nup358 in the inner nuclear membrane and the nuclear pore complex, respectively. HeLa cells were fixed and subjected to indirect immunofluorescence using goat anti-Nup358. Three-dimensional height profiles of the inner (top) and outer (bottom) nuclear membrane of a typical HeLa cell nucleus, as determined by MIET imaging. The outer nuclear membrane roughly follows the profile of the inner nuclear membrane

has shown that optical microscopy allows one not only to measure the distance between the outer and inner nuclear membrane but also to reconstruct its 3D profile over the whole basal area.

Recently, dual-color MIET was combined with Förster resonance energy transfer (FRET) for studies of cytoskeletal elements and adhesions in human mesenchymal

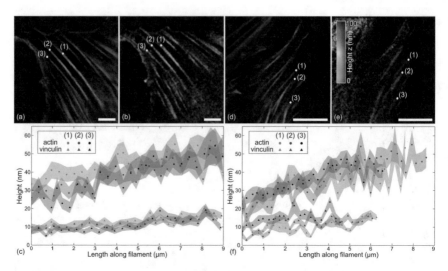

Fig. 8.9 3D architecture of stress fibers at focal adhesions changes from 12 to 24 h. Height profiles along actin filaments and vinculin complexes after 12 and 24 h. Images a and b correspond to intensity-weighted ensemble heights of actin and vinculin, respectively, for a cell fixed 12 h after seeding. Images d and e correspond to intensity-weighted ensemble heights of actin and vinculin, respectively, for a cell fixed 24 h after seeding. White points (1), (2), and (3) on the intensity-weighted height images indicate the starting points of the height profiles shown in images (c) and (f). They show the height of actin filaments (circles) and vinculin clusters (triangles) at the same focal adhesion. The shaded areas mark the 1σ-regions of the height values. Scale bar is $10\,\mu$m

stem cells [16]. In addition to resolving nanometric structural details along the z-axis using MIET, FRET was used to measure the distance between actin and vinculin at focal adhesions. The analysis of the temporal evolution of actin heights shows that the actin filaments move closer to the surface while the cell is spreading and firmly adhering (Fig. 8.9). Although the fibers are distributed over a broad height range during an early phase (1–6 h), their distance to the surface reduces around 12 h and later time points to 40 nm. On the other hand, during maturation of focal adhesion complexes, vinculin aggregates grow larger as indicated by an increase in height, and the mean height of the actin bundles above the surface is decreasing. The nanometer-precise height information along the fibers and of the vinculin clusters Fig. 8.9 gives a detailed picture of stress fibers anchoring at focal adhesions and spanning the cell at a slight inclination of below 1°.

Use of single photon counting detectors for MIET measurements allow one not only to achieve nanometer resolution of sub-cellular structures with high labeling density, but also to do nanometer axial localization of single molecules. The proof of principle study was done by Karedla et al., where the authors determined the height of dye molecules deposited on a dielectric spacer of a known thickness [17]. By varying the thickness of the spacer, the authors showed that the axial position of molecules can be determined with accuracy better than 2.5 nm. The excellent agreement between the known thickness and the height values that were obtained using MIET showed its applicability for single molecule studies with accuracy that unachievable with conventional microscopy techniques.

Isbaner et al. used MIET for colocalizing two single fluorescent emitters along the optical axis with nanometer accuracy [18]. For this purpose, the authors used stepwise photobleaching to find the fluorescence lifetime values of each emitter on one DNA origami pillar, which allowed them to determine their individual heights from the surface and thus their mutual axial distance. The determined distance of 32 ± 11 nm is in excellent agreement with the design value of 32 nm.

8.5 Conclusions

The review of applications of MIET shows its versatility and potential for numerous application in live cell imaging and single molecule localization. The unique combination of its technical simplicity and nanometer axial resolution makes it widely applicable for numerous studies in life science or nanotechnology. The distance range covered by MIET nicely bridges (and complements) the realm of conventional FRET and all the recently developed super-resolution imaging techniques. It opens new perspectives for nanometer localization of sub-cellular structures in cell focal adhesion complexes. Since MIET keeps all the key advantages of conventional fluorescence microscopy, it allows to do simultaneous multi-color imaging of various sub-cellular structures. Extremely high photo-sensitivity of single photon avalanche diodes that are used as photo-detectors for MIET imaging allow one to singe molecule localization with precision that in unimaginable for conventional optical microscopy. We envision further rapid growth of the number of its applications and technical development for increasing its temporal and spatial resolution.

Acknowledgements We are grateful to the Deutsche Forschungsgemeinschaft (DFG) for financial support via project A05 of the SFB755 and project A14 of the SFB937 as well as through the Cluster of Excellence and DFG Research Center "Nanoscale Microscopy and Molecular Physiology of the Brain."

References

1. Chizhik, A. I.: Super-Resolution Depth Measurements: variable Angle TIRF, Super-Critical Angle Fluorescence, MIET, pp. 175–184. Elsevier (2018). https://doi.org/10.1016/B978-0-12-803581-8.09496-0., https://doi.org/10.1007/4243_2014_77
2. Karedla, N., Ruhlandt, D., Chizhik, A.M., Enderlein, J., Chizhik, A.I.: Metal-Induced Energy Transfer, pp. 265–281. Springer International Publishing (2015). https://doi.org/10.1007/4243_2014_77
3. Purcell, E.M.: Spontaneous emission probabilities at radio frequencies. Phys. Rev. **69**, 681 (1946)

4. Ha, T., Enderle, T., Ogletree, D., Chemla, D.S., Selvin, P.R., Weiss, S.: Probing the interaction between two single molecules: fluorescence resonance energy transfer between a single donor and a single acceptor. Proc. Natl. Acad. Sci. **93**(13), 6264–6268 (1996)

5. Drexhage, K.H.: Iv Interaction of Light with Monomolecular Dye Layers. pp. 163–232. Elsevier (1974). https://doi.org/10.1016/S0079-6638(08)70266-X

6. Chance, R.R., Prock, A., Silbey, R.: Molecular Fluorescence and Energy Transfer Near Interfaces, pp. 1–65. Wiley-Blackwell (2007). https://doi.org/10.1002/9780470142561.ch1

7. Lukosz, W., Kunz, R.E.: Light emission by magnetic and electric dipoles close to a plane interface. i. total radiated power. J. Opt. Soc. Am. **67**(12), 1607–1615 (1977). https://doi.org/10.1364/JOSA.67.001607, http://www.osapublishing.org/abstract.cfm?URI=josa-67-12-1607

8. Enderlein, J.: Theoretical study of single molecule fluorescence in a metallic nanocavity. Appl. Phys. Lett. **80**(2), 315–317 (2002). https://doi.org/10.1063/1.1434314

9. Enderlein, J.: Single-molecule fluorescence near a metal layer. Chem. Phys. **247**(1), 1–9 (1999). https://doi.org/10.1016/S0301-0104(99)00097-X

10. Enderlein, J.: A theoretical investigation of single-molecule fluorescence detection on thin metallic layers. Biophys. J. **78**(4), 2151–2158 (2000). https://doi.org/10.1016/S0006-3495(00)76761-0

11. Enderlein, J., Ruckstuhl, T.: The efficiency of surface-plasmon coupled emission for sensitive fluorescence detection. Opt. Express **13**(22), 8855–8865 (2005). https://doi.org/10.1364/OPEX.13.008855, http://www.opticsexpress.org/abstract.cfm?URI=oe-13-22-8855

12. Enderlein, J., Ruckstuhl, T., Seeger, S.: Highly efficient optical detection of surface-generated fluorescence. Appl. Opt. **38**(4), 724–732 (1999). https://doi.org/10.1364/AO.38.000724, http://ao.osa.org/abstract.cfm?URI=ao-38-4-724

13. Chizhik, A.I., Rother, J., Gregor, I., Janshoff, A., Enderlein, J.: Metal-induced energy transfer for live cell nanoscopy. Nat. Photonics **8**(2), 124 (2014)

14. Baronsky, T., Ruhlandt, D., Brückner, B.R., Schäfer, J., Karedla, N., Isbaner, S., Hühnel, D., Gregor, I., Enderlein, J., Janshoff, A., Chizhik, A.I.: Cell-substrate dynamics of the epithelial-to-mesenchymal transition. Nano Lett. **17**(5), 3320–3326 (2017)

15. Chizhik, A.M., Ruhlandt, D., Pfaff, J., Karedla, N., Chizhik, A.I., Gregor, I., Kehlenbach, R.H., Enderlein, J.: Three-dimensional reconstruction of nuclear envelope architecture using dual-color metal-induced energy transfer imaging. ACS Nano **11**(12), 11839–11846 (2017). https://doi.org/10.1021/acsnano.7b04671. PMID: 28921961

16. Chizhik, A.M., Wollnik, C., Ruhlandt, D., Karedla, N., Chizhik, A.I., Hauke, L., Hähnel, D., Gregor, I., Enderlein, J., Rehfeldt, F.: Dual-color metal-induced and förster resonance energy transfer for cell nanoscopy. Mol. Biol. Cell **29**(7), 846–851 (2018)

17. Karedla, N., Chizhik, A.M., Stein, S.C., Ruhlandt, D., Gregor, I., Chizhik, A.I., Enderlein, J.: Three-dimensional single-molecule localization with nanometer accuracy using metal-induced energy transfer (miet) imaging. J. Chem. Phys. **148**(20), 204,201 (2018)

18. Isbaner, S., Karedla, N., Kaminska, I., Ruhlandt, D., Raab, M., Bohlen, J., Chizhik, A., Gregor, I., Tinnefeld, P., Enderlein, J., et al.: Axial colocalization of single molecules with nanometer accuracy using metal-induced energy transfer. Nano Lett. **18**(4), 2616–2622 (2018)

Chapter 9
Reversibly Switchable Fluorescent Proteins for RESOLFT Nanoscopy

Nickels A. Jensen, Isabelle Jansen, Maria Kamper and Stefan Jakobs

Abstract Diffraction-limited lens-based optical microscopy fails to discern fluorescent features closer than ~200 nm. All super-resolution microscopy (nanoscopy) approaches that fundamentally overcome the diffraction barrier rely on fluorophores that can adopt different states, typically a fluorescent 'on-'state and a dark, non-fluorescent 'off-'state. In reversible saturable optical linear fluorescence transitions (RESOLFT) nanoscopy, light is applied to induce transitions between two states and to switch fluorophores on and off at defined spatial coordinates. RESOLFT nanoscopy relies on metastable reversibly switchable fluorophores. Thereby, it is particularly suited for live-cell imaging, because it requires relatively low light levels to overcome the diffraction barrier. Most implementations of RESOLFT nanoscopy utilize reversibly photoswitchable fluorescent proteins (RSFPs), which are derivatives of proteins from the green fluorescent protein (GFP) family. In recent years, analysis of the molecular mechanisms of the switching processes have paved the way to a rational design of new RSFPs with superior characteristics for super-resolution microscopy. In this chapter, we focus on the newly developed RSFPs, the light-driven switching mechanisms and the use of RSFPs for RESOLFT nanoscopy.

9.1 Overcoming the Diffraction Barrier

Optical fluorescence microscopy allows to discern protein distributions in cells. Over the last century, cell biology experienced countless discoveries based on the microscopic visualization of (sub-)cellular structures and their dynamics. Eukaryotic cells, which generally have a diameter of 10–300 μm, are relatively small objects crowded with proteins. Hence, conventional optical microscopy, whose resolution is fundamentally limited by diffraction to about 200 nm, inescapably faces the fact that the finest details are not resolvable.

N. A. Jensen · I. Jansen · M. Kamper · S. Jakobs (✉)
Department of NanoBiophotonics, Max Planck Institute for Biophysical Chemistry,
Am Faßberg 11, 37077 Göttingen, Germany
e-mail: sjakobs@gwdg.de

© The Author(s) 2020
T. Salditt et al. (eds.), *Nanoscale Photonic Imaging*, Topics in Applied Physics 134,
https://doi.org/10.1007/978-3-030-34413-9_9

Resolution in microscopy means the separation of distinct features, such as fluorescently labelled proteins. In conventional diffraction limited microscopy, such as widefield (epifluorescence) or confocal laser scanning microscopy, all fluorophores in closer proximity to each other than the diffraction limit are excited together, they emit together, and their emissions diffract together and therefore they are detected together [1]. Hence, with these approaches, structures closer than the diffraction limit are inseparable. The key to fundamentally overcome the diffraction barrier is to render adjacent molecules discernible for a short period of time, preventing different molecules within the same diffraction region from being detected together [2]. This separation can be implemented either in a coordinate-targeted or in a coordinate-stochastic way (for reviews see [1, 3]).

In coordinate-targeted super-resolution microscopy techniques such as RESOLFT nanoscopy, a light pattern is used to induce transitions between two states and to switch fluorophores on and off at defined spatial coordinates. In the simplest scenario, a single beam creating one intensity minimum in a doughnut-shaped pattern is used, but also approaches with several or many minima are possible. At these minima (zeros), there is no off-switching, and the fluorophores in the on-state can fluoresce. These light patterns are scanned over the sample, in order to record a full image. Such scanning approaches require multiple on-off-cycles of the fluorophores.

9.2 RSFPs for Live-Cell RESOLFT Nanoscopy

In its initial definition, the term 'RESOLFT nanoscopy' covered all coordinate-targeted nanoscopy approaches relying on two distinct (fluorophore) states including STED nanoscopy [4]. Later, this term was primarily used for coordinate-targeted nanoscopy approaches that rely on metastable fluorophores, most prominently on reversibly (photo-)switchable fluorescent proteins (RSFPs). Compared to other super-resolution microscopy techniques that overcome the diffraction barrier, RESOLFT nanoscopy requires remarkably low light dose to achieve nanoscale resolution. The light intensities used are similar to those applied in live-cell confocal fluorescence microscopy and up to six orders of magnitude lower than those in STED-microscopy. The total light dose deposited on the sample is lower by 3–4 orders of magnitude compared to coordinate-stochastic nanoscopy [5, 6]. As the light intensity is an important factor that determines phototoxicity [7], RESOLFT nanoscopy is particularly suitable for live-cell imaging approaches.

RESOLFT microscopy relies on fluorophores that can be reversibly photoswitched between two metastable states. In the overwhelming majority of applications, these have been reversibly photoswitchable fluorescent proteins. RSFPs are structurally highly similar to the green fluorescent protein (GFP). All GFP-based fluorescent proteins share the same overall structure of 11 β-sheets, forming a β-barrel with an α-helix running through the center. The chromophore is autocatalytically formed out of three amino acids within the α-helix requiring only oxygen as an external cofactor for its maturation (Fig. 9.1) [8]. In GFP, the chromophore consists

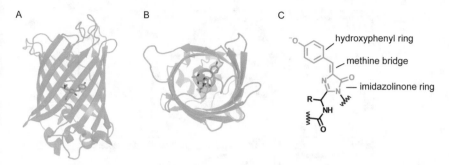

Fig. 9.1 Structure of EGFP and its intrinsic chromophore. **a** Side view of the GFP barrel. **b** Top view of the GFP barrel with the internal chromophore in its center (PDB: 2Y0G). **c** Structure of the GFP chromophore. The chromophore consists of an imidazolinone and a hydroxyphenyl ring connected by a methine bridge

of a hydroxyphenyl ring and an imidazolinone ring connected by a methine bridge. By rotation around the methine bridge, the chromophore can adopt either a *cis* or a *trans* conformation. The hydroxyphenyl ring can be protonated or deprotonated, thereby shifting the absorption spectrum by about 80–120 nm. The β-barrel shields the chromophore in its center, while numerous noncovalent bonds to surrounding amino acid residues and internal water molecules determine the chromophore position within the protein and its protonation state. The spectral properties of a specific RSFP are a result of the chromophore structure as well as of its interactions with the surrounding residues; therefore, those positions are key targets for mutagenesis to modify the properties of RSFPs.

9.3   Photoswitching Mechanisms of RSFPs

Conventional RSFPs can be classified according to their switching mode, i.e. a 'positive' or a 'negative' switching mode (Fig. 9.2) [9]. In negative switching RSFPs, the same wavelength that induces fluorescence also switches the RSFP from the on- to the off-state. In contrast, in positive switching RSFPs, the light that induces fluorescence also transfers the protein from the off- to the on-state. Thereby, in conventional RSFPs, switching and fluorescence excitation are directly interconnected. The mechanistic principles of switching have been initially revealed by spectroscopy and crystallography of the RSFPs asFP595 and Dronpa in their respective on- and off-states [10, 11]. In recent years, numerous further studies [12–16], including detailed molecular dynamics studies, ultrafast spectroscopy and crystallography have led to impressive insights into the details of the light driven switching mechanism [17–20]. The mechanistic key event of the switching of all conventional RSFPs analyzed so far is a light induced *cis-trans* isomerization of the chromophore often combined with a protonation change [13]. This isomerization can be accompanied by shifts of

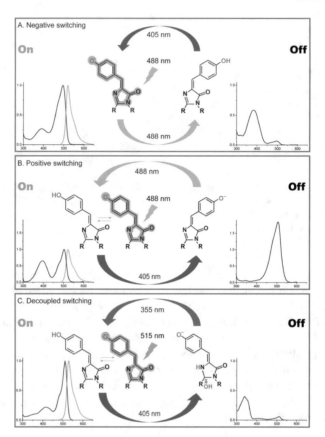

Fig. 9.2 Schematic overview of the different switching mechanisms. **a** Negative switching RSFPs can be switched off with light of the wavelength used for fluorescence excitation. **b** Positive switching RSFPs are switched on by the fluorescence excitation light. **c** Proteins with decoupled switching are excited with light of a different wavelength than used for on- or off-switching. For each switching mechanism, schematic chromophore structures in the on- and the off-state, examples for the switching wavelengths, and the respective absorption and emission spectra are shown

the planarity of the chromophore and/or conformational changes of the chromophore pocket and modifications of the hydrogen-bonding network around the chromophore, influencing the chromophores' ability to fluoresce. Almost all reported crystallographic structures of conventional RSFPs feature a *trans*-conformation of the chromophore in the off-state and a *cis*-conformation in the on-state [21]. However, it should be noted, while the majority of fluorescent proteins contain a chromophore in the *cis*-conformation, bright fluorescent proteins containing a chromophore in the *trans*-state exist [22]. The only reported RSFP so far with a different molecular switching mechanism is the yellow fluorescent RSFP Dreiklang [23] and its descendant Spoon [24] (Fig. 9.2c). In Dreiklang, fluorescence excitation is decoupled from switching, i.e. one wavelength is used for fluorescence excitation (515 nm), another for switching on (405 nm), and a third for switching off (355 nm). In Dreiklang, a light-driven reversible hydration/dehydration of the chromophoric imidazolinone ring causes switching by reversibly disrupting the π-conjugated electron system.

In the following, we will take a closer look at the different switching modes and their application in RESOLFT imaging.

9.3.1 Negative Switching Mode

Negative switching RSFPs are most commonly used for RESOLFT nanoscopy. In the dark state, the chromophore is found in the protonated *trans*-conformation, while in the on-state the chromophore is deprotonated in the *cis*-conformation (Fig. 9.2a). Excitation of the on-state chromophore results in fluorescence (quantum yield typically 0.1–0.9), switching (quantum efficiency typically in the range of 0.01) [25], or to other non-radiative processes. In comparison, the quantum efficiency for switching the protein from the off- to the on-state is generally substantially higher (quantum efficiency typically in the range of 0.1). The wavelength used for switching the RSFP from the off- into the on-state is usually 80–120 nm shorter than the wavelength for switching off and fluorescence excitation. The precise sequence of the intra-molecular events during the switching process, particularly the sequence of isomerization and protonation change, have been controversially discussed [21]. Recent results for Dronpa and IrisFP conclusively suggested that during the off-to-on transition the chromophore isomerization precedes the protonation change [20, 26].

In single-beam scanning RESOLFT nanoscopy, negative switching RSFPs are generally utilized according to the following imaging scheme: First, the proteins are switched to the on-state with a Gaussian beam, then the proteins in the periphery are switched off with a doughnut-shaped beam, and subsequently the remaining fluorescence in the center (whose size is smaller than the diffraction limit) is read out. The on-switching process has generally a higher quantum efficiency than the off-switching process. Thus, the off-switching step is the most time-consuming step, although fast switching proteins have been developed that reduced the dwell time strongly [5]. However, a too high quantum efficiency for switching off is also not desirable, as switching competes with fluorescence excitation in case of RSFPs with a negative switching mode.

9.3.2 Positive Switching Mode

All reported RSFPs with positive switching characteristics, including Padron [9], rsCherry [27] or asFP595 [28], share an Anthozoa heritage. In the off-state, the chromophore of the positive switching RSFPs is generally in the *trans*-state and deprotonated (Fig. 9.2b). Excitation into the absorption band of the deprotonated chromophore induces chromophore isomerization to the *cis*-state. In the *cis*-state, the protein is in an equilibrium between the protonated and the deprotonated state. Excitation of the protonated *cis*-chromophore switches the protein to the *trans* off-state, while excitation of the deprotonated *cis*-chromophore excites fluorescence. The protonation equilibrium of the chromophore in the *cis*- and the *trans*-states is immediately influenced by the amino acids forming the chromophore pocket. Reportedly, the exchange of few amino acids can induce a change in the switching mode of the respective fluorescent protein [9, 12, 27].

Using positive switching RSFPs, the optical system for beam scanning RESOLFT nanoscopy can be reduced to a Gaussian beam for switching on and fluorescence excitation and a doughnut-shaped beam for switching the protein off [29]. A potential advantage of RESOLFT imaging with positive switching fluorescent proteins is that the number of emitted photons during readout is not limited by a competing off-switching process.

9.3.3 Decoupled Switching Mode

The switching of Dreiklang and its descendants differs from the switching of conventional RSFPs by the fact that the excitation of fluorescence does not induce substantial switching [23]. The chromophore of the on-state protein is in an equilibrium between a protonated and a deprotonated state (Fig. 9.2c). Excitation of the deprotonated on-state chromophore at 515 nm results in fluorescence, while excitation of the protonated chromophore at 405 nm leads to a hydration of the C65 atom of the imidazolinone ring, thereby switching the protein to the off-state. More specifically, upon excitation into the absorption band of the protonated chromophore, an ultrafast excited state proton transfer occurs, presumably leading to a charge transfer to the imidizanolinone ring, which subsequently is protonated by Glu222, catalyzing the addition of a water molecule [30]. By this water addition, the π-conjugated electron system is shortened, resulting in the emergence of an absorption band at 395 nm. Excitation of this band results in a water elimination reaction, thereby converting the protein back to the on-state. The positioning of the water molecule reacting with the chromophore and the protonation state of the chromophore are crucial for the efficiency of this switching mechanism. This process is dependent on at least three key amino acids (Gly65, Tyr203, Glu222) [23].

The decoupling of switching from fluorescence excitation enables full control of the state of the protein, which is desirable in RESOLFT nanoscopy. Dreiklang has been used for RESOLFT nanoscopy [23, 31], although it is outperformed by other RSFPs, because some of its other properties (see below) are less favorable for this imaging modality.

9.4   RSFP Properties Important for RESOLFT Nanoscopy

The usefulness of a specific RSFP for live-cell RESOLFT nanoscopy depends on several factors. On the one hand, any usable RSFP must be a suitable fusion tag with a negligible dimerization tendency and a fast maturation rate. On the other hand, it needs to exhibit a combination of favorable photophysical characteristics. The four most important parameters for RESOLFT imaging are the brightness of the protein in its on-state, its switching speed, the residual fluorescence background in the ensemble off-state, and the switching fatigue (Fig. 9.3). These parameters, all of which can be influenced by mutagenesis, are discussed in detail in the following.

9.4.1   Brightness

For a conventional fluorophore, the molecular brightness is defined as the product of the extinction coefficient and the quantum yield. RSFPs can adopt two different states with different quantum yields and in most cases the two states exhibit distinct absorption spectra. Generally, the molecular brightness of the off-state is very small but not necessarily zero. Next to the molecular brightness of the protein in the on-state, two other parameters, namely the effective brightness (of an ensemble of fully matured RSFPs in solution) and the effective cellular brightness (of RSFPs in a living cell) are used to describe and compare RSFPs. In a switching curve (Fig. 9.3a), the effective brightness is measured as the fluorescence intensity of the protein ensemble when switched fully into the on-state. In absolute terms, it is the average molecular

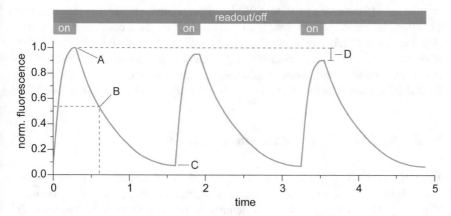

Fig. 9.3 Important switching parameters. Exemplary switching curve of negative switching RSFPs when switched consecutively on and off with light of different wavelengths. Plotted is the emitted fluorescence intensity. Four key parameters can be determined from these switching curves. **a** effective brightness of the RSFP in the on-state. **b** Switching speed. **c** Residual fluorescence in the off-state. **d** Switching fatigue

brightness of the on-switched protein solution. As a RSFP solution may adopt an equilibrium between the on- and the off-state [9, 27], the effective brightness can differ drastically from the brightness at equilibrium.

The effective cellular brightness of a specific RSFP is influenced by a number of factors, such as its maturation time, its turnover-rate, its expression rate, its usability as a tag, temperature, local pH and other difficult to pinpoint factors. As RSFPs are primarily utilized for in vivo imaging, the effective cellular brightness is a very crucial parameter when RSFPs are compared. The cellular brightness can vary strongly between different experimental settings [32, 33] and therefore the data from different studies are difficult to compare.

9.4.2 Ensemble Switching Speed

The switching speed of an ensemble of RSFPs is typically in the microsecond to second range, depending on the light intensity applied [5, 25, 34, 35]. Since RESOLFT is an ensemble method, the ensemble switching speed is a key parameter to describe and compare RSFPs. In a switching curve (Fig. 9.3), this parameter describes the time needed to switch the ensemble from the maximal fluorescence intensity to the minimal fluorescence intensity (from the on- to the off-state) or vice versa (Fig. 9.3b). The speed is dependent on the applied light intensities and wavelengths; applying higher light intensities results in faster switching. As switching is a one photon process in both directions [36, 37], the speed and the applied light intensity for the switching are largely linearly correlated until saturation of the process sets in. The ensemble switching speed is determined primarily by the quantum efficiency for switching, by (often ill-defined) intermediate states, and by a crosstalk between the on- and the off-switching processes. While high quantum efficiencies for switching are beneficial for RESOLFT imaging by shortening the time consuming switching step, the connection between switching and fluorescence readout limits the number of collected photons in a single switching cycle. For point-scanning RESOLFT in living cells, the switching times are generally below 1 ms [5, 6], while in parallelized RESOLFT schemes, switching times of tens of milliseconds are suitable [34, 38].

9.4.3 Residual Fluorescence in the Off-State

The residual fluorescence in the off-state describes the percentage of the fluorescence signal from the off-state compared to the fluorescence signal of the on-state (Fig. 9.3c). The switching contrast is defined as the ratio between the fluorescence signal of the on-state and the fluorescence signal of the off-state measured on an ensemble of proteins.

The switching contrast is determined by the crosstalk between on- and off-switching and the molecular brightness of the on- and the off-state. Furthermore, fast thermal relaxation of the switched protein to the respective equilibrium state, as well as the population of intermediate states, can affect the reachable switching contrast. RSFPs useful for RESOLFT nanoscopy exhibit a contrast higher than 10 (a residual fluorescence below 10%), although smaller values of the residual fluorescence have been reported and are beneficial [25, 39].

9.4.4 Switching Fatigue

With every full switching cycle of an ensemble of RSFPs, a fraction of the proteins is destroyed. This fraction is generally described as the switching fatigue (given as percentage of the effective brightness) (Fig. 9.3d). Presumably, the switching fatigue is mechanistically related to the photostability of a fluorescent protein, which describes its stability while the protein is maintained and excited in the fluorescent on-state.

A low switching fatigue is critical for the usability of RSFPs in RESOLFT microscopy [40]. Photobleaching is highly dependent on the light intensity, as different intensity regimes of photobleaching and nonlinear effects of increasing light intensities have been reported [32, 41]. Therefore, lower light intensities typically result in reduced switching fatigue [39]. Very photostable RSFPs can be switched thousands of times before they are bleached to 50% of their initial brightness [5, 42].

Brightness, switching speed, contrast and switching fatigue are strongly intertwined. The introduction of switching into a conventional fluorescent protein as well as the increase of the switching speed of an existing RSFP by mutagenesis generally lead to a reduction in the fluorescence quantum yield [5, 39, 43, 44]. Furthermore, for negative switching RSFPs a correlation of the switching fatigue with the off-switching speed has been reported [25].

9.5 Overview of RSFPs for RESOLFT Nanoscopy

In the last decade, a number of new RSFPs have been engineered using semi-rational design strategies guided by the crystal structures and insights into the switching mechanism. Thereby, either conventional fluorescent proteins were made switchable, or the characteristics of existing RSFPs were modified. The members of this growing family of RSFPs have been used for a number of applications [45], including intracellular protein tracking [46], photochromic FRET [47], optogenetics [48], optoacoustics [49], and several super-resolution microscopy techniques including coordinate-stochastic approaches [9], SOFI [50, 51], (protected) STED [52, 53] and RESOLFT/NL-SIM [39, 42].

For each application, distinct RSFP properties are crucial and RSFPs have been optimized and adapted for the specific demands. In the following, we provide an

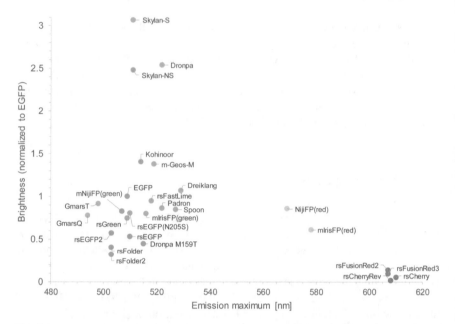

Fig. 9.4 Fluorescence emission wavelength versus molecular brightness of RSFPs. Displayed are the molecular brightness of the on-state (normalized to EGFP) and the fluorescence emission wavelength maximum of selected RSFPs. The selection is not comprehensive. Note that only RSFPs with a (potential) usability for nanoscopy are presented

overview on the published RSFPs emitting in the green, yellow and red regimes of the visible spectrum, which were used in RESOLFT microscopy or have properties beneficial for RESOLFT imaging (Fig. 9.4, Table 9.1).

9.5.1 RSFPs Emitting in the Green

RSFPs emitting in the green have been engineered based on several well-characterized fluorescent proteins isolated from Hydrozoan and Anthozoan species. The switching wavelengths of these green fluorescent RSFPs are at ∼488 and ∼405 nm, corresponding to the absorption bands of the deprotonated and the protonated chromophore. The first experimental implementation of live-cell RESOLFT nanoscopy utilized the negative switching rsEGFP [42], which has been tailored for RESOLFT nanoscopy based on the Hydrozoan EGFP [60]. Derived from rsEGFP, a number of other RSFPs were engineered such as families of rsEGFPs [5, 56], rsGreens [25], and rsFolders [14], with further optimized expression or modified switching characteristics for specific applications.

The EGFP derived RSFPs show very good tagging capabilities, are true monomers, and exhibit fast to very fast switching kinetics. Especially the high resistance of

Table 9.1 Properties of RSFPs for RESOLFT nanoscopy

Protein	Chr. (aa)	λ_{ex} (nm)	λ_{em} (nm)	Switching mode	Switching wavelength (nm) On	Switching wavelength (nm) Off	ϕ_{fluo}	ε $(10^3/(M.cm))$	References
GmarsQ	QYG	470	494	Negative	405	488	0.64	3	[54]
GmarsT	TYG	476	498	Negative	405	488	0.53	55	[35]
rsEGFP2	AYG	478	503	Negative	405	488	0.3	61	[5]
rsFolder	AYG	477	503	Negative	405	488	0.25	52	[14]
rsFolder2	AYG	478	503	Negative	405	488	0.23	44	[14]
NijiFP (green)	HYG	469	507	Negative	405	488	0.64	41	[55]
rsGreen1	TYG	486	509	Negative	405	488	0.42	58	[25]
rsEGFP N205S	TYG	491	510	Negative	405	488	0.45	57	[56]
rsEGFP	TYG	493	510	Negative	405	488	0.36	47	[42]
Skylan-S	SYG	499	511	Negative	405	488	0.64	152	[51]
Skylan-NS	LYG	499	511	Negative	405	488	0.59	134	[39]
Kohinoor	CYG	495	514	Positive	488	405	0.71	63	[29]
mGeos-M	MYG	503	514	Negative	405	488	0.85	52	[57]
Dronpa M159T	CYG	489	515	Negative	405	488	0.23	62	[44]
mIrisFP (green)	HYG	486	516	Negative	405	488	0.54	47	[58]
rsFastLime	CYG	496	518	Negative	405	488	0.77	39	[44]
Dronpa	CYG	503	522	Negative	405	488	0.85	95	[37]
Padron	CYG	503	522	Positive	488	405	0.64	43	[9]
Spoon	GYG	510	527	Decoupled	355	405	0.5	54	[24]
Dreiklang	GYG	511	529	Decoupled	355	405	0.41	83	[23]
NijiFP (red)	HYG	526	569	Negative	450	561	0.65	42	[55]
mIrisFP (red)	HYG	546	578	Negative	450	561	0.59	33	[58]
rsTagRFP	MYG	567	585	Negative	450	561	0.11	37	[47]
asFP595	MYG	572	595	Positive	561	450	n.d.	n.d.	[28]
rsFusionRed2	MYG	580	607	Negative	400–510	592	0.12	36	[34]
rsFusionRed3	MYG	580	607	Negative	400–510	592	0.08	38	[34]
rsCherryRev	MYG	572	608	Negative	450	561	0.005	84	[27]
rsCherryRev1.4	MYG	572	609	Negative	450	592	n.d	n.d.	[59]
rsCherry	MYG	572	610	Positive	561	450	0.02	80	[27]

rsEGFP2 to switching fatigue at high and low light intensities makes it the most widely used RSFP in RESOLFT imaging to date.

The very bright Anthozoa protein Dronpa, which is based on the oligomeric protein 22G from Pectiniidae spec., was the first RSFP to be used in live-cell imaging [46]. Dronpa exhibits comparatively slow switching and very high switching fatigue; Dronpa's crystal structures of the on- and off-state were solved early, guiding the development of models for the switching mechanisms and strategies for semi-rational protein design [10, 44]. Based on Dronpa, several RSFPs with improved switching properties including rsFastLime, Dronpa-M159T [44] and Dronpa-3 [43] were developed. The increase in switching speed of these mutants was always accompanied by a reduction of their respective molecular brightness; still, Dronpa-M159T is well suited for point-scanning RESOLFT [61].

A single-residue exchange at the position 159 (Met159Tyr) is sufficient to reverse the switching mode of rsFastLime, i.e. this mutation converted the negative switching RSFP rsFastLime to the positive switching rsFastLime-M159Y. Additional mutations led to the engineering of the bright, positive switching RSFP Padron, exhibiting a high switching contrast [9]. Further mutagenesis changed the switching characteristics and photostability of Padron and led to the development of Kohinoor [29] opening the door to live-cell RESOLFT with positive switching RSFPs.

In recent years, several RSFPs have been engineered using photoconvertible fluorescent proteins as scaffolds. Conventional photoconvertible fluorescent proteins are irreversibly switched from one emission maximum (generally green) to another one (generally red). The mutation Phe173Ser was shown to introduce negative switching to the photoconvertible Anthozoa proteins Dendra2 [62] and mEos [63], resulting in the multiphotochromic RSFPs NijiFP [55] and (with additional mutations) mIrisFP [64]. NijiFP and mIrisFP are both photoconvertible and negatively switchable in the red and green form.

All reported green to red photoconvertible fluorescent proteins have a histidine at the first amino acid position of their chromophore, which is essential for their photoconvertibility [65]. Exchanging this histidine by another amino acid residue can transform a photoconvertible protein into a negative switching RSFP. This was reported for the Anthozoa protein mEos2 [66]. By exchanging the His for different amino acids (Cys, Glu, Phe, Leu, Met, Ser, partially in combination with the Phe173Ser mutation), a series of negative switching RSFPs (mGeos-X), with switching properties similar to those of various Dronpa variants, was produced [57]. Likewise, the His62Leu mutation of the protein mEos3.1 [67] (a true monomer and bright mEos2 descendant), resulted in the RSFP Skylan-NS [39], which exhibits negative switching characteristics with high brightness and excellent switching contrast.

Following the same strategy, exchange of the chromophoric His in the photoconvertible Anthozoa FP mMaple3 [68], in combination with Met168Ala (corresponding to Met159 in Dronpa), resulted in a series of negative switching RSFPs (GMars-X) displaying various switching kinetics beneficial for different RESOLFT applications, in particular for parallelized RESOLFT [35, 54, 69].

9.5.2 RSFPs Emitting in the Yellow

Based on the yellow fluorescent protein Citrine (a mutant of GFP) [70], Dreiklang [23] and its descendant Spoon [24] were developed. As detailed above, in Dreiklang the switching is decoupled from fluorescence readout and this RSFP has been used for single-beam and parallelized RESOLFT imaging [23, 31, 38]. A limitation of Dreiklang is the comparatively low switching speed, the requirement for two UV light wavelengths for switching and its pronounced switching fatigue. However, in principle, the decoupled switching mechanism should be very beneficial for most RESOLFT applications. Remarkably, no conventional RSFPs have been reported based on a yellow fluorescent protein so far.

9.5.3 RSFPs Emitting in the Red

Even though the positive switching red RSFP asFP595 was the first RSFP used for proof of concept RESOLFT imaging with purified protein solutions [71], its obligate tetramerization and poor performance as a fusion tag hindered its use in live-cell imaging.

Derived from the Anthozoa protein mCherry, the RSFPs rsCherry, rsCherryRev [27] and rsCherryRev1.4 [59] were established. The high photostability and fast switching of rsCherryRev1.4 facilitated point-scanning and parallelized live-cell RESOLFT nanoscopy with this red emitting RSFP [38, 59], but its poor expression in mammalian cells and a tendency for dimerization limit its applicability.

Based on TagRFP [72], the red emitting RSFP rsTagRFP [47] was developed. It shows good contrast and strong changes in the absorption spectra of the on- and off-states. However, a low extinction coefficient and high switching fatigue prevented RESOLFT imaging in living cells. Only recently, a new family of red fluorescent RSFPs based on FusionRed was engineered [34]. The parent protein shows excellent tagging performance and rsFusionRed2 as well as rsFusionRed3 were successfully used in a parallelized RESOLFT approach [34].

In conclusion, a growing family of RSFPs for RESOLFT nanoscopy is available, each member having distinct properties. As many properties of RSFPs are depending on the light intensities used and the (cellular) environment, data recorded in different laboratories may be difficult to compare. Currently, outside the green fluorescence spectral regime, only a few RSFPs are available, underscoring the need for further protein engineering of red and infrared RSFPs.

9.6 Applications of RESOLFT Nanoscopy

Since the first proof of concept demonstration of RESOLFT nanoscopy using purified asFP595 [71], numerous RESOLFT live-cell applications have been reported [45]. It has been used in point-scanning (Fig. 9.5a) [5, 6, 14, 23, 25, 42, 73] as well as in various parallelized implementations (Fig. 9.5b) [34, 35, 38, 54, 56, 69, 74, 75].

Arguably, rsEGFP2 is the most suited protein in the point-scanning mode, as it displays a combination of favorable properties, most prominently its strong resistance against switching fatigue. Amongst others, rsEGFP2 has been used for imaging of fixed cells that had been labeled by a primary antibody against a protein of interest that was detected by a fusion protein consisting of the IgG binding Z-

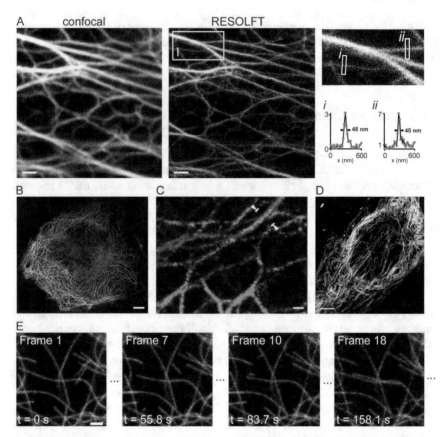

Fig. 9.5 Examples of RESOLFT imaging of living cells. **a** Point-scanning RESOLFT microscopy of intermediate filaments. Keratin19-rsEGFP2 expressed in living Ptk2 cells imaged in the confocal (left) and point-scanning RESOLFT mode (middle) [5]. The graphs show fluorescence intensity line profiles taken at sites indicated in the magnification (right). **b** Parallelized RESOLFT microscopy of intermediate filaments. Keratin19-rsEGFP-N205S expressed in living Ptk2 cells [56]. **c** Two color RESOLFT imaging. Vimentin-rsCherryRev1.4 and Keratin-rsEGFP2 were imaged in HeLa cells [59]. **d** Parallelized RESOLFT microscopy using a red RSFP. Living U2OS cells expressing rsFusionRed3-Vimentin [34]. **e** In vivo RESOLFT microscopy. Time-lapse point-scanning RESOLFT imaging of rsEGFP2-Tubulin facilitates the observation of tubulin dynamics in intact living fly larvae [6]. Images are adapted with permissions from the listed references. Scale bars: **a** 1 μm; **b** 10 μm; **c** 0.25 μm; **d** 5 μm; **e** 1 μm

domain of protein A and rsEGFP2 [73]. Live-cell RESOLFT imaging was established on CRISPR/Cas9 edited cell lines expressing rsEGFP2 fusion proteins at endogenous levels [76]. Likewise, rsEGFP2 was used for RESOLFT nanoscopy in living *Drosophila melanogaster* tissues and even in intact transgenic second instar fly larvae [6] (Fig. 9.5e). Using rsEGFP2, tens of images with a resolution below 50 nm can be taken for small fields of view with a frame rate of a few seconds. To improve specific characteristics, several variants such as rsGreen and rsFolder were generated and used for RESOLFT nanoscopy [14, 25]. The fast switching Dronpa variant Dronpa-M159T has similar switching kinetics as rsEGFP2, and it has also been used in a few RESOLFT applications [61, 77]. Next to these negative-switching RSFPs, the positive-switching RSFP Kohinoor [29] and the decoupled RSFP Dreiklang were used for cellular RESOLFT imaging [23, 31, 38].

Several strategies have been applied to perform dual color live-cell RESOLFT nanoscopy, including an approach that exploited the different switching speeds and fluorescence lifetimes of two RSFPs [78]. Despite the fact that the red fluorescent proteins are outperformed by the green ones, some studies utilized a red fluorescent RSFP in combination with a green [59] or yellow fluorescent RSFP [38] for dual label RESOLFT imaging (Fig. 9.5c).

Because of the rather long dwell times, single beam scanning RESOLFT nanoscopy of large fields of view is rather slow and recording of a single image may take several tens of seconds. To address this limitation, several parallelization strategies have been successfully implemented. The first realization of parallelized RESOLFT relied on incoherently superimposed orthogonal standing light waves that allowed to image whole cells in a few seconds [56], and other strategies were reported subsequently (Fig. 9.5b, d) [38, 74, 79]. RESOLFT nanoscopy was also implemented in a light sheet approach [80] and has been combined with STED nanoscopy [52].

9.6.1   Other Fluorophores for RESOLFT Nanoscopy

Most current realizations of RESOLFT imaging used RSFPs from the GFP family. Only recently, several other photochromic fluorophores have been utilized for RESOLFT-type nanoscopy. This includes rsLov1, an engineered reversibly switchable protein based on the bacterial photoreceptor YtvA from *Bacillus subtilis* that binds flavin mononucleotide as a chromophore [81]. Also, organic fluorophores were successfully used, including carboxylated photoswitchable diarylethenes [82] and Cy3-Alexa647 heterodimers [83]. Currently, however, these novel probes are outperformed by RSFPs based on the GFP backbone, with regard to basically all photophysical parameters important for live-cell RESOLFT nanoscopy.

9.7   Outlook

RESOLFT nanoscopy opens up very flexible imaging schemes. For example, the attained resolution can be adjusted so that the resolution is decreased in favor of reduced phototoxicity, or of an increased recording speed, or in favor of the number of images taken before bleaching. This flexibility, in combination with the relatively low light intensities required, makes RESOLFT nanoscopy an exciting option for live-cell applications.

Recent developments enabling a parallelization of the method allowed to overcome the rather low imaging speed of single-beam solutions. Still, the potential of RESOLFT nanoscopy has not been fully exploited as yet. This is mainly due to the lack of suitable RESOLFT probes emitting in the red and infrared spectral regime, which hinders the development of robust multi-color RESOLFT applications. We predict that this gap will be closed in the near future, as a number of promising templates for the development of such probes are emerging [84].

References

1. Sahl, S.J., Hell, S.W., Jakobs, S.: Fluorescence nanoscopy in cell biology. Nat. Rev. Mol. Cell Biol. **18**(11), 685–701 (2017)
2. Hell, S.W.: Toward fluorescence nanoscopy. Nat. Biotechnol. **21**(11), 1347–1355 (2003)
3. Sigal, Y.M., Zhou, R., Zhuang, X.: Visualizing and discovering cellular structures with super-resolution microscopy. Science **361**(6405), 880–887 (2018)
4. Hell, S.W., Dyba, M., Jakobs, S.: Concepts for nanoscale resolution in fluorescence microscopy. Curr. Opin. Neurobiol. **14**(5), 599–609 (2004)
5. Grotjohann, T., Testa, I., Reuss, M., Brakemann, T., Eggeling, C., Hell, S.W., Jakobs, S.: rsEGFP2 enables fast RESOLFT nanoscopy of living cells. Elife **1**, e00248 (2012)
6. Schnorrenberg, S., Grotjohann, T., Vorbruggen, G., Herzig, A., Hell, S.W., Jakobs, S.: In vivo super-resolution RESOLFT microscopy of drosophila melanogaster. Elife **5** (2016)
7. Waldchen, S., Lehmann, J., Klein, T., van de Linde, S., Sauer, M.: Light-induced cell damage in live-cell super-resolution microscopy. Sci. Rep. **5**, 15348 (2015)
8. Tsien, R.Y.: The green fluorescent protein. Ann. Rev. Biochem. **67**, 509–544 (1998)
9. Andresen, M., Stiel, A.C., Folling, J., Wenzel, D., Schonle, A., Egner, A., Eggeling, C., Hell, S.W., Jakobs, S.: Photoswitchable fluorescent proteins enable monochromatic multilabel imaging and dual color fluorescence nanoscopy. Nat. Biotechnol. **26**(9), 1035–1040 (2008)
10. Andresen, M., Stiel, A.C., Trowitzsch, S., Weber, G., Eggeling, C., Wahl, M.C., Hell, S.W., Jakobs, S.: Structural basis for reversible photoswitching in dronpa. Proc. Natl. Acad. Sci. USA **104**(32), 13005–13009 (2007)
11. Andresen, M., Wahl, M.C., Stiel, A.C., Grater, F., Schafer, L.V., Trowitzsch, S., Weber, G., Eggeling, C., Grubmuller, H., Hell, S.W., Jakobs, S.: Structure and mechanism of the reversible photoswitch of a fluorescent protein. Proc. Natl. Acad. Sci. USA **102**(37), 13070–13074 (2005)
12. Brakemann, T., Weber, G., Andresen, M., Groenhof, G., Stiel, A.C., Trowitzsch, S., Eggeling, C., Grubmuller, H., Hell, S.W., Wahl, M.C., Jakobs, S.: Molecular basis of the light-driven switching of the photochromic fluorescent protein padron. J. Biol. Chem. **285**(19), 14603–14609 (2010)
13. Duan, C., Adam, V., Byrdin, M., Bourgeois, D.: Structural basis of photoswitching in fluorescent proteins. Methods Mol. Biol. **1148**, 177–202 (2014)
14. El Khatib, M., Martins, A., Bourgeois, D., Colletier, J.P., Adam, V.: Rational design of ultra-stable and reversibly photoswitchable fluorescent proteins for super-resolution imaging of the bacterial periplasm. Sci. Rep. **6**, 18459 (2016)
15. Hutchison, C.D.M., Cordon-Preciado, V., Morgan, R.M.L., Nakane, T., Ferreira, J., Dorlhiac, G., Sanchez-Gonzalez, A., Johnson, A.S., Fitzpatrick, A., Fare, C., Marangos, J.P., Yoon, C.H., Hunter, M.S., DePonte, D.P., Boutet, S., Owada, S., Tanaka, R., Tono, K., Iwata, S., van Thor, J.J.: X-ray free electron laser determination of crystal structures of dark and light states of a reversibly photoswitching fluorescent protein at room temperature. Int. J. Mol. Sci. **18**(9) (2017)
16. Pletnev, S., Subach, F.V., Dauter, Z., Wlodawer, A., Verkhusha, V.V.: A structural basis for reversible photoswitching of absorbance spectra in red fluorescent protein rsTagRFP. J. Mol. Biol. **417**(3), 144–151 (2012)

17. Bourgeois, D.: Deciphering structural photophysics of fluorescent proteins by kinetic crystallography. Int. J. Mol. Sci. **18**(6), 1187 (2017)
18. Coquelle, N., Sliwa, M., Woodhouse, J., Schiro, G., Adam, V., Aquila, A., Barends, T.R.M., Boutet, S., Byrdin, M., Carbajo, S., De la Mora, E., Doak, R.B., Feliks, M., Fieschi, F., Foucar, L., Guillon, V., Hilpert, M., Hunter, M.S., Jakobs, S., Koglin, J.E., Kovacsova, G., Lane, T.J., Levy, B., Liang, M., Nass, K., Ridard, J., Robinson, J.S., Roome, C.M., Ruckebusch, C., Seaberg, M., Thepaut, M., Cammarata, M., Demachy, I., Field, M., Shoeman, R.L., Bourgeois, D., Colletier, J.P., Schlichting, I., Weik, M.: Chromophore twisting in the excited state of a photoswitchable fluorescent protein captured by time-resolved serial femtosecond crystallography. Nat. Chem. **10**(1), 31–37 (2018)
19. Warren, M.M., Kaucikas, M., Fitzpatrick, A., Champion, P., Sage, J.T., van Thor, J.J.: Ground-state proton transfer in the photoswitching reactions of the fluorescent protein dronpa. Nat. Commun. **4**, 1461 (2013)
20. Yadav, D., Lacombat, F., Dozova, N., Rappaport, F., Plaza, P., Espagne, A.: Real-time monitoring of chromophore isomerization and deprotonation during the photoactivation of the fluorescent protein dronpa. J. Phys. Chem. B **119**(6), 2404–2414 (2015)
21. Bourgeois, D., Adam, V.: Reversible photoswitching in fluorescent proteins: a mechanistic view. IUBMB Life **64**(6), 482–491 (2012)
22. Petersen, J., Wilmann, P.G., Beddoe, T., Oakley, A.J., Devenish, R.J., Prescott, M., Rossjohn, J.: The 2.0-Å crystal structure of eqFP611, a far red fluorescent protein from the sea anemone entacmaea quadricolor. J. Biol. Chem. **278**(45), 44626–44631 (2003)
23. Brakemann, T., Stiel, A.C., Weber, G., Andresen, M., Testa, I., Grotjohann, T., Leutenegger, M., Plessmann, U., Urlaub, H., Eggeling, C., Wahl, M.C., Hell, S.W., Jakobs, S.: A reversibly photoswitchable GFP-like protein with fluorescence excitation decoupled from switching. Nat. Biotechnol. **29**(10), 942–947 (2011)
24. Arai, Y., Takauchi, H., Ogami, Y., Fujiwara, S., Nakano, M., Matsuda, T., Nagai, T.: Spontaneously blinking fluorescent protein for simple single laser super-resolution live cell imaging. ACS Chem. Biol. **13**(8), 1938–1943 (2018)
25. Duwe, S., De Zitter, E., Gielen, V., Moeyaert, B., Vandenberg, W., Grotjohann, T., Clays, K., Jakobs, S., Van Meervelt, L., Dedecker, P.: Expression-enhanced fluorescent proteins based on enhanced green fluorescent protein for super-resolution microscopy. ACS Nano **9**(10), 9528–9541 (2015)
26. Colletier, J.P., Sliwa, M., Gallat, F.X., Sugahara, M., Guillon, V., Schiro, G., Coquelle, N., Woodhouse, J., Roux, L., Gotthard, G., Royant, A., Uriarte, L.M., Ruckebusch, C., Joti, Y., Byrdin, M., Mizohata, E., Nango, E., Tanaka, T., Tono, K., Yabashi, M., Adam, V., Cammarata, M., Schlichting, I., Bourgeois, D., Weik, M.: Serial femtosecond crystallography and ultrafast absorption spectroscopy of the photoswitchable fluorescent protein IrisFP. J. Phys. Chem. Lett. **7**(5), 882–887 (2016)
27. Stiel, A.C., Andresen, M., Bock, H., Hilbert, M., Schilde, J., Schonle, A., Eggeling, C., Egner, A., Hell, S.W., Jakobs, S.: Generation of monomeric reversibly switchable red fluorescent proteins for far-field fluorescence nanoscopy. Biophys. J. **95**(6), 2989–2997 (2008)
28. Lukyanov, K.A., Fradkov, A.F., Gurskaya, N.G., Matz, M.V., Labas, Y.A., Savitsky, A.P., Markelov, M.L., Zaraisky, A.G., Zhao, X., Fang, Y., Tan, W., Lukyanov, S.A.: Natural animal coloration can be determined by a nonfluorescent green fluorescent protein homolog. J. Biol. Chem. **275**(34), 25879–25882 (2000)
29. Tiwari, D.K., Arai, Y., Yamanaka, M., Matsuda, T., Agetsuma, M., Nakano, M., Fujita, K., Nagai, T.: A fast- and positively photoswitchable fluorescent protein for ultralow-laser-power resolft nanoscopy. Nat. Methods **12**(6), 515–518 (2015)
30. Lacombat, F., Plaza, P., Plamont, M.A., Espagne, A.: Photoinduced chromophore hydration in the fluorescent protein dreiklang is triggered by ultrafast excited-state proton transfer coupled to a low-frequency vibration. J. Phys. Chem. Lett. **8**(7), 1489–1495 (2017)
31. Jensen, N.A., Danzl, J.G., Willig, K.I., Lavoie-Cardinal, F., Brakemann, T., Hell, S.W., Jakobs, S.: Coordinate-targeted and coordinate-stochastic super-resolution microscopy with the reversibly switchable fluorescent protein dreiklang. Chemphyschem **15**(4), 756–762 (2014)

32. Cranfill, P.J., Sell, B.R., Baird, M.A., Allen, J.R., Lavagnino, Z., de Gruiter, H.M., Kremers, G.J., Davidson, M.W., Ustione, A., Piston, D.W.: Quantitative assessment of fluorescent proteins. Nat. Methods 13(7), 557–562 (2016)
33. Heppert, J.K., Dickinson, D.J., Pani, A.M., Higgins, C.D., Steward, A., Ahringer, J., Kuhn, J.R., Goldstein, B.: Comparative assessment of fluorescent proteins for in vivo imaging in an animal model system. Mol. Biol. Cell 27(22), 3385–3394 (2016)
34. Pennacchietti, F., Serebrovskaya, E.O., Faro, A.R., Shemyakina II, Bozhanova, N.G., Kotlobay, A.A., Gurskaya, N.G., Boden, A., Dreier, J., Chudakov, D.M., Lukyanov, K.A., Verkhusha, V.V., Mishin, A.S., Testa, I.: Fast reversibly photoswitching red fluorescent proteins for live-cell resolft nanoscopy. Nat. Methods 15(8), 601–604 (2018)
35. Wang, S., Chen, X., Chang, L., Ding, M., Xue, R., Duan, H., Sun, Y.: Gmars-t enabling multimodal subdiffraction structural and functional fluorescence imaging in live cells. Anal. Chem. 90(11), 6626–6634 (2018)
36. Faro, A.R., Carpentier, P., Jonasson, G., Pompidor, G., Arcizet, D., Demachy, I., Bourgeois, D.: Low-temperature chromophore isomerization reveals the photoswitching mechanism of the fluorescent protein padron. J. Am. Chem. Soc. 133(41), 16362–16365 (2011)
37. Habuchi, S., Ando, R., Dedecker, P., Verheijen, W., Mizuno, H., Miyawaki, A., Hofkens, J.: Reversible single-molecule photoswitching in the gfp-like fluorescent protein dronpa. Proc. Natl. Acad. Sci. USA 102(27), 9511–9516 (2005)
38. Chmyrov, A., Leutenegger, M., Grotjohann, T., Schonle, A., Keller-Findeisen, J., Kastrup, L., Jakobs, S., Donnert, G., Sahl, S.J., Hell, S.W.: Achromatic light patterning and improved image reconstruction for parallelized resolft nanoscopy. Sci. Rep. 7, 44619 (2017)
39. Zhang, X., Zhang, M., Li, D., He, W., Peng, J., Betzig, E., Xu, P.: Highly photostable, reversibly photoswitchable fluorescent protein with high contrast ratio for live-cell superresolution microscopy. Proc. Natl. Acad. Sci. USA 113(37), 10364–10369 (2016)
40. Hell, S.W.: Microscopy and its focal switch. Nat. Methods 6(1), 24–32 (2009)
41. Duan, C., Adam, V., Byrdin, M., Ridard, J., Kieffer-Jaquinod, S., Morlot, C., Arcizet, D., Demachy, I., Bourgeois, D.: Structural evidence for a two-regime photobleaching mechanism in a reversibly switchable fluorescent protein. J. Am. Chem. Soc. 135(42), 15841–15850 (2013)
42. Grotjohann, T., Testa, I., Leutenegger, M., Bock, H., Urban, N.T., Lavoie-Cardinal, F., Willig, K.I., Eggeling, C., Jakobs, S., Hell, S.W.: Diffraction-unlimited all-optical imaging and writing with a photochromic GFP. Nature 478(7368), 204–208 (2011)
43. Ando, R., Flors, C., Mizuno, H., Hofkens, J., Miyawaki, A.: Highlighted generation of fluorescence signals using simultaneous two-color irradiation on dronpa mutants. Biophys. J. 92(12), L97–L99 (2007)
44. Stiel, A.C., Trowitzsch, S., Weber, G., Andresen, M., Eggeling, C., Hell, S.W., Jakobs, S., Wahl, M.C.: 1.8 a bright-state structure of the reversibly switchable fluorescent protein dronpa guides the generation of fast switching variants. Biochem. J. 402(1), 35–42 (2007)
45. Zhou, X.X., Lin, M.Z.: Photoswitchable fluorescent proteins: ten years of colorful chemistry and exciting applications. Curr. Opin. Chem. Biol. 17(4), 682–690 (2013)
46. Ando, R., Mizuno, H., Miyawaki, A.: Regulated fast nucleocytoplasmic shuttling observed by reversible protein highlighting. Science 306(5700), 1370–1373 (2004)
47. Subach, F.V., Zhang, L., Gadella, T.W., Gurskaya, N.G., Lukyanov, K.A., Verkhusha, V.V.: Red fluorescent protein with reversibly photoswitchable absorbance for photochromic fret. Chem. Biol. 17(7), 745–755 (2010)
48. Zhou, X.X., Chung, H.K., Lam, A.J., Lin, M.Z.: Optical control of protein activity by fluorescent protein domains. Science 338(6108), 810–814 (2012)
49. Vetschera, P., Mishra, K., Fuenzalida-Werner, J.P., Chmyrov, A., Ntziachristos, V., Stiel, A.C.: Characterization of reversibly switchable fluorescent proteins in optoacoustic imaging. Anal. Chem. 90(17), 10527–10535 (2018)
50. Geissbuehler, S., Sharipov, A., Godinat, A., Bocchio, N.L., Sandoz, P.A., Huss, A., Jensen, N.A., Jakobs, S., Enderlein, J., Gisou van der Goot, F., Dubikovskaya, E.A., Lasser, T., Leutenegger, M.: Live-cell multiplane three-dimensional super-resolution optical fluctuation imaging. Nat. Commun. 5, 5830 (2014)

51. Zhang, X., Chen, X., Zeng, Z., Zhang, M., Sun, Y., Xi, P., Peng, J., Xu, P.: Development of a reversibly switchable fluorescent protein for super-resolution optical fluctuation imaging (SOFI). ACS Nano 9(3), 2659–2667 (2015)
52. Danzl, J.G., Sidenstein, S.C., Gregor, C., Urban, N.T., Ilgen, P., Jakobs, S., Hell, S.W.: Coordinate-targeted fluorescence nanoscopy with multiple off states. Nat. Photonics 10, 122 (2016)
53. Willig, K.I., Stiel, A.C., Brakemann, T., Jakobs, S., Hell, S.W.: Dual-label sted nanoscopy of living cells using photochromism. Nano Lett. 11(9), 3970–3973 (2011)
54. Wang, S., Chen, X., Chang, L., Xue, R., Duan, H., Sun, Y.: Gmars-Q enables long-term live-cell parallelized reversible saturable optical fluorescence transitions nanoscopy. ACS Nano 10(10), 9136–9144 (2016)
55. Adam, V., Moeyaert, B., David, C.C., Mizuno, H., Lelimousin, M., Dedecker, P., Ando, R., Miyawaki, A., Michiels, J., Engelborghs, Y., Hofkens, J.: Rational design of photoconvertible and biphotochromic fluorescent proteins for advanced microscopy applications. Chem. Biol. 18(10), 1241–1251 (2011)
56. Chmyrov, A., Keller, J., Grotjohann, T., Ratz, M., d'Este, E., Jakobs, S., Eggeling, C., Hell, S.W.: Nanoscopy with more than 100,000 'doughnuts'. Nat. Methods 10(8), 737–740 (2013)
57. Chang, H., Zhang, M., Ji, W., Chen, J., Zhang, Y., Liu, B., Lu, J., Zhang, J., Xu, P., Xu, T.: A unique series of reversibly switchable fluorescent proteins with beneficial properties for various applications. Proc. Natl. Acad. Sci. USA 109(12), 4455–4460 (2012)
58. Fuchs, J., Bohme, S., Oswald, F., Hedde, P.N., Krause, M., Wiedenmann, J., Nienhaus, G.U.: A photoactivatable marker protein for pulse-chase imaging with superresolution. Nat. Methods 7(8), 627–630 (2010)
59. Lavoie-Cardinal, F., Jensen, N.A., Westphal, V., Stiel, A.C., Chmyrov, A., Bierwagen, J., Testa, I., Jakobs, S., Hell, S.W.: Two-color resolft nanoscopy with green and red fluorescent photochromic proteins. Chemphyschem 15(4), 655–663 (2014)
60. Patterson, G.H., Knobel, S.M., Sharif, W.D., Kain, S.R., Piston, D.W.: Use of the green fluorescent protein and its mutants in quantitative fluorescence microscopy. Biophys. J. 73(5), 2782–2790 (1997)
61. Testa, I., Urban, N.T., Jakobs, S., Eggeling, C., Willig, K.I., Hell, S.W.: Nanoscopy of living brain slices with low light levels. Neuron 75(6), 992–1000 (2012)
62. Gurskaya, N.G., Verkhusha, V.V., Shcheglov, A.S., Staroverov, D.B., Chepurnykh, T.V., Fradkov, A.F., Lukyanov, S., Lukyanov, K.A.: Engineering of a monomeric green-to-red photoactivatable fluorescent protein induced by blue light. Nat. Biotechnol. 24(4), 461–465 (2006)
63. Wiedenmann, J., Ivanchenko, S., Oswald, F., Schmitt, F., Rocker, C., Salih, A., Spindler, K.D., Nienhaus, G.U.: EosFP, a fluorescent marker protein with uv-inducible green-to-red fluorescence conversion. Proc. Natl. Acad. Sci. USA 101(45), 15905–15910 (2004)
64. Wiedenmann, J., Gayda, S., Adam, V., Oswald, F., Nienhaus, K., Bourgeois, D., Nienhaus, G.U.: From EosFP to mIrisFP: structure-based development of advanced photoactivatable marker proteins of the GFP-family. J. Biophotonics 4(6), 377–390 (2011)
65. Moeyaert, B., Bich, N.N., De Zitter, E., Rocha, S., Clays, K., Mizuno, H., van Meervelt, L., Hofkens, J., Dedecker, P.: Green-to-red photoconvertible dronpa mutant for multimodal superresolution fluorescence microscopy. ACS Nano 8(2), 1664–1673 (2014)
66. McKinney, S.A., Murphy, C.S., Hazelwood, K.L., Davidson, M.W., Looger, L.L.: A bright and photostable photoconvertible fluorescent protein. Nat. Methods 6(2), 131–133 (2009)
67. Zhang, M., Chang, H., Zhang, Y., Yu, J., Wu, L., Ji, W., Chen, J., Liu, B., Lu, J., Liu, Y., Zhang, J., Xu, P., Xu, T.: Rational design of true monomeric and bright photoactivatable fluorescent proteins. Nat. Methods 9(7), 727–729 (2012)
68. Wang, S., Moffitt, J.R., Dempsey, G.T., Xie, X.S., Zhuang, X.: Characterization and development of photoactivatable fluorescent proteins for single-molecule-based superresolution imaging. Proc. Natl. Acad. Sci. USA 111(23), 8452–8457 (2014)
69. Wang, S., Shuai, Y., Sun, C., Xue, B., Hou, Y., Su, X., Sun, Y.: Lighting up live cells with smart genetically encoded fluorescence probes from gmars family. ACS Sens. 3(11), 2269–2277 (2018)

70. Griesbeck, O., Baird, G.S., Campbell, R.E., Zacharias, D.A., Tsien, R.Y.: Reducing the environmental sensitivity of yellow fluorescent protein. Mechanism and applications. J. Biol. Chem. **276**(31), 29188–29194 (2001)
71. Hofmann, M., Eggeling, C., Jakobs, S., Hell, S.W.: Breaking the diffraction barrier in fluorescence microscopy at low light intensities by using reversibly photoswitchable proteins. Proc. Natl. Acad. Sci. USA **102**(49), 17565–17569 (2005)
72. Merzlyak, E.M., Goedhart, J., Shcherbo, D., Bulina, M.E., Shcheglov, A.S., Fradkov, A.F., Gaintzeva, A., Lukyanov, K.A., Lukyanov, S., Gadella, T.W., Chudakov, D.M.: Bright monomeric red fluorescent protein with an extended fluorescence lifetime. Nat. Methods **4**(7), 555–557 (2007)
73. Ilgen, P., Grotjohann, T., Jans, D.C., Kilisch, M., Hell, S.W., Jakobs, S.: Resolft nanoscopy of fixed cells using a z-domain based fusion protein for labelling. PLoS One **10**(9), e0136233 (2015)
74. Masullo, L.A., Boden, A., Pennacchietti, F., Coceano, G., Ratz, M., Testa, I.: Enhanced photon collection enables four dimensional fluorescence nanoscopy of living systems. Nat. Commun. **9**(1), 3281 (2018)
75. Wang, S., Ding, M., Chen, X., Chang, L., Sun, Y.: Development of bimolecular fluorescence complementation using rsEGFP2 for detection and super-resolution imaging of protein-protein interactions in live cells. Biomed. Opt. Express **8**(6), 3119–3131 (2017)
76. Ratz, M., Testa, I., Hell, S.W., Jakobs, S.: Crispr/cas9-mediated endogenous protein tagging for resolft super-resolution microscopy of living human cells. Sci. Rep. **5**, 9592 (2015)
77. Bohm, U., Hell, S.W., Schmidt, R.: 4Pi-resolft nanoscopy. Nat. Commun. **7**, 10504 (2016)
78. Testa, I., D'Este, E., Urban, N.T., Balzarotti, F., Hell, S.W.: Dual channel resolft nanoscopy by using fluorescent state kinetics. Nano Lett. **15**(1), 103–106 (2015)
79. Li, D., Shao, L., Chen, B.C., Zhang, X., Zhang, M., Moses, B., Milkie, D.E., Beach, J.R., Hammer, J.A. 3rd, Pasham, M., Kirchhausen, T., Baird, M.A., Davidson, M.W., Xu, P., Betzig, E.: Advanced imaging. Extended-resolution structured illumination imaging of endocytic and cytoskeletal dynamics. Science **349**(6251), aab3500 (2015)
80. Hoyer, P., de Medeiros, G., Balazs, B., Norlin, N., Besir, C., Hanne, J., Krausslich, H.G., Engelhardt, J., Sahl, S.J., Hell, S.W., Hufnagel, L.: Breaking the diffraction limit of light-sheet fluorescence microscopy by resolft. Proc. Natl. Acad. Sci. USA **113**(13), 3442–3446 (2016)
81. Gregor, C., Sidenstein, S.C., Andresen, M., Sahl, S.J., Danzl, J.G., Hell, S.W.: Novel reversibly switchable fluorescent proteins for RESOLFT and STED nanoscopy engineered from the bacterial photoreceptor YtvA. Sci. Rep. **8**(1), 2724 (2018)
82. Roubinet, B., Bossi, M.L., Alt, P., Leutenegger, M., Shojaei, H., Schnorrenberg, S., Nizamov, S., Irie, M., Belov, V.N., Hell, S.W.: Carboxylated photoswitchable diarylethenes for biolabeling and super-resolution resolft microscopy. Angew. Chem. Int. Ed. Engl. **55**(49), 15429–15433 (2016)
83. Kwon, J., Hwang, J., Park, J., Han, G.R., Han, K.Y., Kim, S.K.: Resolft nanoscopy with photoswitchable organic fluorophores. Sci. Rep. **5**, 17804 (2015)
84. Lychagov, V.V., Shemetov, A.A., Jimenez, R., Verkhusha, V.V.: Microfluidic system for in-flow reversible photoswitching of near-infrared fluorescent proteins. Anal. Chem. **88**(23), 11821–11829 (2016)

Chapter 10
A Statistical and Biophysical Toolbox to Elucidate Structure and Formation of Stress Fibers

Benjamin Eltzner, Lara Hauke, Stephan Huckemann, Florian Rehfeldt and Carina Wollnik

Abstract We are concerned with statistically validated early mechanically guided differentiation of human mesenchymal stem cells (hMSCs). This chapter reviews and extends methods of fixed and live imaging of hMSCs, automated reliable and unbiased near real-time filament extraction and digitization for massive data via the FilamentSensor, suitable aggregation of simple (area, mean orientation, aspect ratio and order parameter) and advanced (orientation mode persistence and orientation fields) data descriptors and methods of their non-euclidean inferential statistics. Exemplary, we study the morphology of stress fibers in fixed and live hMSCs within 24 h post seeding on elastic matrices exhibiting Young's moduli of 1 kPa (soft, brain-like elasticity), 11 kPa (intermediate, muscle-like stiffness) and 30 kPa (hard, pre-calcified bone rigidity). The combination of these methods constitutes a novel integrated toolbox, where for instance, statistical insight may be used to guide experimental design.

2010 Mathematics Subject Classification: 62H 35 · 62 H 11 · 62 H 30 · 65D18

PACS Subject Classification: 87.16.-b, 87.80.Rb, 87.18.Hf, 87.68.+z

B. Eltzner (✉) · S. Huckemann
Felix-Bernstein-Institute for Mathematical Statistics in the Biosciences, Universität Göttingen, Göttingen, Germany
e-mail: benjamin.eltzner@mathematik.uni-goettingen.de

S. Huckemann
e-mail: stephan.huckemann@mathematik.uni-goettingen.de

L. Hauke · F. Rehfeldt · C. Wollnik
Third Institut of Physics - Biophysics, Universität Göttingen, Göttingen, Germany
e-mail: lara.hauke@phys.uni-goettingen.de

F. Rehfeldt
e-mail: rehfeldt@physik3.gwdg.de

C. Wollnik
e-mail: carina.wollnik@gmx.de

© The Author(s) 2020
T. Saldit et al. (eds.), *Nanoscale Photonic Imaging*, Topics in Applied Physics 134,
https://doi.org/10.1007/978-3-030-34413-9_10

10.1 Introduction

During the last two decades it has become evident that the mechanical properties of the cellular micro-environment are as important for cellular behavior and home-ostasis as traditionally investigated biochemical cues [1, 2]. Especially striking was the finding that differentiation of human mesenchymal stem cells (hMSCs) can be mechanically induced by culturing them on elastic substrates of different Young's moduli E [3]. While upregulation of specific differentiation markers is typically observed after five or more days, fundamental mechanical interactions between cells and the substrates take place immediately after adhesion on the substrate. Interest-ingly, during this early stage (within the first 24 h) of this mechano-guided differen-tiation process in hMSCs, the structure and polarization of actin-myosin *stress fibers* as quantified by an order parameter S depend critically on Young's modulus E [4].

Stress fibers are contractile structures mainly composed of actin *filaments*, myosin motor mini filaments (in particular non-muscle myosin II isoforms) and distinct types of actin cross-linking proteins (e.g. α-actinin, fascin, etc.). They play the role of 'cellular muscles' generating contractile forces and connecting to the *extracellular matrix* (ECM) via focal adhesions, thereby also transmitting forces to the ECM [5]. Acto-myosin filaments are also considered to be the principal force sensors of the cell that translate mechanical cues from the surroundings into biochemical signaling, eventually leading to cell differentiation [2, 6]. Previous experiments with fixed cells revealed the important role of acto-myosin cytoskeleton structure formation for the mechanically induced differentiation of hMSCs [4, 7].

Building statistically validated models and theories linking substrate elasticity to early hMSCs differentiation, the filament structure has to be visualized over time, binarized in an unbiased fashion, aggregated into descriptors and analyzed, possibly within a feedback loop. Due to high biological diversity, large amounts of data are required for statistical power. In turn, such massive data require near real-time processing.

In order to visualize these filaments selectively, fluorescence microscopy proves useful and typical images of acto-myosin stress fibers of different quality in fixed cells are displayed in Fig. 10.1. One of the main differences, however, of experimental visualization of the acto-myosin cytoskeleton in fixed cells at particular time points and life cell imaging is given by the signal to noise ratio of the microscope images. Fixed cells can be stained with many different methods and allow for saturation with fluorescent dyes that typically lead to nice and crisp images (see top row of Fig. 10.1). In contrast, life cell imaging, as detailed in Sect. 10.2, relies on transfection with fluorescent fusion proteins that need to be expressed in the cell and typically leads to worse signal to noise ratios (see bottom row of Fig. 10.1), that are challenging for subsequent image processing.

Once these images are obtained, the challenge consists in extracting the underlying filament structure in near real-time in an automated and unbiased fashion. To this end, we review the *FilamentSensor* (FS) from [8] in Sect. 10.3. It integrates general and specifically tailored preprocessing with an elaborate binarization routine, to identify

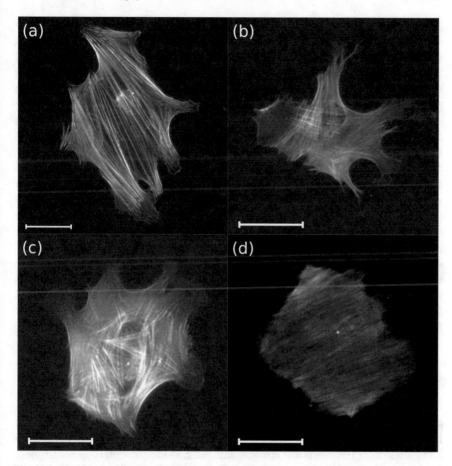

Fig. 10.1 Varying quality (top fixed, bottom live) filament expressions of fluorescence micrographs of human mesenchymal stem cells with scale bars at 50 μm. Subfigures: **a** *good quality* image of a fluorescently stained fixed cell of large size with clearly visible stress fibers on a substrate with a Young's modulus $E = 10$ kPa; **b** *medium quality* image of a fluorescently stained fixed cell of moderate size with inhomogeneous brightness and slight blur on glass; **c** *poor quality* image of a live cell of moderate size with considerable noise and excessive brightness due to overexposure on glass; **d** *very poor quality* image of a live cell of moderate size with very low contrast due to bleaching, considerable blur and hardly discernible stress fibers

filaments of varying widths, lengths and angles. We describe a novel algorithm able to detect slightly bent filaments. Since the FS is modular and open source, it can be easily extended to suit related image analysis tasks. Guaranteeing unbiasedness, tunable parameters can be learned, among others, from the *benchmark data set* (BDS) in [8] that has been manually labeled by specialists.

From the binarized filament structure various morphological descriptors can be extracted. Simple summary statistics are (weighted) mean orientation, area, aspect ratios of principal components (PCs). More subtle is the *order parameter* derived from the angle between extrinsic mean orientation and first PC, quantifying the anisotropy of the acto-myosin cytoskeleton, and [4] linked statistically significant substrate elasticity to stem cell differentiation. A more sophisticated analysis [9] links number and persistence (under smoothing) of modes (i.e. dominating directions) of the distribution of weighted filament orientations to substrate elasticity, and, in particular, this requires development of a causal circular scale space theory in [9].

The above descriptors coarsely describe common filament orientations. In a finer approach, the concept of different single *orientation fields* (OFs) described in Sect. 10.4, pays credit to the tendency of filament orientations to change in a spatially smooth way. In order to simultaneously analyze statistically different moments of their distribution (e.g. jointly the intrinsic mean on a first geodesics PC together with the first geodesic PC), *backward nested descriptor* (BND) analysis is applied in Sect. 10.4.2 along with its asymptotic theory from [10]. This allows to elucidate fundamental differences between fixed and live cell analysis, with consequences for experimental design.

In Sect. 10.5 we conclude with an outlook how our biostatistical toolbox can be used, in various combinations, to tackle problems that have arisen through this research and problems currently high interest, for example tracking of individual filament dynamics and defining and analyzing corresponding descriptors such as filament life times.

10.2   Live Cell Imaging-Opportunities and Challenges

As mentioned above, novel insights into mechanisms of the complex mechanical interplay between cells and the extracellular matrix require the analysis of the dynamics of stress fiber formation and arrangement. Such an experimental approach differs fundamentally from fluorescence microscopy of fixed cells. In experiments using chemically fixed cells, these can be stained with a variety of fluorescent dyes using either antibodies or other small molecules that selectively bind to the protein of choice. Living cells however need to be modified genetically to express a fluorescent protein fused to the protein of interest or a respective binding partner. While both methods allow for fluorescently labeled cellular structures a significant difference is the homogeneity of the fluorescent intensity: fixed cells will overall show similar intensities that depend on the staining method and the cellular concentration of the protein of choice; living cells that are transiently expressing the fluorescent marker will show a broad distribution of intensities (and even dark cells that do not express any fluorescent marker) due to the stochastic nature of the transfection process. In addition, it is essential to monitor many ($N > 30$) cells in parallel leading to an enormous amount of microscopic images. These issues pose distinct challenges to the image segmentation and analysis algorithm that could not be resolved with our traditional approach of a pixel-based eLoG orientation analysis [4] mainly due the varying image quality (as illustrated e.g. in Fig. 10.1) and also computational time.

In the case of fixed cells, we are using a rigorous protocol to ensure the unbiased microscopic analysis of single cells. Due to the intrinsic variation of cellular morphologies it is of paramount importance to exclude any human bias. First, cells are searched in the fluorescence channel of the nucleus dye to find an isolated cells, whose nucleus looks normal and has no direct neighbors. Here, it is critical to exclude deformed nuclei, nuclei with a doubled set of DNA (high intensity), or any nuclei that are within the division process. Next, the fluorescent channel of interest (e.g. actin, non-muscle myosin IIa) is recorded regardless of the morphology, except for cases where now neighboring cells are observed that might interfere.

In contrast to the above described protocol for fixed cells, the situation is more subtle for parallel live cell imaging. Firstly, cells need to be transfected with a fluorescent marker that tags the protein of choice (in our case actin). This can be done using a fluorescent fusion protein (e.g. GFP-actin) with the immediate drawback of over-expression of that protein, differences in assembly kinetics, and potential problems with incorporation in distinct actin structures [11]. Most of these issues can be avoided using LifeAct [12], a short amino-acid sequence that binds to actin and is fused with a fluorescent protein. However, direct comparison also leads to minor differences between this visualization and staining later fixed cells with a phalloidin dye [11].

To avoid influence from neighboring untransfected cells, it is advisable to also record always an image in phase contrast or brightfield mode. That way any unwanted additional interactions can be ruled out. However, during time lapse microscopy recordings several incidents can occur that might affect the statistical analysis of the cell population. Cells might migrate out of the field-of-view, an aspect that we will address by smart repositioning the sample with real-time analysis of the microscopic pictures. Therefore, the full recording will lack a subset of very motile cells. Cells might undergo apoptosis and exclude themselves from further analysis reducing the number of samples. Cells might divide and therefore make the analysis of their cytoskeletal dynamics very complicated, even precluding it in case of not thoroughly separating cells. Cells might interact during the time period with neighboring cells that will also affect their cell-matrix mechanical interplay and acto-myosin structure. Altogether, it becomes clear that the population subsets of microscopy analysis of fixed cells and living cells can differ significantly and appropriate measures and controls need to be developed to fully understand its impact on the statistical analysis.

10.3   Automated Unbiased Binarization of Filament Structure

The present section is heavily based on the authors' previous publication [8], which is published under an open license (CC BY 4.0). The text and contents from said publication is reproduced here to achieve a self-contained description of the FilamentSensor.

10.3.1   Related Work

There is an impressive body of techniques for image processing and in particular, for line detection (for an overview e.g. [13], Chap. 4). Previous to our development of the FS, however, methods for the detection of filaments in cell images were often ad hoc, required manual processing to a considerable extent and were computationally rather time consuming, e.g. [4, 14, 15]. The latter two issues are particularly unfavorable considering the large number of images to be processed in live cell imaging.

Moreover, there is a large number of algorithms focusing on analyzing networks of strongly curved microtubuli (this property is not shared by single filaments), such as line thinning by [16], active contours by [17] and the constrained inverse diffusion (CID) method by [14]. These methods, however, detect only a skeletal filament network structure, they leave out filament orientation, length, and width. They aim at identifying thin microfilaments and not wide stress fibers as we are interested in.

There are methods which aim at extraction not only of filament pixel position but also of local orientation such as the FiberScore algorithm by [18], elongated Laplacians of Gaussians (eLoGs) by [4] and gradient based methods, e.g. [15, 19].

The eLoG method, like the gradient method is geared towards the detection not only of filament pixels but also of their orientations. Although filament width and length are not extracted by these methods, counting the number of pixels per orientation, they yield histograms of cumulated filament length per orientation angle and these histograms are then further analyzed [4, 15].

Local orientation and centerline images are produced by the FiberScore program [18] which provides global information on accumulated line length and average width. Line objects are not produced, however. For our cell images, we tried out the methods applied in FiberScore, but did not yield optimal results [8]. A fundamental drawback for applying FiberScore, however, is that neither the program's nor its framework's source codes are freely available. Even though the original developer has been very helpful and supportive to make the program run, FiberScore could not be tailored to our needs.

With the FS we have developed an image processing tool that returns stress fiber structures from live cell images, as well as from fixed cells images applicable to the use case where images vary widely in brightness, contrast, sharpness and homogeneity of fluorescence, cf. Fig. 10.1. Typically, in our setup of live cell imaging, we observe 30 cells over a period of 24 h, taking an image every 10 min. As we aim at real time processing, this leaves about 20 s process time per image.

The FS thus developed can be used to binarize filament structures for any (sets of) images containing fiber features. Applications in a wide range of use cases come to mind, in particular in the context of actin fiber structures, e.g. [20, 21], but also for more general contexts in medical imaging, biology, and in the material science. As the FS is modular and easily extensible, several authors, including [22, 23], have built on it after it was first published.

Notably, there is also a rising demand to address the task of tracing and tracking stress fibers, both over space and over time. We mention studies on migrating cells which display a variety of stress fiber types (dorsal, ventral, arc) that appear at different loci inside a migrating cell [21, 24–28]. Their exact cellular function is still in the dark, it could be clarified, however, using live cell filament digitization. Indeed, when stress fiber dynamics are followed over time, this may give further insight into formation and function of filament structure. A novel method to analyze traction force microscopy data, so called model-based traction force microscopy has been recently described by [29]. In this context, it is necessary to detect and mark the stress fibers in a cell in order to link forces to fiber location and develop deeper insight into cellular force generation and transmission to the substrate. As mentioned before, ideally, corresponding live cell experiments are performed simultaneously for many cells in order to arrive at statistics that are sufficiently significant. This requires that algorithms for fiber analysis perform tracing and tracking in (nearly) real-time, ideally.

10.3.2 The FilamentSensor and the Benchmark Dataset

To obtain the full information of the stress fibers in cells, namely location, length, width, and orientation, from repeated observations of living cells under widely varying conditions in near real time the FS has to extract

(I) fast and unsupervised
(II) robustly
(III) all filament features: location, length, width and orientation;

where (II) implies dealing with several specific problems illustrated in Fig. 10.2

(IIa) detecting darker lines crossing bright lines,
(IIb) dealing with image inhomogeneities and
(IIc) dealing with image blur and noise.

The FS is specifically designed to meet these challenges. Dealing with image inhomogeneity calls for the application of local image processing tools. Blurring effects will be mitigated by line enhancement through direction sensitive methods. Crossings of lines of varying intensities can be successfully detected by what we call *line Gaussians* which utilize oriented thin masks. After local binarization, an adaption of the semilocal line sensor approach to fingerprint analysis [30] is applied to extract all filament features. As the FS is modularized, employs local and orientation dependent image analysis methods and outputs the entire filament data, expert knowledge such as detecting fewer filaments in specific low variance areas, say, can be easily incorporated.

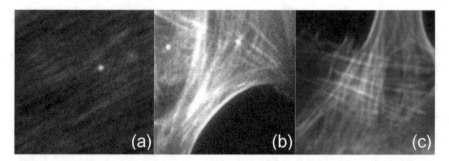

Fig. 10.2 Challenges for filament extraction. a blur (detail from Fig. 10.1d). The overall contrast of the cell body is very low and lines are hardly discernible. **b** overexposure and noise (Fig. 10.1c). The extensive regions of maximal brightness hide any structure that may be present in those regions. Salt and pepper noise is clearly visible as dark spots in bright areas and bright spots in dark areas. **c** filament crossings (Fig. 10.1b). A bundle of roughly vertical filaments of varying brightness crosses a bundle of roughly horizontal filaments with varying brightness

10.3.3 Detecting Slightly Bent Filaments

After preprocessing and binarization, as described in [8], filament data is extracted from the white pixels of the binarized image. Visual inspection of fluorescence microscopic images reveals that actin stress fibers can be slightly bent. To take this into account, we have adjusted the FS to follow slightly curved lines on a piecewise linear path. Line detection is performed by the following algorithm.

1. Every white pixel (x, y) is assigned a *width*, $W(x, y)$. This is done by taking circular pixel neighborhoods of the pixel (cf. Fig. 10.3) with increasing diameter. A diameter is accepted, if the ratio of white pixels of the binary image is above an adjustable tolerance (default 95%). If a diameter was accepted, the next larger diameter is proposed until a diameter is not accepted. The width $W(x, y)$ at the pixel is then given by the largest accepted diameter at the pixel. In particular, this gives a range of widths $1 \leq w_1 < \ldots < w_k = \max W(x, y)$ attained by pixels.

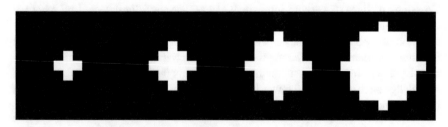

Fig. 10.3 Some circle masks. These are examples of the circular masks used by the segment sensor algorithm to determine line width. The circles displayed here correspond to diameters of 2, 4, 6 and 8 pixels. The masks are squares with an odd number of pixels as they are centered at a unique pixel

A temporary List L, the *filament data set* $\mathcal{F}$, and the *orientation field* $\mathcal{O}$ are each initialized by the empty set.

2. For every white pixel, starting with the highest width value and continuing with decreasing width value, apply the *CurveTracer (CT)*. The curves are represented piecewise linearly. The user specifies four parameters, namely the length of linear pieces l_{lin}, the direction step size in degrees ϕ_{step}, the maximal number of angle steps between two adjacent linear pieces n_{step}, leading to maximal angle of $\phi_{step} \cdot n_{step}$ between adjacent line pieces, and a minimal line length l_{min}.

   a. For each white pixel (x, y) the CT probes into a number of directions (by default $\phi_{step} = 3°$; this corresponds to 60 orientations $3°, 6°, \ldots, 360°$). For each direction the CT follows a straight line from (x, y) for a number of $2l_{lin}$ pixels. For each of these directions, the average width value is calculated and two almost opposite directions (with a relative angle in $[180° - n_{step}\phi_{step}, 180° + n_{step}\phi_{step}]$) with the largest combined average width are selected. For each of these two directions the CT now proceeds separately as follows, using a point list P containing the starting point.

      i. Move l_{lin} pixels in the current direction ϕ_c and add the end point p to P. If the average width along this line is below 1, remove all pixels with width 0 from the end of the line and then proceed removing pixels (possibly with width greater than 0) from the end until the average width is at least 1. Then add the new final point p of the line to P and stop.

      ii. From p probe for $2l_{lin}$ pixels into the $2n_{step} + 1$ directions $\{\phi_c - n_{step}\phi_{step}, \phi_c - (n_{step} - 1)\phi_{step}, \ldots, \phi_c + n_{step}\phi_{step}\}$ and calculate average width values for every direction. Set the new ϕ_c to the direction of highest average width.

      iii. Return to step 1.

   b. When the CT searches in both directions have reached their end points, the combined length of the line pieces is determined and if it is larger than l_{min} the list of points from both pieces is stored to L. The CT is illustrated in Fig. 10.4.

   c. In the next step, segments in L are called in the order of their length, long segments first. For every segment, the orientation field $\mathcal{O}$ (which is empty when first called) is looked up for every pixel on the segment. If less than 30 % of the segment's pixels have a *conflicting orientation* entry in $\mathcal{O}$,—i.e. the entry in $\mathcal{O}$ differs by less than an adjustable tolerance angle (per default 20°) from the segment's orientation—the segment is accepted as valid. For every pixel within a circular neighborhood with diameter $w_j + 2$ pixels (in order to avoid duplicate lines in case the CT does not perfectly follow the path of maximal width) of a segment pixel, the segment's orientation is stored to $\mathcal{O}$ overwriting possible previous entries. The segment is then also added to $\mathcal{F}$. If at least 30 % of the pixels on a segment have a conflicting orientation, we have the following cases.

      i. If $\mathcal{O}$ does not carry a conflicting orientation for any of the endpoints, the segment is discarded.

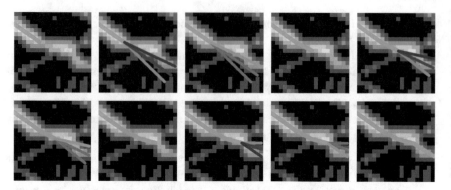

Fig. 10.4 Illustration of the CT algorithm. The probing distance is twice as far as the step length of the CT. Per default, the CT probes three directions at each step where the number of directions as well as the angle between them can be adjusted by the user. The probing directions are visualized in three shades of blue. The chosen direction is then marked green and the others are marked red

ii. Otherwise, the endpoints with conflicting orientations are iteratively removed from the segment until the remaining segment's endpoints no longer have a conflicting orientation. If the resultant segment length is above the threshold of minimal filament length, this new segment is added back to L and the original one is removed. The new segment is revisited when its length is called.

As lines are blurred due to scattering and as the preprocessing usually enhances line width, the FS tends to find greater line width than a human expert (cf. benchmark data set in [8]).

10.4 Orientation Fields

We make use of the individual line data provided by the FS to identify local orientation fields of fibers. These provide a more detailed picture of the cytoskeleton than simple summary statistics or orientation histograms, since they take local features into account. Orientation fields are contiguous regions in the cell filled with filaments of similar orientation. The local orientation of the field may change slowly over the cell, so as to encompass the case of a curved cell, where stress fibers follow the cell shape. As input we use an *orientation map* denoting the line orientations determined by the filament sensor at each pixel. Our algorithm uses relaxation labeling, first described by [31].

As a first step, the image is covered with a rectangular equidistant grid of pixels with spacing a. The grid points define *blocks* in the following. The minimal spacing of the grid is 5 pixels. While the number of blocks b covering the cell area in the image exceeds 500, the spacing is successively increased by steps of two pixels. As a result, the maximal number of non-trivial blocks is 500. In this way, the algorithm can deal with images of different magnification and cells of different size.

For every block, we use a square isotropic Gaussian mask with $\sigma = \max(15, \lfloor 1.2a \rfloor)$ centered at the grid point to assign weights to nearby pixels. The length of the mask from the center is $l = \lfloor 2.5\sigma \rfloor$ pixels in every direction, after which it is truncated. Using the orientation map and these weights, we get an orientation histogram for each block. This orientation histogram is then smoothed with a wrapped Gaussian kernel with $\sigma = 6°$. Of the smoothed histogram, all maxima are stored as local orientations of the block. If several neighboring bins of the smoothed histogram have the same, locally maximal, value, the leftmost bin, corresponding to the smallest angle, is used.

To perform relaxation labeling, it is necessary to have a number of blocks with fixed orientations that can be used as a seed. In order to achieve reliable results, it is desirable to have as many seed blocks as possible. Therefore, we try to determine at least $s = 0.05b$ seed blocks. For this we collect all blocks with only one orientation and apply the following *cleanup procedure*:

1. Keep only the largest contiguous region.
2. Make a histogram of the block orientation smoothed with a Gaussian kernel with $\sigma = 6°$. And determine the global maximum $\phi_{\max}$.
3. Starting with $k = 0$, we determine the largest contiguous region of blocks whose orientation ϕ satisfies $\phi - \phi_{\max} \leq k$. We then increase k by steps of 1, until we reach $k = 6$ or until the largest contiguous region of blocks reaches or exceeds the number of $s = 0.05b$ blocks.

If the number of seed blocks is smaller than s, we repeat the cleanup procedure for all blocks using all orientations of every block in step 2.

Once we have a set of seed blocks with seed orientations, we fix these orientations and perform a relaxation labeling over all orientations of all non-seed blocks. For the relaxation labeling, we use a von Mises type *compatibility function*

$$f(\phi) = C + B \exp(A \cos(2\phi))$$
$$\text{with} \quad f(0) = 1, \quad f(90) = -1, \quad f(\sigma) = 0$$

where we start with $\sigma = 15°$. If the largest field contains less than $2/3$ of line pixels or less than 85% of the blocks, we repeat the relaxation labeling, increasing σ by steps of 5° to a maximum of $\sigma = 25°$.

To preclude too large changes of orientation at medium range, we do not use only nearest neighbors for the relaxation, but every block reacts with isotropic Gaussian weights with surrounding blocks, where the standard deviation of the Gaussian is $\Sigma = 5$ blocks. Every block also has a dummy orientation, whose probability will slowly grow, if none of the block orientations match their neighborhood. Blocks neighboring on seed blocks and having reached a probability of 0.999 on one of their orientations will be turned into seed blocks, so the field will gradually "freeze" to ensure convergence.

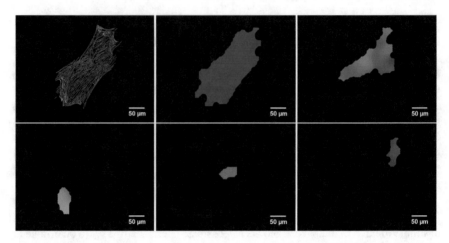

Fig. 10.5 Orientation field detection. Lines segments extracted by the FS (upper left panel) colored by local orientation, giving five dominating orientation fields (their range and local orientations displayed in the other panels). Indeed, colors of the fields, indicating their local orientation, vary only slightly

When the relaxation has converged, the resulting field is saved and its corresponding orientations are removed from the blocks. If the remaining nontrivial blocks do not form a contiguous region, all regions below minimal size s are removed. The procedure is then repeated until no non-trivial blocks remain. Finally, every filament is sorted into the orientation field whose local orientation best matches its own orientation (Fig. 10.5). If the filament orientation diverges by more than $15°$ from all local orientation fields, it is not associated to any field.

10.4.1 Orientation Field Evolution

In most cells on substrates with stiffness 10 and 30 kPa a single orientation field emerges over 24 h, which contains more than 80% of stress fiber length. In order to illustrate the evolution of orientation fields in time, we represent the orientation field at each time point by a gray circle, whose gray level displays the relative amount of fiber length represented by that field, such that the circle is black when all fibers are included in the field. The standard deviation of fiber orientations in the field are displayed by error bars for each circle. The evolution of orientation fields for a typical cell on an intermediate or stiff gel is displayed in Fig. 10.6.

In Fig. 10.7 we show a typical orientation field evolution for a cell on a soft substrate with stiffness 1 kPa. In cells on such soft substrates the cytoskeleton is much less ordered which is reflected by a large number of small orientation fields which are found over time. In a cell where fibers are not ordered, orientation fields

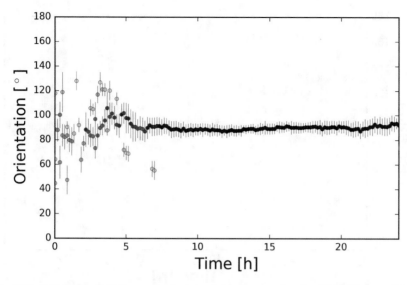

Fig. 10.6 **Orientation field time series for a typical cell on a substrate with stiffness** ≥ 10 kPa, at each time point represented by a gray circle, whose gray level displays the relative amount of fiber length represented by that field (the circle is black when all fibers are included in the field). The standard deviation of fiber orientations in the field are displayed by error bars for each circle. As is typical for a cell on a stiff substrate, a single main orientation field emerges after a few hours and remains stable throughout imaging time

often appear only for few images. While such fields can be considered spurious, they still serve to illustrate the disorder of the actin cytoskeleton.

In many cells on all gels the cytoskeleton is not fully described by just one orientation field but is partially ordered. A frequently observed evolution starts out with an almost chaotic cytoskeleton where the short lived small orientation field converge into a main orientation field over time, as in Figs. 10.6 and 10.7. This process can take between 4 and 20 h and can even be unfinished at the end of the 24 h observation span. However, there are also cases, where a main orientation field, which has remained stable for many hours suddenly disperses as the cell starts to move (corresponding to Figs. 10.6 and 10.7 with inverted time axis).

A behavior, which is observed in less than 5 % of cells, is illustrated in Fig. 10.8. In this case, a stable main orientation field exists, when at some point in time a new orientation field begins to form and the original main field starts to dissolve. This behavior requires more thorough investigation, both into the underlying cell dynamics and into adequate statistical representation of cytoskeleton order. Elucidating this behavior is left for future research.

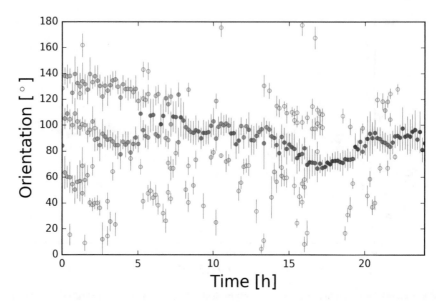

Fig. 10.7 Orientation field time series for a typical cell on a substrate with stiffness 1 kPa (notation as in Fig. 10.6) with several local orientation fields indicating a less ordered cytoskeleton. While over time a dominating orientation field emerges, smaller orientation fields pop out until the end

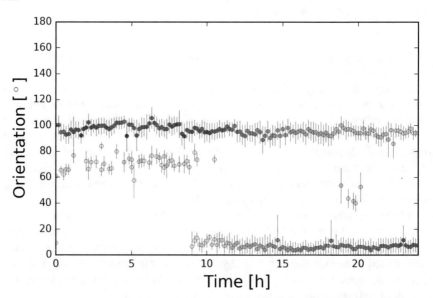

Fig. 10.8 Orientation field time series for a cell on a substrate with stiffness 10 kPa (notation as in Fig. 10.6). The initial main orientation field starts to decay after 10 h and is superseded by a new orientation field which is almost perpendicular to it

10.4.2   Backward Nested Descriptor Analysis

In order to use orientation fields for quantitative analysis, we devised the simple low dimensional orientation fields representation (10.1) in [10]. Denote by M the number of all filament pixels in a cell image, m_1 denotes the number of filament pixels in the largest orientation field and m_2 the number of pixels in all smaller fields combined. $M - m_1 - m_2$ then enumerates the pixels from "rogue" filaments which do not fit into any of the orientation fields due to strongly deviating orientation. In order to compare relative diameters rather than relative areas we observe the quantity

$$x = (x_1, x_2, x_3) := (\sqrt{m_1/M}, \sqrt{m_2/M}, \sqrt{1 - (m_1 + m_2)/M})^T \in S^2 \ (10.1)$$

on a two-sphere. In this representation, the spherical data lie in the first octant. We observe that points tend to accumulate close to the $x_2 = 0$ plane, representing cells with one single orientation field, cf. Fig. 10.9.

In order to interpret live cell observations, we compare these observations to time series of cells which were fixed after different times on a gel. We have taken images of fixed cells for each gel rigidity in intervals of 4 h of time on the gel. The sample sizes are displayed in Table 10.1. These cells were stained with phalloidin as opposed to the live cells, which were transfected with liveAct. To compare the live cell experiment to the fixed cell experiment, we only consider images from live cell movies corresponding to the fixation times. Since we have between 50 and 60 movies on each gel, we can expect a higher data variance for the live cells compared to fixed cells. Some of the investigated samples are displayed in Fig. 10.9.

We analyze the samples on S^2 by applying dimension reduction via *principal nested great spheres* from [32]. This means we first identify the great circle which fits the data best (in terms of accumulated squared spherical distance), then orthogonally (along great circles) project data to this great circle and determine their Fréchet mean on this great circle, called the *nested mean*. Jointly, the two give our *backward nested data descriptor*. To estimate its variance, i.e. the variance of the great circle and the nested mean, we use $B = 1000$ bootstrap replicates from the data. Figure 10.10 displays the backward nested descriptor and bootstrapped means illustrating the spread of the nested mean estimator.

The lower sample size leads to considerably higher variance for live cells, as expected. However, fixed and live cells on the same substrates are strikingly different from each other.

While live cells on the soft ($E = 1$ kPa) gels show little temporal evolution, the fixed cells exhibit a development towards a dominant main field with few smaller fields and eventually less rogue filaments. For the cells on stiffer gels, while the fixed cells strengthen their main field, mainly at the expense of rogue filaments, for live cells this effect is stronger, leading to fewer rogue filaments over time as well. Remarkably, for the fixed cells we observe an increased number of rogue filaments after 16 and 20 h, which does not occur for live cells. A T^2 hypothesis test developed in [10] for backward nested descriptors verifies that this effect is significant.

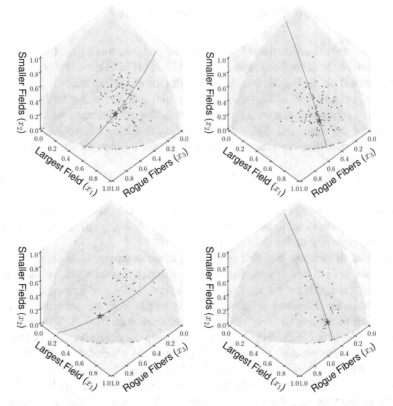

Fig. 10.9 Visualization of the orientation fields representatives on S^2 *(blue points) from* (10.1) with backward nested data descriptor given by the best approximating great circle (blue line) with nested mean (red star). All images correspond to 16 h on the gel. Upper left: 135 fixed cells on a 1 kPa gel; upper right: 127 fixed cells on a 10 kPa gel; lower left: 59 live cells on a 1 kPa gel; lover right: 53 live cells on a 30 kPa gel

Table 10.1 Sample sizes of hMSC skeleton images over varying Young's moduli and cultivation time, left for fixed cells, right for live cells

Time	1 kPa	10 kPa	30 kPa		1 kPa	10 kPa	30 kPa
4h	159	168	153		59	54	53
8h	163	164	153				
12h	176	171	173				
16h	135	127	147				
20h	138	126	127				
24h	166	152	152				

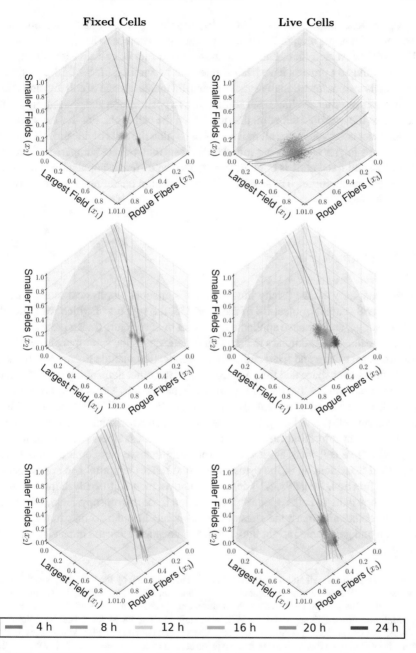

Fig. 10.10 Best fitting great circles with nested means. For every time and sample, nested means of $B = 1000$ bootstrap samples are displayed to illustrate variance of the mean. Rows from top to bottom: 1 kPa gel, 10 kPa gel, 30 kPa gel

Upon closer inspection, as noted in [10], describing cells by their orientation field decomposition, the temporal evolution of fixed cells comes to a stop and at roughly 12 h, often reversed then, hinting to an increased rate of cell division, albeit cells near the division process have been singled out previously, cf. Sect. 10.2. Although these cells are not intentionally synchronized, an increased rate of cell division after a particular time after seeding is not surprising as trypsinization and re-seeding slightly decreases the isotropic temporal distribution on the cell cycle. For live cells, such an effect is not observed, since all movies during which the cells divide are left out. The statistical analysis confirms our initial hypothesis that a direct comparison of results from live and fixed cells is complicated due to the different population subsets and points out the importance of careful experiment design including proper controls for future studies.

10.5 Outlook

In this chapter we have briefly illustrated the statistical biophysical toolbox developed over 8 years in project B8 of the SFB 755, the support of which we gratefully acknowledge, and we have applied a typical set of descriptors (mean great circle and nested mean on it), exemplary to highlight differences in the actin-myosin cytoskeleton structure of live and fixed cells. Future application include tracing and tracking stress fibers over space and time and usage in many other demanding research areas. For example, studies on migrating cells indicate various stress fiber types classically described as (dorsal, ventral, and arcs) appearing at different locations inside a migrating cell [21, 24–28]. Following the filament dynamics over time will give further insight into the formation and function of stress fibers. Using our toolbox applied to live cell imaging, it seems promising that we can come to an unbiased statistical classification of the cytoskeleton that relates temporal and spatial persistence to function. Recently, [29] described a novel method to analyze traction force microscopy data, so called model-based traction force microscopy. Here, it is imperative to detect and mark the stress fibers of a cell in order to link forces to fiber location and gain more insight into cellular force generation and transmission to the substrate.

Acknowledgements All authors gratefully acknowledge funding by the Deutsche Forschungsgemeinschaft (DFG) within the collaborative research center SFB 755 "Nanoscale Photonic Imaging" project B08 and the Open Access Publication Funds of the University of Göttingen. SH also gratefully acknowledges support of the Felix-Bernstein-Institute for Mathematical Statistics in the Biosciences and the Niedersachsen Vorab of the Volkswagen Foundation. FR acknowledges funding from the Volkswagen foundation within the Niedersachsen Israel framework (MWK-VWZN2722). The authors acknowledge Julian Rüger's contribution to an earlier version of the FS.

References

1. Discher, D.E., Janmey, P., Wang, Y.L.: Tissue cells feel and respond to the stiffness of their substrate. Science **310**(5751), 1139–1143 (2005)
2. Rehfeldt, F., Engler, A.J., Eckhardt, A., Ahmed, F., Discher, D.E.: Cell responses to the mechanochemical microenvironment-implications for regenerative medicine and drug delivery. Adv. Drug Deliv. Rev. **59**(13), 1329–1339 (2007)
3. Engler, A.J., Sen, S., Sweeney, H.L., Discher, D.E.: Matrix elasticity directs stem cell lineage specification. Cell **126**(4), 677–689 (2006)
4. Zemel, A., Rehfeldt, F., Brown, A.E.X., Discher, D.E., Safran, S.A.: Optimal matrix rigidity for stress-fibre polarization in stem cells. Nat Phys **6**(6), 468–473 (2010)
5. Wen, J.H., Vincent, L.G., Fuhrmann, A., Choi, Y.S., Hribar, K.C., Taylor-Weiner, H., Chen, S., Engler, A.J.: Interplay of matrix stiffness and protein tethering in stem cell differentiation. Nat. Mater. **13**(10), 979–987 (2014)
6. Swift, J., Ivanovska, I.L., Buxboim, A., Harada, T., Dingal, P.C.D.P., Pinter, J., Pajerowski, J.D., Spinler, K.R., Shin, J.W., Tewari, M., Rehfeldt, F., Speicher, D.W., Discher, D.E.: Nuclear lamin-a scales with tissue stiffness and enhances matrix-directed differentiation. Science **341**(6149) (2013)
7. Zemel, A., Rehfeldt, F., Brown, A.E.X., Discher, D.E., Safran, S.A.: Cell shape, spreading symmetry, and the polarization of stress-fibers in cells. J. Phys. Condens. Matter **22**(19), 194,110 (2010)
8. Eltzner, B., Wollnik, C., Gottschlich, C., Huckemann, S., Rehfeldt, F.: The filament sensor for near real-time detection of cytoskeletal fiber structures. PloS one **10**(5), e0126,346 (2015)
9. Huckemann, S., Kim, K.R., Munk, A., Rehfeldt, F., Sommerfeld, M., Weickert, J., Wollnik, C.: The circular sizer, inferred persistence of shape parameters and application to early stem cell differentiation. Bernoulli **22**(4), 2113–2142 (2016)
10. Huckemann, S.F., Eltzner, B.: Backward nested descriptors asymptotics with inference on stem cell differentiation. Ann. Stat. **46**(5), 1994–2019 (2018). ArXiv:1609.00814
11. Belin, B.J., Goins, L.M., Mullins, R.D.: Comparative analysis of tools for live cell imaging of actin network architecture. BioArchitecture **4**(6), 189–202 (2014)
12. Riedl, J., Crevenna, A.H., Kessenbrock, K., Yu, J.H., Neukirchen, D., Bista, M., Bradke, F., Jenne, D., Holak, T.A., Werb, Z., Sixt, M., Wedlich-Soldner, R.: Lifeact: a versatile marker to visualize f-actin. Nat. Methods **5**, 605 (2008)
13. Szeliski, R.: Computer vision: algorithms and applications. Springer (2010)
14. Basu, S., Dahl, K., Rohde, G.: Localizing and extracting filament distributions from microscopy images. J. Microsc. **250**, 57–67 (2013). Retracted due to image copyright issues
15. Faust, U., Hampe, N., Rubner, W., Kirchgeßner, N., Safran, S., Hoffmann, B., Merkel, R.: Cyclic stress at mHz frequencies aligns fibroblasts in direction of zero strain. PLoS ONE **6**(12), e28963 (2011)
16. Chang, S., Kulikowski, C., Dunn, S., Levy, S.: Biomedical image skeletonization: a novel method applied to fibrin network structures. Medinfo **84**(2), 901–906 (2001)
17. Dormann, D., Libotte, T., Weijer, C.J., Bretschneider, T.: Simultaneous quantification of cell motility and protein-membrane-association using active contours. Cell Motil Cytoskelet. **52**(4), 221–230 (2002)
18. Lichtenstein, N., Geiger, B., Kam, Z.: Quantitative analysis of cytoskeletal organization by digital fluorescent microscopy. Cytometry A **54**(1), 8–18 (2003)
19. Herberich, G., Würflinger, T., Sechi, A., Windoffer, R., Leube, R., Aach, T.: Fluorescence microscopic imaging and image analysis of the cytoskeleton. In: Conference record of the forty fourth Asilomar conference on signals, systems and computers (ASILOMAR), pp. 1359–1363 (2010)
20. Sanchez, T., Chen, D.T.N., DeCamp, S.J., Heymann, M., Dogic, Z.: Spontaneous motion in hierarchically assembled active matter. Nature **491**(7424), 431–434 (2012)
21. Tojkander, S., Gateva, G., Lappalainen, P.: Actin stress fibers-assembly, dynamics and biological roles. J. Cell Sci. **125**(8), 1855–1864 (2012)

22. Alioscha-Perez, M., Benadiba, C., Goossens, K., Kasas, S., Dietler, G., Willaert, R., Sahli, H.: A robust actin filaments image analysis framework. PLOS Computational Biol. **12**(8), 1–23 (2016)
23. Rogge, H., Artelt, N., Endlich, N., Endlich, K.: Automated segmentation and quantification of actin stress fibres undergoing experimentally induced changes. J. Microsc. **268**(2), 129–140 (2017)
24. Ciobanasu, C., Faivre, B., Le Clainche, C.: Actin dynamics associated with focal adhesions. Int. J. Cell Biol. **2012**, 941,292 (2012)
25. Hotulainen, P., Lappalainen, P.: Stress fibers are generated by two distinct actin assembly mechanisms in motile cells. J. Cell Biol. **173**(3), 383–394 (2006)
26. Naumanen, P., Lappalainen, P., Hotulainen, P.: Mechanisms of actin stress fibre assembly. J. Microsc. **231**, 446–454 (2008)
27. Pellegrin, S., Mellor, H.: Actin stress fibres. J. Cell Sci. **120**, 3491–3499 (2007)
28. Vallenius, T.: Actin stress fibre subtypes in mesenchymal-migrating cells. Open Biol. **3**, 13,001 (2013)
29. Soiné, J.R.D., Brand, C.A., Stricker, J., Oakes, P.W., Gardel, M.L., Schwarz, U.S., Vikram, D.: Model-based traction force microscopy reveals differential tension in cellular actin bundles. PLoS Comput Biol **11**(3), e1004,076 (2015)
30. Gottschlich, C., Mihailescu, P., Munk, A.: Robust orientation field estimation and extrapolation using semilocal line sensors. IEEE Trans. Inf. Forensics Secur. **4**(4), 802–811 (2009). https://doi.org/10.1109/TIFS.2009.2033219
31. Rosenfeld, A., Hummel, R.A., Zucker, S.W.: Scene labeling by relaxation operations. IEEE Trans. Syst. Man Cybern. **6**(6), 420–433 (1976)
32. Jung, S., Dryden, I.L., Marron, J.S.: Analysis of principal nested spheres. Biometrika **99**(3), 551–568 (2012)

Chapter 11
Photonic Imaging with Statistical Guarantees: From Multiscale Testing to Multiscale Estimation

Axel Munk, Katharina Proksch, Housen Li and Frank Werner

Abstract   In this chapter we discuss how to obtain statistical guarantees in photonic imaging. We start with an introduction to hypothesis testing in the context of imaging, more precisely we describe how to test if there is signal in a specific region of interest (RoI) or just noise. Afterwards we extend this approach to a family of RoIs and examine the occurring problems such as inflation of type I error and dependency issues. We discuss how to control the family-wise error rate by different modifications, and provide a connection to extreme value theory. Afterwards we present possible extension to inverse problems. Moving from testing to estimation, we finally introduce a method which constructs an estimator of the desired quantity of interest with automatic smoothness guarantees.

2010 Mathematics Subject Classification: Primary 62-01, 62G10 · Secondary 62G20, 62F17

A. Munk (✉) · K. Proksch · H. Li · F. Werner
Institute for Mathematical Stochastics, Universität Göttingen,
Goldschmidtstr. 7, 37077 Göttingen, Germany
e-mail: munk@math.uni-goettingen.de

K. Proksch
e-mail: kproksc@uni-goettingen.de

H. Li
e-mail: housen.li@mathematik.uni-goettingen.de

F. Werner
e-mail: f.werner@math.uni-goettingen.de

A. Munk · F. Werner
Max Planck Institute for Biophysical Chemistry, Am Faßberg 11, 37077 Göttingen, Germany

© The Author(s) 2020
T. Salditt et al. (eds.), *Nanoscale Photonic Imaging*, Topics in Applied Physics 134,
https://doi.org/10.1007/978-3-030-34413-9_11

11.1 Introduction

The analysis of a photonic image typically involves a reconstruction of the measured object of interest which becomes the subject of further evaluation. This approach is frequently employed in photonic image analysis, though it can be quite problematic for several reasons.

1. As the image is noisy and often inherently random, a full reconstruction relies on the choice of a regularisation functional and corresponding a priori assumptions on the image, often implicitly hidden in a reconstruction algorithm. Related to this, the reconstruction relies on the choice of one or several tuning parameters. A proper choice is a sensible task, in particular when the noise-level is high and/or inhomogeneous.
2. The sizes of the objects might be below the resolution of the optical device which further hinders a full reconstruction.
3. As the resolution increases, the object to be recovered becomes random in itself as its fine structure then depends on, e.g., the conformational states of a protein and the interpretation of the recovered object might be an issue.

It is the aim of this chapter to provide a careful discussion of such issues and to address the analysis of photonic images with statistical guarantees. This will be done in two steps. In Sect. 11.2 we survey some recent methodology, which circumvents a full recovery of the image, to extract certain relevant information in such difficult situations mentioned above. Based on this (see Sect. 11.3), we will extend such methods also to situations in which a full reconstruction is reasonable, but still a difficult task, e.g., when the multiscale nature of the object has to be recovered. In both scenarios we will put a particular emphasis on statistical guarantees for the provided methods.

An example where a full recovery of the object of interest is typically not a valid task is depicted in the centre of Fig. 11.1 where a detail of a much larger image is shown (see Fig. 1 in [1] for the full image). The investigated specimen consists of DNA origami which have been designed in such a way that each of the signal clusters contains up to 24 fluorescent markers, arrayed in two strands of up to 12, having a distance of 71 nanometers (nm) (see left panel of Fig. 11.1 for a sketch of such a DNA origami). As the ground truth is basically known, this serves as a real world phantom.

Data were recorded with a STED (STimulated Emission Depletion) microscope at the lab of Stefan Hell of the Department of NanoBiophotonics of the Max Planck Institute for Biophysical Chemistry. In contrast to classical fluorescence microscopy, the resolution in STED microscopy is in theory not limited and can be enhanced by increasing the intensity of the depletion laser [2]. However, this increase comes at the price of a decrease in intensity of the focal spot, which bounds the resolution in practice. Therefore a convolution of the underlying signal with the PSF of the STED microscope is unavoidable and a full reconstruction of the DNA origami (or the shape of the markers) appears to be difficult. However, for most purposes this is also

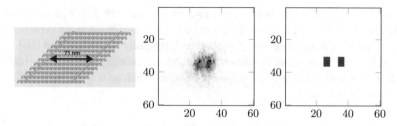

Fig. 11.1 (Detail of Fig. 1 in [1]) Left: Sketch of single DNA origami, middle: detail of image of randomly distributed DNA origami, right: detected strands of markers

not relevant. Instead, less ambitious tasks will provide still important information, e.g., the location of these fluorescent markers. This can be done via a statistical test, which is presumably a much simpler task than reconstruction (estimation in statistical terminology) and it can be tailored towards answering particular questions "How many strands of markers are there?" and "Where are the DNA origamis located?". The right panel of Fig. 11.1 shows the locations of markers as found by such a statistical test (from the data in the middle panel in Fig. 11.1) which will be introduced later on.

11.2 Statistical Hypothesis Testing

11.2.1 Introduction

We will see that proper testing in the above example (Fig. 11.1) is already a complex task. Therefore, in this section, we first introduce the concept of statistical testing in a basic setting. The first step in statistical hypothesis testing is to define the so-called *null hypothesis*, H, and the *alternative hypothesis*, K:

$$H : \text{"Hypothesis to be disproven"}$$
$$K : \text{"Hypothesis to be sustantiated"}.$$

For example, H might correspond to the hypothesis that no marker is contained in a certain given region of the image, K corresponds to the contrary that there is at least one marker in this region. A *statistical significance test* is a decision rule which, based on given data, allows to discriminate between H and K. If a certain criterion is met, H is rejected and K is assumed. If not, H cannot be rejected. For instance, the photon count in a certain given region of a noisy image gives rise to the believe that at least one marker is contained therein. This could be tested, for example by checking whether the total number of photons detected in this region is larger than a certain threshold. However, due to the involved randomness of photon emissions and background noise such a finding is associated with a (certain) risk of being incorrect. A statistical test aims to control this risk. Hence, prior to performing a statistical test, a tolerable risk α is specified, typically in the range of 0.01 up to 0.1, corresponding to accepting the error rate that, on average, in at most $\alpha \cdot 100\%$ of the cases the null-hypothesis H is falsely rejected. Such an α is called *significance level*. This is written as

$$\mathbb{P}\left(\text{"}H \text{ is rejected although H is true"}\right) \leq \alpha. \tag{11.1}$$

Here, $\mathbb{P}$ stands generically for all possible distributions under H and $\mathbb{P}(A)$ denotes the probability[1] of an event A. If the test criterion is chosen such that (11.1) holds, the corresponding test is called a *level-α-test*. The ability of a test to correctly reject H is called *detection power*. If H corresponds to the hypothesis that no marker is located in a certain given region, the test (i.e., the data based decision procedure) is then constructed in such a way that the probability α to falsely detect a marker in an empty region is controlled. H and K are chosen in such a way that the false rejection of H is to be considered the more serious error and controlled in advance. In our scenario, this means that we consider wrong detection of a fluorophore as the more serious error than missing a fluorophore.

11.2.2 A Simple Example

To demonstrate this concept more rigorously, we now consider a very simple Gaussian model, which can be seen as a proxy for more complicated models. Assume that one observes data

$$Y_i = \mu_i + \varepsilon_i, \quad i = 1, \ldots, n, \tag{11.2}$$

where $\mu_i \geq 0$ denote possible "signals" hidden in observations Y_i, and $\varepsilon_i \sim \mathcal{N}(0, 1)$ are independent normal random variables with variances $\sigma^2 = 1$ (for simplicity). Assume for the moment that all signals have the same strength, $\mu_i \equiv \mu \geq 0$. The interest lies in establishing that $\mu > 0$, i.e., presence of such signal in the data. Hence, we set

$$H : \mu = 0 \text{ (to be disproven)} \quad \text{vs.} \quad K : \mu > 0 \text{ (to be substantiated).} \tag{11.3}$$

The goal is now to find a suitable criterion which, given $Y_1, \ldots, Y_n$, allows to decide in favour or against H in such a way that the error to wrongly reject H is controlled by α. From a statistical perspective the aim is to infer about the mean of

---

[1] More formally, (11.1) is meant as $\mathbb{P}\left(\text{"}H \text{ is rejected although it holds"}\right) \leq \alpha$ under all possible configurations under H. Only where necessary this will be made explicit in the following by an additional subscript.

the Y_i which should be close to the empirical mean $\bar{Y} = \frac{1}{n} \sum_{i=1}^{n} Y_i$ of the data. An intuitive decision rule would be to check whether $\bar{Y}$ is "clearly" larger than zero, $\bar{Y} > \gamma_\alpha$, say, for a suitable threshold $\gamma_\alpha > 0$. We consider the *normalized* (i.e. with unit variance) sum

$$T(Y) := \frac{1}{\sqrt{n}} \sum_{i=1}^{n} Y_i$$

and choose, for prescribed $\alpha > 0$, the threshold γ_α such that we have equality in (11.1). As under the assumption H we have that $\mu = 0$, this gives[2]

$$\mathbb{P}\left(H \text{ is falsely rejected}\right) = \mathbb{P}\left(\frac{1}{\sqrt{n}} \sum_{i=1}^{n} \varepsilon_i \geq \gamma_\alpha\right) = \mathbb{P}\left(\mathcal{N}(0, 1) \geq \gamma_\alpha\right)$$

$$= 1 - \Phi(\gamma_\alpha) \overset{!}{=} \alpha, \tag{11.4}$$

since $\frac{1}{\sqrt{n}} \sum_{i=1}^{n} \varepsilon_i \sim \mathcal{N}(0, 1)$. Here Φ denotes the cumulative distribution function of a standard normal random variable: $\Phi(x) = \frac{1}{\sqrt{2\pi}} \int_{-\infty}^{x} e^{-\frac{y^2}{2}} \, dy$. If H holds true, i.e., $\mu = 0$, (11.4) holds if we choose $\gamma_\alpha = z_{1-\alpha}$, where $z_{1-\alpha}$ is the $(1 - \alpha)$-quantile of the standard normal distribution, e.g., $z_{1-\alpha} = 1.6449$, when $\alpha = 0.05$. The statistical test that rejects H whenever $T(Y) > z_{1-\alpha}$ is called *Z-test* and is a level-α test. Furthermore, if a signal is present, i.e., $\mu_i \equiv \mu > 0$ we have that

$$\mathbb{P}_{\mu_i \equiv \mu}(H \text{ is correctly rejected}) = \mathbb{P}_{\mu_i \equiv \mu}\left(\frac{1}{n} \sum_{i=1}^{n} (\mu + \varepsilon_i) \geq \gamma_\alpha\right)$$

$$= 1 - \mathbb{P}\left(\mathcal{N}(0, 1) > \mu\sqrt{n} - z_{1-\alpha}\right) = 1 - \Phi(\mu\sqrt{n} - z_{1-\alpha}).$$

Since $1 - \Phi(x) \leq \exp(-\frac{1}{2}x^2)$ for $x \geq 1/\sqrt{2\pi}$ (see, e.g., [3], inequality (1.8)), we obtain

$$\mathbb{P}_{\mu_i \equiv \mu}(H \text{ is correctly rejected}) \geq 1 - \exp\left(-\frac{1}{2}n\left(\mu - \frac{z_{1-\alpha}}{\sqrt{n}}\right)^2\right)$$

$$\geq 1 - \exp\left(-\frac{1}{4}n\mu^2\right),$$

for sufficiently large n. This means that, if the number n of data points grows, the *detection power* (the case when $\mu > 0$) of the Z-test converges to 1 exponentially fast. This test has been derived in an intuitive way but it can be proven that it is a *uniformly most powerful* (UMP) test (see [4], Chap. 3.4). This means that for all $\mu > 0$ (i.e. the alternative K holds) the detection power is maximized among all

---

[2]Here $\mathbb{P}$ corresponds to only one configuration of distributions when all $\mu_i \equiv \mu = 0$.

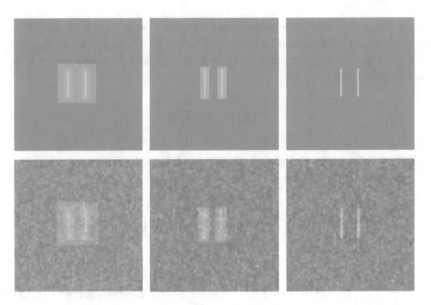

Fig. 11.2 Three different signals (upper row) and noisy signals (lower row)

level-α-tests, i.e., all possible decision rules one might think of which satisfy (11.1) in our set up based on the data $Y_1, \ldots, Y_n$.

Z-test

Comparison of the normalized empirical mean of the set of measurements to a given threshold to assess difference in location to a given constant μ_0. When $\mu_0 = 0$ the Z-test rejects $H : \mu = \mu_0 = 0$ in favor of $K : \mu > 0$ if

$$\frac{1}{\sqrt{n}} \sum_{i=1}^{n} Y_i > z_{1-\alpha}.$$

This is the best possible test at level α if the data $Y_1, \ldots, Y_n$ are independent and $\mathcal{N}(\mu, 1)$ distributed.

11.2.3 Testing on an Image

Subsequently, we consider three illustrative synthetic images of size 60×60, shown in Fig. 11.2 (see the upper panel for a noise-less version and the lower panel for a noisy image). These serve the purpose of explaining how to extend the above simple Z-test to detect a signal in an image, which is a more complex task. To illustrate, we assume for the moment that in these images the intensity on each pixel $Y_i, i = 1, \ldots, n$, follows a $\mathcal{N}(\mu_i, 1)$ distribution, where each μ_i takes one of the four values 0, 2, 3.5 and 5 (see Fig. 11.2). Now, our goal is to segment the image into regions with signal and empty regions while maintaining statistical error guarantees. Note that we do not aim to recover the exact value of each μ_i, only whether it is positive or not (no signal). To this end we will perform many "local" statistical Z-tests on different (and possibly overlapping) regions of this image. We will discuss several approaches (Scenarios 1–5) which provide a step-by-step derivation of our final solution (Scenario 5). As it turns out, the crucial issue will be to control the statistical error of wrong decisions of all these tests simultaneously (overall error).

Scenario 1 (Known position, one test for central 20 × 20 square) *Assume for now that we are only interested whether there is some signal in the central* 20 × 20 *square (framed in blue in the upper row of Fig. 11.3), i.e. we fix the location to be investigated. For this task, we now perform a Z-test at level* $\alpha = 0.05$ *for the central square with* $n = 20 \times 20 = 400$ *pixels, i.e., the test statistic*

$$T_{\text{central } 20 \times 20 \text{ square}}(Y) := \frac{1}{20} \sum_{\text{central } 20 \times 20 \text{ square}} Y_i \tag{11.5}$$

is compared to $z_{1-\alpha} = 1.6449$. *The test allows for exactly two outcomes: rejection (of the hypothesis* H : *no signal in the* 20 × 20 *square) or no rejection. In the second row of Fig. 11.3 the results are depicted. In each of the three test images, the Z-test correctly recognizes that there is signal in the central square, and to visualize this, the square is marked in green. The test decision is correct, however, we cannot draw more (localized) information from this test. Nevertheless, this gives us a first guide how to obtain a segmentation into regions, our final task. Note, that the Z-test, as we derived it in Sect. 11.2.2, is still applicable although we did not assume the alternative that all signals have the same strengths (recall Sect. 11.2.2). This will only affect the power. Crucial is that the test controls the error at level* α *correctly under the assumption that all signals* $\mu_i = 0$.

Given a region of interest (RoI), performing one test on the whole region, as done in the previous scenario, only allows to infer on the entire RoI, i.e., the largest scale there is, finer details cannot be discerned. In the following step we consider the finest possible scales, i.e., tests on single pixels, hoping that we can extract more detailed information on different parts of the image, simultaneously.

Scenario 2 (Known position, pixel-wise tests in 20 × 20 square) *Assume again that we are only interested in testing within the central* 20 × 20 *square. We now perform*

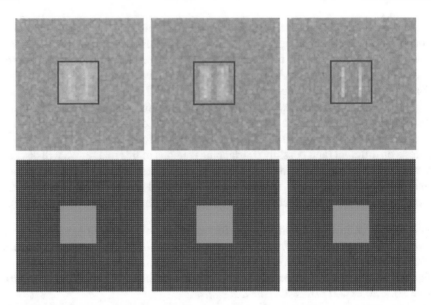

Fig. 11.3 Noisy signals (upper row) and test results from Scenario 1 (lower row). The square is marked in green to show that the test was significant for all three images

a test for each entry in the 20×20 *central region separately, in total 400 tests. The test statistics* $T_{i-\text{th pixel}}(Y)$ *are given by the pixel values. For simplicity, we consider tests for the presence of a signal at pixel "i" which are only based on the observation* Y_i *at pixel i, i.e.*

$$T_{i-\text{th pixel}}(Y) = Y_i, \tag{11.6}$$

and are compared to $z_{1-\alpha} = 1.6449$. *Again, each test allows for two outcomes: rejection or no rejection. In the second row of Fig. 11.4, exemplary results are depicted (all pixels, for which positive test decisions have been made are marked green).*

It is obvious that Scenario 2 gives more detailed information on the signal, but at the expense of several false detections. This is an important issue and will be discussed in more detail in the following section. It is also obvious that parts of the weak signal are missed (see Fig. 11.4: Only 71.25% of the active pixels are detected in the left test image and 85% in the second one). This is due to the fact that the local tests do not take into account neighboring information (surrounding data) from which they could borrow detection strength. This will also be refined in the subsequent sections.

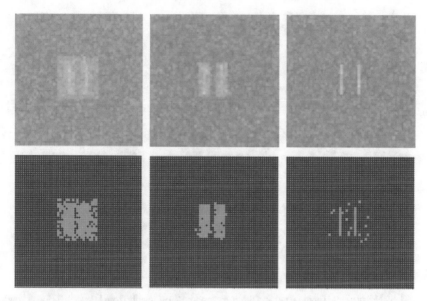

Fig. 11.4 Noisy signals (upper row) and the corresponding test results from Scenario 2 (lower row). The significant pixels are marked green, insignificant pixels are blue

---

False test decisions

There are two kinds of possible false test decisions:

1. Type I error (probability of its occurrence is controlled by α).
   Here: Selection of a RoI although it does not contain any signal (see lower right panel of Fig. 11.4).
2. Type II error (a missed rejection, not controlled).
   Here: Missing to select a RoI that contains signal (see lower left panel of Fig. 11.4).

11.2.4 Testing Multiple Hypotheses

In Scenario 2 in the previous section we applied 400 single Z-tests in the central square of the synthetic image. It is obvious from Fig. 11.4 that this approach suffers from many false detections, in particular when the signal gets sparser (see lower right plot in Fig. 11.4). This issue becomes even more severe if the number of tests increases, as the following test scenario illustrates.

Scenario 3 (Unknown position, pixel-wise tests, whole image) *If we do not have prior information on the particular region which we should investigate, we need*

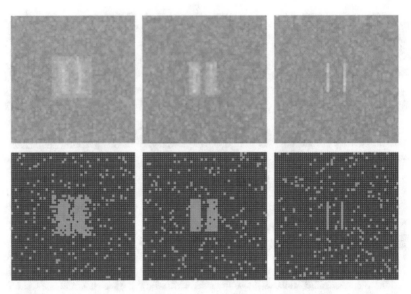

Fig. 11.5 Noisy signal (upper row) and the corresponding test results from Scenario 3 (lower row). The significant pixels are marked green

to scan the entire image. In generalization of Scenario 2 (the RoI is now the full image) to the case of unknown signal position, all single pixels of the entire image are tested. This results in 3600 tests. The results are shown in the second row of Fig. 11.5. Obviously, the number of false rejections increases with the number of tests. In fact, this did not just randomly happen, it is a systematic flaw which we encounter when we naively perform many tests on the same image, simultaneously.

11.2.4.1 Number of False Rejections

The statistical control of false rejections is a general problem one encounters in multiple testing (i.e., testing many hypotheses simultaneously on the same data). The increase of false rejections with increasing number of tests is denoted as *multiplicity effect*.

Figure 11.6 shows the probabilities that out of n independent Z-tests, at least 1 (solid line), 10 (dashed line), 75 (dotted line) and 150 (dash-dotted line) false rejections occur. The curves suggest that in the situation of Scenario 3 we need to expect at least 150 false detections. In fact, the probability that many wrong rejections are made within N tests, each at level α, performed on a data set converges to 1 exponentially fast.

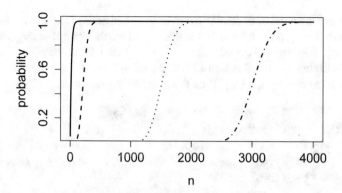

Fig. 11.6   Exact probabilities (y-axis) that out of n (x-axis) independent Z-tests with $\alpha = 0.05$, at least 1 (solid line), 10 (dashed line), 75 (dotted line) and 150 (dash-dotted line) false rejections occur. Here, $n = 1, 2, \ldots, 4000$, where the respective probabilities are zero as long as n is smaller than 10, 75 or 150, respectively

Lemma 11.1 *If* $0 < \alpha \leq 1/2$, $N \geq 2$ *and* $k \leq N \log(1 + \alpha)/\log(N)$, *we have that*

$$\mathbb{P}\,(\textit{at least } k \textit{ out of } N \textit{ false rejections}) \geq 1 - (1 - \alpha^2)^N.$$

Proof The random variables $I\{i - \text{th test rejects}\}$, where I denotes the indicator function, follow a Bernoulli distribution with parameter α. Therefore, if $\alpha \leq 1/2$, we can estimate the probability that out of $N \geq 2$ tests k false rejections are made, as

$$\mathbb{P}\,(\text{at least } k \text{ out of } N \text{ false rejections}) = 1 - \sum_{j=0}^{k-1} \mathbb{P}\,(\text{exactly } k \text{ false rejections})$$

$$= 1 - \sum_{j=0}^{k-1} \binom{N}{j}(1 - \alpha)^{N-j}\alpha^j \geq 1 - (1 - \alpha)^N \sum_{j=0}^{k-1} \binom{N}{j}.$$

It follows, e.g., by induction over k for any $N \geq 2$, that $\sum_{j=0}^{k-1} \binom{N}{j} \leq N^k$, which implies

$$\mathbb{P}\,(\text{at least } k \text{ out of } N \text{ false rejections}) \geq 1 - (1 - \alpha)^N N^k.$$

For $k \leq N \log(1 + \alpha)/\log(N)$ we thus find

$$\mathbb{P}\,(\text{at least } k \text{ out of } N \text{ false rejections}) \geq 1 - (1 - \alpha^2)^N.$$

Hence, the probability of making at least k out of N false rejections converges to 1 exponentially fast, as $N \to \infty$.   □

To reduce the number of false detections, so-called *multiplicity adjustments* have to be made. Two general approaches in this regard are the control of the *family wise error rate* (FWER) and of the *false discovery rate* (FDR). Here, we will mainly focus on the FWER but will briefly discuss FDR control in Sect. 11.2.9. For further reading we refer to the monograph by [5] and the references given there.

> **Multiplicity effect**
>
> If multiple tests are performed without accounting for multiplicity, the chances of making many type I errors are quite large if the false null hypotheses are sparse (see Fig. 11.5).

11.2.4.2 Control of FWER

One possible way to deal with multiplicity is to control the family wise error rate (FWER), that is, controlling the probability of making *any* wrong decision in all tests that are performed. Assume model (11.2) and denote by $\boldsymbol{\mu} = (\mu_1, \ldots, \mu_n)$ the vector of all true means and by $\mathbb{P}_{\boldsymbol{\mu}}$ the probability under configuration $\boldsymbol{\mu}$. In the previous example of imaging, the sample size n corresponded to the number of pixels. Scenarios 2 and 3 were based on many single tests (on many single pixels). Such single tests will be referred to as *local tests* in the sequel. Each of the N (say) local tests corresponds to its own (local) hypotheses H_i versus K_i. For example, in the setup of Scenario 3, a local hypothesis is $H_i : \mu_i = 0$ versus the local alternative $K_i : \mu_i > 0$, for some $i = 1, \ldots, n$. In this case $n = N$ when all local hypotheses are tested. If only a few are tested, then $N \ll n$. If in addition all 2×2 RoIs are tested a total of $N \approx 2n$ tests are performed.

Assume now that all local tests H_i vs. K_i are performed, each at error level α/N. Then the risk of making *any* wrong rejection is controlled at level α, that is, the FWER is controlled.

Theorem 11.1 (Bonferroni correction) *Given N testing problems H_i vs. K_i, $i = 1, \ldots, N$ and local tests at level α/N, we have for any configuration $\boldsymbol{\mu}$*

$$\mathbb{P}_{\boldsymbol{\mu}} \left(\text{``at least one wrong rejection''} \right) \leq \alpha.$$

Proof

$$\mathbb{P}_{\boldsymbol{\mu}} \left(\text{``at least one wrong rejection''} \right) \leq \sum_{i=1}^{N} \mathbb{P}_{\boldsymbol{\mu}} \left( \text{``}i - \text{th test falsely rejects''} \right)$$

$$\leq \sum_{i=1}^{N} \mathbb{P}_{H_i} \left(\text{``}i - \text{th test falsely rejects''} \right) \leq \sum_{i=1}^{N} \frac{\alpha}{N} = \alpha. \tag{11.7}$$

Since the right hand side is independent of μ we say that the FWER is controlled *in the strong sense*. As a consequence, each finding can be considered α-significant and hence can be used as a segment for the final segmentation. Performing tests at an adjusted level such as α/N instead of α is called *level adjusted testing* and the multiple test "reject those H_i which are significant at the adjusted level α/N" is called *Bonferroni procedure*. We stress that although Theorem 11.1 was formulated for the special case of signal detection in independent Gaussian noise, the Bonferroni procedure strongly controls the FWER in much more generality and in particular without any assumptions on the dependency structure between different tests [5, see, e.g., Chap. 3.1.1, for a more detailed discussion].

Scenario 4 (Unknown position, pixel-wise, Bonferroni adjustment) *In the situation of Scenario 3, we now perform a Bonferroni procedure for the entire image, i.e., for all $60 \times 60 = 3600$ entries (see Fig. 11.7). The local testing problems are*

$$H_i : \text{"No signal in } i\text{-th pixel"}, \text{i.e., } \mu_i = 0 \quad vs. \quad K_i : \mu_i > 0.$$

Now $n = N = 3600$ and $\alpha/N \approx 1.3889 \times 10^{-5}$ for $\alpha = 0.05$. In this scenario all single entries are compared to $z_{1-\frac{0.05}{3600}} \approx 4.19096$. (Recall that in Scenarios 2 and 3 we compared each entry to the much smaller threshold 1.6449 and note that any level adjustment corresponds to an increase of the threshold for testing.) The result is shown in the second row of Fig. 11.7. While no false findings were provided by any of these tests, too few detections have been made at all as only parts of the signal have been detected.

Bonferroni multiplicity adjustment

Adjustment (increase) of the thresholds when multiple tests are performed simultaneously to *control the overall type I error*, i.e., the FWER. This is a very general but also a conservative method (in particular if the signal is not sparse).

11.2.5 Connection to Extreme Value Theory

There is a close connection between the control of the FWER in the situation of Scenario 3 and extreme value theory. Recall that the aim is to control $\mathbb{P}_\mu$ ("at least one wrong rejection") for any configuration μ. By monotonicity, we have that

$$\mathbb{P}_\mu \text{ ("at least one wrong rejection")} \leq \mathbb{P}_{\mu=0} \text{ ("at least one wrong rejection")},$$

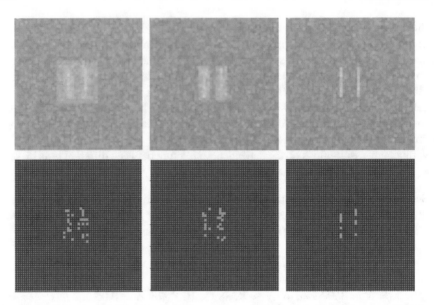

Fig. 11.7 Noisy signals (upper row) and the corresponding test results from Scenario 4 (FWER-controlled, lower row). The significant pixels are marked green, insignificant pixels are blue

which implies that the FWER is controlled if we choose the threshold q for our multiple tests such that

$$\mathbb{P}_{\mu=0} \text{ ("at least one wrong rejection")} = \mathbb{P}\left(\exists\, i \in \{1, \ldots, N\} : \varepsilon_i > q\right) \le \alpha. \tag{11.8}$$

Now, since $\mathbb{P}\left(\exists\, i \in \{1, \ldots, N\} : \varepsilon_i > q\right) = \mathbb{P}\left(\max\{\varepsilon_1, \ldots, \varepsilon_N\} > q\right)$, q can be chosen as the $(1-\alpha)$-quantile of $\max\{\varepsilon_1, \ldots, \varepsilon_N\}$ under the global null hypothesis, H,

$$H = \bigcap_{i=1}^{3600} H_i : \text{"No signal at all"}, \tag{11.9}$$

i.e., $\mu_i = 0$ for all $i = 1, \ldots, 3600$. In this case we have equality in (11.8). Note that

$$
\begin{aligned}
\mathbb{P}\left(\max\{\varepsilon_1, \ldots, \varepsilon_N\} > q\right) &= 1 - \mathbb{P}\left(\max\{\varepsilon_1, \ldots, \varepsilon_N\} \le q\right) \\
&= 1 - \mathbb{P}\left(\varepsilon_1 \le q \text{ and } \varepsilon_2 \le q \text{ and } \ldots \text{ and } \varepsilon_N \le q\right) \\
&= 1 - \left(\mathbb{P}\left(\varepsilon_1 \le q\right)\right)^N = 1 - \left(\Phi\left(q\right)\right)^N.
\end{aligned}
$$

Therefore,

$$\mathbb{P}\left(\max\{\varepsilon_1, \ldots, \varepsilon_N\} > q\right) = \alpha \; \Leftrightarrow \; \left(\Phi(q)\right)^N = 1 - \alpha \; \Leftrightarrow \; \Phi(q) = (1-\alpha)^{\frac{1}{N}},$$

which yields $q = z_{(1-\alpha)^{1/N}}$. Since

$$(1 - \alpha)^{1/N} < 1 - \alpha/N,$$

by Bernoulli's inequality, strict monotonicity of Φ^{-1} implies the same ordering of the thresholds, i.e., $z_{(1-\alpha)^{1/N}} < z_{1-\alpha/N}$. However, it is easy to show that for $N \geq 1$ and $\alpha \leq 1/2$

$$1 - \frac{\alpha + \alpha^2}{N} < (1 - \alpha)^{1/N}$$

and therefore the difference between $z_{(1-\alpha)^{1/N}}$ and $z_{1-\alpha/N}$ is quite small, e.g. $z_{(1-0.05)^{1/3600}} \approx 4.18516$ and $z_{1-0.05/3600} \approx 4.19096$.

The following lemma shows that $z_{1-\alpha/N} \approx \sqrt{2 \log(N)}$ (and therefore also $z_{(1-\alpha)^{1/N}} \approx \sqrt{2 \log(N)}$).

Lemma 11.2 *There exists $N_0 \in \mathbb{N}$ such that for all $N \geq N_0$*

$$\sqrt{2 \log(N)} - \frac{\log \log(N)}{\sqrt{2 \log(N)}} \leq z_{1-\alpha/N} \leq \sqrt{2 \log(N)}$$

Proof To bound the normal quantiles from above and below, we use

$$\frac{\varphi(x)}{x + \frac{1}{x}} < 1 - \Phi(x) < \frac{\varphi(x)}{x},$$

[6, see inequality (10)], where φ and Φ denote the density and cdf of the standard normal distribution, respectively. Since, for sufficiently large N,

$$1 - \Phi(\sqrt{2 \log(N)}) \leq \frac{\varphi(\sqrt{2 \log(N)})}{\sqrt{2 \log(N)}} = \frac{1}{N\sqrt{4\pi \log(N)}} \leq \frac{\alpha}{N},$$

and therefore

$$1 - \frac{\alpha}{N} \leq \Phi(\sqrt{2 \log(N)}) \quad \Leftrightarrow \quad z_{1-\frac{\alpha}{N}} = \Phi^{-1}(1 - \frac{\alpha}{N}) \leq \sqrt{2 \log(N)},$$

the right hand side follows. We further have that

$$1 - \Phi\left(\sqrt{2 \log(N)} - \frac{\log \log(N)}{\sqrt{2 \log(N)}}\right) \geq \frac{\varphi\left(\sqrt{2 \log(N)} - \frac{\log \log(N)}{\sqrt{2 \log(N)}}\right)}{\sqrt{2 \log(N)} - \frac{\log \log(N)}{\sqrt{2 \log(N)}} + \frac{1}{\sqrt{2 \log(N)} - \frac{\log \log(N)}{\sqrt{2 \log(N)}}}}$$

$$\geq \frac{\log(N)}{N\sqrt{2\pi}} \exp\left(-\frac{\log \log(N)}{4 \log(N)}\right) \geq \frac{\alpha}{N},$$

for sufficiently large N, and the left hand side follows.

11.2.5.1 Towards Better Detection Properties

The Bonferroni approach is valid in most generality. Nevertheless, as we have seen in Fig. 11.7, if applied pixel-wise the level adjustment (and the resulting increase of the threshold) is (much) too strict for our purposes. This is not caused by the Bonferroni-adjustment per se, as it can be shown that the detection power of the Bonferroni approach cannot be considerably improved in general [7, Sect. 1.4.1]. The issue is that we have only considered each single pixel as input for our local tests. Therefore, we will extend this from single pixels to larger systems of RoIs, which allow to "borrow strength from neighbouring pixels". This makes sense as soon as the signal has some structure, e.g., whenever signal appears in (small) clusters or filament-like structures. To see this, suppose that for $k > 1$ we have $\mu_1 = \mu_2 = \ldots = \mu_k = \mu$. An uncorrected pixel-wise Z-test would compare each Y_i to the threshold $z_{1-\alpha}$, i.e., signal in a pixel would be detected if

$$Y_i = \underbrace{(Y_i - \mu)}_{\mathcal{N}(0,1)} + \mu > z_{1-\alpha}.$$

This is almost impossible if μ is too small or the noise takes a negative value and becomes even worse if a multiplicity adjustment is performed. If we instead group the first k pixels together and perform a grouped Z-test, i.e., compare $\frac{1}{\sqrt{k}} \sum_{j=1}^{k} Y_j$ to $z_{1-\alpha}$, a signal would be detected if

$$\sqrt{k}\mu + \mathcal{N}(0, 1) > z_{1-\alpha}.$$

This way, the signal is "magnified" by a factor $\sqrt{k}$. Unfortunately, performing, for any k, every test that groups k pixels together and thereby incorporating the fact that positions i and numbers k of relevant pixels are in general not known in advance, is infeasible.[3] However, if the data is clustered spatially we can construct a reasonable test procedure that follows a similar path. Instead of performing all tests that group any configurations of k pixels, we perform *all tests* that merge all pixels in a $k \times k$ square, for many different values of k and "scan" the image for signal in such regions in a computationally and statistically feasible way. Now the local tests become (locally highly) correlated (see Sect. 11.2.6) and a simple Bonferroni adjustment does not provide the best detection power any more, although (11.7) is still valid. This will be the topic of Sects. 11.2.6 and 11.2.7.

---

[3]One issue is computational limitation. Additionally, this has a systematic statistical burden as then tests have to be performed over all possible subsets of the image. For n pixels, these are of size 2^n, which is a collection of sets such that the resulting error probabilities can no longer be controlled in a reasonable way.

> **Amplification of the signal strength by aggregation**
> If a signal is spatially grouped in clusters, cluster-wise tests can increase its detectability. The average of all signal strengths inside the test region is magnified by a factor of $\sqrt{\text{size of cluster}}$.

11.2.6   Scanning

In a way, the two approaches of aggregating data over the entire image (Scenario 1) and performing pixel-wise tests (as done in Scenarios 2–4) are the most extreme scenarios. As a rule of thumb, aggregation makes detection easier at the cost of losing spatial precision whereas pixel-wise testing provides the highest possible spatial precision but makes detection more difficult (after Bonferroni level adjustment as we have seen in Scenario 4. Recall that since the tests are independent we know that there is no substantially better way to control the FWER). In a next step we will combine both ideas. We test on various squares of different sizes to achieve accuracy (small regions) where possible and gain detection power (larger regions) where the signal is not strong enough to be detected pixel-wise, i.e., on small spatial scales. As the system of all subsquares of an image consists of many overlapping squares, we have to deal with locally highly dependent test statistics. Table 11.1 illustrates this effect presenting simulated values of the family wise error rate, based on 1000 simulation runs each, with preassigned value $\alpha = 0.05$. Squares of size $h \times h, h \in \{1, 2, 3, 4, 5\}$ in an image of 60×60 are considered. The parameter h is denoted as a spatial scale. The results of this small simulation study demonstrate that the Bonferroni correction is much too strict if we aggregate data in larger squares. The following scenario is tailored towards dealing with this specific type of dependency structure and is called *multiscale scanning*. Here, the level adjustment is made in an optimal spatially adaptive way, i.e., such that the thresholds are both, large enough so that the FWER is controlled but on the other hand so small that smaller thresholds can no longer guarantee the control of the FWER. The key is now to exploit that the system of all $h \times h$ squares fitting into the $n \times n$ image is highly redundant. For instance, if a square is shifted one pixel to the right, say, both squares share most of their pixels and their contents should not be treated as independent. We discussed in

Table 11.1   Simulated values of FWER at nominal level $\alpha = 0.05$ for a matrix of local averages of $h \times h$ pixels

$h \times h$	1×1	2×2	3×3	4×4	5×5	10×10
Observed error rate	0.049	0.046	0.043	0.028	0.025	0.016

Sect. 11.2.5 that instead of the Bonferroni threshold $z_{1-\frac{\alpha}{N}}$ the $(1-\alpha)$-quantile of the distribution of the maximum of N independent standard normal random variables under the global null hypothesis, H, could be used as a threshold as well. This idea can be transferred to this setting by using the $(1-\alpha)$-quantile of

$$\max_h w(h)(T_{h\times h\,\text{square}}(Y) - w(h)),$$

where $w(h)$ is a size-dependent correction term, given by

$$w(h) := \sqrt{2\ln\frac{N}{h^2}} + 7\ln\left(\sqrt{2\ln\frac{N}{h^2}}\right)\Big/\sqrt{2\ln\frac{N}{h^2}}. \tag{11.10}$$

Under H, the quantiles can be simulated as described in Algorithm 1. Recall that in Lemma 11.2 it was shown that the quantile $z_{1-\frac{\alpha}{N}}$ and therefore also the quantile of the maximum, $z_{(1-\alpha)^N}$, are approximately of size $\sqrt{2\log(N)}$. When pixels are aggregated over $h \times h$ squares, the corresponding quantiles can be shown to be of first asymptotic order $\sqrt{2\log(N/h^2)}$ (the leading term of $w(h)$ in (11.10), see Theorem 11.2 for details), which corresponds to the case of N/h^2 independent tests. This is incorporated into the construction of the thresholds as described in Algorithm 1.

Algorithm 1: Simulation of the thresholds

Parameters : Number of Monte-Carlo runs $M \in \mathbb{N}$, largest size $h_{\max} \in \mathbb{N}$, significance
level $\alpha \in (0, 1)$

1 **for** $n = 1, 2, \ldots, M$ **do**
2   Draw i.i.d. data $Y_i \sim \mathcal{N}(0, 1)$ for $1 \le i \le n$;
3   **for** $1 \le h \le h_{\max}$ **do**
4   Compute all test statistics $T_{h\times h\,\text{square}}(Y)$;
5   Compute all $w(h)(T_{h\times h\,\text{square}}(Y) - w(h))$;
6   Save their maximal value in q_h;
7   Set $t_i := \max_{1\le h\le h_{\max}} q_h$;
8 Sort the values t_i such that $t_1 \le \ldots \le t_M$;
9 Choose $j \in \{1, \ldots, M\}$ such that $j/M \le \alpha < (j+1)/M$;
10 Set $q_{1-\alpha}^h = t_j/w(h) + w(h)$;

In line 12 of Algorithm 1, the size-dependent thresholds $q_{1-\alpha}^h = t_j/w(h) + w(h)$ are defined. Comparing each $T_{h\times h\,\text{square}}(Y)$ to $q_{1-\alpha}^h$ yields a multiplicity adjusted multiple test procedure. Note that in Algorithm 1 the quantile of the maximum over all, locally correlated, test statistics under the global null hypothesis is approximated. This way, the dependence structure is taken into account precisely.

Scenario 5 (Unknown position, multiscale scanning) *We now aggregate test results for several different scanning tests. We consider testing each pixel, as well as testing each 2×2, 3×3, 4×4 and 5×5 square. In total these are 16.830 tests. We now*

Table 11.2 Scale dependent quantiles for the scanning test with windows of variable sizes

α	$q^1_{1-\alpha}$	$q^2_{1-\alpha}$	$q^3_{1-\alpha}$	$q^4_{1-\alpha}$	$q^5_{1-\alpha}$	Bonferroni for 16.830 tests
0.1	5.115	4.760	4.531	4.345	4.208	4.380
0.05	5.267	4.921	4.698	4.527	4.385	4.528
0.01	5.581	5.2538	5.043	4.883	4.750	4.875

adjust the level in a way that accounts for local correlations. We fix $\alpha = 0.05$ and calculate all test statistics $T_{h \times h \text{ square}}(Y)$ (see (11.5)). The local hypotheses $H_{h \times h \text{ square}}$ are

$$H_{h \times h \text{ square}} : \text{``}\mu_i \equiv 0 \text{ in } h \times h \text{ square.''} \qquad (11.11)$$

Each $T_{h \times h \text{ square}}(Y)$ is compared to the size-dependent thresholds $q^h_{1-\alpha}$, which have been generated according to Algorithm 1 and are listed in Table 11.2. We reject the local hypotheses that there is no signal in a particular $h \times h$ square if the corresponding test statistic is larger than the threshold, that is, if

$$T_{h \times h \text{ square}}(Y) = \frac{1}{h} \sum_{h \times h \text{ square}} Y_i > q^h_{1-\alpha}. \qquad (11.12)$$

All significant squares are stored and finally, after all square-wise comparisons have been made, for each pixel, the smallest square that was significant is plotted. Findings for the different sizes are color-coded and for each pixel the color corresponding to the smallest square in which signal was detected is plotted. The results are shown in Fig. 11.8. One big advantage of this approach is that also the weak signal is now completely included in the segmentation in contrast to even the unadjusted approach of Scenario 2 (compare the lower left plots of Figs. 11.4 and 11.8). Also, the color-coding visualizes regions of strong signal and therefore contains "structural information" on the data.

The procedure in Scenario 5 is such that the FWER is still controlled in a strong sense, although the thresholds can be chosen smaller than in a Bonferroni approach. This is much more so if N and h get larger, but is visible starting from $h = 4$, which matches the values given in Table 11.1. This was possible due to the strong local correlations between tests. Roughly speaking, for each size of the moving window a Bonferroni-type adjustment is made for the (maximum) number of non-overlapping squares of that size which is a considerable relaxation. Remarkably, the prize for including many different sizes is extremely small. More theoretical details can be found in Sect. 11.2.7.

To conclude this section, it should be stressed that in many situations, we do not encounter rectangular signals, however, small rectangles can be considered as building blocks for more complex structures. If specific shape information is available,

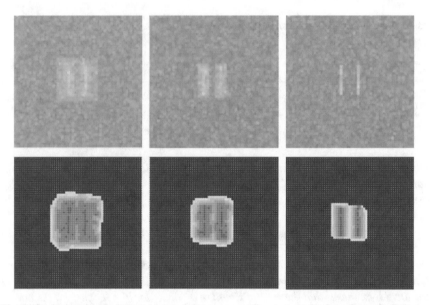

Fig. 11.8 Noisy signals (upper row) and the corresponding test results from Scenario 5 (lower row). Significant 5 × 5 squares are plotted in yellow. Significant 1 × 1 – 4 × 4 squares are plotted in green with increasing brightness. For each pixel, the smallest square which was found significant was plotted. Insignificant regions are coloured in blue

this can be incorporated into the testing procedure as long as the regions are not too irregular and the set of regions satisfies a Vapnik-Cervonenkis-type complexity condition (see [8] for more details). The literature on multiscale scanning methods is vast. In the particular context of imaging, the reader may also consult [9–12] for related ideas.

Multiscale Scanning

With probability guarantee of $1 - \alpha$ all of the RoIs chosen in the multiscale scanning procedure described in Scenario 5, are valid. Hence, we obtain localized RoIs where the signal is sufficiently strong and profit from aggregation, as described in Sect. 11.2.5.1, where the signal is weak and point-wise detection is too difficult.

11.2.7   Theory for the Multiscale Scanning Test

The following theorem is the theoretical foundation for Scenario 5.

Theorem 11.2 *Assume that an $n \times n$ array of independent $\mathcal{N}(\mu_i, 1)$ variables is observed and $\mathcal{H} \subset \{1, \ldots, n\}$ is a set of side lengths of squares. Denote for $h \in \mathcal{H}$*

by $\mathcal{S}(h)$ *the set of all* $h \times h$-*squares. Let* $N = n^2$, $w(h)$ *as defined in* (11.10) *and let further* $\widetilde{q}_{1-\alpha}$ *denote the* $1 - \alpha$-*quantile of*

$$\max_{h \in \mathcal{H}} \max_{S \in \mathcal{S}(h)} w(h)\big(T_S(Y) - w(h)\big) \tag{11.13}$$

under the global hypothesis H : *"no signal in any of the squares". Reject each hypothesis* $H_{h \times h \text{ square}}$ (*see* (11.11)) *for which*

$$T(R) \geq \frac{\widetilde{q}_{1-\alpha}}{w(h)} + w(h). \tag{11.14}$$

(i) *This yields a multiple test for which the FWER at level* α *is controlled asymptotically (as* $|\mathcal{H}|/n \to 0$, $n \to \infty$) *in the strong sense.*

(ii) *This test is minimax optimal in detecting sparse rectangular regions of the signal.*

Claims (i) and (ii) follow from Theorems 7 and 2 in [1]. Roughly speaking, the essence of the previous theorem is that we only need multiplicity control for approximately n^2/h^2 (corresponding to the number of independent) tests instead of $(n - h + 1)^2$ (corresponding to the actual number of all) tests. Control of the FWER in the strong sense means that all significant squares can be used in the final segmentation (lower row of Fig. 11.8).

In this chapter we mainly focused on control of the FWER, however weaker means of error control are of interest as well. A very prominent one is the *false discovery rate* (FDR, [13]), which we briefly discuss in Sect. 11.2.9.

11.2.8 Deconvolution and Scanning

In photonic imaging additional difficulties arise. Firstly, we have to deal with non-Gaussian and non i.i.d. data (see Chap. 4), e.g., following a Poisson distribution with inhomogeneous intensities λ_i. Then, as long as the intensity is not too small, a Gaussian approximation validates model (11.2) as a reasonable proxy for such situations. A formal justification for the corresponding multiscale tests is based on recent results by [14], for details see [1]. The price to pay for such an approximation is a lower bound on the sizes of testing regions that can be used, due to the fact that several data points (of logarithmic order in n) need to be aggregated so that a Gaussian approximation is valid. For ease of notation, we only discussed the Gaussian case in Sect. 11.2.7, generalizations to other distributions can be found in [8].

Secondly, convolution with the PSF of the imaging device induces blur. The first row of Fig. 11.9 shows the convolved synthetic images that were shown in the upper row of Fig. 11.2, where the images in the central row are noisy versions of these convolved images. Note, that some structures are no longer identifiable by eye after convolution. When applying the multiscale scanning approach in Scenario 5 naively

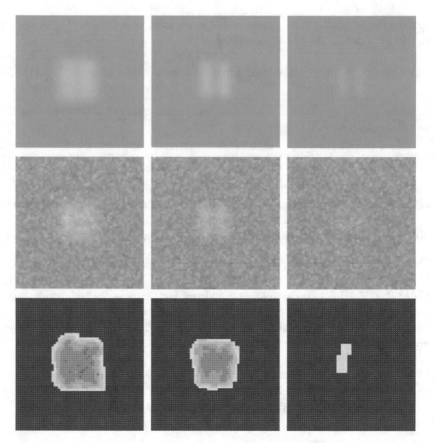

Fig. 11.9 Signals after convolution (upper row), noisy version (central row) and the corresponding test results from Scenario 5, naively applied to the convolved data (lower row). Significant 5×5 squares are plotted in yellow. Significant $1 \times 1 - 4 \times 4$ squares are plotted in green with increasing brightness. For each pixel, the smallest square which was found significant was plotted. Insignificant regions are coloured in blue

to the convolved data (central row of Fig. 11.9). The result (lower row of Fig. 11.11) demonstrates that this is indeed not a competitive strategy and it strongly suggests to take the convolution into account.

We now briefly sketch how to adapt the multiscale scanning procedure (Scenario 5) to the convolution setting. Notice that in the case of data (11.2), we can write the test statistic (11.5) for a particular square S as

$$T_S(Y) = \langle \mathcal{I}_S, Y \rangle ,$$

where $Y = (Y_1, \ldots, Y_n)$ denotes the data vector and $\mathcal{I}_S$ denotes the scaled indicator function on S, i.e., $\mathcal{I}_S(j) = 1/\sqrt{|S|}$ if $j \in S$ and 0 else. Now, the indicator functions are considered as a system of probe functions, which are tested on the data Y. In case of convolution with the PSF k (e.g. of the microscope), model (11.2) turns into

$$Y_i^* = (\mu * k)_i + \varepsilon_i, \quad i = 1, \ldots, n \tag{11.15}$$

where "$*$" denotes convolution. The goal is to find a probe function, acting on the convolved data, denoted as $\mathcal{I}_S^*$ such that

$$\langle \mathcal{I}_S^*, Y^* \rangle \approx \langle \mathcal{I}_S, Y \rangle ,$$

that is, $\mathcal{I}_S^*$ should locally deconvolve. Let $\mu = (\mu_1, \ldots \mu_n)$. Then, if $\mathcal{F}$ denotes the discrete Fourier transform, by Plancherel isometry and the convolution theorem

$$\langle \mathcal{I}_S, \mu \rangle = \left\langle \mathcal{F}^{-1} \left(\frac{\mathcal{F} \mathcal{I}_S}{\mathcal{F} k} \right), \mu * k \right\rangle .$$

This means that (provided $\mathcal{F} k \neq 0$)

$$\mathcal{I}_S^* = \mathcal{F}^{-1} \left(\frac{\mathcal{F} \mathcal{I}_S}{\mathcal{F} k} \right) \tag{11.16}$$

is a reasonable choice of a probe system for the data (11.15) and a statistic that adapts to the convolution is given by

$$T_S^*(Y^*) = \langle \mathcal{I}_S^*, Y^* \rangle .$$

Scenario 5 can now be performed, following Algorithm 1 to derive suitable thresholds, replacing $\mathcal{I}_S$ by $\mathcal{I}_S^*$ and the FWER is controlled. More precisely, it can be shown that Theorem 11.2 also applies in this scenario (see [1] for details). Figure 11.10 d shows the result of this adapted test procedure (MISCAT) applied to our original data (Fig. 11.10 a). As a comparison, we also applied Scenario 5 naively to the data set (Fig. 11.10f). Analogously to [15], $\mathcal{I}_S$ can be chosen such that MISCAT with $\mathcal{I}_S^*$ performs optimally in terms of detection power.

Deconvolution and scanning

In convolution problems sums of pixel values over spatial regions (e.g. squares) will be replaced by probe functionals over the pixels (weighted sums) which can be designed in an optimal way for a given convolution K. The resulting multiscale test scans over all probe functionals which results in substantially more precise segmentation results (for a direct comparison see lower left and lower right panel of Fig. 11.10). It still controls the FWER.

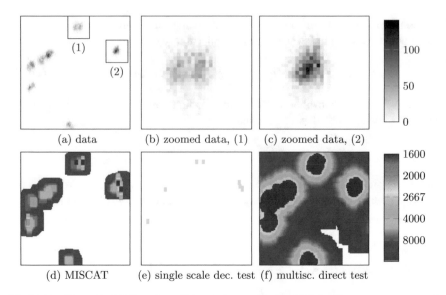

Fig. 11.10 (Figure 2 in [1]) Experimental data and corresponding 90% significance maps computed by different tests. The color-coding of the significance maps always show the size of smallest significance in nm^2, cf. the main text. **a–c** data and zoomed regions, **d** MISCAT, **e** a single scale test with deconvolution, **f** a multiscale scanning test without deconvolution

11.2.9 FDR Control

As discussed in the previous sections of this chapter, as the sample size increases (and therefore the number of tests), the control of the FWER becomes more difficult and thus this may result in low detection power, e.g., in three dimensional imaging. Therefore, a strategy to obtain less conservative procedures of error control is to relax the FWER. The most prominent relaxation is the *false discovery rate* (FDR [13]), defined as

$$\text{FDR} = \mathbb{E}\left[\frac{\#\text{false rejections}}{\max\{\#\text{all rejections, 1}\}}\right],$$

that is, the average proportion of false rejections among all rejections. Hence, in contrast to the FWER this criterion scales with the number of rejections. The control of the FDR is a weaker requirement than the control of the FWER in general. Procedures that control the FDR are often written in terms of *p-values*. In the situation of the Z-test with test statistics $T_{h \times h \, \text{square}}(Y)$ as in (11.12) the p-values are given as

$$p_{h \times h \, \text{square}} = 1 - \Phi\left(T_{h \times h \, \text{square}}(Y)\right), \tag{11.17}$$

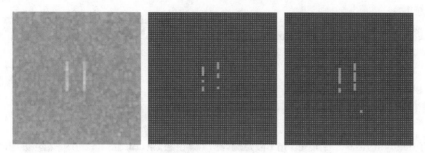

Fig. 11.11 Noisy signal (left) test result of pixel-wise tests after Bonferroni adjustment (middle) and test results from Scenario 6 (right) with FDR control. In both multiple testing procedures $\alpha = 0.05$. Significant pixels are marked green, insignificant regions are coloured in blue

where Φ denotes the cumulative distribution function of the standard normal distribution. The smaller the p-value, the stronger the evidence that the null hypothesis should be rejected.

Benjamini-Hochberg Procedure ([13]) *Consider a multiple testing procedure consisting of independent tests with p-values $p_1, \ldots, p_N$. Sort the p-values increasingly, $p_{(1)} \leq p_{(2)} \cdots \leq p_{(N)}$, and reject all null hypotheses for which $p_i \leq \alpha \frac{\hat{k}}{N}$, where $\hat{k} = \max\{k \mid p_{(k)} \leq \alpha k/N\}$.*

Reference [16] already proposed the above procedure but pointed out that this approach lacks a theoretical justification, which has been given by [13], who showed that FDR $\leq \frac{N_0}{N}\alpha$, where N_0 denotes the number of true null hypotheses.

Scenario 6 (Benjamini-Hochberg (BH) Procedure) *In the situation of Scenario 3, we also performed a BH procedure for all $60 \times 60 = 3600$ entries of the third test image (see left panel of Fig. 11.11). The result is displayed in the right most panel of Fig. 11.11, while in the centre, for a comparison, the result of the Bonferroni procedure on the same data set is displayed. Obviously, more parts of the signal have been found, however, still several positives are missed and a false discovery is included.*

There is a vast literature on FDR control and many generalizations have been proposed. For instance, if $\frac{N_0}{N}$ is much smaller than 1, corresponding to the case of a non-sparse signal, the procedure controls the FDR at much smaller level than α and refined versions of the BH procedure in which N_0/N is estimated from the data have been proposed (see, e.g., [17, 18] and the references given therein).

While the BH procedure grants control of the FDR in test Scenarios 2 and 3 due to independence between pixels, the situation in Scenario 5 is more delicate due to the strong local correlations, in particular in the presence of convolution, where a suitable FDR-procedure is still an open problem and currently investigated by the authors. We stress that while FDR-control under specific dependency structures has been investigated by many authors, e.g., [19, 20] and the references given therein.

Non of the existing methods provide a procedure tailored to deconvolution problems as they occur in photonic imaging. The construction of such adjusted methods is a worthwhile focus for future research.

11.3   Statistical Multiscale Estimation

If one is further interested in the recovery (estimation) of the unknown signal, the multiscale testing procedure developed in Sects. 11.2.6 and 11.2.8 actually provides a collection of feasible candidates for this task in the sense that all signals which fall in the acceptance region of one of the afore-mentioned tests can be considered as "likely" as they cannot be rejected by such a scanning test. More precisely, if we assume model (11.15), any signal $\tilde{\boldsymbol{\mu}}$ which satisfies

$$\max_{h \in \mathcal{H}} \max_{S \in \mathcal{S}(h)} w(h) \left(\left\langle \mathcal{I}_S^*, Y^* - \tilde{\boldsymbol{\mu}} * k \right\rangle - w(h) \right) \le q_{1-\alpha}^*, \qquad (11.18)$$

cannot be rejected. Here $\mathcal{H}$ and $\mathcal{S}(h)$ are defined in Theorem 11.2, and $q_{1-\alpha}^*$ is the $(1 - \alpha)$-quantile of the left hand side of (11.18) with $(Y^* - \tilde{\boldsymbol{\mu}} * k)$ being replaced by noise ε, $w(h)$ is the scale correction term given in (11.10), and $\mathcal{I}_S^*$ is as in (11.16). Among all the candidates $\tilde{\boldsymbol{\mu}}$ lying in (11.18), we will pick the most regular estimate. This is done by means of a (convex) functional $\mathcal{S}(\cdot)$, defined on a domain $\mathcal{D}$ for $\boldsymbol{\mu}$, which encodes prior information about the unknown signal, e.g. sparsity or smoothness. Thus, the final estimator $\hat{\boldsymbol{\mu}}$ is defined as

$$\hat{\boldsymbol{\mu}} \in \arg\min_{\tilde{\boldsymbol{\mu}}} \mathcal{S}(\tilde{\boldsymbol{\mu}}) \quad \text{subject to} \quad \tilde{\boldsymbol{\mu}} \text{ satisfies (11.18)}. \qquad (11.19)$$

Because of the choice of $q_{1-\alpha}^*$ we readily obtain the *regularity guarantee*

$$\mathbb{P}_{\boldsymbol{\mu}} \left(\mathcal{S}(\hat{\boldsymbol{\mu}}) \le \mathcal{S}(\boldsymbol{\mu}) \right) \ge 1 - \alpha \qquad \text{uniformly over all } \boldsymbol{\mu} \in \mathcal{D},$$

i.e., the resulting estimator is at least as regular as the truth with probability $1 - \alpha$, whatever the configuration $\boldsymbol{\mu}$ of the truth is. Furthermore, the remaining residuum $Y^* - \hat{\boldsymbol{\mu}} * k$ is accepted as pure noise by the multiscale procedure described in Sect. 11.2.8.

Before we discuss possible ways how to solve the minimization problem (11.19), note that $\langle \mathcal{I}_S^*, Y - \tilde{\boldsymbol{\mu}} * k \rangle = \langle \mathcal{I}_S^*, Y \rangle - \langle \mathcal{I}_S, \tilde{\boldsymbol{\mu}} \rangle$ in (11.18), and hence the computation can be sped up by avoiding convolutions between $\tilde{\boldsymbol{\mu}}$ and k. Next we emphasize that the discretization of (11.19) has the form

$$\arg\min_{\tilde{\boldsymbol{\mu}}} \mathcal{S}(\tilde{\boldsymbol{\mu}}) \quad \text{subject to} \quad \underline{\lambda} \le K\tilde{\boldsymbol{\mu}} \le \bar{\lambda}, \qquad (11.20)$$

where $\underline{\lambda}, \bar{\lambda}$ are vectors, K a matrix, and "$\leq$" acts element-wise. Thus, whenever S is convex, the whole problem is convex (but, however, non-smooth) and can be solved by many popular methods. In Algorithm 2 we give one possibility which arises from applying the primal-dual hybrid gradient method [21] to an equivalent reformulation of the first order optimality conditions of (11.20) (which are necessary and sufficient by convexity).

Algorithm 2: Primal dual hybrid gradient method for (11.20)

Parameters : Set $\sigma, \tau > 0$ s.t. $\sigma\tau\|K\|^2 < 1$, and $\theta \in [0, 1]$
Initialization: Set $\bar{\mu}_0 = \mu_0 \in \mathcal{D}(K)$ and $\nu_0 \in \mathcal{R}(K)$
1 **for** $n = 1, 2, \dots$ **do**
2 $\quad$ $\nu_n = \nu_{n-1} + \sigma K\bar{\mu}_{n-1}$;
3 $\quad$ $\nu_n = \max\{\nu_n - \sigma\bar{\lambda}, 0\} + \min\{\nu_n - \sigma\underline{\lambda}, 0\}$;
4 $\quad$ $\mu_n = \arg\min_{\tilde{\mu}} \frac{1}{2\tau}\|\tilde{\mu} - (\mu_{n-1} - \tau K^*\nu_n)\|^2 + S(\tilde{\mu})$;
5 $\quad$ $\bar{\mu}_n = \mu_n + \theta(\mu_n - \mu_{n-1})$;

Algorithm 2 relies on efficient computations of the so-called proximal operator of S, see line 4. In most cases, it has either an analytic form if S is ℓ^p-norm ($1 \leq p \leq \infty$), or an efficient solver if S is the total variation semi-norm [22].

One alternative to Algorithm 2 is the alternating direction method of multipliers (ADMM), which can be applied directly to (11.20) and is compatible with any convex functional S [23]. However, Algorithm 2 avoids the projection onto the intersection of convex sets, and turns out to be much faster in practice if step 4 in Algorithm 2 can be efficiently computed. For further algorithms relevant for this problem, see Chaps. 6 and 12.

We stress that a crucial part of the estimator $\hat{\mu}$ in (11.18)–(11.19) is the choice of probe functionals $\mathcal{I}_S^*$ from Sect. 11.2.8. In Fig. 11.12, this estimator $\hat{\mu}$ is referred to as MiScan(short for multiscale image scanning), whereas MrScan(short for multiscale residual scanning) denotes the estimator of a similar form as $\hat{\mu}$ but with $\mathcal{I}_S^*$ being replaced by $\mathcal{I}_S$ see [23–26] i.e., the convolution is not explicitly taken into account in the probe functional. MiScan recovers significantly more features over a range of scales (i.e., various sizes) compared to MrScan.

There is good theoretical understanding on the estimator $\hat{\mu}$ by (11.18)–(11.19) for the regression model (11.2), that is, $k = \delta_0$, the Dirac delta function, in model (11.15). In case of S being Sobolev norms, [27] shows the minimax optimality of $\hat{\mu}$ for Sobolev functions for fixed smoothness, and [28] further show the optimality over Sobolev functions with varying smoothness (adaptation). In case of S being the total variation semi-norm, [29] show the minimax optimality of such an estimator for functions with bounded variation. All the results above are established for L^p-risks ($1 \leq p \leq \infty$). For the more general model (11.15), [30] provide some asymptotic analysis

| Truth | Data | MiScan | MrScan |

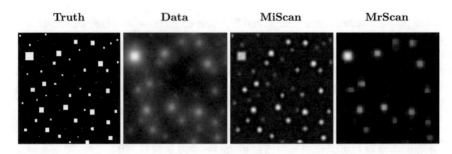

Fig. 11.12 Comparison on a deconvolution problem (SNR $= 100$, and the convolution kernel k satisfying $\mathcal{F}k = 1/(1 + 0.09\| \cdot \|^2)$). MiScanis defined by (11.18)–(11.19); MrScanis similar to MiScanbut with $\mathcal{I}_S^*$ replaced by $\mathcal{I}_S$; For both methods, the regularization functional $\mathcal{S}$ is chosen as the total variation semi-norm

with respect to a relatively weak error measure, the Bregman divergence. A detailed analysis of MiScan exploring the probe functionals in (11.16) in a convolution model is still open and currently investigated by the authors.

References

1. Proksch, K., Werner, F., Munk, A.: Multiscale scanning in inverse problems. Ann. Statist. **46**(6B), 3569–3602 (2018). https://doi.org/10.1214/17-AOS1669
2. Hell, S.W.: Far-field optical nanoscopy. Science **316**, 1153–1158 (2007)
3. Feller, W.: An introduction to probability theory and its applications. Vol. I, 2nd ed, John Wiley and Sons, Inc., New York; Chapman and Hall, Ltd., London (1957)
4. Lehmann, E.L., Romano, J.P.: Testing Statistical Hypotheses. Springer Texts in Statistics, 3rd edn. Springer, New York (2005)
5. Dickhaus, T.: Simultaneous statistical inference. Springer, Heidelberg (2014). https://doi.org/10.1007/978-3-642-45182-9. With applications in the life sciences
6. Gordon, R.D.: Values of Mills' ratio of area to bounding ordinate and of the normal probability integral for large values of the argument. Ann. Math. Stat. **12**, 364–366 (1941)
7. Donoho, D., Jin, J.: Higher criticism for detecting sparse heterogeneous mixtures. Ann. Statist. **32**(3), 962–994 (2004). https://doi.org/10.1214/009053604000000265
8. König, C., Munk, A., Werner, F.: Multidimensional multiscale scanning in exponential families: limit theory and statistical consequences (2018+). Ann. Statist., to appear
9. Arias-Castro, E., Donoho, D.L., Huo, X.: Near-optimal detection of geometric objects by fast multiscale methods. IEEE Trans. Inform. Theory **51**(7), 2402–2425 (2005). https://doi.org/10.1109/TIT.2005.850056
10. Donoho, D.L., Huo, X.: Beamlets and multiscale image analysis. In: Multiscale and Multiresolution Methods, Lect. Notes Comput. Sci. Eng., vol. 20, pp. 149–196. Springer, Berlin (2002). https://doi.org/10.1007/978-3-642-56205-1_3
11. Sharpnack, J., Arias-Castro, E.: Exact asymptotics for the scan statistic and fast alternatives. Electron. J. Stat. **10**(2), 2641–2684 (2016). https://doi.org/10.1214/16-EJS1188
12. Siegmund, D., Yakir, B.: Tail probabilities for the null distribution of scanning statistics. Bernoulli **6**(2), 191–213 (2000). https://doi.org/10.2307/3318574
13. Benjamini, Y., Hochberg, Y.: Controlling the false discovery rate: a practical and powerful approach to multiple testing. J. Roy. Statist. Soc. Ser. B **57**(1), 289–300 (1995). http://links.jstor.org/sici?sici=0035-9246(1995)57:1<289:CTFDRA>2.0.CO;2-E&origin=MSN

14. Chernozhukov, V., Chetverikov, D., Kato, K.: Gaussian approximation of suprema of empirical processes. Ann. Statist. **42**(4), 1564–1597 (2014). https://doi.org/10.1214/14-AOS1230
15. Schmidt-Hieber, J., Munk, A., Dümbgen, L.: Multiscale methods for shape constraints in deconvolution: confidence statements for qualitative features. Ann. Statist. **41**(3), 1299–1328 (2013). https://doi.org/10.1214/13-AOS1089
16. Simes, R.J.: An improved Bonferroni procedure for multiple tests of significance. Biometrika **73**(3), 751–754 (1986). https://doi.org/10.1093/biomet/73.3.751
17. Kumar Patra, R., Sen, B.: Estimation of a two-component mixture model with applications to multiple testing. J. Roy. Statist. Soc. Ser. B **78**(4), 869–893 (2016). https://doi.org/10.1111/rssb.12148
18. Storey, J.D., Taylor, J.E., Siegmund, D.: Strong control, conservative point estimation and simultaneous conservative consistency of false discovery rates: a unified approach. J. Roy. Statist. Soc. Ser. B **66**(1), 187–205 (2004). https://doi.org/10.1111/j.1467-9868.2004.00439.x
19. Benjamini, Y., Yekutieli, D.: The control of the false discovery rate in multiple testing under dependency. Ann. Statist. **29**(4), 1165–1188 (2001). https://doi.org/10.1214/aos/1013699998
20. Finner, H., Dickhaus, T., Roters, M.: Dependency and false discovery rate: asymptotics. Ann. Statist. **35**(4), 1432–1455 (2007). https://doi.org/10.1214/009053607000000046
21. Chambolle, A., Pock, T.: A first-order primal-dual algorithm for convex problems with applications to imaging. J. Math. Imaging Vision **40**(1), 120–145 (2011). https://doi.org/10.1007/s10851-010-0251-1
22. Chambolle, A.: An algorithm for total variation minimization and applications. J. Math. Imaging Vision **20**(1-2), 89–97 (2004). https://doi.org/10.1023/B:JMIV.0000011320.81911.38. Special issue on mathematics and image analysis
23. Frick, K., Marnitz, P., Munk, A.: Statistical multiresolution Dantzig estimation in imaging: fundamental concepts and algorithmic framework. Electron. J. Stat. **6**, 231–268 (2012). https://doi.org/10.1214/12-EJS671
24. Aspelmeier, T., Egner, A., Munk, A.: Modern statistical challenges in high-resolution fluorescence microscopy. Annu. Rev. Stat. Appl. **2**, 163–202 (2015)
25. Frick, K., Marnitz, P., Munk, A.: Statistical multiresolution estimation for variational imaging: with an application in Poisson-biophotonics. J. Math. Imaging Vision **46**(3), 370–387 (2013). https://doi.org/10.1007/s10851-012-0368-5
26. Li, H.: Variational estimators in statistical multiscale analysis. Ph.D. thesis, Georg-August-Universität Göttingen (2016)
27. Nemirovski, A.: Nonparametric estimation of smooth regression functions. Izv. Akad. Nauk. SSR Teckhn. Kibernet. (in Russian) **3**, 50–60 (1985). J. Comput. System Sci., 23:1–11, 1986 (in English)
28. Grasmair, M., Li, H., Munk, A.: Variational multiscale nonparametric regression: smooth functions. Ann. Inst. Henri Poincaré Probab. Stat. **54**(2), 1058–1097 (2018). https://doi.org/10.1214/17-AIHP832
29. del Álamo, M., Li, H., Munk, A.: Frame-constrained total variation regularization for white noise regression (2018). arXiv preprint arXiv:1807.02038
30. Frick, K., Marnitz, P., Munk, A.: Shape-constrained regularization by statistical multiresolution for inverse problems: asymptotic analysis. Inverse Probl. **28**(6), 065,006, 31 (2012). https://doi.org/10.1088/0266-5611/28/6/065006

Chapter 12
Efficient, Quantitative Numerical Methods for Statistical Image Deconvolution and Denoising

D. Russell Luke, C. Charitha, Ron Shefi and Yura Malitsky

Abstract We review the development of efficient numerical methods for statistical multi-resolution estimation of optical imaging experiments. In principle, this involves constrained linear deconvolution and denoising, and so these types of problems can be formulated as convex constrained, or even unconstrained, optimization. We address two main challenges: first of these is to quantify convergence of iterative algorithms; the second challenge is to develop efficient methods for these large-scale problems without sacrificing the quantification of convergence. We review the state of the art for these challenges.

2010 Mathematics Subject Classification: Primary 49J52 · 49M20 · 90C26 · Secondary 15A29 · 47H09 · 65K05 · 65K10 · 94A08.

12.1 Introduction

In this chapter we review progress towards addressing two main challenges in scientific image processing. The first of these is to quantify convergence of iterative algorithms for image processing to solutions (as opposed to optimal values) to the underlying variational problem. The second challenge is to develop efficient methods for these large-scale problems without sacrificing the quantification of convergence.

D. R. Luke (✉) · R. Shefi · Y. Malitsky
Institute for Numerical and Applied Mathematics, Universität Göttingen,
37083 Göttingen, Germany
e-mail: r.luke@math.uni-goettingen.de

R. Shefi
e-mail: ronshefi@gmail.com

Y. Malitsky
e-mail: y.malitskyi@math.uni-goettingen.de

C. Charitha
Indian Institute of Technology Indore, Indore, India
e-mail: charithac@iiti.ac.in

© The Author(s) 2020
T. Salditt et al. (eds.), *Nanoscale Photonic Imaging*, Topics in Applied Physics 134,
https://doi.org/10.1007/978-3-030-34413-9_12

The techniques surveyed here were studied in [1–3]. We present only the main results from these studies, in the context that hindsight provides.

Scientific images are often processed with software that accomplishes a number of tasks like registration, denoising and deblurring. Implicit in the processing is that some systematic error is being corrected to bring the image closer to the truth. This presumption is more complicated for denoising and deblurring. These are often accomplished by filtering or by solving some variational problem such as minimizing the variance of an image. For applications requiring speedy processing, such as audio and video communication, this is sufficient. But the recent development of nanoscale photonic imaging modalities such as STED and RESOLFT featured in Chaps. 1, 7 and 9 has shifted the focus of image denoising and deconvolution from qualitative to quantitative models.

Quantitative approaches to image processing are the subject of Chap. 11 where statistical multiscale estimation is discussed (see Sect. 11.3). Here, the recovered image comes with statistical statements about how far the processed image is, in some statistical sense, from the truth. The estimators are almost exclusively variational, that is, they can be characterized as the solution to an optimization problem. It is important to emphasize that the value of the optimization problem is meaningless. This stands in stark contrast to many conventional applications in economics and operations research, where the value of the optimal solution is related to profit or cost, and so is of principal interest.

The focus on optimal solutions rather than optimal values places heavy demands on the structure of model formulations and the algorithms for solving them. Unless the numerical method allows one to state how far a computed iterate is to the solution of the underlying variational problem, then the scientific significance of the iterate is lost.

The leading computational approaches for solving imaging problems with multi-resolution statistical estimation criterion are based on iterated proximal operators. Most of the analysis for first-order iterative proximal methods is limited to statements about rates of convergence of function values, if rates are discussed at all (see for instance [4–7]). First-order methods have slow convergence in the worst case scenario. A common assumption to guarantee linear convergence of the iterates is strong convexity, but this is far more than is necessary, and in particular it is not satisfied for the Huber function (12.35). It was shown in [8] that metric subregularity is necessary for local linear convergence. Aspelmeier, Charitha and Luke [1] showed that the popular alternating directions method of multipliers algorithm (ADMM) applied to optimization problems with piecewise linear-quadratic objective functions (e.g. the Huber function), together with linear inequality constraints generically satisfies metric subregularity at isolated critical points; hence linear convergence of the iterates for this algorithm can be expected without further ado. More recently, in [3] it was shown that the primal iterates of a modification of the PAPC algorithm (Algorithm 2) converge R-linearly for any *quadratically supportable* objective function (for instance, the Huber function). Conventional results without metric subregularity obtain a convergence rate of $O(1/k)$ with respect to the function values. In settings like qualitative image processing or machine learning, such results are acceptable, but in the setting of statistical image processing these statements do not contain any scientific content. We present in this chapter efficient iterative first-order methods that offer some hope of quantitative guarantees about the distance of the iterates to optimal solutions.

12.2 Problem Formulation

We limit our scope to the real vector space $\mathbb{R}^n$ with the norm generated from the inner product. The closed unit ball centered on the point $y \in \mathbb{R}^n$ is denoted by $\mathbb{B}(y)$. The positive orthant (resp. negative orthant) in $\mathbb{R}^n$ is denoted by $\mathbb{R}^n_+$ (resp. $\mathbb{R}^n_-$). The domain of an extended real-valued function $\varphi : \mathbb{R}^n \to \overline{\mathbb{R}} \equiv (-\infty, +\infty]$ is $\operatorname{dom}(\varphi) \equiv \{z \in \mathbb{R}^n : \varphi(z) < +\infty\}$. The *Fenchel conjugate* of φ is denoted by φ^* and is defined by $\varphi^*(u) \equiv \sup_{z \in \mathbb{R}^n}\{\langle z, u\rangle - \varphi(z)\}$. The set of symmetric $n \times n$ positive (semi)-definite matrices is denoted by $\mathbb{S}^n_{++}$ ($\mathbb{S}^n_+$). The notation $A \succ 0$ ($A \succeq 0$) denotes a positive (semi)definite matrix A. For any $z \in \mathbb{R}^n$ and any $A \in \mathbb{S}^n_+$, we denote the semi-norm $\|z\|_A^2 := \langle z, Az\rangle$. The operator norm is defined by $\|A\| = \max_{u \in \mathbb{R}^n}\{\|Au\| : \|u\| = 1\}$ and coincides with the spectral radius of A whenever A is symmetric. If $A \neq 0$, $\sigma_{min}(A)$ denotes its smallest nonzero singular value. For a sequence $\{z^k\}_{k \in \mathbb{N}}$ converging to z^*, we say the convergence is Q-linear if there exists $c \in (0, 1)$ such that $\frac{\|z^{k+1}-z^*\|}{\|z^k-z^*\|} \leq c$ for all k; convergence is R-linear if there exists a sequence η_k such that $\|z^k - z^*\| \leq \eta_k$ and $\eta_k \to 0$ Q-linearly [9, Chap. 9].

We limit our discussion to proper (nowhere equal to $-\infty$ and finite at some point), lower semi-continuous (lsc), extended-valued (can take the value $+\infty$) functions. We will, in fact, limit our discussion to *convex* functions, but convexity is not the central property governing quantitative convergence estimates. By the *subdifferential* of a function φ, denoted $\partial \varphi$, we mean the collection of all *subgradients* that can be written as limits of sequences of *Fréchet subgradients* at nearby points; a vector v is a *(Fréchet) subgradient* of φ at y, written $v \in \widehat{\partial}\varphi(y)$, if

$$\liminf_{x \to y,\ x \neq y} \frac{\varphi(x) - \varphi(y) - \langle v, x - y\rangle}{\|x - y\|} \geq 0. \tag{12.1}$$

The functions of interest for us are *subdifferentially regular* on their domains, that is, the epigraphs of the functions are *Clarke regular* at points where they are finite [10, Definition 7.25]. For our purposes it suffices to note that, for a function φ that is subdifferentially regular at a point y, the subdifferential is nonempty and all subgradients are Fréchet subgradients, that is, $\partial \varphi(y) = \widehat{\partial}\varphi(y) \neq \emptyset$. Convex functions, in particular, are subdifferentially regular on their domains and the subdifferential has the particularly simple representation as the set of all vectors v where

$$\varphi(x) - \varphi(y) - \langle v, x - y\rangle \geq 0 \quad \forall x. \tag{12.2}$$

A mapping $\Phi : \mathbb{R}^n \rightrightarrows \mathbb{R}^n$ is said to be β-*inverse strongly monotone* [10, Corollary 12.55] if for all $x, x' \in \mathbb{R}^n$

$$\langle v - v', x - x' \rangle \geq \beta \|v - v'\|^2, \quad \text{whenever} \quad v \in \Phi(x), v' \in \Phi(x'). \quad (12.3)$$

The mapping Φ is said to be *polyhedral* (or piecewise polyhedral [10]) if its graph is the union of finitely many sets that are polyhedral convex in $\mathbb{R}^n \times \mathbb{R}^n$ [11]. Polyhedral mappings are generated by the subdifferential of piecewise linear- quadratic functions (see Proposition 12.9).

Definition 12.1 (*piecewise linear-quadratic (plq) functions*) A function $f : \mathbb{R}^n \rightarrow [-\infty, +\infty]$ is called *piecewise linear-quadratic* if $dom f$ can be represented as the union of finitely many polyhedral sets, relative to each of which $f(x)$ is given by an expression of the form $\frac{1}{2}\langle x, Ax \rangle + \langle a, x \rangle + \alpha$ for some scalar $\alpha \in \mathbb{R}$ vector $a \in \mathbb{R}^n$, and symmetric matrix $A \in \mathbb{R}^{n \times n}$.

Closely related to plq functions is quadratically supportable functions.

Definition 12.2 (*pointwise quadratically supportable (pqs)*) A proper, extended-valued function $\varphi : \mathbb{R}^n \rightarrow \mathbb{R} \cup \{+\infty\}$ is said to be *pointwise quadratically supportable at y* if it is subdifferentially regular there and there exists a neighborhood V of y and a constant $\mu > 0$ such that

$$(\forall v \in \partial\varphi(y)) \quad \varphi(x) \geq \varphi(y) + \langle v, x - y \rangle + \frac{\mu}{2}\|x - y\|^2, \quad \forall x \in V. \quad (12.4)$$

If for each bounded neighborhood V of y there exists a constant $\mu > 0$ such that (12.4) holds, then the function φ is said to be *pointwise quadratically supportable at y on bounded sets*. If (12.4) holds with one and the same constant $\mu > 0$ on all neighborhoods V, then φ is said to be *uniformly* pointwise quadratically supportable at y.

For more on the relationship between pointwise quadratic supportability, coercivity, strong monotonicity and strong convexity see [3].

We denote the *resolvent* of Φ by $\mathcal{J}_\Phi \equiv (\text{Id} + \Phi)^{-1}$ where Id denotes the identity mapping and the inverse is defined as

$$\Phi^{-1}(y) \equiv \{x \in \mathbb{R}^n \mid y \in \Phi(x)\}. \quad (12.5)$$

The corresponding *reflector* is defined by $R_{\eta\Phi} \equiv 2\mathcal{J}_{\eta\Phi} - \text{Id}$. One of the more prevalent examples of resolvents is the *proximal map*. For $\varphi : \mathbb{R}^n \rightarrow (-\infty, \infty]$ a proper, lsc and convex function and for any $u \in \mathbb{R}^n$ and $Q \in \mathbb{S}^n_{++}$, the proximal map associated with φ with respect to the weighted Euclidean norm is uniquely defined by:

$$\text{prox}_{Q,\varphi}(u) = \text{argmin}_z\{\varphi(z) + \frac{1}{2}\|z - u\|^2_Q : z \in \mathbb{R}^n\}.$$

When $Q = c^{-1}\mathrm{Id}$, $c > 0$, we simply use the notation $\mathrm{prox}_{c,\varphi}(u)$. We also recall the fundamental Moreau proximal identity [12], that is, for any $z \in \mathbb{R}^n$

$$z = \mathrm{prox}_{Q,\varphi}(z) + Q\mathrm{prox}_{Q^{-1},\varphi^*}(Q^{-1}(z)), \tag{12.6}$$

where Q^{-1} is the inverse of $Q \in \mathbb{S}^n_{++}$.

Notions of *continuity* of set-valued mappings have been thoroughly developed over the last 40 years. Readers are referred to the monographs [10, 11, 13] for basic results. A key property of set-valued mappings that we will rely on is *metric subregularity*, which can be understood as the property corresponding to a Lipschitz-like continuity of the inverse mapping relative to a specific point. It is a weaker property than *metric regularity* which, in the case of an $n \times m$ matrix for instance, is equivalent to surjectivity. Our definition follows the characterization of this property given in [11, Exercise 3H.4].

Definition 12.3 (*metric subregularity*) The mapping $\Phi : \mathbb{R}^n \rightrightarrows \mathbb{R}^m$ is called *metrically subregular at $\overline{x}$ for $\overline{y}$ relative to $W \subset \mathbb{R}^n$* if $(\overline{x}, \overline{y}) \in \mathrm{gph}\Phi$ and there is a constant $c > 0$ and neighborhoods $\mathcal{O}$ of $\overline{x}$ such that

$$\mathrm{dist}(x, \Phi^{-1}(\overline{y}) \cap W) \leq c\mathrm{dist}(\overline{y}, \Phi(x)) \ \forall \, x \in \mathcal{O} \cap W. \tag{12.7}$$

The constant c measures the stability under perturbations of inclusion $\overline{y} \in \Phi(\overline{x})$. An important instance where metric subregularity comes for free is for polyhedral mappings.

Proposition 12.4 (polyhedrality implies metric subregularity) *Let $W \subset V$ be an affine subspace and $T : W \rightrightarrows W$. If T is polyhedral and $\mathsf{Fix}\, T \cap W$ is an isolated point, $\{\overline{x}\}$, then $T - \mathrm{Id} : W \rightrightarrows (W - \overline{x})$ is metrically subregular at $\overline{x}$ for 0 relative to W.*

Proof Polyhedrality and isolated fixed points in fact imply *strong* metric subregularity. See [11, Propositions 3I.1 and 3I.2]. $\quad\square$

A notion related to metric regularity is that of *weak-sharp solutions*. This will be used in the development of error bounds (Theorem 12.6).

Definition 12.5 (*weak sharp minimum* [14]) The solution set $\mathrm{argmin}\,\{f(x) \mid x \in \Omega\}$ for a nonempty closed convex set Ω, is weakly sharp if, for $\overline{p} = \inf_{\Omega} f$, there exists a positive number α (sharpness constant) such that

$$f(x) \geq \overline{p} + \alpha\,\mathrm{dist}(x, S_f) \quad \forall x \in \Omega.$$

Similarly, the solution set S_f is weakly sharp of order $\nu > 0$ if there exists a positive number α (sharpness constant) such that, for each $x \in \Omega$,

$$f(x) \geq \overline{p} + \alpha\,\mathrm{dist}(x, S_f)^{\nu} \quad \forall x \in \Omega.$$

12.2.1 Abstract Problem

The generic problem in which we are interested is

$$\begin{array}{ll} \underset{x \in \mathbb{R}^n}{\text{minimize}} & f(x) \\ \text{subject to} & g_i(A_i x) \leq 0 \quad (i = 1, \ldots M) \end{array} \tag{$\mathcal{P}_0$}$$

The following blanket assumptions on the problem's data hold throughout:

Assumption 1

(i) *The set of optimal solutions for problem $(\mathcal{P}_0)$, denoted S^*, is nonempty.*

(ii) *the function $f : \mathbb{R}^n \to \mathbb{R}$ is proper lsc and convex and coercive, and for $i = 1, \ldots, M$ the functions $g_i : \mathbb{R}^{m_i} \to (-\infty, +\infty]$ are proper, lsc, and convex;*

(iii) *the mappings $A_i : \mathbb{R}^n \to \mathbb{R}^{m_i}$, $i = 1, \ldots, M$ ($m_i \leq n$) are linear and full rank, that is, $\sigma^2_{min}(A_i) = \lambda_{min}(A_i A_i^T) > 0$.*

Assumption (i) implies that the optimal value of $(\mathcal{P}_0)$ is finite. Assumption (ii) implies that the constraint structure is convex. Assumption (iii) implies that the mapping $\mathcal{A} : \mathbb{R}^n \to \mathbb{R}^m$ is linear and full rank, where

$$\mathcal{A} \equiv [A_0^T, A_1^T, \ldots, A_M^T]^T \in \mathbb{R}^m \times \mathbb{R}^n$$

so that $\mathcal{A}x = y$ where $y = (y_1, \ldots, y_M) \in \mathbb{R}^m$ for $m = \sum_{i=1}^M m_i$. The challenge of statistical multi-resolution estimation lies in the feature that the dimension of the constraint structure, m, is much greater than the dimension of the unknowns, n, and grows superlinearly with respect to the number of unknowns.

The above constrained optimization problem is often formulated as an unconstrained-looking problem via the introduction of a (nonsmooth) penalty term enforcing the constraints:

$$\min_{x \in \mathbb{R}^n} \quad f(x) + g(\mathcal{A}x) \tag{$\mathcal{P}$}$$

where

$$g : \mathbb{R}^m \to \overline{\overline{\mathbb{R}}} \equiv \rho\theta \tag{12.8}$$

for ρ a positive scalar and

$\theta : \mathbb{R}^m \to [0, +\infty]$ proper, lsc and convex with $\theta(y) = 0$ if and only if $y \in \mathcal{C}$ (12.9)

where

$$\mathcal{C} \equiv \{y = (y_1, y_2, \ldots, y_M) \mid y \in \text{range } \mathcal{A} \text{ and } g_i(y_i) \leq 0, \ i = 1, 2, \ldots, M\}. \tag{12.10}$$

The requirements on the function θ align this penalty term with *exact* penalization [15], that is, a relaxation of the constraints where, for all parameters ρ large enough, the constraints are exactly satisfied.

The following assumptions are used to guarantee the exact correspondence between solutions to $(\mathcal{P}_0)$ and $(\mathcal{P})$.

Assumption 2

(i) The set $\mathcal{S}_\infty \equiv \operatorname{argmin}\{\theta(Ax) \mid x \in \mathbb{R}^n\}$ is nonempty and weakly sharp (Definition 12.5) of order $\nu \geq 1$.

(ii) The lower level set $\operatorname{lev}_{\leq \alpha} f$ is bounded for each $\alpha \in \mathbb{R}$ and $\inf_{x \in \mathbb{R}^n} f > -\infty$.

Proposition 12.6 *Suppose Assumption 2 holds. Then the set of solutions to $(\mathcal{P})$ with g defined by (12.8) is bounded and for all ρ large enough, the solutions to $(\mathcal{P}_0)$ and $(\mathcal{P})$ coincide.*

Proof This is a distillation of [1, Theorem 3.4].

In $(\mathcal{P}_0)$ and $(\mathcal{P})$ the function f is often smooth, but not prox friendly. In applications it is most often a smooth regularization or a fidelity term. For the ADMM/DR method reviewed in Sect. 12.3 smoothness is not required.

It is assumed that the functions g_i ($i = 1, 2, \ldots, M$) are prox friendly and that they enjoy some structure that makes g also prox friendly. For instance, if the constraints are *separable*, then the function

$$g(Ax) \equiv \rho \sum_{i=1}^{M} g_i(A_i x) \tag{12.11}$$

is also prox-friendly as is the function

$$g(Ax) \equiv \rho \max\{g_1(A_1 x), g_2(A_2 x), \ldots, g_M(A_M x), 0\}. \tag{12.12}$$

The functions $g_i \circ A_i$ can be regularizing functions (like total variation) or hard inequality constraints. For example, hard inequality constraints are modeled by the use of indicator functions for g_i in $(\mathcal{P}_0)$: $g_i(A_i x) = \iota_{\gamma_i \mathbb{B}}(A_i x - b_i)$ where, for a subset $\Omega \subset \mathbb{R}^n$,

$$\iota_\Omega(x) \equiv \begin{cases} 0 & x \in \Omega \\ +\infty & \text{else.} \end{cases}$$

12.2.2 Saddle Point and Dual Formulations

The saddle point formulation is derived by viewing the function g in $(\mathcal{P})$ as the *image* of a function g^* under Fenchel conjugation, that is, $g(x) = (g^*)^*$. Writing this explicitly into $(\mathcal{P})$ yields

$$\min_{x \in \mathbb{R}^n} \left\{ \max_{y \in \mathbb{R}^m} \ f(x) + \langle Ax, y \rangle - g^*(y) \right\}. \tag{$\mathcal{S}$}$$

The bifunction in the saddle point formulation is

$$L(x, y) := f(x) + \langle Ax, y \rangle - g^*(y). \tag{12.13}$$

Contrast this with the Lagrangian for the extended problem

$$\operatorname*{minimize}_{(x,y) \in \mathbb{R}^n \times \mathbb{R}^m} \ f(x) + g(y), \quad \text{subject to } Ax = y. \tag{$\mathcal{P}_{\mathcal{L}}$}$$

The Lagrangian is

$$\mathcal{L}(x, y, z) = f(x) + g(y) + \langle z, Ax - y \rangle, \tag{12.14}$$

and the augmented Lagrangian $\widetilde{\mathcal{L}}$ for $(\mathcal{P}_{\mathcal{L}})$ is given by

$$\widetilde{\mathcal{L}}(x, y, z) = f(x) + g(y) + \langle z, Ax - y \rangle + \tfrac{\eta}{2} \|Ax - y\|^2, \tag{12.15}$$

where $z \in \mathbb{R}^m$, and $\eta > 0$ is a fixed penalty parameter.

Assumption 1(i) guarantees that the mapping $L(\cdot, \cdot)$ has a saddle point, that is, there exists $(\hat{x}, \hat{y}) \in \mathbb{R}^n \times \mathbb{R}^m$ such that

$$L(\hat{x}, y) \leq L(\hat{x}, \hat{y}) \leq L(x, \hat{y}) \quad \forall x \in \mathbb{R}^n, y \in \mathbb{R}^m.$$

The existence of a saddle point corresponds to zero duality gap for the induced optimization problems

$$p(x) = \sup_{y}\{L(x, y) : y \in \mathbb{R}^m\} \quad q(y) = \inf_{x}\{L(x, y) : x \in \mathbb{R}^n\}.$$

By weak duality, we have $\inf_{x \in \mathbb{R}^n} p(x) \geq \sup_{y \in \mathbb{R}^m} q(y)$.

This can be viewed as a *partial dual* to problem $(\mathcal{P})$. The full dual problem involves the Fenchel conjugate of the entire objective function. For $(\mathcal{P})$ the dual problem is

$$\sup_{y \in \mathbb{R}^m} -f^*(A^T y) - g^*(-y). \tag{$\mathcal{D}'$}$$

Instead of working with this dual, it is more convenient to work with the following equivalent formulation via the change of variable $y \to -y$:

$$\inf_{y \in \mathbb{R}^m} f^*(-A^T y) + g^*(y). \tag{$\mathcal{D}$}$$

Under standard constraint qualifications (e.g., [16, Theorem 2.3.4]), $(\hat{x}, \hat{y})$ is a saddle point of L if and only if $\hat{x}$ is an optimal solution of the primal problem $(\mathcal{P}_0)$, and $\hat{y}$ is an optimal solution of the dual problem ($\mathcal{D}$). The following two inclusions characterize the solutions of the problems ($\mathcal{P}_0$) and ($\mathcal{D}$) respectively:

$$0 \in \partial f(\overline{x}) + \partial(g \circ \mathcal{A})(\overline{x});$$

$$0 \in \partial \left(f^* \circ (-\mathcal{A}^T)\right)(\overline{y}) + \partial g^*(\overline{y}).$$

In both cases, one has to solve an inclusion of the form

$$0 \in (B + D)(x), \tag{12.16}$$

for general set-valued mappings B and D.

12.2.3 Statistical Multi-resolution Estimation

Statistical multi-resolution estimation (SMRE) discussed in Sect. 11.2.7 of Chap. 11 is specialized here for the case of imaging systems with Gaussian noise. Let $\mathcal{G} = \{1, \ldots, N\} \times \{1, \ldots, N\}$ denote a grid of $N^2 = n$ points. Denote by $\mathcal{V}$ a collection of subsets of $\mathcal{G}$. Suppose there are M such subsets, that is $M = |\mathcal{V}|$, each of these subsets V_i consisting of $m_i \leq n$ grid points with $\sum_{i=1}^{M} m_i = m \geq n$.

The variational model for statistical multi-resolution estimation with Gaussian noise takes the form

$$\begin{aligned} &\underset{x \in \mathbb{R}^n}{\text{minimize}} && f(x) \\ &\text{subject to} && \left|\sum_{j \in V_i} w_i(j) \left((Ax)_j - b_j\right)\right| \leq \gamma_i && \forall i = 1, 2, \ldots, M. \end{aligned}$$
$$(\mathcal{P}_{SMRE})$$

Here $f : \mathbb{R}^n \to \mathbb{R}$ is a regularization functional, which incorporates a priori knowledge about the unknown signal $\widehat{x}$ such as smoothness, w_i is a weighting function for the grid points in the subset V_i, and $A : \mathbb{R}^n \to \mathbb{R}^n$ is the linear imaging operator that models the experiment. The constant γ_i has an interpretation in terms of the *quantile* of the estimator.

In the context of the general model ($\mathcal{P}_0$),

$$g_i(A_i x) = \left|\sum_{j \in V_i} w_i(j) \left((Ax)_j - b_j\right)\right| - \gamma_i, \quad i = 1, 2, \ldots, M. \tag{12.17}$$

Here the affine mapping $A_i x \equiv \sum_{j \in V_i} w_i(j) \left((Ax)_j - b_j\right)$ is an averaging operator that accounts for sampling at different resolutions of the image. Note that the observation b need not be in the range of the imaging operator A - all that is assumed is that this mapping is injective, not surjective. This means that, in applications, practitioners need to be careful not to make the constraint γ_i too small, otherwise the optimization problem might be *infeasible*. If the algorithms presented below appear to be diverging for a particular instance of ($\mathcal{P}_{SMRE}$), it is because the problem is infeasible; increasing the constants γ_i should solve the problem.

12.3   Alternating Directions Method of Multipliers and Douglas Rachford

In this section we survey the main results (without proofs) from [1]. For proofs of the statements, readers are referred to that article. A starting point for most of the main approaches to solving ($\mathcal{P}_0$) is the alternating directions method of multipliers (ADMM) (primary sources include [17–21]). This method is one of many *splitting methods* which are the principal approach to handling the computational burden of large-scale, separable problems [22]. ADMM belongs to a class of *augmented Lagrangian methods* whose original motivation was to regularize Lagrangian formulations of constrained optimization problems. The ADMM algorithm for solving ($\mathcal{P}_{\mathcal{L}}$) follows. The penalty parameter η need not be a constant, and indeed evidence indicates that the choice of η can greatly impact the complexity of the algorithm. For simplicity we keep this parameter fixed.

Algorithm 1: ADMM for ($\mathcal{P}_{\mathcal{L}}$) as in [1, Algorithm 2.1].

Parameters : Set $\eta > 0$
Initialization: Set $(x^0, y^0, z^0) \in \mathbb{R}^n \times \mathbb{R}^m \times \mathbb{R}^m$
1 **for** $k = 1, 2, \ldots$ **do**
2 $\quad$ $x^{k+1} \in \operatorname{argmin}_x \left\{ f(x) + \langle z^k, \mathcal{A}x \rangle + \frac{\eta}{2} \| \mathcal{A}x - y^k \|^2 \right\}$;
3 $\quad$ $y^{k+1} \in \operatorname{argmin}_y \left\{ g(y) - \langle z^k, y \rangle + \frac{\eta}{2} \| \mathcal{A}x^{k+1} - y \|^2 \right\}$;
4 $\quad$ $z^{k+1} = z^k + \eta(\mathcal{A}x^{k+1} - y^{k+1})$.

We do not specify how the argmin in Algorithm 1 should be calculated, and indeed, the analysis that follows assumes that these can be computed exactly.[1] One problem that should be immediately apparent is that this algorithm operates on a space of dimension $n + 2m$. Since one of the two challenges we address is high dimension, this expansion in the dimension of the problem formulation should be troubling. Nevertheless, we show with this algorithm how the first challenge, namely *quantification* of convergence is achieved.

The connection between the ADMM algorithm and the Douglas–Rachford algorithm introduced in Chap. 6, (6.30) was first discovered by Gabay [19] (see also the thesis of Eckstein [17]). For any $\eta > 0$, the Douglas–Rachford algorithm [23, 24] for solving (12.16) is given by

---

[1]This is not true in practice and remains an unresolved issue in numerical variational analysis.

$$z^{k+1} \in T'z^k \quad (k \in \mathbb{N}), \tag{12.18}$$

$$\text{for} \quad T' \equiv \mathcal{J}_{\eta D}\left(\mathcal{J}_{\eta B}(\text{Id} - \eta D) + \eta D\right), \tag{12.19}$$

where $\mathcal{J}_{\eta D} \equiv (\text{Id} + \eta D)^{-1}$ and $\mathcal{J}_{\eta B} \,(\text{Id} + \eta B)^{-1}$ are the *resolvents* of ηD and ηB respectively.

Given z^0 and $y^0 \in Dz^0$, following [25], define the new variable $\xi^0 \equiv z^0 + \eta y^0$ so that $z^0 = \mathcal{J}_{\eta D}\xi^0$. We thus arrive at an alternative formulation of the Douglas–Rachford algorithm (12.18):

$$\xi^{k+1} \in T\xi^k \quad (k \in \mathbb{N}), \tag{12.20}$$

$$\text{for} \quad T \equiv \tfrac{1}{2}(R_{\eta B}R_{\eta D} + \text{Id}) = \mathcal{J}_{\eta B}(2\mathcal{J}_{\eta D} - \text{Id}) + (\text{Id} - \mathcal{J}_{\eta D}), \tag{12.21}$$

where $R_{\eta D}$ and $R_{\eta B}$ are the reflectors of the respective resolvents. This is the form of Douglas–Rachford considered in [26].

Specializing this to our application yields

$$B \equiv \partial\left(f^* \circ (-A^T)\right) \quad \text{and} \quad D \equiv \partial g^*, \tag{12.22}$$

and so the resolvent mappings are the proximal mappings of the convex functions $\left(f^* \circ (-A^T)\right)$ and g^* respectively, and hence the resolvent mappings and corresponding fixed point operator T are single-valued [12].

Proposition 12.7 (Proposition 2.2 [1]) *Let $f : \mathbb{R}^n \to \overline{\mathbb{R}}$ and $g : \mathbb{R}^m \to \overline{\mathbb{R}}$ be proper, lsc and convex. Let $A : \mathbb{R}^n \to \mathbb{R}^m$ be linear and suppose there exists a solution to $0 \in (B + D)(x)$ for B and D defined by (12.22). For fixed $\eta > 0$, given any initial points ξ^0 and $\left(y^0, z^0\right) \in \text{gph}\,D$ such that $\xi^0 = y^0 + \eta z^0$, the sequences $\left(z^k\right)_{k \in \mathbb{N}}$, $\left(\xi^k\right)_{k \in \mathbb{N}}$ and $\left(y^k\right)_{k \in \mathbb{N}}$ defined respectively by (12.18), (12.20) and $y^k \equiv \tfrac{1}{\eta}\left(\xi^k - z^k\right)$ converge to points $\overline{z} \in \text{Fix}\,T'$, $\overline{\xi} \in \text{Fix}\,T$ and $\overline{y} \in D\left(\text{Fix}\,T'\right)$. The point $\overline{z} = \mathcal{J}_{\eta D}\overline{\xi}$ is a solution to (D), and $\overline{y} = \tfrac{1}{\eta}\left(\overline{\xi} - \overline{z}\right) \in D\overline{z}$. If, in addition, A has full column rank, then the sequence $\left(y^k, z^k\right)_{k \in \mathbb{N}}$ corresponds exactly to the sequence of points generated in steps 2–3 of Algorithm 1 and the sequence $\left(\xi^{k+1}\right)_{k \in \mathbb{N}}$ converges to $\overline{\xi}$, a solution to $(\mathcal{P}_0)$.*

The correspondence between Douglas-Rachford and ADMM in the proposition above means that if quantitative convergence can be established for one of the algorithms, it is automatically established for the other. Linear convergence of Douglas-Rachford under the assumption of strong convexity and Lipschitz continuity of f was already established by Lions and Mercier [26]. Recent published work in this direction includes [27, 28]. Local linear convergence of the iterates to a *solution* was established in [29] for linear and quadratic programs using spectral analysis. In Proposition 12.8, two conditions are given that guarantee linear convergence of the ADMM iterates to a solution. The first condition is classical and follows Lions and Mercier [26]. The second condition, based on [30], is much more prevalent in applications and generalizes the results of [29].

Proposition 12.8 (Theorem 2.3 of [1]) *Let $f : \mathbb{R}^n \to \overline{\mathbb{R}}$ and $g : \mathbb{R}^m \to \overline{\mathbb{R}}$ be proper, lsc and convex. Suppose there exists a solution to $0 \in (B + D)(x)$ for B and D defined by (12.22) where $\mathcal{A} : \mathbb{R}^n \to \mathbb{R}^m$ is an injective linear mapping. Let $\widehat{\xi} \in$ Fix T for T defined by (12.21). For fixed $\eta > 0$ and any given triplet of points (ξ^0, y^0, z^0) satisfying $\xi^0 \equiv z^0 + \eta y^0$, with $y^0 \in Dz^0$, generate the sequence $(y^k, z^k)_{k \in \mathbb{N}}$ by Steps 2–3 of Algorithm 1 and the sequence $(\xi^k)_{k \in \mathbb{N}}$ by (12.20).*

(i) *Let $\mathcal{O} \subset \mathbb{R}^n$ be a neighborhood of $\widehat{\xi}$ on which g is strongly convex with constant μ and ∂g is β-inverse strongly monotone for some $\beta > 0$. Then, for any $(\xi^0, y^0, z^0) \in \mathcal{O}$ satisfying $\xi^0 \equiv z^0 + \eta y^0 \in \mathcal{O}$, the sequences $(\xi^k)_{k \in \mathbb{N}}$ and $(y^k, z^k)_{k \in \mathbb{N}}$ converge linearly to the respective points $\overline{\xi} \in$ Fix T and $(\overline{y}, \overline{z})$ with*

    rate at least $K = \left(1 - \frac{2\eta\beta\mu^2}{(\mu+\eta)^2}\right)^{\frac{1}{2}} < 1$.

(ii) *Suppose that $T : W \to W$ for some affine subspace $W \subset \mathbb{R}^n$ with $\widehat{\xi} \in W$. On the neighborhood $\mathcal{O}$ of $\widehat{\xi}$ relative to W, that is $\mathcal{O} \cap W$, suppose there is a constant $\kappa > 0$ such that*

$$\|\xi - \xi^+\| \geq \sqrt{\kappa}\text{dist}(\xi, \text{Fix } T) \quad \forall \xi \in \mathcal{O} \cap W, \ \forall \xi^+ \in T\xi. \qquad (12.23)$$

Then the sequences $(\xi^k)_{k \in \mathbb{N}}$ and $(y^k, z^k)_{k \in \mathbb{N}}$ converge linearly to the respective points $\overline{\xi} \in$ Fix $T \cap W$ and $(\overline{y}, \overline{z})$ with rate bounded above by $\sqrt{1 - \kappa}$.

In either case, the limit point $\overline{z} = \mathcal{J}_{\eta D}\overline{\xi}$ is a solution to $(\mathcal{D})$, $\overline{y} \in D\overline{z}$ and the sequence $\left(x^k\right)_{k \in \mathbb{N}}$ of Step 1 of Algorithm 1 converges to $\overline{x}$, a solution of $(\mathcal{P}_0)$.

The strong convexity assumption (i) of Theorem 12.8 fails in many applications of interest, and in particular for feasibility problems (minimizing the sum of indicator functions). By [31, Theorem 2.2], case (ii) of Theorem 12.8, in contrast, holds in general for mappings T for which $T - \text{Id}$ is *metrically subregular* and the fixed point sets are *isolated points* with respect to an affine subspace to which the iterates are confined. The restriction to the affine subspace W is a natural generalization for the Douglas–Rachford algorithm, where the iterates are known to stay confined to affine subspaces orthogonal to the fixed point set [32, 33]. We show that metric subregularity with respect to this affine subspace holds in many applications.

Proposition 12.9 (polyhedrality of the Douglas–Rachford operator) *Let $f : \mathbb{R}^n \to \overline{\mathbb{R}}$ and $g : \mathbb{R}^m \to \overline{\mathbb{R}}$ be proper, lsc and convex. Suppose, in addition, that f and g are piecewise linear-quadratic. The operator $T : \mathbb{R}^m \to \mathbb{R}^m$ defined by (12.21) with $\eta > 0$ fixed, is polyhedral for B and D given by (12.22) where $\mathcal{A} : \mathbb{R}^n \to \mathbb{R}^m$ is a linear mapping.*

Proof This is Proposition 2.6 of [1].

Proposition 12.10 (local linear convergence, Theorem 2.7 of [1]) *Let $f : \mathbb{R}^n \to \overline{\mathbb{R}}$ and $g : \mathbb{R}^m \to \overline{\mathbb{R}}$ be proper, lsc, convex, piecewise linear-quadratic functions (see Definition 12.1). Define the operator $T : \mathbb{R}^m \to \mathbb{R}^m$ by (12.21) with $\eta > 0$ fixed and B and D given by (12.22) where $\mathcal{A} : \mathbb{R}^n \to \mathbb{R}^m$ is a linear mapping. Suppose that* *there exists a solution to $0 \in (B + D)(x)$, that $T : W \to W$ for W some affine sub-space of $\mathbb{R}^m$ and that* Fix $T \cap W$ *is an isolated point* $\{\bar{\xi}\}$. *Then there is a neighborhood $\mathcal{O}$ of $\bar{\xi}$ such that, for all starting points (ξ^0, y^0, z^0) with $\xi^0 \equiv z^0 + \eta y^0 \in \mathcal{O} \cap W$ for $y^0 \in D(z^0)$ so that $\mathcal{J}_{\eta D}\xi^0 = z^0$, the sequence $(\xi^k)_{k \in \mathbb{N}}$ generated by (12.20) converges Q-linearly to $\bar{\xi}$ where $\bar{z} \equiv \mathcal{J}_{\eta D}\bar{\xi}$ is a solution to (D). The rate of linear convergence is bounded above by $\sqrt{1 - \kappa}$, where $\kappa = c^{-2} > 0$, for c a constant of metric subregularity of $T -$ Id at $\bar{\xi}$ for the neighborhood $\mathcal{O}$. Moreover, the sequence $\left(y^k, z^k\right)_{k \in \mathbb{N}}$ generated by steps 2–3 of Algorithm 1 converges linearly to $(\bar{y}, \bar{z})$ with $\bar{y} = \frac{1}{\eta}(\bar{x} - \bar{z})$, and the sequence $\left(x^k\right)_{k \in \mathbb{N}}$ defined by Step 1 of Algorithm 1 converges to a solution to $(\mathcal{P}_0)$.*

12.3.1   ADMM for Statisitcal Multi-resolution Estimation of STED Images

The theoretical results above are demonstrated with an image $b \in \mathbb{R}^n$ (Fig. 12.1) generated from a Stimulated Emission Depletion (STED) microscopy experiment [34, 35] conducted at the Laser-Laboratorium Göttingen examining tubulin, represented as the "object" $x \in \mathbb{R}^n$. The imaging model is simple linear convolution. The measurement b, shown in Fig. 12.1, is *noisy* or otherwise inexact, and thus an exact solution is not desirable. Although the noise in such images is usually modeled by

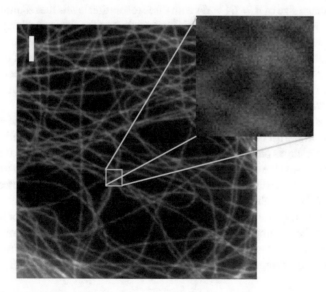

Fig. 12.1 Recorded STED image used in [1]. Inset: area to be processed ($640 \times 640\,\text{nm}^2$). The scale bar (white) is $1\,\mu\text{m}$

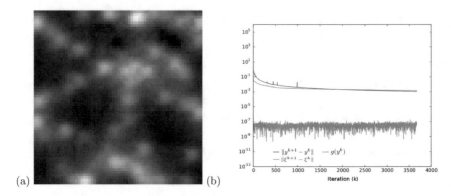

(a)   (b)

Fig. 12.2 **a** Numerical reconstruction via Algorithm 1 from the imaging data shown in Fig. 12.1 for $\rho = 4096$. **b** Iterates of Algorithm 1: primal and dual step sizes, as well as constraint violation $g(\mathcal{A}x)$ given by (12.12) with $\rho = 4096$. Reproduced from [1]

Poisson noise, a Gaussian noise model with constant variance suffices as the photon counts are of the order of 100 per pixel and do not vary significantly across the image. We calculate the numerically reconstructed tubulin density $\overline{x}$ shown in Fig. 12.2a via Algorithm 1 for the problem $(\mathcal{P})$ with the qualitative regularization

$$f(x) \equiv \alpha ||x||^2 \qquad (12.24)$$

and exact penalty $g(\mathcal{A}x)$ given by (12.12) with g_i given by (12.17).

For an image size of $n = 64 \times 64$ with three resolution levels the resulting number of constraints is $m = 12161$ (that is, 64^2 constraints at the finest resolution, $4 * 32^2$ constraints at the next resolution and $9 * 21^2$ constraints at the lowest resolution). The constant $\alpha = 0.01$ in (12.24) is used to balance the contributions of the individual terms to make the most of limited numerical accuracy (double precision). The constant γ_i is chosen so that the model solution would be no more than 3 standard deviations from the noisy data on each interval of each scale.

Since this is experimental data, there is no "truth" for comparison - the constraint, together with the error bounds on the numerical solution to the model solution provide statistical guarantees on the numerical reconstruction [36]. In Fig. 12.2b the iteration is shown with the value of $\rho = 4096$ for which the constraints are exactly satisfied (to within machine precision), indicating the correspondence of the computed solution of problem $(\mathcal{P})$ to a solution to the exact model problem $(\mathcal{P}_0)$.

The only assumption from Proposition 12.10 that cannot be verified for this implementation is the assumption that the algorithm fixed point is a singleton; all other assumptions are satisfied *automatically* by the problem structure. We *observe*, however, starting from around iteration 1500 in Fig. 12.2b, behavior that is *consistent* with (i.e. does not contradict) linear convergence. From this, the observed convergence rate is $c = 0.9997$, which yields an a posteriori upper estimate of the pixel-wise error of about $8.9062e^{-4}$, or 3 digits of accuracy at each pixel.[2]

12.4 Primal-Dual Methods

The ADMM method presented above suffers from the extreme computational cost of computing the prox-operator in step 1. The results of the previous section required several days of cpu time on a 2016-era laptop. In this section we present a method studied in [3] that can achieve results in about 30 s on the same computer architecture. In this section we survey the main results (without proofs) from [1]. There is one subtle difference in the present survey over [3] that has major implications for the application and implementation of the main Algorithm 2.

In this section we consider exclusively functions g in problem $(\mathcal{P})$ of the form (12.11). The algorithm we revisit is the proximal alternating predictor-corrector (PAPC) algorithm proposed in [37] for solving $(\mathcal{S})$. It consists of a predictor-corrector gradient step for handling the smooth part of L in (12.13) and a proximal step for handling the nonsmooth part.

Algorithm 2: Extended Proximal Alternating Predictor-Corrector (EPAPC) for $(\mathcal{S})$.

Parameters : Set $\eta > 0$ and choose the parameters τ and σ to satisfy
$$\tau \in \left(0, \tfrac{1}{K_f}\right), \quad 0 < \tau\sigma < \tfrac{1}{\|\mathcal{A}\,\mathcal{A}\|}, \quad \text{and}$$
$$(\forall w \in (\ker \mathcal{A}^T)^\perp)(\forall w_0 \in \ker \mathcal{A}^T) \; \tfrac{1}{\sigma}\|w_0\|^2 \geq g^*(w) - g^*(w + w_0).$$

Initialization: Choose $(x^0, y^0) \in \mathbb{R}^n \times (\ker \mathcal{A}^T)^\perp$

1 **for** $k = 1, 2, \ldots$ **do**

2 $\quad p^k = x^{k-1} - \tau(\nabla f(x^{k-1}) + \mathcal{A}^T y^{k-1})$;

3 $\quad$ **for** $i = 1, \ldots, M$ **do**

4 $\quad\quad w_i^k = \operatorname{argmax}_{w_i \in \mathbb{R}^{m_i}} \{\langle A_i p^k, w_i\rangle - g_i^*(w_i) - (1/2\sigma)\left\| w_i - y_i^{k-1}\right\|^2\}$

5 $\quad\quad \equiv \operatorname{prox}_{\sigma, g_i^*}(y_i^{k-1} + \sigma A_i p^k)$;

6 $\quad y^k = P_{(\ker \mathcal{A}^T)^\perp} w^k$;

7 $\quad x^k = x^{k-1} - \tau(\nabla f(x^{k-1}) + \mathcal{A}^T y^k)$.

At each iteration the algorithm computes one gradient and a prox-mapping corresponding to the nonsmooth function, both of which are assumed to efficiently implementable. We suppose these can be evaluated exactly, though this does not take into account finite precision arithmetic. The dimension of the iterates of the EPAPC

---

[2]This statement does not take into account finite precision evaluation of the steps in Algorithm 1.

algorithm is on the same order of magnitude as with the ADMM/Douglas-Rachford method, but the individual steps can be run in parallel and, with the exception of the projection in Step 6, are much less computationally intensive to execute.

For quantitative convergence guarantees of primal-dual methods, additional assumptions are required.

Assumption 3

(i) *The function $f : \mathbb{R}^n \to \mathbb{R}$ is convex and continuously differentiable with Lipschitz continuous gradient ∇f (constant K_f), that is for all $x, x' \in \mathbb{R}^n$, we have*

$$\|\nabla f(x) - \nabla f(x')\| \leq K_f \|x - x'\|; \tag{12.25}$$

(ii) *The function $f : \mathbb{R}^n \to \mathbb{R}$ is pointwise quadratically supportable (Definition 12.2) at each $\hat{x}$ in the solution set $\mathcal{S}^*$.*

(iii) *There exists a $\sigma > 0$ such that*

$$(\forall w \in (\ker \mathcal{A}^T)^\perp)(\forall w_0 \in \ker \mathcal{A}^T) \ \tfrac{1}{\sigma}\|w_0\|^2 \geq g^*(w) - g^*(w + w_0).$$

The assumption of Lipschitz continuous gradients Assumption 3(i), is standard, but stronger than one might desire in general. The assumption is included mainly to guarantee boundedness of the iterates. Lipschitz continuity of the gradients is enough for our purposes, however. By the standing Assumption 1 the mapping $\mathcal{A}$ is injective and when $m \leq n$, then $\mathcal{A}$ has full row rank, and $\mathcal{A}\mathcal{A}^T$ is invertible. When $m > n$, $\mathcal{A}$ is still injective but $\mathcal{A}\mathcal{A}^T$ has a nontrivial kernel and care must be taken that the conjugate function g^* does not decrease too fast in the direction of the kernel of $\mathcal{A}^T$. This is assured by Assumption 3(iii). This assumption comes into play in Lemma 12.1.

Step (3) of Algorithm 2 can be written more compactly when $g(w) := g(w_1, \ldots, w_M) = \sum_{i=1}^M g_i(w_i)$. In this case, the convex conjugate of a separable sum of functions is the sum of the individual conjugates: $g^*(w) := \sum_{i=1}^M g_i^*(w_i)$. Defining the matrix $S = \sigma^{-1} I_m$ we immediately get that for any point $\zeta_i \in \mathbb{R}^{m_i}$, $i = 1, \ldots, M$,

$$\text{prox}_{S,g^*}(\zeta) = (\text{prox}_{\sigma, g_1^*}(\zeta_1), \text{prox}_{\sigma, g_2^*}(\zeta_2), \ldots, \text{prox}_{\sigma, g_M^*}(\zeta_M)).$$

Thus Step (3) of Algorithm 2 can be written in vector notation by $w^k = \text{prox}_{S,g}(y^{k-1} + \sigma \mathcal{A}p^k)$. It is possible to use different proximal step constants σ_i, $i = 1 \ldots, M$, see details in [37]. The choice $\sigma_i = \sigma$ for $i = 1, \ldots, M$ is purely for simplicity of exposition. The projection onto $(\ker \mathcal{A}^T)^\perp$ in (6) is carried out by applying the pseudo inverse:

$$P_{(\ker \mathcal{A}^T)^\perp} = \mathcal{A} \left(\mathcal{A}^T \mathcal{A}\right)^{-1} \mathcal{A}^T.$$

When $m \leq n$ and A_i is full rank for all $i = 1, 2, \ldots, M$, then $\ker \mathcal{A}^T = \{0\}$ and the above operation is not needed. But an unavoidable feature of multi-resolution analysis, our motivating application, is that $m > n$, so some thought must be given to efficient computation of $(A^T A)^{-1}$.

The next technical result, which is new, establishes a crucial upper bound on the growth of the Lagrangian with respect to the primal variables.

Lemma 12.1 *Let Assumption 3 hold and let* $((p^k, y^k, x^k))_{k \in \mathbb{N}}$ *be the sequence generated by the EPAPC algorithm. Then for every $k \in \mathbb{N}$ and every $(x, y) \in \mathbb{R}^n \times \mathbb{R}^m$,*

$$L(x^k, y^k) - L(x, y^k) \le \frac{1}{2\tau} \left(\| x - x^{k-1} \|^2 - \| x - x^k \|^2 \right)$$
$$- \frac{1}{2} \left(\frac{1}{\tau} - K_f \right) \| y^k - y^{k-1} \|^2. \tag{12.26}$$

and

$$L(x^k, y) - L(x^k, y^k) \le \frac{1}{2} \left(\| y - y^{k-1} \|_G^2 - \| y - y^k \|_G^2 - \| y^k - y^{k-1} \|_G^2 \right) \tag{12.27}$$

where

$$G := \sigma^{-1} I_m - \tau A A^T. \tag{12.28}$$

Note that for the choice of τ given in the parameter initialization of Algorithm 2, $G \succ 0$.

Proof The proof of (12.26) follows exactly as in the proof of [37, Lemma 3.1(i)]. The second inequality (12.27) follows exactly as the proof of [37, Lemma 3.1(ii)] in the case that $\ker A^T = \{0\}$. The case where $\ker A^T$ is nontrivial requires more care. The proof of this is suppressed in the interest of brevity.

The next intermediate result establishes pointwise quadratic supportability (Definition 12.2) on bounded sets at all saddle points under Assumptions 1 and 3.

Proposition 12.11 ([3]) *Let* $((p^k, y^k, x^k))_{k \in \mathbb{N}}$ *be the sequence generated by the EPAPC algorithm. If Assumptions 1 and 3 are satisfied, then for any primal solution $\hat{x}$ to the saddle point problem (S), there exists a $\mu > 0$ such that*

$$f(x^k) \ge f(\hat{x}) + \langle \nabla f(\hat{x}), x^k - \hat{x} \rangle + \tfrac{1}{2}\mu \|x^k - \hat{x}\|^2 \quad \forall k. \tag{12.29}$$

The constant μ in Proposition 12.11 depends on the choice of (x^0, y^0) and so depends implicitly on the distance of the initial guess to the point in the set of saddle point solutions.

Convergence of the primal-dual sequence is with respect to a weighted norm on the primal-dual product space built on G in (12.28).

$$u = \begin{pmatrix} x \\ y \end{pmatrix}, \qquad H = \begin{pmatrix} \tau^{-1} I_n & 0 \\ 0 & G \end{pmatrix} \tag{12.30}$$

where by the assumptions on the choice of τ given in Algorithm 2, $G \succ 0$. We can then define an associated norm using the positive definite matrix H, $\|u\|_H^2 := \frac{1}{\tau}\|x\|^2 + \|y\|_G^2$.

We are now ready to state the main result and corollaries, whose proofs can be found in [3].

Theorem 12.4.1 *Let* $\left((p^k, x^k, y^k)\right)_{k\in\mathbb{N}}$ *be the sequence generated by the EPAPC algorithm. Let* $\lambda_{min+}(\mathcal{A}\mathcal{A}^T)$ *denote the smallest nonzero eigenvalue of* $\mathcal{A}\mathcal{A}^T$. *If Assumptions 1 and 3 are satisfied, then there exists a saddle point solution for* $L(\cdot, \cdot)$, *the pair* $\hat{u} = (\hat{x}, \hat{y})$, *with* $\hat{y} \in (\ker \mathcal{A}^T)^\perp$, *and for any* $\alpha > 1$ *and for all* $k \geq 1$, *the sequence* $\left(u^k = (x^k, y^k)\right)_{k\in\mathbb{N}}$ *satisfies*

$$\left\|u^k - \hat{u}\right\|_H^2 \leq \frac{1}{1+\delta}\left\|u^{k-1} - \hat{u}\right\|_H^2, \tag{12.31}$$

where

$$\delta = \min\left\{\frac{(\alpha-1)\tau\sigma(1-\tau L_f)\lambda_{min+}(\mathcal{A}\mathcal{A}^T)}{\alpha}, \frac{\mu\tau\sigma\lambda_{min+}(\mathcal{A}\mathcal{A}^T)}{\alpha\tau L_f^2 + \sigma\lambda_{min+}(\mathcal{A}\mathcal{A}^T)}\right\} \tag{12.32}$$

is positive and $\mu > 0$ *is the constant of pointwise quadratic supportability of* f *at* $\hat{x}$ *depending on the distance of the initial guess to the point* $(\hat{x}, \hat{y})$ *in the solution set* S^*. *In particular,* $\left((x^k, y^k)\right)_{k\in\mathbb{N}}$ *is Q-linearly convergent with respect to the H-norm to a saddle-point solution.*

An unexpected corollary of the result above is that the set of saddle points is a singleton.

Corollary 12.12 (Unique saddle point) *Under Assumptions 1 and 3 the solution set* S^* *is a singleton.*

The above theorem yields the following estimate on the number of iterations required to achieve a specified distance to a saddle point.

Corollary 12.13 *Under Assumptions 1 and 3, let* $\bar{u} = (\bar{x}, \bar{y})$ *be the limit point of the sequence generated by the EPAPC algorithm. In order obtain*

$$\|x^k - \bar{x}\| \leq \epsilon \quad (\text{resp. } \|y^k - \bar{y}\|_G \leq \epsilon), \tag{12.33}$$

it suffices to compute k iterations, with

$$k \geq \frac{2\log\left(\frac{C}{L_f\epsilon}\right)}{\delta} \quad \left(\text{resp. } k \geq \frac{2\log\left(\frac{C}{\epsilon}\right)}{\delta}\right), \tag{12.34}$$

where $C = \|u^0 - \bar{u}\|_H = \left(\frac{1}{\tau}\|x^0 - \bar{x}\|^2 + \|y^0 - \bar{y}\|_G^2\right)^{1/2}$, *and* δ *is given in* (12.32).

12.4.1   EPAPC for Statisitcal Multi-resolution Estimation of STED Images

An efficient computational strategy for evaluating or at least approximating the projection $P_{(\ker \mathcal{A}^T)^\perp}$ in Step 6 of Algorithm 2 has not yet been established. We report here preliminary computational results of Algorithm 2 *without* computing Step 6. Our results show that the method is promising, though error bounds to the solution to $(\mathcal{S})$ are not justified without computation of $P_{(\ker \mathcal{A}^T)^\perp}$.

In our numerical experiments, the constraint penalty in $(\mathcal{S})$ takes the form $g(y) = \sum_i \iota_{\mathcal{C}_i}(y)$ where each $\mathcal{C}_i = \{y : | \sum_{j \in V_i} \omega_i(y_j - b_j)| \leq \gamma_i\}$. This is an exact penalty function, and so solutions to $(\mathcal{S})$ correspond to solutions to $(\mathcal{P}_0)$. Using Moreau's identity (12.6), the prox-mapping is evaluated explicitly in (6) for each constraint by

$$
\begin{aligned}
y_i^k &= \text{prox}_{\sigma, \iota_{\mathcal{C}_i}^*} (y_i^{k-1} + \sigma A_i p^k) \\
&= y_i^{k-1} + \sigma A_i p^k - \sigma P_{\mathcal{C}_i}\left(\frac{y_i^{k-1} + \sigma A_i p^k}{\sigma}\right), \qquad i = 1, \ldots, M.
\end{aligned}
$$

The proximal parameter is a function of τ and given by $\sigma = 1/(\tau\|\mathcal{A}\mathcal{A}^T\|_2)$. More details in [37, Sect. 4.1].

Here, we also consider the smooth approximation of the L^1-norm as the qualitative objective. The L^1-norm is non-smooth at the origin, thus in order to make the derivative-based methods possible we consider a smoothed approximation of this, known as the Huber approximation.

The Huber loss function is defined as follows:

$$
\|x\|_{1,\alpha} = \sum_{i,j} \phi_\alpha(x_{i,j}), \qquad \phi_\alpha(t) = \begin{cases} \frac{t^2}{2\alpha} & \text{if } |t| \leq \alpha \\ |t| - \frac{\alpha}{2} & \text{if } |t| > \alpha, \end{cases} \tag{12.35}
$$

where $\alpha > 0$ is a small parameter defining the trade-off between quadratic regularization (for small values) and L^1 regularization (for larger values). The function ϕ is smooth with $\frac{1}{\alpha}$-Lipschitz continuous derivative and its derivative is given by

$$
\phi_\alpha'(t) = \begin{cases} \frac{t}{\alpha} & \text{if } |t| \leq \alpha \\ \text{sgn}(t) & \text{if } |t| > \alpha. \end{cases} \tag{12.36}
$$

Pointwise quadratically supportability of this function at solutions is not unreasonable but still must be assumed.

We demonstrate our reconstruction of the image inset shown in Fig. 12.1 of size $n = 64^2$ with the same SMRE model as the demonstration in Sect. 12.3.1. The confidence level γ_i was set to $0.25 * i$ at each resolution level ($i = 1, 2, 3$). Figure 12.3(top left) shows the reconstruction with the L^2 function $f(x) = 0.01\|x\|^2$ (compare to Fig. 12.2). Figure 12.3(top right) shows the reconstruction with the Huber function

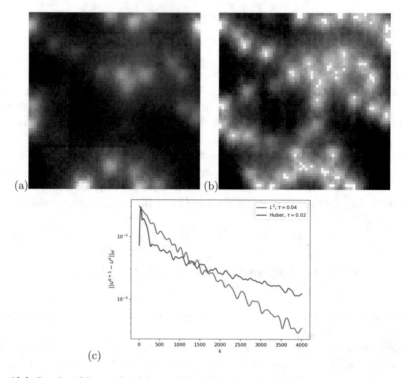

Fig. 12.3 Results of image denoising and deconvolution via Algorithm 2 for statistical multi-resolution estimation with three levels at iteration 4000 with **a** $f(x) = 0.01 * \|x^2\|^2$ and **b** $f(x) = \|x\|_{1,0.25}$, the Huber loss function. Frame **c** shows the step length $\|u^{k+1} - u^k\|_H$ where $u^k = (x^k, y^k)$ of the associated sequence as function of the number of iterations

$\|x\|_{1,\alpha}$, where $\alpha = 0.25$. Figure 12.3(bottom) shows the step size of the primal-dual pair for each of these regularized problems as a function of iteration. The model with quadratic regularization achieves a better average rate of convergence, but for both objective functions the algorithm appears to exhibit R-linear convergence (not Q-linear). What is not evident from these experiments is the computational effort required per iteration. Without computation of the pseudo-inverse in step 6, the EPAPC algorithm computes these results in about 30 s on a 2018-era laptop, compared to several days for the results shown in Fig. 12.2.

12.5 Randomized Block-Coordinate Primal-Dual Methods

The previous sections reviewed numerical strategies and structures that yield quantitative estimates of the distance of an iterate to the solution of the underlying variational problem. In this section we examine *implementation* strategies for dealing with high dimensional problems. These are implementation strategies because they do not involve changing the optimization model. Instead, we select at random a smaller subset of variables or constraints in the computation of an update in the full-dimensional iterative procedure. This is the principal strategy for handling problems that, due to their size, must be distributed across many processing and storage units (see for instance [26, 38–40] and references therein). We survey here a randomized primal-dual technique proposed and analyzed in [2]. The main theoretical question to resolve with such approaches is whether, and in what sense, iterates converge to a solution to the original problem. We can determine whether the iterates converge, but obtaining an estimate of the distance to the solution remains an open problem.

The algorithm below is a primal-dual method like the algorithms reviewed above, with the exception that it solves an extension of the dual problem $(\mathcal{D})$:

$$
\begin{array}{ll}
\underset{(x,y)\in\mathbb{R}^n\times\mathbb{R}^m}{\text{minimize}} & f^*(x) + g^*(y) \\
\text{subject to} & x = -\mathcal{A}^T y
\end{array}
\qquad (\widetilde{\mathcal{D}})
$$

The main prox operation is computed on the dual objective in $(\mathcal{D})$, that is $f^*(x) + g^*(y)$ with respect to the variables $(x, y) \in \mathbb{R}^n \times \mathbb{R}^m$. The dimension of the basic operations is unchanged from the previous approaches, but the structure of the sum of functions allows for efficient evaluation of the prox mapping. Implicit in this is that the function f is prox friendly. In the algorithm description below it is convenient to use the convention $f \equiv g_0$, $A_0 \equiv \text{Id}$. The algorithm is based in part on [39].

Algorithm 3: Random Block-coordinate Primal-Dual Algorithm (RBPD) for $(\widetilde{\mathcal{D}})$

Parameters : Choose $\tau > 0$, $\sigma = (\sigma_0, \sigma_1 \ldots, \sigma_M) \in \mathbb{R}_{++}^M$.
Initialization: Choose $y^1 = 0 \in \mathbb{R}^m$, and set $x^1 = u^1 = -\mathcal{A}^T y^1 = 0 \in \mathbb{R}^n$.

1 **for** $k = 1, 2, \ldots$ **do**
2 $\quad$ Choose $i \in \{0, 1, 2, \ldots, M\}$ uniformly at random;
3 $\quad$ $y_i^{k+1} = \text{prox}_{\sigma_i, g_i^*}(y_i^k - \sigma_i A_i x^k)$;
4 $\quad$ $\delta^{k+1} = A_i^T(y_i^{k+1} - y_i^k)$;
5 $\quad$ $u^{k+1} = u^k + \frac{\tau}{m}\delta^{k+1}$;
6 $\quad$ $x^{k+1} = u^{k+1} + x^k + \tau\delta^{k+1}$;

Notice that each iteration of Algorithm 3 requires only two small matrix-vector multiplications: $A_i(\cdot)$ and $A_i^T(\cdot)$. The methods of the previous sections, in contrast, worked with full matrix $\mathcal{A} = [A_1^T, \ldots, A_m^T]^T$. This means that all iterations involve full vector operations. For some applications this might be not feasible, at least on standard desktop computers due to the size of problems. Algorithm 3 uses only blocks A_i of $\mathcal{A}$, therefore each iteration requires fewer floating point operations, at the cost of having less information available for choosing the next step. This reduction in the effectiveness of the step is compensated for through larger block-wise steps. Computation of the step size is particularly simple. This follows the same hybrid step-length approach developed in [41] for the nonlinear problem of blind ptychography. In particular, we use step sizes adjusted to each block A_i:

$$\tau \sigma_i \lambda_{\max}(A_i^T A_i) < 1 \quad \forall i = 0, 1, \ldots, m.$$

Proposition 12.14 (Theorem 1 of [2]) *Suppose Assumption 1 holds and let* $\tau \sigma_i \|A_i\|^2 < 1$. *Then* (x^k, y^k) *generated by Algorithm 3 converges to a solution to* $(\widetilde{\mathcal{D}})$. *In particular, the sequence* (x^k) *converges almost surely to a solution to* $(\mathcal{P})$.

The statement above concerns part (i) of Theorem 1 of [2]. No smoothness is required of the qualitative regularization f. Instead, it is assumed that this function is prox-friendly. This opens up the possibility of using the 1-norm as a regularize, promoting, in some sense, sparsity in the image. No claim is made on the *rate* of convergence, though the numerical experiments below indicate that, for regular enough functions f, convergence might be locally linear. This remains to be proved.

12.5.1 RBPD for Statisitcal Multi-resolution Estimation of STED Images

Despite many open questions regarding convergence, random methods offer a way to handle extremely large problems. To make a comparison with the deterministic approaches above, cycles of the RBPD Algorithm 3 are counted in terms of *epochs*. An epoch contains the number of passes through steps 1–6 of Algorithm 3 required before each block is chosen at least once. After k epochs, therefore, the i-th coordinate of x will be updated, on average, the same number of times for the randomized algorithm as for the deterministic methods. In other words, an epoch for a randomized block-wise method is comparable to an iteration of a deterministic method. As the RBPD updates only one block per *iteration*, each iteration is less computationally intensive than the the deterministic counterparts. However, in our case this efficient iteration still requires one to evaluate two (possibly) expensive convolution products (embedded in $A_i x$ and $A_i^T y$). Thus, if these operations are relatively expensive, the efficiency gain will be marginal. Nevertheless, because of the ability to operate on smaller blocks, the randomized method requires, per epoch, approximately half the time required per iteration of the deterministic methods. Although the quantitative convergence analysis remains open, our numerical experiments indicate that the method achieves a comparable step-residual to the EPAPC Algorithm 2 after the same number of epochs/iterations.

As with the experiments in the previous Sections, we use three resolutions, which results in one block at the highest resolution, four blocks at the next resolution (four possible shifts of 2×2 pixels), and nine blocks at the third resolution (nine different shifts of 3×3 pixels). We applied Algorithm 3 with different regularization f in $(\mathcal{P}_0)$: the 1-norm $f(x) = \|x\|_1$, Huber loss function $f(x) = \|x\|_{1,\alpha}$ given by (12.35) ($\alpha = 0.25$) and the squared Euclidean norm. As with the EPAPC experiments, the

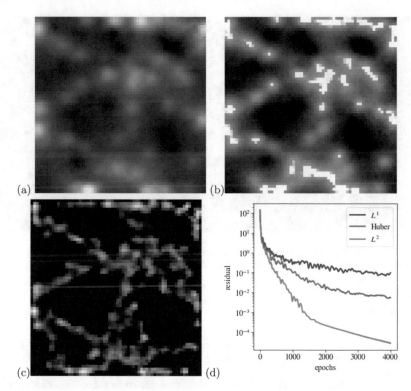

Fig. 12.4 Results of image denoising and deconvolution using RBPD (Algorithm 3) for statistical multi-resolution estimation of a 64×64 pixel image with three levels at epoch 4000 with **a** $f(x) = 0.01 * \|x\|^2$, **b** $f(x) = \|x\|_{1,0.25}$, the Huber loss function and **c** $f(x) = \|x\|_1$. **d** The step length $\|u^{k+1} - u^k\|_H$ where $u^k = (x^k, y^k)$ of the associated sequence as function of the number of iterations

function g is given by (12.11) with g_i given by (12.17) for the parameter $\gamma_i = 0.25 * i$ for $i = 1, 2, 3$. All of these functions are prox-friendly and have closed-form Fenchel conjugates. The gain in efficiency over the deterministic EPAPC method proposed above (without computation of the pseudo-inverse) is a factor of 2.

Figure 12.4a–c shows the reconstructions on the same 64×64 image data used in the previous sections. The numerical performance of the algorithm is shown in Fig. 12.4(d). What the more efficient randomization strategy enables is for the full 976×976 pixel image to be processed. The result for regularization with the 1-norm is shown in Fig. 12.5.

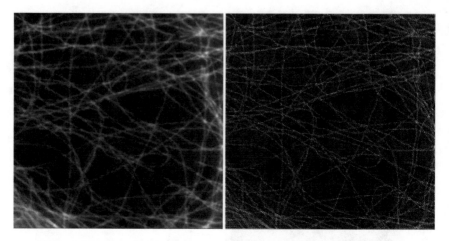

Fig. 12.5 Left: original image. Right: image denoising and deconvolution using RBPD (Algorithm 3) for statistical multi-resolution estimation with three levels at epoch 4000 with $f(x) = \|x\|_1$

References

1. Aspelmeier, T., Charitha, C., Luke, D.R.: Local linear convergence of the ADMM/Douglas-Rachford algorithms without strong convexity and application to statistical imaging. SIAM J. Imaging Sci. **9**(2), 842–868 (2016)
2. Luke, D.R., Malitsky, Y.: Block-coordinate primal-dual method for the nonsmooth minimization over linear constraints. In: Giselsson, P., Rantzer, A. (eds.) Distributed and Large-Scale Optimization. Springer (2018)
3. Luke, D.R., Shefi, R.: A globally linearly convergent method for pointwise quadratically supportable convex-concave saddle point problems. J. Math. Anal. Appl. **457**(2), 1568–1590 (2018). https://doi.org/10.1016/j.jmaa.2017.02.068
4. Burger, M., Sawatzky, A., Steidl, G.: First order algorithms in variational image processing. In: Splitting Methods in Communication, Imaging, Science, and Engineering, pp. 345–407. Springer (2016)
5. Chambolle, A., Pock, T.: An introduction to continuous optimization for imaging. Acta Numerica **25**, 161–319 (2016)
6. Combettes, P.L., Pesquet, J.C.: Proximal splitting methods in signal processing. In: Fixed Point Algorithms for Inverse Problems in Science and Engineering, pp. 185–212 (2011)
7. Komodakis, N., Pesquet, J.C.: Playing with duality: an overview of recent primal? Dual approaches for solving large-scale optimization problems. IEEE Signal Proc. Mag. **32**(6), 31–54 (2015)
8. Luke, D.R., Teboulle, M., Thao, N.H.: Necessary conditions for linear convergence of iterated expansive, set-valued mappings. Math. Program. A **180**, 1–31 (2020). https://doi.org/10.1007/s10107-018-1343-8
9. Ortega, J.M., Rheinboldt, W.C.: Iterative Solution of Nonlinear Equations in Several Variables. Academic Press, New York (1970)
10. Rockafellar, R.T., Wets, R.J.: Variational Analysis. Grundlehren Math. Wiss. Springer, Berlin (1998)
11. Dontchev, A.L., Rockafellar, R.T.: Implicit Functions and Solution Mapppings, 2nd edn. Springer, New York (2014)
12. Moreau, J.J.: Proximité et dualité dans un espace Hilbertian. Bull. de la Soc. Math. de France **93**(3), 273–299 (1965)

13. Aubin, J.P., Frankowska, H.: Set-valued Analysis. Birkhäuser, Boston (1990)
14. Burke, J.V., Ferris, M.C.: Weak sharp minima in mathematical programming. SIAM J. Control. Optim. **31**, 1340–1359 (1993)
15. Friedlander, M.P., Tseng, P.: Exact regularization of convex programs. SIAM J. Optim. **18**(4), 1326–1350 (2007). https://doi.org/10.1137/060675320
16. Borwein, J., Vanderwerff, J.: Convex Functions: Constructions, Characterizations and Counterexamples, Encyclopedias in Mathematics, vol. 109. Cambridge University Press, New York (2010)
17. Eckstein, J.: Splitting methods for monotone operators with applications to parallel optimization. Ph.D. thesis, MIT, Cambridge, MA (1989)
18. Eckstein, J., Bertsekas, D.P.: On the Douglas-Rachford splitting method and the proximal point algorithm for maximal monotone operators. Math. Program. **55**, 293–318 (1992)
19. Gabay, D.: Applications of the method of multipliers to variational inequalities. In: Augmented Lagrangian Methods: Applications to the Solution of Boundary-Value Problems, pp. 299–331. North-Holland, Amsterdam (1983)
20. Glowinski, R., Marroco, A.: Sur l'approximation, par elements finis d'ordre un, et las resolution, par penalisation-dualitè, d'une classe de problemes de dirichlet non lineares. Revue Francais d'Automatique, Informatique et Recherche Opérationnelle **9**(R-2), 41–76 (1975)
21. Rockafellar, R.T.: Monotone operators and the proximal point algorithm. SIAM J. Control. Optim. **14**, 877–898 (1976)
22. Boyd, S., Parikh, N., Chu, E., Peleato, B., Eckstein, J.: Distributed optimization and statistical learning via the alternating direction method of multipliers. Found. Trends Mach. Learn. **3**(1), 1–122 (2011)
23. Douglas Jr., J., Rachford Jr., H.H.: On the numerical solution of heat conduction problems in two or three space variables. Trans. Am. Math. Soc. **82**, 421–439 (1956)
24. Lions, P.L., Mercier, B.: Splitting algorithms for the sum of two nonlinear operators. SIAM J. Numer. Anal. **16**, 964–979 (1979)
25. Svaiter, B.F.: On weak convergence of the Douglas-Rachford method. SIAM J. Control. Optim. **49**(1), 280–287 (2011)
26. Strohmer, T., Vershynin, R.: A randomized kaczmarz algorithm with exponential convergence. J. Fourier Anal. Appl. **15**(2), 262–278 (2009)
27. He, B., Yuan, X.: On non-ergodic convergence rate of Douglas-Rachford alternating direction method of multipliers. Numerische Mathematik **130**(3), 567–577 (2015)
28. He, B., Yuan, X.: On the o(1/n) convergence rate of the Douglas-Rachford alternating direction method. SIAM J. Numer. Anal. **50**(2), 700–709 (2012)
29. Boley, D.: Local linear convergence of the alternating direction method of multipliers on quadratic or linear programs. SIAM J. Optim. **23**(4), 2183–2207 (2013). https://doi.org/10.1137/120878951
30. Hesse, R., Luke, D.R.: Nonconvex notions of regularity and convergence of fundamental algorithms for feasibility problems. SIAM J. Optim. **23**(4), 2397–2419 (2013)
31. Luke, D.R., Thao, N.H., Tam, M.K.: Quantitative convergence analysis of iterated expansive, set-valued mappings. Math. Oper. Res. **43**(4), 1143–1176 (2018). https://doi.org/10.1287/moor.2017.0898
32. Hesse, R., Luke, D.R., Neumann, P.: Alternating projections and Douglas-Rachford for sparse affine feasibility. IEEE Trans. Signal. Process. **62**(18), 4868–4881 (2014). https://doi.org/10.1109/TSP.2014.2339801
33. Phan, H.: Linear convergence of the Douglas-Rachford method for two closed sets. Optimization **65**, 369–385 (2016)
34. Hell, S.W., Wichmann, J.: Breaking the diffraction resolution limit by stimulated emission: stimulated-emission-depletion fluorescence microscopy. Opt. Lett. **19**(11), 780–782 (1994). https://doi.org/10.1364/OL.19.000780
35. Klar, T.A., Jakobs, S., Dyba, M., Egner, A., Hell, S.W.: Fluorescence microscopy with diffraction resolution barrier broken by stimulated emission. Proc. Natl. Acad. Sci. **97**(15), 8206–8210 (2000). https://doi.org/10.1073/pnas.97.15.8206

36. Frick, K., Marnitz, P., Munk, A.: Statistical multiresolution dantzig estimation in imaging: fundamental concepts and algorithmic framework. Electron. J. Stat. **6**, 231–268 (2012)
37. Drori, Y., Sabach, S., Teboulle, M.: A simple algorithm for a class of nonsmooth convex-concave saddle-point problems. Oper. Res. Lett. **43**(2), 209–214 (2015)
38. Defazio, A., Bach, F., Lacoste-Julien, S.: SAGA: A fast incremental gradient method with support for non-strongly convex composite objectives. In: Advances in Neural Information Processing Systems, pp. 1646–1654 (2014)
39. Fercoq, O., Richtárik, P.: Accelerated, parallel, and proximal coordinate descent. SIAM J. Optim. **25**(4), 1997–2023 (2015)
40. Johnson, R., Zhang, T.: Accelerating stochastic gradient descent using predictive variance reduction. In: Advances in Neural Information Processing Systems, pp. 315–323 (2013)
41. Hesse, R., Luke, D.R., Sabach, S., Tam, M.: The proximal heterogeneous block implicit-explicit method and application to blind ptychographic imaging. SIAM J. Imaging Sci. **8**(1), 426–457 (2015)

Chapter 13
Holographic Imaging and Tomography of Biological Cells and Tissues

Tim Salditt and Mareike Töpperwien

Abstract This chapter reviews recent progress in propagation-based phase-contrast imaging and tomography of biological matter. We include both inhouse μ-CT results recorded in the direct-contrast regime of propagation imaging (large Fresnel numbers F), as well as nanoscale phase contrast in the holographic regime with synchrotron radiation. The current imaging capabilities starting from the cellular level all the way to small animal imaging are illustrated by recent examples of our group, with an emphasis on 3D histology.

13.1 Propagation-Based Phase-Contrast Tomography

For the high photon energies E of hard X-rays which are needed to penetrate bulk samples, absorption contrast becomes negligible and phase contrast prevails. This is simply the result of the energy dependence of the X-ray index of refraction $n(E, \mathbf{r}) = 1 - \delta(E, \mathbf{r}) + i\beta(E, \mathbf{r})$, which is well suited to describe the propagation of X-rays in matter as long as the continuum approximation holds, i.e., if scattering angles are small. One of the advantages of hard X-ray imaging is that the spatial distribution of the real part of the refractive index is proportional to the corresponding electron density $\rho(\mathbf{r})$. The predominance of phase interaction over absorption is quantified by the ratio $\delta/\beta \gg 1$, which is particularly large for the low-Z elements of soft (unmineralized) biological tissues. The resulting phase shift and amplitude decrement of a beam traversing a resolution element (voxel of side length a) are $\Delta\phi = -k\delta a$ and $\Delta A = \exp(-k\beta a)$, respectively. Even for materials where absorption contrast is still sufficient at large length scales, it will become impossible to distinguish structural details if the side length a of the voxel is decreased. Hence, high resolution imaging with hard X-rays always requires phase contrast. It is therefore not surprising that

T. Salditt (✉) · M. Töpperwien
Institute for X-ray Physics, Universität Göttingen, Friedrich-Hund-Platz 1,
37077 Göttingen, Germany
e-mail: tsaldit@gwdg.de

M. Töpperwien
e-mail: mtoeppe@gwdg.de

© The Author(s) 2020   339
T. Salditt et al. (eds.), *Nanoscale Photonic Imaging*, Topics in Applied Physics 134,
https://doi.org/10.1007/978-3-030-34413-9_13

suitable implementations of phase-contrast radiography and tomography have been a major research goal over the last two decades, after X-ray sources with sufficient partial coherence had become available.

All phase-contrast methods rely on wave-optical transformations of the phase shifts, which the sample induces in the (partially) coherent wavefront, into measurable intensities patterns by ways of interference, see also [1, 2] for a review. The particular geometries and mechanisms by which waves are brought to interfere can be quite different. The methods can be classified according to the order of phase contrast. Zero-order methods such as a Bonse-Hart interferometry [3] are capable to measure the absolute phase shift $\Delta\phi$ between an object and an empty reference beam. First-order methods such as crystal [4] or grating (Talbot) interferometers [5, 6] are sensitive to the first spatial derivative of the phase $\nabla\phi$ in the object plane. Finally, second-order methods are based on contrast formation proportional to $\nabla^2\phi$. This is the case for propagation-based phase contrast, where the self-interference of the diffracted beam behind the object and the unattenuated or weakly attenuated primary beam interfere to form a defocused 'image'. As a second-order phase-contrast method, propagation imaging is particularly well suited for high spatial resolution. Further, it does not require any optical components acting as an 'analyzer'. First-order techniques make use of optical components which are scanned or rotated during data acquisition, such as in crystal interferometers (diffraction enhanced imaging) [4], or in grating (Talbot) interferometers [5, 6]. A particular advantage of Talbot interferometry is that along with the phase information it also generates an additional and completely separated darkfield image [5]. Furthermore, phase sensitivity is very high. A related first-order phase-contrast technique uses edge illumination or coded apertures [7, 8]. As in Talbot interferometry, the beam is structured in the object plane by a periodic array creating many beamlets. In contrast to a phase grating, these are spatially separated and small angular changes in their directions induced by the object are recorded downstream by a detector with sharp absorbing edges, without interference between the different beamlets. Hence, this technique is also applicable in the case of an incoherent source. Finally, in a more recent variant, the Talbot grating is replaced by a random speckle producing pattern [9].

A major disadvantage of all first-order techniques is the fact that the resolution is limited, e.g., by the grating period, aperture size or speckle grain. At the same time, the number of images to be acquired is fairly high and can pose a serious challenge in terms of acquisition time and dose. Note that tomography already requires the acquisition of hundreds or even thousands of projection images. It is therefore important to keep the acquisitions per projection angle at minimum and to record a sufficient number of resolution elements in parallel, i.e., to use detectors with a large number of pixels, and an optical setup where resolution and magnification are matched. For high-resolution tomography of biological samples, phase contrast by free space propagation thus remains the method of choice.

A central challenge in propagation-based imaging has been the formulation of accurate and efficient phase-retrieval schemes [1, 10, 11]. This has been particularly difficult in the holographic regime (small Fresnel number F), where the wave diffracted from any point in the object reaches all or a large set of detector pixels. However, high geometric magnification results in small F and hence one always ends up in the holographic regime, since the effective pixel size in the object plane enters quadratically [12]. At the same time, this holographic regime offers highest sensitivity to small phase shifts. As we review in this chapter, quantitative phase retrieval can be accomplished with a set of four measurements per projection. The key concept as introduced by Peter Cloetens and coworkers is to record images at four different F_i [10], e.g, by varying the sample/detector position or the photon energy. In the meantime, the initial limitation to objects with weakly varying phase can be overcome by iterative algorithms [13, 14]. In special cases, additional constraints such as compact support [15], sparsity [16], or range and dimensionality constraints [17] can reduce this set to a single acquisition.

It is also instructive to briefly compare propagation-based phase contrast to coherent diffractive imaging (CDI), which is typically carried out in the optical far field. Strictly speaking, far-field diffraction can also be counted as a 'phase-contrast' technique, since it is dominated by the Fourier transform of $\delta(\mathbf{r})$, but the term 'phase contrast' is mostly used only in full-field radiography, and not in diffractive imaging. How then, do CDI and propagation imaging compare, in particular if both record diffraction patterns at small F? Importantly, holographic imaging exploits the interference terms $2 \cdot \mathrm{Re}[\psi_s \psi_0^*]$ between a scattered wave ψ_s and a reference wave ψ_0 (enlarged primary wave), while CDI uses the far-field diffraction pattern $\psi_s \psi_s^*$ [18–23], without additional mixing with a reference wave. This important difference changes the way in which phase information is encoded in the intensity images, including the mathematical nature of the phase problem [24]. Furthermore, in near-field imaging, a weak scattered signal can be amplified high above background signals of residual scatter. These differences could in principle also affect the dose-resolution relationship, as we will further discuss later on. In this chapter, we show how propagation-based phase-contrast imaging has now matured to a powerful tool for 3D imaging of biological matter. We both include inhouse μ-CT results, where partial coherence has become sufficient to observe edge enhancement in the direct-contrast regime of propagation imaging (large Fresnel numbers F), and high-resolution phase contrast in the holographic regime by synchrotron radiation. Recent studies on biological cells and tissues performed by our group serve as examples for the current capabilities of phase-contrast imaging and tomography.

13.2 Nano-CT Using Synchrotron Radiation: Optics, Instrumentation and Phase Retrieval

13.2.1 Cone-Beam Holography

X-ray propagation imaging with nanoscale resolution requires a correspondingly small focus of the X-ray beam, which is today only possible using synchrotron radiation. The divergent cone-beam behind the focus is then exploited for illumination and recordings in the holographic regime. We therefore use the term 'holography' or 'holographic X-ray imaging' synonymously for this type of high-resolution propagation imaging. The properties of the beam, which is also denoted as *the probe* in the field of coherent X-ray imaging, are hence essential for holographic imaging. High resolution and quantitative phase contrast can only be achieved by efficient nano-focusing optics, sufficient coherence and smooth wavefronts [25–28]. These properties are certainly beneficial for all coherent X-ray techniques, however, to a different degree. Coherent diffractive imaging (CDI) in the far field, for example, is less sensitive to wavefront errors, but requires higher coherence [29].

In holography, aberration-free image formation relies on the quasi-spherical nature of wavefronts, since in data treatment one tacitly assumes spherical wavefronts in order to apply the Fresnel scaling theorem [2]. A small focus in a coherent probe also warrants a large numerical aperture in the illuminating wavefront. This directly affects the spatial resolution and also facilitates high geometric magnifications M, and hence small effective pixel sizes, in the case of a limited detector distance. Typical values for the focus-to-sample distance in high-resolution holography range between a few millimeters and several centimeters. Unfortunately, X-ray nano-focusing is associated with significant wavefront distortions, see also [30] for an overview of different X-ray focusing optics. These wavefront artifacts violate the idealizing assumptions made on the probe in the course of image reconstruction, such as point-source emission or distortion-free wavefront. The validity of these assumptions has recently been investigated, showing that they lead to reduced resolution and image quality [31, 32]. To avoid this, additional optical filtering and wavefront cleaning can be used. Alternatively, phase-retrieval schemes have to be generalized to non-ideal illumination conditions. In short, either *hardware* or *software* solutions, or a combination of both, are required.

For the latter case, the ptychographic concept of simultaneous probe and object reconstruction was recently generalized to near-field (propagation) imaging [26, 33–36]. Ptychographic algorithms were initially formulated only for confined probes (but extended objects), a setting which is typical for far-field coherent diffractive imaging [23, 37–41]. A generalization to extendeded illumination wavefronts was given in [36, 42], but a wavefront diffuser was required in order to increase the diversity of the probe. Thus, only the artificially modified wavefront and not the 'natural' probe could be recovered. This limitation was lifted by introducing longitudinal scanning of the object in [34], as well as the combination of lateral and longitudinal scans [33] to generate diversity in the data. Reconstructions of the illumination produced by a set of Kirkpatrick-Baez (KB) mirrors in the imaging plane were presented in [35] for the upgraded beamline ID16a of the European Synchrotron Radiation Facility (ESRF) in Grenoble [43], and in [26] for the holography endstation GINIX (Göttingen Instrument for Nano-Imaging with X-rays, cf. Fig. 13.1a) [44] at the P10 beamline of PETRAIII (DESY, Hamburg). As an alternative to shifting the sample, probe reconstruction was also achieved without any object in the beam by translating the detector [26], using an improved multiple magnitude projections (MMP) scheme [31, 45, 46]. The disadvantage of these approaches is that multiple images with different sample translations have to be recorded for each tomographic projection. Further, they only work if the empty beam remains temporally stable.

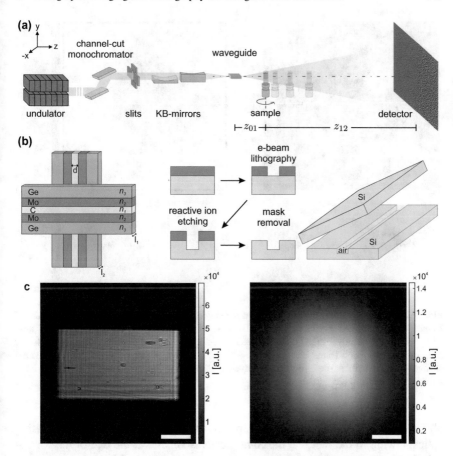

Fig. 13.1 Experimental realization of holographic imaging at the GINIX setup installed at the P10 beamline at PETRAIII (DESY, Hamburg). **a** The X-rays are generated in an undulator and monochromatized by a double-crystal Si(111) channel-cut monochromator. Subsequently, they are focused by a set of Kirkpatrick-Baez (KB) mirrors. An X-ray waveguide is placed in the KB focal plane as a coherence filter and to increase the numerical aperture for holographic imaging. The sample is mounted on a fully motorized sample tower at a distance z_{01} behind the waveguide and the evolving intensity distributions are recorded in the detection plane at distance z_{12} behind the sample. **b** A waveguide can be either realized by combining two 1d devices, consisting of a multilayer structure with carbon as guiding layer, or by etching a channel into a silicon wafer and bonding a second wafer on top, leading to a closed channel with air as guiding layer. **c** By introducing these waveguides into the setup, the disturbed illumination (left), caused by small sub-nanometer irregularities on the surface of the KB-mirrors, is spatially filtered, resulting in a smooth illumination in which high-frequency variations are suppressed (right). Scale bars: 0.5 mrad. Adapted from [47]

13.2.2 Waveguide Optics and Imaging

A hardware solution to avoid artifacts due to a non-ideal illumination is given by additional optical elements for coherence and wavefront filtering. This can be accom-

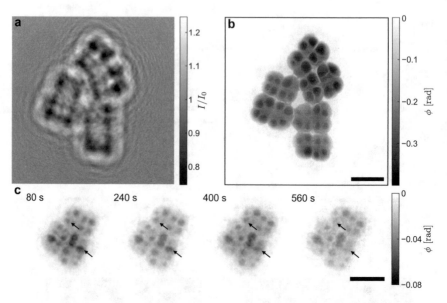

Fig. 13.2 Waveguide holographic imaging at the single cellular level. **a** Normalized hologram of *Deinococcus radiodurans* cells (freeze-dried), obtained in a single recording with 8 s dwell time along with **b** the iterative mHIO phase reconstruction. **c** mHIO reconstruction of (initially) living cells in solution. Each frame was recorded during 80 s (every other frame is shown). Gradual changes in the densities (see arrow) are observed in response to successive irradiation. Scale bars: 4 μm. Adapted from [25]

plished by X-ray waveguides [48–52], positioned in the focal plane of the focusing optics. Owing to the fact that the X-ray focus is smaller at the exit of the waveguide than in the front, this also offers the important advantage of increased numerical aperture. The significant challenges in fabricating two-dimensionally confining X-ray waveguides of suitable quality were solved by crossing two planar one-dimensional waveguides [53, 54], or by advanced lithographic techniques and wafer bonding [55, 56], cf. Fig. 13.1b, including advanced schemes with tapered [57] or curved waveguides [58]. As shown in [12, 14, 15, 25], X-ray waveguides provide highly coherent, well controlled, smooth and quasi point-like illumination for nanoscale X-ray imaging. A comparison between the illumination in the imaging plane provided by a waveguide and a set of KB mirrors is depicted in Fig. 13.1c. The 2D imaging capabilities of this filtered wavefront for biological tissues are demonstrated in Fig. 13.2 at the example of freeze-dried *Deinococcus radiodurans* cells. The first tomography application using waveguides was demonstrated for bacterial cells in [59]. Significantly increased 3D image quality has been achieved in the meantime due to various improvements on different levels, starting from the waveguide optics, the alignment and image processing procedure, the recording and detection scheme, and finally the phase retrieval [14]. The current state-of-the art for tomography of biological cells is reported in [14] and for biological tissues in [60, 61].

13.2.3 Dose-Resolution Relationship

The resolution values as obtained for waveguide-based holography have reached the range of 50 nm (half-period resolution). Fitting of line cuts through lithographic test patterns has given FWHM values down to 25 nm, but based on limited numerical aperture twice this value seems a more realistic resolution estimate. This is still significantly lower than typical values for ptychography, e.g. 10–15 nm for the same test pattern imaged in [62, 63]. Note that resolution determination of near-field holographic imaging deserves a careful consideration [64], and shows particularities not known from far-field imaging, such as a dependence of the maximum theoretical resolution on the object position in the field of view. The analysis in [64] also indicates the importance of the numerical aperture in holographic imaging. From an experimental point of view, benchmark experiments on realistic biological samples are required, beyond the typical demonstrations for specially designed test charts, which are highly contrasted and for which high resolution can be much more easily achieved.

For this reason, ptychographic and holographic reconstructions have been compared for the same objects, namely *Deinococcus radiodurans* bacterial cells. This bacterium had served early on as a first demonstration that ptychographic imaging with hard X-rays is possible for low contrast biological samples [66]. 3D reconstructions by ptycho-tomography were presented in [62], holographic imaging in [25], and an early holo-tomography result in [59]. In these studies the point was made that X-ray imaging yields quantitative electron density contrast. Therefore, the long debated (mass) density in bacterial nucleoids could be addressed in quantitative terms [25, 62, 66, 67].

These studies also provided a starting point for considerations of the dose-resolution relationship. Surprisingly, they pointed to an advantage of holography in terms of dose with respect to (far-field) ptychography. Since experiments are never

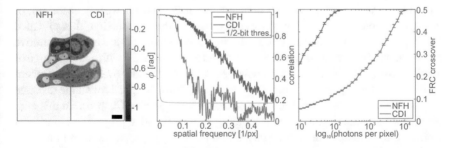

Fig. 13.3 Dose-resolution relationship in holographic and coherent diffractive imaging. **a** Example reconstructions for 200 photons per pixel after 200 iterations of RAAR ($\beta_0 = 0.99$, $\beta_{max} = 0.75$ and $\beta_s = 150$ iterations) for near-field holography (NFH) and coherent diffractive imaging (CDI) using a support and pure phase object constraint. **b** Fourier ring correlation of the reconstructions. The intersection with the 1/2-bit threshold curve determines the maximum frequency that can be resolved. **c** Resolution as a function of dose. Scale bar: 50 pixels. Adapted from [65]

completely free of uncontrolled parameters, this issue was studied also by analytical and numerical work. Starting with random binary bitmap patterns, the information content of near-field and far-field diffraction patterns was compared, as well as reconstructibility as a function of fluence (see [68] for a precise definition of 'reconstructibility'), based on a maximum likelihood approach. Earlier work had already used random bitmap pattern and mutual information theory for a one-dimensional model [69]. More realistic phantoms and reconstruction algorithms were used in [65] for numerical simulations of the dose-resolution relationship for near- and far-field coherent diffractive imaging (cf. Fig. 13.3). In this study, a dose advantage for near field phase retrieval over CDI was found. This conclusion can, however, only me made for the particular reconstruction algorithms which have been used. Other authors who compared holographic and ptychographic reconstruction did not find any considerable differences [70]. Numerical simulations have also considered the effect of finite partial coherence, multi-modal wavefield reconstructions [71], as well as the coherence-resolution relationship [29]. Analytical scaling of the dose-resolution curves as well as experimental results on the resolution limits of biological objects due to radiation damage are given in [72, 73].

13.2.4 Phase Retrieval Algorithms

In many cases, phase-contrast experiments are carried out in the direct-contrast regime where phase-contrast effects are visible as edge enhancement and phase reconstruction can be performed by linearization of the transport of intensity equation (TIE) along the propagation direction z [74, 75]. In the holographic regime, however, where the phase contrast transfer and hence phase sensitivity is highest, these reconstruction algorithms fail. An approach for the inversion of phase-contrast effects in this regime based on the contrast transfer function (CTF) was proposed 20 years ago by Cloetens and coworkers [10]. It relies on several different measurement planes with varying Fresnel numbers and it is valid for weakly absorbing or pure-phase objects with a slowly varying phase. In the next section, this so-called CTF approach and its limits will be discussed in detail, and a particular iterative approach will be presented which is well suited to replace CTF with a larger range of applicability [13]. Before we do so, however, we give a broader overview over recent work on iterative algorithms, including those which are designed for single distance acquisition or are based on different constraint sets (see also Chap. 6). To this end, we already assume a general understanding of how iterative projection algorithms work, see for example the tutorial chapter on basic X-ray propagation and imaging.

We first want to mention the so-called Holo-TIE algorithm [12], which is a one step direct reconstruction scheme operating on two images recorded at slightly different (defocus) distances z from the source. By Holo-TIE, the TIE approach [75, 76] is extended towards arbitrary defocus distances z, including the holographic regime addressed here. This is important since TIE phase retrieval enjoys much success in the direct-contrast regime, and can thus be seamlessly extended, given at least two recordings. Importantly, Holo-TIE does not rely on assumptions on material composition nor on linearization of the specimen's optical constants, as it is typically the case in conventional non-iterative reconstruction schemes. An extension of the initial Holo-TIE reconstruction was presented in [14], operating on four measurements recorded at different, well-chosen distances (or Fresnel numbers). In order to further improve reconstruction quality, if necessary, Holo-TIE reconstructions can also be used to initialize iterative algorithms.

Next, we consider iterative algorithms for single-distance acquisitions. For this case, the so-called modified hybrid-input-output (mHIO) algorithm [15] was proposed as an iterative reconstruction for (single distance) X-ray holograms. The designation 'modified' refers to the fact that the HIO was well established in CDI, and was modified to the near-field case. The mHIO uses a support estimate from the holographic reconstruction to slowly push the phase outside the support to zero. Importantly, it was shown that this algorithm can fill in the lost information due to the zero crossings of the oscillatory CTF. Hence, samples of arbitrary composition can be phased, overcoming the common assumption that the $\delta(\mathbf{r})$ and $\beta(\mathbf{r})$ components of the complex index of refraction $n(\mathbf{r}) = 1 - \delta(\mathbf{r}) + i\beta(\mathbf{r})$ are coupled, which strictly is true only for samples consisting of a single material. Support estimation in mHIO was demonstrated in a fully automated manner for tomography in [59]. In [77], the scheme of mHIO was compared to RAAR, again using the same single distance holographic data and constraints. Significant improvements were provided by an iteratively regularized Gauss-Newton (IRGN) method, reaching higher resolution and image quality than mHIO for noisy data [78, 79]. In [14], different iterative phase-retrieval techniques were compared for the holographic regime, using both numerical simulation and experimental data of biological cells.

Finally, ptychographic algorithms should be mentioned, which offer a solution for cases where the separation between object and probe is challenging (for example if intensity minima occur in the illumination wavefront) or if the constraint set is insufficient. This can be the case, e.g., if no support constraint is available as the object covers the entire field of view and further object constraints, such as sparsity, cannot be applied either. In this case, additional data is required, for example by translating the object with respect to the probe. For such a scan series, ptychographic algorithms can exploit the constraint of separability, which typically offers high reconstruction quality for object and probe. This, however, comes at the prize of a significantly increased number of acquisitions, by a factor on the order of 10 or more. This is impractical for tomography, in particular with large fields of view, as these also lead to a high number of projection angles in order to fulfill the sampling criterion.

13.3   CTF-based Reconstruction and Its Limits

While the assumption of a weakly varying phase in CTF-based phase retrieval is a reasonable approximation for unstained biological tissues [47], samples with a larger

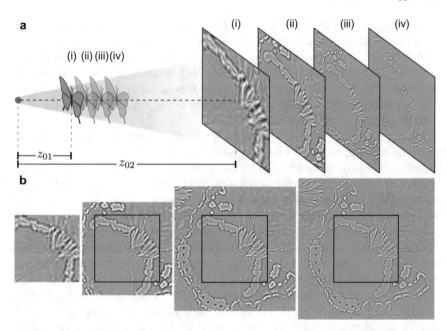

Fig. 13.4 Procedure for the alignment of projection images in multi-distance cone-beam phase-contrast imaging using a butterfly as test object. **a** By acquiring images at varying source-to-object distance z_{01}, while the source-to-detector distance z_{02} is kept constant, different Fresnel numbers can be reached in the resulting projections. The simultaneously changing magnification, however, also alters the effective pixel size and field of view of the single images. **b** To account for a varying magnification, all images are scaled to the effective pixel size of the projection with the highest magnification. Subsequently, the images are aligned to each other in Fourier space to identify overlapping regions. By subsequently cutting all projections accordingly, projection images with the same field of view and effective pixel size but varying Fresnel numbers can be obtained, well suited for the application of multi-distance phase-retrieval approaches

variance in electron density lead to artifacts in the reconstructed phase distributions. An alternative phase-retrieval approach is given by iterative algorithms, in which the phase distribution is reconstructed by alternately propagating between the object and measurement plane and applying according constraints as, e.g., a compact support [80]. As many samples are, however, not compactly supported, no suitable phase-retrieval algorithms exist.

In [13], a simple approach of iterative alternating projections is introduced which can fill this gap, providing superior image quality for extended samples which do not obey the assumption of a slowly varying phase. The input data corresponds to the same set of measurements which is typically used in CTF-based phase retrieval with images acquired in $N = 4$ measurement planes. Due to the cone-beam geometry of the setup, these projection images are recorded at varying magnification and hence fields of view and have to be aligned to each other prior to the phase-retrieval step. This can be implemented according to the scheme presented in Fig. 13.4. In a first step, the different projections are scaled to the effective pixel size of the projection which was recorded at highest magnification. Subsequently, matching fields of view are determined via a cross-correlation of the corresponding projection and its predecessor in Fourier space [81]. As the variation of the source-to-sample distance also results in a variation of the geometric magnification, images are acquired at different Fresnel numbers and hence, the occurring interference fringes vary in all projections. Since this can affect the quality of image alignment, it can be of advantage to use single-step CTF reconstructions instead of the raw projections for alignment. In a last step, all images are cropped to the same field of view, leading to projection images with the same effective pixel size but varying Fresnel numbers, which are well suited for the application of multi-distance phase-retrieval approaches.

In Fig. 13.5, the results of the CTF approach for homogeneous objects [82] as well as the iterative phase retrieval are shown on a 2D projection of polystyrene spheres with a diameter of 15 μm and a 3D reconstruction of an epon-embedded Golgi-Cox stained brain slice of a wild type mouse hippocampus [60]. In both examples, the CTF approach results in severe artifacts due to the violation of the underlying assumptions whereas the iterative approach leads to superior quality in the reconstructed phase distribution, especially in the 2D case of the polystyrene spheres.

So far, the computation time of iterative algorithms impeded the application of iterative algorithms on large data sets. However, with the advent of new computational hardware, especially GPUs, this limitation no longer applies, making iterative reconstructions feasible for a large variety of measurements in which the assumptions of the CTF approach are violated.

13.4 Laboratory μ-CT: Instrumentation and Phase Retrieval

Phase-contrast imaging based on free-space propagation between object and detector requires a high degree of spatial coherence. Therefore, it was long considered to be an imaging technique which is only applicable at large scale synchrotron facilities. With the development of microfocus X-ray sources, however, the degree of spatial coherence could be considerably increased, as the lateral coherence length $L_\perp = \lambda z_{01}/s$ is proportional to the source-to-sample distance z_{01}, and inversely proportional to the source size s. In order to observe phase-contrast effects for object features of spatial length scale d, we must have $L_\perp \geq d$ in the source plane. The flux density on the sample $I \propto s/z_{01}^2$ is directly proportional to the source size, since the power loading of the target is proportional to the linear dimension s of the source (and not its area!). Hence, intensity is maximized if the coherence condition is just fulfilled, i.e., if the object is moved to the minimum distance which still fulfills the coherence constraint $z_{01} = sd/\lambda$. Inserting this distance in the flux density expression gives $I \propto \lambda^2/(sd^2)$, showing that the source size ought to be reduced in order to maximize the lateral coherence (even if this reduces the power) and that the wavelength should be increased to the maximum value compatible with object transmission.

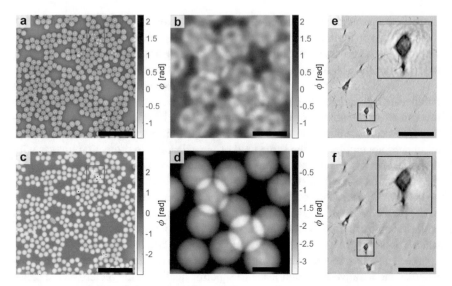

Fig. 13.5 Comparison between CTF-based and iterative phase retrieval in 2D and 3D. **a, b** Phase distribution of a layer of polystyrene spheres with a diameter of 15 μm obtained by the CTF approach. The region marked by the red rectangle in (**a**) is shown at higher magnification in (**b**). **c, d** Iteratively obtained reconstruction of the same data set. The magnified part in (**d**) shows the unwrapped phase obtained by Matlab's `unwrap` function. **e, f** Virtual slice through the density of a Golgi-Cox stained mouse hippocampus obtained from projections reconstructed according to the CTF (**e**) as well as iterative approach (**f**). The insets show regions marked by rectangles at higher magnification. Scale bars: 100 μm (**a, c, d, f**) and 15 μm (**b, e**). Adapted from [13]

The coherence length, however, is not the only factor which has to be taken into account. In [83] it was shown that diffraction effects behind the object have to be constrained to sufficiently small angles, such that the coherent wave scattered by an object feature of size d does not scatter 'out' of the coherent radiation cone. This sets a limit for the scattering angle α, or equivalently the 'shearing length' $L_{shear} = z_{12}\alpha$. This finally results in the condition $L_{shear}/L_{\perp} = (M-1)s/(Md) < 1$ (with the geometrical magnification $M = z_{02}/z_{01}$). Along with $L_{\perp}/d \geq 1$ this must be fulfilled in order for the associated phase-contrast effects to become measurable. Further, visibility increases towards smaller values of $L_{shear}/L_{\perp}$ and larger $L_{\perp}/d$. In Fig. 13.6a, the ratio $L_{shear}/L_{\perp}$ is plotted as a function of the geometrical magnification M and the object feature size d for a constant source size FWHM$_{src}$ = 10 μm. For feature sizes below 10 μm, the lateral coherence length is thus insufficient, even if the object was illuminated coherently, since the diffraction angle would be too large, and the signal would not interfere coherently with the primary wave in the detection plane. Only when using the 'inverse' geometry at low M, the shearing condition $L_{shear}/L_{\perp} < 1$ can be met for feature sizes smaller than 10 μm.

While a small source size s warrants sufficient partial coherence for the realization of propagation-based phase-contrast, it also compromises the flux, as the power loading of the anode cannot be increased without melting of the target material. To

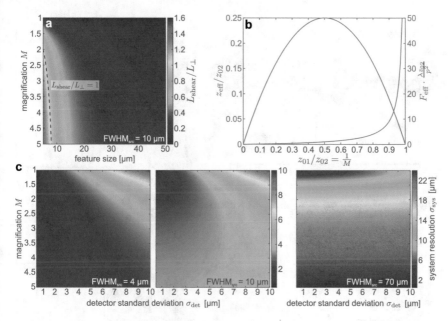

Fig. 13.6 Considerations of partial spatial coherence and resolution in a microfocus setup. **a** Effect of coherence on phase contrast. The ratio $L_{shear}/L_\perp$, as defined by [83], is an indicator for the visibility of phase-contrast effects, with enhanced visibility for smaller values of $L_{shear}/L_\perp$. For $L_{shear}/L_\perp \geq 1$, the partial coherence is not sufficient for phase-contrast effects to be measurable. The ratio depends on the source size (here: 10 µm), the geometrical magnification M as well as the size of features of interest. Note that other important factors as the detector resolution are not considered. **b** Effective propagation distance as a function of the inverse magnification $1/M$. For a given source-to-detector distance z_{02}, the maximum effective propagation distance is $z_{eff,max} = 0.25 \cdot z_{02}$, reached at a geometrical magnification of $M = 2$. While it decreases symmetrically for larger and smaller values of the inverse magnification, the Fresnel number increases monotonically according to $F_{eff} \lambda z_{02}/p^2 = 1/(M-1)$. **c** System resolution given by (13.1) as a function of detector standard deviation and geometrical magnification for different typical source sizes

overcome this limitation, more elaborate anode schemes are required, such as rotating anodes (cf. Fig. 13.7a) [84] or liquid-metal jets consisting of the alloy Galinstan which is liquid at room temperature (cf. Fig. 13.7c) [85]. In the first case, a higher photon flux is enabled by the decrease in stationary heat development as the interaction point between electron focus and anode material is constantly changing. The minimum spot size, however, lies in the range of ∼70 µm. In the second case, the advantage of the heat transfer is combined with an anode that is already in a liquid state, so that the electron power is no longer limited by the melting of the anode material, and which is continuously regenerated, allowing for electron power densities that can vaporize the anode material. Additionally, electron spot sizes well below 10 µm can be reached.

The divergent beam emanating from the quasi-point sources and the associated geometrical magnification M result in an effective pixel size $p_{eff} = p/M$ in the object plane, as well as an effective propagation distance $z_{eff} = (z_{02} - z_{01})/M$ and effec-

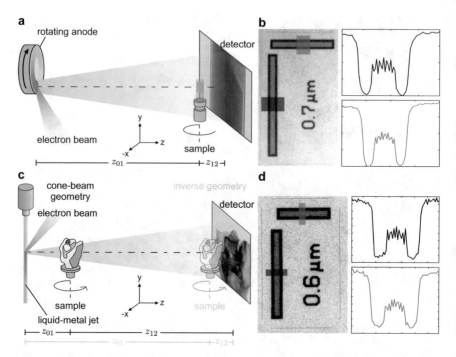

Fig. 13.7 Sketch of the laboratory setups. **a** The X-rays are generated by a microfocus source with a rotating copper anode. In a distance z_{01} behind the source, the sample is positioned on a fully motorized sample stage and the intensity distributions are recorded in a distance $z_{12} = z_{02} - z_{01}$ behind the sample. Due to the comparably large source diameter, this setup is only operated in the inverse geometry. **b** In this geometry, the 700 nm lines and spaces can be resolved in both directions, as revealed in the profiles averaged over the indicated lines shown on the right. Note that in horizontal direction also the 600 nm lines and spaces can be recognized (not shown) due to the smaller source spot in this direction. **c** In a different setup, the X-rays are generated by a liquid-metal jet source with Galinstan as anode material. Depending on the desired resolution and field of view, the setup can be either operated in a cone-beam ($z_{01} \ll z_{12}$) or inverse geometry ($z_{01} \gg z_{12}$). **c** In the inverse geometry, the 600 nm lines and spaces of a test pattern can be resolved in 2D

tive Fresnel number $F_{\text{eff}} = p_{\text{eff}}^2/(z_{\text{eff}}\lambda)$, based on the Fresnel scaling theorem [2]. Interestingly, the maximum effective propagation distance is given by $0.25 \cdot z_{02}$, reached at a magnification of $M = 2$, while for larger or smaller values of the magnification, the effective propagation distance decreases symmetrically, see Fig. 13.6b. As the effective Fresnel number, on the other hand, also depends on the effective pixel size, it deviates from the symmetric behavior of the effective propagation distance, and hence, can be freely adjusted by changing the geometric magnification.

Laboratory setups are often implemented in geometries which correspond to either of two limiting cases: (cf. Fig. 13.7c). In the 'cone-beam geometry', the source-to-sample distance z_{01} is small compared to the sample-to-detector distance z_{02}, leading to a large magnification $M > 1$, while in the 'inverse geometry', the the sample is moved close to the detector, resulting in a small magnification $M \simeq 1$. The resolution of the imaging system as a function of the source standard deviation σ_{src}, corresponding to a Gaussian source distribution, and detector standard deviation σ_{det} is given by [86]

$$\sigma_{sys} = \sqrt{(M-1)^2 M^{-2} \sigma_{src}^2 + M^{-2} \sigma_{det}^2}. \tag{13.1}$$

Hence, in the cone-beam geometry, the resolution is limited by the source size, while in the inverse geometry, the resolution of the detector is the limiting factor. In Fig. 13.6c the system resolution as a function of geometrical magnification and detector standard deviation is shown for typical source sizes of a liquid-metal jet source ($FWHM_{src} = 4$ μm and $FWHM_{src} = 10$ μm) and a source with a rotating anode ($FWHM_{src} = 70$ μm). It is evident that for the smaller source sizes, resolutions well below 5 μm can be reached in both geometries, while for the larger source size, only the inverse geometry provides resolutions sufficient to resolve features smaller than 10 μm. Provided that a detector with a point-spread-function (PSF) of standard deviation in the range of ∼1 μm is available, the highest resolution for both types of X-ray sources can be reached in inverse geometry, in the same order of magnitude as the detector resolution. This could be experimentally validated by imaging an absorbing test pattern with the XSight Micron (Rigaku, Czech Republic) [47, 87], a lens-coupled high-resolution detector, showing that half-period resolutions well below 1 μm are possible (cf. Fig. 13.7b, d). Note that the constraints of shearing length and system resolution result in a similar expression for the maximum magnification.

In order to allow for tomographic imaging at the laboratory, the setup should comprise a fully motorized sample tower, containing one rotational axis for the tomographic scans, three translations above the rotation axis for the alignment of the sample in the field of view and one translation perpendicular to the optical axis for alignment of the rotation axis. Additionally, a further translation along the optical axis can be used for varying the source-to-sample distance z_{01}. By also enabling the motion of the detector perpendicular to the optical axis, all relevant degrees of freedom for a proper alignment of the rotation axis are given [47, 90, 91].

Data analysis starts by phase retrieval on the individual projections, followed by tomographic reconstruction of the 3D volume in which the cone-beam geometry of the setup has to be taken into account, e.g., by using the implementation of the algorithm by Feldkamp, Davis and Kress (FDK) [92] provided by the ASTRA tomography toolbox [93, 94]. As shown in [89], a suitable phase-reconstruction strategy follows the Bronnikov-aided correction (BAC) [88], since this approach is robust with respect to the non-ideal beam conditions at compact X-ray sources such as low spatial coherence and large bandpass, providing sharp and quantitative reconstructions of the sample's 3D density distribution (cf. Fig. 13.8). More examples of the results that can be obtained at the laboratory, both in cone-beam and inverse geometry, can be found in Sect. 13.6.

13.5   Novel Tomography Approaches

13.5.1   Combined Phase Retrieval and Tomographic Reconstruction

In the classical phase-contrast tomography approach, data reconstruction is performed in a sequential manner, i.e., it starts with phase retrieval on the individual projections, followed by tomographic reconstruction. The main challenge in this reconstruction scheme is the phase-retrieval step and many methods have been developed relying on, e.g., the linearization of the transport of intensity equation [74, 75] or an analytic form of the free-space contrast transfer functions [10, 82]. However, in most cases, phase-retrieval techniques require additional assumptions like negligible absorption, slowly varying phase, or known compact support as well as measurements from several measurement planes. In [17], a combination of phase retrieval and tomographic reconstruction was introduced, called 'iterative reprojection phase retrieval (IRP)', which can overcome these limitations, providing reconstructions of

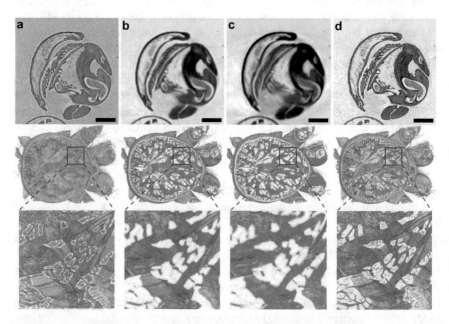

Fig. 13.8 Comparison of different phase-retrieval approaches. **a** Results from the raw projections. Top: Virtual slice through the reconstructed pedipalp of an iodine stained cobweb spider. Bottom: Volume rendering of its thorax, showing individual muscle strands that appear hollow due to the edge-enhancement effects caused by free-space propagation between the object and the detector. After the application of the **b** MBA [74], **c** SMO [75] and **d** BAC approach [88], quantitative gray values can be reconstructed at high signal-to-noise ratio, though at the cost of resolution in the case of the MBA and SMO. Only the BAC provides a reconstruction with a resolution comparable to the raw data. Scale bars: 100 μm. Adapted from [89]

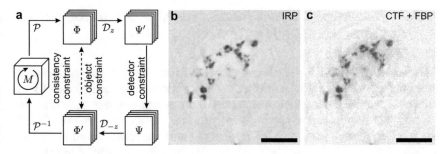

Fig. 13.9 Combined phase retrieval and tomographic reconstruction. **a** Sketch of the combined iterative reconstruction algorithm called 'iterative reprojection phase retrieval (IRP)' [17]. $\mathcal{P}$ denotes the tomographic projection, whereas $\mathcal{D}$ stands for Fresnel propagation. **b** Virtual slice through the reconstruction obtained by the IRP scheme. **c** Virtual slice through the reconstruction obtained after application of the CTF approach for phase retrieval and subsequent filtered backprojection (FBP). Compared to the IRP approach, a lower signal-to-noise ratio can be observed and twin image artifacts occur due to the imperfect phase retrieval. Scale bars: 5 μm. Adapted from [14, 17]

the object from projections acquired at a single reconstruction plane without further assumptions on the phase-shifting and absorption properties of the object or its support. It relies on the Helgason-Ludwig consistency which states that tomographic projections are not independent from each other as the finite size of the object imposes systematic correlations. This helps to phase in particular low spatial frequencies, which pose a significant challenge in single-distance phase retrieval. The general scheme of the IRP algorithm is shown in Fig. 13.9a.

Reconstruction starts with an initial guess for the 3D distribution of the refractive index decrement $\delta(\mathbf{r})$ as well as for $\beta(\mathbf{r})$. By forward projection, a first iteration of the phase and amplitude distribution of the exit wave $\Phi_\alpha(x, y) \propto \exp\left(-k \int_z [i\delta_\alpha(\mathbf{r}) + \beta_\alpha(\mathbf{r})] dz\right)$, i.e., the wave field directly behind the object, is obtained for each tomographic angle α. Subsequently, the exit wave for each angle is propagated to the detection plane (Fresnel propagator $\mathcal{D}_z$) and the magnitude constraint is enforced. Back propagation then yields a modified exit wave Φ_α'. In the last step, the 3D distributions of $\delta(\mathbf{r})$ and $\beta(\mathbf{r})$ are reconstructed from these modified exit waves in a similar fashion as in the algebraic reconstruction technique (ART). By also enforcing positivity of the electron density, corresponding to $\delta(\mathbf{r}) > 0$, as well as positivity for $\beta(\mathbf{r})$ (no generation of X-rays), additional unrestrictive constraints can be implemented for the object. This basic sequence of propagators and projectors is iterated M times, leading to a consistent tomographic reconstruction. The result of the IRP algorithm for the example of a barium-stained macrophage is shown in Fig. 13.9b [14]. A higher signal-to-noise ratio can be reached and twin image artifacts are reduced, compared to the standard sequential scheme (with CTF-based reconstruction of all projections followed by a filtered backprojection), which is depicted in (c).

13.5.2 Tomographic Reconstruction Based on the 3D Radon Transform (3DRT)

As discussed in Sect. 13.4, a significant challenge in laboratory-based phase-contrast tomography is to reach sufficient brilliance with a laboratory source. To this end, small source sizes are required, which often means insufficient photon flux. In [95], a novel tomographic reconstruction approach based on the 3D Radon transform (3DRT) has been introduced, instead of the 2D Radon transform (2DRT) used in classical tomography. The 3DRT could help to solve this intensity/coherence dilemma, as it allows for relaxations of the source size in one of the two source dimensions, while exploiting the smaller dimension for resolution and coherence.

To this end, it was shown that by proper extension of the data recording scheme, in particular rotation around two instead of one tomographic axes, an experimental realization of the area integrals required for 3DRT becomes possible. Within this scheme, the recorded projections are integrated along the 'low-resolution direction' in which the source spot is elongated. At the same time, the resolution and contrast of the entire 3D object reconstruction are determined by the perpendicular 'high-resolution direction'. The 3DRT filtered backprojection is performed analogously to the 2D case by filtering the 1D absorption profiles and subsequent backprojection or 'smearing' into the 3D space. Note, however, that the filter function in Fourier space is given by k^2 as opposed to $|k|$ in the 2D case. Figure 13.10 presents an example of a 3DRT reconstruction applied to experimental data of a gerbil cochlea, which was recorded with anisotropic source conditions [96]. In the recorded projection in (a), the blurring in horizontal direction is clearly visible. After reconstruction with the 3DRT, the numerical reprojection under the same angle in (b) as well as the virtual slice in (c) show an isotropically sharp representation of the cochlea. Note that the sampling scheme is chosen such that the full 3D Fourier space is equidistantly sampled. However, in the practical implementation, the computation

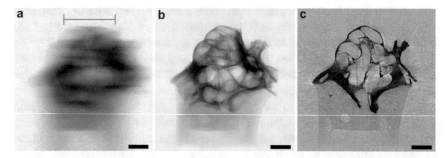

Fig. 13.10 Tomography with extended sources based on the 3D Radon transform. **a** Empty-beam corrected projection of a gerbil cochlea prior to tomographic reconstruction. The bar at the top indicates the effective width of the source. **b** Numerical reprojection of the reconstructed volume, showing a sharp image of the projected cochlea. **c** Slice through the reconstructed volume using the 3D Radon transform. Scale bars: 1 mm. Adapted from [96]

of additional 1D profiles from each of the anisotropically blurred projections in a sector of $\pm\Delta\theta$ around the high resolution direction yielded better results. As this leads to a nonuniform sampling of the unit sphere, partitioning of the hemisphere into Voronoi regions was used for normalization in the 3DRT reconstruction step.

The 3DRT can also be used for phase-contrast imaging based on, e.g., grating interferometry, edge illumination or free space propagation. This was demonstrated in [95] for the case of propagation imaging, taking the example of a common match. Phase retrieval according to the BAC algorithm [88] was performed prior to the tomographic reconstruction. This phase retrieval step was carried out on the sinogram and hence on the 1D projections acquired after integration along the low-resolution direction. Note that only in the high-resolution direction of the anisotropic source, the required spatial coherence was provided.

One particular motivation to develop reconstruction based on the 3DRT is related to local or region-of-interest tomography. It can be shown mathematically that the reconstruction only depends on the local values of the Radon transformed object function, so that artifacts introduced by object components outside the reconstruction volume, which often affect the image quality in local tomography of standard 2DRT, should in principle be suppressed. This, however, could not be confirmed by numerical simulations.

13.6 Tomography of Biological Tissues: Applications and Benchmarks

As is well known from classical histology, physiological function is enabled by the underlying tissue structure, and conversely, alterations lead to different pathological states, e.g., in neurodegenerative diseases. Deciphering the 3D tissue structure from the the the whole organ down to the cellular scale enables the quantification of these relations and the underlying mechanisms. Conventional approaches as histological sectioning or electron microscopy (EM), are associated with serial sectioning, staining and subsequent investigation under a light or electron microscope. They provide excellent results on single 2D sections, but the 3D anatomy can only be determined after aligning the individual sections, leading to a non-isotropic resolution within the tissue. Apart from possible artifacts due to the slicing or staining procedure, they are labor-intensive and time-consuming techniques, impeding the visualization of large fields of view, e.g. entire organs, even at moderate resolution.

To this end, phase-contrast tomography based on free propagation offers a unique capability for high resolution imaging of soft tissues over a cross section of several mm, and with a geometric zoom capability to visualize selected regions of interest down to 20–50 nm voxel sizes. Zoom-tomography is enabled by variation of the focus-to-sample distance, yielding 3D reconstructions at selectable magnification, resolution and field of view (FOV). The zoom capability and the dose efficiency are particularly pronounced if highly divergent and highly coherent beams with low wavefront distortions are available. Such wavefronts are provided by optimized X-ray waveguide optics, as presented in Sect. 13.2. In combination with suitable phase-retrieval algorithms, challenging radiation sensitive and low-contrast samples can be reconstructed with minimal artifacts.

In this section, we review phase-contrast X-ray tomography of biological tissues, presenting examples and benchmark studies for two different cases. First, phase-contrast μ-CT using (in-house) laboratory sources, and second, nano-CT using synchrotron radiation (SR). The first case is illustrated by tomography on the scale of small animal organs, notably cochlea [97, 98], as well as tomography at the small animal scale [99]. For the second case, we present tomography of nerves from mouse (optic nerve, sciatic nerve), showing each axon in the nerve with details such as the node of Ranvier and Schmidt-Lantermann incisures [100], lung tissue for asthma and control mice [27], and finally high-resolution reconstructions of human cerebellum, yielding the precise locations of neurons in the molecular and granular layer [61]. The last example comprises both synchrotron nano-CT and laboratory μ-CT.

13.6.1 *3D Structure of Cochlea*

Imaging of the delicate and complex anatomy of the cochlea in small animal models is required to understand malformations caused by genetic defects, to guide new treatments and to develop cochlear implants [101, 102], including novel optogenetic approaches [103]. Cochlea imaging is perfectly suited to illustrate the particular advantages of phase-contrast X-ray imaging, since soft tissues and membranes have to be visualized while surrounded by bone. Phase-contrast tomography of cochleae using synchrotron radiation can overcome the limitations of imaging approaches such as classical histology or magnetic resonance imaging [102, 104, 105]. Reaching sufficient contrast and resolution at laboratory sources, however, poses a much larger challenge. In [97], a well chosen combination of a liquid-metal jet anode (cf. Sect. 13.4), high resolution detectors, an optimized geometry and reconstruction algorithms was used to achieve sufficient contrast and resolution down to 2 μm, enabling the visualization of thin membranes and nerve fibers surrounded by bone. Importantly, the high data quality allowed for automatic histogram-based segmentation between bone and soft tissue. Figure 13.11 illustrates the achieved contrast, data quality and resolution for the visualization of thin membranes and nerve fibers within the cochlea.

The presented results show that polychromatic illumination of laboratory X-ray sources does not *per se* impede high data quality. However, the reconstructed grey values can by no means be regarded as quantitatively correct. Neither does the phase-retrieval approach by [88] properly separate phase from amplitude, nor is phase or amplitude well defined in the case of a broad bandpass. Effects of beam hardening make even effective values for (mean) photon energy extremely questionable. Such

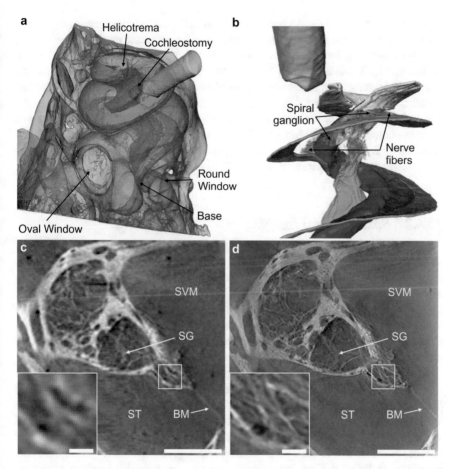

Fig. 13.11 Results of in-house phase-contrast tomography on cochlea. **a** 3D visualization of a mouse cochlea with bone (brown, semi-transparent), basilar membrane (green), Reissner's membrane (yellow), Rosenthal's canal (blue) and optical fiber (gray). **b** Magnified view showing nerve tissue (orange). The nerve fibers of the spiral ganglion pass out between the two layers of the lamina spiralis ossea (bottom layer shown in magenta). **c, d** Slices through regions of interest within the cochlea, revealing scala tympani (ST), basilar membrane (BM), scala vestibuli et media (SVM) and spiral ganglion (SG) for **c** cone-beam and **d** inverse geometry. Finer nerve fibers are resolved in the inverse geometry (see inset). Scale bars: 200 μm and 20 μm (insets). Adapted from [97]

problems are often particularly noticeable if strongly absorbing materials as metal are present in the object. Cochlea implants with wires and electric components in the vicinity of soft tissues fall into this category of multiple material objects with strong differences in the δ/β-ratio. To find a solution for such applications, a new class of narrow-band and compact radiation sources was evaluated in [98], based on the interaction of accelerated electrons and laser photons (inverse Compton effect) [106]. As a prototype of such sources, the Munich Compact Light Source (MuCLS) generates narrow-band X-ray photons within a continuously tunable energy spectrum [107–109], providing a very useful source for phase-contrast imaging, and closing a performance gap between conventional laboratory instruments and synchrotron facilities. MuCLS data enabled high quality reconstruction of the functional soft tissue within guinea pig and marmoset cochleae even in the presence of an electrical cochlear implant with metallic components. Figure 13.12 illustrates imaging of a guinea pig cochlea, at a resolution in the range of 10 μm [98]. The higher and tunable photon energy and in particular the narrow bandpass of the MuCLS allows in principle for more quantitative reconstruction values (grey levels) than possible with conventional laboratory microfocus X-ray sources.

13.6.2 Small Animal Imaging

In the next example, we show that not only excised organs, but entire small animals are amenable to propagation-based phase-contrast tomography even at compact laboratory sources. This is important since synchrotron radiation sources are rarely in direct vicinity of small animal and biomedical research facilities, and beamtime scheduling constraints easily interfere with the requirements of small animal studies. Contrarily, a much wider range of premedical research applications can be addressed after translation of phase-contrast tomography to the laboratory scale.

The chosen example is concerned with in situ 3D lung imaging of small animals. Phase contrast had been demonstrated earlier for this application by grating-based phase contrast [110, 111], which does not, however, achieve the resolution to resolve small features in tissue. With the advent of improved sources, instrumentation and analysis, 2D [112] and 3D [99] propagation-based phase-contrast imaging has now become practical also at the level of small animals. A suitable strategy, demonstrated in [99], is as follows: First, large overview scans are recorded in absorption contrast. Subsequently, by changing according geometric parameters and increasing the magnification of the setup, a phase-contrast data set is acquired in local tomography mode. As shown in [99] and illustrated in Fig. 13.13, fine terminal airways and thousands of small alveoli of the lung can be resolved at a resolution of about 5 μm, despite the rather thick and absorbing surrounding tissue.

The results required an optimization of the energetic spectrum by pre-hardening. Hence, for future experiments it would be useful to enrich the liquid jet alloys with indium or to replace it with a suitable silver alloy. By further improvements resulting in a decrease of the acquisition time by a factor of ten, live animal phase-contrast imaging would become possible. The high availability of laboratory sources would thus enable longitudinal studies, as well as the necessary statistical power.

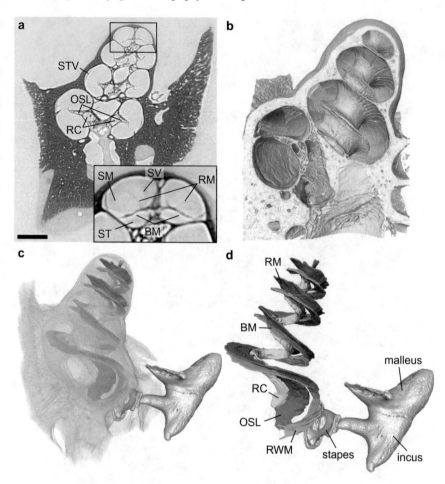

Fig. 13.12 Reconstruction results for a guinea pig cochlea measured at the MuCLS. **a** Virtual slice through the reconstructed 3D volume, showing the typical anatomical features of the cochlea in high detail without artifacts like beam hardening. In particular, the Rosenthal's canal (RC), the osseous spiral lamina (OSL) and the stria vascularis (STV) can be recognized. In the inset, in which contrast was optimized for the soft tissue components, also the basilar membrane (BM) and the Reissner's membrane (RM) are visible as well as the corresponding chambers separated by these membranes, the scala tympani (ST), scala media (SM) and the scala vestibuli (SV). **b** 3D rendering of part of the volume with a cut revealing the inner structure of the cochlea in high detail. **c, d** Segmentation of typical anatomical features of the cochlea together with a volume rendering displayed semi-transparently to put it in context. Note that due to rupturing, the theoretical shape of the membranes was derived from the position of typical landmarks in the volume. The segmentation includes the ossicles (malleus, incus and stapes), the round window membrane (RWM) as well as the OSL, RC, RM and BM. Scale bar: 1 mm. Adapted from [98]

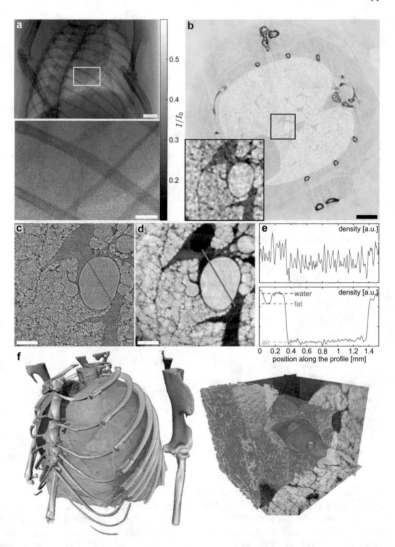

Fig. 13.13 **a** Projection of the large-FOV data set that covers the thorax of the whole mouse (top), showing mostly absorption contrast, and a phase-contrast projection obtained by zooming into the region marked by the rectangle (bottom). **b** Virtual slice through the large-FOV measurement. The inset shows a zoom into the lung area marked by the rectangle with adjusted contrast. **c** Virtual slice obtained in the zoom configuration setting, where the sample-detector distance was increased to obtain phase contrast. **d** The same slice after the application of the phase retrieval approach proposed by Paganin et al. [75]. The signal-to-noise ratio is increased while simultaneously the gray values get more quantitative. **e** Profiles along the 6 pixel wide lines indicated in (**c**) and (**d**), respectively. The positive effect of phase retrieval is clearly visible. **f** 3D rendering of the large-FOV measurement (left), containing automatically segmented bones (gray), the heart (red) and lung tissue (pink). On the right, a 3D rendering of the zoomed reconstruction volume is shown, with orthogonal slices through the volume and a rendering of the soft-tissue structure. Scale bars: 2 mm (**a**, top and **b**), 400 μm (**a**, bottom) and 500 μm (**c**, **d**). Adapted from [99]

13.6.3 3D Virtual Histology of Nerves

A fast conduction of action potentials in specific nerves in the peripheral or central nervous system (PNS/CNS) is enabled by myelin sheaths which surround the corresponding parallel arranged axons within the nerve, leading to an electrical insulation against the surrounding fluids. These myelin sheaths comprise myelin segments with a length of 150–200 μm in the CNS and up to 1 mm in the PNS [113, 114], followed by myelin-free gaps called *nodes of Ranvier (RN)*. This segmental structure results in saltatory conduction, with the action potentials propagating from one node of Ranvier to the next, where a large sodium influx leads to a regeneration of the signal [115]. In more or less regular distances within the myelin segments of the PNS, its compact structure is interrupted by clefts, the so-called Schmidt-Lanterman (SL) incisures.

3D virtual histology by phase-contrast tomography offers a unique access to probe the spatial organization of the axon bundles within the nerve and to answer questions of axon organization and size distribution as well as correlations of RN and SL between different axons. In [100], entire (uncut) optic, saphenous and sciatic nerves were prepared from mouse using high pressure freezing, and scanned using the nanofocus KB optics at legacy beamline ID22NI of ESRF. In subsequent work, the recent ESRF upgrade beamline ID16A [43], as well as the upgraded GINIX endstation at beamline P10/PETRAIII [44], have been used to demonstrate the suitability of these novel setups for nerve tomography [116].

It was found that intrinsic electron density without additional labeling or staining is sufficient to identify axonal structures. However, to specifically image the myelin sheath surrounding the axon, labeling by an osmium tetroxide stain was required. By placing the nerve at different defocus positions in the diverging waveguide, both overview scans of entire sciatic nerves, as well as zoom tomograms of relevant sub-structures as nodes of Ranvier and Schmidt-Lanterman incisures were recorded (cf. Fig. 13.14). The reconstructions were found to be very consistent with histology sections and EM micrographs, but offered the clear advantage of probing much larger volumes that could be visualized with isotropic 3D resolution.

13.6.4 Macrophages in Lung Tissue

The lung is the primary organ of the respiratory system in air-breathing vertebrates. It enables the oxygen exchange between the inhaled oxygen-rich air and the blood in the cardiovascular system of the body. The air is transported through the trachea, which branches into many bronchi and bronchioles that eventually end in the alveoli. This anatomical structure leads to a continuous surface enlargement, enabling a fast exchange of oxygen between the alveoli and surrounding blood vessels. One of the major diseases associated with the lung is asthma, with typical symptomps as coughing or shortness of breath, the cause and progression of which is still not fully understood [117]. It leads to a chronic inflammation of the respiratory tract, especially

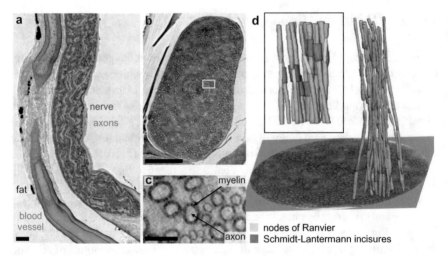

Fig. 13.14 Phase-contrast tomography of nerves. **a** Three dimensional visualization of a mouse saphenus nerve stained with osmium and embedded in agarose (voxel size 430 nm). The nerve is rendered in blue, while an adjacent blood vessel is depicted in red. Additionally, a longitudinal virtual slice is shown, revealing the single axons within the nerve due to the high electron density of the osmium-stained myelin sheath. **b** Virtual slice through an EPON-embedded osmium stained mouse sciatic nerve measured with a voxel size of 430 nm. **c** Virtual slice through a zoom-tomogram of the same nerve (50 nm voxel size), recorded in the region marked in (**b**). **d** 3D rendering of the same nerve measured with a voxel size of 100 nm. 20 axons (turquoise) are shown along with a virtual slice through the reconstructed volume. Nodes of Ranvier are rendered yellow, Schmidt-Lanterman incisures red. An additional rendering of 13 axons (black box) suggests a correlation between the positions of these nodes and incisures of neighbouring axons. Scale bars 50 μm (**a**), 100 μm (**b**) and 10 μm (**c**). Adapted from [100]

the bronchi and bronchioles. Macrophages, which are part of the immune system, are a special kind of phagocytes, protecting the organism by ingesting harmful pathogens and other foreign substances. They are known to be involved in processes of allergic inflammation [118], but their precise role in asthma and the underlying mechanisms are still debated [119], including in particular their migration properties [120].

High resolution X-ray phase-contrast tomography is a promising tool to visualize the 3D distribution of macrophages in situ. In [27], tissue slices from lungs of mice were imaged at the legacy beamline ID22NI of ESRF as well as the GINIX setup with voxel sizes in the range of 50–430 nm. In this study, the intricate three-dimensional (3D) structure of lung tissue was visualized, with its system of the bronchial tree, alveoli, and blood vessels (see Fig. 13.15). In addition, the distribution of macrophages and their migration properties within the lung were investigated by 3D visualization with high resolution and contrast. Precise tracking of alveolar macrophages in relation to anatomical structures was enabled by barium-labeling [27, 121]. It was shown that the intratracheally applied macrophages (MH-S cell line [122]) localize predominantly on alveoli and are able to penetrate the epithelial layer between the airway lumen and parenchyma.

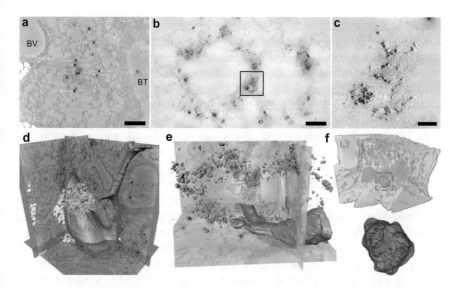

Fig. 13.15 Phase-contrast tomography on lung tissue. **a** Virtual slice through the reconstructed asthmatic lung tissue obtained at ID22NI (ESRF) with a voxel size of $p = 430$ nm. Barium sulphate particles (black) and fat (white) show a strong density contrast compared to soft tissue. A blood vessel (BV) and a bronchial tube (BT) can be identified based on their different wall morphologies. **b**, **c** Virtual slices through the reconstructed lung tissue from a healthy control measured at the GINIX setup (DESY) both in an overview scan (**b**, voxel size: 245 nm) and a zoom configuration (**c**, voxel size: 52 nm). The position of the zoom scan is indicated by a rectangle in (**b**). **d** 3D rendering of the ID22NI data set together with barium clusters (green), alveolar walls in a small ROI (yellow) and part of a blood vessel (purple). **e**, **f** 3D visualization of the tomography results obtained at the GINIX setup. The 3D renderings show barium aggregates in macrophages (green), part of a blood vessel (purple), the bronchial wall (yellow) and the outline of a single macrophage (blue). This cell is additionally shown at higher magnification in (**f**, bottom). Scale bars: 100 μm (**a**), 50 μm (**b**) and 10 μm (**c**). Adapted from [27]

13.6.5 Neuron Locations in Human Cerebellum

The cerebellum, which is among other things important for the maintenance of upright posture and synergy of movements [123], is located at the back of the brain of mammals. Compared to the largest part of the brain, the cerebral cortex, it has a significantly higher cell density and contains 80% of the total number of neurons within the human brain, despite its relatively small weight of $\sim$10% of the total brain mass [124]. The cerebellum generally consists of the tightly folded cerebellar cortex, comprising three distinct layers, the cell-rich granular layer, the low-cell molecular layer and the intermediary mono-cellular Purkinje cell layer, located above white matter with a large amount of axon bundles.

Studying the cytoarchitecture of the cerebellum via propagation-based phase-contrast tomography requires an additional contrast enhancement as hydrated unstained tissue does not allow for an unambiguous identification of all cells in the densely packed granular layer [47]. Contrary to radiocontrast agents, which specifically increase contrast in certain features of the sample, e.g., the myelin sheaths in mouse nerves or macrophages in the mouse lung, a global contrast enhancement can be reached by exchanging the surrounding medium with a medium with lower electron density. This makes it possible to examine conventional neuropathological samples as human brain obtained during routine autopsy since these are usually embedded in paraffin after fixation.

In [61], propagation-based phase-contrast tomography was performed on paraffin-embedded human cerebellum both at the GINIX endstation and at the laboratory setup in inverse geometry, providing insights into the 3D cytoarchitecture at sub-cellular level (cf. Fig. 13.16). In order to fully exploit the potential of this 3D virtual histology, a workflow was developed in order to automatically locate the small cells in the molecular and granular layer, leading to the segmentation of several ten thousands (GINIX) to $\sim$1.8 million of cells (laboratory). This has enabled the analysis of spatial organization of neurons in the granular layer, e.g., based on local density estimations or pair correlation functions, pointing towards a strong short-range order of these cells visible as local clustering (cf. Fig. 13.17). Moreover, the availability of the exact cell positions in 3D allows for the precise quantification of cellular distributions, revealing an anisotropy in the arrangement of nearest neighbors within the granular layer which is governed by the principle directions of the Purkinje cell layer, a result which would not have been accessible by conventional 2D histology.

13.6.6 Outlook: Time-Resolved Phase-Contrast Tomography

Together with the progress in detector technology, the current upgrades of synchrotron sources to a multi-bend achromat lattice and the corresponding increase in brilliance will offer unique opportunities for time-resolved (dynamic) tomography. In other words, the data acquisition rate f could become high enough to observe dynamic processes in biological matter on the micro-scale and in some cases even on sub-micron scales. The phase-contrast imaging capabilities presented in this chapter could thus be (at least partially) extended from 3D to 4D (time & space) imaging. The first question to be asked concerns the temporal sampling required to probe the dynamic process of the object. In [125], e.g, phase-contrast tomography was performed in vivo on *Xenopus laevis* embryos, revealing new aspects of their gastrulation over time. The time scale of this development was long enough for the single tomographic scans to be considered as static, enabling data reconstruction following the classical approach via simple filtered backprojection, and the development of the gastrulation was monitored by recording several tomograms with a time lapse of $\sim$10 min.

For faster processes, more elaborate and generalized recording or analysis schemes have to developed to meet the challenges of dynamic tomography. In the special case of cyclic processes such as a beating heart, acquisitions can be gated or trig-

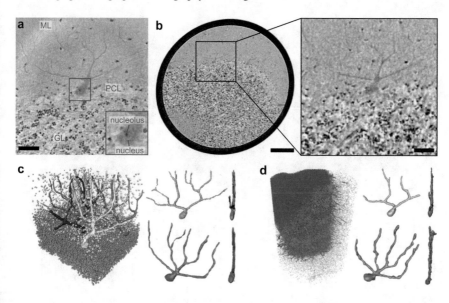

Fig. 13.16 Phase-contrast tomography of tissue from human cerebellum, showing reconstructions obtained both at the GINIX endstation and the laboratory setup. **a** The virtual slice through the reconstructed volume of the synchrotron data set reveals the interface between the low-cell molecular (ML) and cell-rich granular layer (GL), including a cell of the mono-cellular Purkinje cell layer (PCL). **b** Corresponding slice of the laboratory data set, showing the larger volume accessible by the laboratory setup while maintaining the resolution required for single cell identification. A magnified view of the region marked by the rectangle is shown on the right, corresponding to the FOV of the synchrotron data set in (**a**). **c** Segmentation of the cells in the granular layer (dark red), the molecular layer (light red) and the Purkinje cell layer (shades of gray) with two exemplary Purkinje cells shown separately, from front and side view. The segmentation for the granular and molecular layer was performed automatically whereas for the Purkinje cell layer, a semi-automatic approach was used. **d** The same segmentation for the laboratory data set. Note that the individual Purkinje cells are the same as for the synchrotron data set and that the thick branches of the dendritic tree can already be resolved with the laboratory setup. Scale bars: 50 μm (**a** and **b**, right) and 200 μm (**b**, left). Adapted from (a) [47] and (b–d) [61]

gered (hardware or *a posteriori* software) to cover different phases of the considered motion. By combining projections which are recorded at the same state of the motion but at different rotation angles, static solutions can be generated for the different time points within one cycle, unraveling, e.g., the complex muscle movement during insect flight [126]. However, such approaches fail for non-cyclic processes.

In [127], an approach for the reconstruction of time-resolved processes based on filtered backprojection along dynamically curved paths was introduced. It can account for non-affine and non-cyclic motion on time scales shorter than the time needed for an entire tomogram, provided that the motion model can be estimated. The workflow is depicted in Fig. 13.18 for the example of a burning match. In order to monitor the burning process, 47 single tomograms with 401 projections each were recorded at the TOMCAT beamline (SLS, Villigen, Switzerland) while the match

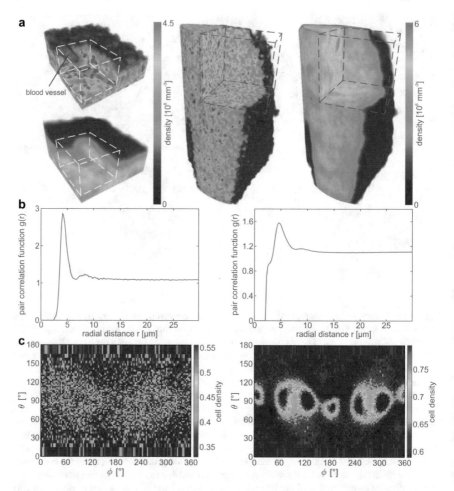

Fig. 13.17 Analysis of cell distributions in the granular layer of human cerebellum, as obtained at the synchrotron (left column) and the laboratory setup (right column). **a** Local cell density distribution within the granular layer of the cerebellum. By considering small volumes for the computation of the local density, a clustering of cells within this layer can be clearly recognized as hotspots in the density distribution. With increasing volume, the differences in cell density are vanishing, leading to an almost uniform density distribution within the granular layer. **b** Angular averaged pair correlation function of the cells in the granular layer, revealing two distinct peaks at approximately once and twice the cell diameter, which indicates a local clustering of the cells. **c** Angular distribution of nearest neighbors in the granular layer. Note that the data sets were aligned with respect to the Purkinje cell layer such that the dendritic tree lies approximately in the xy-plane and hence at $\theta \simeq 90°$. The majority of nearest neighbors are clearly distributed in parallel to the dendritic tree of the Purkinje cells as hotspots in the angular distribution are visible at $\theta \simeq 90°$. Adapted from [61]

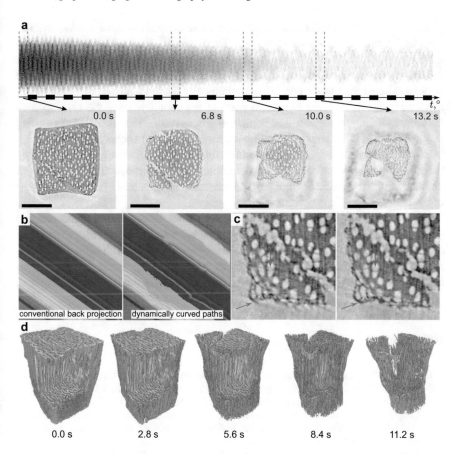

Fig. 13.18 Four dimensional movie of a burning match. **a** Exemplary sinogram extracted from all 18800 recorded projections. The bottom row depicts tomographic reconstructions of the highlighted sinogram segments, each consisting of 401 equidistant projections. While the shape, features of the wooden structure and the stages of the burning process can be clearly identified, the reconstructions show motion artifacts such as 'streaks' or non-closed shapes. **b** By estimating the motion between successive time points via optical flow, a better tomographic reconstruction can be carried out by backprojection on dynamically curved paths (right). Note that the motion amplitude was increased by the factor of 3 for better visibility. **c** The improved reconstruction quality can be observed in fine object details which can be resolved in the image on the right (red arrows), compared to the result of a conventional direct filtered backprojection, shown on the left. **d** Rendered 3D structure of the burning match at different time points. Scale bars: 1 mm. Adapted from [127]

was continuously rotated at a rate of 1.25 Hz. In the exemplary sinogram in (a) at the top, the shrinking of the structure as well as a decrease of signal intensity due to the burning process can be observed. This process can be approximately depicted in 3D by selecting intervals of 401 projections from the sinogram and performing standard filtered backprojections, leading to the reconstructions shown in the lower row. The shape and features of the wooden structure and the stages of the burning process can be clearly identified. However, the reconstructions show motion artifacts such as 'streaks' or non-closed shapes. These can be reduced by estimating the motion perpendicular to the rotation axis between each subsequent pair of slices in the time series via optical flow analysis [128]. This motion model is then used to perform filtered backprojection along dynamically curved paths, as depicted in (b), accounting for the motion of certain parts of the sample during the time span of the corresponding tomogram. The comparison between the reconstructed slice using this approach and a standard filtered backprojection in (c) shows that artifacts can be significantly reduced, enabling the investigation of the dynamics of the burning process at high temporal and spatial resolution in the μm-range. The 4D nature of the data is illustrated in (d), showing the rendered 3D structure of the wooden part of the match at 5 different points in time.

As this simple example of the burning match shows, time-resolved phase-contrast tomography based on advanced reconstruction schemes allows us to observe dynamic processes in the interior of biomaterials and biological matter. With further improvements, 4D reconstructions of such processes in the interior of multi-cellular assemblies and tissues up to the level of entire organs and small animals can be anticipated.

References

1. Nugent, K.A.: Coherent methods in the X-ray sciences. Adv. Phys. **59**(1), 1–99 (2010)
2. Paganin, D.M.: Coherent X-ray Optics. Oxford University, New York (2006)
3. Bonse, U., Hart, M.: An X-ray interferometer. Appl. Phys. Lett. **6**(8), 155–156 (1965)
4. Chapman, D., Thomlinson, W., Johnston, R.E., Washburn, D., Pisano, E., Gmür, N., Zhong, Z., Menk, R., Arfelli, F., Sayers, D.: Diffraction enhanced X-ray imaging. Phys. Med. Biol. **42**(11), 2015 (1997)
5. Pfeiffer, F., Bech, M., Bunk, O., Kraft, P., Eikenberry, E.F., Brönnimann, C., Grünzweig, C., David, C.: Hard-X-ray dark-field imaging using a grating interferometer. Nat. Mater. **7**, 134–137 (2008)
6. Weitkamp, T., Diaz, A., David, C., Pfeiffer, F., Stampanoni, M., Cloetens, P., Ziegler, E.: X-ray phase imaging with a grating interferometer. Opt. Express **13**(16), 6296–6304 (2005)
7. Munro, P.R.T., Ignatyev, K., Speller, R.D., Olivo, A.: Phase and absorption retrieval using incoherent x-ray sources. PNAS **109**(35), 13922–13927 (2012)
8. Olivo, A., Speller, R.: A coded-aperture technique allowing X-ray phase contrast imaging with conventional sources. Appl. Phys. Lett. **91**(7), 074106 (2007)
9. Zanette, I., Zhou, T., Burvall, A., Lundström, U., Larsson, D.H., Zdora, M., Thibault, P., Pfeiffer, F., Hertz, H.M.: Speckle-based X-ray phase-contrast and dark-field imaging with a laboratory source. Phys. Rev. Lett. **112**(25), 253903 (2014)
10. Cloetens, P., Ludwig, W., Baruchel, J., Van Dyck, D., Van Landuyt, J., Guigay, J.P., Schlenker, M.: Holotomography: quantitative phase tomography with micrometer resolution using hard synchrotron radiation X-rays. Appl. Phys. Lett. **75**(19), 2912–2914 (1999)
11. Gureyev, T.E., Davis, T.J., Pogany, A., Mayo, S.C., Wilkins, S.W.: Optical phase retrieval by use of first Born- and Rytov-Type approximations. Appl. Opt. **43**(12), 2418–2430 (2004)
12. Krenkel, M., Bartels, M., Saldit, T.: Transport of intensity phase reconstruction to solve the twin image problem in holographic X-ray imaging. Opt. Express **21**(2), 2220–2235 (2013)
13. Hagemann, J., Töpperwien, M., Saldit, T.: Phase retrieval for near-field X-ray imaging beyond linearisation or compact support. Appl. Phys. Lett. **113**(4), 041109 (2018)

14. Krenkel, M., Toepperwien, M., Alves, F., Salditt, T.: Three-dimensional single-cell imaging with X-ray waveguides in the holographic regime. Acta Crystallogr. A **73**(4), 282–292 (2017)
15. Giewekemeyer, K., Krüger, S.P., Kalbfleisch, S., Bartels, M., Beta, C., Salditt, T.: X-ray propagation microscopy of biological cells using waveguides as a quasipoint source. Phys. Rev. A **83**(2), 023804 (2011)
16. Pein, A., Loock, S., Plonka, G., Salditt, T.: Using sparsity information for iterative phase retrieval in X-ray propagation imaging. Opt. Express **24**(8), 8332–8343 (2016)
17. Ruhlandt, A., Krenkel, M., Bartels, M., Salditt, T.: Three-dimensional phase retrieval in propagation-based phase-contrast imaging. Phys. Rev. A **89**, 033847 (2014)
18. Chapman, H.N., Nugent, K.A.: Coherent lensless X-ray imaging. Nat. Photon. **4**(12), 833–839 (2010)
19. Fienup, J.R.: Reconstruction of a complex-valued object from the modulus of its Fourier transform using a support constraint. J. Opt. Soc. Am. A **4**(1), 118–123 (1987)
20. Marchesini, S.: Invited article: a unified evaluation of iterative projection algorithms for phase retrieval. Rev. Sci. Instrum. **78**(1), 011301 (2007)
21. Miao, J., Charalambous, P., Kirz, J., Sayre, D.: Extending the methodology of X-ray crystallography to allow imaging of micrometre-sized non-crystalline specimens. Nature **400**(6742), 342–344 (1999)
22. Schroer, C.G., Boye, P., Feldkamp, J.M., Patommel, J., Schropp, A., Schwab, A., Stephan, S., Burghammer, M., Schoder, S., Riekel, C.: Coherent X-ray diffraction imaging with nanofocused illumination. Phys. Rev. Lett. **101**(9), 090801 (2008)
23. Thibault, P., Dierolf, M., Bunk, O., Menzel, A., Pfeiffer, F.: Probe retrieval in ptychographic coherent diffractive imaging. Ultramicroscopy **109**(4), 338–343 (2009)
24. Maretzke, S.: A uniqueness result for propagation-based phase contrast imaging from a single measurement. Inverse Probl. **31**(6), 065003 (2015)
25. Bartels, M., Krenkel, M., Haber, J., Wilke, R.N., Salditt, T.: X-Ray holographic imaging of hydrated biological cells in solution. Phys. Rev. Lett. **114**, 048103 (2015)
26. Hagemann, J., Robisch, A.L., Osterhoff, M., Salditt, T.: Probe reconstruction for holographic X-ray imaging. J. Synchrotron Rad. **24**(2), 498–505 (2017)
27. Krenkel, M., Markus, A., Bartels, M., Dullin, C., Alves, F., Salditt, T.: Phase-contrast zoom tomography reveals precise locations of macrophages in mouse lungs. Sci. Rep. **5**, 09973 (2015)
28. Mokso, R., Cloetens, P., Maire, E., Ludwig, W., Buffiere, J.: Nanoscale zoom tomography with hard X-rays using Kirkpatrick-Baez optics. Appl. Phys. Lett. **90**, 144104 (2007)
29. Hagemann, J., Salditt, T.: Coherence-resolution relationship in holographic and coherent diffractive imaging. Opt. Express **26**(1), 242 (2018)
30. Stangl, J., Mocuta, C., Chamard, V., Carbone, D.: Nanobeam X-ray Scattering: Probing Matter at the Nanoscale. Wiley (2013)
31. Hagemann, J., Robisch, A.L., Luke, D.R., Homann, C., Hohage, T., Cloetens, P., Suhonen, H., Salditt, T.: Reconstruction of wave front and object for inline holography from a set of detection planes. Opt. Express **22**(10), 11552–11569 (2014)
32. Homann, C., Hohage, T., Hagemann, J., Robisch, A.L., Salditt, T.: Validity of the empty-beam correction in near-field imaging. Phys. Rev. A **91**, 013821 (2015)
33. Robisch, A.L., Kröger, K., Rack, A., Salditt, T.: Near-field ptychography using lateral and longitudinal shifts. New J. Phys. **17**(7), 073033 (2015)
34. Robisch, A.L., Salditt, T.: Phase retrieval for object and probe using a series of defocus near-field images. Opt. Express **21**(20), 23345–23357 (2013)
35. Robisch, A.L., Wallentin, J., Pacureanu, A., Cloetens, P., Salditt, T.: Holographic imaging with a hard X-ray nanoprobe: ptychographic versus conventional phase retrieval. Opt. Lett. **41**(23), 5519–5522 (2016)
36. Stockmar, M., Cloetens, P., Zanette, I., Enders, B., Dierolf, M., Pfeiffer, F., Thibault, P.: Near-field ptychography: phase retrieval for inline holography using a structured illumination. Sci. Rep. **3**, 1927 (2013)

37. Kewish, C.M., Guizar-Sicairos, M., Liu, C., Qian, J., Shi, B., Benson, C., Khounsary, A.M., Vila-Comamala, J., Bunk, O., Fienup, J.R., Macrander, A.T., Assoufid, L.: Reconstruction of an astigmatic hard X-ray beam and alignment of K-B mirrors from ptychographic coherent diffraction data. Opt. Express **18**(22), 23420–23427 (2010)
38. Maiden, A.M., Rodenburg, J.M.: An improved ptychographical phase retrieval algorithm for diffractive imaging. Ultramicroscopy **109**(10), 1256–1262 (2009)
39. Marchesini, S., Krishnan, H., Shapiro, D.A., Perciano, T., Sethian, J.A., Daurer, B.J., Maia, F.R.N.C.: SHARP: a distributed, GPU-based ptychographic solver. arXiv **1602.01448** (2016)
40. Schropp, A., Boye, P., Feldkamp, J.M., Hoppe, R., Patommel, J., Samberg, D., Stephan, S., Giewekemeyer, K., Wilke, R.N., Salditt, T., Gulden, J., Mancuso, A.P., Vartanyants, I.A., Weckert, E., Schoder, S., Burghammer, M., Schroer, C.G.: Hard X-ray nanobeam characterization by coherent diffraction microscopy. Appl. Phys. Lett. **96**(9), 091102 (2010)
41. Vine, D.J., Williams, G.J., Abbey, B., Pfeifer, M.A., Clark, J.N., de Jonge, M.D., McNulty, I., Peele, A.G., Nugent, K.A.: Ptychographic Fresnel coherent diffractive imaging. Phys. Rev. A **80**(6), 063823 (2009)
42. Stockmar, M., Zanette, I., Dierolf, M., Enders, B., Clare, R., Pfeiffer, F., Cloetens, P., Bonnin, A., Thibault, P.: X-ray near-field ptychography for optically thick specimens. Phys. Rev. Appl. **3**, 014005 (2015)
43. Morawe, C., Barrett, R., Cloetens, P., Lantelme, B., Peffen, J.C., Vivo, A.: Graded multilayers for figured Kirkpatrick-Baez mirrors on the new ESRF end station ID16A. Proc. SPIE **9588**, 958803 (2015)
44. Salditt, T., Osterhoff, M., Krenkel, M., Wilke, R.N., Priebe, M., Bartels, M., Kalbfleisch, S., Sprung, M.: Compound focusing mirror and X-ray waveguide optics for coherent imaging and nano-diffraction. J. Synchrotron Rad. **22**(4), 867–878 (2015)
45. Allen, L.J., Oxley, M.P.: Phase retrieval from series of images obtained by defocus variation. Opt. Commun. **199**, 65–75 (2001)
46. Loetgering, L., Hammoud, R., Juschkin, L., Wilhein, T.: A phase retrieval algorithm based on three-dimensionally translated diffraction patterns. Europhys. Lett. **111**(6), 64002 (2015)
47. Töpperwien, M.: 3d virtual histology of neuronal tissue by propagation-based X-ray phase-contrast tomography. Ph.D. thesis, Universität Göttingen (2018)
48. Fuhse, C., Ollinger, C., Salditt, T.: Waveguide-based Off-Axis holography with hard X-rays. Phys. Rev. Lett. **97**(25), 254801 (2006)
49. Jarre, A., Fuhse, C., Ollinger, C., Seeger, J., Tucoulou, R., Salditt, T.: Two-dimensional hard X-ray beam compression by combined focusing and waveguide optics. Phys. Rev. Lett. **94**(7), 074801 (2005)
50. Osterhoff, M., Salditt, T.: Coherence filtering of X-ray waveguides: analytical and numerical approach. New J. Phys. **13**(10), 103026 (2011)
51. Pfeiffer, F., David, C., Burghammer, M., Riekel, C., Salditt, T.: Two-dimensional X-ray waveguides and point sources. Science **297**(6), 230 (2002)
52. Salditt, T., Kruger, S.P., Fuhse, C., Bahtz, C.: High-transmission planar X-ray Waveguides. Phys. Rev. Lett. **100**(18), 184801–184804 (2008)
53. Krüger, S.P., Giewekemeyer, K., Kalbfleisch, S., Bartels, M., Neubauer, H., Salditt, T.: Sub-15 nm beam confinement by two crossed X-ray waveguides. Opt. Express **18**(13), 13492–13501 (2010)
54. Krüger, S.P., Neubauer, H., Bartels, M., Kalbfleisch, S., Giewekemeyer, K., Wilbrandt, P.J., Sprung, M., Salditt, T.: Sub-10 nm beam confinement by X-ray waveguides: design, fabrication and characterization of optical properties. J. Synchrotron Rad. **19**(2), 227–236 (2012)
55. Hoffmann-Urlaub, S., Höhne, P., Kanbach, M., Salditt, T.: Advances in fabrication of X-ray waveguides. Microelectron. Eng. **164**, 135–138 (2016)
56. Neubauer, H., Hoffmann, S., Kanbach, M., Haber, J., Kalbfleisch, S., Krüger, S.P., Salditt, T.: High aspect ratio X-ray waveguide channels fabricated by e-beam lithography and wafer bonding. J. Appl. Phys. **115**(21), 214305 (2014)
57. Chen, H.Y., Hoffmann, S., Salditt, T.: X-ray beam compression by tapered waveguides. Appl. Phys. Lett. **106**(19), 194105 (2015)

58. Salditt, T., Hoffmann, S., Vassholz, M., Haber, J., Osterhoff, M., Hilhorst, J.: X-ray optics on a chip: guiding X-rays in curved channels. Phys. Rev. Lett. **115**, 203902 (2015)
59. Bartels, M., Priebe, M., Wilke, R.N., Krüger, S., Giewekemeyer, K., Kalbfleisch, S., Olendrowitz, C., Sprung, M., Salditt, T.: Low-dose three-dimensional hard X-ray imaging of bacterial cells. Opt. Nanoscopy **1**(1), 10 (2012)
60. Töpperwien, M., Krenkel, M., Müller, K., Salditt, T.: Phase-contrast tomography of neuronal tissues: from laboratory-to high resolution synchrotron CT. Proc. SPIE **9967**, 99670T (2016)
61. Töpperwien, M., van der Meer, F., Stadelmann, C., Salditt, T.: Three-dimensional virtual histology of human cerebellum by X-ray phase-contrast tomography. PNAS **115**(27), 6940–6945 (2018)
62. Wilke, R.N., Priebe, M., Bartels, M., Giewekemeyer, K., Diaz, A., Karvinen, P., Salditt, T.: Hard X-ray imaging of bacterial cells: nano-diffraction and ptychographic reconstruction. Opt. Express **20**(17), 19232–19254 (2012)
63. Wilke, R.N., Vassholz, M., Salditt, T.: Semi-transparent central stop in high-resolution X-ray ptychography using Kirkpatrick-Baez focusing. Acta Crystallogr. A **69**(5), 490–497 (2013)
64. Maretzke, S.: Locality estimates for fresnel-wave-propagation and stability of near-field X-ray propagation imaging with finite detectors. arXiv preprint arXiv:1805.06185 (2018)
65. Hagemann, J., Salditt, T.: The fluence-resolution relationship in holographic and coherent diffractive imaging. J. Appl. Crystallogr. **50**(2), 531–538 (2017)
66. Giewekemeyer, K., Thibault, P., Kalbfleisch, S., Beerlink, A., Kewish, C.M., Dierolf, M., Pfeiffer, F., Salditt, T.: Quantitative biological imaging by ptychographic X-ray diffraction microscopy. PNAS **107**(2), 529–534 (2010)
67. Wilke, R.N.: Coherent X-ray diffractive imaging on the single-cell-level of microbial samples: ptychography, tomography, Nano-diffraction and waveguide-imaging. Ph.D. thesis, Universität Göttingen (2014)
68. Jahn, T., Wilke, R.N., Chushkin, Y., Salditt, T.: How many photons are needed to reconstruct random objects in coherent X-ray diffractive imaging? Acta Crystallogr. A **73**(1) (2017)
69. Elser, V., Eisebitt, S.: Uniqueness transition in noisy phase retrieval. New J. Phys. **13**(2), 023001 (2011)
70. Du, M., Gursoy, D., Jacobsen, C.: Near, far, wherever you are: simulations on the dose efficiency of holographic and ptychographic coherent imaging. arXiv preprint arXiv:1908.06770 (2019)
71. Hagemann, J., Salditt, T.: Reconstructing mode mixtures in the optical near-field. Opt. Express **25**(13), 13969–13973 (2017)
72. Howells, M.R., Beetz, T., Chapman, H.N., Cui, C., Holton, J.M., Jacobsen, C.J., Kirz, J., Lima, E., Marchesini, S., Miao, H., Sayre, D., Shapiro, D.A., Spence, J.C.H., Starodub, D.: An assessment of the resolution limitation due to radiation-damage in X-ray diffraction microscopy. J. Electron Spectros. Relat. Phenomena **170**(1–3), 4–12 (2009)
73. Huang, X., Miao, H., Steinbrener, J., Nelson, J., Shapiro, D., Stewart, A., Turner, J., Jacobsen, C.: Signal-to-noise and radiation exposure considerations in conventional and diffraction X-ray microscopy. Opt. Express **17**(16), 13541–13553 (2009)
74. Groso, A., Stampanoni, M., Abela, R., Schneider, P., Linga, S., Müller, R.: Phase contrast tomography: an alternative approach. Appl. Phys. Lett. **88**, 214104 (2006)
75. Paganin, D., Mayo, S.C., Gureyev, T.E., Miller, P.R., Wilkins, S.W.: Simultaneous phase and amplitude extraction from a single defocused image of a homogeneous object. J. Microsc. **206**(Pt 1), 33–40 (2002)
76. Nugent, K.A., Gureyev, T.E., Cookson, D.F., Paganin, D., Barnea, Z.: Quantitative phase imaging using Hard X-rays. Phys. Rev. Lett. **77**(14), 2961–2964 (1996)
77. Hagemann, J.: X-ray near-field holography: beyond idealized assumptions of the probe. Ph.D. thesis, Universität Göttingen (2017)
78. Maretzke, S.: Regularized Newton methods for simultaneous Radon inversion and phase retrieval in phase contrast tomography. arXiv preprint arXiv:1502.05073 (2015)
79. Maretzke, S., Bartels, M., Krenkel, M., Salditt, T., Hohage, T.: Regularized Newton methods for X-ray phase contrast and general imaging problems. Opt. Express **24**(6), 6490–6506 (2016)

80. Fienup, J.R.: Phase retrieval algorithms: a comparison. Appl. Opt. **21**(15), 2758–2769 (1982)
81. Guizar-Sicairos, M., Thurman, S.T., Fienup, J.R.: Efficient subpixel image registration algorithms. Opt. Lett. **33**(2), 156–158 (2008)
82. Turner, L., Dhal, B., Hayes, J., Mancuso, A., Nugent, K., Paterson, D., Scholten, R., Tran, C., Peele, A.: X-ray phase imaging: demonstration of extended conditions for homogeneous objects. Opt. Express **12**(13), 2960–2965 (2004)
83. Wu, X., Liu, H.: Clarification of aspects in in-line phase-sensitive X-ray imaging. Med. Phys. **34**(2), 737–743 (2007)
84. Rigaku: Microfocus rotating anode X-ray generator (2018). https://www.rigaku.com/en/products/protein/micromax007
85. Hemberg, O., Otendal, M., Hertz, H.M.: Liquid-metal-jet anode electron-impact X-ray source. Appl. Phys. Lett. **83**(7), 1483–1485 (2003)
86. Gureyev T.E., Nesterets, Y.I., Stevenson, A.W., Miller, P.R., Pogany, A., Wilkins, S.W.: Some simple rules for contrast, signal-to-noise and resolution in in-line X-ray phase-contrast imaging. Opt. Express **16**(5), 3223–3241 (2008)
87. Reichardt, M., Frohn, J., Töpperwien, M., Nicolas, J.D., Markus, A., Alves, F., Salditt, T.: Nanoscale holographic tomography of heart tissue with X-ray waveguide optics. Proc. SPIE **10391**, 1039105 (2017)
88. Witte, Y.D., Boone, M., Vlassenbroeck, J., Dierick, M., Hoorebeke, L.V.: Bronnikov-aided correction for x-ray computed tomography. J. Opt. Soc. Am. A **26**(4), 890–894 (2009)
89. Töpperwien, M., Krenkel, M., Quade, F., Salditt, T.: Laboratory-based X-ray phase-contrast tomography enables 3D virtual histology. Proc. SPIE **9964**, 99640I (2016)
90. Bartels, M.: Cone-beam X-ray phase contrast tomography of biological samples: optimization of contrast, resolution and field of view. Ph.D. thesis, Universität Göttingen (2013)
91. Krenkel, M.: Cone-beam x-ray phase-contrast tomography for the observation of single cells in whole organs. Ph.D. thesis, Universität Göttingen (2015)
92. Feldkamp, L.A., Davis, L.C., Kress, J.W.: Practical cone-beam algorithm. J. Opt. Soc. Am. A **1**(6), 612–619 (1984)
93. van Aarle, W., Palenstijn, W.J., Cant, J., Janssens, E., Bleichrodt, F., Dabravolski, A., De Beenhouwer, J., Batenburg, K.J., Sijbers, J.: Fast and flexible X-ray tomography using the ASTRA toolbox. Opt. Express **24**(22), 25129–25147 (2016)
94. van Aarle, W., Palenstijn, W.J., De Beenhouwer, J., Altantzis, T., Bals, S., Batenburg, K.J., Sijbers, J.: The ASTRA Toolbox: a platform for advanced algorithm development in electron tomography. Ultramicroscopy **157**, 35–47 (2015)
95. Vassholz, M., Koberstein-Schwarz, B., Ruhlandt, A., Krenkel, M., Salditt, T.: New X-ray tomography method based on the 3D radon transform compatible with anisotropic sources. Phys. Rev. Lett. **116**, 088101 (2016)
96. Lohse, L.M., Vassholz, M., Salditt, T.: Tomography with extended sources: theory, error estimates, and a reconstruction algorithm. Phys. Rev. A **96**(6), 063804 (2017)
97. Bartels, M., Hernandez, V.H., Krenkel, M., Moser, T., Salditt, T.: Phase contrast tomography of the mouse cochlea at microfocus X-ray sources. Appl. Phys. Lett. **103**(8), 083703 (2013)
98. Töpperwien, M., Gradl, R., Keppeler, D., Vassholz, M., Meyer, A., Hessler, R., Achterhold, K., Gleich, B., Dierolf, M., Pfeiffer, F., Moser, T., Salditt, T.: Propagation-based phase-contrast X-ray tomography of cochlea using a compact synchrotron source. Sci. Rep. **8**, 4922 (2018)
99. Krenkel, M., Töpperwien, M., Dullin, C., Alves, F., Salditt, T.: Propagation-based phase-contrast tomography for high-resolution lung imaging with laboratory sources. AIP Adv. **6**(3), 035007 (2016)
100. Bartels, M., Krenkel, M., Cloetens, P., Möbius, W., Salditt, T.: Myelinated mouse nerves studied by X-ray phase contrast zoom tomography. J. Struct. Biol., 561–568 (2015)
101. Lareida, A., Beckmann, F., Schrott-Fischer, A., Glueckert, R., Freysinger, W., Müller, B.: High-resolution X-ray tomography of the human inner ear: synchrotron radiation-based study of nerve fibre bundles, membranes and ganglion cells. J. Microsc. **234**(1), 95–102 (2009)
102. Rau, C., Robinson, I.K., Richter, C.P.: Visualizing soft tissue in the mammalian cochlea with coherent hard X-rays. Microsc. Res. Tech. **69**(8), 660–665 (2006)

103. Hernandez, V.H., Gehrt, A., Reuter, K., Jing, Z., Jeschke, M., Schulz, A.M., Hoch, G., Bartels, M., Vogt, G., Garnham, C.W., et al.: Optogenetic stimulation of the auditory pathway. J. Clin. Investig. **124**(3), 1114–1129 (2014)
104. Rau, C., Hwang, M., Lee, W.K., Richter, C.P.: Quantitative X-ray tomography of the mouse cochlea. PLoS ONE **7**(4), e33568 (2012)
105. Richter, C.P., Shintani-Smith, S., Fishman, A., David, C., Robinson, I., Rau, C.: Imaging of cochlear tissue with a grating interferometer and hard X-rays. Microsc. Res. Tech. **72**(12), 902–907 (2009)
106. Huang, Z., Ruth, R.D.: Laser-electron storage ring. Phys. Rev. Lett. **80**, 976–979 (1998)
107. Achterhold, K., Bech, M., Schleede, S., Potdevin, G., Ruth, R., Loewen, R., Pfeiffer, F.: Monochromatic computed tomography with a compact laser-driven X-ray source. Sci. Rep. **3**, 1313 (2013)
108. Eggl, E., Schleede, S., Bech, M., Achterhold, K., Loewen, R., Ruth, R.D., Pfeiffer, F.: X-ray phase-contrast tomography with a compact laser-driven synchrotron source. PNAS **112**(18), 5567–5572 (2015)
109. Gradl, R., Dierolf, M., Hehn, L., Günther, B., Yildirim, A.Ö., Gleich, B., Achterhold, K., Pfeiffer, F., Morgan, K.S.: Propagation-based Phase-contrast X-ray imaging at a compact light source. Sci. Rep. **7**, 4908 (2017)
110. Schleede, S., Meinel, F.G., Bech, M., Herzen, J., Achterhold, K., Potdevin, G., Malecki, A., Adam-Neumair, S., Thieme, S.F., Bamberg, F., Nikolaou, K., Bohla, A., Yildirim, A.Ö., Loewen, R., Gifford, M., Ruth, R., Eickelberg, O., Reiser, M., Pfeiffer, F.: Emphysema diagnosis using X-ray dark-field imaging at a laser-driven compact synchrotron light source. PNAS **109**(44), 17880–17885 (2012)
111. Yaroshenko, A., Meinel, F.G., Bech, M., Tapfer, A., Velroyen, A., Schleede, S., Auweter, S., Bohla, A., Yildirim, A.Ö., Nikolaou, K., Bamberg, F., Eickelberg, O., Reiser, M.F., Pfeiffer, F.: Pulmonary emphysema diagnosis with a preclinical small-animal X-ray dark-field scatter-contrast scanner. Radiology **269**(2), 427–433 (2013)
112. Larsson, D.H., Lundström, U., Westermark, U.K., Arsenian Henriksson, M., Burvall, A., Hertz, H.M.: First application of liquid-metal-jet sources for small-animal imaging: high-resolution CT and phase-contrast tumor demarcation. Med. Phys. **40**(2), 021909 (2013)
113. Kirschner, D.A., Blaurock, A.E.: Organization, Phylogenetic Variations, and Dynamic Transitions of Myelin. CRC Press (1992)
114. Siegel, G.J.: Basic Neurochemistry: Molecular, Cellular and Medical Aspects. Elsevier Academic Press (2006)
115. Baumann, N., Pham-Dinh, D.: Biology of oligodendrocyte and myelin in the mammalian central nervous system. Physiol. Rev. **81**, 871–927 (2001)
116. Töpperwien, M., Krenkel, M., Ruhwedel, T., Möbius, W., Pacureanu, A., Cloetens, P., Salditt, T.: Phase-contrast tomography of sciatic nerves: image quality and experimental parameters. J. Phys: Conf. Ser. **849**(1), 012001 (2017)
117. Martinez, F.D.: Genes, environments, development and asthma: a reappraisal. Eur. Respir. J. **29**(1), 179–184 (2007)
118. Moreira, A.P., Hogaboam, C.M.: Macrophages in allergic asthma: fine-tuning their pro-and anti-inflammatory actions for disease resolution. J. Interf. Cytok. Res. **31**(6), 485–491 (2011)
119. Balhara, J., Gounni, A.S.: The alveolar macrophages in asthma: a double-edged sword. Mucosal Immunol. **5**(6), 605 (2012)
120. Mizue, Y., Ghani, S., Leng, L., McDonald, C., Kong, P., Baugh, J., Lane, S.J., Craft, J., Nishihira, J., Donnelly, S.C., et al.: Role for macrophage migration inhibitory factor in asthma. PNAS **102**(40), 14410–14415 (2005)
121. Dullin, C., dal Monego, S., Larsson, E., Mohammadi, S., Krenkel, M., Garrovo, C., Biffi, S., Lorenzon, A., Markus, A., Napp, J., Salditt, T., Accardo, A., Alves, F., Tromba, G.: Functionalized synchrotron in-line phase-contrast computed tomography: a novel approach for simultaneous quantification of structural alterations and localization of barium-labelled alveolar macrophages within mouse lung samples. J. Synchrotron Rad. **22**(1), 143–155 (2015)

122. Mbawuike, I.N., Herscowitz, H.B.: MH-S, a murine alveolar macrophage cell line: morphological, cytochemical, and functional characteristics. J. Leukoc. Biol. **46**(2), 119–127 (1989)
123. Siegel, A., Sapru, H.N.: Essential Neuroscience. Point (Lippincott Williams & Wilkins). Wolters Kluwer Health/Lippincott Williams & Wilkins (2011)
124. Azevedo, F.A.C., Carvalho, L.R.B., Grinberg, L.T., Farfel, J.M., Ferretti, R.E.L., Leite, R.E.P., Lent, R., Herculano-Houzel, S., et al.: Equal numbers of neuronal and nonneuronal cells make the human brain an isometrically scaled-up primate brain. J. Comp. Neurol. **513**(5), 532–541 (2009)
125. Moosmann, J., Ershov, A., Altapova, V., Baumbach, T., Prasad, M.S., LaBonne, C., Xiao, X., Kashef, J., Hofmann, R.: X-ray phase-contrast in vivo microtomography probes new aspects of Xenopus gastrulation. Nature **497**(497), 374–377 (2013)
126. Walker, S.M., Schwyn, D.A., Mokso, R., Wicklein, M., Müller, T., Doube, M., Stampanoni, M., Krapp, H.G., Taylor, G.K.: In Vivo time-resolved microtomography reveals the mechanics of the blowfly flight motor. PLoS Biol. **12**(3), e1001823 (2014)
127. Ruhlandt, A., Töpperwien, M., Krenkel, M., Mokso, R., Salditt, T.: Four dimensional material movies: high speed phase-contrast tomography by backprojection along dynamically curved paths. Sci. Rep. **7**(1), 6487 (2017)
128. Liu, C., et al.: Beyond pixels: exploring new representations and applications for motion analysis. Ph.D. thesis, Massachusetts Institute of Technology (2009)

Chapter 14
Constrained Reconstructions in X-ray Phase Contrast Imaging: Uniqueness, Stability and Algorithms

Simon Maretzke and Thorsten Hohage

Abstract This chapter considers the inverse problem of X-ray phase contrast imaging (XPCI), as introduced in Chap. 2. It is analyzed how physical a priori knowledge, e.g. of the approximate size of the imaged sample (support knowledge), affects the inverse problem: uniqueness and—for a linearized model—even well-posedness are shown to hold under support constraints, ensuring stability of reconstruction from real-world noisy data. In order to exploit these theoretical insights, regularized Newton methods are proposed as a class of reconstruction algorithms that flexibly incorporate constraints and account for the inherent nonlinearity of XPCI. A Kaczmarz-type variant of the approach is considered for 3D image-recovery in tomographic XPCI, which remains applicable for large-scale data. The relevance of constraints and the capabilities of the proposed algorithms are demonstrated by numerical reconstruction examples.

2010 Mathematics Subject Classification: 65R10 · 65R32 · 78A45 · 78A46 · 92C55 · 94A08

14.1 Forward Models

We aim to describe (propagation-based) X-ray phase contrast imaging (XPCI) in the language of inverse problems. To this end, we deduce *forward operators* $F : X \to Y$, that model the dependence of the measured near-field diffraction patterns (called *holograms*) $I \in Y$ from the sample-characterizing parameters $f \in X$ (the sought *image*). Different models F are obtained for various settings of practical interest, including X-ray phase contrast tomography (XPCT).

S. Maretzke (✉) · T. Hohage
Institute for Numerical and Applied Mathematics, Universität Göttingen, Lotzestr. 16-18,
37083 Göttingen, Germany
e-mail: s.maretzke@math.uni-goettingen.de

T. Hohage
e-mail: hohage@math.uni-goettingen.de

© The Author(s) 2020
T. Salditt et al. (eds.), *Nanoscale Photonic Imaging*, Topics in Applied Physics 134,
https://doi.org/10.1007/978-3-030-34413-9_14

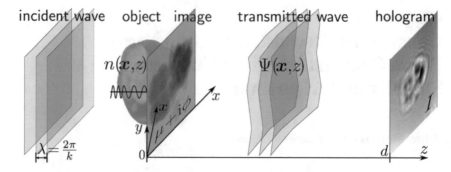

Fig. 14.1 Basic physical model of XPCI: incident plane waves scatter on the imaged object, that is parametrized by a spatially varying refractive index n. The resulting diffraction-pattern (hologram) is recorded in the optical near-field at some distance behind the sample

14.1.1 Physical Model and Preliminaries

The basic physical model of XPCI is detailed in Chap. 2 and summarized by Fig. 14.1: incident monochromatic X-rays, modeled by plane waves, are scattered by the imaged sample, that is parametrized by its spatially varying refractive index $n(x, z) = 1 - \delta(x, z) + \mathrm{i}\beta(x, z)$ (δ, β: refractive- and absorption decrement). By the scattering-interaction, a perturbation (the image) is imprinted upon the transmitted X-ray wave-field. The intensity I of the perturbed wave-field is recorded by a detector placed at a finite distance $d > 0$ behind the sample.

As derived in Sect. 2.1, the dependence of the hologram-intensities I from the sample-parameters δ and β is given by

$$I(x) = |\Psi(x, d)|^2 = |\mathcal{D}\left(\exp\left(-\mu - \mathrm{i}\phi\right)\right)(x)|^2 \quad \text{for all} \quad x \in \mathbb{R}^2,$$

$$\phi(x) = k \int_{\mathbb{R}} \delta(x, z)\,\mathrm{d}z, \qquad \mu(x) := k \int_{\mathbb{R}} \beta(x, z)\,\mathrm{d}z. \tag{14.1}$$

The *phase-* and *absorption-images* ϕ and μ are 2D-*projections* of the 3D-densities δ and β (k: X-ray wavenumber) along the incident z-direction. The *Fresnel-propagator* $\mathcal{D}$, modeling free-space propagation of the X-rays between object and detector, is defined by

$$\mathcal{D}(f) \mapsto \mathcal{F}\left(m_{\mathfrak{f}} \cdot \mathcal{F}^{-1}(f)\right) \quad \text{with} \quad m_{\mathfrak{f}}(\xi) := \exp\left(-\mathrm{i}\xi^2/(2\mathfrak{f})\right). \tag{14.2}$$

Here, $\mathcal{F}(f)(\xi) := (2\pi)^{-m/2} \int_{\mathbb{R}^m} \exp(-\mathrm{i}\xi \cdot x) f(x)\,\mathrm{d}x$ is the Fourier-transform and $\mathfrak{f} = kb^2/d > 0$ is the modified[1] Fresnel-number associated with the physical length b, that is identified with 1 in the chosen dimensionless coordinates.

---

[1] The classical Fresnel-number is given by $\mathfrak{f}_{(b)} := b^2/(\lambda d) = \mathfrak{f}/(2\pi)$. However, using the parameter $\mathfrak{f}$ is notationally more convenient as it avoids excessive occurence of 2π-factors.

XPCI-experiments provide intensity data I of the form (14.1) (up to data errors), whereas the images ϕ, μ are the quantities of interest. Hence, the following principal *inverse problem* has to be solved:

Inverse Problem 1 (XPCI) *For some set A, reconstruct a 2D-image $h = \mu + i\phi \in A$ from measured holograms I of the form (14.1).*

By rotating the object in Fig. 14.1, holograms I_{θ_j} may be acquired for different *incident directions* $\boldsymbol{\theta}_j \in \mathbb{S}^2 = \{\boldsymbol{x} \in \mathbb{R}^3 : |\boldsymbol{x}| = 1\}$ of the X-rays onto the sample (in Fig. 14.1, the incident direction coincides with the z-axis). This is the setting of X-ray phase contrast *tomography* (XPCT). A mathematical model will be provided in Sect. 14.1.3. XPCT allows to probe *3D-variations* of the parameters δ, β beyond mere projections ϕ, μ.

Inverse Problem 2 (XPCT) *For some set A, recover a 3D-image $f = k\beta + ik\delta \in A$ from holograms $\{I_{\theta_j}\}$ measured under different incident directions $\{\boldsymbol{\theta}_j\} \subset \mathbb{S}^2$.*

14.1.1.1 A Priori Constraints

The set of admissible images A in inverse Problems 1 and 2 is highly relevant. In order to facilitate and stabilize image reconstruction, the set A should be restricted as far as possible by available physical a priori knowledge:

- *Support constraints*: real-world samples are of finite size. This implies that the functions $f \in \{\phi, \mu, \delta, \beta\} : \mathbb{R}^m \to \mathbb{R}$ have a *compact support*, i.e. are identically zero outside some bounded *object-domain* $\Omega \subset \mathbb{R}^m$.
- *Non-negativity*: by the physics of hard X-rays, the decrements δ, β—and thus also ϕ, μ—are always non-negative.
- *Pure phase object*: especially for biological samples, β and μ are typically orders of magnitude smaller than δ and ϕ. Assuming a purely shifting-, i.e. *non-absorbing* object β, $\mu = 0$, is then a good approximation.
- *Homogeneous objects*: as is rigorously true for samples composed of a single material, proportionality of δ and β [ϕ and μ] may often be assumed.
- *Regularity*: realistic images ϕ, μ, δ, β are not arbitrarily singular functions, but typically have some characteristic smoothness properties.
- *Tomographic consistency*: Images ϕ and μ that arise as tomographic projections of one object under different incident directions are correlated.

Focussing on support-knowledge, we study the role of such *constraints* on inverse Problems 1 and 2 and outline how to exploit them algorithmically.

14.1.1.2   Additional Notation

We study inverse Problems 1 and 2 in spaces of square-integrable functions:

$$L^2(\mathbb{R}^m) = \{f : \mathbb{R}^m \to \mathbb{C} : \|f\|_{L^2} < \infty\}, \quad \|f\|_{L^2}^2 := \int_{\mathbb{R}^m} |f(x)|^2 \, dx \qquad (14.3)$$

The focus lies on functions $f \in L^2(\mathbb{R}^m)$ that have *compact support* supp(f), i.e. that vanish outside some bounded domain $\Omega \subset \mathbb{R}^m$:

$$\text{supp}(f) \subset \Omega \;\Leftrightarrow\; f|_{\mathbb{R}^m \setminus \Omega} = 0, \qquad (14.4)$$

$f|_B$ denotes the *restriction* of f to $B \subset \mathbb{R}^m$, defined by $f|_B(x) = f(x)$ if $x \in B$ and $f|_B(x) = 0$ otherwise. For $\Omega \subset \mathbb{R}^m$, we write

$$L^2(\Omega) = \{f \in L^2(\mathbb{R}^m) : \text{supp}(f) \subset \Omega\}. \qquad (14.5)$$

Furthermore, we define spaces of *real-valued* L^2-functions:

$$L^2(\Omega, \mathbb{R}) = \{f \in L^2(\Omega) : \text{Im}(f) = 0\}, \qquad (14.6)$$

where $\text{Re}(\cdot)$, $\text{Im}(\cdot)$ denote the real- and imaginary parts, respectively.

14.1.2   Forward Operators for XPCI

Based on Sect. 14.1.1, we introduce forward maps $F : X \to Y$ modeling different settings of XPCI. Note that we define the maps in arbitrary dimensions $m \in \{1, 2, 3, \ldots\}$ although the natural case are images and holograms in $m = 2$ dimensions. The benefit of this will be seen in Sect. 14.3.4.1.

14.1.2.1   General Nonlinear Forward Operator

The most general (and most challenging) XPCI-setting is the reconstruction of both phase ϕ and absorption μ from a single hologram. According to (14.1), this setting is modeled by the forward map

$$\mathcal{N}(h) = I - 1 = |\mathcal{D}(\exp(-h))|^2 - 1, \qquad (14.7)$$

for *complex-valued* images $h = \mu + i\phi$. Note that the constant background intensity 1 has been subtracted, such that $\mathcal{N}(0) = 0$. As a benefit, $\mathcal{N}$ can be analyzed as an operator on L^2-spaces: for any bounded $\Omega \subset \mathbb{R}^m$,

$$\mathcal{N} : L^2(\Omega) \to L^2(\mathbb{R}^m) \tag{14.8}$$

is essentially[2] a well-defined, *nonlinear* operator. Moreover, it can be shown [1, 2] that $\mathcal{N}$ is continuously Fréchet-differentiable, i.e. sufficiently smooth to admit local linear approximations. The derivative is given by

$$\mathcal{N}'[f]h = -2\mathrm{Re}\big(\overline{\mathcal{D}\left(\exp\left(-f\right)\right)} \cdot \mathcal{D}\left(\exp\left(-f\right) \cdot h\right)\big). \tag{14.9}$$

14.1.2.2 Linearized Forward Map and Contrast-Transfer-Functions

The nonlinearity of the forward map $\mathcal{N}$ causes difficulties in both analysis and practical image reconstruction. It is therefore standard [3–7] to resort to a *linearization* valid for weakly scattering samples (see e.g. [7] for details on the regime-of-validity): the idea is that the image f is sufficiently "small" so that higher-order terms are negligible:

$$\mathcal{N}(h) = \mathcal{T}(h) + \mathcal{O}(h^2) \approx \mathcal{T}(h) \quad \text{with} \quad \mathcal{T}(h) := -2\mathrm{Re}\left(\mathcal{D}(h)\right). \tag{14.10}$$

The linearized forward map $\mathcal{T} = \mathcal{N}'[0]$ is also known as the *contrast-transfer-function- (CTF-)model*, which refers to the following alternate form (compare with Sect. 2.2):

$$\mathcal{T}(-\mu - i\phi) = -2\mathcal{F}^{-1}\bigg(\underbrace{\sin\left(\frac{|\xi|^2}{2\mathfrak{f}}\right)}_{=:s_0(\xi)} \mathcal{F}(\phi) + \underbrace{\cos\left(\frac{|\xi|^2}{2\mathfrak{f}}\right)}_{=:c_0(\xi)} \mathcal{F}(\mu)\bigg) \tag{14.11}$$

According to (14.11), the linearized contrast in Fourier-space is given by a superposition of the Fourier-transforms of phase- and absorption-image ϕ, μ modulated by the oscillatory CTFs s_0 and c_0, respectively.

As $|s_0(\xi)|, |c_0(\xi)| \leq 1$ for all $\xi \in \mathbb{R}^m$, $\mathcal{T} : L^2(\mathbb{R}^m) \to L^2(\mathbb{R}^m)$ is a bounded $(\mathbb{R})$-linear operator with $\|\mathcal{T}(h)\|_{L^2} \leq 2\|h\|_{L^2}$ for all $h \in L^2(\mathbb{R}^m)$.

14.1.2.3 Homogeneous Objects and Pure Phase Objects

The cases of *homogeneous objects* and *pure phase objects*, see Sect. 14.1.1.1, may be treated in a unified manner, by expressing the complex-valued image $h = \mu + i\phi = ie^{-i\nu}\varphi$ in terms of a single *real-valued* function φ and a parameter $\nu = \arctan(\beta/\delta) \in [0; \pi/2)$ ($\nu = 0$: pure phase object).

Such a *homogeneity-constraint* may be incorporated into the general forward model, via a modified forward map

---

[2]To ensure well-definedness on the whole space $L^2(\Omega)$, the exponential has to be suitably truncated for the physically irrelevant case of negative absorption $\mathrm{Re}(f) = \mu < 0$.

$$\mathscr{N}_\nu : L^2(\Omega, \mathbb{R}) \to L^2(\mathbb{R}^m);$$

$$\varphi \mapsto \mathscr{N}(\mathrm{ie}^{-\mathrm{i}\nu}\varphi) = \left| \mathcal{D}\left(\exp(-\mathrm{ie}^{-\mathrm{i}\nu}\varphi)\right) \right|^2 - 1. \tag{14.12}$$

The linearized model under a homogeneity constraint may be expressed via a single CTF $s_\nu(\boldsymbol{\xi}) := \sin\left(|\boldsymbol{\xi}|^2/(2\mathfrak{f}) + \nu\right)$: for $\varphi \in L^2(\mathbb{R}^m, \mathbb{R})$, it holds that

$$\mathscr{S}_\nu(\varphi) := -2\mathcal{F}^{-1}\left(s_\nu \cdot \mathcal{F}(\varphi)\right) = \mathscr{T}(\mathrm{ie}^{-\mathrm{i}\nu}\varphi). \tag{14.13}$$

Although (14.13) only holds for real-valued φ, we define $\mathscr{S}_\nu : L^2(\mathbb{R}^m) \to L^2(\mathbb{R}^m)$ ($\|\mathscr{S}_\nu\| = 2$) on general L^2-spaces. For its properties, it is widely irrelevant if real- or complex-valued functions are considered, as $\mathscr{S}_\nu$ commutes with the pointwise real-part: $\mathrm{Re}\left(\mathscr{S}_\nu(h)\right) = \mathscr{S}_\nu(\mathrm{Re}\,(h))$ for all $h \in L^2(\mathbb{R}^m)$.

14.1.2.4 Multiple Holograms

In order to obtain richer data in XPCI, it is standard to acquire multiple holograms $I_1, I_2, \ldots, I_\ell$ at several object-to-detector-distances, corresponding to different Fresnel-numbers $\mathfrak{f}_1, \mathfrak{f}_2, \ldots, \mathfrak{f}_\ell$. This may be modeled by combining the forward maps for the individual holograms $F_j : X \to L^2(\mathbb{R}^m)$; $h \mapsto I_j - 1$, $F_j \in \{\mathscr{N}^{(\mathfrak{f}_j)}, \mathscr{N}_\nu^{(\mathfrak{f}_j)}, \mathscr{T}^{(\mathfrak{f}_j)}, \mathscr{S}_\nu^{(\mathfrak{f}_j)}\}$ to a "vector-valued" operator:

$$F^{(\mathfrak{f}_1, \ldots, \mathfrak{f}_\ell)} : X \to L^2(\mathbb{R}^m)^\ell; \quad h \mapsto (F_1(h), \ldots, F_\ell(h)) \tag{14.14}$$

14.1.3 Forward Operators for XPCT

In X-ray phase contrast tomography (XPCT), holograms are measured under different incident directions $\boldsymbol{\theta} \in \mathbb{S}^2$. According to the basic model (14.1), the resulting intensities I_θ are then given by

$$I_\theta = |\mathcal{D}\left(\exp\left(-\mathscr{P}_\theta(k\beta + \mathrm{i}k\delta)\right)\right)(\boldsymbol{x})|^2, \tag{14.15}$$

where $\mathscr{P}_\theta$ is the parallel-beam projector along $\boldsymbol{\theta}$ ($\boldsymbol{\theta} \perp \boldsymbol{n}_x \perp \boldsymbol{n}_y \perp \boldsymbol{\theta}$):

$$\mathscr{P}_\theta(f)(x, y) := \int_\mathbb{R} f(x\boldsymbol{n}_x + y\boldsymbol{n}_y + z\boldsymbol{\theta})\,\mathrm{d}z, \quad x, y \in \mathbb{R}, \tag{14.16}$$

According to the standard theory of computed tomography, projection-data $\{\mathscr{P}_\theta(f)\}_{\theta \in \Theta}$ for a suitable set of incident-directions Θ allows to reconstruct the underlying 3D-function $f : \mathbb{R}^3 \to \mathbb{C}$. Analogously, the goal of XPCT is to reconstruct 3D-variations of the decrements δ and β of the sample's refractive index from a tomographic series of holograms $\{I_\theta\}_{\theta \in \Theta}$.

Composition of the projectors $\mathscr{P}_{\theta}$ with any of the forward maps $F \in \{\mathscr{N}, \mathscr{N}_{\nu}, \mathscr{T}, \mathscr{S}_{\nu}\}$: $X \to L^2(\mathbb{R}^m)$ from Sect. 14.1.2 induces a corresponding XPCT-model: for $\Theta = \{\theta_1, \ldots, \theta_t\}$, the tomographic hologram-data is modeled by

$$F_{\text{PCT}} : f \mapsto \left(F\left(\mathscr{P}_{\theta}(f)\right)\right)_{\theta \in \Theta} = \left(I_{\theta} - 1\right)_{\theta \in \Theta}. \tag{14.17}$$

14.2 Uniqueness Theory

In practice, it is highly relevant whether the measured intensity data I uniquely determines the sought image $h = \mu + i\phi$ (or $f = k\beta + ik\delta$ in XPCT). Otherwise, it might happen that two structurally different samples are indistinguishable by the imaging method, which is not desirable. (Non-)uniqueness of an inverse problem is equivalent to (non-)injectivity of the governing forward operator $F : X \to Y$. Hence, it depends on different aspects:

1. The richness of the data, i.e. the size of the *data-space* Y: for example, it is commonly argued that measuring several holograms $I_1, I_2, \ldots$ at different Fresnel-numbers (see Sect. 14.1.2.4) helps to ensure uniqueness in XPCI.
2. Available a priori knowledge, i.e. the size of the *object-space* X: the smaller X the more likely it is that any two images $h_1, h_2 \in X$ with $h_1 \neq h_2$ induce distinguishable data $F(h_1) \neq F(h_2)$.

In addition, it may happen that the nonlinear forward model is unique but its linearization is non-unique or vice verser. Accordingly, the different forward models from Sect. 14.1.2 have to be investigated individually.

14.2.1 Preliminary Results and Counter-Examples

We first review some known results on (non-)uniqueness of XPCI. Firstly, image reconstruction from a single hologram is generally *non-unique*:

- Linearized model: $\mathscr{T} : L^2(\mathbb{R}^m) \to L^2(\mathbb{R}^m)$; $h \mapsto -2\mathrm{Re}(h)$ has a huge null-space composed of all h for which $\mathscr{D}(h)$ is purely imaginary-valued:

$$\mathrm{kern}(\mathscr{T}) := \{h \in L^2(\mathbb{R}^m) : \mathscr{T}(h) = 0\} = \mathscr{D}^{-1}\left(iL^2(\mathbb{R}^m, \mathbb{R})\right) \tag{14.18}$$

- Nonlinear model (example from [8]): Images $h_{\pm} : \mathbb{R}^2 \setminus \{0\} \to \mathbb{C}$; $x \mapsto a(|x|) \pm i\nu \arctan2(x)$ for $\nu \in \mathbb{N}$ and smooth functions $a : \mathbb{R}_{\geq 0} \to \mathbb{R}$ give rise to so-called *phase-vortices* in the wave-field. The sign of the vortex is not determined by Fresnel-intensities ($A := \exp(-a)$):

$$|\mathcal{D}(\exp(-h_+))|^2 = |\mathcal{D}(A \cdot \exp(-i\nu\arctan2(\cdot)))|^2$$
$$= |\mathcal{D}(A \cdot \exp(i\nu\arctan2(\cdot)))|^2 = |\mathcal{D}(\exp(-h_-))|^2 \quad (14.19)$$

Based on these negative results, it is typically argued that at least two holograms and/or a homogeneity-constraint are required for uniqueness. Indeed, the situation improves substantially in the latter settings:

- Uniqueness under homogeneity-constraints (linear): the operator $\mathscr{S}_\nu : L^2(\mathbb{R}^m) \to L^2(\mathbb{R}^m)$ from Sect. 14.1.2.3 is *injective*, as the zero-manifolds of the Fourier-multiplier s_ν are sets of the Lebesgue-measure 0 in $\mathbb{R}^m$.
- Uniqueness for two holograms (linear): in [9], it is shown by a similar argument based on the CTF-representation (14.11) that also the operator $\mathscr{T}^{(f_1,f_2)} : L^2(\mathbb{R}^m) \to L^2(\mathbb{R}^m)^2$ (see Sect. 14.1.2.4) is injective for $f_1 \neq f_2$.

Moreover, it is argued in [9] that both results carry over to the nonlinear model, provided that the image h is *compactly supported*. Indeed, a much stronger uniqueness result holds true under such an assumption, as will be seen in the following.

14.2.2 Sources of Non-uniqueness—The Phase Problem

According to the basic physical model (14.1), image-formation mathematically amounts to three operations: pointwise exponential, $h \mapsto \exp(-h)$, Fresnel-propagation, $\exp(-h) \mapsto \mathcal{D}(\exp(-h))$, and computation of the pointwise squared modulus, $\mathcal{D}(\exp(-h)) \mapsto |\mathcal{D}(\exp(-h))|^2$. Among those, $\mathcal{D}$ is an *invertible* operation, i.e. does not destroy information. This is not true for the other two operations, which give rise to different sources of non-uniqueness:

- *Phase-wrapping*: The exponential is 2π-periodic in the imaginary-part of its argument. Hence, the *phase-image* $\phi = \text{Im}(h)$ may only be determined by the data up to increments by multiples of 2π.
- *Phase problem*: The squared modulus, arising from the restriction of X-ray detectors to measuring *intensities*, eliminates the phase-information.

The first aspect is simpler to analyze and often turns out to be of lesser practical impact in XPCI: for moderately strongly scattering samples, ϕ is a priori known to assume values within $[0; 2\pi)$, so that non-uniqueness due to *phase-wrapping* is not an issue. In the following, we therefore focus on possible ambiguities due to the phase problem.

14.2.3 Relation to Classical Phase Retrieval Problems

Up to possibly remaining phase-wrapping ambiguities, the image reconstruction problem in XPCI may be rephrased as follows:

> Given data $I = |\mathcal{D}(O)|^2$, reconstruct the object-transmission-function (OTF) $O := \exp(-h) \in \tilde{A}$ from some admissible set $\tilde{A}$.

Such settings are known as *phase retrieval* problems as recovering O is equivalent to retrieving the missing phase of $\mathcal{D}(O)$ (and then inverting $\mathcal{D}$). Uniqueness of phase retrieval has been extensively studied ever since the pioneering works of Walther [10] and Akutowicz [11, 12], primarily for the case where $\mathcal{D}$ is replaced by the Fourier-transform $\mathcal{F}$, i.e. for the reconstruction from phaseless Fourier-data. We refer to [13–17] for reviews.

Indeed, Fresnel-data may be readily reduced to the classical Fourier-setting, by rewriting the Fresnel-propagator in the form

$$\mathcal{D}(f)(x) = u_0 \mathfrak{f}^{\frac{m}{2}} n_{\mathfrak{f}}(x) \cdot \mathcal{F}\left(n_{\mathfrak{f}} \cdot f\right)(\mathfrak{f}x) \quad \text{for all} \quad x \in \mathbb{R}^m \quad (14.20)$$

with $n_{\mathfrak{f}}(x) = \exp(\mathrm{i}\mathfrak{f}x^2/2)$ and $u_0 = \exp(-\mathrm{i}m\pi/4)$. Hence, if we define $\tilde{O} := n_{\mathfrak{f}} \cdot O$, then the holograms in XPCI provide Fourier-data for $\tilde{O}$:

$$I(\xi/\mathfrak{f}) = |\mathcal{D}(O)(\xi/\mathfrak{f})|^2 = \mathfrak{f}^m \mathcal{F}(\tilde{O})(\xi) \quad \text{for all} \quad \xi \in \mathbb{R}^m. \quad (14.21)$$

Based on the identification in (14.21), uniqueness results for Fourier-phase retrieval may be adapted to the Fresnel-regime. Notably, however, most of such uniqueness theorems assume a compact support of the objective. Importantly, this is *not* justified in the setting of XPCI:

> The OTF O is *not* a compactly supported function in any realistic setting. Only the *contrast* $o := O - 1$ typically has compact support.

14.2.4 Holographic Nature of Phase Retrieval in XPCI

In order to emphasize the structural difference to classical phase retrieval problems, it is illustrative to rewrite the XPCI-model in the form

$$I = |\mathcal{D}(O)|^2 = |\mathcal{D}(1) + \mathcal{D}(o)|^2 = 1 + \underbrace{2\mathrm{Re}(\mathcal{D}(o))}_{=\mathcal{D}(o)+\overline{\mathcal{D}(o)}} + |\mathcal{D}(o)|^2, \quad (14.22)$$

where it has been used that $\mathcal{D}$ maps constant functions onto themselves.[3] According to the physical model from Sect. 14.1.1, the summands on the r.h.s. of (14.22) can be interpreted in terms of the *scattered-* and *transmitted* parts of the X-ray wave-field: the constant 1 is the intensity of the incident plane wave and the last summand that of the waves scattered by the object, whereas the second term describes the *interference* of these two wave-field components on the detector.

Formula (14.22) places the inverse problem of XPCI in the realm of *holographic* phase retrieval problems, i.e. reconstruction in the presence of a *reference signal*— here provided by the unscattered part of the incident X-rays. Several theoretical and practical works have shown that such a holographic reference facilitates phase retrieval, see e.g. [18–21].

14.2.5 General Uniqueness Under Support Constraints

According to (14.22), image reconstruction in XPCI is equivalent to retrieving $o = \exp(-h) - 1$ from data of the form (14.22) (up to possible phase-wrapping). By invertibility of the Fresnel-propagator $\mathcal{D}$, uniqueness thus holds if it is possible to disentangle the summands $\mathcal{D}(o)$, $\overline{\mathcal{D}(o)}$, and $|\mathcal{D}(o)|^2$. As shown in [22] using the theory of entire functions, the latter is indeed possible whenever o is known to have compact support, which is true for any sample of finite size. The principal result reads as follows:

Theorem 14.1 (Uniqueness of XPCI [22]) *Let o $(= \exp(-h) - 1)$ be a compactly supported function (or distribution).*

Then o is uniquely determined by XPCI-data $I = |\mathcal{D}(1 + o)|^2$. Furthermore, uniqueness is retained if only restricted data $I|_K$ is available, measured for any detection-domain $K \subset \mathbb{R}^m$ that contains an open set.

For any such K and $\Omega \subset \mathbb{R}^m$ bounded, $\mathcal{N}_K : h \mapsto \mathcal{N}(h)|_K$ is injective up to phase-wrapping: if $\mathcal{N}(h_1)|_K = \mathcal{N}(h_2)|_K$ for $h_1, h_2 \in L^2(\Omega)$, then

$$h_1(x) - h_2(x) \in 2\pi i \mathbb{Z} \quad \text{for almost all} \quad x \in \mathbb{R}^m. \tag{14.23}$$

Importantly, Theorem 14.1 establishes uniqueness in the most challenging setting of XPCI: single hologram, no homogeneity-constraint. The result trivially extends to every less difficult case with more data or additional constraints. However, note that the extension of uniqueness to restricted measurements $I|_K$ is based on analytic continuation of the data—a very *unstable* procedure in practice.

---

[3]Note that this behavior of $\mathcal{D}$ is fundamentally different from that of the Fourier-transform $\mathcal{F}$, which maps constants to Dirac-deltas centered at the origin.

14.2.5.1 Uniqueness for the Linearized Model

Uniqueness for the *linearized* XPCI-model has to be shown individually. According to Sect. 14.1.2.2, it corresponds to data of the form $I_{\text{lin}} = 1 - \mathcal{D}(h) - \overline{\mathcal{D}(h)}$. Compared to (14.22), merely the quadratic term $|\mathcal{D}(o)|^2$ is omitted and $o = \exp(-h) - 1$ is replaced by $-h$ (note that this rules out phase-wrapping!). Hence, the principal uniqueness argument from [22] remains valid: the summands $\mathcal{D}(h)$ and $\overline{\mathcal{D}(h)}$ may be disentangled owing to their different "finger-prints" as entire functions:

Corollary 14.1 (Uniqueness of linearized XPCI [22]) *For any bounded domain $\Omega \subset \mathbb{R}^m$ and any $K \subset \mathbb{R}^m$ that contains an open set, the linearized forward operator $\mathscr{T}_K : L^2(\Omega) \to L^2(\mathbb{R}^m); \ h \mapsto \mathscr{T}(h)|_K$ is injective.*

14.2.5.2 Uniqueness for XPCT

By combining with standard results on uniqueness of tomographic reconstruction described by the theory of the Radon transform, the uniqueness theorems may be easily extended to XPCT. We refer to [22] for details.

14.3 Stability Theory

The uniqueness results of the preceding Sect. 14.2, suprisingly strong though they are, do not guarantee that accurate images may actually be reconstructed from holograms acquired in real-world XPCI-setups. Experimental data always contains errors due to noise and/or inaccuracies of the physical model. As detailed in Chap. 5 such data errors may lead to arbitrarily strongly corrupted images due to the phenomenon of *ill-posedness*: even if a forward model $F : X \to Y$ is injective, its inverse $F^{-1} : F(X) \to X$ may be *discontinuous* such that small perturbations in the data $g^{\text{obs}} = F(f) + \epsilon$ may be arbitrarily amplified in the reconstruction $F^{-1}(F(f) + \epsilon)$.

The aim of this section is thus to supplement the uniqueness results with an analysis of *stability*, exploring how susceptible image reconstruction is to data errors. Thereby, it sheds a light on the question which reconstructions are feasible in practice. Due to difficulties arising from nonlinearity, the stability analysis is restricted to the linearized forward models.

14.3.1 Lipschitz-Stability and its Meaning

Although other (weaker) concepts of stability are common in the field of inverse problems, the notion of *Lipschitz-stability* turns out to be most suitable for XPCI: a forward map $F : X \to Y$ between normed spaces X, Y is said to be *Lipschitz-stable* if a stability estimate of the form

$$\|F(f_1) - F(f_2)\|_Y \geq C_{\text{stab}} \|f_1 - f_2\|_X \quad \text{for all} \quad f_1, f_2 \in X \quad (14.24)$$

holds for some constant $C_{\text{stab}} > 0$. In this case, F has a *Lipschitz-continuous* inverse F^{-1}: $\|F^{-1}(g_1) - F^{-1}(g_2)\|_X \leq C_{\text{stab}}^{-1} \|g_1 - g_2\|_Y$ for all $g_1, g_2 \in F(X)$. Notably, this implies robustness to data errors: given measurements $g_\epsilon = F(f^\dagger) + \epsilon \in F(X)$, the resulting reconstruction-error is bounded by

$$\left\| f^\dagger - F^{-1}(g_\epsilon) \right\|_X = \left\| F^{-1}(F(f^\dagger)) - F^{-1}(F(g_\epsilon)) \right\|_X \leq C_{\text{stab}}^{-1} \|\epsilon\|_Y. \quad (14.25)$$

The bound (14.25) states that data errors manifest at most amplified by a finite factor C_{stab}^{-1} in the recovered object. Therefore C_{stab} should be as large as possible: if $C_{\text{stab}} \ll 1$, the error-amplification predicted by (14.25) may be too large to guarantee accurate reconstructions at realistic noise-levels $\|\epsilon\|_Y$.

Notably, for *linear* forward models $F : X \rightarrow Y$, (14.24) is equivalent to

$$C_{\text{stab}} = \inf_{f \in X, \|f\|_X = 1} \|F(h)\|_Y > 0. \quad (14.26)$$

Moreover, a linear inverse problem is *well-posed* if and only if (14.26) holds.

14.3.2   Stability for General Objects and one Hologram

Firstly, we consider the most challenging setting of reconstructing arbitrary phase- and absorption-images ϕ, μ from a single (linearized) hologram $I \approx 1 + \mathscr{T}(h)$. Stable inversion of the forward map $\mathscr{T}$ is commonly argued to be infeasible. Indeed, as seen in Sect. 14.2.1, the forward model is not even unique for general images ϕ, $\mu \in L^2(\mathbb{R}^m)$, but only if ϕ, μ are compactly supported. Accordingly, we assume a *support contraint* in the following:

$$h = \mu + i\phi \in L^2(\Omega) \quad \text{for some} \quad \Omega \subset \mathbb{R}^m \quad \text{bounded.} \quad (14.27)$$

14.3.2.1   Analytical Approach

Our approach to analyzing stability is ultimately based on the principle of holographic reconstruction [23], that earned DENNIS GABOR the Nobel Prize in physics in 1971. The idea is to rewrite the forward map in the form

$$-\mathscr{T}(h) = 2\text{Re}\,(\mathcal{D}(h)) = \mathcal{D}(h) + \overline{\mathcal{D}(h)} = \mathcal{D}(h) + \mathcal{D}^{-1}(\overline{h}), \quad (14.28)$$

which reveals linearized XPCI data to be a superposition of a propagated image $\mathcal{D}(h)$ and the *back*-propagated *twin-image* $\overline{h}$. Applying the Fresnel-propagator $\mathcal{D}$ to

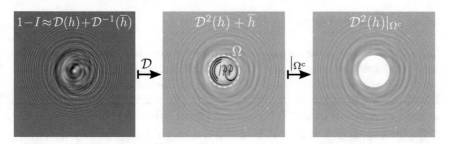

Fig. 14.2 Idea of the stability analysis of $\mathscr{T}$ [24]: by applying the propagator $\mathcal{D}$ to linear XPCI-data $I - 1 \approx \mathscr{T}(h)$, the *twin-image* $\overline{h}$ becomes sharp, see logo in central panel. By restricting to the complement of $\Omega \supset \mathrm{supp}(\overline{h})$, $\overline{h}$ is eliminated and incomplete Fresnel data $\mathcal{D}^2(h)|_{\Omega^c}$ is obtained (right panel). Images show real parts of images computed from a hologram (left panel) acquired at GINIX [25, 26], P10-beamline, DESY

a hologram thus recovers the twin-image $\overline{h}$, perturbed by a fringe-pattern originating from the doubly propagated image $\mathcal{D}^2(h)$:

$$-\mathcal{D}\big(\mathscr{T}(h)\big) = \mathcal{D}^2(h) + \overline{h}, \tag{14.29}$$

This is Gabor's original idea of *holographic reconstruction*, which is illustrated in the first and second panel of Fig. 14.2 for a real-data example.

For stability analysis, we use the idea in a converse manner. By the constraint $h \in L^2(\Omega)$, the sharp twin-image (the valuable part ins Gabor's eyes!) can be eliminated from (14.29) by restricting to the complement Ω^c of Ω:

$$-\mathcal{D}\big(\mathscr{T}(h)\big)|_{\Omega^c} = \mathcal{D}^2(h)|_{\Omega^c} + \overline{h}|_{\Omega^c} \stackrel{\mathrm{supp}(\overline{h}) \subset \Omega}{=} \mathcal{D}^2(h)|_{\Omega^c}. \tag{14.30}$$

By (14.30), we are left with *incomplete* (but phased!) Fresnel-data $\mathcal{D}^2(h)|_{\Omega^c}$. Notably, to this point, only *stable* operations have been applied to the XPCI-data, which do not amplify data errors in L^2-norm: for any $g \in L^2(\mathbb{R}^m)$, it holds that $\|\mathcal{D}(g)|_{\Omega^c}\|_{L^2} \leq \|g\|_{L^2}$. When applied to (14.30) this bound yields

$$\|\mathscr{T}(h)\|_{L^2} \geq \big\|\mathcal{D}\big(\mathscr{T}(h)\big)|_{\Omega^c}\big\|_{L^2} = \big\|\mathcal{D}^2(h)|_{\Omega^c}\big\|_{L^2}. \tag{14.31}$$

Finally, by employing the alternate form (14.20) of $\mathcal{D}$, the Fresnel-data on the r.h.s. of (14.31) may be identified with incomplete *Fourier*-data:

$$\|\mathscr{T}(h)\|_{L^2} \geq \big\|\mathcal{D}^2(h)|_{\Omega^c}\big\|_{L^2} = \big\|\mathcal{F}(\tilde{h})|_{\Omega_{\mathfrak{f}}^c}\big\|_{L^2} \tag{14.32}$$

with $\tilde{h} = n_{\mathfrak{f}/2} \cdot h$ and $\Omega_{\mathfrak{f}} := \{x \in \mathbb{R}^m : (2/\mathfrak{f}) \cdot x \in \Omega\}$.

14.3.2.2 Stability Bound

Since $\|\tilde{h}\|_{L^2} = \|h\|_{L^2}$, the bound (14.32) can be regarded as a *relative* stability esti-mate: recovering an image $h \in L^2(\Omega)$ from XPCI-data $\mathcal{T}(h)$ is *at least as sta-ble* as the reconstruction of $\tilde{h} \in L^2(\Omega)$ from Fourier-data outside the domain $\Omega_{\mathfrak{f}}$. Reconstruction from incomplete Fourier-data in turn is a well-studied problem: an *uncertainty principle* from [27] implies that Lipschitz-stability holds, $\||\mathcal{F}(\tilde{h})|_{\Omega_{\mathfrak{f}}^c}\| \geq C_{stab}^{gen}\|\tilde{h}\|$ for some $C_{stab}^{gen} > 0$, provided that Ω is bounded along at least one dimension. For rectangular domains Ω, the stability-constant C_{stab}^{gen} may be expressed in terms of the principal eigenvalue of a compact selfadjoint operator, for which asymptotics are derived in [28]. Via (14.32), these results yield stability estimates for linearized XPCI:

Theorem 14.2 (Stability estimate for general images [24]) *Let* $\Omega = [-\frac{1}{2}; \frac{1}{2}]^m$. *Then*

$$C_{stab}^{gen}(\Omega, \mathfrak{f}) := \inf_{h \in L^2(\Omega), \|h\|=1} \|\mathcal{T}(h)\| > 0 \qquad (14.33)$$

i.e. the reconstruction of images h with support in Ω from linearized XPCI-data is well-posed. For $\mathfrak{f} \to \infty$, the stability constant satisfies the bound

$$C_{stab}^{gen}(\Omega, \mathfrak{f}) \geq m^{\frac{1}{2}} (2\pi\mathfrak{f})^{\frac{1}{4}} \left(1 - \frac{3}{8\mathfrak{f}} + \mathcal{O}\left(\mathfrak{f}^{-2}\right)\right) \exp(-\mathfrak{f}/8). \qquad (14.34)$$

While Theorem 14.2 only gives a *worst-case* bound on the data-contrast $\|\mathcal{T}(h)\|/\|h\|$ over all images h, the result may be sharpened considerably, as detailed in [24]: for any $h \in L^2(\Omega)$, an individual lower bound for $\|\mathcal{T}(h)\|$ may be given based on the eigenvalues from [28] and the images that minimize $\|\mathcal{T}(h)\|/\|h\|$ may be characterized in terms of the associated eigenmodes.

14.3.2.3 Stability in a Practical Sense?

Numerical computations in [24] indicate that the bound (14.34) is quite sharp. While this is good news for a (pure) mathematician, it is bad news from an applied perspec-tive: the predicted (quasi) exponential decay $C_{stab}^{gen}(\Omega, \mathfrak{f}) \sim \exp(\mathfrak{f}/8)$ implies that the constant quickly becomes very small for larger values of $\mathfrak{f}$, e.g. $C_{stab}^{gen}(\Omega, \mathfrak{f}) \lesssim 10^{-5}$, for $\mathfrak{f} \gtrsim 100$. Notably, $\mathfrak{f} = kb^2/d$ is the modified Fresnel-number associated with the width of the support-domain Ω, i.e. with the diameter of the imaged sample.[4] In typi-cal XPCI-experiments at synchrotrons, one has $10^2 \lesssim \mathfrak{f} \lesssim 10^5$, so that Theorem 14.2 only guarantees stability *in practice* for imaging settings at the lower end of typical Fresnel-numbers.

---

[4]The lateral lengthscale b associated with $\mathfrak{f}$ is implicitly fixed to the width of Ω by assuming the latter to be 1 in Theorem 14.2, as will also be done in all subsequent results.

Notably, this is in line with empirics: after all, independent reconstruction of phase- and absorption-image ϕ and μ from a single hologram, as analyzed here, is widely considered as infeasible by practioners. It is thus highly surprising in the first place that the problem is technically *well-posed* at all.

14.3.2.4 Extension to Other Domains

Theorem 14.2 seemingly only applies to a very particular choice of the domain $\Omega \subset \mathbb{R}^m$. Yet, it may be readily generalized via the following properties:

- *Translation- and rotation-invariance*: As the map $\mathscr{T}$ is invariant under shifts and/or rotations of the coordinates, it holds that $C_{\text{stab}}^{\text{gen}}(\tilde{\Omega}, \mathfrak{f}) = C_{\text{stab}}^{\text{gen}}(\Omega, \mathfrak{f})$ whenever $\tilde{\Omega}$ is a shifted and/or rotated version of $\Omega \subset \mathbb{R}^m$.
- *Monotonicity*: $C_{\text{stab}}^{\text{gen}}(\Omega_1, \mathfrak{f}) \geq C_{\text{stab}}^{\text{gen}}(\Omega_2, \mathfrak{f})$ for any $\Omega_1 \subset \Omega_2 \subset \mathbb{R}^m$.
- *Scaling*: $C_{\text{stab}}^{\text{gen}}(r \cdot \Omega, \mathfrak{f}) \geq C_{\text{stab}}^{\text{gen}}(\Omega, r^2 \mathfrak{f})$ for any $\Omega \subset \mathbb{R}^m$ and $r > 0$.

Analogous properties hold for the stability constants in Sect. 14.3.3.

14.3.3 Homogeneous Objects and Multiple Holograms

In most practical works, one aims to stabilize image reconstruction in XPCI by one of the following approaches (often both, actually):

1. Impose a homogeneity-constraint, e.g. assuming a pure phase object $h = i\phi$ if absorption is negligible ($\mu \approx 0$), see Sect. 14.1.2.3.
2. Reconstruct from more than one hologram, see Sect. 14.1.2.4.

According to Sect. 14.2.1, uniqueness then also holds without support constraints, but image reconstruction is still *ill-posed* in general: the associated forward maps $\mathscr{S}_\nu : L^2(\mathbb{R}^m) \to L^2(\mathbb{R}^m)$ and $\mathscr{T}^{(\mathfrak{f}_1, \dots, \mathfrak{f}_\ell)} : L^2(\mathbb{R}^m) \to L^2(\mathbb{R}^m)^\ell$ do not have a bounded inverse due to zeros of the CTFs.

When both homogeneity- and support constraints can be assumed, well-posedness holds true with an improved stability constant compared to (14.34):

Theorem 14.3 (Stability estimate for homogeneous objects [24]) *Let $\Omega \subset \mathbb{R}^m$ be a ball of diameter 1, w.l.o.g. $\Omega = \{x \in \mathbb{R}^m : |x| \leq \frac{1}{2}\}$. Then*

$$C_{\text{stab}}^{hom}(\Omega, \mathfrak{f}, \nu) := \inf_{\varphi \in L^2(\Omega), \|\varphi\|_{L^2}=1} \|\mathscr{S}_\nu(\varphi)\|_{L^2}$$

$$\geq \max\left\{\min\left\{c_1, c_2\mathfrak{f}^{-1}\right\}, \min\left\{c_3\nu, c_4\mathfrak{f}^{-\frac{1}{2}}\right\}\right\} \quad (14.35)$$

for some constants $c_j > 0$ that depend only on m.

By Theorem 14.3, the original decay $C_{stab}^{gen} \sim \exp(-\mathfrak{f}/8)$ of the stability constant as $\mathfrak{f} \to \infty$ improves to $C_{stab}^{hom}(\Omega, \mathfrak{f}, \nu) \sim \mathfrak{f}^{-\gamma}$ with $\gamma = 1$ for $\nu = 0$ and $\gamma = 1/2$ for $\nu > 0$. This ensures *practical* stability also at larger Fresnel-numbers.

A similar improvement applies for the reconstruction of general objects (no homogeneity-constraint) from *two* holograms:

Theorem 14.4 (Stability estimate for two holograms [24]) *Let $\mathscr{T}^{(\mathfrak{f}_1, \mathfrak{f}_2)} : h \mapsto (\mathscr{T}^{(\mathfrak{f}_1)}(h), \mathscr{T}^{(\mathfrak{f}_2)}(h))$ denote the linearized XPCI-model for two holograms at Fresnel-numbers $\mathfrak{f}_1 \neq \mathfrak{f}_2$ (see Sect. 14.1.2.4). Let $\mathscr{S}_0^{(\mathfrak{f}-)}$ be the forward map from Theorem 14.3 for $\nu = 0$ and $\mathfrak{f} = \mathfrak{f}_- := |\mathfrak{f}_1^{-1} - \mathfrak{f}_2^{-1}|^{-1}$. Then*

$$\left\| \mathscr{T}^{(\mathfrak{f}_1, \mathfrak{f}_2)}(h) \right\|_{L^2} \geq 2^{-\frac{1}{2}} \left\| \mathscr{S}_0^{(\mathfrak{f}-)}(h) \right\|_{L^2} \quad \text{for all} \quad h \in L^2(\mathbb{R}^m). \tag{14.36}$$

In particular, for any support-domain $\Omega \subset \mathbb{R}^m$, the following stability estimate holds true:

$$C_{stab}^{two}(\Omega, \mathfrak{f}_1, \mathfrak{f}_2) := \inf_{\substack{h \in L^2(\Omega) \\ \|h\|_{L^2} = 1}} \left\| \mathscr{T}^{(\mathfrak{f}_1, \mathfrak{f}_2)}(h) \right\|_{L^2} \geq 2^{-\frac{1}{2}} C_{stab}^{hom}(\Omega, \mathfrak{f}_-, 0). \tag{14.37}$$

Note that the r.h.s. of the stability bound (14.37) increases with the difference $\mathfrak{f}_-^{-1}$ between the reciprocal Fresnel-numbers $\mathfrak{f}_1^{-1}, \mathfrak{f}_2^{-1}$. Improved stability is thus guaranteed only if $\mathfrak{f}_1$ and $\mathfrak{f}_2$ differ strongly, i.e. if the two holograms are acquired in significantly different experimental setups.

14.3.3.1 Order-Optimality

For $\nu = 0$, it can be shown that the $\sim \mathfrak{f}^{-1}$ order of the decay in Theorem 14.3 cannot be improved: for a fixed bounded domain $\Omega \subset \mathbb{R}^m$ with non-empty interior, there exists a constant $c_{max}(\Omega) > 0$ such that

$$C_{stab}^{hom}(\Omega, \mathfrak{f}, 0) \leq c_{max}(\Omega) \mathfrak{f}^{-1}. \tag{14.38}$$

This is a consequence of the bound $\|\mathscr{S}_0(\varphi)\|_{L^2} \leq 1/\mathfrak{f} \|\Delta \varphi\|_{L^2}$ where Δ is the Laplacian, which in turn follows from $|s_0(\xi)| \leq \xi^2/(2\mathfrak{f})$ for all $\xi \in \mathbb{R}^m$.

Notably, better rates do not even hold in a setting with multiple holograms: for any Fresnel-numbers $\mathfrak{f}_1, \ldots, \mathfrak{f}_\ell$, it holds by a similar argument that

$$\inf_{\varphi \in L^2(\Omega), \|\varphi\|_{L^2} = 1} \|\mathscr{S}_0^{(\mathfrak{f}_1, \ldots, \mathfrak{f}_\ell)}(\varphi)\|_{L^2} \leq c_{max}(\Omega) \left(\sum_{i=1}^{\ell} \mathfrak{f}_i^{-2} \right)^{\frac{1}{2}}. \tag{14.39}$$

The reason for this surprising negative result on the benefit of multiple holograms is that the CTFs $s_0^{(\mathfrak{f}_i)}(\xi) = \sin(\xi^2/(2\mathfrak{f}_i))$ all share a second order zero at $\xi = 0$. This corresponds to the well-known *low-frequency instability* of XPCI that gives rise to the proven f^{-1}-rates of the stability constant.

14.3.3.2   Numerical Stability Computations

Other than for the setting in Theorem 14.2, the prediction (14.35) for the stability constant $C_{\text{stab}}^{\text{hom}}$ (and thus for $C_{\text{stab}}^{\text{two}}$) is far from sharp if the analytical bounds on the constants c_j from [24] are inserted. Sharp values of $C_{\text{stab}}^{\text{hom}}$ may however be computed numerically by approximating the minimum singular value of the operator $\mathscr{S}_\nu$ via techniques presented in [29, Sect. 3.4].

14.3.4   Extensions

14.3.4.1   Phase Contrast Tomography

Although the physical setting of XPCI corresponds to $m = 2$ dimensions, the stability results in Theorems 14.2 to 14.4 have been formulated for arbitrary m. As a benefit, stability may be readily extended to XPCT: for the considered linearized forward models, XPCT data is of the form $I_\theta - 1 = T (\mathscr{P}_\theta(f))$ for $T \in \{\mathscr{T}, \mathscr{S}_\nu\}$ and incident directions $\theta \in \Theta \subset \mathbb{S}^2$, compare Sect. 14.1.3. As noted in [30, 31], the order of the projector $\mathscr{P}_\theta$ and T may be interchanged:

$$I_\theta - 1 = T (\mathscr{P}_\theta(f)) = \mathscr{P}_\theta \left(T^{(3d)}(f) \right), \tag{14.40}$$

where $T^{(3d)} \in \{\mathscr{T}^{(3d)}, \mathscr{S}_\nu^{(3d)}\}$ is the equivalent of T in $m = 3$ dimensions.

As detailed in [29, Sect. 3.3], the relation (14.40) allows to express stability of linearized XPCT via known results for tomographic reconstruction, combined with stability bounds for $\mathscr{T}^{(3d)}, \mathscr{S}_\nu^{(3d)} : L^2(\Omega) \to L^2(\mathbb{R}^3)$ where $\Omega \subset \mathbb{R}^3$. Stability then depends on a *three-dimensional* support constraint $\text{supp}(\beta + i\delta) \subset \Omega \subset \mathbb{R}^3$ for the imaged sample's refractive index.

14.3.4.2   Imaging with Finite Detectors

There are a number of idealizing assumptions underlying to the obtained stability estimates: in addition to the neglected nonlinearity and idealizations in the basic physical model such as full coherence, it has also been assumed that the hologram I is measured in the whole detector-plane in Fig. 14.1. Due to the finite size of real-world detectors (and—more fundamentally—the finite width of illuminating X-ray beams), however, only restrictions $I|_K$ to some bounded domain K are available in practice.

According to Theorem 14.1, such restricted data has *no* impact on *uniqueness* (if K contains an open set). The situation is quite different in terms of *stability*, as analyzed in [32]: for any bounded $K \subset \mathbb{R}^m$—however large—the inverse problem of XPCI becomes *severely ill-posed*, i.e. Lipschitz-stability is lost so that data errors may severely corrupt the reconstructed images. Yet, it is also proven in [32] that the situation may be repaired by restricting to images $h = \mu + i\phi$ of *finite resolution* (smoothness constraint in the sense of Sect. 14.1.1.1): by imposing that the h are B-splines on a Cartesian grid of sufficiently large spacing $r(\Omega, K, \mathfrak{f}) > 0$ (i.e. *pixelated* images in some sense), Lipschitz-stability can be restored in the finite-detector setting. Physically, the necessity of such a restriction corresponds to a *resolution limit* that arises due to the finite numerical aperture associated with the detector size.

14.4   Regularized Newton Methods for XPCI

The following section considers regularized Newton-type methods for image reconstruction in XPCI. The proposed algorithm is motivated by the theoretical insights gained from Sects. 14.2 and 14.3.

14.4.1   Motivation

14.4.1.1   Significance of Constraints

The stability results of Sect. 14.3 heavily rely on *support constraints*—without such, XPCI is ill-posed or even non-unique. To guarantee stability in practice, image reconstruction methods must thus be able to exploit support-knowledge. Also other types of a priori knowledge (see Sect. 14.1.1.1) are known to be beneficial. In particular, imposing *non-negativity* often has a similar stabilizing effect as support constraints.

14.4.1.2   Necessity of Iterative Methods

By far the most commonly used reconstruction method for XPCI at synchrotrons is direct CTF-inversion, as presented in Sect. 2.3. Within the notation of this chapter, the approach corresponds to quadratic Tikhonov regularization applied to the linearized forward maps $\mathscr{S}_\nu^{(\mathfrak{f}_1,...,\mathfrak{f}_\ell)}$ or $\mathscr{T}^{(\mathfrak{f}_1,...,\mathfrak{f}_\ell)}$. Owing to the *linearity* and *translation-invariance* of these maps, the reconstruction may be implemented via a multiplication in Fourier-space (deconvolution), which renders the approach computationally fast.

However, direct CTF-inversion is incompatible with the above constraints:

- Support constraints $\mathrm{supp}(h) \subset \Omega$ for $\Omega \subsetneq \mathbb{R}^m$ break translation-invariance
- Non-negativity is a *nonlinear* constraint: any reconstruction imposing it depends nonlinearly on the data $I - 1$—even for linear forward models!

In either case, reconstruction may thus no longer be achieved by deconvolution. Thus, *iterative* algorithms have to be applied to impose support- and/or nonnegativity-constraints in lack of efficient direct reconstruction formulas.

14.4.1.3   XPCI Beyond Linear Models

Although the linear CTF-model of XPCI has a surprisingly large regime-of-validity, there are settings where linear image reconstruction induces severe artifacts arising from the neglected nonlinearity, as demonstrated in Sect. 13.3. Reconstruction algorithms based on the full nonlinear XPCI-model are thus preferable in principle. The main obstacle in using such is that direct inversion formulas for the nonlinear model are not known. However, when iterative methods are needed anyway (Sect. 14.4.1.2), nonlinear forward maps cause little additional difficulty.

14.4.2   Reconstruction Method

In the following, we propose a reconstruction algorithm that meets the requirements discussed in Sect. 14.4.1. Details can be found in [33].

By choosing $F : X \to L^2(\mathbb{R}^m) \in \{\mathcal{N}, \mathcal{N}_\nu\}$ with $X = L^2(\Omega, (\mathbb{R}))$ for $\Omega \subset \mathbb{R}^m$, optional homogeneity- and/or support constraints are incorporated in the forward operator F. Consequently, such constraints are imposed *automatically* if image reconstruction in XPCI is performed by inverting F via any generic regularization method for inverse problems, see Chap. 5. In order to exploit Fréchet-differentiability of $\mathcal{N}$ (see Sect. 14.1.2.1) and the comparably moderate nonlinearity of XPCI, we choose regularized Newton methods as introduced in Chap. 5:

$$h_{k+1} \in \mathrm{argmin}_{h \in X} \left\| F(h_k) + F'[h_k](h - h_k) - (I^{\mathrm{obs}} - 1) \right\|_{L^2}^2$$
$$+ \alpha_k \| h - h_0 \|_{H^s}^2 + \mathcal{R}_{\geq 0}(h, h_k). \tag{14.41}$$

for $k = 0, 1, \ldots, k_{\mathrm{stop}}$, with initial guess $h_0 \in X$ (usually $h_0 = 0$), observed (noisy) hologram(s) I^{obs} and regularization parameters $\alpha_k > 0$.

Note that we use a standard squared L^2-norm as a *data-fidelity* term in (14.41), in lack of an accurate model for the data error statistics in flat-field corrected holograms. The squared Sobolev-term $\alpha_k \| h - h_0 \|_{H^s}^2$ ($\| f \|_{H^s}^2 := \| (1 + \xi^2)^{s/2} \cdot \mathcal{F}(f) \|_{H^s}^2$) imposes tunable (by the choice of $s \geq 0$) smoothness of the iterates h_k and acts as a regularizer. Finally, $\mathcal{R}_{\geq 0}(h, h_k)$ is a quadratic penalty term that is designed to correct negative values of $\text{Re}(h_k)$ or $\text{Im}(h_k)$ in the subsequent iterate h_{k+1}, see [33] for details.

In the numerical algorithm, a discretized analogue of the quadratic minimization problem in (14.41) is solved for images $\boldsymbol{h}_* \in \mathbb{C}^N$, data $\boldsymbol{I}^{\text{obs}} \in \mathbb{R}^M$ and forward map $F_{\text{dis}} : \mathbb{C}^N \to \mathbb{R}^M$, via a conjugate-gradient method. The α_k and k_{stop} are chosen in a widely automated fashion, as detailed in [33].

14.4.3 Reconstruction Example

We assess the capabilities of the proposed method by reconstructing phase ϕ and absorption μ as independent parameters from a single simulated noisy hologram, which is shown in Fig. 14.3a. The considered test case is detailed in [33], where also a real-data example is considered for an analogous setting.

The true phase-image ϕ (Fig. 14.3b) is given by a bulk disk of magnitude 0.2, whereas the true absorption-image $0 \leq \mu \leq 0.02$ shows a logo-structure (Fig. 14.3c). Accordingly, *no* homogeneity-constraint is applicable so that the test-case is situated in the most challenging, unstable setting of XPCI, which has been analyzed in Sect. 14.3.2. In particular, recall that image reconstruction is non-unique without exploiting further constraints.

The data is reconstructed using the regularized Newton method from Sect. 14.4.2, imposing non-negativity of ϕ and μ as well as support constraint, allowing nonzero values of ϕ, μ only within the circular region marked by the blue dashed line in Fig. 14.3b, c. The reconstructed images in Fig. 14.3d, e show that the proposed method correctly attributes the disk-structure to the phase-image ϕ and the logo-pattern to μ, without visible signs of "mixing things up". The overall lower reconstruction-quality in μ compared to ϕ is due to the lower signal-to-noise in this parameter, as a realistically low absorption-refraction-ratio $\beta/\delta \leq 0.1$ has been assumed in the test case.

Now why does reconstruction of both ϕ and μ from a single hologram work here, contrary to the usual experience? The diameter of the circular support corresponds to a relatively low (modified) Fresnel-number $\mathfrak{f} \approx 87$. According to the analysis in Sect. 14.3.2, this ensures stability of image reconstruction, as is discussed to greater detail in [33] and [24, Sect. 6]. By its ability to impose support constraints (and non-negativity), the proposed Newton-type method allows to exploit this theoretical stability in practice.

14.5 Regularized Newton-Kaczmarz-SART for XPCT

In the final section, we present a Newton-type reconstruction method for X-ray phase contrast *tomography* (XPCT) that is a compromise between flexibility w.r.t. a priori constraints and computational performance. We note that the method is an

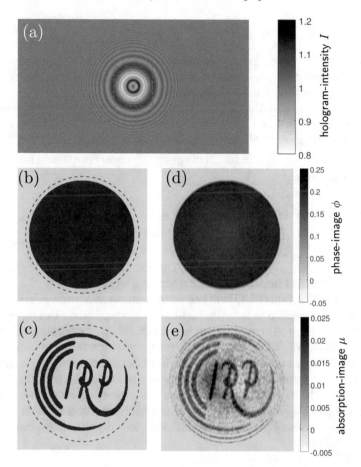

Fig. 14.3 Reconstruction of a general image $h = \mu + i\phi$ from a single simulated hologram by a regularized Newton method (test case from [33]). **a** Hologram of size 1920×1080. **b, c** True images ϕ and μ (zooms to the relevant region, that is marked by a red-dashed line in (**a**)). **d, e** Reconstructed images ϕ and μ, obtained by imposing non-negativity and *support* of ϕ, μ within the circular region bounded by the blue-dashed line in (**b**), (**c**)

all-at-once approach, as also proposed in [30, 31, 34]: the 3D-object parameters δ, β are recovered directly from the full tomographic hologram-series, instead of first reconstructing 2D-images ϕ, μ for each hologram individually. Thereby, *tomographic consistency* is imposed as an additional constraint in image reconstruction, compare Sect. 14.1.1.1.

By replacing $F \in \{\mathcal{N}, \mathcal{N}_\nu\}$ with the corresponding *tomographic* forward operator from Sect. 14.1.3, $F_{\mathrm{PCT}} : X \to L^2(\mathbb{R}^2)^t; \ f \mapsto (F(\mathscr{P}_{\theta_j}(f)))_{j=1}^t$ with $X = L^2(\Omega, (\mathbb{R}))$, $\Omega \subset \mathbb{R}^3$, the regularized Newton method from Sect. 14.4.2 may be readily adapted to solve the inverse problem of XPCT:

$$F_{\text{PCT}}(f) \approx (I_{\theta_j}^{\text{obs}} - 1)_{j=1}^t \qquad \text{for} \qquad f = k\beta + ik\delta \in X. \tag{14.42}$$

This is done in [2]. Yet, typical problem-sizes in XPCT with $\sim 10^9$ dimensions of the discretized object- and data-space, are too large for this approach to be competitive in terms of computation times and memory requirements.

As a remedy, we supplement the approach with a *Kaczmarz-type* strategy that exploits the block-structure of the XPCT-problem (14.42). The idea is to cyclically perform regularized Newton-steps w.r.t. the small sub-problems $I_{\theta_j}^{\text{obs}} - 1 \approx F(\mathscr{P}_{\theta_j}(f))$ defined by the measured holograms $I_{\theta_j}^{\text{obs}}$ under the different tomographic incident directions θ_j:

$$f_{k+1} \in \operatorname{argmin}_{f \in X} \left\| F(\mathscr{P}_{\theta_{j_k}}(f_k)) + F'[\mathscr{P}_{\theta_{j_k}}(f_k)]\mathscr{P}_{\theta_{j_k}}(f - f_k) - (I_{\theta_{j_k}}^{\text{obs}} - 1) \right\|^2$$
$$+ \alpha\big((1 - \gamma) \|f - f_k\|_{L^2}^2 + \gamma \|\nabla(f - f_k)\|_{L^2}^2\big) \tag{14.43}$$

for $k = 0, 1, \ldots, tn_{\text{stop}} - 1$ with $n_{\text{stop}} \in \mathbb{N}$. The parameters $\alpha > 0, 0 \leq \gamma \leq 1$ control the regularization and smoothing w.r.t. the preceding iterate f_k.

Iterations of the form (14.43) are known as *regularized Newton-Kaczmarz* [35]. The advantage compared to *bulk* (i.e. non-Kaczmarz-)methods is that the operator-blocks $f \mapsto F(\mathscr{P}_{\theta_j}(f))$ require much less computations to evaluate than the total XPCT operator F_{PCT}, which permits efficient computation of the iterates (14.41). Moreover, Kaczmarz-type methods often exhibit fast initial convergence, typically reaching a good reconstruction already after one or two *cycles* over the data, i.e. for $n_{\text{stop}} \in \{1, 2\}$. To promote convergence, the processing order $\{j_1, j_2, \ldots\} \subset \{1, \ldots, t\}$ of the data-blocks should be chosen such that subsequently fitted directions $\theta_{j_k}, \theta_{j_{k+1}}$ differ as strongly as possible, which we achieve by following a "multi-level-scheme" from [36].

14.5.1 Efficient Computation by Generalized SART

Although the processed *data*-size is reduced by the Kaczmarz-strategy, the iterates (14.41) still involve a minimization problem on a high-dimensional space of 3D-objects f. Moreover, if the minimization is performed iteratively, each iteration requires evaluations of the (discretized) projector $\mathscr{P}_{\theta_{j_k}}$ and its adjoint $\mathscr{P}_{\theta_{j_k}}^*$, the *back*-projector, both of which typically amount to much higher computational costs than evaluating the XPCI forward map F.[5]

Both computational issues can be resolved by computing the iterates (14.41) via a *generalized SART*[6] (GenSART-) scheme, as introduced in [38] for a much more general class of tomographic Kaczmarz-iterations:

---

[5] For images of size $N \times N$, the discretized forward maps $F = \mathscr{N}^{(\mathfrak{f}_1, \ldots, \mathfrak{f}_\ell)}$ may be evaluated in $\mathcal{O}(\ell N^2 \log N)$ operations, while (back-)projecting 3D-arrays of size $N \times N \times N$ is $\mathcal{O}(N^3)$.

[6] "SART" refers to the simultaneous algebraic reconstruction technique from [37].

GenSART for Newton-Kaczmarz-iterations:

1. Forward-projection: $p_k := \mathscr{P}_{\theta_{j_k}}(f_k)$
2. Optimization in *projection-space* ($u_j = \mathscr{P}_{\theta_j}(\mathbf{1}_\Omega)$):

$$\Delta p_k \in \text{argmin}_{p \in L^2(\mathbb{R}^2,(\mathbb{R}))} \| F(p_k) + F'[p_k](u_{j_k} \cdot p) - (I_{\theta_{j_k}}^{\text{obs}} - 1)\|_{L^2}^2$$
$$+ \alpha\big((1-\gamma)\|u_{j_k}^{1/2} \cdot p\|_{L^2}^2 + \gamma\|u_{j_k}^{1/2} \cdot \nabla p\|_{L^2}^2\big) \qquad (14.44)$$

3. Back-projection update: $f_{k+1} = f_k + \mathscr{P}_{\theta_{j_k}}^*(\Delta p_k)$

The main benefit of the approach is that the required minimization is cast to *projection-space*, i.e. no longer needs to be solved on a high-dimensional space of 3D-objects but merely on 2D-*images*. Moreover, the whole scheme requires only a single evaluation of $\mathscr{P}_{\theta_{j_k}}$ (1.) and its adjoint $\mathscr{P}_{\theta_{j_k}}^*$ (3.), whereas the optimization (2.) does not involve any of these costly operations anymore.

As is standard for Kaczmarz-type methods, non-negativity of the iterates f_{k+1} (in real- and imaginary part) may be imposed by adding a final step to the GenSART-scheme: $f_{k+1, \geq 0} = \max\{0, \text{Re}(f_{k+1})\} + \text{i} \max\{0, \text{Im}(f_{k+1})\}$.

14.5.2   Parallelization and Large-Scale Implementation

Regularized Newton-Kaczmarz, computed via GenSART-schemes, is well-suited for large-scale computations and can be efficiently implemented in a parallelized manner. While we refer to [29, Sect. 6.3] for a detailed discussion, we mention the most important aspects here:

- *Low memory requirements*: if the back-projection update (3.) (as well as the optional non-negativity projection) is implemented as an *in-place* operation, only a single 3D-array (storing $f_0, f_1, (f_{1, \geq 0},)f_2, \ldots$) needs to be kept in memory throughout the whole Newton-Kaczmarz-reconstruction.
- *Parallelized optimization*: as the optimization-step (2.) works on *2D*-images only, its memory-requirements are low enough to be performed on a single graphical processing unit (GPU) even for large-scale data. This permits efficient parallized implementation of this step.
- *Parallelized 3D-computations*: The only operations on the 3D-objects f_k are forward- and back-projections $\mathscr{P}_{\theta_{j_k}}, \mathscr{P}_{\theta_{j_k}}^*$ and pointwise arithmetics. All of these can be easily parallelized at low communication requirements between the different processors. In fact, it is possible to implement GenSART-schemes in a distributed manner: the object-iterates f_k may be split into chunks, that are stored

and managed by dedicated machines throughout the whole reconstruction. This property allows to run Newton-Kaczmarz reconstructions efficiently on multiple GPUs.

14.5.3 Reconstruction Example

We assess the Newton-Kaczmarz method for XPCT-data of freeze-dried *Deinococcus radiodurans* bacteria. The experimental data set, acquired with the GINIX setup from Chap. 3, is composed of 641 holograms of size 2048×2048 at tomographic incident angles $\theta = 0°, 0.25°, \ldots, 119°, 139°, 139.25°, \ldots, 180°$ (one hologram per angle). 2D orthoslices of the 3D tomographic data (two spatial and one angular dimension) are shown in Fig. 14.4a–c, emphasizing the missing data between $\theta = 119°$ and $\theta = 139°$.

The biological sample constitutes a *pure phase object* to good approximation, i.e. vanishing absorption $\beta = 0$ may be assumed. Moreover, the sample is localized in a small subdomain of the imaged 2048×2048-sized field-of-view, as can be seen from Fig. 14.4a–c, i.e. support constraints may be imposed.

For comparison, we reconstruct the XPCT-data with different methods:

1. CTF+FBP: direct CTF-inversion for each hologram, followed by filtered back-projection applied to the recovered projections of δ.
2. Linear Kaczmarz: reconstruction by (14.43) over a single cycle $n_{stop} = 1$, using the *linearized* XPCI-model $F = \mathscr{S}_0$. Non-negativity of the reconstructed δ and support in a centered cube of 512^3 voxels is imposed.
3. Newton-Kaczmarz: same as (2.), but with the *nonlinear* model $F = \mathscr{N}_0$.

2D orthoslices through the reconstructed $512 \times 512 \times 512$ volumes are plotted in Fig. 14.4d–l. We note the following observations:

- The additional constraints exploited in "Linear Kaczmarz" compared to "CTF+FBP" widely eliminate low-frequency background-artifacts (compare Fig. 14.4e–h) and thereby enable *quantitatively* correct reconstructions δ.
- Though the sample-induced phase shifts are moderate, $\phi_\theta = k\mathscr{P}_\theta(\delta) \lesssim 1$, going over to the nonlinear XPCI-model has significant effects: especially in Fig. 14.4h, it can be seen that using the linearized model causes artificial distortions in the recovered object-density compared to the nonlinear Newton-Kaczmarz-reconstruction in Fig. 14.4i–l.

Accordingly, both the nonlinearity and the ability to exploit a priori constraints of the proposed Newton-Kaczmarz method turn out to be vital here to accurately reconstruct the anticipated 3D structure of the imaged bacteria[7]: cytoplasm with blob-shaped inclusions containing the DNA, where each of the two compounds is of approximately uniform density.

---

[7]The additional object in the top-left of Fig. 14.4e, h, k is a contaminant particle.

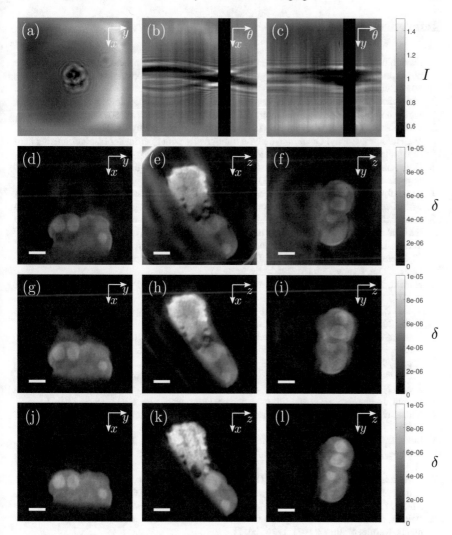

Fig. 14.4 XPCT-reconstruction of *Deinococcus radiodurans* bacteria with different algorithms. Rows show 2D orthoslices for: **a–c** the stack of 641 holograms of 2048 × 2048 pixels each (x, y: detector-coordinates, θ: tomographic incident angle) **d–l** reconstructed object-volumes with different methods (**d–f** CTF-inversion followed by FBP-reconstruction, **g–i** Linear Kaczmarz, **j–l** Newton-Kaczmarz). The tomographic axis is the y-axis. Scale bars: 1 μm. For details, see text

References

1. Davidoiu, V., Sixou, B., Langer, M., Peyrin, F.: Nonlinear approaches for the single-distance phase retrieval problem involving regularizations with sparsity constraints. Appl. Opt. **52**(17), 3977–3986 (2013)
2. Maretzke, S.: Regularized Newton methods for simultaneous Radon inversion and phase retrieval in phase contrast tomography (2015). arXiv preprint arXiv:1502.05073
3. Cloetens, P., Ludwig, W., Baruchel, J., Van Dyck, D., Van Landuyt, J., Guigay, J., Schlenker, M.: Holotomography: Quantitative phase tomography with micrometer resolution using hard synchrotron radiation X-rays. Appl. Phys. Lett. **75**(19), 2912–2914 (1999)
4. Hofmann, R., Moosmann, J., Baumbach, T.: Criticality in single-distance phase retrieval. Opt. Express **19**(27), 25881–25890 (2011)
5. Krenkel, M., Toepperwien, M., Alves, F., Salditt, T.: Three-dimensional single-cell imaging with X-ray waveguides in the holographic regime. Acta Crystallogr. A **73**(4), 282–292 (2017)
6. Langer, M., Cloetens, P., Guigay, J.P., Peyrin, F.: Quantitative comparison of direct phase retrieval algorithms in in-line phase tomography. Med. Phys. **35**(10), 4556–4566 (2008)
7. Turner, L., Dhal, B., Hayes, J., Mancuso, A., Nugent, K., Paterson, D., Scholten, R., Tran, C., Peele, A.: X-ray phase imaging: Demonstration of extended conditions for homogeneous objects. Opt. Express **12**(13), 2960–2965 (2004)
8. Nugent, K.A.: X-ray noninterferometric phase imaging: a unified picture. J. Opt. Soc. Am. A **24**(2), 536–547 (2007)
9. Jonas, P., Louis, A.: Phase contrast tomography using holographic measurements. Inverse Probl. **20**(1), 75 (2004)
10. Walther, A.: The question of phase retrieval in optics. J. Mod. Opt. **10**(1), 41–49 (1963)
11. Akutowicz, E.J.: On the determination of the phase of a Fourier integral, i. Proc. Am. Math. Soc., 179–192 (1956)
12. Akutowicz, E.J.: On the determination of the phase of a Fourier integral, ii. Proc. Am. Math. Soc. **8**(2), 234–238 (1957)
13. Fienup, J.: Phase retrieval algorithms: a personal tour. Appl. Opt. **52**(1), 45–56 (2013)
14. Klibanov, M.V., Sacks, P.E., Tikhonravov, A.V.: The phase retrieval problem. Inverse Probl. **11**(1), 1 (1995)
15. Luke, D.R.: Phase retrieval, what's new. SIAG/OPT Views News **25**(1), 1–5 (2017)
16. Millane, R.: Phase retrieval in crystallography and optics. J. Opt. Soc. Am. A **7**(3), 394–411 (1990)
17. Shechtman, Y., Eldar, Y.C., Cohen, O., Chapman, H.N., Miao, J., Segev, M.: Phase retrieval with application to optical imaging: a contemporary overview. IEEE Signal Proc. Mag. **32**(3), 87–109 (2015)
18. Beinert, R.: One-dimensional phase retrieval with additional interference intensity measurements. Results Math. **72**(1–2), 1–24 (2017)
19. Bendory, T., Beinert, R., Eldar, Y.C.: Fourier phase retrieval: Uniqueness and algorithms. In: Compressed Sensing and its Applications, pp. 55–91. Springer (2017)
20. Leshem, B., Xu, R., Dallal, Y., Miao, J., Nadler, B., Oron, D., Dudovich, N., Raz, O.: Direct single-shot phase retrieval from the diffraction pattern of separated objects. Nat. Commun. **7**, 10,820 (2016)
21. Raz, O., Leshem, B., Miao, J., Nadler, B., Oron, D., Dudovich, N.: Direct phase retrieval in double blind Fourier holography. Opt. Express **22**(21), 24935–24950 (2014)
22. Maretzke, S.: A uniqueness result for propagation-based phase contrast imaging from a single measurement. Inverse Probl. **31**, 065,003 (2015)
23. Gabor, D., et al.: A new microscopic principle. Nature **161**(4098), 777–778 (1948)
24. Maretzke, S., Hohage, T.: Stability estimates for linearized near-field phase retrieval in X-ray phase contrast imaging. SIAM J. Appl. Math. **77**, 384–408 (2017)
25. Kalbfleisch, S., Neubauer, H., Krüger, S., Bartels, M., Osterhoff, M., Mai, D., Giewekemeyer, K., Hartmann, B., Sprung, M., Salditt, T.: The göttingen holography endstation of beamline p10 at petra iii/desy. In: AIP Conference Proceedings, Vol. 1365, pp. 96–99. AIP (2011)

26. Salditt, T., Osterhoff, M., Krenkel, M., Wilke, R.N., Priebe, M., Bartels, M., Kalbfleisch, S., Sprung, M.: Compound focusing mirror and x-ray waveguide optics for coherent imaging and nano-diffraction. J. Synchrotron Rad. **22**(4), 867–878 (2015)
27. Havin, V., Jöricke, B.: The Uncertainty Principle in Harmonic Analysis. Springer, Berlin (1994)
28. Slepian, D., Sonnenblick, E.: Eigenvalues associated with prolate spheroidal wave functions of zero order. Bell Syst. Tech. J. **44**(8), 1745–1759 (1965)
29. Maretzke, S.: Inverse problems in propagation-based X-ray phase contrast imaging and tomography: stability analysis and reconstruction methods. eDiss Uni Göttingen (2019)
30. Kostenko, A., Batenburg, K.J., King, A., Offerman, S.E., van Vliet, L.J.: Total variation minimization approach in in-line X-ray phase-contrast tomography. Opt. Express **21**(10), 12185–12196 (2013)
31. Ruhlandt, A., Salditt, T.: Three-dimensional propagation in near-field tomographic X-ray phase retrieval. Acta Crystallogr. A **72**(2) (2016)
32. Maretzke, S.: Locality estimates for Fresnel-wave-propagation and stability of near-field X-ray propagation imaging with finite detectors. Inverse Probl. **34**(12), 124,004 (2018). https://doi.org/10.1088/1361-6420/aae78f
33. Maretzke, S., Bartels, M., Krenkel, M., Salditt, T., Hohage, T.: Regularized Newton methods for X-ray phase contrast and general imaging problems. Opt. Express **24**(6), 6490–6506 (2016)
34. Ruhlandt, A., Krenkel, M., Bartels, M., Salditt, T.: Three-dimensional phase retrieval in propagation-based phase-contrast imaging. Phys. Rev. A **89**(3), 033,847 (2014)
35. Burger, M., Kaltenbacher, B.: Regularizing Newton-Kaczmarz methods for nonlinear ill-posed problems. SIAM J. Numer. Anal. **44**(1), 153–182 (2006)
36. Guan, H., Gordon, R.: A projection access order for speedy convergence of ART (algebraic reconstruction technique): a multilevel scheme for computed tomography. Phys. Med. Biol. **39**(11), 2005 (1994)
37. Andersen, A.H., Kak, A.C.: Simultaneous algebraic reconstruction technique (SART): a superior implementation of the ART algorithm. Ultrason. Imaging **6**(1), 81–94 (1984)
38. Maretzke, S.: Generalized SART-methods for tomographic imaging. arXiv preprint p. arXiv:1803.04726 (2018)

Chapter 15
Scanning Small-Angle X-ray Scattering and Coherent X-ray Imaging of Cells

Tim Salditt and Sarah Köster

Abstract In this chapter we review recent work towards high resolution imaging of unstained biological cells in the hydrated and living state, using synchrotron radiation (SR) and free electron laser (FEL) radiation. Specifically, we discuss the approaches of scanning small-angle X-ray scattering (scanning SAXS) and coherent diffractive X-ray imaging (CDI) of cells.

15.1 X-ray Structure Analysis of Biological Cells: A Brief Overview

The desire to probe the three-dimensional (3D) structure of biological cells and tissues at high resolution and under hydrated conditions has motivated a continuous and long-lasting effort to develop suitable high resolution microscopy techniques. Fluorescence light microscopy provides an excellent tool to label specific biomolecules and organelles. As the last three decades have shown, an ever increasing number of imaging problems can be addressed by this specific labeling approach. However, not only the strength but also the limitation of this microscopy technique is linked to the selective imaging of a few components within the cells. Firstly, fluorescence microscopy of living cells can typically not be applied when transfection with fluorescent proteins is not possible or too invasive. Secondly, some questions in biology and biophysics cannot be answered from mapping selected macromolecular components, but necessitate the visualization of the entire mass density distribution in the cell. In these cases high resolution images with quantitative mass or electron density

T. Salditt (✉) · S. Köster
Institute for X-ray Physics, Universität Göttingen, Friedrich-Hund-Platz 1,
37077 Göttingen, Germany
e-mail: tsalditt@gwdg.de

S. Köster
e-mail: sarah.koester@physik.uni-goettingen.de

© The Author(s) 2020

405

T. Salditt et al. (eds.), *Nanoscale Photonic Imaging*, Topics in Applied Physics 134,
https://doi.org/10.1007/978-3-030-34413-9_15

contrast are needed, rather than the distribution of a selected label. Hard X-rays with multi-keV photon energies can contribute exactly this contrast mechanism related to the native electron density distributions in biological matter.

Apart from this aspect, other specific advantages of X-rays are: (i) a scalable resolution down to the X-ray wavelength of Å to nm, (ii) a kinematic nature of the scattering process enabling quantitative image analysis unaffected by multiple scattering, (iii) element specific contrast variation exploiting anomalous effects at absorption edges, (iv) compatibility with unsliced (three-dimensionally extended), unstained and hydrated specimens due to the large penetration depth. In this chapter we review recent studies of biological cells with hard X-rays. We focus on proof-of-concept experiments with micro- and nano-focused X-ray beams which have extended classical small-angle X-ray scattering (SAXS) to cellular imaging, combining real and reciprocal space. Classical SAXS is known as a structural technique for soft and biological matter, biomaterials and proteins which does *not* offer any real space information and can hardly be used on systems which are as heterogeneous as a biological cell. We also include coherent diffractive X-ray imaging (CDI) techniques as an X-ray imaging modality, which can complement scanning SAXS. We do no include, however, X-ray fluorescence microscopy of cells, which is by now quite well established, see [1].

The development of X-ray microscopy and imaging techniques has always been closely related to the availability of high brilliance radiation, provided by synchrotron radiation sources, and recently also by X-ray free electron lasers (FEL). Ultra-short and high brilliance FEL pulses may offer sharp still images of structure even of living cells, since the signal is recorded before structural changes occur by radiation damage. However, these very recent opportunities should not lead us to believe that imaging of cells with X-rays is an entirely new research topic. It has, in fact, already started with the pioneering work in the eighties both by the Göttingen group of G. Schmahl [2] and the Brookhaven group led by Kirz [3]. Biological microscopy with Fresnel zone plates in the so-called water window spectral range is by now a mature technique [4, 5]. In this chapter we restrict ourselves to the more recent developments of hard X-ray microscopy (i.e. photon energies above 5 keV).

In scanning SAXS, resolution in real space and reciprocal space is combined in a hybrid manner. This differs from other approaches which either reach high resolution in reciprocal space—based on diffraction averaging over a large ensemble such as in SAXS—or in real space based on inverting the diffraction pattern, e. g. by CDI. In fact, scanning SAXS with nano- or microbeams combines high resolution in reciprocal space (by analysis of the diffraction patterns and accounting for the available q-range) with resolution in real space on the order of the beam size. The method can hence probe local structures (in reciprocal space) in a range smaller than the beamsize down to the length scale given by the signal-to-noise cut-off. This cut-off depends on the degree of order in the sample and is typically intermediate between length scales of the organelle and the molecular constituents. At the same time, the resolution in real space is limited by the focal spot size. Scanning SAXS has been first demonstrated on biomaterial specimen such as wood and bone [6, 7], and then also for various tissue samples [8], with typical real space resolution values in the range of several microns. More recently, nano-focusing techniques (see Chap. 3) have made it possible to reach spot sizes well below 100 nm, based on reflective (mirrors, waveguides), refractive (compound refractive lenses) or diffractive optics (Fresnel zone plates, mutlilayer zone plates). Nano-focusing with X-rays is reviewed in [9], and also treated in advanced textbooks [10], while biological materials and cells imaged by diffraction and scattering have been previously reviewed in [11].

Direct imaging in real space, with a resolution below the X-ray spot size, is enabled by (far-field) CDI or (near-field) holographic techniques, see Chap. 2. For extended and non-compact objects such as cells, ptychographic CDI, or multi-plane holographic recordings are well suited to solve the phase problem, since the support constraint cannot be used. As in scanning SAXS, contrast is based on the native electron density distribution in hydrated biological cells. However, in contrast to scanning SAXS the specimen is directly imaged in real space. This is possible without any labeling, fixation, or staining. Within certain dose restrictions and for a short time span, CDI is also amenable to living cells [12]. The dose values to observe a time series on the same cell, however, are prohibitively high. For static images, a resolution below 100 nm is possible at synchrotron sources, while a range below 10 nm may be reachable by single ultra-short X-ray pulse using FEL radiation. In fact, the first ptychographic imaging of a cell already achieved 85 nm resolution on low contrast (unstained) bacterial cells at a fluence of 10^7 photons/μm^2 [13, 14]. Extrapolating from these results and assuming a $I \propto q^{-4}$ power law decay for the scattering intensity FEL pulses, delivering 10^{13} photons/μm^2 in a time span below 50 fs, results in a resolution better than 3 nm. Actual experiments, however, are still about a factor of ten above this estimate, see for example [15], who have reported 2D reconstructions of projected electron density with 37 nm (half period) resolution for living bacterial cells.

As we will review here, recent work has now brought scanning SAXS and CDI of biological cells to the level where they can complement optical fluorescence and electron microscopy. In particular, they can 'shed X-ray light' on unlabeled cellular structures in cells by providing an electron density based contrast. Biomolecular assemblies can hence be investigated without slicing and staining, in fixed cells and—with restrictions—also in living cells [16–18]. This new "contrast mechanism" can possibly be useful for a very diverse range of problems. Here we name just a few examples, which may simply be closer to our perspective than others:

- protein network architecture and the impact on cellular mechanics
- protein filament bundling by cross-linking
- force generation in cellular locomotion and muscle contraction
- DNA compaction in the nucleus
- amyloid aggregation.

After the following section, which addresses requirements of beam preparation and sample environments, we will first review scanning SAXS, followed by a section on direct imaging by ptychography and holography. We then address the topic of cellular imaging with FEL, and close with a section discussing multi-scale imaging, from the cellular to tissue level.

15.2   Methods: X-ray Optics and Sample Environment

15.2.1   Focusing Optics and Imaging Modalities

Imaging of cells with hard X-rays has been enabled only by the recent progress in X-ray optics and focusing, required to concentrate photons on a single cell or to specific regions within the cell. Since scattering of biological matter is typically weak, due to low-Z elements involved, beam preparation, cleaning of the beam path by apertures, efficient detection, and background subtraction are major issues. Finally, in situ optical microscopy is required to select scanning regions, perform alignments, and to monitor the cell with respect to radiation damage. The work reviewed here has been performed at synchrotron beamlines which combine these functionalities, notably ID13 of the European Synchrotron Radiation Facility (ESRF) in Grenoble, the cSAXS beamline of the Swiss Light Source, and the Göttingen Instrument for Nano-Imaging with X-rays (GINIX), installed at the P10 coherence beamline of the PETRA III storage ring at DESY in Hamburg. GINIX has been specifically designed for imaging of cells and tissues by holography and scanning SAXS [19]. Furthermore, it is fully compatible with tomography and also correlative optical microscopy [20].

As an example, Fig. 15.1 illustrates the different beam configurations and imaging modalities offered by the modular compound nano-focus optical system of GINIX. Similar modalities have also been realized at other beamlines. The optical system is composed of a high gain fixed curvature Kirkpatrick-Baez (KB) mirror and a probe filtering module, based on cleaning apertures and/or X-ray waveguides. Three different imaging modalities are sketched:

(a) **Scanning SAXS**, or more generally nano-diffraction in the small angle or wide-angle regime depending on detector position. For cells without mineralized or crystallized components, only SAXS signals are observed. Diffraction data are recorded for each scan point, forming a tensor product with two reciprocal space dimensions and two real space dimensions. As in conventional diffraction, a beamstop is required, sampling in real and reciprocal space is not very constrained, and coherence can be very low. The analysis is largely based on models and fitting of diffraction patterns in reciprocal space.

(b) **Ptychography**, i.e. far-field CDI with ptychographic phase retrieval. The slits in front of the KB are closed to achieve full coherence, and the anti-scatter apertures of (a) are replaced by pinholes to compactify the probe, i. e. to absorb the tails of the KB in focal space [21]. The sample is then scanned laterally behind the pinhole with partial overlap between exposures. Oversampling in the detector plane is required, and the beamstop must be sufficiently small or semi-transparent [22, 23] to recover low spatial frequencies.

(c) **Holography**, i.e. near-field phase contrast imaging in a diverging spherical wave, emitted from the exit of an X-ray waveguide. Due to the smaller confinement in the waveguide's guiding channel, the divergence of the exit beam increases with respect to the KB beam, resulting in higher numerical aperture. The waveguide also results

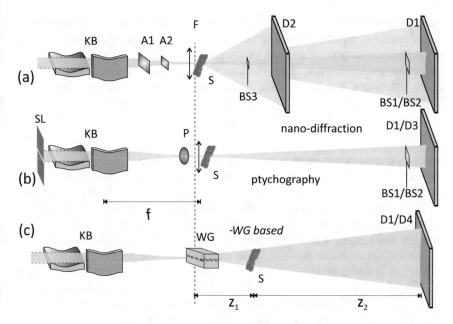

Fig. 15.1 Schematic of different imaging modalities for biological cells, for the example of the GINIX endstation of the P10 beamline (PETRAIII storage ring, DESY). **a** For scanning SAXS the beam is focused by Kikpatrick-Baez (KB) mirrors and cleaned by two successive soft-edge apertures (A1,A2) to cut the KB tails, and the diffraction is recorded by SAXS (D1) and WAXS (D2) detectors, each with respective beam stops (BS). **b** For ptychography, the KB-beam is made fully coherent by closing the entrance slits (SL), and the probe can be compactified by pinholes (P) if necessary for sampling. **c** For holographic imaging, the sample is moved to a defocus position, and after alignment with the pixel detector (D1), a high resolution detector (D4) is used to record the hologram. To increase the numerical aperture and hence the resolution, and to filter the wavefront, an X-ray waveguide (WG) is place into the focal plane of the KB. In this way, artifacts related to the typical wavefront distortions of a KB beam can be avoided. Adapted from [19]

in wavefield and coherence filtering [24]. The holographic pattern is recorded by a high resolution detector, and is treated by phase retrieval in the optical near-field, as discussed in Chaps. 2 and 13. One advantage of (c) over (b) and (a) is that it is a full field technique and that images can be recorded without scanning. Both modalities (b, c) yield the projected electron density of the object.

15.2.2   X-ray Compatible Microfluidic Sample Environments for Cells

X-ray experiments on fixed hydrated and on living cells require a suitable sample environment, which closely mimics physiological conditions and does not further deteriorate the already low signal-to-noise ratios. For living cells, it is indispensable to provide nutrients and control metabolites by a continuous exchange media or buffer. Depending on the experiment, one may choose between cultivation of cells directly in the X-ray chamber or a suitable transfer strategy. Finally, a high throughput of suspended cells or the ability to scan a large number of adhering cells is important to obtain statistically relevant data sets. Another challenge to be faced when studying biological matter, and in particular living cells, is radiation damage. The dose is defined as

$$D = \frac{\mu I_0 T h\nu}{\rho\sigma},\tag{15.1}$$

where I_0 is the primary beam intensity in photons per time, T the exposure time, ρ the mass density of the sample, and σ the exposed area per scan point. Fast scanning, for example, helps to decrease the dose and protect the samples from deterioration.

To this end, the advent of microfluidic devices fabricated by soft lithography has been an enabling event for this research field. Even before X-ray experiments on cells had become a topic of interest, solution SAXS and structure analysis of suspended biomolecular assemblies had already been augmented by the possibility to observe in situ structural dynamics by making use of hydrodynamic focusing in microfluidic devices, as reviewed in [27, 28]. This approach paved the way to then adapt the fabrication processes and to develop X-ray compatible cell culture chambers [18, 25]. The most important requirements for flow chambers that are compatible with both X-ray studies and cell culture are: radiation stability, low absorption, low background scattering, control over the degree of cell adhesion to the materials, biological compatibility, and ease of fabrication. A method to custom-build flow chambers from UV curable adhesive as channel defining material, Kapton® film as radiation resistant window material and optional silicon-rich nitride (SiRN) windows as a substrate for cell growth is shown in Fig. 15.2a step-by-step and the resulting flow chambers are shown in photographs in Fig. 15.2b [25]. In these devices, the cells are constantly supplied with nutrients during the experiments and waste products of the cells are flushed away. Another advantage of the constant flow of liquid is that free radicals produced due to the radiation are flushed away, the sample is permanently cooled and air bubbles are reduced. The device design is very flexible since the fabrication is based on photolithography. Therefore, virtually any channel geometry can be realized and tested.

In addition to the home-built flow chambers, one can use also commercially available microfluidic cell chambers (e.g. ibidi®, Germany) as routinely used for optical microscopy. For X-ray microscopy, modified versions with adapted window materials (coated and uncoated SiRN) are preferred, see Fig. 15.2c. Finally, simple home-built chambers with windows made of cover slides may also do for fixed hydrated cells, if an absorption of X-ray photons by $\leq 200\,\mu m$ glass can be tolerated. In principle, a large set of window materials can be used for the X-ray measurements. The constraints differ for the imaging modality, since 'low background' in SAXS versus holography is associated with different material properties. For example, the residual phase shifts induced by traversing a few mms of Zeonor-® were found to be small enough to still allow for phase contrast imaging of attached cells in the channel by holographic full-field imaging [12]. For scanning SAXS using a nanobeam, on the

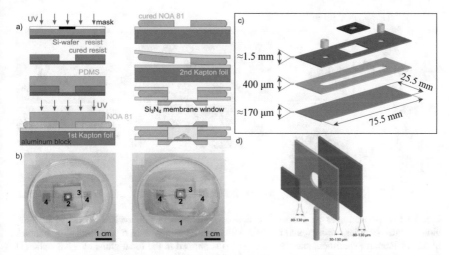

Fig. 15.2 **a** Step-by-step sketch for fabricating X-ray compatible microfluidic devices for cell imaging. The insertion of the SiRN window is optional. **b** Photographs (back/front) of the flow chamber as described in (**a**). 1: PDMS (polydimethylsiloxane) support; 2: SiRN window; 3: flow channel; 4: inlet/outlet. **c** Flow chambers based on commercial microscopy slides with further insertion of a SiRN window. **d** Simple chambers with coverslip windows. **a**, **b** reproduced from [25] with permission from The Royal Society of Chemistry, **c**, **d** from [26]

other hand, the excess polymer material resulted in an increased background level, which was reduced by inserting SiRN windows. As also compiled in a number of recent monographs [29–33], it is now fairly well known which window materials are compatible (and at which thickness) with a given imaging modality. Different window materials (glass, cyclic olefin copolymers [34], polypropylene) are now available for cellular growth.

An entirely different challenge is to create the sample environment for single pulse experiments by FEL. As a new sample has to be delivered to the 'interaction zone' for a subsequent 'hit', tailored microfluidic platforms are required. The strength of microfluidics is the high level of controllability. Flows are typically laminar, enabling exact control of important experimental parameters such as buffer/media conditions, temperature, induction period of reagents to the cells and so on. However, in contrast to synchrotron experiments, window-less flow chambers are necessary, since a single pulse would already destroy window materials. Free jets or microfluidic channels with holes offer of minimum background scatter and full compatibility with the X-ray beam propagating in ultra high vacuum.

Figure 15.3a shows the example of a free microfluidic jet with laminar flow beneath a nozzle with 13 μm exit, which has been used for diffraction on suspensions of biomolecules with micro-focus synchrotron radiation [35], but is also fully compatible with vacuum injection and therefore well suited for FEL experiments, see also [38]. The large speed of the jet, however, results in an elongational shear stress, which would certainly harm most eukaryotic cells, whereas it has been shown

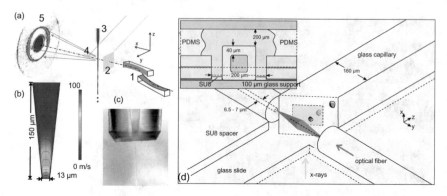

Fig. 15.3 a Microfluidic jet for sample delivery. Jets offer a windowless access to hydrated biomolecular samples, viruses, and bacteria with continuous high-rate replenishment of samples. In the diffraction experiment sketched, the jet (3) is aligned in the focal plane of a KB optic (1), behind a cleanup aperture (2), and the far-field diffraction is recorded by a 2D detector (5), with a miniaturized beamstop (4) blocking the primary beam directly behind the jet. **b** Finite element simulation of the flow field for a jet with diameter of 13 μm and a break-up length of a few mms. High flow verlocity can be used for orientational alignment of biomelecular assemblies, for example of a membrane suspension [35], but elongational and shear strain must be reduced for delivery of cells. **c** Photograph of the nozzle and the jet. **d** Optical stretcher. The sketch shows the integration of the laser system on the microfluidic chip, and the optical axis of the X-ray beam. The capillary is half-cut to give an inside view. The inset shows the central x-y-cross-section of the system. When a cell enters the trap, a highly anisotropic stress profile on the cell contour results in its trapping and stretching [36]. **a–c** adapted from [35], **d** from [37]

that bacteria can indeed be delivered by these jets, withstanding the hydrodynamic stresses [35]. In the meantime, jet technology has advanced tremendously, including electrostatic control, gas confinement and focusing layers [39]. Further, dedicated aerosole injectors have been designed for FEL sample delivery [40]. However, while some of these devices have been demonstrated to be suitable for viruses and smaller particles, the sample delivery for cells is still in its infancy. Living cells kept in a micro-liquid enclosure array, with each element used only once, as presented in [15], are one possible option. Jets with moderate hydrodynamic stress coupled out of microfluidic devices may offer more flexibility and control parameters. Finally, microfluidic channels with micro-sized holes drilled into their enclosure materials may be a third attractive route for further development.

Manipulation of cells in the beam and probing at controlled application of force is an entirely different challenge for the sample environment. In Fig. 15.3d we show the example of an optical stretcher, which was recently used to trap macrophages, and to rotate them in the beam for a tomographic scan [37]. In the stretcher, two opposing and divergent laser beams are used to trap and to stretch cells. The stretcher has been developed by Guck et al. [41] as a tool to studying the elastic properties of biological cells based on video microscopy of their deformed shape functions, since high deforming forces can be applied to biological objects such as cells. In this respect, stretchers offer an added functionality complementary to optical tweezers, which are commonly used to micro-manipulate micron and sub-micron-sized particles. In [37] the experimental capabilities of an optical stretcher as a potential sample delivery system for X-ray diffraction and imaging studies was explored. Even in a non-optimized configuration based on a commercially available optical stretcher system, X-ray holograms could be recorded from different views on a biological cell and the three-dimensional phase of the cell could be reconstructed. By means of high throughput screening, the optical stretcher could possibly become a useful tool, both for SR and FEL studies.

15.3 Scanning Small-Angle X-ray Scattering of Cells

In the past, we have focused our efforts to image cellular components mostly on, first, the cytoskeleton, due to particularly well ordered structures (see, e.g., Figs. 15.4a and 15.7a, b). Second, we have investigated the packing and (de-)compaction of DNA in the nucleus of eukaryotes (see Fig. 15.6a) and nucleoids of bacteria (Fig. 15.9d). The cytoskeleton determines the shape, motility, viscoelastic properties and generated forces of eukaryotic cells and is also a fascinating active soft matter system. It consist mainly of three distinct filament systems, along with associated binding proteins and molecular motors. Actin filaments play an essential role for directed cellular motion via polymerization and depolarization in lamellipodia and filopodia, form the stabilizing actin cortex and, together with myosin motors, stress fibers as the basic building blocks enabling the contractability of muscle cells. Microtubules contribute a system of 'tracks' within the cell, along which motor proteins transport cargo. They are also instrumental in cell division as they pull the chromosomes into the daughter cells. Intermediate filaments, such as vimentin or keratin, finally contribute greatly to passive mechanical properties of cells and protect the cell from destruction by heavy impact.

The DNA, which contains the genetic information of an organism, is densely packed in the nucleus or nucleoid and needs to be unpacked in a highly controlled manner for protein production. This apparent contradiction is solved by nature using a strictly hierarchical way of packing. To study the architecture and interactions of biomolecular assemblies, biophysicists have since long used SAXS. As the model systems get more and more complex, however, SAXS data interpretation is often impeded, in particular in the presence of large heterogeneity and the lack of molecule specific information available. Scanning SAXS, in particular when combined with visible light microsopy, addresses this challenge by providing additional real space constraints. Since scanning SAXS patterns average only locally over a small volume, a complete powder average is no longer obtained. For this reason, the anisotropy of the SAXS pattern can be measured, providing further clues. Of course, the indirect modeling of diffraction signals could be made obsolete altogether, if direct inversion of a coherent diffraction pattern was possible. However, this is to date not possible at the resolution which is typically achieved by SAXS. Apart from coherence and

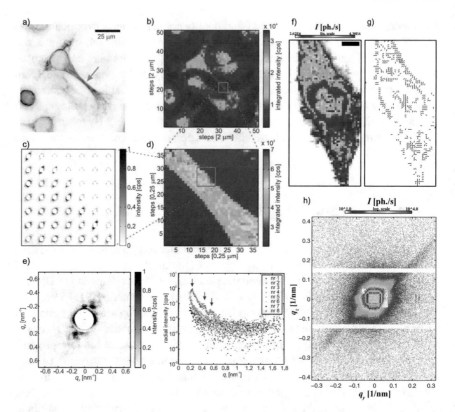

Fig. 15.4 Scanning SAXS with a nano-focused beam. **a** Inverted gray scale fluorescence micrograph of a SW13 cell with keratin K8/K18; the red arrow points to a particularly dense keratin bundle structure. **b** Corresponding X-ray dark field image overview image. **c** Composite image of individual SAXS patterns corresponding to the region in the red box in **b**. **d** Detailed X-ray dark field image of ROI (red box in **b**) recorded at smaller step size. **e** Single 2D diffraction pattern which clearly shows the orientation of the keratin bundles by the anisotropy of the signal (left) and integrated 1D intensity curve (right). Segments 1 and 5 align with the anisotropy and show distinct modulations of the signal. **f** X-ray darkfield scan of muscle induced human mesenchymal stem cell (hMSC), along with **g** the corresponding analysis of the anisotropy of the diffraction pattern revealing for example local orientations of networks of cytoskeletal components. **h** Single diffraction pattern of the region marked in (**g**). **a–e** adapted from [17], **f–h** from [26]

sampling issues which have to be met, inversion of data at higher momentum transfer and at low signal-to-noise level is much more difficult than just fitting a decay. Most importantly, model-based interpretation of SAXS data makes sense despite the fact that only one 2D projection is available. Contrarily, the information of the 2D projected electron density in real space, obtained for a single projection angle becomes useless at small scales without 3D tomography (which is often times technically not feasible).

For this reason, we regard scanning SAXS as an indispensable tool in X-ray microscopy of biological cells, even if the first SAXS studies of individual cells (disregarding tissues and biomaterials) were only published in 2012 [17, 42]. One of the two papers used SAXS only as a complement to a ptychographic experiment to circumvent the stringent oversampling conditions, in search for structural clues on the condensation of DNA in the bacterial nucleoid [42]. *Deinococcus radiodurans* was chosen as a model system for the debated mechanisms of radiation damage repair [14, 42, 43]. Cells were prepared by rapid vitrification followed by freeze drying, and the SAXS signal was found to obey a power law decay with $q^{-\nu}$ with $\nu \simeq 3.2 - 3.7$, depending on the scan. The cross-over of the power law decay to the noise floor, which can serve as a resolution criterion, was observed at $q = 0.188 \text{Å}^{-1}$. The first eukaryotic cells were studied at the same time by experiments dedicated to the development of scanning SAXS on cells. For this purpose, SW13 epithelial cells with a pronounced keratin K8/K18 network were chosen in order to highlight one of the three components of the cytoskeleton, see Fig. 15.4a–e. The cells were grown on SiRN windows, which are excellent substrates for cell growth and virtually transparent for X-rays, and subsequently chemically fixed, plunge-frozen and freeze-dried. This ensured a high electron density contrast between the cellular material and the surrounding air, which facilitated the recording in the starting phase. Cells were then imaged by scanning SAXS with a small beam ($140 \times 110 \, \text{nm}^2$ and $200 \times 125 \, \text{nm}^2$) and with small step sizes (50–2000 nm). For illustration, an X-ray dark field image, where in each position all scattering is integrated and plotted on a color scale, is shown in Fig. 15.4b, a detail in Fig. 15.4d. Example diffraction patterns are shown in Fig. 15.4c and at larger magnification in Fig. 15.4e. The signal was then integrated azimuthally to obtain 1D $I(q)$ curves and, despite the small sample volume probed (beam size multiplied by sample thickness), distinct diffraction peaks were observed [17].

Pronounced streaks in diffraction patterns of *Dictyostelium discoideum* were explained with formation of fiber bundles in the acto-myosin contractile ring, based on comparing the ring like occurance of these features in SAXS scan to the typical contractile ring observed in fluorescence microscopy [16]. Modulations in the streaks were modeled based on a simple fiber scattering model [16, 32]. The strength and anisotropy of diffraction patterns were compared between naive human mesenchymal stem cells and differentiated stem cells, indicating that order of the cytoskeleton of stem cells increases during the differentiation process [26]. Modulations in the azimuthally integrated SAXS data could even be quantified by rendering the qualitative models into fitting procedures. Thus, the packing geometry, filament diameter and center-to-center distance and bundle diameter could be determined for keratin bundles in intact cells. The results correspond exactly to those derived from scanning electron microscopy, albeit without the need for slicing the cell [44].

Yet another highly ordered state of actin, namely hair cell stereocilia from the inner ear, were studied in [45]. These cell protrusions act as force sensors and enable hearing by transforming mechanical bending into a neuronal signal. They are filled with numerous parallel, densely packed actin filaments. Stereocilia were 'stamped' on coated SiRN windows, chemically fixed, stained for actin, and freeze-dried. Scan-

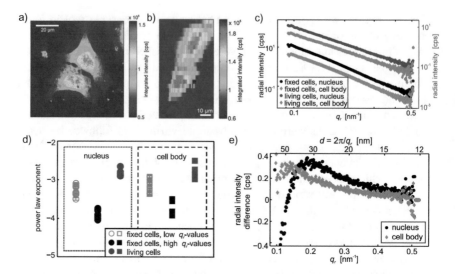

Fig. 15.5 **a** Darkfield image of a chemically fixed, hydrated cell and **b** of a living cell. Note that in (**b**) several scan lines were skipped in order to account for the severe effects of radiation damage in this case. **c** 1D radial intensities plotted against the q values; note that the data for living cells (blue) are scaled by a factor of 10 for better visibility. **d** Power law exponents of chemically fixed (black) and living (blue) cells. The latter are consistently larger, indicating nanoscale changes. **e** Difference signal; data for fixed cells subtracted from data for living cells. Thus, values below 0 show structures emerging upon fixation, whereas values above 0 hint at structures that were destroyed by the fixation process. Reprinted with permission from [18], Copyright (2014) by the American Physical Society

ning SAXS experiments revealed an extremely high order within the stereocilia, but also considerable spatial heterogeneity in the translational and orientational structure.

After these early studies proved successful regarding interpretation and signal level, the next step was to investigate chemically fixed and living cells in hydrated state. Interestingly, distinct differences in the power-spectra of living and chemically fixed cells (see Fig. 15.5) were observed, pinpointing to both emerging and destroyed nanostructures upon chemical fixation [18]. This study illustrated very clearly the advantage of X-ray nano-imaging over other imaging methods. So far, X-ray imaging is the only way to directly compare fixed (or labeled, stained) and untreated whole cells since extensive sample preparation is not *necessary* for imaging. The results become particularly important now that fluorescence microscopy, where chemical fixation is a routine method, reaches well into the affected length scales (see top axis in Fig. 15.4e). In further studies, scanning SAXS was carried out on cryogenically preserved (i. e. vitrified) cells [16], which offers a larger range of dose before damage is observed. A direct comparison of the power law exponents for the decaying 1D SAXS curves shows that the results from cryo-preserved samples can be reproduced at room temperature, however, the primary beam needs to be attentuated and the exposure time decreased [46].

Following the first proof-of-concepts of scanning SAXS for biological cells, the optimization of imaging and analysis capabilities became important. To this end, several issues of optics, instrumentation and analysis were addressed: (i) choice of focus and influence of focal size on the SAXS data quality, (ii) rapid scanning using continuous movements and synchronized detector read-out, (iii) improvements of alignment procedures and tools based on *in situ* optical microscopy, (iv) optimized pixel detector read out and control software [47], (v) optimized beam preparation and suppression of tail scattering, for example, of the KB mirror system, (vi) specially fabricated semi-transparent beam stops [22], as well as (vii) the completion of a versatile toolbox for scanning SAXS which was made publically available [48, 49]. In this way, it now has become possible to compute structural observables in a fast and automated way, based on empirical data descriptors. Algorithms for semi-automatic quantification of the diffraction patterns include analysis of anisotropy parameters by automized fitting of ellipsoids [17], decomposition into principal components [26], the automized power law fitting [18, 44, 50, 51] and the computation of cumulants [50] to describe the azimuthally averaged structure factors for different regions-of-interest within the cells. In this way, empirical analysis has become possible also for extremely large data sets. At the same time, model based analysis of diffraction patterns, based on fiber models with free fitting parameters [16, 44], or sarcomere models for muscle cells [49] has been used. The effects of different sample preparation methods on the detected structure factor, including the freeze-dried, chemically fixed, frozen hydrated and living states was investigated more systematically. Radiation dose effects were also studied, and dose was precisely quantified by including ptychographic reconstruction of the probing beam [42]. In several studies, scanning SAXS and CDI was combined on the same cell, including combinations of SAXS with holography [51], and SAXS with ptychography [42, 44, 45].

It is not possible (yet) to record "movies" of living cells using scanning SAXS. Thus, to obtain temporally resolved information, an indirect approach by recording "snap shots" of cell has to be taken. An example is shown in Fig. 15.6, where the stage in the cell division cycle was determined by visible light bright field microscopy and the nuclei of these cells were recorded by scanning SAXS. Thus, it is possible to relate the SAXS signal and the exponents derived to the cell cycle time point (Fig. 15.6b) and derive information of the compaction and decompaction of the DNA.

The best choice of real space and reciprocal space resolution in scanning SAXS deserves some special consideration, and a compromise has to be found for the specific application at hand. For example, a tight nano-focus leads to higher divergence and often also pronounced beam tails or streaks which can easily compromise the low-q signal in SAXS pattern. If this is to be prevented, the beam size must not be too small, typically in range of a few microns rather than in the range of a few hundred nanometers. Figures 15.7 and 15.8 show examples of micro-SAXS for cardiac tissue cell, where a clean recording of the sarcomere diffraction pattern at low-q required relaxation of real-space resolution to the micron range. Figure 15.7 shows that the corresponding real space images of neonatal rat cardiac tissue cells can still be recognized and correlated to the fluorescence micrographs, while at the same time the quality of the SAXS signal was improved, enabling automated decomposition of

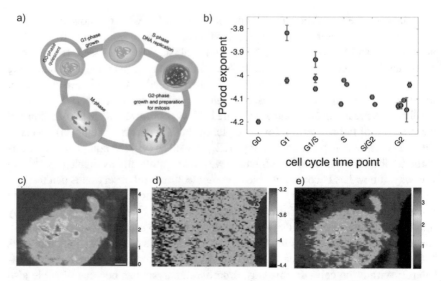

Fig. 15.6 a Schematic of the cell and nucleus division cycle of a eukaryotic cell. **b** Porod exponents of data taken from cells in different stages of the cell dividing cycle. An increase of the exponent towards 3.8 coincides with the decompaction of the DNA as the cell grows. The subsequent decrease is related to the compaction as the DNA is duplicated. Spatial representation of the DNA compaction state, **c** dark field image, **d** Porod exponent and **e** Porod constant. Interestingly, the Porod exponent is fairly homogeneous throughout nucleus and cytoplasm indicating a similar degree of heterogeneity in the structures. The Porod constant, by contrast, highlights dense reagins in the nucleus, presumably nucleoli. Reprinted with permission from [46], Copyright (2016) by the American Chemical Society

scattering patterns into principal components [26]. Using the micro-SAXS approach and hydrated adult cardiomyocytes, the myofibril diffraction signal reflecting inter-filament distances of thick filaments (myosin) and thin filaments (actin) could indeed be observed from selected regions within a single cell, see Fig. 15.8.

15.4 Coherent X-ray Imaging of Cells

15.4.1 Ptychography

In scanning SAXS, the real-space resolution is determined by the focal spot size. Contrarily, super-resolution below spot size can be achieved by CDI, where an over-sampled far-field diffraction pattern is inverted by solving the phase problem. This was first achieved only under the restrictive assumption of a compact object with known support, fully coherent and plane wave illumination [54, 55]. The restriction to a 'compact object' was lifted by ptychographic CDI (pCDI), which uses a compact probe and partial overlap between illuminations of adjacent scan points

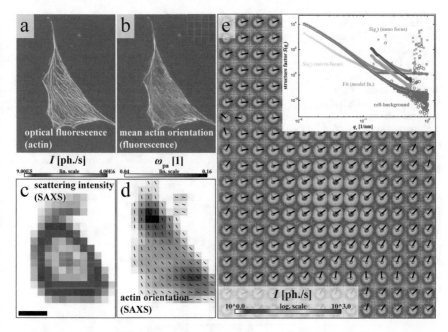

Fig. 15.7 Scanning SAXS with a micro-focused beam. **a, b** Visible light fluorescence micrographs of a freeze-dried neonatal rat cardiac tissue cell with labeled actin cytoskeleton. The most significant filaments were identified by the filament sensor algorithm [52], see (**a**) (yellow lines), and a mean orientation angle was computed for all segmenting blocks with a blocksize equal to the stepsize of the SAXS scan, see (**b**) (red lines). **c** X-ray dark field of the SAXS scan. **d** PCA results on each diffraction pattern showing the fiber orientation as black lines. The degree of anisotropy was quantified by a unitless order parameter ω_{PA} as further detailed in [26, 50]. **e** Composite image showing the diffraction patterns with respect to their relative recording position. Black and gray arrows show the direction of the major and minor principal component axis. Diffraction patterns can be integrated azimuthally, see (**e**, inset), indicating the local structure factors, which can then be described mathematically by different fitting models (pink lines). Scale bar: 40 μm. From [50]

to phase the diffraction pattern [56–58]. Exploiting the constraint of separability for phasing, both an unknown object o and an unknown probe p can be reconstructed [58], so that pCDI became applicable not only to extended samples, but also to non-idealized illumination functions (probes), including partial coherence [59]. A first application of ptychography to biological cells was shown in [14] for the gram-positive bacterium *Deinococcus radiodurans* [14]. Images of the projected mass density in freeze-dried preparations were reconstructed with a phase resolution below 0.01 rad up to a half-period resolution of 85 nm, at a relatively low fluence of 6.6×10^7 photons/μm^2. Subsequent studies for the same bacteria, with substantially improved optics and phase retrieval extended this to 3D tomographic imaging [42, 43], see Fig. 15.9, and to cryogenically fixed cells [32]. By poviding mass density values, the results contributed to the long standing debate on DNA compactification in bacterial nucleoids [43]. In [60], malaria infected red blood cells were imaged by

Fig. 15.8 Micro-SAXS results obtained for adult murine cardiomyocytes (CM). **a** Isolated adult CM from rat (optical brightfield image). **b** The diffraction signal recorded in each scan point reflects the acto-myosin structure of myofibrils. **c** The extracted d-spacing of the acto-myosin lattice, and **d** the filament orientation, determined by automated fitting. **e** Different physiological conditions of the heart and fixatives yield different interfilament periods. The distribution functions (histograms) of the structural parameters is reflected by the violin plots. Adapted from [53]

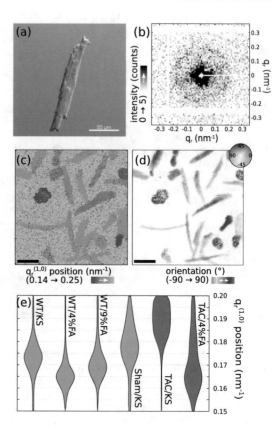

ptychographic tomography. In [61], quantitative phase and amplitude pCDI reconstructions were used to segment features and to localize mitochondrial constriction sites in mouse embryonic fibroblasts. Frozen-hydrated cells, which tolerate more dose than room temperature samples, and thus offer higher resolution, were imaged in 2D [62, 63], and in 3D by X-ray nano-tomography in [64]. Correlative cellular ptychography with functionalized Fe nanoparticles was shown in [65].

15.4.2   Holography

The downsides of both original CDI and ptychography are: (i) stringent oversampling and coherence conditions, (ii) comparatively slow convergence of the reconstruction, (iii) rather long data acquisition for large field of view (no zooming out), and (iv) the high radiation dose. A major problem of applying CDI and pCDI to biological cells is in fact the low contrast in the hydrated state, and the correspondingly high radiation dose and associated radiation damage [66]. In fact, the typical radiation doses of CDI are in the range of $10^7 - 10^9$ Gy, creating a need for cryogenic conditions to warrant

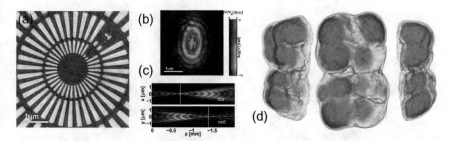

Fig. 15.9 Ptychographic reconstructions with simultaneous reconstruction of probe and object. **a** Phase image of a Siemens star test pattern with 50 nm lines and spaces, and **b** complex-valued field of the probe (focused beam) in the sample position. **c** Cuts along the optical axis of the propagated probe indicating significant astigmatism in the X-ray optics (Fresnel zone plate with upstream focusing). The results show that even with an astigmatic probe ptychography can yield faithful object reconstructions. **d** Tomographic reconstruction of *Deinococcus radiodurans* bacteria. Dense regions are attributed to DNA-rich bacterial nucleoids, obtained with the same setup at 6.2 keV photon energy (cSAXS beamline, Swiss Light Source). From [42]

structure preservation. The theoretical dose-resolution curve is characterized by an already very steep algebraic increase with an exponent $3 \leq \gamma \leq 4$, as derived for the case of Fraunhofer far-field diffraction [66]. As an example, CDI was applied to hydrated bacterial cells ('wet' CDI), demonstrated at about 30 nm resolution (stated as half-period throughout this work), but at the 'cost' of 10^8 Gy [67].

These shortcomings could possibly be solved by holographic X-ray imaging, employing a highly coherent divergent cone for illumination and geometric adjustable magnification, and providing a much directed encoding of phase information. As shown in [12], X-ray waveguides can provide the clean wavefronts required for high resolution inline holography. While this approach has not yet reached the resolution of pCDI or CDI, it was proven experimentally to be very dose efficient. A dose advantage over CDI was also found by numerical simulations [68]. Furthermore, by variation of the object position, it becomes easily possible to zoom in an out and thus record multiple magnifications of a sample. Combined studies by holography and ptychography with scanning SAXS were published in [51] and [42, 44, 45], respectively. Holography, in particular, allows for low dose overviews recorded before identification of regions-of-interest for scanning SAXS. Figure 15.10 illustrates the state-of-art for holographic tomography of cells [69].

By now the initially rather exotic area of CDI of cells has become more and more established. However, even ptychography, which is already more common than holography, has not yet spread into the biological community, for which many forms of microscopy are available. Contrarily, for tomography of biological materials and tissues it has found convincing use and for this purpose, the holographic approach has become more widespread than ptychography, see [70] for a state-of-the art application to human neuronal tissue. One very promising application of X-ray imaging on cells is the combination with other microscopy methods, as decribed in the next section.

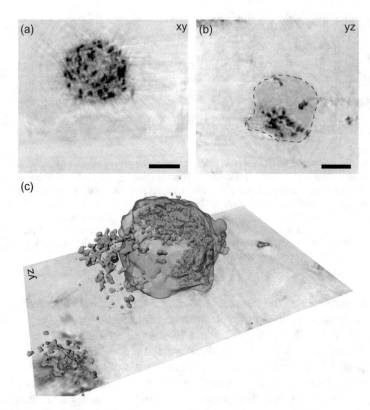

Fig. 15.10 Holo-Tomography reconstruction of a macrophage, labeled with barium sulfate and osmium tetroxide. **a** Virtual slice through a plane perpendicular to the tomographic axis. The cell with its internal contrast particles, the Petri dish and the resin used for embedding can be distinguished based on grey shades which are proportional to electron density. **b** Virtual slice coplanar to a projection direction. **c** 3D rendering of the dataset, showing barium particles (green) and the cell volume (half-transparent blue). Phase retrieval was based on the CTF approach with 4 distances. Scalebars denote $5\,\mu$m. From [69]

15.5 Correlative Microscopy

Correlative X-ray and optical microscopy may help to overcome some of the persisting challenges in X-ray data analysis of scanning SAXS and will at the same time provide information not accessible by employing just one or the other method. In particular, correlative optical fluorescence microscopy can help to formulate models and constrain parameters, by providing additional information on specifically labeled biomolecules. In the absence of such information, previous studies of biological cells by scanning SAXS were mostly analyzed in terms in empirical, model-free data analysis, with only few exceptions [16, 44]. As shown in Sect. 15.3, already without a scattering model, a wealth of parameters can be extracted from the diffraction patterns in an automated manner, for example total diffraction intensity (darkfield), differential phase contrast, second moments of the scattering distribution, power-law exponents, or anisotropy parameters based on fitting of the 2D scattering patterns or principal component analysis (PCA) [17, 50]. By inspection of the diffraction patterns and the real-space maps, it seems plausible to attribute diffraction signals in some locations to the presence of filamentous proteins of the cytoskeleton or DNA in the nucleus. However, such conclusions need confirmation by optical fluorescence microscopy, at highest possible resolution. With this information at hand, the local diffraction patterns can be interpreted and analyzed, providing in the end much more information than either the optical image or the X-ray data alone.

In [20] a correlative microscopy approach for biological cells and tissues was proposed, which combines holographic X-ray imaging, X-ray scanning diffraction, and stimulated emission depletion (STED)-microscopy as a super-resolution optical fluorescence technique. All three imaging modalities were integrated into the same dedicated synchrotron nano-focus endstation GINIX at the P10 beamline of the PETRAIII storage ring (DESY, Hamburg). With this setup, both labeled and unlabeled biomolecular components in the cell can be imaged in a quasi-simultaneous scheme, exploiting the complementary contrast mechanisms of X-ray microscopy and optical fluorescence. This was demonstrated for heart tissue cells with a fluorescently labeled actin cytoskeleton. Micrographs of all three modalities were registered. The principal directions of the anisotropic diffraction patterns were found to coincide to a certain extent with the actin fiber directions. Further, actin filaments bundles were also recognizable in the phase map reconstructed from holographic recordings. We expect that the co-localization constraints provided by such correlative microscopy approaches will be instrumental for the formulation of advanced diffraction models, to fully exploit the data which is becoming available (Fig. 15.11).

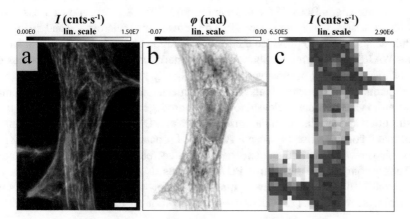

Fig. 15.11 Correlative microscopy. Neonatal cardiac tissue cell with labeled actin, imaged in three different modalities. **a** STED micrograph. Scale bar: 5 μm. **b** X-ray phase reconstruction. **c** X-ray dark field map of the cell obtained by scanning SAXS. From [20]

15.6 From Cells to Tissues

For many biological functions it is important to integrate structural aspects on scales ranging from the single cell to the entire organ. Heart contractility as one of the most important physiological functions is a perfect example of how function relies on an intricate molecular and cellular architecture. The classical research field which addresses the way that cells form a functional tissue is histology, which combines sophisticated sectioning with optical or electron microscopy. While the cytoarchitecture can thus be imaged in 2D, conventional histology lacks the capability to probe the tissue structure in full 3D. Furthermore, the high resolution molecular structure as revealed by electron microscopy can only be carried out in very small volumes, and structural variations within the tissues. Important and functionally relevant structural properties, such as for example the 3D vector field of myofibril orientation in heart, cannot be suitably assessed by conventional histology. Furthermore, different regions within the heart exhibit variations of the intrinsic sarcomere structure. For example, the acto-myosin lattice spacing near the ventrical wall may differ from the outer perimeter of the heart, as observed for mouse heart [49], see Fig. 15.12. Using the scanning SAXS approach it becomes possible to probe molecular orientation of heart tissue, combining the required real space resolution with molecular sensitivity by diffraction [49]. Extending this to a series of slices, or - as an alternative - to X-ray darkfield tomography [71], one could possibly probe the entire 3D assembly of myofibrils. In this way, the multi-scale challenge of mapping molecular structures and orientation over length scales of an entire heart may become possible in future.

Scanning small and wide-angle X-ray scattering (SAXS/WAXS) and X-ray fluorescence (XRF) with micro-focused synchrotron radiation have also been used in [72] to study histological sections from human brain tissue, notably of the midbrain and of substantia nigra. Both XRF and scanning SAXS/WAXS were shown to visualize tissue properties, which are inaccessible by conventional microscopy and histology. While scanning SAXS provided the local orientation and ordering of myelin structure, WAXS provide the distribution of cholesterol crystallites, and XRF maps of transition metals. All observables were intrinsically registered (aligned) since they were acquired in the same scan. Transition metals and more generally elemental distribution has become a relevant topic for neurodegeneration, for example the iron distribution and speciation in Parkinson's disease (PD). In [72], variations in transition metal concentration between a PD and CTR patient were observed. The XRF analysis showed increased amounts of iron and decreased amounts of copper in the PD tissue compared to the control. PD tissue scans also exhibited increased amounts of crystallized cholesterol. However, as only tissues from one PD patient and one control were available, [72] can only serve as a proof-of-feasibility.

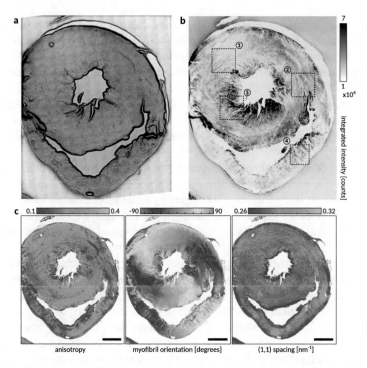

Fig. 15.12 Scanning SAXS of mouse cardiac muscle. **a** Optical micrograph of a histological section. **b** X-ray darkfield image, i.e. the integrated scattering intensity. **c** Multiple scattering parameters extracted in a fully automated manner from the diffraction patterns of the scan: (left) anisotropy of the scattering resulting from the $(1, 1)$ reflection from the acto-myosin lattice, (center) the corresponding myofibril orientation, and (right) the mean position of the reflection along q_r, as obtained from a Gaussian fit to the structure factor $I(q_r)$ with a background model. Scale bar: 1mm. From [49]

15.7 Outlook: FEL Studies of Cells

The advent of highly brilliant pulsed X-ray radiation from free electron lasers (FEL) has opened up a novel route to high resolution imaging by short femto-second (fs) pulses, before radiation damage takes place [73–75]. Ultra short pulses not only enable highest temporal resolution for example for pump-probe experiments, but also static single pulse imaging unaffected by any (Brownian) motion. If the structure is recorded by ultra-fast elastic scattering *before* changes occur due to multiple ionization, this holds promise to record sharp still images of extremely high resolution, unaffected by structure deterioration due to radiation damage. This so-called "diffract and destroy" principle was initially coined for single molecule CDI envisioned for FEL, but can also be applied to colloids, viruses or (small) entire cells. To this end, feasibility of imaging living cells by ultrafast CDI was discussed in [76], based on numerical simulations of the interaction of FEL pulses (10–100 fs) with biomolecular matter. It was concluded that subnanometer resolutions could be reached on micron-sized cells at fluences of 10^{11}–10^{12} photons/μm^2. For mimivirus [77] and small bacteria of *microbacterium lacticum* [15], single pulse CDI has been indeed been demonstrated by now. However, resolution was much lower, i.e. 37 nm (half-period) for the bacterial cells.

Furthermore, single-shot CDI of large extended objects such as eukaryotes is in practice impeded by oversampling restrictions, the beam stop induced missing data problem [22], and the lack of a priori information (support). This is well illustrated by the FEL experiments on freeze dried cells presented below in Fig. 15.13. Notably, current pixel detector technology restricts the field of view to around $d \simeq 1\mu m$ for hard X-rays, which is prohibitive for eukaryotic cells. At the same time, FEL imaging of cells is limited to 2D. 3D imaging by serial shots with randomly sampled projections is possible for identical particles, but not for most cell types. A serial implementation of cellular imaging with high throughput of cells would nevertheless give a useful distribution of 2D views for a given cell type and state.

How can the maximum support cross section d resulting from support/ oversampling constraints be increased? As we have $d = \lambda z/2p$, where p is the detector pixel size, z the distance, and λ the wavelength, it seems reasonable to increase the wavelength for eukaryotic cells, which in addition also increases the scattering intensity. To this end, single pulse CDI experiments on freeze dried cells were also carried out in the water window spectral range, using FEL radiation of $\lambda = 8$ mn at the FLASH facility (DESY, Hamburg). While the signal and hit rate were sufficient, see Fig. 15.13, several restrictive conditions have impeded reconstruction: (1) Insufficient degree of coherence: The global degree of coherence of the third harmonic was determined to be only around 0.4 [78]. (2) Insufficient data at low q: The missing data due to overly sized beam stops and beam stop holders leads to unconstrained low frequencies. (3) Insufficient sampling: The large fields of view needed for the adherent cells were still not compatible with the oversampling conditions. From this attempt and other examples, one can learn that the strategy to collect and analyze data for eukaryotic cells at FEL has to be revised.

The solution to this challenge could be a hybrid approach based on combining two separate paradigms of structural analysis: imaging and diffraction. Low and medium resolution projected electron density should be assessed in real space, based on near-field (holographic) phase retrieval, while high resolution (molecular) Fourier components should be assessed by model based analysis of diffraction data, relaxing the restrictive conditions for inversions. Importantly, by holographic (near-field) imaging—with the sample in the defocus position of the nano-focused FEL—even extended objects (without support constraints) can be reconstructed from single shot data. To this end, iterative algorithms using mild constraints can be readily employed in the near-field regime, such as negativity of the phase, unit amplitude (pure phase object) or sparsity, as well as combinations thereof. This hybrid approach could be implemented for example at the MID instrument at XFEL, using a compound refrac-

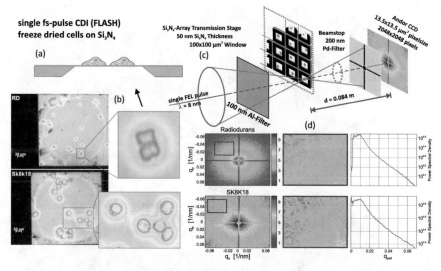

Fig. 15.13 Single pulse CDI experiments on cells, using soft X-ray FEL radiation at BL2/FLASH, DESY ($\lambda = 8$ mn), at experimental parameters as reported in [78–80]. **a** Schematic of freeze dried cells attached to the multi-window SiRN array (576 100 × 100 μm squared windows) with membrane thickness of 100 nm. The system is designed such that a single shot at a given micro-chamber will leave the other chambers with cells intact. **b** Optical microscopy of the *D. radiodurans* (top) and SK8K18 (human epithelial) cells (bottom). **c** Setup with the focused beam, multi-window sample holder, and the thin foils used to suppress the 3rd harmonic at $\lambda_{3rd} = 2.66$ nm by a 200 nm free-standing Pd-filter, installed directly in front of the CCD detector. To avoid radiation damage on the CCD detector, a beam stop blocks the central pixels. **d** Single pulse diffraction patterns of *D. radiodurans* bacteria and human epithelial cell SK8K18. The magnification shows that the oversampling criterion is fulfilled at this wavelength. The power spectral density shows a dynamic range of more than three orders of magnitude, and hence a sufficient signal. However, the existence of a beam stop and the lack of compact support impedes object reconstruction. From D.-D. Mai et al., unpublished

tive lens system focusing to 50 nm. At the same time, and in addition to the high resolution CCD or CMOS detectors needed for this imaging modality, a wide angle pixel detectors could be used to record the far field scattering intensity outside of the central diverging radiation cone, i.e. covering Fourier components corresponding to scales below 50 nm. Figure 15.14 shows a sketch of an FEL holography experiment. The sample is placed in controlled defocus position z_1 behind the focal plane of the CRL. The direct beam traverses a pixel detector with a hole, and reaches the high resolution detector at large distance where the in-line hologram is recorded.

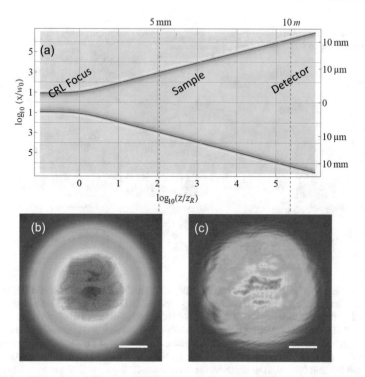

Fig. 15.14 **a** Schematic of divergent FEL beam, calculated in natural units (Rayleigh length z_R, waist w_0) for 8 keV photon energy, and beam focusing by 50 compound refractive lenses (CRL) made of Beryllium ($R = 50\,\mu m$, $f = 94.6\,mm$, $w_0 = 45.9\,nm$); parameters are adapted to MID instrument of XFEL. Data can be recorded in two ways: (i) single shot far-field diffraction patterns recorded by a pixel detector (AGIPD) at 10 m distance, and (ii) holographic recordings by a high resolution sCMOS detector with the sample in a defocus position. **b** Simulation of the beam intensity at $z_1 = 5\,mm$ behind the CRL focus, overlaid with the cell phantom. Scale bar 2.5 μm. **c** Simulated hologram at 10 m distance. Scale bar 1 mm

References

1. de Jonge, M.D., Vogt, S.: Hard X-ray fluorescence tomography—an emerging tool for structural visualization. Curr. Opin. Struct. Biol. **20**(5), 606–614 (2010). https://doi.org/10.1016/j.sbi. 2010.09.002. https://doi.org/10.1016/j.sbi.2010.09.002
2. Schmahl, G., Rudolph, D., Niemann, B., Christ, O.: X-ray microscopy of biological specimens with a zone plate microscope. Ann. NY Acad. Sci. **342**(1), 368–386 (1980)
3. Jacobsen, C., Kenney, J., Kirz, J., Rosser, R., Cinotti, F., Rarback, H., Pine, J.: Quantitative imaging and microanalysis with a scanning soft X-ray microscope. Phys. Med. Biol **32**(4), 431 (1987)
4. Larabell, C.A., Nugent, K.A.: Imaging cellular architecture with X-rays. Curr. Opin. Struct. Biol. **20**(5), 623–631 (2010)
5. Schneider, G., Guttmann, P., Heim, S., Rehbein, S., Mueller, F., Nagashima, K., Heymann, J.B., Müller, W.G., McNally, J.G.: Three-dimensional cellular ultrastructure resolved by X-ray microscopy. Nat. Meth. **7**(12), 985 (2010)

6. Lichtenegger, H., Müller, M., Paris, O., Riekel, C., Fratzl, P.: Imaging of the helical arrangement of cellulose fibrils in wood by synchrotron X-ray microdiffraction. J. Appl. Cryst. **32**(6), 1127–1133 (1999)

7. Rinnerthaler, S., Roschger, P., Jakob, H., Nader, A., Klaushofer, K., Fratzl, P.: Scanning small angle X-ray scattering analysis of human bone sections. Calcified Tissue Int. **64**(5), 422–429 (1999)

8. Bunk, O., Bech, M., Jensen, T., Feidenhans, R., Binderup, T., Menzel, A., Pfeiffer, F.: Multimodal X-ray scatter imaging. New. J. Phys. **11**(12), 123,016 (2009)

9. Attwood, D., Sakdinawat, A.: X-rays and Extreme Ultraviolet Radiation: Principles and Applications. Cambridge University Press (2017)

10. Stangl, J., Mocuta, C., Chamard, V., Carbone, D.: Nanobeam X-ray Scattering: Probing Matter at the Nanoscale. Wiley (2013)

11. Hémonnot, C.Y.J., Köster, S.: Imaging of biological materials and cells by X-ray scattering and diffraction. ACS Nano **11**(9), 8542–8559 (2017). https://doi.org/10.1021/acsnano.7b03447. https://doi.org/10.1021/acsnano.7b03447

12. Bartels, M., Krenkel, M., Haber, J., Wilke, R., Salditt, T.: X-ray holographic imaging of hydrated biological cells in solution. Phys. Rev. Lett. **114**(4), 048,103 (2015)

13. Giewekemeyer, K., Neubauer, H., Kalbfleisch, S., Krüger, S.P., Salditt, T.: Holographic and diffractive X-ray imaging using waveguides as quasi-point sources. New J. Phys. **12**(3), 035,008 (2010)

14. Giewekemeyer, K., Thibault, P., Kalbfleisch, S., Beerlink, A., Kewish, C., Dierolf, M., Pfeiffer, F., Salditt, T.: Quantitative biological imaging by ptychographic X-ray diffraction microscopy. Proc. Natl. Acad. Sci. U.S.A. **107**, 529 (2010)

15. Kimura, T., Joti, Y., Shibuya, A., Song, C., Kim, S., Tono, K., Yabashi, M., Tamakoshi, M., Moriya, T., Oshima, T., et al.: Imaging live cell in micro-liquid enclosure by X-ray laser diffraction. Nat. Commun. **5**, 3052 (2014)

16. Priebe, M., Bernhardt, M., Blum, C., Tarantola, M., Bodenschatz, E., Salditt, T.: Scanning X-ray nanodiffraction on dictyostelium discoideum. Biophys. J. **107**(11), 2662–2673 (2014)

17. Weinhausen, B., Nolting, J.F., Olendrowitz, C., Langfahl-Klabes, J., Reynolds, M., Salditt, T., Köster, S.: X-ray nano-diffraction on cytoskeletal networks. New J. Phys. **14**(8), 085,013 (2012)

18. Weinhausen, B., Saldanha, O., Wilke, R.N., Dammann, C., Priebe, M., Burghammer, M., Sprung, M., Köster, S.: Scanning X-ray nanodiffraction on living eukaryotic cells in microfluidic environments. Phys. Rev. Lett. **112**(8), 088,102 (2014). https://doi.org/10.1103/PhysRevLett.112.088102

19. Salditt, T., Osterhoff, M., Krenkel, M., Wilke, R.N., Priebe, M., Bartels, M., Kalbfleisch, S., Sprung, M.: Compound focusing mirror and X-ray waveguide optics for coherent imaging and nano-diffraction. J. Synchrotron Radiat. **22**(4), 867–878 (2015)

20. Bernhardt, M., Nicolas, J.D., Osterhoff, M., Mittelstädt, H., Reuss, M., Harke, B., Wittmeier, A., Sprung, M., Köster, S., Salditt, T.: Correlative microscopy approach for biology using X-ray holography, X-ray scanning diffraction and sted microscopy. Nat. Commun. 3641 (2018)

21. Giewekemeyer, K., Wilke, R.N., Osterhoff, M., Bartels, M., Kalbfleisch, S., Salditt, T.: Versatility of a hard X-ray kirkpatrick-baez focus characterized by ptychography. J. Synchr. Radiat. **20**(3), 490–497 (2013)

22. Wilke, R., Vassholz, M., Salditt, T.: Semi-transparent central stop in high-resolution X-ray ptychography using kirkpatrick-baez focusing. Acta Cryst. A **69**(5), 490–497 (2013)

23. Wilke, R., Wallentin, J., Osterhoff, M., Pennicard, D., Zozulya, A., Sprung, M., Salditt, T.: High-flux ptychographic imaging using the new 55 μm-pixel detectorlambda'based on the medipix3 readout chip. Acta Cryst. A **70**(6), 552–562 (2014)

24. Osterhoff, M., Salditt, T.: Coherence filtering of x-ray waveguides: analytical and numerical approach. New J. Phys. **13**(10), 103,026 (2011)

25. Weinhausen, B., Köster, S.: Microfluidic devices for x-ray studies on hydrated cells. Lab Chip (2013). https://doi.org/10.1039/c2lc41014a

26. Bernhardt, M., Priebe, M., Osterhoff, M., Wollnik, C., Diaz, A., Salditt, T., Rehfeldt, F.: X-ray micro- and nanodiffraction imaging on human mesenchymal stem cells and differentiated cells. Biophys. J. **110**(3), 680–690 (2016)
27. Ghazal, A., Lafleur, J.P., Mortensen, K., Kutter, J.P., Arleth, L., Jensen, G.V.: Recent advances in x-ray compatible microfluidics for applications in soft materials and life sciences. Lab Chip **16**(22), 4263–4295 (2016). https://doi.org/10.1039/c6lc00888g. https://doi.org/10.1039/c6lc00888g
28. Köster, S., Pfohl, T.: X-ray studies of biological matter in microfluidic environments. Mod. Phys. Lett B **26**(26), 1230,018 (2012)
29. Bernhardt, M.: X-ray micro- and nano-diffraction imaging on human mesenchymal stem cells and differentiated cells. Ph.D. thesis, Georg-August-Universität Göttingen (2017)
30. Hemonnot, C.: Investigating cellular nanoscale with x-rays. Ph.D. thesis, Georg-August-Universität Göttingen (2016)
31. Nicolas, J.D.: Mulitscale x-ray analyis of biological cells and tissues by scanning diffraction and coherent imaging. Ph.D. thesis, Georg-August-Universität Göttingen (2018)
32. Priebe, M.: Scanning x-ray nanodiffraction on dictyostelium discoideum. Ph.D. thesis, Georg-August-Universität Göttingen (2015)
33. Weinhausen, B.: Scanning x-ray nano-diffraction on eukaryotic cells: From freeze-dried to living cells. Ph.D. thesis, Georg-August-Universität Göttingen (2013)
34. Denz, M., Brehm, G., Hémonnot, C.Y.J., Spears, H., Wittmeier, A., Cassini, C., Saldanha, O., Perego, E., Diaz, A., Burghammer, M., Köster, S.: Cyclic olefin copolymer as an x-ray compatible material for microfluidic devices. Lab Chip **18**(1), 171–178 (2018). https://doi.org/10.1039/c7lc00824d. https://doi.org/10.1039/c7lc00824d
35. Priebe, M., Kalbfleisch, S., Tolkiehn, M., Köster, S., Abel, B., Davies, R., Salditt, T.: Orientation of biomolecular assemblies in a microfluidic jet. New J. Phys. **12**(4), 043,056 (2010)
36. Lincoln, B., Wottawah, F., Schinkinger, S., Ebert, S., Guck, J.: High-throughput rheological measurements with an optical stretcher. Meth. Cell Biol. **83**, 397–423 (2007)
37. Nicolas, J.D., Hagemann, J., Sprung, M., Salditt, T.: The optical stretcher as a tool for single-particle x-ray imaging and diffraction. J. Synchrotron Radiat. **25**(4) (2018)
38. DePonte, D., Weierstall, U., Schmidt, K., Warner, J., Starodub, D., Spence, J., Doak, R.: Gas dynamic virtual nozzle for generation of microscopic droplet streams. J. Phys. D: Appl. Phys. **41**(19), 195,505 (2008)
39. Chavas, L., Gumprecht, L., Chapman, H.: Possibilities for serial femtosecond crystallography sample delivery at future light sources. Struct. Dynam. **2**(4), 041,709 (2015)
40. Kirian, R., Awel, S., Eckerskorn, N., Fleckenstein, H., Wiedorn, M., Adriano, L., Bajt, S., Barthelmess, M., Bean, R., Beyerlein, K., et al.: Simple convergent-nozzle aerosol injector for single-particle diffractive imaging with x-ray free-electron lasers. Struct. Dynam. **2**(4), 041,717 (2015)
41. Guck, J., Ananthakrishnan, R., Mahmood, H., Moon, T.J., Cunningham, C.C., Käs, J.: The optical stretcher: a novel laser tool to micromanipulate cells. Biophys. J. **81**(2), 767–784 (2001)
42. Wilke, R., Priebe, M., Bartels, M., Giewekemeyer, K., Diaz, A., Karvinen, P., Salditt, T.: Hard x-ray imaging of bacterial cells: nano-diffraction and ptychographic reconstruction. Opt. Express **20**(17), 19232–19254 (2012)
43. Wilke, R.: Coherent x-ray diffractive imaging on the single-cell-level of microbial samples: ptychography, tomography, nano-diffraction and waveguide-imaging. Ph.D. thesis, Georg-August-Universität Göttingen (2014)
44. Hemonnot, C.Y., Reinhardt, J., Saldanha, O., Patommel, J., Graceffa, R., Weinhausen, B., Burghammer, M., Schroer, C.G., Köster, S.: X-rays reveal the internal structure of keratin bundles in whole cells. ACS Nano **10**(3), 3553–3561 (2016)
45. Piazza, V., Weinhausen, B., Diaz, A., Dammann, C., Maurer, C., Reynolds, M., Burghammer, M., Köster, S.: Revealing the structure of stereociliary actin by x-ray nanoimaging. ACS Nano **8**(12), 12228–12237 (2014)

46. Hémonnot, C.Y.J., Ranke, C., Saldanha, O., Graceffa, R., Hagemann, J., Köster, S.: Following DNA compaction during the cell cycle by x-ray nanodiffraction. ACS Nano **10**(12), 10661–10670 (2016). https://doi.org/10.1021/acsnano.6b05034. https://doi.org/10.1021/acsnano.6b05034
47. Osterhoff, M.: dada–a web-based 2d detector analysis tool. J. Phys. Conf. Ser. **849**, 012059 (2017) (IOP Publishing)
48. https://irpgoe.github.io/nanodiffraction/
49. Nicolas, J.D., Bernhardt, M., Markus, A., Alves, F., Burghammer, M., Salditt, T.: Scanning x-ray diffraction on cardiac tissue: automatized data analysis and processing. J. Synchrotron Rad. **24**(6), 1163–1172 (2017). https://doi.org/10.1107/S1600577517011936. https://doi.org/10.1107/S1600577517011936
50. Bernhardt, M., Nicolas, J.D., Eckermann, M., Eltzner, B., Rehfeldt, F., Salditt, T.: Anisotropic x-ray scattering and orientation fields in cardiac tissue cells. New J. Phys. **19**(1), 013,012 (2017)
51. Nicolas, J.D., Bernhardt, M., Krenkel, M., Richter, C., Luther, S., Salditt, T.: Combined scanning x-ray diffraction and holographic imaging of cardiomyocytes. J. Appl. Cryst. **50**(2), 612–620 (2017). http://journals.iucr.org/j/issues/2017/02/00/rg5124/rg5124.pdf
52. Eltzner, B., Wollnik, C., Gottschlich, C., Huckemann, S., Rehfeldt, F.: The filament sensor for near real-time detection of cytoskeletal fiber structures. PLoS ONE **10**(5), e0126,346 (2015)
53. Nicolas, J.D., Bernhardt, M., Schlick, S.F., Tiburcy, M., Zimmermann, W.H., Khan, A., Markus, A., Alves, F., Toischer, K., Salditt, T.: X-ray diffraction imaging of cardiac cells and tissue. Prog. Biophys. Mol. Biol. (2018)
54. Chapman, H.N., Barty, A., Marchesini, S., Noy, A., Hau-Riege, S.P., Cui, C., Howells, M.R., Rosen, R., He, H., Spence, J.C., et al.: High-resolution ab initio three-dimensional x-ray diffraction microscopy. JOSA A **23**(5), 1179–1200 (2006)
55. Miao, J., Charalambous, P., Kirz, J., Sayre, D.: Extending the methodology of x-ray crystallography to allow imaging of micrometre-sized non-crystalline specimens. Nature **400**(6742), 342 (1999)
56. Rodenburg, J., Hurst, A., Cullis, A., Dobson, B., Pfeiffer, F., Bunk, O., David, C., Jefimovs, K., Johnson, I.: Hard-x-ray lensless imaging of extended objects. Phys. Rev. Lett. **98**(3), 034,801 (2007)
57. Rodenburg, J.M.: Ptychography and related diffractive imaging methods. Adv. Imag. Electr. Phys. **150**, 87–184 (2008)
58. Thibault, P., Dierolf, M., Menzel, A., Bunk, O., David, C., Pfeiffer, F.: High-resolution scanning x-ray diffraction microscopy. Science **321**(5887), 379–382 (2008)
59. Thibault, P., Menzel, A.: Reconstructing state mixtures from diffraction measurements. Nature **494**(7435), 68 (2013)
60. Jones, M.W., Van Riessen, G.A., Abbey, B., Putkunz, C.T., Junker, M.D., Balaur, E., Vine, D.J., McNulty, I., Chen, B., Arhatari, B.D., et al.: Whole-cell phase contrast imaging at the nanoscale using fresnel coherent diffractive imaging tomography. Sci. Rep. **3**, 2288 (2013)
61. Jones, M.W., Elgass, K., Junker, M.D., Luu, M.B., Ryan, M.T., Peele, A.G., Van Riessen, G.A.: Mapping biological composition through quantitative phase and absorption x-ray ptychography. Sci. Rep. **4**, 6796 (2014)
62. Deng, J., Vine, D.J., Chen, S., Nashed, Y.S., Jin, Q., Phillips, N.W., Peterka, T., Ross, R., Vogt, S., Jacobsen, C.J.: Simultaneous cryo x-ray ptychographic and fluorescence microscopy of green algae. Proc. Natl. Acad. Sci. U.S.A. p. 201413003 (2015)
63. Lima, E., Diaz, A., Guizar-Sicairos, M., Gorelick, S., Pernot, P., Schleier, T., Menzel, A.: Cryo-scanning x-ray diffraction microscopy of frozen-hydrated yeast. J. Microsc. **249**(1), 1–7 (2013)
64. Diaz, A., Malkova, B., Holler, M., Guizar-Sicairos, M., Lima, E., Panneels, V., Pigino, G., Bittermann, A.G., Wettstein, L., Tomizaki, T., et al.: Three-dimensional mass density mapping of cellular ultrastructure by ptychographic x-ray nanotomography. J. Struct. Biol. **192**(3), 461–469 (2015)
65. Gallagher-Jones, M., Dias, C.S.B., Pryor, A., Bouchmella, K., Zhao, L., Lo, Y.H., Cardoso, M.B., Shapiro, D., Rodriguez, J., Miao, J.: Correlative cellular ptychography with functionalized nanoparticles at the fe l-edge. Sci. Rep. **7**(1), 4757 (2017)

66. Howells, M.R., Beetz, T., Chapman, H.N., Cui, C., Holton, J., Jacobsen, C., Kirz, J., Lima, E., Marchesini, S., Miao, H., et al.: An assessment of the resolution limitation due to radiation-damage in x-ray diffraction microscopy. J. Electron Spectrosc. **170**(1–3), 4–12 (2009)
67. Nam, D., Park, J., Gallagher-Jones, M., Kim, S., Kim, S., Kohmura, Y., Naitow, H., Kunishima, N., Yoshida, T., Ishikawa, T., et al.: Imaging fully hydrated whole cells by coherent x-ray diffraction microscopy. Phys. Rev. Lett. **110**(9), 098,103 (2013)
68. Hagemann, J., Salditt, T.: The fluence-resolution relationship in holographic and coherent diffractive imaging. J. Appl. Cryst. **50**(2), 531–538 (2017)
69. Krenkel, M., Toepperwien, M., Alves, F., Salditt, T.: Three-dimensional single-cell imaging with x-ray waveguides in the holographic regime. Acta Cryst. A **73**(4), 282–292 (2017)
70. Töpperwien, M., van der Meer, F., Stadelmann, C., Salditt, T.: Three-dimensional virtual histology of human cerebellum by x-ray phase-contrast tomography. Proc. Natl. Acad. Sci. U.S.A. p. 201801678 (2018)
71. Liebi, M., Georgiadis, M., Menzel, A., Schneider, P., Kohlbrecher, J., Bunk, O., Guizar-Sicairos, M.: Nanostructure surveys of macroscopic specimens by small-angle scattering tensor tomography. Nature **527**(7578), 349–352 (2015). https://doi.org/10.1038/nature16056. https://doi.org/10.1038/nature16056
72. Carboni, E., Nicolas, J.D., Töpperwien, M., Stadelmann-Nessler, C., Lingor, P., Salditt, T.: Imaging of neuronal tissues by x-ray diffraction and x-ray fluorescence microscopy: evaluation of contrast and biomarkers for neurodegenerative diseases. Biomed. Opt. Express **8**(10), 4331 (2017). https://doi.org/10.1364/boe.8.004331. https://doi.org/10.1364/boe.8.004331
73. Chapman, H.N., Fromme, P., Barty, A., White, T.A., Kirian, R.A., Aquila, A., Hunter, M.S., Schulz, J., DePonte, D.P., Weierstall, U., et al.: Femtosecond x-ray protein nanocrystallography. Nature **470**(7332), 73 (2011)
74. Gaffney, K., Chapman, H.: Imaging atomic structure and dynamics with ultrafast x-ray scattering. Science **316**(5830), 1444–1448 (2007)
75. Neutze, R., Wouts, R., van der Spoel, D., Weckert, E., Hajdu, J.: Potential for biomolecular imaging with femtosecond x-ray pulses. Nature **406**(6797), 752 (2000)
76. Bergh, M., Huldt, G., Timneanu, N., Maia, F.R., Hajdu, J.: Feasibility of imaging living cells at subnanometer resolutions by ultrafast x-ray diffraction. Q. Rev. Biophys. **41**(3–4), 181–204 (2008)
77. Seibert, M.M., Ekeberg, T., Maia, F.R., Svenda, M., Andreasson, J., Jönsson, O., Odić, D., Iwan, B., Rocker, A., Westphal, D., et al.: Single mimivirus particles intercepted and imaged with an x-ray laser. Nature **470**(7332), 78 (2011)
78. Mai, D., Hallmann, J., Reusch, T., Osterhoff, M., Düsterer, S., Treusch, R., Singer, A., Beckers, M., Gorniak, T., Senkbeil, T., et al.: Single pulse coherence measurements in the water window at the free-electron laser flash. Opt. Express **21**(11), 13005–13017 (2013)
79. Dronyak, R., Gulden, J., Yefanov, O., Singer, A., Gorniak, T., Senkbeil, T., Meijer, J.M., Al-Shemmary, A., Hallmann, J., Mai, D., et al.: Dynamics of colloidal crystals studied by pump-probe experiments at FLASH. Phys. Rev. B **86**(6), 064,303 (2012)
80. Singer, A., Sorgenfrei, F., Mancuso, A., Gerasimova, N., Yefanov, O., Gulden, J., Gorniak, T., Senkbeil, T., Sakdinawat, A., Liu, Y., et al.: Spatial and temporal coherence properties of single free-electron laser pulses. Opt. Express **20**(16), 17480–17495 (2012)

Chapter 16
Single Particle Imaging with FEL Using Photon Correlations

Benjamin von Ardenne and Helmut Grubmüller

Abstract Scattering experiments with femtosecond high-intensity free-electron laser pulses provide a new route to macromolecular structure determination without the need for crystallization at low material usage. In these experiments, the X-ray pulses are scattered with high repetition on a stream of identical single biomolecules and the scattered photons are recorded on a pixelized detector. The main challenges are the unknown random orientation of the molecule in each shot and the extremely low signal to noise ratio due to the very low expected photon count per scattering image, typically well below the number of over 100 photons required by available analysis methods. The latter currently limits the scattering experiments to nanocrystals or larger virus particles, but the ultimate goal remains to retrieve the atomic structure of single biomolecules. Here, we use photon correlations to overcome the issue with low photon counts and present an approach that can determine the molecular structure *de novo* from as few as three coherently scattered photons per image. We further validate the method with a small protein (46 residues), show that near-atomic resolution of 3.3 Å is within experimental reach and demonstrate structure determination in the presence of isotropic noise from various sources, indicating that the number of disordered solvent molecules attached to the macromolecular surface should be kept at a minimum. Our correlation method allows to infer structure from images containing multiple particles, potentially opening the method to other types of experiments such as fluctuation X-ray scattering (FXS).

B. von Ardenne (✉) · H. Grubmüller
Department of Theoretical and Computational Biophysics, Max Planck Institute
for Biophysical Chemistry Göttingen, Am Fassberg 11, 37077 Göttingen, Germany
e-mail: Benjamin.von.Ardenne@gmail.com

H. Grubmüller
e-mail: hgrubmu@mpibpc.mpg.de

T. Salditt et al. (eds.), *Nanoscale Photonic Imaging*, Topics in Applied Physics 134,
https://doi.org/10.1007/978-3-030-34413-9_16

16.1   The Single Molecule Scattering Experiment

Despite the great effort in biomolecular structure determination, the structures of less than 1% (~160,000) of the more than 21 million transcribed proteins [1] have been determined to high resolution [2]. Over the past years existing structure determination methods such as X-ray crystallography and NMR have been stagnating, leaving room for novel methods that can extend the knowledge of biomolecular structures. To this end, X-ray scattering experiments with single biomolecules have been proposed by Neutze et al. as a new *de novo* structure determination approach for proteins without the need for crystallization [3–7]. Single molecule X-ray imaging becomes possible due to newly-developed free electron laser that produce very high-intensity femtosecond-short X-ray pulses with a focus size of down to 100 nm.

As illustrated in Fig. 16.1, in the experiment, a stream of (typically) hydrated and randomly oriented proteins enters the pulsed X-ray beam at a rate of one molecule per pulse. Despite the high photon flux of the incident beam, only a few photons are scattered by the molecules and recorded on the pixelized detector.

Sample delivery is non-trivial due to the nanoscopic size of the biomolecules and several solutions have been proposed, e.g., using electrospraying techniques [8], gas focused liquid jets [9], oil/water droplet immersion jets [10] or embedding the molecules into polymers (lipidic cubic phase injector) to save material [11]. In each

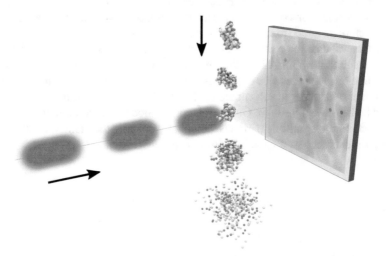

Fig. 16.1 Experimental setup of single molecule scattering imaging. A stream of randomly-oriented particles is injected into the high-intensity short-pulsed FEL beam, hit sequentially by femtosecond X-ray pulses, and the few coherently scattered photons (red dots) are recorded on the pixel detector. The spatial distribution of the photons follows the Fourier intensity of the molecule which is depicted here in light blue in the background of the photon pattern. After illumination, ionization effects charge the molecules and the resulting Coulomb forces quickly disintegrate the molecule

sample delivery method, it is important that the single molecules stay in their physiological environment in order to observe the their natural conformations.

In the scattering process, ionization (Auger decay) charges the atoms in the molecule and leads to Coulomb explosion, coining the method as a "diffract and destroy" experiment. In fact, only 10% of all photons are scattered coherently, all others are absorbed due to the photo-electric effect and expelled shortly after from the molecules at lower energies. However, the short pulses, usually less than 100 fs long, outrun the severe radiation damage because the molecular motion in response to the changed electronic configuration is estimated to take longer than 100 fs [7, 12] and the incident photons are scattered by the unperturbed structure before the molecule disintegrates.

Like in conventional X-ray crystallography, only the intensities and not the phases are measured. In the absence of crystals, the measured signal is the continuous Fourier transformation of the molecule, rendering the phase problem accessible to established *ab initio* phase-retrieval methods [13].

Whereas previous X-ray sources, including synchrotron sources, have primarily engaged in studies of static structures, X-ray FELs are by their nature suited for studying dynamic systems at the time and length scales of atomic interactions. In contrast to methods that measure a structure ensemble (NMR, SAXS, FRET), this method gives access to single molecule images and, with a seed model, the images could be e.g., sorted probabilistically to distinguish between different native conformations. Further, similar to nano-crystallography, in systems where reactions can be easily induced, e.g., by light, a sequence of structures at different reaction times may be recorded which opens the window to molecular movies as a long-standing dream [14]. Even without sorting, the variance of the native conformations can be assessed via the variance of the determined electron density in which flexible regions would be smeared out more than rigid protein motifs.

16.2   Structure Determination Using Few Photons

Single molecule scattering images sample spherical dissections (Ewald sphere) of the continuous 3D Fourier intensity, $I(\mathbf{k}) = |\mathcal{F}[\rho(\mathbf{x})]|^2$ and the orientation of the dissection depends on the orientation of the molecule at the time of illumination. The structure determination from these single molecule images faces two major challenges. First, the orientation of the molecule at the time of illumination is unknown and hard to control because it is usually injected into the "reaction chamber" via electro-spraying in which the molecules tumble inside a solvent bubble. Second, only a low number of photons is coherently scattered (as a statistical Poisson process following the Fourier intensity) and the additional background noise from, e.g., inelastic scattering, the photo-electric effect or background radiation leads to very low signal-to-noise levels. In fact, we estimated that a rather small protein (46 residues) scatters only 20 photons coherently at realistic beam parameters of the next generation European XFEL which add an additional layer of complexity to the structure determination problem due to the additional Poisson noise (shot noise).

Over the past years, several structure determination methods have been proposed and demonstrated which mainly fall into two major classes. The first class of methods predicts the orientation of the molecules at the time of illumination for each scattering image either explicitly or implicitly e.g., through statistical similarities between images or by using a coarse seed model. Images that belong to the same orientation are averaged and these averages are assembled into the 3D intensity similar to cryo-EM. However, almost all of the orientation classification methods are limited to scattering datasets with usually many more than 100 average photons per image.

The second class of methods forgoes the classification of orientations by using photon correlations as an averaged summary statistics of the entire image dataset that is independent of the individual orientations and will be covered in this Chapter. Previous attempts have focused on extracting as much as possible information from two correlated photons using additional knowledge such as symmetry or molecular rotations around a fixed axis. From early work by Kam on electron micrograph images, it is known that two-photon correlations do not carry sufficient information to retrieve the full 3D intensity ab initio [15, 16]. Motivated by these observation, we suspected and eventually validated the claim that three photon suffice and therefore developed a method method that allows for *de novo* structure determination from as few as three coherently scattered photons per single molecule X-ray scattering image. The main idea is to determine the molecule's intensity $I(\mathbf{k})$ from the *full* three-photon correlation $t(k_1, k_2, k_3, \alpha, \beta)$ which is accumulated from all photon triplets in the recorded scattering images, independent of the respective molecular orientations and therefore free of errors associated with the classification of the orientations.

16.2.1 Theoretical Background on Three-Photon Correlations

A single photon triplet is characterized by the angles α and β between the photons and the distances of the photons to the detector center (Fig. 16.2). Each triplet is comprised of three correlated doublets $(k_1, k_2, \alpha,)$, (k_2, k_3, β) and $(k_1, k_3, \alpha + \beta)$ and the angles are chosen as the minimum difference between the pairs, $\alpha, \beta \in [0, \pi]$. The probability of observing a coherently scattered photon at pixel position $\mathbf{k}^\star$ is proportional to the intensity $I(\mathbf{k}^\star)$ at this pixel which lies on the projection of the intensity $I(\mathbf{k})$ on the Ewald sphere in 3D Fourier space. The full three-photon correlation $t(k_1, k_2, k_3, \alpha, \beta)$ is the sum over all possible triplets which is equivalent to the orientational average $\langle \rangle_\omega$ of the product between three intensities $I(\mathbf{k})$ that lie on the intersection between the Ewald sphere and the 3D Fourier density,

$$t(k_1, k_2, k_3, \alpha, \beta)_{I(\mathbf{k})} = \left\langle I_\omega\left(\mathbf{k}_1^\star(k_1, 0)\right) \cdot I_\omega\left(\mathbf{k}_2^\star(k_2, \alpha)\right) \cdot I_\omega^*\left(\mathbf{k}_3^\star(k_3, \beta)\right)\right\rangle_\omega . \quad (16.1)$$

Fig. 16.2 Schematic depiction of the three-photon correlation using an exemplary synthetic single molecule scattering image of Crambin with only coherently scattered photons. In the detector plane $k_x k_y$ the recorded photons are grouped into triplets, each of which is characterized by distances k_1, k_2, k_3 to the detector center (orange lines) and the angles α and β between the respective photons (orange circular arcs)

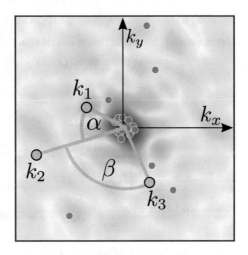

Here, without loss of generality, the three vectors $\mathbf{k_1}^\star$, $\mathbf{k_1}^\star$ and $\mathbf{k_1}^\star$ are the projection onto the Ewald sphere of the three photon positions $\mathbf{k_1} = (k_1, 0, 0)$, $\mathbf{k_2} = k_2(\cos\alpha, \sin\alpha, 0)$ and $\mathbf{k_3} = k_3(\cos\beta, \sin\beta, 0)$ in the detector plane. These positions are chosen as one arbitrary realization of the tuple $(k_1, k_2, k_3, \alpha, \beta)$.

For the orientational average $\langle\rangle_\omega$ it is assumed that in the experiment the orientation of the molecule is unknown and uniformly sampled. Note that the orientational average can either be expressed as an average over all rotations of $I_\omega(\mathbf{k})$ for fixed $\mathbf{k}_{1,2,3}$ (our approach) or as an average over all rotations of the vectors $\mathbf{k}_{1,2,3,\omega}$ for a fixed $I(\mathbf{k})$.

The orientational integral over all possible triple products of 3D intensities $I(\mathbf{k})$ in 16.1 is challenging to calculate and may be simplified by decomposing $I(\mathbf{k})$ into spherical shells with radius k and by expanding each shell using a spherical harmonics basis [17],

$$I(\mathbf{k}) = \sum_{lm} A_{lm}(k) Y_{lm}(\theta, \varphi). \qquad (16.2)$$

The coefficients $A_{lm}(k)$ describe the intensity function on the respective shells and are non-zero only for even $l \in \{0, 2, 4, ..., L\}$ because of the symmetry of $I(\mathbf{k}) = I(-\mathbf{k})$ (Friedel's law). In this description, a 3D Euler rotation ω of $I(\mathbf{k})$ is expressed by transforming the spherical harmonics coefficients according to $A_{lm}^{rot}(k) = \sum_{mm'} D_{mm'}^l A_{lm'}^{unrot}(k)$, using the rotation operators $D_{m'm}^l$ which are composed of elements of the Wigner D-matrix as defined, e.g., in [17], yielding the rotated intensity,

$$I_\omega(\mathbf{k}) = \sum_{lmm'} A_{lm}(k) Y_{lm'}(\theta, \varphi) D_{m'm}^l(\omega). \qquad (16.3)$$

Inserting the spherical harmonics expansion of the rotated intensity $I_\omega(\mathbf{k})$, evaluated at positions $\mathbf{k}_1^*$, $\mathbf{k}_2^*$ and $\mathbf{k}_3^*$ on the Ewald sphere ($\theta_i = \cos^{-1}(\frac{k_i \lambda}{4\pi})$), into the expression for the three-photon correlation, (16.1), yields

$$t(k_1, k_2, k_3, \alpha, \beta)_{\{A_{lm}(k)\}} =$$

$$\sum_{l_1 l_2 l_3 m_1 m_2 m_3 m_1' m_2' m_3'} \sum \sum A_{l_1 m_1}(k_1) A_{l_2 m_2}(k_2) A^*_{l_3 m_3}(k_3)$$

$$Y_{l_1 m_1'}(\theta_1(k_1), 0) \cdot Y_{l_2 m_2'}(\theta_2(k_2), \alpha) \cdot Y^*_{l_3 m_3'}(\theta_3(k_3), \beta)$$

$$\left\langle D^{l_1}_{m_1 m_1'} \cdot D^{l_2}_{m_2 m_2'} \cdot D^{l_3*}_{m_3 m_3'} \right\rangle_\omega, \tag{16.4}$$

such that the orientational average only involves the elements of the Wigner D-matrix $D^l_{mm'}$.

Using the Wigner-3j symbols $\begin{pmatrix} l_1 & l_2 & L \\ m_1 & m_2 & -M \end{pmatrix}$ [18], the product of two rotation elements $D^l_{mm'}$ reads

$$D^{l_1}_{m_1 m_1'} D^{l_2}_{m_2 m_2'} = \sum_{L=|l_1-l_2|}^{l_1+l_2} \sum_{MM'} (2L+1)(-1)^{M-M'} \tag{16.5}$$

$$\begin{pmatrix} l_1 & l_2 & L \\ m_1 & m_2 & -M \end{pmatrix}$$

$$\begin{pmatrix} l_1 & l_2 & L \\ m_1' & m_2' & -M' \end{pmatrix} D^L_{MM'}.$$

With the orthogonality theorem for orientational averages of the product of two Wigner D operators,

$$\left\langle D^L_{MM'} D^{l_3*}_{m_3 m_3'} \right\rangle_\omega = \frac{1}{2L+1} \delta_{l_3 L} \delta_{m_3 M} \delta_{m_3' M'}, \tag{16.6}$$

the three-photon correlation finally reads

$$t(k_1, k_2, k_3, \alpha, \beta)_{\{A_{lm}(k)\}} = \sum_{l_1 l_2 l_3 m_1 m_2 m_3} \sum A_{l_1 m_1}(k_1) A_{l_2 m_2}(k_2) A^*_{l_3 m_3}(k_3) \tag{16.7}$$

$$\begin{pmatrix} l_1 & l_2 & l_3 \\ m_1 & m_2 & -m_3 \end{pmatrix}$$

$$\sum_{m_1' m_2' m_3'} (-1)^{m_3 - m_3'} \begin{pmatrix} l_1 & l_2 & l_3 \\ m_1' & m_2' & -m_3' \end{pmatrix}$$

$$Y_{l_1 m_1'}(\theta_1(k_1), 0) Y_{l_2 m_2'}(\theta_2(k_2), \alpha) Y^*_{l_3 m_3'}(\theta_3(k_3), \beta).$$

This expression only involves sums of products of three spherical harmonics coefficients $A_{lm}(k)$ with known Wigner-3j symbols and spherical harmonics basis functions $Y_{lm}(\theta, \varphi)$. The numerical calculation of the three photon correlation (forward model) is the computationally limiting step in the structure determination approach. The correlations, expressed in spherical harmonics terms, are faster to calculate than e.g., the numerical integration, and they allow for adapting the number $K(L^2 + 3L + 2)/2$ of spherical harmonics basis functions to the target resolution via the largest considered wave number k_{cut}, the number K of used shells between $0...k_{cut}$, and the expansion order L. The hierarchical properties of spherical harmonics basis functions further allow to determine the structure first with low angular resolution and then to successively refine it to higher resolutions and higher expansion limits, respectively.

16.2.2 Bayesian Structure Determination

Currently no analytic inversion of the three-photon correlation in (16.7) is known, and the number of unknowns (e.g., 4940 for $K = 26$, $L = 18$) is too large for a straightforward numeric solution. Instead we have developed a probabilistic approach [19] in which we asked which intensity $I(\mathbf{k})$ is most likely to have generated the complete set of measured scattering images and triplets, respectively. To this end, we considered the Bayesian probability p (with uniform prior) that a given intensity $I(\mathbf{k})$, expressed in spherical harmonics by $\{A_{lm}(k)\}$, generated the set of triplets, $\left\{k_1^i, k_2^i, k_3^i, \alpha^i, \beta^i\right\}_{i=1...T}$,

$$p\left(\left\{k_1^i, k_2^i, k_3^i, \alpha^i, \beta^i\right\}_{i=1...T}\middle| \{A_{lm}(k)\}\right) = \prod_{i=1}^{T} \tilde{t}(k_1^i, k_2^i, k_3^i, \alpha^i, \beta^i)_{\{A_{lm}(k)\}}. \quad (16.8)$$

Due to the statistical independence of the triplets, this probability p is a product over the probabilities $\tilde{t}(k_1^i, k_2^i, k_3^i, \alpha^i, \beta^i)$ of observing the individual triplets i which is given by the normalized three-photon correlation $\tilde{t}(k_1, k_2, k_3, \alpha, \beta)$. Here, $\tilde{t}(k_1, k_2, k_3, \alpha, \beta)$ is calculated using (16.7) for varying intensity coefficients $\{A_{lm}(k)\}$ and the coefficients that maximized $p\left(\left\{k_1^i, k_2^i, k_3^i, \alpha^i, \beta^i\right\}\right)$ are determined using a Monte Carlo scheme as discussed in Sect. 16.2.4.

In contrast to the direct inversion, the probabilistic approach has the benefit of fully accounting for the Poissonian shot noise implied by the limited number of photon triplets that are extracted from the given scattering images. We note that this approach also circumvents the limitation faced in previous works on degenerate three photons correlations by Kam [16], where only triples are considered, in which two photons are recorded at the same detector position. Because all other triples had to be discarded, Kam's approach is limited to very high beam intensities, and cannot be applied in the present extreme Poisson regime.

Calculating the probability from (16.8) (and energy in the Monte Carlo scheme) is computationally expensive due to the typically large number of triples T. We therefore approximated this product by grouping triplets with similar α, β angles and distances k into bins and calculated the function $t(k_1, k_2, k_3, \alpha, \beta)$ for each bin only once, denoted $t_{k_1,k_2,k_3,\alpha,\beta}$, thus markedly reducing the number of function evaluations to the number of bins. To improve the statistics for each bin, the intrinsic symmetry of the triple correlation function was also used. In particular, all triplets were mapped into the sub-region of the triple correlation that satisfies $k_1 \geq k_2 \geq k_3$. In this mapping, special care was taken to correct for the fact that triplets with $k_1 = k_2 \neq k_3$ or $k_1 \neq k_2 = k_3$ or $k_1 = k_3 \neq k_2$ occur 3 times more often than $k_1 = k_2 = k_3$ and triplets with $k_1 \neq k_2 \neq k_3$ occur 6 times more often. To compensate for different binsizes, each bin was normalized by $k_1 k_2 k_3$.

16.2.3 Reduction of Search Space Using Two-Photon Correlations

The high-dimensional search space may be reduced by utilizing the structural information contained within the two-photon correlation. In analogy to the three-photon correlation, the two photon-correlation is expressed as a sum over products of spherical harmonics coefficients $A_{lm}(k)$ weighted with Legendre polynomials P_l [16, 20],

$$c_{k_1,k_2,\alpha} = \sum_l P_l(\cos(\alpha)) \sum_m A_{lm}(k_1)(\omega) A_{lm}^*(k_2). \tag{16.9}$$

Please note that the α which is seen on the detector is different from the angle $\alpha^\star = \cos^{-1}(\sin(\theta_1)\sin(\theta_2)\cos(\alpha) + \cos(\theta_1)\cos(\theta_2))$ between the two points in 3D intensity space due to the Ewald curvature ($\theta = \cos^{-1}(k\lambda/4\pi)$).

The inversion yields coefficient vectors $\mathbf{A}_l^0(k) = (A_{l-m}^0, ..., A_{lm}^0)$ for all $l \leq L \leq K_{max}/2$ and $-l < m < l$, as first demonstrated by Kam [16]. However, all rotations in the $2l + 1$-dimensional coefficient eigenspaces of $\mathbf{A}_l^0(k)$ by $\mathbf{U}_l$ are also solutions,

$$\mathbf{A}_l(k) = \mathbf{U}_l \mathbf{A}_l^0(k). \tag{16.10}$$

The result implies that the inversion only gives a degenerate solution for the coefficients and the intensity cannot be determined solely from two photons. Note that the maximum L, corresponding to the angular resolution of the intensity model, scales with the number of shells K_{max} (or the inverse of the shell spacing Δk respectively) used for the two-photon inversion.

16.2.4 Optimizing the Probability Using Monte Carlo

In our method, we decided to maximize the probability p from (16.8) with a Monte Carlo/simulated annealing approach on the 'energy' function

$$
\begin{aligned}
E\left(\{k_1^i, k_2^i, k_3^i, \alpha^i, \beta^i\} \mid \{A_{lm}(k)\}\right) \\
= -\log p\left(\{k_1^i, k_2^i, k_3^i, \alpha^i, \beta^i\} \mid \{A_{lm}(k)\}\right) \\
= -\sum_i \log \tilde{t}(k_1^i, k_2^i, k_3^i, \alpha^i, \beta^i)_{\{A_{lm}(k)\}},
\end{aligned}
\tag{16.11}
$$

in the space of all rotations $\mathbf{U}_l$ given by the inversion of the two-photon correlation discussed in the previous Section.

Each Monte Carlo run is initialized with a random set of rotations $\{\mathbf{U}_l\}$ and the set of unaligned coefficients $\{\mathbf{A}_l^0\}$. In each Monte Carlo step j, all rotations $\mathbf{U}_l^j$ are varied by small random rotations $\mathbf{\Delta}_l(\beta_l)$ such that the updated rotations for each l $(l \leq L)$ read $\mathbf{U}_l^{j+1} = \mathbf{\Delta}_l(\beta_l) \cdot \mathbf{U}_l^j$ using stepsizes β_l. In order to escape local minima, a simulated annealing is performed using an exponentially decaying temperature protocol, $T(j) = T_{\text{init}} \exp(j/\tau)$. Steps with an increased energy were also accepted according to the Boltzmann factor $\exp(-\Delta E/T)$. We further used adaptive stepsizes such that all $\beta(l)$ were increased or decreased by a factor μ when accepting or rejecting the proposed steps, respectively. Convergence was improved by using a hierarchical approach in which the intensity was first determined with low angular resolution and further increased to high resolution. To this end, the variations of low-resolution features were "frozen out" faster than the variations of high-resolution features.

The random rotations $\{\mathbf{U}_l \in R^{2l+1 \times 2l+1}\}$ were generated using QR decompositions of matrices whose entries were drawn from a normal distribution as described by Mezzadri [21]. The rotational variations $\mathbf{\Delta}_l(\beta)$ were calculated via the basis transformation

$$
\mathbf{\Delta}_l(\beta) = \mathbf{R}_l \mathbf{S}_l(\beta) \mathbf{R}_l^{-1}
\tag{16.12}
$$

with

$$
\mathbf{S}_l(\beta) =
\begin{pmatrix}
\cos(\beta) & -\sin(\beta) & 0 & \dots & 0 \\
\sin(\beta) & \cos(\beta) & 0 & \dots & 0 \\
0 & 0 & I_{2l+1-2} & & \\
\dots & \dots & & & \\
0 & 0 & & &
\end{pmatrix}
\tag{16.13}
$$

and random rotation matrices $\mathbf{R}_l$ [22]. Here, sub-matrix $\mathbf{I}_{2l-1}$ in $\mathbf{S}_l$ is a $2l - 1$-dimensional unity matrix.

By using the small rotational variations $\mathbf{\Delta}_l(\beta)$, the SO(n) is sampled ergodically. Approximately $[1/(2 - 2\cos(\beta))]n \cdot \log(n)$ steps are necessary to achieve sufficient sampling aaccording to [22]. For the largest search space of $L = 18$ with a rotation dimension of $n = 37$ ($n = 2L + 1$) and a minimum stepsize of $\beta = 0.025$ rad,

213,777 steps are required to sample rotations in $SO(37)$ sufficiently dense. To ensure that the search space is exhaustively explored, we aimed at an optimization length of over 200,000 Monte Carlo steps. To this end, a time constant for the temperature decrease of $\tau = 50000$ steps was chosen. The initial temperature T_{init} was calculated as 10% of the standard deviation of the energy within 50 random steps away from the starting structure using the initial stepsizes. Further, we used a factor $\mu = 1.01$ for the adaptive stepsizes. The hierarchical approach was implemented by distributing the initial stepsizes according to $\beta(l) = (l - 1)\pi$ such that spherical harmonics coefficients with larger expansion orders l are always varied with a larger stepsize $\beta(l)$ than coefficients with lower orders.

16.3 Method Validation

Currently, experimental single molecule scattering data is only available for very large icosahedral viruses and in the absence of single molecule scattering images of smaller bio-molcules such as proteins, we have resorted to synthetic scattering experiments to validate our method. Thus, we have tested the method with a Crambin molecule for which we have estimated approx. 20 coherently scattered photons per image at realistic beam parameters. To stay below the estimate of approximately 20 photons per image, we generated up to 3.3×10^9 synthetic scattering images with only 10 photons on average, totalling up to 3.3×10^{10} recorded photons. With an expected XFEL repetition rate of up to 27 kHz [23], and assuming a hit-rate of 10%, this data can be collected within a few days. However, the data acquisition time substantially decreases to e.g., approx. 30 min when on average 100 photons per image are recorded, reducing the total number of required photons by a factor 100 to 3.3×10^8 (and reducing the number of images by a factor 1000 to 3.3×10^6).

For the synthetic image generation, we approximated the 3D electron density $\rho(\mathbf{x})$ by a sum of Gaussian functions centered at the atomic positions $\mathbf{x}_i$,

$$\rho(\mathbf{x}) = \sum_{i=1}^{N_{atoms}} N_i \exp^{-(\mathbf{x}-\mathbf{x}_i)^2/(2\sigma_i^2)} . \tag{16.14}$$

The heights and variances of the Gaussian spheres depend on the type of atom i. The variances σ_i correspond to the size of the atoms with respect to their scattering cross-section and the height is determined by N_i, the number of electrons which are the potential targets for scattering.

The absolute square of the electron densities' Fourier transformation $I(\mathbf{k}) = |\mathcal{F}[\rho(\mathbf{x})]|^2$ was used to generate the images. In each synthetic scattering experiment, In each shot, the molecule, and thus also $I(\mathbf{k})$, was randomly oriented and on average P photons per image were generated according to the distribution given by the dissection of the randomly oriented Ewald sphere and the intensity $I_\omega(\mathbf{K})$.

To generate the distributions numerically, first, a random set of N_{pos} positions $\{\mathbf{K}_i\}$ in the $k_x k_y$-plane was generated according to a 2D Gaussian distribution $G(\mathbf{K})$ with width $\sigma = 1.05\,\text{Å}^{-1}$ (specific to the Crambin intensity). Given a random 3D rotation $\mathbf{U}$, *rejection sampling* was used to accept or reject each position according to $\xi < I_\omega(\mathbf{U} \cdot \mathbf{K}_i)/(M \cdot G(\mathbf{K}_i))$ using uniformly-distributed random numbers $\xi \in [0, 1]$ each. Here, the constant M was chosen as $I_{max} \cdot \max(G(\mathbf{K}))$ such that the ratio $I_\omega(\mathbf{U} \cdot \mathbf{K}_i)/(M \cdot G(\mathbf{K}_i))$ is below 1 for all $\mathbf{K}$.

In accordance with our most conservative estimate, the number of positions N_{pos} was chosen such that on average 10 scattered photons were generated. For assessing the dependency of the resolution on the number of scattered photons, additional image sets with 25, 50 or 100 scattered photons were also generated (see Sect. 16.3.2).

16.3.1 Resolution Scaling with Photon Counts

Starting from the histograms obtained from 3.3×10^9 synthetic scattering images with 10 photons, we performed 20 independent structure determination runs. For all runs we used an expansion order $L = 18$, $K = 26$ shells and a cutoff $k_{cut} = 2.15\,\text{Å}^{-1}$, thus setting the maximum achievable resolution to 2.9 Å. To assess the achievable resolution of the determined Fourier intensities, we calculated 20 real space electron density maps using the relaxed averaged alternating reflections (RAAR) iterative phase retrieval algorithm by Luke [13]. Figure 16.3 compares the average of the 20 retrieved densities (a, green shaded structure) with the the reference electron density (b, blue shaded structure) which has been calculated from the Fourier density (including phases) with same cutoff k_{cut} as (a). The cross-correlation between the two densities is 0.9.

The resolution of the phased electron densities was characterized by the Fourier shell correlation (FSC),

$$\text{FSC}(k) = \frac{\sum\limits_{k_i \in k} F_1(k_i) \cdot F_2(k_i)^*}{\sqrt{\sum\limits_{k_i \in k} |F_1(k_i)|^2 \cdot \sum\limits_{k_i \in k} |F_2(k_i)|^2}}. \tag{16.15}$$

We have adopted the common definition of the resolution from cryo-EM [24] for cases in which the reference density is known. The resolution is then defined as the scattering angle k_{res} at which $\text{FSC}(k) = 0.5$, yielding a radial resolution $\Delta r = 2\pi/k_{res}$. In cases where the two densities in the FSC come from densities retrieved from independent image-sets (cross-validation), a lower cut-off $\text{FSC}(k) = 0.143$ is typically used. Here, we have achieved a near-atomic resolution of 3.3 Å from the correlation derived from 3.3×10^9 images.

Next, we have determined the structure from increasing number of images to asses how the resolution scales with the total number of observed photons and,

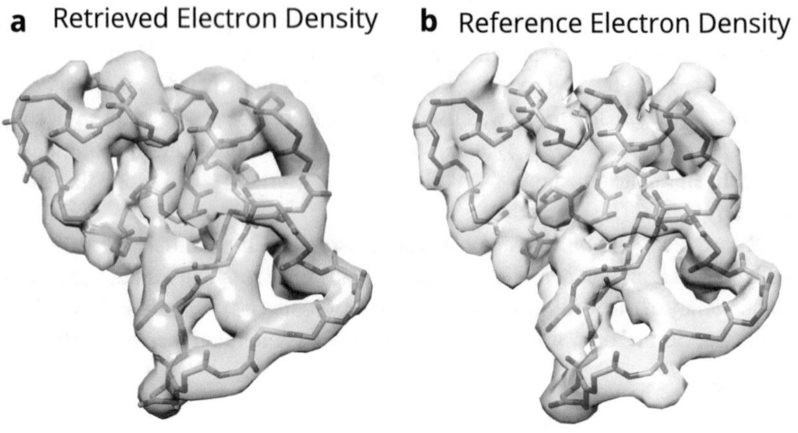

a Retrieved Electron Density   **b** Reference Electron Density

Fig. 16.3 Comparison of the retrieved electron density (**a**) and the reference electron density (**b**). The reference density (**b**) was calculated from the known Fourier density using the same cutoff $k_{cut} = 2.15\,\text{Å}^{-1}$ in reciprocal space as (**a**). The resolution of the retrieved density is 3.3 Å, the resolution of the reference density is 2.9 Å and the cross-correlation between the two densities is 0.9

hence, the number of recorded images. To this end, electron densities were calculated and averaged as above starting from 1.3×10^6 and going up to 3.3×10^9 images (4.7×10^8 up to 1.2×10^{12} triplets).

Figure 16.4 shows the FSC curves of all retrieved (averaged) densities along with the 0.5 cutoff (vertical dashed line) and the corresponding resolutions (inset). In Fig. 16.5 visualizes how the resolution improves with the increasing number of detected photons by comparing four electron densities that were retrieved from histograms with 2.0×10^8 to 3.3×10^{10} photons.

As mentioned before, the best electron density was retrieved with a near-atomic resolution of 3.3 Å (Fig. 16.5a) from the histograms that was derived from a total of 3.3×10^{10} photons. Decreasing the number of photons by a factor of 10 decreased the resolution only slightly by 0.4–3.7 Å (Fig. 16.5c), which indicates that very likely fewer than 3.3×10^{10} photons suffice to achieve near-atomic resolution. If much fewer photons are recorded, e.g. 2.0×10^8, the resolution decreased markedly to 7.8 Å (Fig. 16.5a) and even 14 Å resolution for 1.3×10^7 photons. For comparison, the diameter of Crambin is 17 Å.

To address the question how much further the resolution can be increased, we mimicked an experiment with infinite number of photons by determining the intensity from the analytically calculated three-photon correlation. As can be seen in Fig. 16.4 (purple line), the resolution only slightly improved by 0.1 Å to about 3.2 Å indicating that at this point either the expansion order L or insufficient convergence of the Monte Carlo based structure search became resolution limiting. To distinguish between these two possible causes, we phased the electron density directly from the reference intensity, using the same expansion order $L = 18$ as in the other experiments.

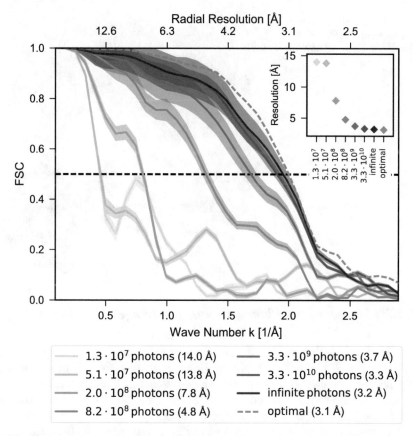

Fig. 16.4 Fourier shell correlations (FSC) of densities retrieved from 1.3×10^7 to 3.3×10^{10} photons (4.7×10^8–1.2×10^{12} triplets) and infinite photon number. As a reference, the "optimal" FSC is shown (dashed grey), which was calculated directly from the known intensity using the same expansion parameters. The inset shows the corresponding resolutions estimated from $FSC(k_{res}) = 0.5$

The reference intensity is free from convergence issues of the Monte Carlo structure determination and the resulting electron density only includes the phasing errors introduced by the limited angular resolution of the spherical harmonics expansion in Fourier space. The FSC curve of the "optimal phasing" (grey dashed) shows only a minor increase in resolution to 3.1 Å indicating that the Monte Carlo search decreases the resolution by 0.1 Å. The remaining 0.2 Å difference to the optimal resolution of 2.9 Å at the given k_{cut} (not shown) is attributed to the finite expansion order L and the corresponding phasing errors.

We have also independently assessed the overall phasing error by calculating the intensity shell correlation (ISC) between the intensities of the phased electron densities $I_{phased} = |\mathcal{F}[\rho_{retrieved}]|^2$ and the intensities before phasing $I_{retrieved}$. The phasing method does not markedly deteriorate the structures.

Fig. 16.5 Electron densities retrieved from **a** 2.0×10^8, **b** 8.2×10^8, **c** 3.3×10^9 and **d** 3.3×10^{10} photons

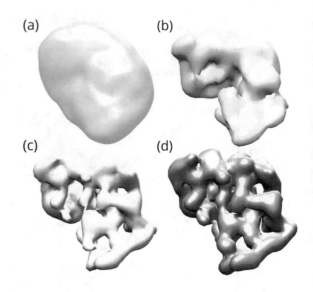

16.3.2 Impact of the Photon Counts per Image

The maximum number of triplets T that can be collected from an image with P photons is $T = P \cdot (P - 1) \cdot (P - 2)/6$. However, these triplets are not all statistically independent; instead, starting from 3 photons, each additional photon adds only two real numbers to the triple correlation: a new angle β (with respect to another photon) and a new distance k to the detector center.

The sampling of the three-photon correlation is improved by either collecting more photons per image P or by collecting more images I. However, because for each image, the orientation (3 Euler angles) needs to be inferred, the total amount of information that remains available for structure determination increases with the number of photons per image. Therefore, for every structure determination method, including ours, increasing P is preferred over increasing I, especially at low photon counts. For larger photon counts, the ratio between the 3 Euler angles and P becomes small and hence also the information asymmetry between P and I.

To assess this effect, we asked how the resolution depends on the number of images I and the photons per image P and therefore carried out additional synthetic experiments using image sets with 10, 25, 50 and 100 average photons P per shot at different image counts yielding different total number of photons. In Fig. 16.6, the achieved resolutions are shown as a function of the number of collected photons for four different $P = [10, 25, 50, 100]$. For the best achievable resolution of 3.3 Å, e.g., the total number of required photons decreases by a factor of 100 from 3.3×10^{10} to 3.3×10^8 photons (and the number of images decreased by a factor of 1000 from 3.3×10^9 to 3.3×10^6 images) when increasing the photons per image from 10 to 100, thus substantially decreasing the data acquisition time from over 20.000 min to only 30 min.

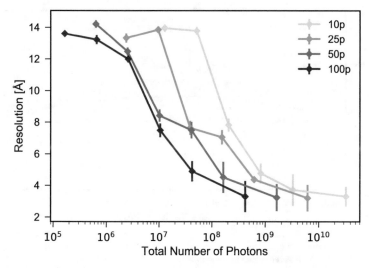

Fig. 16.6 The resolution as a function of the total number of photons collected from images with 10, 25, 50 and 100 photons on average

16.3.3 Structure Results in the Presence of Non-Poissonian Noise

To asses how additional noise (beyond the Poisson noise due to low photon counts) affects the achievable resolution, we have carried out synthetic scattering experiments including Gaussian distributed photons, $G(\mathbf{k}, \sigma) = (2\pi\sigma^2)^{-1/2} \exp\left(-|\mathbf{k}|^2/2\sigma^2\right)$ (see Fig. 16.7), as a simple noise model. From the generated scattering images, intensities $S(\mathbf{k})$ were determined with the discussed structure determination scheme.

Assuming that the noise is independent of the molecular structure, the obtained intensities $S(\mathbf{k}) = I(\mathbf{k}) + \gamma N(\mathbf{k})$ are a linear superposition of the molecules' intensity $I(\mathbf{k})$ and the intensity of the unknown noise $N(\mathbf{k})$. Accordingly, the noise was subtracted from $S(\mathbf{k})$ in 3D Fourier space using our noise model $N(\mathbf{k}) = G(\mathbf{k}, \sigma)$ and the estimated signal to noise ratio γ. Since the spherical harmonics expansion of a Gaussian distribution is described by a single coefficient $G_{l=0,m=0}(k) = G(k, \sigma)$ on each shell k, the noise subtraction simplified to $A_{l=0,m=0}^{\text{noise-free}}(k) = A_{l=0,m=0}^{\text{noisy}}(k) - \gamma G(k, \sigma)$.

As discussed in the main text, we assessed the effect of noise for different Gaussian widths ($\sigma = [0.5, 0.75, 1.125, 2.5]\,\text{Å}^{-1}$ and several signal to noise ratios $\gamma \in [10\%, ..., 50\%]$. Figure 16.7 compares the Crambin intensity (green) with the different Gaussian distributions (puples shades, black) at signal to noise ratio of $\gamma = 100\%$.

The Figure also shows the noise expected from Compton scattering (grey), which was estimated using the Klein-Nishina differential cross-section [25].

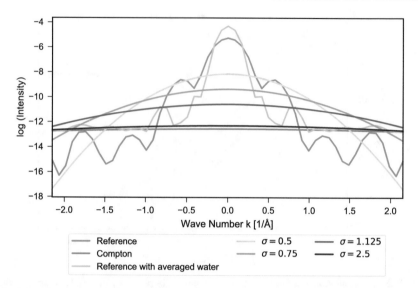

Fig. 16.7 Comparison of linear cuts through the normalized intensities of noise distributed according to Gaussian functions with widths $\sigma = [0.5, 0.75, 1.125, 2.5]\,\text{Å}^{-1}$ (purple shades and black), noise from Compton scattering (grey) and noise from the a disordered water shell of 5 Å thickness (aqua). A cut through the Crambin intensity without noise (green) is given for reference. Note that, due to the normalization in 3D, the noise intensities are shown at a signal to noise ratio $\gamma = 100\%$; at different signal to noise ratios, the noise intensities are shifted vertically with respect to the Crambin intensity

$$d\sigma = \frac{1}{2}\frac{\alpha^2}{m^2}\left(\frac{E'}{E}\right)^2\left[\frac{E'}{E} + \frac{E}{E'} - \sin^2\theta\right]d\Omega, \qquad (16.16)$$

with the scattering angle θ, the energy of the incoming photons E, the energy of the scattered photon $E' = E/(1 + \frac{E}{m}(1 - \cos\theta))$, the fine structure constant $\alpha = 1/137.04$ and the electron resting mass $m_e = 511\,\text{keV}/c^2$. As can be seen, the noise from Compton scattering (grey) is described well by a Gaussian distributions with width $\sigma = 2.5\,\text{Å}^{-1}$ (black), and thus was used to approximate incoherent scattering.

Finally, we also estimated the noise from the disordered fraction of the water shell by averaging the intensities of 100 Crambin structures with different 5 Å-thick water shells. The resulting intensity (aqua) is similar to the reference intensity with fewer signal in the intermediate regions ($0.2\,\text{Å}^{-1} < k < 1.0\,\text{Å}^{-1}$) and more signal in the center and the high-resolution regions ($k > 1.0\,\text{Å}^{-1}$). Since the noise of the water shell depends on the structure of the biomolecule, potentially combined with ordered water molecules, it is unlikely to be well described by our simple Gaussian model. Therefore, simple noise subtraction will be challenging, and more advanced iterative techniques will be required.

In Fig. 16.8, the electron densities from the discussed runs are compared to each other.

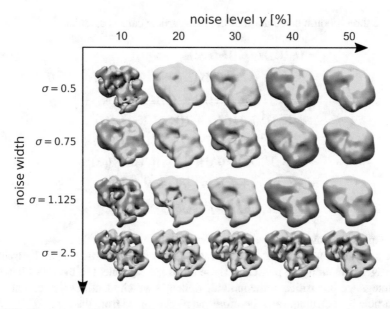

noise level γ [%]

10   20   30   40   50

noise width

$\sigma = 0.5$

$\sigma = 0.75$

$\sigma = 1.125$

$\sigma = 2.5$

Fig. 16.8 Comparison of the electron densities retrieved from images containing noise of different levels $\gamma \in [10\%, ..., 50\%]$ and widths $\sigma \in [0.5, 0.75, 1.125, 2.5]$

16.4   Structure Determination from Multi-Particle Images

Structure determination approaches are usually limited by the total number of single molecule shots that can be recorded. Remarkably, our method can process images with multiple illuminated particles because the two- and three-photon correlations of these images are connected to the correlations of the single particle shots. In order to show this relation, here, we derived the connection for the two-particle case.

The intensity of an image containing two randomly oriented particles $I_2(\mathbf{k})$ is the superposition of the the individual particle intensities' with the relative orientation being random,

$$I_2(\mathbf{k}) = \langle I(\mathbf{k}) + I_\omega(\mathbf{k}) \rangle_\omega \qquad (16.17)$$
$$= I(\mathbf{k}) + \langle I_\omega(\mathbf{k}) \rangle$$
$$= I(\mathbf{k}) + I^1(k).$$

The two-photon correlation then reads,

$$c^{(2)}_{k_1,k_2,\alpha} = \langle I_2(\mathbf{K}_1) I_2(\mathbf{K}_2) \rangle >_\omega \qquad (16.18)$$
$$= \langle I(\mathbf{K}_1) I(\mathbf{K}_2) + I(\mathbf{K}_1) I^1(k_2) + I^1(k_1) I(\mathbf{K}_2) + I^1(k_1) I^1(k_2) \rangle >_\omega$$
$$= c^{(1)}_{k_1,k_2,\alpha} + 3 I^1(k_1) I^1(k_2)$$

and the three-photon correlation of the two-particle case is calculated as,

$$
\begin{aligned}
t^{(2)}_{k_1,k_2,k_3,\alpha,\beta} &= \langle I_2(\mathbf{K}_1) I_2(\mathbf{K}_2) I_2(\mathbf{K}_3) \rangle_\omega \\
&= \langle (I(\mathbf{K}_1) + I^1(k_1))(I(\mathbf{K}_2) + I^1(k_2))(I(\mathbf{K}_3) + I^1(k_3)) \rangle_\omega \\
&= \langle I(\mathbf{K}_1) I(\mathbf{K}_2) I(\mathbf{K}_3) + I^1(k_1) I(\mathbf{K}_2) I(\mathbf{K}_3) + \\
&\quad I(\mathbf{K}_1) I^1(k_2) I(\mathbf{K}_3) + I(\mathbf{K}_1) I(\mathbf{K}_2) I^1(k_3) + \\
&\quad I^1(k_1) I^1(k_2) I(\mathbf{K}_3) + I^1(k_1) I(\mathbf{K}_2) I^1(k_3) + \\
&\quad I(\mathbf{K}_1) I^1(k_2) I^1(k_3) + I^1(k_1) I^1(k_2) I^1(k_3) \rangle_\omega \\
&= t^{(2)}_{k_1,k_2,k_3,\alpha,\beta} + I^1(k_2) c^{(1)}_{k_1,k_3,\beta} + I^1(k_1) c^{(1)}_{k_2,k_3,(\alpha-\beta)} + \\
&\quad I^1(k_3) c^{(1)}_{k_1,k_2,\alpha} + 4 I^1(k_1) I^1(k_2) I^1(k_3)
\end{aligned}
\tag{16.19}
$$

The expressions above are readily generalized to the N-particle case and the only remaining unknowns are the mixture ratios γ_i for the N_i-particles, i.e. the fraction of images containing N_i particles. These ratios are equivalent to the ratios between the integrated intensities of the individual images which identifies the total number of particle in each image and therefore can be calculated from the experimental data without additional effort.

The robustness of the two- and three-photon correlation in the presence of multiple particles in the beam potentially makes our method also interesting for other types of experiments such as fluctuation X-ray scattering (FXS) [26, 27] which is similar to solution scattering. In conventional solution scattering, the orientational averaging that occurs during the X-ray illumination results in signal which carries only 1-dimensional (radial) intensity information and all angular information is averaged out. In FXS experiments, however, the X-ray pulses from synchronous or free electron lasers are much shorter than the orientational diffusion times of the molecules such that they appear to be fixed in space. In each image multiple particles with different orientations are recorded and as a result speckle patterns emerge from which angular correlations can be calculated as described above.

References

1. Pruitt, K.D., Tatusova, T., Brown, G.R., Maglott, D.R.: NCBI reference sequences (RefSeq): current status, new features and genome annotation policy. Nucl. Acid. Res. **40**(D1), D130–D135 (2012). https://doi.org/10.1093/nar/gkr1079
2. Berman, H.M., Westbrook, J., Feng, Z., Gilliland, G., Bhat, T.N., Weissig, H., Shindyalov, I.N., Bourne, P.E.: The protein data bank. Nucl. Acid. Res. **28**(1), 235–242 (2000). https://doi.org/10.1093/nar/28.1.235, http://www.ncbi.nlm.nih.gov/pubmed/10592235http://www.pubmedcentral.nih.gov/articlerender.fcgi?artid=PMC102472
3. Gaffney, K.J., Chapman, H.N.: Imaging atomic structure and dynamics with ultrafast X-ray scattering. Science **316**(5830), 1444–1449 (2007). https://doi.org/10.1126/science.1135923, http://www.ncbi.nlm.nih.gov/pubmed/17556577

4. Hajdu, J.: Single-molecule X-ray diffraction. Curr. Opin. Struct. Biol. **10**(5), 569–573 (2000).   https://doi.org/10.1016/S0959-440X(00)00133-0,   http://www.google.de/search?client=safari&rls=10_7_4&q=Single+molecule+X+ray+diffraction&ie=UTF-8&oe=UTF-8&gws_rd=cr&ei=HFPVUo72FeWu4ATfsYCYCwpapers3://publication/uuid/813EAD12-8319-4370-9E9E-3B5B3E04F224

5. Huldt, G., Szoke, A., Hajdu, J.: Diffraction imaging of single particles and biomolecules. J. Struct. Biol. **144**(1–2), 219–227 (2003). https://doi.org/10.1016/j.jsb.2003.09.025,   http://ac.els-cdn.com/S1047847703001825/1-s2.0-S1047847703001825-main.pdf?_tid=6e21c252-3aec-11e7-9dc3-00000aab0f26&acdnat=1495017459_e37a85365e7eb5cfe1c99eb78e4121be

6. Miao, J., Ishikawa, T., Robinson, I.K., Murnane, M.M.: Beyond crystallography: diffractive imaging using coherent X-ray light sources. Science **348**(6234), 530–535 (2015). https://doi.org/10.1126/science.aaa1394

7. Neutze, R., Wouts, R., van der Spoel, D., Weckert, E., Hajdu, J.: Potential for biomolecular imaging with femtosecond X-ray pulses. Nature **406**(6797), 752–757 (2000). https://doi.org/10.1038/35021099,   http://www.google.de/search?client=safari&rls=10_7_4&q=Potential+for+biomolecular+imaging+with+femtosecond+X+ray+pulses&ie=UTF-8&oe=UTF-8&gws_rd=cr&ei=UVPVUpL1O8np4gSh8oGYAgpapers3://publication/uuid/FEE86AEB-32C4-4EBD-9E75-6A590DEA4E35

8. Bogan, M.J., Benner, W.H., Boulet, S., Rohner, U., Frank, M., Barty, A., Marvin Seibert, M., Maia, F., Marchesini, S., Bajt, S., Woods, B., Riot, V., Hau-Riege, S.P., Svenda, M., Marklund, E., Spiller, E., Hajdu, J., Chapman, H.N.: Single particle X-ray diffractive imaging. Nano Lett. **8**(1), 310–316 (2008). https://doi.org/10.1021/nl072728k

9. James, D.: Injection methods and instrumentation for serial X-ray free electron laser experiments. Ph.D. thesis (2015).   https://doi.org/10.1007/s13398-014-0173-7.2,   http://search.proquest.com/openview/993954d4328d83f473768bdf86a13b47/1?pq-origsite=gscholar&cbl=18750&diss=y

10. Nelson, G.: Sample injector fabrication and delivery method development for serial crystallography using synchrotrons and X-ray free electron lasers. Ph.D. thesis (2015). http://search.proquest.com/openview/fbeec51fc04b16d830c362b05954da27/1?pq-origsite=gscholar&cbl=18750&diss=y

11. Weierstall, U., James, D., Wang, C., White, T.A., Wang, D., Liu, W., Spence, J.C.H., Bruce Doak, R., Nelson, G., Fromme, P., Fromme, R., Grotjohann, I., Kupitz, C., Zatsepin, N.A., Liu, H., Basu, S., Wacker, D., Han, G.W., Katritch, V., Boutet, S., Messerschmidt, M., Williams, G.J., Koglin, J.E., Marvin Seibert, M., Klinker, M., Gati, C., Shoeman, R.L., Barty, A., Chapman, H.N., Kirian, R.A., Beyerlein, K.R., Stevens, R.C., Li, D., Shah, S.T.A., Howe, N., Caffrey, M., Cherezov, V.: Lipidic cubic phase injector facilitates membrane protein serial femtosecond crystallography. Nat. Commun. **5**, 3309 (2014). https://doi.org/10.1038/ncomms4309, http://www.ncbi.nlm.nih.gov/pubmed/24525480,   http://www.pubmedcentral.nih.gov/articlerender.fcgi?artid=PMC4061911,   http://www.pubmedcentral.nih.gov/articlerender.fcgi?artid=4061911&tool=pmcentrez&rendertype=abstract

12. Inhester, L., Groenhof, G., Grubmueller, H.: Auger spectrum of a water molecule after single and double core-ionization by intense X-ray radiation. Biophys. J. **102**(3), 392A–392A (2012).   http://gateway.webofknowledge.com/gateway/Gateway.cgi?GWVersion=2&SrcAuth=mekentosj&SrcApp=Papers&DestLinkType=FullRecord&DestApp=WOS&KeyUT=000321561202570papers3://publication/uuid/8FCEB065-61AB-4E2E-9B4E-0147C14B1415

13. Luke, D.R.: Relaxed averaged alternating reflections for diffraction imaging. Inverse Probl. **37**(1), 13 (2004). https://doi.org/10.1088/0266-5611/21/1/004, http://arxiv.org/abs/math/0405208

14. Schoenlein, R.: New science opportunities enabled by LCLS-II X-ray lasers. Technical report (2015).   https://portal.slac.stanford.edu/sites/lcls_public/Documents/LCLS-IIScienceOpportunities_final.pdf

15. Kam, Z.: Determination of macromolecular structure in solution by spatial correlation of scattering fluctuations. Macromolecules **10**(5), 927–934 (1977). https://doi.org/10.1021/ma60059a009

16. Kam, Z.: The reconstruction of structure from electron micrographs of randomly oriented particles. J. Theor. Biol. **82**(1), 15–39 (1980). https://doi.org/10.1016/0022-5193(80)90088-0, http://www.google.de/search?client=safari&rls=10_7_4&q=The+reconstruction+of+structure+from+electron+micrographs+of+randomly+oriented+particles&ie=UTF-8&oe=UTF-8&gws_rd=cr&ei=PlPVUuG4Jumo4ASH3oDwCApapers3://publication/uuid/619E60E4-19CE-4532-80D1-0134D6

17. Baddour, N.: Operational and convolution properties of three-dimensional Fourier transforms in spherical polar coordinates. J. Opt. Soc. Am. A **27**(10), 2144 (2010). https://doi.org/10.1364/JOSAA.27.002144, http://josaa.osa.org/abstract.cfm?URI=josaa-27-10-2144

18. Wigner, E.P.: On the Matrices Which Reduce the Kronecker Products of Representations of S. R. Groups. In: Quantum Theory of Angular Momentum, pp. 87–133. Springer, Berlin, Heidelberg (1965). http://link.springer.com/10.1007/978-3-662-02781-3_42

19. von Ardenne, B., Mechelke, M., Grubmüller, H.: Structure determination from single molecule X-ray scattering with three photons per image. Nat. Commun. **9**(1), 2375 (2018). https://doi.org/10.1038/s41467-018-04830-4

20. Saldin, D.K., Shneerson, V.L., Fung, R., Ourmazd, A.: Structure of isolated biomolecules obtained from ultrashort X-ray pulses: exploiting the symmetry of random orientations. J. Phys. Condens. matter Inst. Phys. J **21**(13), 134,014 (2009). https://doi.org/10.1088/0953-8984/21/13/134014, http://www.google.de/search?client=safari&rls=10_7_4&q=Structure+of+isolated+biomolecules+obtained+from+ultrashort+x+ray+pulses+exploiting+the+symmetry+of+random+orientations&ie=UTF-8&oe=UTF-8&gfe_rd=cr&ei=yFPVUoC7FeWG8Qfh9IGICgpapers3://publication/uuid

21. Mezzadri, F.: How to generate random matrices from the classical compact groups. arXiv:math-ph/0609050 (2006)

22. Rosenthal, J.S.: Random rotations: characters and random walks on SO(N). Ann. Probab. **22**(1), 398–423 (1994). https://doi.org/10.1214/aop/1176988864, http://www.jstor.org/stable/2244511

23. Barty, A., Küpper, J., Chapman, H.N.: Molecular imaging using X-ray free-electron lasers. Annu. Rev. Phys. Chem. **64**(1), 415–435 (2013). 10.1146/annurev-physchem-032511-143708

24. Van Heel, M., Schatz, M.: Fourier shell correlation threshold criteria. J. Struct. Biol. **151**(3), 250–262 (2005). https://doi.org/10.1016/j.jsb.2005.05.009, http://ac.els-cdn.com/S1047847705001292/1-s2.0-S1047847705001292-main.pdf?_tid=e1dad656-2a67-11e7-a019-00000aacb35d&acdnat=1493201311_8120e3e22f5ad836cf2fac5a2533828a

25. Klein, O., Nishina, T.: Über die Streuung von Strahlung durch freie Elektronen nach der neuen relativistischen Quantendynamik von Dirac. Zeitschrift für Physik **52**(11–12), 853–868 (1929). https://doi.org/10.1007/BF01366453

26. Donatelli, J.J., Zwart, P.H., Sethian, J.A.: Iterative phasing for fluctuation X-ray scattering. Proc. Natl. Acad. Sci. **112**(33), 10286–10291 (2015). https://doi.org/10.1073/pnas.1513738112. http://www.ncbi.nlm.nih.gov/pubmed/26240348, http://www.pubmedcentral.nih.gov/articlerender.fcgi?artid=PMC4547282, http://www.pnas.org/lookup/doi/10.1073/pnas.1513738112

27. Kirian, R.A.: Structure determination through correlated fluctuations in X-ray scattering. J. Phys. B At. Mol. Opt. Phys. **45**(22), 223,001 (2012). https://doi.org/10.1088/0953-4075/45/22/223001, http://stacks.iop.org/0953-4075/45/i=22/a=223001?key=crossref.66de6e4d5e36e0c0aa6803c8da621c4d

Chapter 17
Development of Ultrafast X-ray Free Electron Laser Tools in (Bio)Chemical Research

Simone Techert, Sreevidya Thekku Veedu and Sadia Bari

Abstract The chapter will focus on fundamental aspects and methodological challenges of X-ray free electron laser research and recent developments in the related field of ultrafast X-ray science. Selected examples proving "molecular movie capabilities" of Free-electron laser radiation investigating gas phase chemistry, chemistry in liquids and transformations in the solid state will be introduced. They will be discussed in the context of ultrafast X-ray studies of complex biochemical research, and time-resolved X-ray characterisation of energy storage materials and energy bionics.

17.1 Introduction

After a preparation phase of almost twenty years—from the first vision of a common research effort in synchrotron radiation in Europe (1975) up to the first electron beam entering the first high-brilliant, third generation X-ray synchrotron ring in 1992 (at the European Synchrotron Radiation Facility)—synchrotron radiation has proven its unique X-ray photons characteristics in brilliance, coherence, pulsed properties and high-repetition frequencies—allowing for the development of various synchrotron-typical experiments such as high-resolution X-ray crystallography and anomalous X-ray scattering, the various types of X-ray spectroscopy techniques, X-ray time-resolved methods and coherent and incoherent X-ray diffraction techniques (to name some examples). The techniques have been adapted for and applied to material research, chemistry, solid state and biophysics, earth and planetary research, astrophysics etc.

S. Techert (✉) · S. Thekku Veedu · S. Bari
FS-Strukturdynamik (bio)chemischer Systeme, Deutsches Elektronen-Synchrotron DESY, Notkestraße 85, 22607 Hamburg, Germany
e-mail: simone.techert@desy.de

T. Salditt et al. (eds.), *Nanoscale Photonic Imaging*, Topics in Applied Physics 134,
https://doi.org/10.1007/978-3-030-34413-9_17

In the begin of the 2nd millennium, synchrotron researchers faced a similar situation as in begin of 1980s—again with a 20 years agenda from 2000 on—entering their next research "quantum step" by extending the novel technique of the so-called Free-Electron Laser (FEL) principle [1–6] from short arrangements of water-cooled magnets to very long arrangements of water-cooled (and now even superconducting) magnets [1–6]. Since in free-electron lasers high-speed electrons freely move in a magnetic field [1, 2], being accelerated and decelerated in the field, and by thus generating radiation, changing the properties of the magnetic fields as explained allow for moving from the FEL-generation of micro-waves [2, 4] to the ones of X-ray radiation [4–6]. Free-electron lasers are therefore fully tunable and they have the widest frequency range of any laser type.

By developing the FEL-principle to the hard X-ray regime [4–7]—a gap between the "traditional" synchrotron radiation world and the FEL radiation world has been closed—one of the reasons why FEL radiation is also sometimes called "synchrotrons of the 4th generation" (at least from the perspective of synchrotron researchers). As a consequence, the properties of the novel FEL sources and the properties of the produced X-ray radiation have been amplified by many orders of magnitudes compared to current synchrotron sources of the 3rd generation.

Milestones in the FEL development agenda were the first soft X-ray free-electron laser in operation—2005 the *Free-Electron LASer* FLASH at DESY in Hamburg, the first hard X-ray free-electron laser in operation—2009 the *Linear Coherent Light Source* LCLS at SLAC in Stanford (which is a so-called 1st generation of FELs based on a low-repetition frequency approach), and the first high-repetition frequency hard X-ray FEL in operation—2017 the *European X-ray Free Electron Laser* European XFEL at DESY and in Schenefeld. How these "quantum jumps" in X-ray properties can be made used for novel approaches of research in chemistry will be topic of this chapter.

In order to reflect milestones developments from 2006 on, the chapter reflects recent research summaries [8–10] as well as novel developments.

If one reads about chemical reactions and the measurement of real-time responses in chemical reactions, one very quickly ends up with the concept of the resolution of measurements on ultrafast time scales, which are characteristic for the movement of atoms in molecules, i.e. femtoseconds. A femtosecond is defined as the millionth of a billionth of a second. Life-relevant motions, however, can be as slow as seconds or even up to minutes' or hours' time scales. The origin of these time scale differences is based on the complexity of the coordination space of a proceeding reaction [8–14].

In a classical kinetic scheme, following the explanatory approach, the gradients on the potential energy hypersurface define the molecular dynamics. Statistically population-weighted, they compose to the kinetics of a chemical reaction. The coordinates describing the dynamics and kinetics of a chemical reaction are the reaction coordinate and the energy. The reaction coordinate is defined as a one-dimensional projection of the reactant's and product's normal coordinates, which span the potential energy hypersurface of reactant and product and the potential energy hypersurfaces of their transitions (Fig. 17.1). The energy gradient along a reaction coordinate

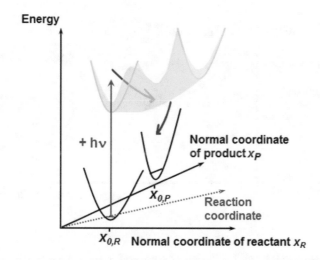

Energy

+ hν

Normal coordinate of product x_P

$x_{0,P}$

Reaction coordinate

$x_{0,R}$ **Normal coordinate of reactant** x_R

Fig. 17.1 References [11–16]: Potential energy hyper-surface of a chemical reaction. The reaction coordinate (abscissa) of a chemical reaction is defined as a one-dimensional projection of the normal coordinates spanning the reactant's potential energy hypersurface versus the normal coordinates spanning the product's potential energy hypersurface, and their corresponding transitions [11, 14]. The figure is adapted from the [8–10] and references therein

is defined as reaction dynamics, the energy gradient along the normal coordinates as molecular dynamics.

Commonly, the potential energy is presented as the characteristic curve or hyper-surface in the graph, and the ordinate axis presents the sum of the potential and kinetic energy of the nuclei involved in the chemical reaction (Fig. 17.2). Potential and kinetic energy of molecules can be detangled through the projection of the potential energy onto the total energy axis. The activated complex and transition state (according to Eyring) includes an imaginary vibrational mode [11–14, 16].

From a chemical physicist's point of view, one would like to understand which elementary chemical processes happen at which time scales, and how these time scales are inter-connected. To what extend do structural motifs "freeze in" time and dynamics information of chemical reactions? Which type of apparatus needs to be built and which kind of methods need to be developed for investigating the created femtosecond "time stamps" in the structure of complex matter during a chemical or biochemical reaction?

Ultrafast X-ray methods bear the potential of determining the complexity of chemical reactions—*during their reactions*, in particular in the bulk, with techniques which utilize the specific characteristics of X-ray/matter interaction: in well-ordered systems, X-ray crystallography as a Thomson scattering process allows for element-specific determinations such as electron densities (redox states) and high-precision spatial resolution determination of atoms in lattices from which hydrogen bonding, chemical bonding or van der Waals stacks can be derived. In less-ordered and disordered systems, X-rays deliver element-specific information of the investi-

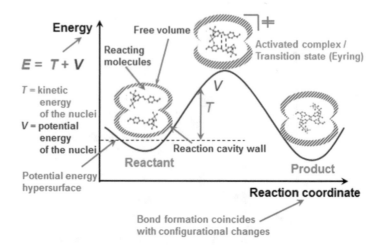

Fig. 17.2 References [11–16]: Total energy of a chemical system as the sum of potential energy (V) and kinetic energy (T) of the molecules (ordinate). Inside the graph, only the contribution of the potential energy is plotted; its projection onto the ordinate allows determining the kinetic energy [12, 16]. The figure is adapted from the [8–10] and references therein

gated molecules utilizing, for example, X-ray spectroscopy, X-ray absorption, photo-induced electron cascades, or X-ray and electron emission properties. Site-specific information is obtained by elastic and inelastic scattering processes such as multidimensional X-ray spectroscopy, X-ray diffraction, or X-ray scattering. The chemical consequences of X-ray/matter interactions leading to fragmentation can be characterized by X-ray mass spectrometry.

Characteristic for the X-ray photons generated in synchrotrons and free-electron lasers are their

1. energy tunability (allowing for excitation-energy-sensitive methods like X-ray spectroscopy or advanced X-ray diffraction methods)
2. pulsed structure (allowing for in-situ and time-resolved X-ray methods)
3. defined polarization (allowing for advances in X-ray spectroscopy)
4. coherence (allowing for X-ray imaging or correlation spectroscopy methods)
5. high flux (allowing for high-resolution X-ray experiments in all experimental domains)

Fourth generation accelerator-based light sources (free-electron lasers, FELs) in the VUV or X-ray regime deliver ultra-brilliant coherent radiation in very short pulses (10^{12} to 10^{13} photons/bunch/10–100 fs). In order to fully exploit their unique photon capabilities, novel instrumentation is required based on single-shot (collection) schemes. Moreover, hundreds up to trillions of fragment particles, ions, electrons or scattered photons can emerge when a single light flash impinges on matter with intensities up to (predicted for the XFEL) 10^{22} W cm^{-2}. In order to meet these challenges, in the starting time of FLASH (Free-Electron Laser in Hamburg [16a]) and the LCLS (Linac Coherent Light Source [16b]), various experimental chambers and endstations have been designed. Further constructions, in particular for relevance for this chapter have been developed and are under construction at the European XFEL [16e], as shortly been explained in the following.

Starting from this introduction as a comprehensive summary of the basic principles, in the following, various FEL methods developed so far will be summarised, including the concept of *filming chemical reactions in real time utilizing ultrafast high-flux X-ray sources* [11–16], *X-ray Diffraction and Crystallography for Condensed State Chemistry Studies—Crystallography with Ultrahigh Temporal and Ultrahigh Spatial Resolution* [14, 17–52] and *Applications in organic electronics and energy research* [14, 53–59]; from *From Local to Global: Ultrafast Multidimensional Soft X-ray Spectroscopy and Ultrafast X-ray Diffraction Shake Their Hands* [60–76] and *Applications in Bimolecular Reaction Studies and Photocatalysis* [77–80] to *Applications in Unimolecular Liquid Phase Reaction Dynamics* [81–89] and *Applications in Bioelectronics, Aqueous and Prebiotics Reaction Dynamics* [90–108] to *Applications in Biophysics and Gas Phase Biomolecules* [109–121]; from *Ultrafast Imaging of Gas Phase Reactions* [122–127] and *Applications in Nanoscience and Multiphoton-Ionisations* [128–150] to *Applications in Unimolecular Gas Phase Dynamics* [122, 151–163] and *Outlook*.

17.2 The Concept: Filming Chemical Reactions in Real Time Utilizing Ultrafast High-Flux X-ray Sources

In a proof-of-principle experiment at the white beam beamline at the ESRF (The European Synchrotron Research Facility) in 2001, it has been demonstrated that high-flux, pulsed X-rays—as created with synchrotrons of the 3rd generation—can act as the "photons of choice" for studying the dynamics and kinetics of small chemical systems on their complex reaction landscape [1]. These studies have been used to define various expectation values for time-resolved experiments at free-electron lasers and saddling the ground for ultrafast X-ray experiments at these sources. Since then, also the phrase of *recording the molecular movie* has been born (Fig. 17.3) [2–6b].

Figure 17.3 summarizes the principles of such a "molecular movie" approach: after the initialization of a chemical reaction with a short laser pulse, ultrafast X-ray FEL snapshots take photographs of the X-ray spectroscopic or X-ray diffraction signal. By varying the time delay between laser pump and X-ray probe pulse, information about the structural changes as a function of time are collected.

Time-wise, the criterion for "recording the molecular movie" is given when the time resolution of the pump and probe sources meet the time scales of the *structural dynamics* investigated. The resolution criterion for *structural dynamics* studies is fulfilled in chemistry, when the refined structure allows for determining the electron

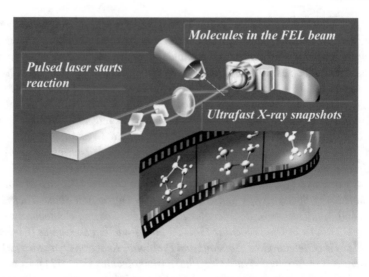

Fig. 17.3 References [11–16]: Principle of the "molecular movie". After the initialization of a chemical reaction by a pulsed trigger laser, ultrafast X-ray snapshots and photographs of the X-ray pulses are collected as a function of time (courtesy by DESY, MPIbpC and European XFEL)

density or the charge densities (which are equal in the redox state) around a moving atom. High-resolution X-ray crystallography studies contains such precision [2, 8–16, 21, 23, 25, 35, 44, 46, 51].

17.3 X-ray Diffraction and Crystallography for Condensed State Chemistry Studies—Crystallography with Ultrahigh Temporal and Ultrahigh Spatial Resolution

Crystallography with ultrahigh temporal and spatial resolution allows studying photochemical reactions beyond conventional quantum chemical approaches. Far beyond any present laboratory technique, time-resolved synchrotron (picosecond time resolution) and FEL (femtosecond time resolution) experiments emphasize the uniqueness of the pulsed, ultrafast, high brilliant and coherent X-ray methods and metrology. For chemical bond breaking and bond formation, the criterion for spatial resolution is met in periodic systems (crystallography) when 0.01–0.001Å resolution diffractograms yield high precision structural information [8–16, 21–23, 25, 35, 44, 46, 51].

Figure 17.4 reflects the changes of X-ray synchrotron beam characteristics when evolving from synchrotrons of the 2nd generation towards hard X-ray free electron lasers. The diffractograms have been collected on a molecular crystal of the same species, the same crystal quality and the same orientation. Utilizing broadband

Average Brilliance / ph/(sec mrad² mm² 0.1% bw)

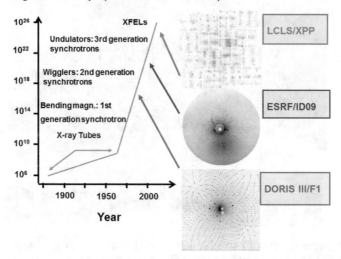

Fig. 17.4 References [17–52]: Evolution of the average brilliance of synchrotron radiation light of synchrotrons of the 2nd generation (DORIS) though the ESRF towards LCLS free-electron laser—characterized at the high-resolution diffraction pattern of the same molecular crystal system with same crystal quality (courtesy by DESY, ESRF and LCLS). The figure is adapted from the [8–10] and references therein

wiggler radiation in 2nd generation synchrotrons (F1/DORIS) Laue diffraction pattern have been collected. Taped undulator radiation of synchrotrons of the 3rd generation yields in quasi Pink-Laue diffractograms (ID09/ESRF).

Compared to the Pink Laue white beam at XPP as well as CXI beamline of LCLS, the FEL radiation is about one to two orders of magnitude smaller in bandwidth, allowing only the investigation of a statistical number of Bragg reflections for small molecular crystals when the crystal is rotated.

Small molecular crystallography at free-electron lasers could be quantitatively utilized for structure determination by the combination of traditional Laue crystallography and FEL- specific serial crystallography techniques with rapid sample exchange and based on a single shot data collection strategy analogous to the developments of time-resolved Laue diffraction at synchrotrons of the 3rd generation.

In contrast to conventional Laue crystallography, for normalization purposes during FEL experiments, every diffractogram needs to be associated to an online collected X-ray spectrum. Utilizing high X-ray energies well above 15 keV (use of the third harmonic and smaller X-ray/atom cross section) with no monochromatization in pink Laue modus reduces radiation damage so that with a monotonically running spindle and randomly changing X-ray wavelength with known X-ray spectral characteristics, various orientations under defined X-ray conditions can be collected. They are sufficient for determining the orientation matrix of small molecules and hence following the indexing of the collected diffractograms.

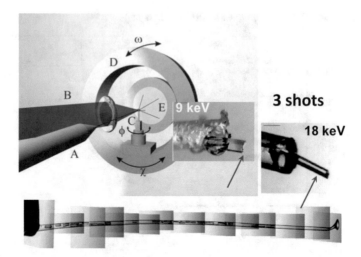

Fig. 17.5 References [17–52]: Ultrafast time-resolved small molecule crystallography at LCLS, XPP and CXI beamlines (interim set-up installed during the first in-house test phase of LCLS). Left side, top: used three circle diffractometer. Right side, top: organic crystals exposed to 9 keV X-ray radiation (single shot) inducing severe radiation damage and 18 keV radiation allowing for the collection of various quasi Laue diffractograms with the Laue pink FEL beam under different orientations. Bottom: small molecule crystals lined up for a three-shot-serial type of crystallography experiment. The figure is adapted from the [8–10] and references therein

Since the studied materials are normally compounds of small amounts, in the presented approach highest quality crystals are stacked behind each other in a capillary (or other type of sample target holder) and high energy X-ray radiation is utilized for collecting high resolution diffraction patterns and minimizing accumulative radiation damage. Due to the monochromaticity of FEL X-ray beams even in its pink Laue mode, quasi Pink Laue diffraction pattern will be recorded for various orientations—allowing a precise determination of the orientation matrix of the small molecule crystals. Additionally, on a single shot base, the Bragg peak intensities are wavelength and X-ray intensity normalized (Fig. 17.5).

In the current example, unravelling the mechanisms of electron-transfer induced structural changes (ET), and the interplay of charge-transfer (CT) and structural reorganisations in complex molecular systems is of prime importance in both chemical and biological processes and has gained considerable importance over decades. ET/CT and the associated structural changes are pervasive, path to inevitably diverse processes occurring in nature, including light harvesting systems like photosynthesis to technological innovations such as solar energy converters and optical devices. It is challenging to trap the fleeting excited species or transient species and gain insight into the unprecedented dynamics which occurs in ultrafast time scale in the range of as to ps. Emerging scientific innovations in many synchrotrons and FELs, with high brilliance and extremely high photon flux have wide opened a new domain of research in investigating many ultrafast processes. Information about what happens

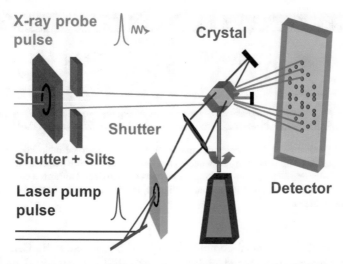

Fig. 17.6 References [17–52]: Optical laser pump/X-ray probe set-up for ultrafast X-ray crystallography. By varying the time-delay between optical laser pump and X-ray probe beam for different crystallographic orientations, X-ray diffraction snapshots as a function of time are collected. Adding the diffraction signals for one time-delay into one Ewald sphere allows an analysis of structural evolution as a function of time

immediately after the triggering of the reaction is satisfactory but to evaluate what happens next is crucial. In order to tackle these questions, we must know how fast the processes happen and what are the time scales of each processes we are interested in.

By employing time-resolved X-ray diffraction (TR-XRD) technique, the CT/ET reaction has been initiated by a laser pump pulse from the ground state to the excited state or the CT/ET state. The photo-induced subsequent structural changes have been monitored utilizing the pulsed structure of the X-ray beam by varying the time delay between the optical laser pump and X-ray probe pulses (pump/X-ray probe scheme) as been shown in Fig. 17.6.

In the following examples of structural dynamics of organic and metallo-organic systems by trapping charge transfer states using high brilliance Pink Laue time-resolved X-ray diffraction are presented. The photo-induced dynamics has been studied by looking from simple organic donor acceptor molecules (Fig. 17.7) to complex inorganic single crystalline systems. In the former the structural changes are induced by an ET/CT of an organic donor to an organic acceptor moiety, and in the later it is a induced by a metal to ligand charge transfer (MLCT) (see following chapter). Due to their high photo-conversion efficiencies and lifetimes, first are highly suited for solar cell developments, and last are highly suited as photo-catalysts.

The experiments have been performed combining high resolution static single crystal XRD studies utilizing a home-source (Bruker Apex II) system and P11 beamline of PETRA III at DESY. To map out the details of the potential landscapes, ultrafast, light-induced Laue diffraction studies of the pure organic

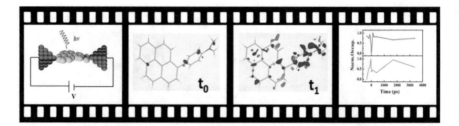

Fig. 17.7 References [17–52]: Molecular movie of a non-conventional molecular diode which cannot be described within the Born-Oppenheimer approximation. Within femtoseconds and nearly immediately after optical photo-absorption the electrons are rearranged towards a conducting state meanwhile the structure knocks into a tilted configuration. The figure is adapted from the [8–10] and references therein

system as well as the metal-complex have been performed at BioCARS, 14-ID beamline/Argonne (picosecond pink-Laue), ID09/ESRF (picosecond pink-Laue), XPP/CXI/LCLS (femtosecond pink-Laue) and P11/PETRA-III (picosecond, monochromatic). In the pure organic system within few hundreds of picoseconds electron density migration occurs and the structural changes are clearly visible from the photo difference map with prominent differences in the electron density maps at the electron donor and electron acceptor system (Fig. 17.7). In the metal-complex, it has been oberseved that upon photo-excitation the electron migration or the charge migration is mostly on the proximal atoms of the organic ligands from the metal center (not shown here).

Figure 17.7 depicts the refined result of such an experiment- the femtosecond structural dynamics or "molecular movie" of a molecular crystal, which consist of only light elements (carbon, nitrogen). The patented system has the most efficient optical light/electron transfer rate possible (100%), by utilizing quantum effects such as electron and structural dynamics pathways which cannot be described through the conventional Born-Oppenheimer approximation.

17.4   Applications in Energy Research

Understanding this "beyond Born-Oppenheimer" behaviour and combining the properties of this type of system with smart semiconducting plastic types of compounds, it is possible to build flexible, solid, plastic-type of solar cells and organic light emitting diodes with very high efficiency (Fig. 17.8).

Both as a consequence of successive technical developments, and based on the chemical rules and time laws derived during the various method development process, it has become possible to optimize functional performances of solar cell organic materials and devices (Fig. 17.8).

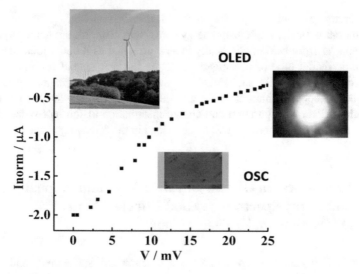

Fig. 17.8 References [53–59]: Once structural relaxation processes are understood the derived structural and dynamical properties can be utilized to improve functional dynamics performance allowing developing new classes of organic solar cells (OSC) (see current-voltage (I–V)curve) or organic light emitting diodes (OLED) based on plastic (see inset). The figure is adapted from the [8–10] and references therein

By moving the fundamental method developments into the application regime (again as a proof for the methods' developments), the circle of promises closes that 3rd and 4th generation X-ray sources may help to optimize strategies for modern material performances. In order to test whether the structural dynamics laws derived through the "molecular movie" approach can have direct consequences for applications of the molecular system, electronic devices such as organic solar cells and organic light emitting diodes have been built and Fig. 17.8 presents such efforts.

As can be seen on the current voltage (I–V) curve, when combining with semi-conducting plastic material (which by itself is not photo active), efficient organic solar cells with high mechanic flexibility and cheap in production can be designed.

Since the time-resolved X-ray methods allow for detangling "local-to-global" and "global-to-local" structural responses, desired functional actions of a device like energy storage can be distinguished from "energy-eating" processes based on non-desired heating and energy quenching processes. Figure 17.7 presents the structural mechanisms underlying an optimized all-over organic solar cell. Small atomic changes on the light-absorbing chromophore unit lead to a complete switching of its functional dynamics—from light absorbing solar cell devices [53–59] to a light emitting organic diode [54]. In another example the understanding of the crystallization processes of organic material out of time-resolved X-ray diffraction (TR-XRD) studies has been influenced by the optimization of the recycling process of molten PET bottles to ultra-hard polyethylene [56]. Such ultra-hard plastic material is currently used in every 2nd wind craft machine produced world-wide.

The current examples emphasize, however, that the real world, functional materials, pharmaceuticals, catalysts or energy converting materials are not always crystalline and far from being periodically ideally arranged as it could look like when performing model type of investigations. If the intrinsic spatial resolution of the system does not allow for such detailed investigations, a combination of ultrafast X-ray spectroscopy and ultrafast X-ray diffraction or scattering as the "local to global approach" deliver configuration and charge information of the molecules studied [17–52, 60–76]. This approach will be described in the following chapter.

17.5   From Local to Global: Ultrafast Multidimensional Soft X-ray Spectroscopy and Ultrafast X-ray Diffraction Shake Their Hands

As ultrafast X-ray spectroscopy and X-ray diffraction can "shake their hands" in X-ray Free-electron laser science, they open up new ways to study complex chemical reactions.

X-ray diffraction [17–59] and X-ray spectroscopy [60–76] are complementary techniques. X-ray spectroscopy allows probing the electronic properties in an element-, orbital- and site-specific way, for example bonding or oxidation state changes [63]. Since one gets local information for the system under investigation the method is referenced as the *local approach*. In X-ray diffraction the structural changes of the whole bulk are probed which is termed as the *global approach*. With these approaches it is possible to get the overall structural properties of the target system. Furthermore, by applying both methods more complete information can be obtained. However, to follow whole reaction pathways or reaction intermediates between the start and end of a reaction, the experimental approaches have to be extended using the time-resolved method, i.e. the pump-probe scheme, as described in the previous section and shown in Fig. 17.9.

With the availability of the first X-ray FEL, FLASH at DESY in Hamburg, Germany, it became possible to implement X-ray spectroscopy and diffraction for investigating ultrafast chemical reaction processes. In the following, we present two pioneering soft X-ray experiments performed at the FEL facilities FLASH and LCLS.

The first example is a time-resolved X-ray diffraction measurement on silver-behenate [60–76] with an outlook towards combination of the liquid jet with X-ray spectroscopy, the second example is a femtosecond time-resolved X-ray spectroscopy experiment on iron-pentacarbonyl ($Fe(CO)_5$) [78–81]. Figure 17.10a shows the apparatus which has been developed for such experiments at FELs. The ultrafast X-ray diffraction of a silver-containing redox system embedded in a supramolecular organic structure has been studied (Fig. 17.10b). The experiment is a proof-of-principle, utilizing FEL radiation for ultrafast X-ray diffraction of chemical systems in real-time. By investigating the time-evolution of the Bragg reflections (Fig. 17.10c) complex

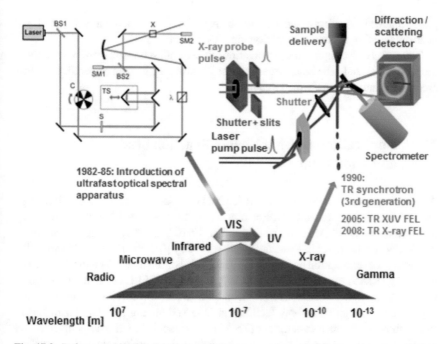

Fig. 17.9 References [60–76]: Principle of various types of pump-probe experiments applied to the optical region (left) and the corresponding type of experiments in the X-ray region (right). An optical laser pump initiates a chemical reaction and an optical (left) or X-ray laser pulse (right) probes the proceeding reaction pathways. Right: Both the X-ray spectroscopy and X-ray diffraction signal can be recorded, yielding complementary information. The figure is adapted from the [8–10] and references therein

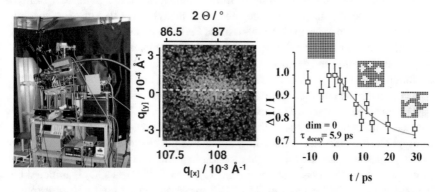

Fig. 17.10 References [60–76]: **a** The X-ray photon endstation at the FEL, FLASH at The modular built-up endstation has been used for the experiment presented in (**b**) and (**c**). **b** Bragg diffraction peak (110) of silver-behenate studied by single-shot FEL pulses. **c** The time-dependent behavior of the Bragg peak after photo-excitation. The figure is adapted from the [8–10] and references therein

photo-induced transformation kinetics of partially photo-chemically induced and heat-propagation-influenced reaction kinetics has been found.

The decay curve depicts the propagation of the ultrafast transformation throughout the whole material. The black areas present the non-transformed material, the white areas the island of transformed material.

17.6 Applications in Bimolecular Reaction Studies and Photocatalysis

Another experiment in the field of soft X-ray spectroscopy has recently been worked out for studying photocatalysis. For pioneering time-resolved X-ray spectroscopy measurements the FLASH end station in Fig. 17.10 has been modified by adding a soft X-ray spectrometer in Rowland circle geometry [63, 66, 77–80]. This so called Liquid Jet (LJ) end station has then been successfully used in resonant inelastic X-ray scattering (RIXS) experiments on Fe(CO)$_5$ at the LCLS-FEL at SLAC in Stanford, CA, USA. Similar end stations have been built to perform soft X-ray spectroscopy experiments at high flux X-ray facilities at DESY/Hamburg [15] and HZB/Berlin. In the above mentioned experiment [77–80], the dissociation of iron-pentacarbonyl (Fe(CO)$_5$) in ethanol has been studied in real-time. After optical excitation with 266 nm photons, Fig. 17.11a, Fe(CO)$_5$ dissociates into iron-tetracarbonyl (Fe(CO)$_4$) and carbon monoxide (CO) under solvent-assistance. For every time-delay, the incident monochromatic FEL photon energy has been scanned over the Fe $2p$ edge, and the X-ray emission spectra have been recorded.

The time evolution was deduced from differences of the pumped (positive time-delay) and the un-pumped (negative time-delay) Fe(CO)$_5$ emission spectra. Figure 17.11 summarizes the ethanol-assisted Fe(CO)$_5$ photo-dissociation pathways and simulations are compared to the experiment. The complexity of reaction increases due to the formation and decay of a triplet state, where Fig. 17.11b–e summarises the evolution of the valence-electronic structure of Fe(CO)$_4$ in ethanol upon femtosecond spin crossover and complex formation.

The involved orbitals are assigned according to the Fe $2p$ and $3d$, or ligand $2p$ characteristics, and according to the symmetry along the Fe–CO bonds. The star (*) in Fig. 17.11b, marks the antibonding orbitals of the electron configuration of the Fe(CO)$_5$ ground state and the single-electron transitions of the laser pump—X-ray probe processes. The dissociation of (Fe(CO)$_5$ $\longrightarrow$ Fe(CO)$_4$ + CO) is initiated by the optical $d_\pi \rightarrow 2\pi^*$ excitation. The RIXS measurements at the Fe L$_3$-absorption edge to final valence-excited states involves the probing of the $d_\pi \rightarrow d_\sigma^*$ transition. In Fig. 17.11(c, top), the difference RIXS spectra (RIXS intensity of incident photon energy versus energy transfer in eV) of the summed pumped and un-pumped sample is illustrated. Figure 17.11(c, bottom) displays the time structure in the regions (1–4) in comparison to simulated populations of the excited (E) and triplet (T) states and ligated complex (L). Figure 17.11d shows the measured RIXS spectra for Fe(CO)$_5$

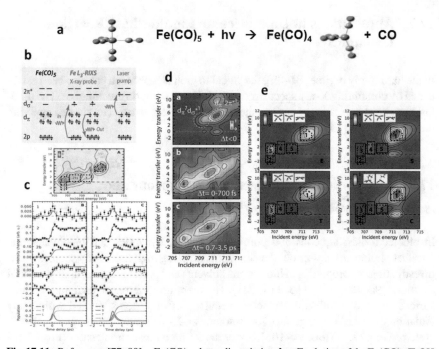

Fig. 17.11 References [77–80]: **a** Fe(CO)$_5$ photo-dissociation. **b–e** Evolution of the Fe(CO)$_4$/EtOH valence-electronic structure upon femtosecond spin crossover and complex formation. Term schema of the laser pump—X-ray probe experiment (**b**), measured (**c**, **d**) and simulated (**e**) RIXS scattering spectra, with and without included kinetic model on the excited state (E), triplet state (T) and ligand complex (L), Fe(CO)$_4$. **d**: The experimental Fe L$_3$-RIXS intensities (encoded in color) versus energy transfer and incident photon energy. (**d**, top) Fe(CO)$_5$ ground state (negative delays, probe before pump, scattering to d$_\pi^7$d$_\sigma^{*1}$ and d$_\pi^7$2π^{*1} marked by circles). (**d**, middle and bottom) Difference intensities for 0–700 fs and 0.7–3.5 ps time delays, respectively. **e** Calculated Fe L$_3$-RIXS intensities (color coded as in (**d**) and molecular-orbital diagrams of Fe(CO)$_5$ (ground state and hot), excited, triplet, and singlet Fe(CO)$_4$ (all three non-complexed) and solvent complexed singlet Fe(CO)$_4$–EtOH. The figure is adapted from [8–10] and references therein

at negative time delays and the difference after delay intervals of 0–700 fs and 0.7–3.5 ps respectively (color-encoded). By subtracting the negative time delay with a weight of 0.9, the pumped contributions are isolated (top: Fe(CO)$_5$, ground state (negative delays, probe before pump, scattering to d$_\pi^7$d$_\sigma^{*1}$ and d$_\pi^7$2π^{*1} marked by middle/bottom: difference intensities for 0–700 fs and 0.7–3.5 ps time delays). The experimental data is compared to RIXS calculations displayed in Fig. 17.11e. Simulated Fe L$_3$-RIXS intensities and molecular-orbital diagrams of Fe(CO)$_5$ in the ground and hot state and Fe(CO)$_4$ in various valence excited states are depicted.

The 2p → LUMO resonance positions and d$_\pi$ → d$_\sigma^*$ RIXS transitions are marked by arrows. The RIXS pattern at early time evolution can only be reproduced when a complex between the Fe(CO)$_4$ with the solvent is taken into account.

17.7 Applications in Unimolecular Liquid Phase Reaction Dynamics

In the hard X-ray regime [81–89], the "local to global" approach has also been established by combining X-ray spectroscopy with X-ray scattering techniques. The ultrafast structural dynamics of various metal organic systems has been studied applying these techniques [84, 88, 89].

17.8 Applications in Bioelectronics, Aqueous and Prebiotics Reaction Dynamics

In bioelectronics, aqueous and prebiotics reaction dynamics and biophysics, the ultrafast photon-in/photon-out developments utilizing high flux X-ray sources allow investigating the properties of biorelevant solvents and proteins during their structural reactions [90–108, 115, 116, 118, 121]. The literature-referenced examples include various types of small up to macromolecular model systems studied with X-ray radiation of ultrashort, high flux X-ray sources—X-ray scattering on quasi-periodic systems (Fig. 17.12) [90, 94–103], X-ray scattering on macromolecules (Fig. 17.13) and (Fig. 17.14) [90, 115, 116, 118, 121] and X-ray spectroscopy (Fig. 17.15) [91–93, 104–108].

Two examples of dynamical studies of soft condensed matter via high-flux time-resolved X-ray scattering will be given [90–108, 115, 116, 118, 121]: first, the studies of the dynamics of a phase transition of a liquid crystal-to-microemulsion system (Fig. 17.12) [95], and second, a real-time study of complex protein dynamics (Fig. 17.13/Fig. 17.14) [115, 116] will be given.

In liquid crystal type of soft condensed matter systems (Fig. 17.12), a manifold of phase-ordering transitions relevant to chemical and biological systems occur, ranging from liquids to self-assembled soft solids (like membranes or liquid crystals). In the present case the dynamics of the driving forces (activation energy and entropy) of a liquid crystal-to-microemulsion phase transition has been studied (Fig. 17.12 (left)). The purpose of this work was to clarify the influence of concentration effects of the amphiphilic molecules on the nature of these self-assembly processes.

By photosensitization of the model system (polyalkylglycolether ($C_{10}E_4$), water, decane, and cyclohexane) with laser dyes, the phase transition could effectively be photo-induced and controlled through the absorption of optical photons (as a photo-induced phase transition, PIPT). The photo transformation conditions were chosen in such a way that the system was in thermal equilibrium as starting conditions.

By applying time-resolved photo small-angle X-ray scattering it has been found that the conversion process depends on the surfactant concentration and the activation energy, which is observable through the length of the induction time (Fig. 17.12 (right)). The phase-transition, though photo-triggered, is still diffusion controlled in the rate-determining steps.

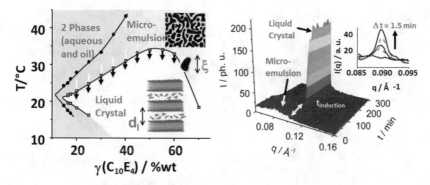

Fig. 17.12 References [90–95, 95–108, 115, 116, 118, 121]: (Left) Phase diagram of the system $C_{10}E_4$, water, *n*-decane, and cyclohexane at equal volume fractions of water and oil. Cyan represents an area where two phases coexist, dark gray the ME phase, and white the LC phase. The dots represent points of the phase boundaries. Arrows indicate the concentrations of the PIPT experiment. Inset: schematical drawings of the corresponding phases with the assignment of characteristic length scales like correlation length ξ (ME) and distance d between the lamellae (LC). (Right) Photoinduced ME-LC phase transition investigated by TR-SAXS. The decrease of the ME and the increase of lamellar phase resemble in a concomitant increase of the discovered Bragg peak. The monoexponential nucleation and growth period is preceded by a concentration dependent induction time. Inset: Traces of the LC scattering. The figure is adapted from [8–10] and references therein

In the second example, X-ray scattering techniques, comprising of small-angle/wide-angle X-ray scattering (SAXS/WAXS) techniques are increasingly used to characterize the structure and interactions of biological macromolecules and their complexes in solution (Figs. 17.13 and 17.14) [90, 94–103, 115, 116, 118, 121]. It is a method of choice to characterize flexible, partially folded and unfolded protein systems. X-ray scattering is the last resort for proteins that cannot be investigated by crystallography or NMR and acts as a complementary technique with different biophysical techniques to answer challenging scientific questions. The marriage of the X-ray scattering technique with the fourth dimension "time" yields structural dynamics and kinetics information for protein motions in hierarchical timescales from picoseconds to days (Fig. 17.13).

In the "fourth-dimension – time" of X-ray scattering technique (Fig. 17.13) the timescales accessible range from hours down to femtoseconds, even for investigating non-photoactive protein dynamics with different time-resolved X-ray scattering techniques. Various time-resolved and in-situ X-ray scattering techniques listed are complementary to the photon-triggered time-resolved X-ray scattering techniques and can be adaped to structural dynamics investigations of ubiquitous non-photoactive proteins.

Depending on the time scale of the system studied, it has been demonstrated that it is furthermore possible to merge X-ray scattering techniques like diffuse X-ray scattering with pressure jump, temperature jump, electric field modulations, and structural freezing methods or, on the chemical modulation side, with, rapid mixing or photo-switching methods.

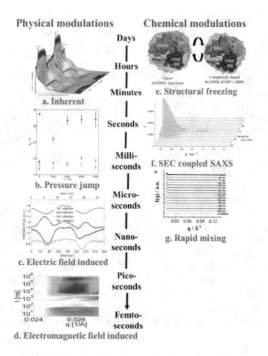

Fig. 17.13 References [90–108, 115, 116, 118, 121]: The "fourth-dimension – time" of X-ray scattering technique. The timescales accessible to investigate non-photoactive protein dynamics by using different time-resolved X-ray scattering techniques and the typical signal changes have been presented with representative published literature. **a** Inherent, **b** pressure-jump, **c** electric-field modulations, **d** electromagnetic field modulations, **e** structural freezing. **f** SAXS coupled with size-exclusion chromatography (SEC) technique and **g** rapid-mixing with the help of microfluidic devices. The figure is adapted from the [8–10] and references therein

Figure 17.13 summarises typical X-ray scattering signal changes—inherent (a), as pressure-jump (b), in electric-field modulations (c), as electromagnetic field modulations (d), as structural freezing (e) and in small angle X-ray scattering experiments coupled with size-exclusion chromatography (SEC) technique (f). Special attention is given to rapid-mixing approaches with the help of microfluidic devices (g) and (Fig. 17.14). Various techniques listed in the summary figure have independently been developed in neighbor projects of the collaborative research center (see "references" and references therein).

Figure 17.14 presents a scheme for a rapid-mixing time-resolved small angle X-ray scattering experiments for the investigation of the ubiquitin unfolding process as model system and model process. Figure 17.14a presents the 20-microchannel continuous-flow microfluidic device and Fig. 17.14b the time-resolved small angle X-ray scattering (SAXS) setup for the kinetic ubiquitin unfolding studies (proving the secondary structure unfolding on the millisecond time scale), followed by

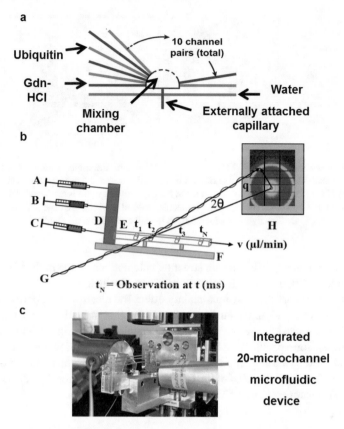

Fig. 17.14 References [91–93, 104–108]: Rapid-mixing TR-SAXS experiments to investigate the ubiquitin unfolding process. **a** 20-microchannel continuous-flow microfluidic device. **b** TR-SAXS setup for the kinetic ubiquitin unfolding studies (A: ubiquitin solution, B: 8 M guanidinium-HCl solution, C: water, D: 20-microchannel microfluidics device, E: external glass capillary (X-ray probing area), F: capillary holder, G: X-ray beam, H: detector). **c** The integrated rapid mixing SAXS experimental setup at the cSAXS beamline, PSI Switzerland. The figure is adapted from [8–10] and references therein

the integrated rapid mixing SAXS experimental setup at the cSAXS beamline, PSI Switzerland.

The third example (of [91–93, 104–108]) concentrates on the use of multidimensional X-ray spectroscopy for studying the dynamics in soft condensed matter, here the dynamics of ions in aqueous solutions (Fig. 17.15). Hydration shells around ions are crucial for many fundamental biological and chemical processes. Their local physicochemical properties are quite different from those of bulk water and hard to probe experimentally.

By combining high-resolution soft X-ray spectroscopy using liquid jet technology as core hole clock spectroscopy method (Fig. 17.15 (left)) with molecular dynamics

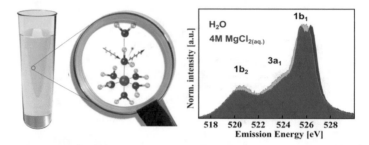

Fig. 17.15 References [90–108, 115, 116, 118, 121]: (Left) Principle of core-hole-clock spectroscopy/resonant inelastic X-ray scattering for investigating soft condensed matter reaction dynamics. (Right) The difference spectrum between 4 M MgCl$_2$ and pure water is displayed. The difference represents the first hydration shell of water molecules around Mg^{2+} on the time scale of the oxygen core-hole-clock (5 fs). The figure is adapted from [8–10] and references therein

simulations and ab initio electronic structure calculations, at the molecular level the water-ion interaction in MgCl$_2$ solution has been elucidated.

The results reveal that salt ions mainly affect the electronic properties of water molecules in close vicinity. Furthermore, in the first solvation shell the oxygen K-edge X-ray emission spectrum of water molecules differs significantly from that of bulk water. Ion-specific effects are identified by fingerprint features in the water X-ray emission spectra. While Mg^{2+} ions cause a bathochromic shift of the water lone pair orbital, the $3p$ orbital of the Cl$^-$ ions causes an additional peak in the water emission spectrum at around 528 eV (Fig. 17.15 (right)).

17.9 Applications in Biophysics and Gas Phase Biomolecules

Prolongating from the already presented techniques of combining X-ray scattering techniques like diffuse X-ray scattering with pressure jump, temperature jump, electric field modulations, and structural freezing methods or, on the chemical modulation side, with, rapid mixing or photo-switching methods and their successful application to biophysical questions [90, 94–103, 115, 116, 118, 121] or the application of core hole clock and multidimensional X-ray spectroscopy techniques towards aqueous and pre-biotics questions [91–93, 104–108] or novel liquid jet developments for biophysical research [120], in X-ray spectroscopy various other coupling techniques have been successfully proven.

The combination of synchrotrons or free-electron laser radiation with techniques, such as electrospray ionization mass spectrometry, allows deriving entirely novel experimental techniques for investigating macromolecules [109–121]. E.g. a mass spectrometric study on gas-phase ubiquitin at FLASH has revealed a fast local struc-

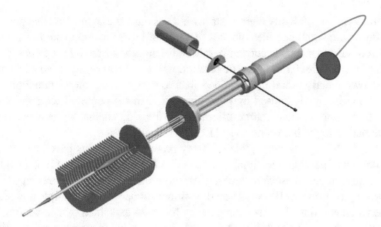

Fig. 17.16 References [109, 111–114, 119]: The mobile SMIS setup combines an electrospray ionization source, radio frequency (RF) mass filters, an RF trap and a mass spectrometer, which as a unity can be interfaced with different photon sources. The figure is adapted from [8–10, 114] and references therein

tural response, leading to small fragments with yields increasing linearly with photon intensity [109, 119].

Investigating the interaction of light with biologically relevant molecules has gained interest for a wide variety of research fields including photochemical reactions such as light harvesting as well as radiation damage in proteins and DNA related to cutting-edge cancer treatment techniques. However, in the condensed phase, disentangling direct and indirect radiation effects is often difficult [109–121].

In the beginning, studies on isolated polyatomic molecules in the gas phase were limited to small molecules, which are stable against thermal decomposition. In order to advance to more complex biomolecular systems, a novel apparatus has been designed. This mobile setup combines an electrospray ionization source, radio frequency (RF) mass filters, an RF trap and a mass spectrometer, which as a unity can be interfaced with different photon sources (see Fig. 17.16). Electrospray ionization introduces biomolecular ions from solution into the gas phase, allowing for studies of molecular systems in a well-defined state [109, 111–114, 119].

The coupling of electrospray ionization sources with synchrotrons [109, 111–114] and free-electron lasers [119] opens the way to the investigation of the electronic structure of biomolecular systems and of a fine description of their relaxation mechanisms in the VUV and soft X-ray energy range. The wide-ranging photon energy available at the synchrotrons enables systematic studies of ionization and dissociation as a function of the photon energy. Inner-shell excitations provide a localized site of energy deposition.

The extremely high photon flux and fs pulse duration offered by free-electron lasers allow studying the molecular properties in intense fields. Furthermore, using the assets of free-electron lasers in a pump-probe scheme enables the study of the dynamics of charge migration and charge transfer within gas-phase biomolecules.

In the VUV to soft X-ray regime, strong indications have been found that photoabsorption within small peptides induces fast loss of aromatic amino acid side chains through repulsive states, occurring before redistribution of the internal energy. The very fast loss of side chains thus efficiently cools the residual peptide, and may enable the survival of early functional peptide substructures under photon irradiation. This dissociation is mainly caused by photoabsorption in the peptide backbone, which seems to trigger ultrafast charge migration to the side chains, without leading to fragmentation to the backbone itself [111–114, 119].

In the following, X-ray/peptide or X-ray/protein interactions have been studied as a function of peptide size. Small peptides like leucine enkephalin [109, 112] (five amino acid residues) dissociate into small immonium fragments, whereas bigger proteins like cytochrome c (106 amino acids residues) undergo only multiple ionizations and no fragmentation. The size dependency has been explained by considering that the average excess energy left in the peptide is redistributed over the ro-vibrational modes of the molecule and averaged over the various degrees of freedom. Therefore the larger proteins can handle much more energy due to more degrees of freedom to redistribute the energy.

First studies of a gas-phase protein (ubiquitin, 76 residues) at the free-electron laser in Hamburg, FLASH, has revealed two different photoabsorption regimes: non-dissociative ionization in the few-photon regime (pulse energy of $0.1\,\mu J$), in contrast to side-chain losses in the multi-photon regime (pulse energy of $2.3\,\mu J$) [119]. The yields of these side-chain fragments increase linearly with the number of photons in the pulses. No region has been found where intermediate fragments due to backbone scission prevail. These effects suggest that in the XUV multiphoton regime, proteins seem to react as an ensemble of small peptides losing the side chains in fast local fragmentation processes.

Near edge X-ray absorption fine structure spectroscopy (NEXAFS) probes transitions between atomic core levels and orbitals of the molecular bonding states of intra-molecular neighbours. Therefore, NEXAFS is a powerful structural tool that provides information on the electronic structure. Data taken by near edge X-ray absorption mass spectrometry (NEXAMS) of gas-phase oligonucleotides [114], peptides [109, 112] (and references therein) and proteins [114c,118,119] show (π^*, σ^*) transitions and Rydberg states similar to conventional NEXAFS spectra of thin films and liquids. Additional structural and dissociation dynamics within the molecules can be extracted from the NEXAMS information of the different individual ionization and dissociation products. Moreover, the gas-phase NEXAMS method can be sensitive to the secondary structure of proteins as shown in the study of melittin [113].

In the photoexcitation case of the protein melittin, the molecules end up singly ionized after excitation of the electron and singly Auger emission. The yields of the singly ionized parent peak of different initial charge states of melittin against the photon energy around the carbon $1s$ edge are shown in Fig. 17.17 (top). In the photoionization case, the molecules end up doubly ionized after ionization of the electron into the continuum and singly Auger electron emission. The yields of the doubly ionized parent peak are shown in Fig. 17.17 (bottom).

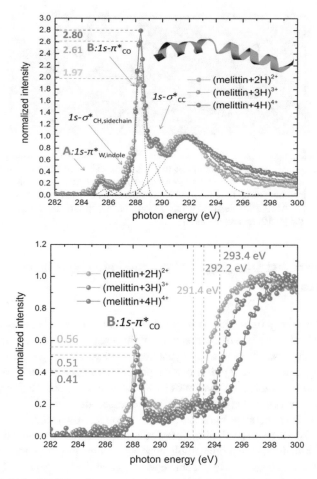

Fig. 17.17 References [109, 111–114, 119]: Yields/NEXAMS of different initial charge states of melittin against the photon energy. (top) NEXAMS spectra of the singly ionized parent peak around the carbon 1s edge. (bottom) NEXAMS spectra of doubly ionized melittin after ionizations of the electron into the continuum and singly Auger electron emission

A decrease of the resonance $(1s - \pi^*_{CO})$ with increasing charge state is observed, an opposite trend as in the photoexcitation case. The NEXAMS properties and photoionisation behaviour are explained as additional ionization of the molecule by secondary electrons and the probability of the molecules being hit by these electrons depending on their geometries. Consequently, a helical and more compact structure with the charge state 2+ the molecule has a higher probability to be hit again by an escaping electron compared to linear structures with four charges.

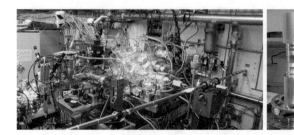

Fig. 17.18 References [122, 122–127, 127–163]: FLASH/CFEL-ASG MultiPurpose (CAMP) endstation at the AMO beamline at the LCLS in 2012. Since 2014, CAMP is a permanent endstation at FLASH/BL1 at DESY in Hamburg. Various experimental techniques can be employed in the instrument, including novel time-resolved X-ray studies of chemical reactions in the gas phase. On the right side, CAMP's double-sided velocity-map-imaging spectrometer is shown. The figure is adapted from the [8–10] and references therein

17.10   Ultrafast Imaging of Unimolecular Gas-Phase Reactions

The advent of X-ray free-electron lasers enabled not only novel studies in the condensed phase, as described in the first parts of this chapter, but also brought forward unprecedented possibilities to study dynamical processes in the gas phase. The very short and very intense X-ray pulses made it possible, for the first time, to probe ultrafast photo-induced molecular dynamics by electron or ion momentum spectroscopy following multiple, element-specific inner-shell absorption.

These new 4th generation light sources also called for novel, dedicated instrumentation (Fig. 17.18). To this end, different endstations have been developed, initially in particular for the use at the atomic, molecular, and optical physics (AMO) beamline, which was the first beamline to become operational at the LCLS [122, 122–163] in 2009. Several of the early, pioneering experiments from 2009 to 2012 have been performed at the AMO beamline in the CFEL-ASG MultiPurpose (CAMP) endstation [125, 127], see Fig. 17.18, which was developed within the Max-Planck Advanced Study Group (ASG) at the Center for Free-Electron Laser Science (CFEL) in Hamburg. Experiments in this instrument range from X-ray imaging of biomolecules, nanocrystals, and clusters [122, 125, 127–150] to (time-resolved) ion and electron spectroscopy on atoms and small gas-phase molecules [122, 122–127, 151–163]. Since 2014, CAMP is a permanent user endstation at FLASH/BL1 at DESY in Hamburg [124], and its successor, LAMP, has become operational at the AMO beamline at the LCLS [16b].

17.11 Applications in Nanoscience and Multiphoton-Ionisation

Moreover, the high-field physics (HFP) instrument has been developed at the LCLS [128–150], housing an ion momentum spectrometer and several electron time-of-flight spectrometers mounted under different angles. It also offers a pulsed, supersonic molecular beam, delivering gas-phase molecules to the interaction region. The examples given in the following, as well as multiple other pioneering gas-phase studies have been conducted in the HFP instrument [122–150].

17.12 Applications in Unimolecular Gas Phase Dynamics

Gas-phase FEL experiments allow questioning fundamental definitions in chemistry, for example the investigation of processes beyond the Born-Oppenheimer approximation. As an example, a UV-pump, X-ray probe study of two complementary halomethane molecules with different photochemistry is shown here [122, 122–124, 126, 151–163].

In Fig. 17.19, the schematic potential energy curves (PECs) of iodomethane (CH_3I) and fluoromethane (CH_3F) are displayed, illustrating that the different halogen species give rise to qualitatively different PECs [122, 152–155, 161, 163]. One reason for this is the considerably different electronegativity of iodine and fluorine, stabilizing the C–F bond in contrast to the C–I bond. Upon absorption of one 267 nm UV photon, CH_3I dissociates into two neutral fragments, $CH_3 + I$, whereas in CH_3F, no PEC is resonantly accessible at 4.6 eV. In the latter case, absorption of at least three UV photons in the same molecule populates several higher-lying ionic PECs, also resulting in dissociation of the molecules.

After a tunable time delay, an intense X-ray pulse (727 eV, 1 mJ) probes the dissociating system by ionizing predominantly the iodine ($3d$) or the fluorine ($1s$) level, respectively, because of their large absorption cross section (3.3 Mb for I and 0.4 Mb for F, compared to 0.1 Mb for CH_3), resulting in a localized positive charge on the halogen [122, 152–155, 161, 163].

At these very high X-ray intensities, a single molecule can absorb many photons, such that very highly charged ions up to I^{21+}/F^{4+} and C^{4+} are created.

As the charge is initially created locally at the halogen atom though, the fact that highly charged carbon ions are also detected already shows that the charge rearranges within the molecule before or during the fragmentation.

In Fig. 17.20, the calculated electrostatic potentials of an $I^{6+} + CH_3$ are plotted for three different internuclear distances between the two fragments, together with the binding energy of the highest occupied orbital. In the intact molecule (a), the electrons are delocalized, but as the fragments move apart the potential barrier rises, until at a certain critical distance (b), it reaches the electron binding energy.

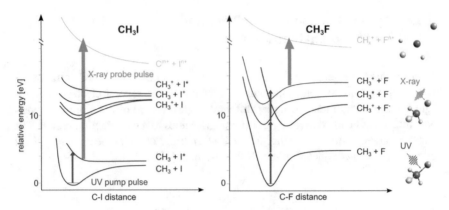

Fig. 17.19 References [122, 122–127, 151–163]: Schematic potential energy curves for iodomethane (left) and fluoromethane (right). Absorption of one 267 nm photon in CH₃I leads to resonant population of a repulsive neutral state, whereas multi-photon UV absorption in CH₃F populates to several higher-lying ionic states. After a given time delay, these states are probed by Coulomb explosion following inner-shell ionization of the respective halogen atom by one or several X-ray photons, probing the transition from a molecule to isolated atoms. The figure is adapted from the [8–10] and references therein

Therefore, for larger distances (c) the electrons can classically be regarded as localized at one of the two fragments. It is this transition from a bound molecule to isolated atoms that is probed by time-resolved ion spectroscopy [122, 152–155, 161, 163].

The delay-dependent time-of-flight peaks of selected ions of iodomethane and fluoromethane are shown in Fig. 17.21. It is evident that the fragmentation patterns of the two molecules are qualitatively different. For iodine charge states $\geq I^{4+}$, the appearance of low-energy ions at positive delays is clearly visible (channel 3 in Fig. 17.21a).

Signatures of long-distance intramolecular electron transfer have been observed for both, CH₃I and CH₃F, and the reconstructed critical distances (up to 15 Å for I^{21+}) are in good agreement with a classical over-the-barrier model.

These ions originate from the pump-probe process as indicated for iodomethane in Fig. 17.21, and can be used to extract the critical internuclear distance, up to which electron transfer from methyl to iodine is classically allowed for a given charge state [122, 152–155, 161, 163].

Two other channels can be seen in Fig. 17.21 that correspond to Coulomb explosion of intact molecules by only the FEL (1) and to ionic dissociation induced by multi-photon UV absorption (2), as illustrated for fluoromethane in Fig. 17.21, which also occurs with a lower probability in CH₃I. The low-energy channel is absent in the fluorine ions.

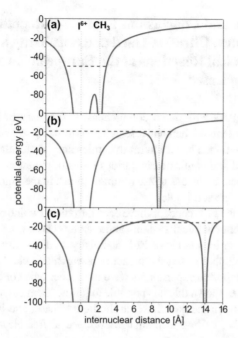

Fig. 17.20 References [122, 122–127, 151–163]: Calculated Coulomb potentials for an I^{6+} atom and a neutral methyl radical for **a** the equilibrium distance, **b** the cricital distance (see text) and **c** for isolated atoms (in a classical picture). The dashed blue line indicates the energy of the electron in the highest occupied orbital. The figure is adapted from the [8–10] and references therein

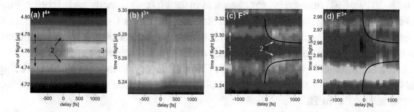

Fig. 17.21 References [122, 122–127, 151–163]: **a–b** Time-of-flight spectra as a function of the pump-probe delay for selected fragments of iodomethane and fluoromethane. Different fragmentation channels are indicated by 1, 2, 3 (see text). Additionally, in (**c**) and (**d**), calculated delay-dependent time-of-flight curves are overlaid with the data, corresponding to an asymptotic kinetic energy of 0.4 eV. Positive delays correspond to the UV pulse arriving before the X-ray pulse. The figure is adapted from the [8–10] and references therein

17.13 Outlook and Conclusion: First High-Repetition Frequency, Ultrafast Hard and Soft X-ray Studies of Chemical Reactions at the European X-ray Free Electron Laser

The current chapter tried to give a comprehensive summary of the very first and pioneering developments of chemical research tools utilizing soft and hard X-ray Free Electron Laser radiation. The structural dynamics and "molecular movie" concept has been introduced and confirmed in various states of matter and in different type of chemical reactions: in the gas phase (unimolecular), in the liquid phase (uni- and bimolecular) and in the solid state.

Novel experimental FEL concepts have been introduced which needed to be developed for the different chemical systems and reaction classes, and their pioneering applications in the various fields of FEL chemistry studies—from molecular reaction studies, through photocatalysis to energy research and biophysics—have been demonstrated. All reported experiments were of pioneer type concerning its kind. It has successfully been shown that it is possible to follow chemical reactions with FEL radiation in real time in the gas phase, in the liquid phase, and in the solid phase. Comparisons to other exiting techniques developed at free-electron laser sources have been given.

As due to the coincidence between the time, where FELs in the soft and hard X-ray regime came first into operational mode (2005–2018), and the start and running time of the Collaborative Research Center SFB755 *Nanoscale Photonic Imaging* (2007–2019), FEL milestones for chemical research have directly and right-on-time been reported within the SFB755. From our side, therefore, the SFB covers the most precious "life-time" of our experimental efforts so far, making it most important, essential and special for us as intellectual exchange and scientific support platform, and as a common effort of the Göttingen Campus. Every further development being sketched in the last paragraph can be built on the scientific construction platform supported by the SFB.

Since end 2017 the European X-ray Free-electron Laser as the first FEL of the 2nd generation is operationable. It unifies MHz repetition frequency capabilities with highest brilliance, stable time-synchronisation (allowing for view femtosecond time-resolved experiments), highest degree of coherence, tunability etc. allowing for experiments in an evolution to pioneering experiments of 1st generation FELs but also towards novel experimental strategies employing multidimensional X-ray spectroscopy or ultrafast inelastic scattering techniques. In the "molecular movie" field first feasibility and demonstration experiments have been performed and are discussed in the context of existing X-ray FEL chemistry studies. In the next 10 years novel superconducting LINAC technologies will come in operation (i.e. in LCLS-II) allowing for expanding existing techniques and a shift of experimental paradigm from proof-of-concept studies towards sample-curiosity driven X-ray experiments.

17.14 Acknowledgements

The Helmholtz Society is thanked for continuous financial support and DESY for preferred access to X-ray instrumentation. DESY staff is thanked for their competent help in large scale facility technical and computer support. Up to 2017, the Max Planck Society is acknowledged for continuous financial support. The workshops and chemical facility of the Max Planck Institute for Biophysical Chemistry is thanked for their competent help in chemical synthesis, analysis and technical design of the micro apparatus and computer support.

When the SFB755 "Nanoscale Photonic Imaging" started, in 2006, synchrotron researchers faced their next challenge in developing and proving X-ray Free Electron Laser Science. Throughout and fully, over the last 12 years research work of method developments utilizing X-ray free-electron lasers and synchrotron research (method development) as been presented in this chapter have been financially supported by the Deutsche Forschungsgemeinschaft within the Collaborative Research Center SFB755 "Nanoscale Photonic Imaging" (project B03 and project B10). Systems- and sample-related research work also presented in this chapter have been funded through the Collaborative Research Center SFB1073 "Atomic-scale Control of Energy Conversion" (project B06 and C02). Furthermore, the German Academic Exchange Service DAAD, the Alexander von Humboldt Foundation AvH, the Fonts of the Chemical Industry FCI, the Aventis Foundation, "Niedersachsen Vorab", the Peter-Paul-Ewald Fellowship program of the Volkswagen Foundation (No 87008) are thanked for financial support.

The independent young investigator groups of Daniel Rolles ("Watching Chemistry in Action" (VH-NG-904, 2012–2017), now Kansas State U) and Sadia Bari ("Structure and Dynamics of Gas Phase Biomolecules" (VH-NG-1104, 2016-now) have additionally financially been supported by the project B03 of the SFB755 (see "references").

The Advanced Study Group of the Max Planck Society is thanked for financial support 2006–2011, the Initiative and Networking Fund of the Helmholtz Association is thanked from 2012 on (programs: VH-NG-1104, NV-NG-904, HG-recruitment, HG-Innovation "ECRAPS", HG-Innovation "HESEB/SESAME", DSF, DASHH, DGP, CMWS).

The German-Israeli Foundation for Scientific Research and Development (GIF) and the Technion Society/BMBF are thanked for support in education and research.

Over the last decade, beamline staff of the Helmholtz Center Berlin (HZB—Prof. A. Foehlisch and his team) is thanked for the fruitful collaborations, experimental support and hospitality. Beamline responsibles of the European Synchrotron Radiation Facility (ID09, ID11, ESRF—Prof. M. Wulff and his team), the Advanced Photon Source (ID14, APS—V. Srajier and the team), the Swiss Light Source (cSAXS (A. Menzel), SLS), the Stanford Synchrotron Radiation Facility (SPEAR, SSRL), PETRA III (P01 (H. Yavas), P03 (S. Roth), P04 (J. Viefhaus), P10 (M. Stumpf) and P11 (A. Burkhardt)) and the Advanced Light Source (ALS) are thanked for their fruitful collaborations, experimental support and hospitality.

Staff and beamline scientists of the free-electron laser FLASH (BL1-BL3, CAMP (B. Erk); E. Ploenjes, B. Manschwetus, R. Treusch, M. Kuhlmann and the teams) @ DESY, staff and beamline scientists of the free-electron laser LCLS (AMO (J. Bozek, C. Bostedt), SXR (B. Schlotter, J. Turner), XPP (D. Fritz) and CXI (M. Messerschmidt, S. Boutet)) @ SLAC, staff and beamline scientists of the Free Electron Laser SACLA @ SPRING-8/Riken and staff and—finally—the European X-ray free-electron laser European XFEL in Schenefeld and their beamline staff (FXE (C. Bressler), SPB (A. Mancuso), SCS (A. Scherz) and SQS (M. Meyer)) are deeply acknowledged for their collaborations, support and hospitality with the DESY/FS-SCS group members.

Use of the Linac Coherent Light Source (LCLS), SLAC National Accelerator Laboratory, is supported by the U.S. Department of Energy, Office of Science, Office of Basic Energy Sciences under Contract No. DE-AC02-76SF00515.

References

1. Madey, J.M.J.: Stimulated emission of bremsstrahlung in a periodic magnetic field. J. Appl. Phys. **42**(5), 1906–1913 (1971). https://doi.org/10.1063/1.1660466
2. Benson, S., Deacon, D.A.G., Eckstein, J.N., Madey, J.M.J., Robinson, K., Smith, T.I., Taber, R.: Review of recent experimental results from the stanford 3 μ m free-electron laser. Le J. de Physique Colloques **44**(C1), C1–353–C1–362 (1983). https://doi.org/10.1051/jphyscol:1983128
3. Halbach, K.: Permanent magnet undulators. Le J. de Physique Colloques **44**(C1), C1–211–C1–216 (1983). https://doi.org/10.1051/jphyscol:1983120
4. Kondratenko, A.M., Saldin, E.L.: Linear theory of free-electron lasers with fabry-perot resonators. Sov. Phys.-Tech. Phys.(Engl. Transl.);(United States) **27**(2) (1982)
5. Luccio, A., Pellegrini, C.: One dimensional numerical simulation of a free-electron laser–storage ring system. Le J. de Physique Colloques **44**(C1), C1–373–C1–374 (1983). https://doi.org/10.1051/jphyscol:1983133
6. Madey, J.M.J.: Application of transverse gradient wigglers in high efficiency storage ring fel's. Le J. de Physique Colloques **44**(C1), C1–169–C1–178 (1983). https://doi.org/10.1051/jphyscol:1983116
7. Sprangle, P., Smith, R.A.: Theory of free-electron lasers. Phys. Rev. A **21**(1), 293–301 (1980). https://doi.org/10.1103/physreva.21.293
8. Bari, S., Boll, R., Idzik, K., Kubicek, K., Raiser, D., Veedu, S.T., Yin, Z., Techert, S.: Ultrafast time structure imprints in complex chemical and biochemical reactions. R. Soc. Chem. (2017). https://doi.org/10.1039/9781782624097
9. Techert, S.: Handbook for using free-electron lasers in chemical research. Struct. Dyn. (2019). In press
10. Techert, S.: Tutorial lecture, iff spring school (2019). https://www.fz-juelich.de/SharedDocs/Termine/PGI/EN/2019/2019-03-11-IFFSpringSchool-50th.html
11. Schmatz, S.: Four-mode calculations of resonance states of intermediate complexes in the SN2 reaction $Cl^- + CH_3Cl$ to $ClCH_3 + Cl^-$. J. Chem. Phys. **118**(1–3), 4499–4516 (2003). https://doi.org/10.1063/1.1541626
12. Techert, S.: First-, second- and third-order correlation function in time-resolved X-ray diffraction experiments. J. Appl. Crystallogr. **37**(3), 445–450 (2004). https://doi.org/10.1107/s0021889804007381
13. Techert, S., Schmatz, S.: Time-resolved X-ray scattering and the observation of intramolecular reaction dynamics in liquids. Zeitschrift für Physikalische Chem. **216**(10–12), 575–583 (2002). https://doi.org/10.1515/zpch-2015-0610

14. Techert, S., Schotte, F., Wulff, M.: Picosecond X-ray diffraction probed transient structural changes in organic solids. Phys. Rev. Lett. **86**(10), 2030–2033 (2001). https://doi.org/10.1103/physrevlett.86.2030
15. (a) flash; (b) lcls, (c) sacla, (d) swissfel, (e) european xfel
16. Debnarova, A., Techert, S., Schmatz, S.: Ab initio treatment of time-resolved X-ray scattering: application to the photoisomerization of stilbene. J. Chem. Phys. **125**(22), 224,101–224,108 (2006). https://doi.org/10.1063/1.2400231
17. See Ref. [28] pp. 335–337
18. Altarelli, M., Brinkmann, R., Chergui, M., Decking, W., Dobson, B., Düsterer, S., Grübel, G., Graeff, W., Graafsma, H., Hajdu, J., Marangos, J., Pflüger, J., Redlin, H., Riley, D., Robinson, I., Rossbach, J., Schwarz, A., Tiedtke, K., Tschentscher, T., Vartaniants, I., Wabnitz, H., Weise, R.W., Witte, H.K., Wolf, A., Wulff, M., Yurkov, M.: The european X-ray free-electron laser (2007)
19. Blome, C., Tschentscher, T., Davaasambuu, J., Durand, P., Techert, S.: Femtosecond time-resolved powder diffraction experiments using hard X-ray free-electron lasers. J. Synchrotron Radiat. **12**(6), 812–819 (2005). https://doi.org/10.1107/s0909049505026464
20. Blome, C., Tschentscher, T., Davaasambuu, J., Durand, P., Techert, S.: Ultrafast time-resolved powder diffraction using free-electron laser radiation. In: AIP Conference Proceedings. AIP (2007). https://doi.org/10.1063/1.2436292
21. Busse, G., Tschentscher, T., Plech, A., Wulff, M., Frederichs, B., Techert, S.: First investigations of the kinetics of the topochemical reaction of p-formyl-trans-cinnamic acid by time-resolved X-ray diffraction. Faraday Discuss. **122**, 105–117 (2002). https://doi.org/10.1039/b202831j
22. Cavalieri, A.L., Fritz, D.M., Lee, S.H., Bucksbaum, P.H., Reis, D.A., Rudati, J., Mills, D.M., Fuoss, P.H., Stephenson, G.B., Kao, C.C., Siddons, D.P., Lowney, D.P., MacPhee, A.G., Weinstein, D., Falcone, R.W., Pahl, R., Als-Nielsen, J., Blome, C., Düsterer, S., Ischebeck, R., Schlarb, H., Schulte-Schrepping, H., Tschentscher, T., Schneider, J., Hignette, O., Sette, F., Sokolowski-Tinten, K., Chapman, H.N., Lee, R.W., Hansen, T.N., Synnergren, O., Larsson, J., Techert, S., Sheppard, J., Wark, J.S., Bergh, M., Caleman, C., Huldt, G., van der Spoel, D., Timneanu, N., Hajdu, J., Akre, R.A., Bong, E., Emma, P., Krejcik, P., Arthur, J., Brennan, S., Gaffney, K.J., Lindenberg, A.M., Luening, K., Hastings, J.B.: Clocking femtosecond X-rays. Phys. Rev. Lett. **94**, 11 (2005). https://doi.org/10.1103/PhysRevLett.94.114801
23. Collet, E., Lemee-Cailleau, M.H., Cointe, M.B.L., Cailleau, H., Wulff, M., Luty, T., Koshihara, S.Y., Meyer, M., Toupet, L., Rabiller, P., Techert, S.: Laser-induced ferroelectric structural order in an organic charge-transfer crystal. Science **300**(5619), 612–615 (2003). https://doi.org/10.1126/science.1082001
24. Davaasambuu, J., Busse, G., Techert, S.: Aspects of the photodimerization mechanism of 2, 4-dichlorocinnamic acid studied by kinetic photocrystallography? J. Phys. Chem. A **110**(9), 3261–3265 (2006). https://doi.org/10.1021/jp054723m
25. Davaasambuu, J., Durand, P., Techert, S.: Experimental requirements for light-induced reactions in powders investigated by time-resolved x-ray diffraction. J. Synchrotron Radiat. **11**(6), 483–489 (2004). https://doi.org/10.1107/s090904950402463x
26. Davaasambuu, J., Wright, J., Soerensen, H.O., Schmidt, S., Poulsen, H.F., Techert, S.: An application of multigrain approaches to the structural solution of grains from polycrystalline samples. Solid State Phenom. **288**, 119–123 (2019). https://doi.org/10.4028/www.scientific.net/SSP.288.119
27. Debnarova, A., Techert, S., Schmatz, S.: Contribution of coulomb explosion to form factors and mosaicity spread in single particle X-ray scattering. Phys. Chem. Chem. Phys. **16**(2), 792–798 (2014). https://doi.org/10.1039/c3cp54011a
28. Fuest, H.: Built-up of a time correlated single photon counting apparatus for optical control in photo crystallography (2012)
29. Hallmann, J.: Photoinduced solid-state reactions—spectroscopy and X-ray diffraction analysis (2010)

30. Hallmann, J., More, R., Morgenroth, W., Paulmann, C., Kong, Q., Wulff, M., Techert, S.: Evidence for point transformations in photoactive molecular crystals by the photoinduced creation of diffuse diffraction patterns. J. Phys. Chem. B **116**(36), 10996–11003 (2012). https://doi.org/10.1021/jp3020832

31. Hallmann, J., Morgenroth, W., Paulmann, C., Davaasambuu, J., Kong, Q., Wulff, M., Techert, S.: Time-resolved X-ray diffraction of the photochromic α-styrylpyrylium trifluoromethane-sulfonate crystal films reveals ultrafast structural switching. J. Am. Chem. Soc. **131**(41), 15018–15025 (2009). https://doi.org/10.1021/ja905484u

32. Hallmann, J., Techert, S.: Photoluminescence properties of a molecular organic switching system. J. Phys. Chem. Lett. **1**(6), 959–961 (2010). https://doi.org/10.1021/jz100115k

33. Krasniqi, F., Zhong, Y.P., Reis, D.A., Scholz, M., Hartmann, R., Hartmann, A., Rolles, D., Rudenko, A., Epp, S.W., Foucar, L., Trigo, M., Fuchs, M., Fritz, D.M., Cammarata, M., Zhu, D., Lemke, H., Braune, M., Ilchen, M., Larsson, J., Techert, S., Strüder, L., Schlichting, I., Ullrich, J.: Resonant X-ray emission spectroscopy with free- electron lasers: nonequilibrium electron dynamics in highly excited polar semiconductors. In: Research in Optical Sciences. OSA (2012). https://doi.org/10.1364/icusd.2012.iw1d.2

34. Kubiček, K., Veedu, S.T., Storozhuk, D., Kia, R., Techert, S.: Geometric and electronic properties in a series of phosphorescent heteroleptic cu(i) complexes: crystallographic and computational studies. Polyhedron **124**, 166–176 (2017). https://doi.org/10.1016/j.poly.2016.12.035

35. Lindenberg, A.M., Larsson, J., Sokolowski-Tinten, K., Gaffney, K., Blome, C., Synnergren, O., Sheppard, J., Caleman, C., MacPhee, A.G., Weinstein, D., Lowney, D.P., Allison, T., Matthews, T., Falcone, R.W., Cavalieri, A.L., Fritz, D.M., Lee, S.H., Bucksbaum, P.H., Reis, D.A., Rudati, J., Mills, D.M., Fuoss, P.H., Stephenson, G.B., Kao, C.C., Siddons, D.P., Pahl, R., Als-Nielsen, J., Dusterer, S., Ischebeck, R., Schlarb, H., Shulte-Schrepping, H., Tschentscher, T., Schneider, J., Hignette, O., Sette, F., Chapman, H.N., Lee, R.W., Hansen, T.N., Techert, S., Wark, J.S., Bergh, M., Huldt, G., van der Spoel, D., Timneanu, M., Hajdu, J., von der Linde, D., Akre, R.A., Bong, E., Emma, P., Krejcik, P., Arthur, J., Brennan, S., Luening, K., Hastings, J.B.: Atomic-scale visualization of inertial dynamics. Science **308**(5720), 392–395 (2005). https://doi.org/10.1126/science.1107996

36. Messerschmidt, M., Tschentscher, T., Cammarata, M., Meents, A., Sager, C., Davaasambuu, J., Busse, G., Techert, S.: Ultrafast potential energy surface softening of one-dimensional organic conductors revealed by picosecond time-resolved laue crystallography. J. Phys. Chem. A **114**(29), 7677–7681 (2010). https://doi.org/10.1021/jp104081b

37. More, R.: Untersuchung zur strukturdynamik organischer molekularer halbleiter: Photochemische festkörperreaktionen—protonentransfer in carbonsäuredimeren (2012)

38. Moré, R., Busse, G., Hallmann, J., Paulmann, C., Scholz, M., Techert, S.: Photodimerization of crystalline 9-anthracenecarboxylic acid: a nontopotactic autocatalytic transformation. he J. Phys. Chem. C **114**(9), 4142–4148 (2010). https://doi.org/10.1021/jp909513v

39. Moré, R., Scholz, M., Busse, G., Busse, L., Paulmann, C., Tolkiehn, M., Techert, S.: Hydrogen bond dynamics in crystalline β-9-anthracene carboxylic acid—a combined crystallographic and spectroscopic study. Phys. Chem. Chem. Phys. **14**(29), 10,187 (2012). https://doi.org/10.1039/c2cp40216e

40. Neutze, R., Techert, S.: Watching ultrafast chemistry with FEL radiation. TESLA Tech. Des. Rep.—Part **I**, (2001)

41. Quevedo, W., Busse, G., Hallmann, J., Moré, R., Petri, M., Krasniqi, F., Rudenko, A., Tschentscher, T., Stojanovic, N., Düsterer, S., Treusch, R., Tolkiehn, M., Techert, S., Rajkovic, I.: Ultrafast time dynamics studies of periodic lattices with free electron laser radiation. J. Appl. Phys. **112**(9), 093,519 (2012). https://doi.org/10.1063/1.4764918

42. Rajkovic, I., Busse, G., Hallmann, J., Moré, R., Petri, M., Quevedo, W., Krasniqi, F., Rudenko, A., Tschentscher, T., Stojanovic, N., Düsterer, S., Treusch, R., Tolkiehn, M., Techert, S.: Diffraction properties of periodic lattices under free electron laser radiation. Phys. Rev. Lett. **104**(12) (2010). https://doi.org/10.1103/physrevlett.104.125503

43. Scholz, M.: Ultrafast X-ray diffraction of metalloorganic compounds (2010)

44. Schotte, F., Techert, S., Anfinrud, P., Srajer, V., Moffat, K., Wulff, M.: Recent advantages in the generation of pulsed synchrotron radiation suitable for picosecond time-resolved X-ray studies. Wiley-Interscience (2002)
45. Sørensen, H.O., Schmidt, S., Wright, J.P., Vaughan, G.B.M., Techert, S., Garman, E.F., Oddershede, J., Davaasambuu, J., Paithankar, K.S., Gundlach, C., Poulsen, H.F.: Multigrain crystallography. Zeitschrift für Kristallographie **227**(1), 63–78 (2012). https://doi.org/10.1524/zkri.2012.1438
46. Techert, S.: Current developments in time-resolved X-ray diffraction. Crystallogr. Rev. **12**(1), 25–45 (2006). https://doi.org/10.1080/08893110600688873
47. Techert, S.: Watching chemistry with femtosecond X-rays, scientific case of an energy recovery linac (2006)
48. Techert, S.: Structural changes during the photoswitching of matter. MPIbpC News **10**, 1–8 (2007)
49. Techert, S.: Ultrafast photochemistry. In: Tschentscher, T., Rousse, A. (eds.) Conference Proceedings on X-ray Investigations of Ultrafast Processes, pp. 410–445. Hasylab at DESY, Hamburg (2007)
50. Techert, S.: Concepts of structural dynamics investigations in chemical research, pp. 129–139. Springer Netherlands (2014)
51. Techert, S., Zachariasse, K.A.: Structure determination of the intramolecular charge transfer state in crystalline 4-(diisopropylamino)benzonitrile from picosecond X-ray diffraction. J. Am. Chem. Soc. **126**(17), 5593–5600 (2004). https://doi.org/10.1021/ja0379518
52. Veedu, S.T., Cao, D., Lavy, T., Botoshansky, M., Kaftory, M.: Unexpected molecular flip in solid-state photodimerization. Cryst. Growth & Des. **13**(2), 936–941 (2013). https://doi.org/10.1021/cg3016707
53. Grossmann, P., Rajkovic, I., Moré, R., Norpoth, J., Techert, S., Jooss, C., Mann, K.: Time-resolved near-edge X-ray absorption fine structure spectroscopy on photo-induced phase transitions using a tabletop soft-X-ray spectrometer. Rev. Sci. Instrum. **83**(5), 053,110 (2012). https://doi.org/10.1063/1.4718936
54. Idzik, K.R., Cywiński, P.J., Kuznik, W., Frydel, J., Licha, T., Ratajczyk, T.: The optical properties and quantum chemical calculations of thienyl and furyl derivatives of pyrene. Phys. Chem. Chem. Phys. **17**(35), 22758–22769 (2015). https://doi.org/10.1039/c5cp03013g
55. Mildner, S., Hoffmann, J., Blöchl, P.E., Techert, S., Jooss, C.: Temperature- and doping-dependent optical absorption in the small-polaron systemPr1-xCaxMnO3. Phys. Rev. B **92**(3) (2015). https://doi.org/10.1103/physrevb.92.035145
56. Petri, M.: Private communication (2016)
57. Raiser, D.: Development of ultrafast optical methods for the study of the structural dynamics of complex solids (2017)
58. Raiser, D., Mildner, S., Ifland, B., Sotoudeh, M., Blöchl, P., Techert, S., Jooss, C.: Evolution of hot polaron states with a nanosecond lifetime in a manganite perovskite. Adv. Energy Mater. **7**(12), 1602,174 (2017). https://doi.org/10.1002/aenm.201602174
59. Veedu, S.T., Raiser, D., Kia, R., Scholz, M., Techert, S.: Ultrafast dynamical study of pyrene-n, n-dimethylaniline (PyDMA) as an organic molecular diode in solid state. J. Phys. Chem. B **118**(12), 3291–3297 (2014). https://doi.org/10.1021/jp4121222
60. David, C., Rösner, B., Döring, F., Guzenko, V., Koch, F., Lebugle, M., Marschall, F., Seniutinas, G., Raabe, J., Watts, B., Grolimund, D., Yin, Z., Beye, M., Techert, S., Viefhaus, J., Falkenberg, G., Schroer, C.: Diffractive X-ray optics for synchrotrons and free-electron lasers. Microsc. Microanal. **24**(S2), 268–269 (2018). https://doi.org/10.1017/s1431927618013673
61. Döring, F., Marschall, F., Yin, Z., Rosner, B., Beye, M., Miedema, P., Kubicek, K., Glaser, L., Raiser, D., Soltau, J., Guzenko, V.A., Viefhaus, J., Buck, J., Risch, M., Techert, S., David, C.: ID-full field microscopy of elastic and inelastic scattering with transmission off-axis fresnel zone plates. Microsc. Microanal. **24**(S2), 184–185 (2018). https://doi.org/10.1017/s1431927618013260
62. Hallmann, J., Grübel, S., Rajkovic, I., Quevedo, W., Busse, G., Scholz, M., More, R., Petri, M., Techert, S.: First steps towards probing chemical systems and dynamics with free-electron

laser radiation—case studies at the FLASH facility. J. Phys. B: At. Mol. Opt. Phys. **43**(19), 194,009 (2010). https://doi.org/10.1088/0953-4075/43/19/194009

63. Kunnus, K., Rajkovic, I., Schreck, S., Quevedo, W., Eckert, S., Beye, M., Suljoti, E., Weniger, C., Kalus, C., Grübel, S., Scholz, M., Nordlund, D., Zhang, W., Hartsock, R.W., Gaffney, K.J., Schlotter, W.F., Turner, J.J., Kennedy, B., Hennies, F., Techert, S., Wernet, P., Föhlisch, A.: A setup for resonant inelastic soft X-ray scattering on liquids at free electron laser light sources. Rev. Sci. Instrum. **83**(12), 123,109 (2012). https://doi.org/10.1063/1.4772685

64. Marschall, F., Yin, Z., Rehanek, J., Beye, M., Döring, F., Kubiček, K., Raiser, D., Veedu, S.T., Buck, J., Rothkirch, A., Rösner, B., Guzenko, V.A., Viefhaus, J., David, C., Techert, S.: Transmission zone plates as analyzers for efficient parallel 2d RIXS-mapping. Sci. Rep. **7**(1) (2017). https://doi.org/10.1038/s41598-017-09052-0

65. Rajkovic, I., Grübel, S., Quevedo, W., Techert, S.: Ultrafast pump/probe diffraction and spectroscopy experiments with FEL radiation: setup development from the soft to the hard X-rays with the aim of studying chemical processes. In: Tschentscher, T., Cocco, D. (eds.) Advances in X-ray Free-electron Lasers: Radiation Schemes, X-ray Optics, and Instrumentation. SPIE (2011). https://doi.org/10.1117/12.886763

66. Rajkovic, I., Hallmann, J., Grübel, S., More, R., Quevedo, W., Petri, M., Techert, S.: Development of a multipurpose vacuum chamber for serial optical and diffraction experiments with free-electron laser radiation. Rev. Sci. Instrum. **81**(4), 045,105 (2010). https://doi.org/10.1063/1.3327816

67. Rajkovic, I., Hallmann, J., Grübel, S., More, R., Quevedo, W., Petri, M., Techert, S.: Development of a multipurpose vacuum chamber for serial optical and diffraction experiments with free- electron laser radiation. Virtual J. Ultrafast X-ray Sci. **9**(1), 5 (2010). https://doi.org/10.1063/1.3327816

68. Techert, S.: Choreografie der moleküle—röntgenblitze "filmen" molekulare schalter. MPIbpC News **5**, 1–3 (2010)

69. Techert, S.: Capturing the structural dynamics of biomolecules with X-ray photon-in/photon-out techniques. MPIbpC News **1**, 1–4 (2011)

70. Wernet, P., Kunnus, K., Schreck, S., Quevedo, W., Kurian, R., Techert, S., de Groot, F.M.F., Odelius, M., Föhlisch, A.: Dissecting local atomic and intermolecular interactions of transition-metal ions in solution with selective X-ray spectroscopy. J. Phys. Chem. Lett. **3**(23), 3448–3453 (2012). https://doi.org/10.1021/jz301486u

71. Yin, Z.: X–ray spectroscopy of complex chemical systems in liquid phase (2016)

72. Yin, Z., Löchel, H., Rehanek, J., Goy, C., Kalinin, A., Schottelius, A., Trinter, F., Miedema, P., Jain, A., Valerio, J., Busse, P., Lehmkühler, F., Möller, J., Grübel, G., Madsen, A., Viefhaus, J., Grisenti, R.E., Beye, M., Erko, A., Techert, S.: X-ray spectroscopy with variable line spacing based on reflection zone plate optics. Opt. Lett. **43**(18), 4390 (2018). https://doi.org/10.1364/ol.43.004390

73. Yin, Z., Peters, H.B., Hahn, U., Agåker, M., Hage, A., Reininger, R., Siewert, F., Nordgren, J., Viefhaus, J., Techert, S.: A new compact soft X-ray spectrometer for resonant inelastic X-ray scattering studies at PETRA III. Rev. Sci. Instrum. **86**(9), 093,109 (2015). https://doi.org/10.1063/1.4930968

74. Yin, Z., Peters, H.B., Hahn, U., Gonschior, J., Mierwaldt, D., Rajkovic, I., Viefhaus, J., Jooss, C., Techert, S.: An endstation for resonant inelastic X-ray scattering studies of solid and liquid samples. J. Synchrotron Radiat. **24**(1), 302–306 (2017). https://doi.org/10.1107/s1600577516016611

75. Yin, Z., Rajković, I., Raiser, D., Scholz, M., Techert, S.: Experimental setup for high resolution X-ray spectroscopy of solids and liquid samples. In: Klisnick, A., Menoni, C.S. (eds.) X-Ray Lasers and Coherent X-Ray Sources: Development and Applications X. SPIE (2013). https://doi.org/10.1117/12.2023992

76. Yin, Z., Rehanek, J., Löchel, H., Braig, C., Buck, J., Firsov, A., Viefhaus, J., Erko, A., Techert, S.: Highly efficient soft X-ray spectrometer based on a reflection zone plate for resonant inelastic X-ray scattering measurements. Opt. Express **25**(10), 10,984 (2017). https://doi.org/10.1364/oe.25.010984

77. Kunnus, K., Josefsson, I., Rajkovic, I., Schreck, S., Quevedo, W., Beye, M., Grübel, S., Scholz, M., Nordlund, D., Zhang, W., Hartsock, R., Gaffney, K.J., Schlotter, W.F., Turner, J.J., Kennedy, B., Hennies, F., Techert, S., Wernet, P., Odelius, M., Föhlisch, A.: Anti-stokes resonant X-ray raman scattering for atom specific and excited state selective dynamics. New J. Phys. **18**(10), 103,011 (2016). https://doi.org/10.1088/1367-2630/18/10/103011
78. Kunnus, K., Josefsson, I., Rajkovic, I., Schreck, S., Quevedo, W., Beye, M., Weniger, C., Grübel, S., Scholz, M., Nordlund, D., Zhang, W., Hartsock, R.W., Gaffney, K.J., Schlotter, W.F., Turner, J.J., Kennedy, B., Hennies, F., de Groot, F.M.F., Techert, S., Odelius, M., Wernet, P., Föhlisch, A.: Identification of the dominant photochemical pathways and mechanistic insights to the ultrafast ligand exchange of fe(CO)5 to fe(CO)4etoh. Struct. Dyn. **3**(4), 043,204 (2016). https://doi.org/10.1063/1.4941602
79. Wernet, P., Kunnus, K., Josefsson, I., Rajkovic, I., Quevedo, W., Beye, M., Schreck, S., Grübel, S., Scholz, M., Nordlund, D., Zhang, W., Hartsock, R.W., Schlotter, W.F., Turner, J.J., Kennedy, B., Hennies, F., de Groot, F.M.F., Gaffney, K.J., Techert, S., Odelius, M., Föhlisch, A.: Orbital-specific mapping of the ligand exchange dynamics of fe(CO)5 in solution. Nature **520**(7545), 78–81 (2015). https://doi.org/10.1038/nature14296
80. Yin, Z., Quevedo, W., Rajkovic, I., Wernet, P., Föhlisch, A., Pietsch, A., Techert, S.: Ionic solutions probed by resonant inelastic X-ray scattering. Zeitschrift für Physikalische Chem. **1**(1), 100–120 (2015). https://doi.org/10.1515/zpch-2015-0610
81. Debnarova, A.: Ab-initio studies of chemical reactions under intense X-ray radiation (2009)
82. Debnarova, A., Techert, S., Schmatz, S.: Ab initio studies of ultrafast X-ray scattering of the photodissociation of iodine. J. Chem. Phys. **133**(12), 124,309 (2010). https://doi.org/10.1063/1.3475567
83. Debnarova, A., Techert, S., Schmatz, S.: Computational studies of the X-ray scattering properties of laser aligned stilbene. J. Chem. Phys. **134**(5), 054,302 (2011). https://doi.org/10.1063/1.3523569
84. Kjaer, K.S., Zhang, W., Alonso-Mori, R., Bergmann, U., Chollet, M., Hadt, R.G., Hartsock, R.W., Harlang, T., Kroll, T., Kubicek, K., Lemke, H.T., Liang, H.W., Liu, Y., Nielsen, M.M., Robinson, J.S., Solomon, E.I., Sokaras, D., van Driel, T.B., Weng, T.C., Zhu, D., Persson, P., Warnmark, K., Sundström, V., Gaffney, K.J.: Ligand manipulation of charge transfer excited state relaxation and spin crossover in [fe(2,2ʹ-bipyridine)2(CN)2]. Struct. Dyn. **4**(4), 044,030 (2017). https://doi.org/10.1063/1.4985017
85. Neutze, R., Wouts, R., Techert, S., Kirrander, A., Davidson, J., Kocsis, M., Schotte, F., Wulff, M.: Visualising photo-chemical dynamics in solution through picosecond X-ray scattering. Phys. Rev. Lett. **87**(4), 195,508–195,512 (2001). https://doi.org/10.1103/PhysRevLett.86.2030
86. Petrov, N.K., Gulakov, M.N., Alfimov, M.V., Busse, G., Techert, S.: Solvation-shell effect on the cyanine-dye fluorescence in binary liquid mixtures. Zeitschrift für Physikalische Chem. **221**(4), 537–547 (2007). https://doi.org/10.1524/zpch.2007.221.4.537
87. Techert, S., Neutze, R.: Ultrafast photochemistry, pp. 133–139. TESLA Technical Design Report (2001)
88. Zhang, W., Kjær, K.S., Alonso-Mori, R., Bergmann, U., Chollet, M., Fredin, L.A., Hadt, R.G., Hartsock, R.W., Harlang, T., Kroll, T., Kubicek, K., Lemke, H.T., Liang, H.W., Liu, Y., Nielsen, M.M., Persson, P., Robinson, J.S., Solomon, E.I., Sun, Z., Sokaras, D., van Driel, T.B., Weng, T.C., Zhu, D., Warnmark, K., Sundstrom, V., Gaffney, K.J.: Tracking excited-state charge and spin dynamics in iron coordination complexes. Nature **509**(7500), 345–348 (2014). https://doi.org/10.1038/nature13252
89. Zhang, W., Kjær, K.S., Alonso-Mori, R., Bergmann, U., Chollet, M., Fredin, L.A., Hadt, R.G., Hartsock, R.W., Harlang, T., Kroll, T., Kubicek, K., Lemke, H.T., Liang, H.W., Liu, Y., Nielsen, M.M., Persson, P., Robinson, J.S., Solomon, E.I., Sun, Z., Sokaras, D., van Driel, T.B., Weng, T.C., Zhu, D., Warnmark, K., Sundstrom, V., Gaffney, K.J.: Manipulating charge transfer excited state relaxation and spin crossover in iron coordination complexes with ligand substitution. Chem. Sci. **8**(1), 515–523 (2017). https://doi.org/10.1039/c6sc03070j

90. Debnarova, A., Techert, S., Schmatz, S.: Limitations of high-intensity soft x-ray laser fields for the characterisation of water chemistry: Coulomb explosion of the octamer water cluster. Phys. Chem. Chem. Phys. **14**(27), 9606 (2012). https://doi.org/10.1039/c2cp40598a

91. Jay, R.M., Norell, J., Eckert, S., Hantschmann, M., Beye, M., Kennedy, B., Quevedo, W., Schlotter, W.F., Dakovski, G.L., Minitti, M.P., Hoffmann, M.C., Mitra, A., Moeller, S.P., Nordlund, D., Zhang, W.W.W.L., Kunnus, K., Kubicek, K., Techert, S.A., Lundberg, M., Wernet, P., Gaffney, K., Odelius, M., Föhlisch, A.: Disentangling transient charge density and metal–ligand covalency in photoexcited ferricyanide with femtosecond resonant inelastic soft X-ray scattering. J. Phys. Chem. Lett. **9**(12), 3538–3543 (2018). https://doi.org/10.1021/acs.jpclett.8b01429

92. Kunnus, K., Josefsson, I., Schreck, S., Quevedo, W., Miedema, P.S., Techert, S., de Groot, F.M.F., Föhlisch, A., Odelius, M., Wernet, P.: Quantifying covalent interactions with resonant inelastic soft x-ray scattering: case study of Ni2 aqua complex. Chem. Phys. Lett. **669**, 196–201 (2017). https://doi.org/10.1016/j.cplett.2016.12.046

93. Mitzner, R., Rehanek, J., Kern, J., Gul, S., Hattne, J., Taguchi, T., Alonso-Mori, R., Tran, R., Weniger, C., Schröder, H., Quevedo, W.H.L., Sierra, R.G., Han, G., Lassalle-Kaiser, B., Koroidov, S., Kubicek, K., Schreck, S., Kunnus, K., Brzhezinskaya, M., Firsov, A., Minitti, M.P., Turner, J.J., Moeller, S., Sauter, N.K., Bogan, M.J., Nordlund, D.W.F.S., Messinger, J., Borovik, A., Techert, S., de Groot, F.M.F., Föhlisch, A., Erko, A.U.B., Yachandra, V.K., Wernet, P., Yano, J.: L-edge x-ray absorption spectroscopy of dilute systems relevant to met-alloproteins using an x-ray free-electron laser. J. Phys. Chem. Lett. **4**(21), 3641–3647 (2013). https://doi.org/10.1021/jz401837f

94. Petri, M.: Time-resolved structure determination in low-periodic systems (2011)

95. Petri, M., Busse, G., Quevedo, W., Techert, S.: Photo-induced phase transitions to liquid crystal phases: influence of the chain length from c8e4 to c14e4. Materials **2**(3), 1305–1322 (2009). https://doi.org/10.3390/ma2031305

96. Petri, M., Menzel, A., Bunk, O., Busse, G., Techert, S.: Concentration effects on the dynamics of liquid crystalline self-assembly: time-resolved x-ray scattering studies. J. Phys. Chem. A **115**(11), 2176–2183 (2011). https://doi.org/10.1021/jp1108224

97. Quevedo, W.: Time-resolved investigations in liquid crystals and periodic nanostructures (2010)

98. Quevedo, W., Peth, C., Busse, G., Mann, K., Techert, S.: Nanosecond dynamics of photoex-cited lyotropic liquid crystal structures. J. Phys. Chem. B **114**(26), 8593–8599 (2010). https://doi.org/10.1021/jp101609q

99. Quevedo, W., Peth, C., Busse, G., Scholz, M., Mann, K., Techert, S.: Time-resolved soft x-ray diffraction reveals transient structural distortions of ternary liquid crystals. Int. J. Mol. Sci. **10**(11), 4754–4771 (2009). https://doi.org/10.3390/ijms10114754

100. Quevedo, W., Petri, M., Busse, G., Techert, S.: On the mechanism of photoinduced phase transitions in ternary liquid crystal systems near thermal equilibrium. J. Chem. Phys. **129**(2), 024,502 (2008). https://doi.org/10.1063/1.2943200

101. Quevedo, W., Petri, M., Techert, S.: Home-based time-resolved photo small angle x-ray diffraction and its applications. Zeitschrift für Kristallographie **223**(4-5/2008) (2008). https://doi.org/10.1524/zkri.2008.0031

102. Ramos, A.S.F., Techert, S.: Influence of the water structure on the acetylcholinesterase effi-ciency. Biophys. J. **89**(3), 1990–2003 (2005). https://doi.org/10.1529/biophysj.104.055798

103. Ramos, A.S.F., Techert, S.: Determination of the relative permittivity of acetylcholinesterase. J. Phys. Chem. Lett. **1**(1), 417–419 (2009). https://doi.org/10.1021/jz900261z

104. Schreck, S., Beye, M., Sellberg, J.A., McQueen, T., Laksmono, H., Kennedy, B., Eckert, S., Schlesinger, D., Nordlund, D., Ogasawara, H., Sierra, R.G., Segtnan, V., Kubicek, K., Schlotter, W.F., Dakovski, G.L., Moeller, S., Bergmann, U., Techert, S., Pettersson, L., Wernet, P., Bogan, M.J., Harada, Y., Nilsson, A., Föhlisch, A.: Reabsorption of soft X-ray emission at high X-ray free-electron laser fluences. Phys. Rev. Lett. **113**(15) (2014). https://doi.org/10.1103/physrevlett.113.153002

105. Schreck, S., Pietzsch, A., Kennedy, B., Såthe, C., Miedema, P.S., Techert, S., Strocov, V.N., Schmitt, T., Hennies, F., Rubensson, J.E., Föhlisch, A.: Ground state potential energy surfaces around selected atoms from resonant inelastic X-ray scattering. Sci. Rep. **7**, 20,054 (2016). https://doi.org/10.1038/srep20054

106. Sellberg, J.A., McQueen, T.A., Laksmono, H., Schreck, S., Beye, M., DePonte, D.P., Kennedy, B., Nordlund, D., Sierra, R.G., Schlesinger, D., Tokushima, T., Zhovtobriukh, I., Eckert, S., Segtnan, V.H., Ogasawara, H., Kubicek, K., Techert, S., Bergmann, U., Dakovski, G.L., Schlotter, W.F., Harada, Y., Bogan, M.J., Wernet, P., Föhlisch, A., Pettersson, L.G.M., Nilsson, A.: X-ray emission spectroscopy of bulk liquid water in "no-man's land". J. Chem. Phys. **142**(4), 044,505 (2015). https://doi.org/10.1063/1.4905603

107. Yin, Z., Inhester, L., Veedu, S.T., Quevedo, W., Pietzsch, A., Wernet, P., Groenhof, G., Föhlisch, A., Grubmüller, H., Techert, S.: Cationic and anionic impact on the electronic structure of liquid water. J. Phys. Chem. Lett. **8**(16), 3759–3764 (2017). https://doi.org/10.1021/acs.jpclett.7b01392

108. Yin, Z., Rajkovic, I., Kubicek, K., Quevedo, W., Pietzsch, A., Wernet, P., Föhlisch, A., Techert, S.: Probing the hofmeister effect with ultrafast core-hole spectroscopy. J. Phys. Chem. B **118**(31), 9398–9403 (2014). https://doi.org/10.1021/jp504577a

109. (a) Fenn, J.B.: Angew. Chem. Int. Ed. Engl. **42**, 3871 (2003). (b) Bari, S., Gonzalez-Magaña, O., Reitsma, G., Werner, J., Schippers, S., Hoekstra, R., Schlathölter, T.: J. Chem. Phys. **134**, 024314 (2011). (c) Milosavljević, A.R., Nicolas, C., Lemaire, J., Dehon, C., Thissen, R., Bizau, J.-M., Réfrégiers, M., Nahon, L., Giuliani, A.: Phys. Chem. Chem. Phys. **13**, 15432 (2011). (d) Schwob, L., Lalande, M., Egorov, D., Rangama, J., Hoekstra, R., Vizcaino, V., Schlathölter, T., Poully, J.-C.: Phys. Chem. Chem. Phys. 22895 (2017)

110. Arnlund, D., Johansson, L.C., Wickstrand, C., Barty, A., Williams, G.J., Malmerberg, E., Davidsson, J., Milathianaki, D., DePonte, D.P., Shoeman, R.L., Wang, D., James, D., Katona, G., Westenhoff, S., White, T.A., Aquila, A., Bari, S., Berntsen, P., Bogan, M., van Driel, T.B., Doak, R.B., Kjær, K.S., Frank, M., Fromme, R.I.G., Henning, R., Hunter, M.S., Kirian, R.A., Kosheleva, I., Kupitz, C., Liang, M., Martin, A.V., Nielsen, M.M., Messerschmidt, M., Seibert, M.M.J.S., Stellato, F., Weierstall, U., Zatsepin, N.A., Spence, J.C.H., Fromme, P., Schlichting, I., Boutet, S., Groenhof, G., Chapman, H.N., Neutze, R.: Visualizing a protein quake with time-resolved X-ray scattering at a free-electron laser. Nat. Methods **11**(9), 923–926 (2014). https://doi.org/10.1038/nmeth.3067

111. Bari, S., Egorov, D., Jansen, T.L.C., Boll, R., Hoekstra, R., Techert, S., Zamudio-Bayer, V., Bülow, C., Lindblad, R., Leistner, G., Lawicki, A., Hirsch, K., Miedema, P.S., von Issendorff, B., Lau, T., Schlathölter, T.: Soft X-ray spectroscopy as a probe for gas-phase protein structure: electron impact ionization from within. Chem.—A Eur. J. **24**(30), 7631–7636 (2018). https://doi.org/10.1002/chem.201801440

112. Bari, S., González-Magaña, O., Reitsma, G., Werner, J., Schippers, S., Hoekstra, R., Schlathölter, T.: Photodissociation of protonated leucine-enkephalin in the VUV range of 8–40 eV. J. Chem. Phys. **134**(2), 024,314 (2011). https://doi.org/10.1063/1.3515301

113. Egorov, D., Bari, S., Boll, R., Dörner, S., Deinert, S., Techert, S., Hoekstra, R., Zamudio-Bayer, V., Lindblad, R., Bülow, C., Timm, M., von Issendorff, B., Lau, T., Schlathölter, T.: Near-edge soft X-ray absorption mass spectrometry of protonated melittin. J. Am. Soc. Mass Spectrom. **29**(11), 2138–2151 (2018). https://doi.org/10.1007/s13361-018-2035-6

114. González-Magaña, O., Tiemens, M., Reitsma, G., Boschman, L., Door, M., Bari, S., Lahaie, P.O., Wagner, J.R., Huels, M.A., Hoekstra, R., Schlathölter, T.: Fragmentation of protonated oligonucleotides by energetic photons and cq+ions. Phys. Rev. A **87**(3) (2013). https://doi.org/10.1103/physreva.87.032702

115. Jain, R., Petri, M., Kirschbaum, S., Feindt, H., Steltenkamp, S., Sonnenkalb, S., Becker, S., Griesinger, C., Menzel, A., Burg, T.P., Techert, S.: X-ray scattering experiments with high-flux X-ray source coupled rapid mixing microchannel device and their potential for high-flux neutron scattering investigations. Eur. Phys. J. E **36**(9) (2013). https://doi.org/10.1140/epje/i2013-13109-9

116. Jain, R., Techert, S.: Time-resolved and in-situ X-ray scattering methods beyond photoactivation: utilizing high-flux X-ray sources for the study of ubiquitous non-photoactive proteins. Protein & Pept. Lett. **23**(3), 242–254 (2016). https://doi.org/10.2174/0929866523666160106153847

117. Kupitz, C., Basu, S., Grotjohann, I., Fromme, R., Zatsepin, A.N., Rendek, K.N., Hunter, M.S., Shoeman, R.L., White, T.A., Wang, D., James, D., Yang, J.H., Cobb, D.E., Reeder, B., Sierra, R.G., Liu, H., Barty, A., Aquila, A.L., Deponte, D., Kirian, R.A., Bari, S., Bergkamp, J.J., Beyerlein, K.R., Bogan, M.J., Caleman, C., Chao, T.C., Conrad, C.E., Davis, K.M., Fleckenstein, H., Galli, L., Hau-Riege, S.P., Kassemeyer, S., Laksmono, H.M.L., Lomb, L.S.M., Martin, A.V., Messerschmidt, M., Milathianaki, D., Nass, K., Ros, A., Roy-Chowdhury, S., Schmidt, K., Seibert, M., Steinbrener, J., Stellato, F., Yan, L., Yoon, C., Moore, T.A., Moore, A.L., Pushkar, Y., Williams, G.J., Boutet, S., Doak, R.B., Weierstall, U., Frank, M.H.N.C., Spence, J.C.H., Fromme, P.: Serial time-resolved crystallography of photosystem II using a femtosecond X-ray laser. Nature **513**(7517), 261–265 (2014). https://doi.org/10.1038/nature13453

118. Petri, M., Frey, S., Menzel, A., Görlich, D., Techert, S.: Structural characterization of nanoscale meshworks within a nucleoporin FG hydrogel. Biomacromolecules **13**(6), 1882–1889 (2012). https://doi.org/10.1021/bm300412q

119. Schlathölter, T., Reitsma, G., Egorov, D., González-Magaña, O., Bari, S., Boschman, L., Bodewits, E., Schnorr, K., Schmid, G., Schröter, C.D., Moshammer, R., Hoekstra, R.: Multiple ionization of free ubiquitin molecular ions in extreme ultraviolet free-electron laser pulses. Angew. Chem. Int. Edition **55**(36), 10741–10745 (2016). https://doi.org/10.1002/anie.201605335

120. Schulz, J., Bari, S., Buck, J., Uetrecht, C.: Sample refreshment schemes for high repetition rate FEL experiments. In: Tschentscher, T. Tiedtke, K. (eds.) Advances in X-ray Free-Electron Lasers II: Instrumentation. SPIE (2013). https://doi.org/10.1117/12.2019754

121. Techert, F., Techert, S., Woo, L., Beck, W., Lebsanft, H., Wizemann, V.: High blood flow rates with adjustment of needle diameter do not increase hemolysis during hemodialysis treatment. J. Vasc. Access **8**(4), 252–257 (2007). https://doi.org/10.1177/112972980700800406

122. Boll, R., Anielski, D., Bostedt, C., Bozek, J.D., Christensen, L., Coffee, R., De, S., Decleva, P., Epp, S.W., Erk, B., Foucar, L., Krasniqi, F., Küpper, J., Rouzée, A., Rudek, B., Rudenko, A., Schorb, S., Stapelfeldt, H., Stener, M., Stern, S., Techert, S., Trippel, S., Vrakking, M.J.J., Ullrich, J., Rolles, D.: Femtosecond photoelectron diffraction on laser-aligned molecules: towards time-resolved imaging of molecular structure. Phys. Rev. A **88**(6) (2013). https://doi.org/10.1103/physreva.88.061402

123. Bomme, C., Anielski, D., Savelyev, E., Boll, R., Erk, B., Bari, S., Viefhaus, J., Stener, M., Decleva, P., Rolles, D.: Diffraction effects in the recoil-frame photoelectron angular distributions of halomethanes. J. Phys.: Conf. Ser. **635**(11), 112,020 (2015). https://doi.org/10.1088/1742-6596/635/11/112020

124. Erk, B., Müller, J.P., Bomme, C., Boll, R., Brenner, G., Chapman, H.N., Correa, J., Düsterer, S., Dziarzhytski, S., Eisebitt, S., Graafsma, H., Grunewald, S., Gumprecht, L., Hartmann, R., Hauser, G., Keitel, B., von Korff Schmising, C., Kuhlmann, M., Manschwetus, B., Mercadier, L., Müller, E., Passow, C., Plönjes, E., Ramm, D., Rompotis, D., Rudenko, A., Rupp, D., Sauppe, M., Siewert, F., Schlosser, D., Strüder, L., Swiderski, A., Techert, S., Tiedtke, K., Tilp, T., Treusch, R., Schlichting, I., Ullrich, J., Moshammer, R., Möller, T., Rolles, D.: CAMP@FLASH: an end-station for imaging, electron- and ion-spectroscopy, and pump-probe experiments at the FLASH free-electron laser. J. Synchrotron Radiat. **25**(5), 1529–1540 (2018). https://doi.org/10.1107/s1600577518008585

125. Foucar, L., Barty, A., Coppola, N., Hartmann, R., Holl, P., Hoppe, U., Kassemeyer, S., Kimmel, N., Küpper, J., Scholz, M., Techert, S., White, T.A., Strüder, L., Ullrich, J.: CASS-CFEL-ASG software suite. Comput. Phys. Commun. **183**(10), 2207–2213 (2012). https://doi.org/10.1016/j.cpc.2012.04.023

126. Savelyev, E., Boll, R., Bomme, C., Schirmel, N., Redlin, H., Erk, B., Düsterer, S., Müller, E., Höppner, H., Toleikis, S., Müller, J., Czwalinna, M.K., Treusch, R., Kierspel, T., Mullins, T., Trippel, S., Wiese, J., Küpper, J., Brauße, F., Krecinic, F., Rouzee, A., Rudawski, P., Johnsson,

P., Amini, K., Lauer, A., Burt, M., Brouard, M., Christensen, L., Thørgersen, J., Stapelfeldt, H., Berrah, N., Müller, M., Ulmer, A., Techert, S., Rudenko, A., Rolles, D.: Jitter-correction for IR/UV-XUV pump-probe experiments at the FLASH free-electron laser. New J. Phys. **19**(4), 043,009 (2017). https://doi.org/10.1088/1367-2630/aa652d

127. Strueder, L., Epp, S., Rolles, D., Hartmann, R., Holl, P., Lutz, G., Soltau, H., Eckart, R., Reich, C., Heinzinger, K., Thamm, C., Rudenko, A., Krasniqi, F., Techert, S., Mosshammer, R., Kühnel, K.U., Bauer, C., Schröter, C.D., Miessner, D., Porro, M., Hälker, O., Meidinger, N., Ziegler, N., Hermann, S., Pietsch, U., Walenta, A., Leitenberger, W., Boestedt, C., Möller, T., Rupp, D.M.A., Graafsma, H., Hirsemann, H., Gärtner, K., Richter, R., Foucar, L., Shoeman, R.L., Schlichting, I., Ullrich, J.: Large-format, high-speed, X-ray pnCCDs combined with electron and ion imaging spectrometers in a multipurpose chamber for experiments at 4th generation light sources. Nucl. Instrum. Methods Phys. Res. Sect. A: Accel. Spectrom. Detect. Assoc. Equip. **614**(3), 483–496 (2010). https://doi.org/10.1016/j.nima.2009.12.053

128. Ablikim, U., Bomme, C., Xiong, H., Savelyev, E., Obaid, R., Kaderiya, B., Augustin, S., Schnorr, K., Dumitriu, I., Osipov, T., Bilodeau, R., Kilcoyne, D., Kumarappan, V., Rudenko, A., Berrah, N., Rolles, D.: Identification of absolute geometries of cis and trans molecular isomers by coulomb rxplosion imaging. Sci. Rep. **6**(1), 433 (2016). https://doi.org/10.1038/srep38202

129. Anielski, D., Boll, R., Rolles, D.: Time-resolved photoelectron diffraction on laser-aligned molecules. In: Research in Optical Sciences. OSA (2012). https://doi.org/10.1364/hilas.2012.jt2a.38

130. Gomez, L.F., Ferguson, K.R., Cryan, J.P., Bacellar, C., Tanyag, R.M.P., Jones, C., Schorb, S., Anielski, D., Belkacem, A., Bernando, C., Boll, R., Bozek, J., Carron, S., Chen, G., Delmas, T., Englert, L., Epp, S.W., Erk, B., Foucar, L., Hartmann, R., Hexemer, A., Huth, M., Kwok, J., Leone, S.R., Ma, J.H.S., Maia, F.R.N.C., Malmerberg, E., Marchesini, S., Neumark, D.M., Poon, B., Prell, J., Rolles, D., Rudek, B., Rudenko, A., Seifrid, M., Siefermann, K.R., Sturm, F.P., Swiggers, M., Ullrich, J., Weise, F., Zwart, P., Bostedt, C., Gessner, O., Vilesov, A.F.: Shapes and vorticities of superfluid helium nanodroplets. Science **345**(6199), 906–909 (2014). https://doi.org/10.1126/science.1252395

131. Gorkhover, T., Schorb, S., Coffee, R., Adolph, M., Foucar, L., Rupp, D., Aquila, A., Bozek, J.D., Epp, S.W., Erk, B., Gumprecht, L., Holmegaard, L., Hartmann, A., Hartmann, R.G.H., Holl, P., Hömke, A., Johnsson, P., Kimmel, N., Kühnel, K.U., Messerschmidt, M., Reich, C., Rouzée, A., Rudek, B., Schmidt, C., Schulz, J., Soltau, H., Stern, S., Weidenspointner, G., White, B., Küpper, J., Strüder, L., Schlichting, I., Ullrich, J., Rolles, D., Rudenko, A., Möller, T., Bostedt, C.: Femtosecond and nanometre visualization of structural dynamics in superheated nanoparticles. Nat. Photonics **10**(2), 93–97 (2016). https://doi.org/10.1038/nphoton.2015.264

132. Hädrich, S., Rothhardt, J., Klas, R., Tschernajew, M., Hoffmann, A., Tadesse, G.K., Klenke, A., Gottschall, T., Eidam, T., Limpert, J., Tünnermann, A., Boll, R., Bomme, C., Dachraoui, H., Erk, B., Fraia, M.D., Horke, D., Kierspel, T., Mullins, T., A. Przystawik, E.S., Wiese, J., Laarmann, T.J.K., Rolles, D., Barkowski, M., Sadashivaiah, S., Urbancic, J., Aeschlimann, M., Mathias, S.: High photon flux 70 eV HHG source for applications in molecular and solid state physics. In: High-brightness Sources and Light-Driven Interactions. OSA (2016). https://doi.org/10.1364/hilas.2016.ht1b.2

133. Hau-Riege, S.P., Graf, A., Döppner, T., London, R.A., Krzywinski, J., Fortmann, C., Glenzer, S.H., Frank, M., Sokolowski-Tinten, K., Messerschmidt, M., Bostedt, C., Schorb, S., Bradley, J.A., Lutman, A., Rolles, D., Rudenko, A., Rudek, B.: Ultrafast transitions from solid to liquid and plasma states of graphite induced by X-ray free-electron laser pulses. Phys. Rev. Lett. **108**(21) (2012). https://doi.org/10.1103/physrevlett.108.217402

134. Jones, C.F., Bernando, C., Tanyag, R.M.P., Bacellar, C., Ferguson, K.R., Gomez, F., Anielski, D., Belkacem, A., Boll, R., Bozek, J., Carron, S., Cryan, J., Englert, L., Epp, W., Erk, B., Foucar, L., Hartmann, R.D.M.N., Rolles, D., Rudenko, A., Siefermann, K.R., Weise, F., Rudek, B., Sturm, F.P., Ullrich, J., Bostedt, C., Gessner, O., Vilesov, A.F.: Coupled motion of

xe clusters and quantum vortices in he nanodroplets. Phys. Rev. B **93**(18), 180,510 (2016). https://doi.org/10.1103/physrevb.93.180510

135. Kierspel, T., Wiese, J., Mullins, T., Robinson, J., Aquila, A., Barty, A., Bean, R., Boll, R., Boutet, S., Bucksbaum, P.H., Chapman, H.N., Christensen, L.A.F., Hunter, M., Koglin, J.E., Liang, M., Mariani, V., Morgan, A.A.N., Petrovic, V., Rolles, D.A.R., Schnorr, K., Stapelfeldt, H., Stern, S., Thogersen, J., Yoon, C.H., Wang, F., Trippel, S., Küpper, J.: Strongly aligned gas-phase molecules at free-electron lasers. J. Phys. B: Atom. Mol. Opt. Phys. **48**(20), 204,002 (2015). https://doi.org/10.1088/0953-4075/48/20/204002

136. Liekhus-Schmaltz, C., Tenney, I., Osipov, T., Bucksbaum, P.H., Petrovic, V.S., Belkacem, A., Berrah, N., Boll, R., Bomme, C., Bostedt, C., Bozek, J.D., Carron, S., Coffee, R.N., Devin, J., Erk, B., Fang, L., Field, R.W., Ferguson, K., Foucar, L., Frasinski, L.J., Glownia, J.M., Guehr, M., Kamalov, A., Krzywinski, J., Li, H., Marangos, J.P., Martinez, T., McFarland, B.K., Miyabe, S., Murphy, B.F., Natan, A.A., Rolles, D., Rudenko, A., Sanchez, A., Siano, M., Simpson, E., Spector, L.S., Swiggers, M.L., Walke, D.J., Wang, S., Weber, T.: Mapping the fragmentation of acetylene with femtosecond resolution pump probe at LCLS using 2, 3, and 4 particle coincidences. In: CLEO: 2014. OSA (2014). https://doi.org/10.1364/cleo_at.2014.jth2a.88

137. Liekhus-Schmaltz, C., Tenney, I., Osipov, T., Sanchez-Gonzalez, A., Berrah, N., Boll, R., Bomme, C., Bostedt, C., Bozek, J.D., Carron, S., Coffee, R.N., Devin, J., Erk, B., Ferguson, K.R., Field, R.W.L.F., Frasinski, L.J., Glownia, J.M., Guehr, M., Kamalov, A., Krzywinski, J., Li, H., Marangos, J.P., Martinez, T., McFarland, B.K., Miyabe, S., Murphy, B.F., Natan, A.A., Rolles, D., Rudenko, A., Siano, M., Simpson, E.R., Spector, L., Swiggersa, M., Walke, D.S., Weber, T., Bucksbaum, P.H., Petrovic, V.S.: Ultrafast isomerization initiated by X-ray core ionization. Nat. Commun. **6**(1) (2015). https://doi.org/10.1038/ncomms9199

138. Rolles, D., Boll, R., Erk, B., Rompotis, D., Manschwetus, B.: An experimental protocol for femtosecond nir/uv–xuv pump–probe experiments with free-electron lasers. JoVE **140**(e57055) (2018). https://doi.org/10.3791/57055

139. Rolles, D., Boll, R., Tamrakar, S.R., Anielski, D., Bomme, C.: Femtosecond photoelectron diffraction: a new approach to image molecular structure during photochemical reactions. In: Liu, Z. (ed.) Ultrafast Nonlinear Imaging and Spectroscopy II. SPIE (2014). https://doi.org/10.1117/12.2061783

140. Rothhardt, J., Hädrich, S., Shamir, Y., Tschernajew, M.R.K., Hoffmann, A., Tadesse, G.K., Klenke, A., Gottschall, T., Eidam, T., Limpert, J., Tünnermann, A., Boll, R., Bomme, C., Dachraoui, H., Erk, B., Fraia, M.D., Horke, D.A., Kierspel, T., Mullins, T.A.P., Savelyev, E., Wiese, J.T.L., Küpper, J., Rolles, D.: High-repetition-rate and high-photon-flux 70 eV high-harmonic source for coincidence ion imaging of gas-phase molecules. Opt. Express **24**(16), 18,133 (2016). https://doi.org/10.1364/oe.24.018133

141. Rudek, B., Son, S.K., Foucar, L., Epp, S.W., Erk, B., Hartmann, R., Adolph, M., Andritschke, R., Aquila, A., Berrah, N., Bostedt, C., Bozek, J., Coppola, N., Filsinger, F., Gorke, H., Gorkhover, T., Graafsma, H., Gumprecht, L., Hartmann, A., Hauser, G., Herrmann, S., Hirsemann, H., Holl, P., Hömke, A., Journel, L., Kaiser, C., Kimmel, N., Krasniqi, F., Kühnel, K.U., Matysek, M., Messerschmidt, M., Miesner, D., Möller, T., Moshammer, R., Nagaya, K., Nilsson, B., Potdevin, G., Pietschner, D., Reich, C., Rupp, D., Schaller, G., Schlichting, I., Schmidt, C.F.S., Schorb, S., Schröter, C., Schulz, J., Simon, M., Soltau, H., Strüder, L., Ueda, K., Weidenspointner, G., Santra, R., Ullrich, J., Rudenko, A., Rolles, D.: Ultra-efficient ionization of heavy atoms by intense X-ray free-electron laser pulses. Nat. Photonics **6**(12), 858–865 (2012). https://doi.org/10.1038/nphoton.2012.261

142. Rudenko, A., Rolles, D.: Time-resolved studies with fels. J. Electron Spectrosc. Relat. Phenom. **204**, 228 (2015). https://doi.org/10.1016/j.elspec.2015.07.010

143. Sauppe, M., Rompotis, D., Erk, B., Bari, S., Bischoff, T., Boll, R., Bomme, C., Bostedt, C., Dörner, S., Düsterer, S., Feigl, T., Flückiger, L., Gorkhover, T., Kolatzki, K., Langbehn, B., Monserud, N., Mueller, E., Müller, J.P., Passow, C., Ramm, D., Rolles, D., Schubert, K., Schwob, L., Senfftleben, B., Treusch, R., Ulmer, A., Weigelt, H., Zimbalsko, J., Zimmermann, J., Moeller, T., Rupp, D.: XUV double-pulses with femtosecond to 650 ps separation from a

multilayer-mirror-based split-and-delay unit at FLASH. J. Synchrotron Radiat. **25**(5), 1517–1528 (2018). https://doi.org/10.1107/s1600577518006094

144. Schippers, S., Buhr, T., Mueller, A., Perry-Sassmannshausen, A., Ricz, S., Klumpp, S., Mertens, K., Reinwardt, S., Martins, M., Schubert, K., Bari, S., Waitz, F.M., Jahnke, T., Schöffler, M., Dörner, R.: Pipe: The photon-ion-endstation at petra iii for experimental studies of xuv-photoprocesses in small quantum systems. In: Deutsche Tagung für Forschung mit Synchrotronstrahlung, Neutronen und Ionenstrahlen an Großgeräten (2018)

145. Schippers, S., Martins, M., Beerwerth, R., Bari, S., Holste, K., Schubert, K., Viefhaus, J., Savin, D.W., Fritzsche, S., Mueller, A.: Near l-edge single and multiple photoionization of singly charged iron ions. Astrophys. J. **849**(1), 5 (2017). https://doi.org/10.3847/1538-4357/aa8fcc

146. Schnorr, K., Senftleben, A., Kurka, M., Rudenko, A., Foucar, L., Schmid, G., Broska, A., Pfeifer, T., Meyer, K., Anielski, D., Boll, R., Rolles, D., Kübel, M., Kling, M.F., Jiang, Y.H., Mondal, S., Tachibana, T., Ueda, K., Marchenko, T., Simon, M., Brenner, G., Treusch, R., Scheit, S., Averbukh, V., Ullrich, J., Schröter, C.D., Moshammer, R.: Time-resolved measurement of interatomic coulombic decay in Ne$_2$. Phys. Rev. Lett. **111**(9) (2013). https://doi.org/10.1103/physrevlett.111.093402

147. Schnorr, K., Senftleben, A., Schmid, G., Augustin, S., Kurka, M., Rudenko, A., Foucar, L., Broska, A., Meyer, K., Anielski, D., Boll, R., Rolles, D., Kübel, M., Kling, M.F., Jiang, Y.H., Mondal, S., Tachibana, T., Ueda, K., Marchenko, T., Simon, M., Brenner, G., Treusch, R., Scheit, S., Averbukh, V., Ullrich, J., Pfeifer, T., Schröter, C.D., Moshammer, R.: Time-resolved study of ICD in ne dimers using FEL radiation. J. Electron Spectros. Relat. Phenom. **204**, 245–256 (2015). https://doi.org/10.1016/j.elspec.2015.07.009

148. Schorb, S., Gorkhover, T., Cryan, J.P., Glownia, J.M., Bionta, M.R., Coffee, R.N., Erk, B., Boll, R., Schmidt, C., Rolles, D., Rudenko, A., Rouzee, A., Swiggers, M., Carron, S., Castagna, J.C., Bozek, J.D., Messerschmidt, M., Schlotter, W.F., Bostedt, C.: X-ray-optical cross-correlator for gas-phase experiments at the linac coherent light source free-electron laser. Appl. Phys. Lett. **100**(12), 121,107 (2012). https://doi.org/10.1063/1.3695163

149. Tanyag, R., Bernando, C., Jones, C.F., Bacellar, C., Ferguson, K.R., Anielski, D., Boll, R., Carron, S.J.P.C., Englert, L., Epp, S.W., Erk, B., Foucar, L., Gomez, L.F.R.H., Neumark, D.M., Rolles, D.B.R., Rudenko, A., Siefermann, K.R., Ullrich, J., Weise, F., Bostedt, C., Gessner, O., Vilesov, A.F.: Communication: X-ray coherent diffractive imaging by immersion in nanodroplets. Struct. Dyn. **2**(5), 051,102 (2015). https://doi.org/10.1063/1.4933297

150. Ziaee, F., Rudenko, A., Rolles, D., Savelyev, E., Boll, R., Manschwetus, B., Erk, N., Trippel, S., Wiese, J., Kuepper, J., Amini, K., Lee, J., Brouard, M., Brauße, F., Rouzee, A., Olshin, P., Mereshchenko, A., Lahl, J., Johnsson, P., Simon, M., Marchenko, T., Holland, D., Underwood, J.: Probing ultrafast nuclear dynamics in halomethanes by time-resolved electron and ion imaging. In: APS Division of Atomic and Molecular Physics Meeting (2016)

151. Ablikim, U., Bomme, C., Savelyev, E., Xiong, H., Kushawaha, R., Boll, R., Amini, K., Osipov, T., Kilcoyne, D., Rudenko, A., Berrah, N., Rolles, D.: Isomer-dependent fragmentation dynamics of inner-shell photoionized difluoroiodobenzene. Phys. Chem. Chem. Phys. **19**(21), 13419–13431 (2017). https://doi.org/10.1039/c7cp01379e

152. Allum, F., Burt, M., Amini, K., Boll, R., Köckert, H., Olshin, P.K., Bari, S., Bomme, C., Brauße, F., de Miranda, B.C., Düsterer, S., Erk, B., Géléoc, M., Geneaux, R., Gentleman, A.S., Goldsztejn, G., Guillemin, R., Holland, D.M.P., Ismail, I., Johnsson, P., Journel, L., Küpper, J., Lahl, J., Lee, J.W.L., Maclot, S., Mackenzie, S.R., Manschwetus, B., Mereshchenko, A.S., Mason, R., Palaudoux, J., Piancastelli, M.N., Penent, F., Rompotis, D., Rouzée, A., Ruchon, T., Rudenko, A., Savelyev, E., Simon, M., Schirmel, N., Stapelfeldt, H., Techert, S., Travnikova, O., Trippel, S., Underwood, J.G., Vallance, C., Wiese, J., Ziaee, F., Brouard, M., Marchenko, T., Rolles, D.: Coulomb explosion imaging of CH3i and CH2cli photodissociation dynamics. J. Chem. Phys. **149**(20), 204,313 (2018). https://doi.org/10.1063/1.5041381

153. Amini, K., Boll, R., Lauer, A., Burt, M., Lee, J., Christensen, L., Brauße, F., Mullins, T., Savelyev, E., Ablikim, U., Berrah, N., Bomme, C., Düsterer, S., Erk, B., Höppner, H., Johnsson, P., Kierspel, T., Krecinic, F., Küpper, J., Müller, M., Müller, E., Redlin, H., Rouzée,

A., Schirmel, N., Thøgersen, J., Techert, S., Toleikis, S., Treusch, R., Trippel, S., Ulmer, A., Wiese, J., Vallance, C., Rudenko, A., Stapelfeldt, H., Brouard, M., Rolles, D.: Alignment, orientation, and coulomb explosion of difluoroiodobenzene studied with the pixel imaging mass spectrometry (PImMS) camera. J. Chem. Phys 147(1), 013,933 (2017). https://doi.org/10.1063/1.4982220

154. Amini, K., Savelyev, E., Brauße, F., Berrah, N., Bomme, C., Brouard, M., Burt, M., Christensen, L., Düsterer, S., Erk, B., Höppner, H., Kierspel, T., Krecinic, F., Lauer, A., Lee, J.W.L., Müller, M., Müller, E., Mullins, T., Redlin, H., Schirmel, N., Thøgersen, J., Techert, S., Toleikis, S., Treusch, R., Trippel, S., Ulmer, A., Vallance, C., Wiese, J., Johnsson, P., Küpper, J., Rudenko, A., Rouzée, A., Stapelfeldt, H., Rolles, D., Boll, R.: Photodissociation of aligned CH3i and c6h3f2i molecules probed with time-resolved coulomb explosion imaging by site-selective extreme ultraviolet ionization. Struct. Dyn. 5(1), 014,301 (2018). https://doi.org/10.1063/1.4998648

155. Boll, R., Erk, B., Coffee, R., Trippel, S., Kierspel, T., Bomme, C., Bozek, J.D., Burkett, M., Carron, S., Ferguson, K.R., Foucar, L., Küpper, J., Marchenko, T., Miron, C., Patanen, M., Osipov, T., Schorb, S., Simon, M., Swiggers, M., Techert, S., Ueda, K., Bostedt, C., Rolles, D., Rudenko, A.: Charge transfer in dissociating iodomethane and fluoromethane molecules ionized by intense femtosecond X-ray pulses. Struct. Dyn. 3(4), 043,207 (2016). https://doi.org/10.1063/1.4944344

156. Boll, R., Rouzée, A., Adolph, M., Anielski, D., Aquila, A., Bari, S., Bomme, C., Bostedt, C., Bozek, J.D., Chapman, H.N., Christensen, L., Coffee, R., Coppola, N., De, S., Decleva, P., Epp, S.W., Erk, B., Filsinger, F., Foucar, L., Gorkhover, T., Gumprecht, L., Hömke, A., Holmegaard, L., Johnsson, P., Kienitz, J.S., Kierspel, T., Krasniqi, F., Kühnel, K.U., Maurer, J., Messerschmidt, M., Moshammer, R., Müller, N.L.M., Rudek, B., Savelyev, E., Schlichting, I., Schmidt, C., Scholz, F., Schorb, S., Schulz, J., Seltmann, J., Stener, M., Stern, S., Techert, S., Thøgersen, J., Trippel, S., Viefhaus, J., Vrakking, M., Stapelfeldt, H., Küpper, J., Ullrich, J., Rudenko, A., Rolles, D.: Imaging molecular structure through femtosecond photoelectron diffraction on aligned and oriented gas-phase molecules. Faraday Discuss. 171, 57–80 (2014). https://doi.org/10.1039/c4fd00037d

157. Brauße, F., Goldsztejn, G., Amini, K., Boll, R., Bari, S., Bomme, C., Brouard, M., Burt, M., de Miranda, B.C., Düsterer, S., Erk, B., Géléoc, M., Geneaux, R., Gentleman, A.S., Guillemin, R., Ismail, I., Johnsson, P., Journel, L., Kierspel, T., Köckert, H., Küpper, J., Lablanquie, P., Lahl, J., Lee, J.W.L., Mackenzie, S.R., Maclot, S., Manschwetus, B., Mereshchenko, A.S., Mullins, T., Olshin, P.K., Palaudoux, J., Patchkovskii, S., Penent, F., Piancastelli, M.N., Rompotis, D., Ruchon, T., Rudenko, A., Savelyev, E., Schirmel, N., Techert, S., Travnikova, O., Trippel, S., Underwood, J.G., Vallance, C., Wiese, J., Simon, M., Holland, D.M.P., Marchenko, T., Rouzée, A., Rolles, D.: Time-resolved inner-shell photoelectron spectroscopy: From a bound molecule to an isolated atom. Phys. Rev. A 97(4) (2018). https://doi.org/10.1103/physreva.97.043429

158. Burt, M., Boll, R., Lee, J.W.L., Amini, K., Gentleman, A.S., Klöckert, H., Mackenzie, S.R., Vallance, C., Bari, S., Bomme, C., Dachraoui, H., Düsterer, S., Erk, B., Manschwetus, B., Müller, E., Rompotis, D., Savelyev, E., Schirmel, N., Techert, S., Treusch, R., Küpper, J., Trippel, S., Wiese, J., Stapelfeldt, H., de Miranda, B.C., Guillemin, R., Ismail, I., Journel, L., Marchenko, T., Palaudoux, J., Penent, F., Piancastelli, M.N., Simon, M., Travnikova, O., Brauße, F., Goldsztejn, G., Rouzee, A., Geleoc, M., Geneaux, R., Ruchon, T., Underwood, J., Holland, D.M.P., Mereshchenko, A.S., Olshin, P.K., Johnsson, P., Maclot, S., Lahl, J., Rudenko, A., Ziaee, F., Brouard, M., Rolles, D.: Coulomb-explosion imaging of concurrent CH2bri photodissociation dynamics. Phys. Rev. A 96(4) (2017). https://doi.org/10.1103/physreva.96.043415

159. Erk, B., Boll, R., Trippel, S., Anielski, D., Foucar, L., Rudek, B., Epp, S.W., Coffee, R., Carron, S., Schorb, S., Ferguson, K.R., Swiggers, M., Bozek, J.D., Simon, M., Marchenko, T., Kupper, J., Schlichting, I., Ullrich, J., Bostedt, C., Rolles, D., Rudenko, A.: Imaging charge transfer in iodomethane upon X-ray photoabsorption. Science 345(6194), 288–291 (2014). https://doi.org/10.1126/science.1253607

160. Erk, B., Rolles, D., Foucar, L., Rudek, B., Epp, S.W., Cryle, M., Bostedt, C., Schorb, S., Bozek, J., Rouzee, A., Hundertmark, A., Marchenko, T., Simon, M., Filsinger, F., Christensen, L., De, S., Trippel, S., Küpper, J., Stapelfeldt, H., Wada, S., Ueda, K., Swiggers, M., Messerschmidt, M., Schröter, C.D., Moshammer, R., Schlichting, I., Ullrich, J., Rudenko, A.: Ultrafast charge rearrangement and nuclear dynamics upon inner-shell multiple ionization of small polyatomic molecules. Phys. Rev. Lett. **110**(5) (2013). https://doi.org/10.1103/physrevlett.110.053003
161. Müller, A., Borovik, A., Bari, S., Buhr, T., Holste, K., Martins, M., Perry-Saßmannshausen, A., Phaneuf, R., Reinwardt, S., Ricz, S., Schubert, K., Schippers, S.: Near-k-edge double and triple detachment of the f-negative ion: Observation of direct two-electron ejection by a single photon. Phys. Rev. Lett. **120**(13) (2018). https://doi.org/10.1103/physrevlett.120.133202
162. Rolles, D., Boll, R., Adolph, M., Aquila, A., Bostedt, C., Bozek, J.D., Chapman, H.N., Coffee, R., Coppola, N., Decleva, P., Delmas, T., Epp, S.W., Erk, B., Filsinger, F., Foucar, L., Gumprecht, L., Hömke, A., Gorkhover, T., Holmegaard, L., Johnsson, P., Kaiser, C., Krasniqi, F., Kühnel, K.U., Maurer, J., Messerschmidt, M., Moshammer, R., Quevedo, W., Rajkovic, I., Rouzée, A., Rudek, B., Schlichting, I., Schmidt, C., Schorb, S., Schröter, C.D., Schulz, J., Stapelfeldt, H., Stener, M., Stern, S., Techert, S., Thøgersen, J., Vrakking, M.J.J., Rudenko, A., Küpper, J., Ullrich, J.: Femtosecond X-ray photoelectron diffraction on gas-phase dibromobenzene molecules. J. Phys. B: Atom. Mol. Opt. Phys. **47**(12), 124,035 (2014). https://doi.org/10.1088/0953-4075/47/12/124035
163. Rudenko, A., Inhester, L., Hanasaki, K., Li, X., Robatjazi, S.J., Erk, B., Boll, R., Toyota, K., Hao, Y., Vendrell, O., Bomme, C., Savelyev, E., Rudek, B., Foucar, L., Southworth, S.H., Lehmann, C.S., Kraessig, B., Marchenko, T., Simon, M., Ueda, K., Ferguson, K.R., Bucher, M., Gorkhover, T., Carron, S., Alonso-Mori, R., Koglin, J.E., Correa, J., Williams, G.J., Boutet, S., Young, L., Bostedt, C., Son, S.K., Santra, R., Rolles, D.: Femtosecond response of polyatomic molecules to ultra-intense hard X-ray. Nature **546**(7656), 129–132 (2017). https://doi.org/10.1038/nature22373

Chapter 18
Polarization-Sensitive Coherent Diffractive Imaging Using HHG

Sergey Zayko, Ofer Kfir and Claus Ropers

Abstract High harmonic generation (HHG) from lasers have attractive properties for probing ultrafast dynamics at the nanoscale. The spectral range of high harmonics at the extreme-UV and soft-X-rays ($\lambda \sim 100$ nm–1 nm, $\hbar\omega \sim 10$ eV–1 keV) enables element specificity, the short wavelengths combined with high spatial coherence allows for imaging with nanometric spatial resolution, the extremely short pulse durations provide access to dynamics faster than a femtosecond ($1 \ fs = 10^{-15}$ s), and all that, on a compact system. In this chapter, we focus on experimental aspects of imaging with high harmonics. First, we present the experimental system and the image reconstruction procedure. Second, we show experimental results from the various configurations that were used throughout this project. Finally, we discuss mechanisms that played an important role in this imaging effort, and would contribute to the advancement of nanoscale imaging.

18.1 Experimental Setup

The experimental setup for high harmonic generation is schematically depicted in Fig. 18.1. The laser system delivers up to 8 mJ per pulse (pulse duration of 40 fs) at 1 kHz repetition rate, operating at a central wavelength of 800 nm. Initially, only a fraction of the output power of the laser system was used (between 0.4 and 1 mJ) for the generation of high harmonics in argon. Focusing the laser beam using a lens with a relatively short focal length of 20 cm ensures that the field intensity is sufficient to drive this nonlinear process. A typical Ar pressure needed for efficient harmonic up-conversion is more than 30 mbar. However, already at this pressure,

S. Zayko (✉) · O. Kfir · C. Ropers
IV. Physical Institute - Solids and Nanostructures, Universität Göttingen,
Friedrich-Hund-Platz 1, 37077 Göttingen, Germany
e-mail: szayko@gwdg.de

O. Kfir
e-mail: okfir@uni-goettingen.de

C. Ropers
e-mail: cropers@gwdg.de

T. Salditt et al. (eds.), *Nanoscale Photonic Imaging*, Topics in Applied Physics 134,
https://doi.org/10.1007/978-3-030-34413-9_18

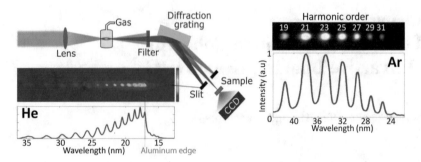

Fig. 18.1 Experimental setup for lensless imaging with high-harmonic radiation. Femtosecond laser pulses are focused in a gas cell filled with Ar or He. Generated harmonics are spatially dispersed with a toroidal diffraction grating and refocused onto a sample. The resulting diffraction patterns are recorded using a charge-coupled device camera. The measured spectra from Ar and He are shown on the left and on the right, respectively

most of the harmonics yield will be reabsorbed within a just a few mm. To mitigate the reabsorbtion of the generated radiation the interaction region is confined within a metallic capillary (diameter of 5–10 mm) with a very small entrance and exit holes that are self-drilled by the laser. Having only small holes is beneficial, since it reduces the overall gas pressure in the generation chamber. Generally, the short absorption length of extreme-UV radiation requires that all experiments are carried out in high-vacuum conditions. After the HHG beam passes through an aluminum filter, a toroidal grating spatially disperses different harmonic orders, and refocuses a selected order onto the sample. When the sample is removed from the beam path all harmonic orders are incident on a charge-coupled device (CCD) camera and the HHG flux can be improved by finding the optimal phase matching conditions with a recursive fine-tuning of the following parameters: gas pressure, laser beam diameter, position of the capillary relative to the laser focus and the laser pulse chirp.

The aluminum filter used in the experimental setup is a 150-nm-thick free-standing foil that separates the generation chamber and imaging system. It prevents oil contamination from the roughing pump of the generation chamber and blocks (by reflecting) any visible radiation, including the fundamental laser beam, from entering the imaging chamber. Depending on the thickness of the aluminum oxide layer, the transmission of the filter can be as high as 50% in the spectral range around 30 nm. The toroidal diffraction grating (550 grooves/mm, focal length 16 cm) spatially disperses the harmonics and refocuses them in its focal plane. In the used configuration the brightest harmonic order from Argon (23rd order, wavelength $\lambda = 34.8$ nm) in the plateau region of the HHG spectrum is selected and isolated by the slit. The slit in front of the sample is used to reduce the unnecessary stray light in the imaging chamber.

The sample is positioned in the focus of the harmonic beam, and the light scattered off the sample forms a diffraction pattern which is recorded downstream at distances ranging from 15 mm to 60 mm with a cooled back-illuminated CCD-camera (20 μm pixel size, 1340×1300 array). When the CCD is placed closer to the sample, the scattered light is acquired at higher numerical aperture (NA) resulting to potentially higher spatial resolution. On the other hand, for the reconstruction procedure to converge, the diffraction pattern must be sufficiently oversampled [1], which requires to place the CCD far enough. In practice, the distance is chosen in a tradeoff between these two requirements.

The experimental results with quasi-binary mask samples are summarized in Fig. 18.2. The samples are prepared by focused ion beam (FIB) etching of silicon nitride membranes coated with a gold layer of different thickness: 150 nm (Fig. 18.2c, f), 200nm (Fig. 18.2b, e) and 460 nm (Fig. 18.2a, d). The left column shows the measured diffraction patterns on a logarithmic scale and the corresponding SEM micrographs of the samples in the insets. When recording at short distances, central spots of diffraction patterns can be overexposed after just a few seconds of exposure. Thus, to record high scattering angles (carrying high spatial resolution information) with sufficient signal-to-noise ratio (SNR) and to increase the dynamical range of the data, several identical diffraction patterns (5–100) of the same sample were captured and averaged. The diffraction patterns are not centro-symmetric, indicating a non-trivial phase-structure of the exit wave. To obtain (reconstruct) real-space images of the samples from the corresponding far-field diffraction patterns methods for coherent diffractive imaging (CDI) were implemented. The approach retrieves scattering phases from measured diffraction intensities. Further general information on phase retrieval can be found in Chap. 6. The magnitudes of the CDI reconstructions, i.e., the amplitude of light field distribution at the exit-surface of the sample are shown in the right column of Fig. 18.2 (in inverse gray colormap).

To investigate the capabilities of lensless imaging with a high-harmonic source we designed various samples with different spatial features. These ranged from a heavily sparse objects (Fig. 18.2b, e) to a structure with a large open area (Fig. 18.2c, f). The diffraction pattern from the latter case requires an extremely high dynamic range since the central spot (mainly direct, unscattered beam) is very intense compared to the high-scattering angle components. This adds complexity to the data acquisition procedure, and typically requires a physical beam stop to block the intense center and consequent stitching of diffraction patterns captured with and without a beam stop. Furthermore, the phase retrieval process in the case of non-sparse object becomes rather challenging, as discussed next.

18.2   Phase Retrieval of Experimental Data

With the available detectors only far-field intensities can be recorded, while the phase information is lost. As discussed in Chap. 5, without this information one can back propagate the measured far-field data from reciprocal to real space using, e.g., Kirchhoff's diffraction formula. Once the recorded far-field intensities are phased, the near-field information is linked by a Fourier transformation in the case of far-field diffraction. The missing-phase problem (see Chap. 5) can be solved with various

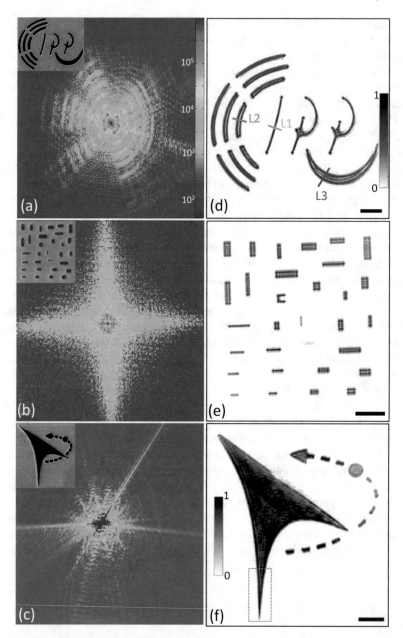

Fig. 18.2 Coherent diffractive imaging results using an illumination wavelength of 35 nm for various samples. Left column **a–c**—the measured diffraction data. Right column **d–f**—CDI reconstructions from the corresponding diffraction pattern. The scale bars of the reconstructions are 1 μm, and the corresponding SEM images are placed as insets

well-established reconstruction algorithms for iterative phase retrieval described in Chap. 6 and [2, 3]. However, in realistic experimental conditions, the diffraction images must first undergo post-processing procedures. Furthermore, the real-space support, which is the necessary a priori knowledge, has to be defined. In our scheme, we used the same procedure for every sample, irrespective of its shape, to post-process diffraction data. The process has several steps: first, the dark counts (signal emerging from the camera itself, irrespective of the illumination) were removed by recording and subtracting an image without an HHG beam, i.e., a dark image. If necessary, the dark image was subtracted from the measured data with an additional constant offset. Second, the center of mass of each data set was used to center the diffraction patterns. Finally, the images were mapped onto an equidistantly-spaced discrete Fourier plane, i.e., the Ewald sphere, to account for distortions from the use of a flat detector [4, 5]. This correction becomes important at high numerical apertures.

To determine the support of the near-field, we start from the autocorrelation of the signal scattered form the object, that is, the Fourier transformation of the measured far-field intensities. A more precise support can be obtained by deconvolution of the autocorrelation function [6]. Depending on the shape of the sample, this deconvolved support can be sufficiently tight, and accurately define the transmissive parts of the object. Having a well-defined support drastically simplifies the phase-retrieval process. Generally, a subsequent refinement of the support can be achieved with methods such as "shrink wrap" [7] or by simply setting a magnitude threshold to the final reconstruction.

To examine the phase retrieval performance and to find the most suitable reconstruction algorithm for the data obtained experimentally, we applied and tested multiple algorithms: ER, DM, HIO, and RAAR, with some modifications for noise resistance [8, 9]. For HIO and RAAR with fixed relaxation β parameter we added additional constraints and an averaging procedure since these methods do not tend to stagnate [10]. The reconstruction process was done for 2000 steps, after initiation from a random guess for the far-field phase. We notice that HIO and RAAR performed significantly better than the other algorithms, whereas ER fails to converge for most of the experimental data. To find successful reconstructions within a run of 2000 iterations for RAAR and HIO, the real space error (sum of counts outside the support) in every step is compared to the errors of the two preceding steps. Alternatively, one can calculate the far-field error by comparing the reconstructed far-field amplitudes with the measured data. If a local minimum of an error is found, the corresponding reconstruction is saved, with the purpose of keeping only ten reconstructions with the smallest error. Once all iterations are completed, the average of the 10 reconstructions corresponding to the 10 minima with lowest errors serve as the final reconstruction [10]. We note that averaging procedure was necessary only for data sets recorded with a relatively low SNR [11].

For the extended (autocorrelation-based) support, employing a positivity constraint in real space and limiting phase variations to less than π was required for consistent convergence of the phase retrieval process. We found it useful to reconstruct an image multiple times, as described above, where a successful reconstruction provides for the tighter support for the next reconstruction procedure. This new support is determined as the reconstructed amplitudes above a certain threshold. A tighter support accelerates and improves the convergence, so the positivity constraint can be relaxed, or removed completely. Alternatively, in the shrink-wrap method [7], the autocorrelation-based support is repeatedly redefined during the first reconstruction run by shrinking the support to include only regions above some threshold every given number of steps. This technique, however, requires a few more fine-tune parameters, especially for non-sparse objects.

If the support is well-defined, we find RAAR (see Chap. 6, 6.25) to be the best method for reconstruction among the tested ones. Starting with a relaxation parameter, β close to 1, we gradually reduced it down to 0.5 after the first few tens of steps. In this case, the algorithm converges consistently to a very similar or an identical solution every time, making a multi-image averaging redundant (see PRTF in Fig. 18.11 of Sect. 18.6) [10].

It is important to note that the above procedure was performed on diffraction data with linear oversampling of 4 and even lower without the need for a higher oversampling ratio to handle noise [12]. Increasing the oversampling ratio by recording at a larger distance, and/or by using a CCD with a smaller pixel size, and/or by imaging smaller samples adds information redundancy to the diffracion data, and thus, CDI becomes more noise tolerant. For the phase-retrieval imaging of smaller structures at a given resolution (inverse numerical aperture) has two additional advantages: It reduces the coherence requirements of the source [13, 14] as well as the dynamical range of the scattered signal. The reduced dynamical range is achieved since a smaller portion of the beam remain un-scattered, thus, saturation effects of the central spot of the diffraction pattern are less severe. For these reasons, the phase retrieval process of a diffraction pattern from a smaller sample may converge to a reasonable solution for the given parameters such as wavelength, CCD pixel size and NA even when the diffraction data have low SNR or insufficient bandwidth. In this regard, ptychography [15] can be, in many cases, an efficient approach. In ptychography, multiple diffraction patterns of the same object are recorded—one for every shift of a confined illumination. A large real-space overlap between the illuminated regions in each acquisition provides for additional redundancy in the data which eases the phase-retrieval process compared to a single diffraction pattern in CDI. Clearly this extra redundancy comes at a price of an increased exposure time, increased requirements for stability and positioning of sample relative to the beam.

The oversampling ratio is a pre-requirement for CDI, but the form of a sample may also drastically affect the phase-retrieval convergence. First of all, sparse objects have less data points with unknown values, i.e., number of pixels with values to be determined *within* the real-space support. Furthermore, a well-defined sparse object provides for an accurate autocorrelation support—a crucial step for a successful reconstruction. This further reduces the number of unknowns due to a tighter support and "forbids" the reconstruction to move within the support. An uncertainty in the position of the reconstruction within the support will lead to a blurred reconstructed image, especially when non-stagnating algorithm is used together with an averaging over multiple reconstructions. In this regard, an object with a cross-correlation term with a delta function (single pixel, or close to that) in the autocorrelation (inverse Fourier transformation of the measured far-field intensitites) gives a significant improvement for the phase retrieval. This feature is demonstrated in Sect. 18.5 with small holographic reference holes drilled in the vicinity of an investigated specimen.

The parameters discussed above also affect the number of steps required for phase retrieval. For instance, in similar experimental conditions, a reconstruction based on RAAR algorithm was accurate after just over 50 steps, (Fig. 18.2b, e data in Sect. 18.5), or required hundreds to thousands of iterations (Fig. 18.2a, c, d, f).

18.3 Experimental Results

The reconstructions in Fig. 18.2 are in a good agreement with the SEM micrographs. However, further inspection of the experimental results reveals field and phase modulations that, at a first glance, may not be expected from a binary opaque transmission mask. Interestingly, such modulations have not been identified or reported in the literature even for a very similar experimental conditions. This might be because the achieved spatial resolution was not sufficient to accurately resolve such small features. In this case, the interpretation of the experimental results as well as an estimate for the achieved spatial resolution might be misleading. In the follwoing we show that the origin of these modulations is associated with the light propagation and multiple scattering within the objet (in this case opaque binary mask) itself.

The reconstructed field—the field profile at the sample's exit surface—relates to a product of the incident field profile and a *non-scalar transmission* function of the object. While Si_3N_4 as well as gold are optically thick media at extreme-UV wavelengths, the light propagating through the removed regions of the sample can be represented as a sum of discrete propagating eigen modes, equivalent to the propagation of electromagnetic wave in a waveguide. These modes propagate through the sample to its exit surface, where they scatter and freely travel to the detector. Clearly, the exit-surface wave is a superposition of these propagating modes and the observed modulation in the reconstructed image resutls from multi-mode interference. In the following, we perform 2D and 3D numerical simulations using finite element modeling and semi-analytical solution to corroborate the experimental findings.

Figure 18.3a shows a 2D numerical simulation of light propagation through slab waveguides of different width with geometries similar to the experimental conditions (marked with three solid lines L1–L3 in Fig. 18.2d). The material properties and wavelengths are as in the experiment as well. The field distributions at the exit surface of the waveguides are plotted with red dashed lines for numerically simulated data in the right column of Fig. 18.3a. The solid blue lines depict the experimental field values obtained from lineouts L1–L3 in CDI reconstruction. Similarly, the structure marked with a red dashed rectangle in Fig. 18.2f can be approximated as 2D slab waveguides of different width and the expected exit surface field distribution

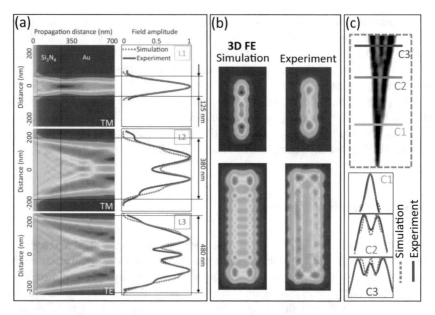

Fig. 18.3 Interpretation of the experimental results: waveguiding at extreme-UV. **a** Numerical simulations of light propagation in slab waveguides for the wavelengths and materials used in the experiment. The waveguide dimensions correspond to the regions of the sample marked with L1, L2 and L3 in (**a**) and C1, C2 and C3 in (**c**). Solid blue lines—measured data, red dashed lines—simulated data. **b** Three dimensional simulation of light propagation in rectangular waveguides of various sizes with our experimental conditions

can be simulated. The blue solid lines and the red dashed lines in the bottom of Fig. 18.3c are the experimental and simulated lineouts for the regions marked as $C1$, $C2$ and $C3$. For the structure shown in Fig. 18.3b, e, aspect ratios of individual features (waveguides) are not as high as in the case of the structures shown in Fig. 18.2d, f. Therefore, here the approximation of the structure with a 2D model as a slab waveguide is not accurate and a 3D simulation is required. Figure 18.3b compares the expected exit surface fields (simulated using a 3D finite-element modeling) on the left with the experimentally measured ones on the right (from reconstruction shown in Fig. 18.2e) for two different waveguides. Again, as in the case with the other samples the reconstructed field distribution is in a close agreement with the simulated data.

Further insights into waveguiding at extreme-UV frequencies and fundamental reasons for mode beating at the exit surface of CDI reconstructions follow from a semi-analytical solution of the mode propagation within the structure using eigenmode expansion. Figure 18.4a demonstrates field distribution (field profile) of the first three eigenmodes in a gold-cladded slab waveguide. Figure 18.4b shows the computed transmission of these modes through a 700-nm-long waveguide as a function of waveguide's width. Here, TE and TM modes correspond to the polarizations parallel and perpendicular to the cladding, respectively. We note that only even order

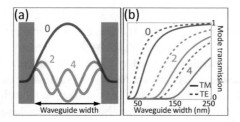

Fig. 18.4 Analytical solution for waveguiding for interpretation of the experimental results shown in Fig. 18.3. **a** Field profiles of first three allowed eigenmodes in a symmetrical slab waveguide. **b** Mode transmission through 700-nm-long gold waveguide with perpendicular (TM) and parallel (TE) incident polarization

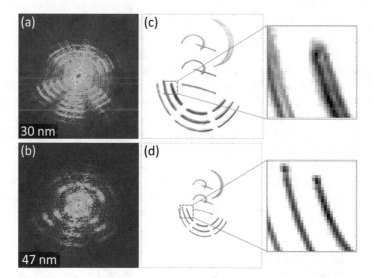

Fig. 18.5 Coherent diffractive imaging using illuminating wavelength of 30 (**a, c**) and 47 nm (**b, d**). The damping of higher order modes is already evident from the far-field diffraction patterns which is accurately reproduced in CDI reconstructions

modes are supported by a symmetrical waveguide. As expected, higher order modes experience stronger damping in narrow waveguides. Thus, the relative intensities of these modes at the exit surface are governed by the waveguide dimensions and the intensity profile resulting from a superposition of these modes at the waveguide's exit can strongly differ for very similar geometries, e.g., a slight width difference.

Similarly, the illuminating wavelength affects the mode distribution at the exit plane. Figure 18.5 illustrates the experimental results of the same object imaged with a wavelength of 30 nm (a, c) and 47 nm (b, d). The zoomed regions (insets) emphasize the difference of the reconstructed fields. The image obtained with the longer wavelength contains dominantly the fundamental mode, whereas the image obtained with 30 nm wavelength exhibit a complex mode-beating profile. Notably, this feature is already evident from the corresponding diffraction patterns, where the maximum spatial frequency (spatial frequency with sufficient SNR) scattered off the sample at 47 nm wavelength is much lower than the ones present on Fig. 18.5a, recorded with 30 nm illuminating wavelength. High-order waveguiding modes for longer wavelength are strongly damped due to a multiple scattering within the object and do not pass through the structure to its exit surface and therefore absent (heavily suppressed) on the far-field image as well as on the reconstruction. Clearly, for such high aspect ratio structures or for the structures comparable to the illuminating wavelength the sharpness of the CDI reconstruction will be limited to the highest Eigen mode transmitted. Therefore, using a knife-edge technique without accounting for wave propagation effects in such structures might lead to a strong overestimation.

We note, however, that a similar mode-beating profile can be observed when the high spatial frequencies are not fully recorded in the far-field, i.e., span beyond the CCD edges. Such a truncation of the far-field intensities determines the upper limit for the spatial resolution of the reconstructed image.

To obtain high spatial resolution one needs to collect data at high NA. However, in most cases the signal-to-noise ratio at high scattering angles might be very weak and this will limit the maximum spatial frequency that can be accurately reconstructed with a phase retrieval algorithm and consequently the achieved spatial resolution. This is demonstrated in Fig. 18.6 where two diffraction patterns of the same structure are recorded with the same wavelength (35 nm) and the same NA (0.5). The top row shows a single diffraction pattern recorded for 5 s and the corresponding reconstruction. The bottom row is an average of 200 individual exposures (HDR data), so the signal-to-noise ratio especially at high scattering angles (c.f. insets) is improved. The effect of higher SNR ratio also imprints to the corresponding CDI

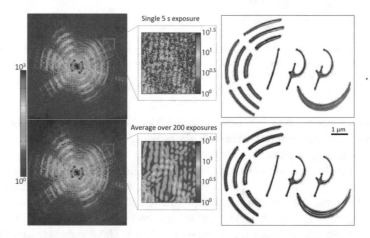

Fig. 18.6 Coherent diffractive imaging with high dynamic range data demonstrates influence of SNR ratio at high scattering angles to the reconstruction quality. Note that fine features of the wavegude mode beating pattern cannot be resolved from the non-HDR (top) diffraction data with low SNR ratio. Adapted from [10]

reconstruction demonstrating a sharper image. While both reconstructions are in a good agreement with the SEM micrographs, the second CDI reconstruction from HDR data contains finer features that cannot be resolved in the first reconstruction. This is because the diffraction signal from smaller features scatters at higher angles where SNR is noticeably lower. Increasing the dynamic range by multi-exposure averaging results in a better SNR at high scattering angles and improves spatial resolution, although both images were recorded at the same NA. Clearly, poor quality of a diffraction pattern and/or non-accurate phase retrieval impede imaging with high resolution irrespective of how high the NA is.

18.4   Polarization Dependence

Figure 18.4b shows a simulation of wave propagation within a gold-cladded waveguide. As discussed in Sect. 18.3 for a relatively narrow waveguide (smaller than 70 nm) higher-order modes can be fully damped whereas the fundamental mode may still be transmitted with relatively low losses. The narrower the slit the stronger the polarization anisotropy, i.e., transmission difference for polarization parallel (TE mode) and perpendicular (TM) to the walls of a waveguide. Higher suppression of the TM polarization can be explained by the fact that perpendicular fields penetrate deeper into the gold cladding where it experiences an exponential decay. Interestingly, this polarization dependent transmission effect is opposite to the one in wire-grid polarizers where perpendicular polarization is transmitted.

To investigate this phenomenon and its effects in nanoscale imaging with extreme-UV radiation, we designed and fabricated a structure with an angular arrangement of identical 50 nm-wide slits, etched in a gold coated Si_3N_4 membrane. The SEM image of the structure is shown in Fig. 18.7a. The diffraction pattern shown in (Fig. 18.7b) was recorded with S-polarized illumination at wavelength of 35 nm. The corresponding CDI reconstruction (exit-field intensity) is shown in Fig. 18.7c. The reconstruction reveals that slits parallel to the field polarization appear noticeably brighter than the ones that are perpendicular to the electric field. Figure 18.7d plots the field intensity transmitted through each slit as a function of the angle between the slit orientation and the polarization, with comparison to Malus' law for an imperfect polarizer. The reconstructed field (red circles) and the measured angular far-field (blue triangles) accurately follow the predicted pattern. The measurement was done for multiple linear-polarization states to verify that the polarization dependence originates solely from the sample, and that the slits are indeed identical. In contrast to far-field pattern where it is impossible to disentangle contributions from parallel slits, intensity information from the reconstructed image provide information for each slit individually. Based on the quantitative information from the CDI reconstruction and the simulated polarization dependence for such a structure, we estimated width of the slits to be 52 nm.

This experiment brought three insights to Lensless imaging with a high-harmonic source:

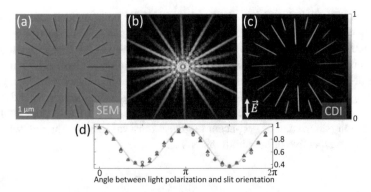

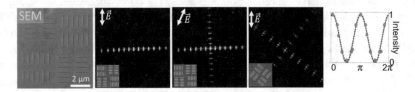

Fig. 18.7 Extreme-UV polarimetry. **a** An SEM image of a structure with nanoscale slits. The width of the slits is identical. **b** The diffraction pattern in logarithmic scale. **c** The reconstructed intensity at the exit surface of the sample obtained by CDI from (**b**). **d** An analysis of the experimental data showing polarization dependent transmission through the nanoscale slits. Experimental data from the CDI reconstruction and from the diffraction pattern (red circles, and blue triangles, respectively). The line is the Malus' law for an imperfect polarizer

Fig. 18.8 Polarizer for Extreme-UV radiation. **a** An SEM image of the structure with nanoscale slits of a typical width of 40 nm. The structure demonstrates polarization dependent transmission with high extinction ratio TE/TM mode

1. In the extreme-UV and soft-X-ray range the scalar projection approximation is not valid. Instead, CDI can be used to accurately and quantitatively map polarization anisotropies and waveguiding effects at nanoscales.
2. A structure with nanoscale slit arrangement provides information on the polarization state of the incident extreme-UV light in a single acquisition measurement compared to a conventional reflection-based polarimeter, where incident polarization can be estimated only from a series of measurements at various angles [16]. This polarization analyzer shown in Fig. 18.7a proved to be very useful in the future experiments where optimization of polarization state of the extreme-UV light was required.
3. A structure with an array of only parallel nanoscale slits can serve as an effective polarizer for extreme-UV radiation as demonstrated in Fig. 18.8. Here, the diffraction patterns contain detectible scattering signal only from the polarization component parallel to the slits. The verification of the polarization anisotropy was done by rotating the sample and rotating the incident polarization.

18.5   Magneto-Optical Imaging Using High-Harmonic Radiation

Recent developments in the generation of high harmonics with arbitrary polarization [17, 18] allows to access X-ray magnetic circular dichroism (XMCD). The M-edge absorption lines in the important 3d-ferromagnets Fe (52 eV), Co (60 eV) and Ni (75 eV) are within the spectral range of a typical HHG source based on a Ti:Sapphire amplified laser [19, 20], and additional materials (e.g. Gd, 145 eV) can be accessed using HHG sources at the soft-X-ray [21]. The circularly polarized HHG are generated by a bi-chromatic laser field in a gas (typically He). The laser field combines the fundamental laser and its second harmonic $\lambda = 800$ nm and $\lambda = 400$ nm, respectively. The driving fields are circularly polarized, with opposing handedness, thus the selection rule for the high harmonics imposes the suppression of every third harmonic order, while the allowed harmonics are circularly polarized [17, 22]. In this project, we implemented two schemes for circularly polarized high harmonic generation, as described in [17, 18]. We find that the latter scheme, namely an in-line MAZEL-TOV (stands for MAch-ZEhnder-Less for Threefold Optical Virgina spiderwort) device, is robust and reliable, due to its inherent transmission geometry. This device offers a drastically simplified alignment process compared to the scheme involving a Mach-Zehnder interferometer. The use of a single quarter-wave retarder to set the polarization of the bi-chromatic field enables a direct control on the laser polarization in the gas, and therefore, on the HHG polarization. For example, a reflection from a tilted surface may result in an unequal amplitude or phase for the incident TE or TM polarizations, thus deteriorating the laser's degree of circular polarization. For the detection of XMCD contrast a direct polarization control is crucial—the helicity of the circularly polarized HHG is flipped from left-handed to right-handed by flipping the quarter-wave plate (QWP) retarder to 45° or to −45°. Furthermore, the MAZEL-TOV apparatus enables a straightforward fine-tuning of the recollision process in a way that allows to access to any harmonic order with circular polarization, even including the typically-suppressed harmonics orders (e.g. 36, 39, . . .) [23].

The experimental scheme for magnetic imaging with high harmonics is depicted in Fig. 18.9. The HHG source is based on the setup described in Sect. 18.3, with some modifications [20, 24]. First, Ar gas was replaced with He, which has a higher ionization potential and, thus, generates higher harmonic orders [25]. Specifically, we can access the 38th harmonic order in He and in Ne, providing for the optimal XMCD contrast at the M-edge of cobalt [26]. Helium was preferred over neon, since the price is significantly lower and the smaller absorption coefficient may results in brighter harmonics with a better beam quality. However the pulse energy (2–3.5 mJ) required to drive HHG process in He is higher than for Ar due to a higher ionization potential and higher phase matching pressures. Second, the need for circularly polarized HHG required the use of a MAZEL-TOV device. This device is inserted after the focusing lens to convert the linearly-polarized driving laser field into a bi-chromatic counter-rotating field. Finally, the interaction region, i.e., the

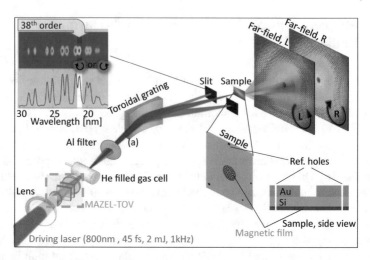

Fig. 18.9 Experimental setup for nanoscale magnetic imaging using high-harmonic radiation. **a** A scheme of the experiment, **b** A measured spectrum generated using a circularly polarized bi-chromatic field tailored with MAZEL-TOV device. The suppression of every third harmonic order (e.g. ..., 30, 33, 36, ...) indicates that the harmonics are circularly polarized. The toroidal grating focuses the harmonics on the sample plane, and a slit selects the 38th harmonic order (59 eV), which provides for the optimal XMCD contrast in Co. To isolate the magneto-optical signal, two diffraction patterns are recorded—one with left- and one with right-handed circularly polarized HHG beam. The sample includes the region of interest (central aperture) and four fully-drilled reference holes provide for a strong reference field that interferes in the far-field with the scattering from the magnetic sample

focus of the fundamental beam, was positioned further away from the diffraction grating. This allowed us to illuminate a larger number of slits on the grating, which improves the spectral dispersion of the toroidal grating. Increasing the dispersion improved the temporal coherence of the beam in the sample region, and thus allows for a higher spatial resolution [1]. We note that, for previous experiments, described in Sect. 18.3 the estimated monochromaticity $^\lambda/_{\Delta\lambda}$ was larger than 500, corresponding to a spatial resolution down to ∼30 nm [1, 12]. For the magneto-optical imaging experiments, the monochromaticity was increased to enable improved resolution—beyond 20 nm. In a more recent development, the sample was illuminated with a replica of the fundamental beam ($\lambda = 800$ nm), through a controlled time delay with respect to the HHG beam. This pump-probe addition would allow to combine imaging experiments with femtosecond time resolution enabling movies of ultrafast magnetic dynamics at the nanoscale.

In order to isolate a magnetic signal from non-magnetic background, two diffraction patterns with opposite helicities were recorded (c.f. *L* and *R* in Figs. 18.9 and 18.10 for the illumination with left- and right-handed circularly polarized 38th harmonic, respectively). The HHG helicity (*L* vs. *R*) is easily done by rotating the quarter wave-plate of the MAZEL-TOV device from 45° to −45°, and vice versa. For each helicity, the diffraction pattern dynamic range was increased by combin-

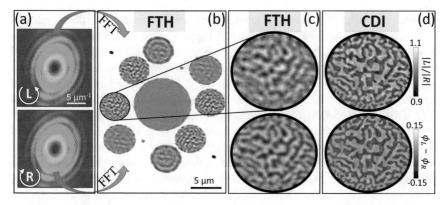

Fig. 18.10 Lensless imaging of nanoscale magnetic domains using high-harmonic radiation. **a** Holographic diffraction patterns recorded with left- (L) and right-handed (R) circularly polarized 38th harmonic order. **b** A full-field Fourier transform holography (FTH) reconstruction. **c** The magneto-optical absorption (top) and phase contrast (bottom) reconstructions as recovered with FTH. The spatial resolution is estimated between 150 nm and 200 nm. **d** the magneto-optical absorption and phase contrast, as reconstructed via CDI. Using CDI, the spatial resolution reaches below 50 nm

ing two exposure times. A long exposure—up to 10 min per helicity—provided the diffraction pattern for the medium and high scattering angles with sufficient SNR. A short exposure time (typically several images of a few seconds each) captured the low scattering angles, without any saturation effects. For example, diffraction patterns shown in Fig. 18.10a are composed of scattering data from an exposure time of 10 min with the average of 24 frames of 5 s exposure each. The diffraction patterns were scaled properly and stitched to form a single diffraction pattern with high dynamic range. Finally, the diffraction patterns were prepared for reconstruction as described in Sect. 18.1.

The phase retrieval process used for the magneto-optical images is similar to the one described in Sect. 18.2. As the real-space support, a single cross-correlation of the diffraction pattern was used. Since for magneto-optical experiments we used reference holes, the cross-correlation includes accurate replicas of the structure, and thus dramatically improves the convergence of the phase retrieval algorithm. For samples with reference holes, the required number of RAAR steps can be reduced to about a 100. The combination of FTH with CDI allows for a direct low-resolution but noise-tolerant reconstruction by a single-step Fourier transformation, and a high resolution CDI reconstruction. The experimental results from the magnetic structures are summarized in Fig. 18.10. First, two diffraction patterns from left- and right-handed circularly polarized illumination are recorded (c.f. L and R in (a), shown in logarithmic scale). Second, a single Fourier transformation of the measured far-field intensities provides for the holographic reconstruction (Fourier transform holography—FTH). Finally, the real-space support is derived from the FTH, thus assisting the iterative phase retrieval to recover the high resolution image. Figure 18.10b shows the magneto-optical amplitude for the holographic reconstructions. Since the sample had four reference holes, the FTH reconstructs eight replicas of the sample—a reconstruction and its complex conjugate for each reference hole. The magneto-optical amplitude (top) and phase (bottom) contrasts originating from the smallest reference hole (approx. 200 nm in diameter) are shown in Fig. 18.10c. The magnetic phase contrast is the phase difference between the reconstructions recorded with left- and right-handed HHG helicity. The corresponding magneto-optical contrast CDI reconstructions based on the RAAR algorithm are shown in Fig. 18.10d. Both the FTH and the CDI reconstructions show the same pattern of magnetic domains. The resolution of the CDI reconstruction is clearly higher (below 50 nm), since it is limited by the high NA recorded far-field, whereas, the resolution of the FTH reconstruction is set by the pre-drilled reference hole. Notably, the CDI reconstruction in this case has a multiple binary-like transitions between up and down magnetization, where the domain transition region is below a single-pixel in a vast region of the image. Since the domain-wall width for this Co/Pd multilayer structure is expected to be 10 nm to 15 nm, it can provide for a test sample for higher resolution imaging, even below the illuminating wavelength ($\lambda_{38\text{th harmonic}}$ = 21 nm) to investigate the capabilities of HHG for magnetic imaging.

In principle, the size of reference holes can be reduced in order to improve the spatial resolution of the Fourier-transform hologram. However, the intensity of the light transmitted through a small reference hole is significantly lower due to propagation effects in the narrow channel (see Sect. 18.3). As a result, longer exposure times would be required. Additionally, the manufacturing of narrow reference holes is a technical challenge, limiting the repeatability of the experiment. A convenient approach in FTH is to use of reference holes of slightly varying sizes. Thus, the achievable spatial resolution (hole size) and an image contrast (hole transmission) can be determined while analyzing the scattering data. In a later section we show that adding large and strongly scattering reference holes to the sample, assists the CDI reconstruction in few ways [20]. First, determining the support from the cross-correlation is much easier and the phase retrieval algorithm converges faster when reference holes are introduced. Second, the interference of weak scattering signals with a strong reference field enhances the weak signal so that it can be detected above the instrumental noise level [27, 28]. In contrast to FTH, the size of reference holes plays a secondary role in determination of resolution in CDI experiments. To demonstrate this, we designed a similar structure with reference holes larger than the sizes of the features to be observed.

Figure 18.11 shows the imaging of worm-like domains in a Co/Pd multilayer stack in a field of view diameter of 4 μm, for which the reference holes in the mask had a diameter of 500–600 nm. These diameters are twice larger than a typical size of the magnetic-domains and an order of magnitude larger than the final resolution for images obtained via CDI reconstructions. Figure 18.11a shows the diffraction pattern recorded with left-handed circular polarization and Fig. 18.11b shows the field magnitude of the corresponding CDI reconstruction. Note that the image is presented with a true-pixel resolution and contain features in the order of a single pixel. Figure 18.11c represents the phase contrast dichroic image, which is the angle

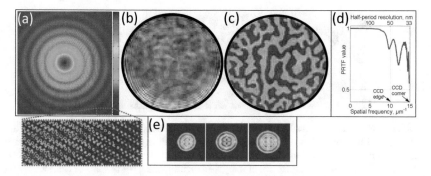

Fig. 18.11 Magnetic imaging using large reference holes. **a** Holographic diffraction pattern recorded with left-handed circularly polarized light. Inset highlights very good speckle visibility at high scattering angles. **b** CDI reconstruction of (**a**) for a single helicity illumination. Image contains magnetic as well as non-magnetic contributions. **c** Magnetic phase-contrast CDI reconstruction, i.e., isolated XMCD signal. **d** PRTF for 20 reconstructions initiated from a random first guess indicating a consistent convergence for all spatial frequencies recorded. **e** Exit fields of the reference holes (interpolated) demonstrating strong intensity modulations due to waveguidng and Fourier-truncation effects

of the ratio of two reconstructions recorded with opposite helicities. Despite the fact that reference holes are too large to resolve individual magnetic domains via FTH, CDI successfully retrieves high-resolution information by finding the far-field phase at high NA. To estimate the achieved spatial resolution, we use the phase retrieval transfer function (PRTF) [29] . Here, PRTF is calculated as the average far-field phase of 20 reconstructions that were initiated from a random first guess phase. Figure 18.11d depicts that the phase retrieval process is consistent throughout the entire far-field. Reconstructed phase reaching beyond 15 inverse μm corresponds to a spatial resolution better than 33 nm. However, since the truncation of the diffraction patterns at the physical edges of the CCD begins just above 10 μm, the spatial resolution that can be claimed is defined by this value to 50 nm (a single pixel resolution).

18.6   Dichroic Imaging

The extrme-UV light propagating in the reference holes undergoes the same waveguiding effects, as described in Sect. 18.3 (c.f. Fig. 18.11e). A similar effect is noted near the edges of the central aperture (c.f. Fig. 18.11d). Generally, wave effects in the illuminating field reduce the image quality and add artefacts [11] that are associated solely with an imaging scheme (a Fresnel number for the used wavelength and sample geometry). In contrast, dichroic imaging provides for a unique type of microscopy, which can eliminate these effects in the reconstructed image. When the magnetization map is obtained from the ratio of two reconstructed images with

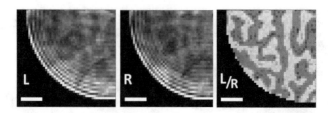

Fig. 18.12 Waveguiding and edge-diffraction artifacts elimination with dichroic imaging. Left and middle images are absorption contrast reconstructions for left (L) and right (R)-handed circularly polarized illumination. Wave modulations from edge diffraction effects are observed and can not be separated from magnetic signal. Right image—a dichroic image—the ratio of L/R isolates the magnetic signal from non-magnetic contributions

opposite helicities (*L* and *R* helicities), the artifacts vanish. The dichroic contrast isolates the magnetic (dichroic) signal from the non-magnetic background.

Figure 18.12 shows a part (left lower quarter) of the central aperture for left- and for right-handed circularly polarized illumination marked as L and R. The image for each helicity exhibits fine intensity modulations in the order of a single pixel, mainly near the edge of the aperture. These fine modulations are real but cancel each other out in the dichroic image (Right image in Fig. 18.12 for the ratio L/R), provided that reconstructions from opposite helicities are accurately overlaid. To consistently position the two reconstructions with a sub-pixel accuracy, it is convenient to match the lower order moments of their far-field phase (i.e. the global phase, and the phase gradients). To do so, we fully reconstruct the data recorded for one helicity, and use its far-field phase as an initial guess for the phase retrieval of the opposite helicity diffraction pattern. Since the dichroic component is only a small component of the far-field, only a few iterations are required for a full phase retrieval, when starting from the far-field phase of the opposing helicity. This provides for a robust and reliable imaging of magnetic samples without any artifacts of smearing effects from reconstruction shifts and waveguiding effects.

18.7 Signal Enhancement Mechanism

A strong reference or an auxiliary wave in the vicinity of the field-of-view has an advantage for recording diffraction data. Specifically, a strong scattering field from multiple reference holes interferes on the CCD detector with the weak magneto-optical scattering signal. Thus, photons carrying magnetic information can be detected above the instrumental noise, and the exposure times can be drastically

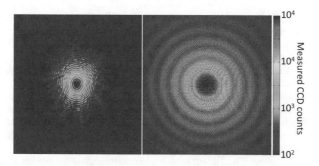

Fig. 18.13 Signal enhancement effect through the interference with a strong reference wave. (left) A diffraction pattern from a sample with very narrow reference holes (sub-100 nm). (right) A diffraction pattern from the same sample with large reference holes (500–600 nm). The larger reference holes result in a high signal-to-noise ratio at higher scattering angles. The weak magneto-optical signal scales as square-root of the strong reference intensity, because it is an interference effect. Since the reference holes allow for the auxiliary field to cover the entire CCD, the weak signal is amplified above the noise throughout the recorded diffraction pattern, and a high-resolution image is reconstructable

reduced. Figure 18.13 shows a comparison of two diffraction patterns recorded from the same structure, albeit with a different intensity of the reference field. When the reference holes were small (left), the signal includes mostly the low-angle scattering from the central aperture, where the high resolution information is buried and lost in the noise. When the reference holes are large (right), a meaningful portion of the light passes them and scatters to higher angles, thus lifting the signal above the level of the instrumental noise of the camera. Thus, this auxiliary field brings the weak magneto-optical scattering above the noise through interference. In this example, the scattering from the waveguides is two orders of magnitude higher when the reference holes are large, which means that through interference, the magneto-optical signal is enhanced by at almost one order of magnitude.

Acknowledgements   We gratefully acknowledge the support from Eike Mönnich for development, implementation and advancement of phase retrieval algorithms used in Sect. 18.2. We thank Felix Schulze for 3D simulations used to interpret the experimental results in Sect. 18.3.

References

1. Miao, J., Ishikawa, T., Anderson, E.H., Hodgson, K.O.: Phase retrieval of diffraction patterns from noncrystalline samples using the oversampling method. Phys. Rev. B **67**(17), 174,104 (2003). https://doi.org/10.1103/PhysRevB.67.174104
2. Fienup, J.R.: Phase retrieval algorithms: a comparison. Appl. Opt. **21**(15), 2758–2769 (1982). https://doi.org/10.1364/AO.21.002758

3. Luke, R.: Relaxed averaged alternating reflections for diffraction imaging. Inverse Probl. **21**(1), 37–50 (2005). https://doi.org/10.1088/0266-5611/21/1/004. http://stacks.iop.org/0266-5611/21/i=1/a=004?key=crossref.68c3009a049400c5a687e8a6894d540d
4. Raines, K.S., Salha, S., Sandberg, R.L., Jiang, H., Rodríguez, J.A., Fahimian, B.P., Kapteyn, H.C., Du, J., Miao, J.: Three-dimensional structure determination from a single view. Nature **463**(7278), 214–217 (2010). https://doi.org/10.1038/nature08705
5. Sandberg, R.L., Raymondson, D.A., La-o vorakiat, C., Paul, A., Raines, K.S., Miao, J., Murnane, M.M., Kapteyn, H.C., Schlotter, W.F.: Tabletop soft-x-ray Fourier transform holography with 50 nm resolution. Opt. Lett. **34**(11), 1618 (2009). https://doi.org/10.1364/OL.34.001618. http://ol.osa.org/abstract.cfm?URI=ol-34-11-1618delimiter026E30Fn. http://www.opticsinfobase.org/abstract.cfm?URI=ol-34-11-1618
6. Fienup, J.R., Crimmins, T.R., Holsztynski, W.: Reconstruction of the support of an object from the support of its autocorrelation. J. Opt. Soc. Am. **72**(5), 610 (1982). https://doi.org/10.1364/JOSA.72.000610
7. Marchesini, S., He, H., Chapman, H.N., Hau-Riege, S.P., Noy, A., Howells, M.R., Weierstall, U., Spence, J.C.H.: X-ray image reconstruction from a diffraction pattern alone **11**(19), 5 (2003). https://doi.org/10.1103/PhysRevB.68.140101. http://arxiv.org/abs/physics/0306174
8. Giewekemeyer, K., Thibault, P., Kalbfleisch, S., Beerlink, A., Kewish, C.M., Dierolf, M., Pfeiffer, F., Salditt, T.: Quantitative biological imaging by ptychographic X-ray diffraction microscopy. Proc. Natl. Acad. Sci. USA **107**(2), 529–534 (2010). https://doi.org/10.1073/pnas.0905846107
9. Martin, A.V., Wang, F., Loh, N.D., Ekeberg, T., Maia, F.R.N.C., Hantke, M., van der Schot, G., Hampton, C.Y., Sierra, R.G., Aquila, A., Bajt, S., Barthelmess, M., Bostedt, C., Bozek, J.D., Coppola, N., Epp, S.W., Erk, B., Fleckenstein, H., Foucar, L., Frank, M., Graafsma, H., Gumprecht, L., Hartmann, A., Hartmann, R., Hauser, G., Hirsemann, H., Holl, P., Kassemeyer, S., Kimmel, N., Liang, M., Lomb, L., Marchesini, S., Nass, K., Pedersoli, E., Reich, C., Rolles, D., Rudek, B., Rudenko, A., Schulz, J., Shoeman, R.L., Soltau, H., Starodub, D., Steinbrener, J., Stellato, F., Strüder, L., Ullrich, J., Weidenspointner, G., White, T., Wunderer, C.B., Barty, A., Schlichting, I., Bogan, M.J., Chapman, H.N.: Noise-robust coherent diffractive imaging with a single diffraction pattern. Opt. Express **20**(15), 16,650 (2012). https://doi.org/10.1364/OE.20.016650. http://pubdb.xfel.eu/record/139753/files/Martin2012-OE-Noise-tolerant-CDI.pdf
10. Zayko, S., Mönnich, E., Sivis, M., Mai, D.D., Salditt, T., Schäfer, S., Ropers, C.: Coherent diffractive imaging beyond the projection approximation: waveguiding at extreme ultraviolet wavelengths. Opt. Express **23**(15), 19,911 (2015). https://doi.org/10.1364/OE.23.019911. https://www.osapublishing.org/abstract.cfm?URI=oe-23-15-19911
11. Zayko, S.: Nanoscale Waveguiding Studied by Lensless Coherent Diffractive Imaging Using EUV High-harmonic Generation Source. PhD Thesis, Georg-August-Universität Göttingen. (2016). http://hdl.handle.net/11858/00-1735-0000-0023-3EB2-E
12. Sandberg, R.L., Paul, A., Raymondson, D.A., Hädrich, S., Gaudiosi, D.M., Holtsnider, J., Tobey, R.I., Cohen, O., Murnane, M.M., Kapteyn, H.C., Song, C., Miao, J., Liu, Y., Salmassi, F.: Lensless diffractive imaging using tabletop coherent high-harmonic soft-X-ray beams. Phys. Rev. Lett. **99**(9), 098,103 (2007). https://doi.org/10.1103/PhysRevLett.99.098103
13. Chen, B., Abbey, B., Dilanian, R., Balaur, E., Van Riessen, G., Junker, M., Tran, C.Q., Jones, M.W.M., Peele, A.G., McNulty, I., Vine, D.J., Putkunz, C.T., Quiney, H.M., Nugent, K.A.: Diffraction imaging: the limits of partial coherence. Phys. Rev. B - Condens. Matter Mater. Phys. **86**(23) (2012). https://doi.org/10.1103/PhysRevB.86.235401
14. Spence, J., Weierstall, U., Howells, M.: Coherence and sampling requirements for diffractive imaging. Ultramicroscopy **101**(2-4), 149–152 (2004). https://doi.org/10.1016/j.ultramic.2004.05.005. http://linkinghub.elsevier.com/retrieve/pii/S0304399104000956
15. Rodenburg, J.M., Hurst, A.C., Cullis, A.G., Dobson, B.R., Pfeiffer, F., Bunk, O., David, C., Jefimovs, K., Johnson, I.: Hard-X-ray lensless imaging of extended objects. Phys. Rev. Lett. **034801**(January), 1–4 (2007). https://doi.org/10.1103/PhysRevLett.98.034801
16. Zayko, S., Sivis, M., Schäfer, S., Ropers, C.: Polarization contrast of nanoscale waveguides in high harmonic imaging. Optica **3**(3), 239–242 (2016). https://doi.org/10.1364/OPTICA.3.000239. http://www.osapublishing.org/optica/abstract.cfm?URI=optica-3-3-239

17. Fleischer, A., Kfir, O., Diskin, T., Sidorenko, P., Cohen, O.: Spin angular momentum and tunable polarization in high-harmonic generation. Nat. Photonics **8**(7), 543–549 (2014). https://doi.org/10.1038/nphoton.2014.108

18. Kfir, O., Bordo, E., Ilan Haham, G., Lahav, O., Fleischer, A., Cohen, O.: In-line production of a bi-circular field for generation of helically polarized high-order harmonics. Appl. Phys. Lett. **108**(21) (2016). https://doi.org/10.1063/1.4952436. http://scitation.aip.org/content/aip/journal/apl/108/21/10.1063/1.4952436

19. Kfir, O., Grychtol, P., Turgut, E., Knut, R., Zusin, D., Popmintchev, D., Popmintchev, T., Nembach, H., Shaw, J.M., Fleischer, A., Kapteyn, H., Murnane, M., Cohen, O.: Generation of bright phase-matched circularly-polarized extreme ultraviolet high harmonics. Nat. Photonics **9**(2), 99–105 (2014). https://doi.org/10.1038/nphoton.2014.293

20. Kfir, O., Zayko, S., Nolte, C., Sivis, M., Möller, M., Hebler, B., Sai, S., Kanth, P., Steil, D., Schäfer, S., Albrecht, M., Cohen, O., Mathias, S., Ropers, C.: Nanoscale magnetic imaging using circularly polarized high-harmonic radiation, pp. 1–6 (2017)

21. Fan, T., Grychtol, P., Knut, R., Hernandez-Garcia, C., Hickstein, D.D., Zusin, D., Gentry, C., Dollar, F.J., Mancuso, C.A., Hogle, C.W., Kfir, O., Legut, D., Carva, K., Ellis, J.L., Dorney, K.M., Chen, C., Shpyrko, O.G., Fullerton, E.E., Cohen, O., Oppeneer, P.M., Milosevic, D.B., Becker, A., Jaron-Becker, A.A., Popmintchev, T., Murnane, M.M., Kapteyn, H.C.: Bright circularly polarized soft X-ray high harmonics for X-ray magnetic circular dichroism. Proc. Natl. Acad. Sci. **112**(46), 14, 206–211 (2015). https://doi.org/10.1073/pnas.1519666112. http://www.pnas.org/content/early/2015/11/02/1519666112

22. Alon, O.E., Averbukh, V., Moiseyev, N.: Selection Rules for the High Harmonic Generation Spectra, pp. 3743–3746 (1998)

23. Kfir, O., Zayko, S., Nolte, C., Mathias, S., Cohen, O., Ropers, C.: Attosecond-precision coherent control of electron recombination in the polarization plane. In: Conference Lasers Electro-Optics, p. FM3D.2. Optical Society of America (2017). https://doi.org/10.1364/CLEO_QELS.2017.FM3D.2. http://www.osapublishing.org/abstract.cfm?URI=CLEO_QELS-2017-FM3D.2

24. Zayko, S., Kfir, O., Nolte, C., Sivis, M., Möller, M., Ganss, F., Hebler, B., Steil, D., Schäfer, S., Albrecht, M., Cohen, O., Mathias, S., Ropers, C.: Nanoscale imaging of magnetic domains using a high-harmonic source. In: Conference Lasers Electro-Optics, p. FW1H.8. Optical Society of America (2017). https://doi.org/10.1364/CLEO_QELS.2017.FW1H.8. http://www.osapublishing.org/abstract.cfm?URI=CLEO_QELS-2017-FW1H.8

25. Popmintchev, T., Chen, M.C., Bahabad, A., Gerrity, M., Sidorenko, P., Cohen, O., Christov, I.P., Murnane, M.M., Kapteyn, H.C.: Phase matching of high harmonic generation in the soft and hard X-ray regions of the spectrum. Proc. Natl. Acad. Sci. USA **106**(26), 10516–10521 (2009). https://doi.org/10.1073/pnas.0903748106

26. Valencia, S., Gaupp, A., Gudat, W., Mertins, H.C., Oppeneer, P.M., Abramsohn, D., Schneider, C.M.: Faraday rotation spectra at shallow core levels: 3p edges of Fe Co, and Ni. New J. Phys. **8**, (2006). https://doi.org/10.1088/1367-2630/8/10/254

27. Noh, D.Y., Kim, C., Kim, Y., Song, C.: Enhancing resolution in coherent X-ray diffraction imaging. https://doi.org/10.1088/0953-8984/28/49/493001

28. Shintake, T.: Possibility of single biomolecule imaging with coherent amplification of weak scattering X-ray photons, pp. 1–9 (2008). https://doi.org/10.1103/PhysRevE.78.041906

29. Chapman, H.N., Barty, A., Marchesini, S., Noy, A., Hau-Riege, S.P., Cui, C., Howells, M.R., Rosen, R., He, H., Spenceet, J.C. al.: High-resolution ab initio three-dimensional X-ray diffraction microscopy. JOSA A **23**(5), 1179–1200 (2006) https://doi.org/10.1364/JOSAA.23.001179

Chapter 19
Nonlinear Light Generation in Localized Fields Using Gases and Tailored Solids

Murat Sivis and Claus Ropers

Abstract In Chap. 18, we demonstrated polarization-sensitive imaging at extreme-ultraviolet (EUV) wavelengths using a gas-phase high-harmonic generation (HHG) source. In a related project, we have investigated new types of gas-phase and solid-state EUV light sources employing field localization in plasmonic nanostructures and structured targets. Whereas our first results indicate that strong field confinement leads to exceedingly inefficient high-harmonic generation in gas-phase targets, for solid-state media efficient high-harmonic generation is possible in localized fields. The latter has great ramifications for new types of high-harmonic generation experiments and technological developments. Therefore, our research efforts aim in two directions: firstly, the development of new types of solid-state sources for high-harmonic generation and, secondly, the application of locally generated solid-state high-harmonic signals for spectroscopy and imaging.

PACS Subject Classification: 42.65.Ky · 81.07.-b · 73.20.Mf

19.1 Plasmonic Enhancement for EUV Light Generation

In this section, we discuss EUV light generation in gas-phase and solid targets using strong laser fields confined in plasmonic nanostructures. In the case of gas-phase targets, we critically revisit the feasibility of high-harmonic generation in resonant nanoantennas and tapered hollow waveguides (see Fig. 19.1a and b), as reported previously [1, 2]. The results of our studies [3–5] show that for gas-phase targets, such as noble gas atoms, the measurable EUV emission exclusively originates from incoherent fluorescence instead of coherent high-harmonic radiation, which is due to an unfavorable conversion efficiency for coherent signals generated in low-density targets

M. Sivis · C. Ropers (✉)
IV. Physical Institute - Solids and Nanostructures, Universität Göttingen,
Friedrich-Hund-Platz 1, 37077 Göttingen, Germany
e-mail: claus.ropers@uni-goettingen.de

M. Sivis
e-mail: murat.sivis@uni-goettingen.de

© The Author(s) 2020   523
T. Salditt et al. (eds.), *Nanoscale Photonic Imaging*, Topics in Applied Physics 134,
https://doi.org/10.1007/978-3-030-34413-9_19

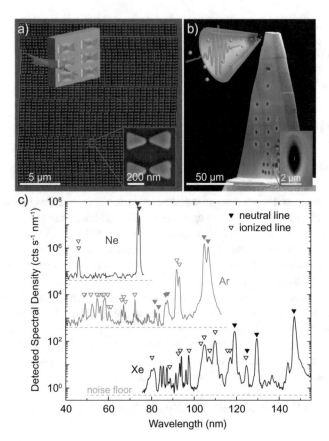

Fig. 19.1 Plasmon-enhanced EUV light generation. **a** Scanning electron micrograph of an array of gold bow-tie nanoantennas on a sapphire substrate. Lower inset: Close-up of two bow-tie antenna pairs. Upper inset: Schematic illustration of the gas excitation in the enhanced fields localized in the gap between the tips of the triangular antenna arms. **b** Scanning electron micrograph of a gold plateau containing a tapered hollow waveguide structure. Lower inset: Close-up of a waveguide, taken from the entrance aperture side. The waveguide has an entrance aperture size of several micrometers and tapers down to an exit aperture of few tens of nanometers in size. Upper inset: Schematic illustration of gas excitation in the locally enhanced field near the waveguide's exit aperture. **c** EUV spectra from xenon (Xe), argon (Ar) and neon (Ne), excited in the waveguide structure shown in (**b**). The argon and neon spectra are up-shifted to avoid overlap. Similar spectra have been observed from xenon and argon gas using bow-tie antennas for field enhancement

in nanometrically confined excitation volumes. After briefly describing our experiments and discussing the implications of the outcome, we highlight recent experiments, which overcome the limitations for plasmon-enhanced HHG by exchanging the gas-phase targets with solid-state materials.

In the experiment, the nanostructures are illuminated under vacuum conditions (pressure below 10^{-6} mbar) with low-energy, few-femtosecond laser pulses centered at 800 nm wavelength from a 78 MHz Ti:sapphire oscillator. The incident laser field is confined in the particular structure leading to local intensities which are enhanced by more than two orders of magnitude compared to the incident intensity. Injected noble gas atoms (backing pressures up to 500 mbar) are excited and ionized in the enhanced field (see upper insets in Fig. 19.1a and b) via multiphoton absorption and strong-field excitation. The radiation emitted by the atoms is collected with an EUV flat-field spectrometer, which spectrally resolves the signal on an imaging microchannel plate detector. For both excitation schemes, the laser radiation is tightly focused to achieve incident intensities in excess of 0.1 TW/cm^2. More detailed information on the experimental setup and conditions can be found elsewhere [3, 5].

Figure 19.1c shows a set of EUV fluorescence spectra obtained from different noble gases excited in a hollow waveguide structure, where all spectral features are identified with emission corresponding to electronic transitions of the respective gas species. The wavelength positions of the fluorescence lines are marked with full and open triangles for emission from neutral and singly ionized atoms, respectively. Similar spectra were measured using bow-tie nanostructures. The reason for the lack of coherent emission in these spectra can be found by considering the absolute conversion efficiency of the HHG process in the localized generation volume. The number of gas atoms in the generation volume is too small for a very efficient coherent build-up of the harmonic radiation. More specifically, the output power of (phase-matched) high-harmonic generation scales quadratically with the pressure-length product, which is drastically reduced in implementations based on surface-plasmon field confinement in nanostructures. Since incoherent fluorescence scales only linearly with the pressure-length product, it can be efficiently generated and therefore is the predominant contribution to the EUV signal.

Estimates suggest that high-harmonic signal levels in plasmon-enhanced scenarios should be several orders of magnitude below that of the simultaneously emitted incoherent fluorescence [5, 6]. Therefore, if possible at all, it will be very difficult to discriminate coherent from incoherent signals.

In a recent publication, the authors of the first studies on plasmon-enhanced gas-phase high-harmonic generation [1, 2] report that a reproduction of their initial experiment resulted in the exclusive observation of incoherent emission and they acknowledge that their interpretation of the previous results in [1] was not fully correct, since they ignored the possibility of any incoherent contributions [7].

As described above, plasmon-enhanced high-harmonic generation in gaseous media is unfeasible under the given experimental conditions. However, the main limitation—the too low gas atom density in the generation volume—can be overcome by exchanging the gas with condensed matter targets. Reports on plasmon-enhanced high-harmonic generation in sapphire [5, 7], silicon [8], and zinc-oxide [9] crystals re-initiate the experimental research on high-harmonic generation in localized fields and, moreover, extend the prospects for solid-state high-harmonic generation.

19.2 High-Harmonic Generation and Imaging in Tailored Semiconductors

Beyond enabling high harmonic generation from localized fields in plasmonic nanostructures, solids have an important advantage over gaseous targets: The structure and chemical composition of solid targets can be tailored via established micro- and nano-engineering methods. This enables novel means to shape the generated high-harmonic wave field in terms of intensity, phase or polarization. In turn, coherent signals generated in structured solids can be used for imaging and spectroscopy with nanometer spatial resolution.

Here, we use locally structured and chemically modified zinc oxide (ZnO) and silicon targets to demonstrate high-harmonic generation and imaging [10]. Figure 19.2a depicts the experimental setup and principle. A zinc oxide crystal with an array of microcones milled to its surface (focused ion beam fabrication, see scanning electron micrograph in Fig. 19.2c) is placed in the focus of an infrared (2.1 μm central wavelength), 70-femtosecond laser beam with 10 kHz pulse repetition rate. At incident peak intensity exceeding 30 GW cm^{-2}, high-harmonic radiation up to the 9th harmonic order of the fundamental frequency (see Fig. 19.2b) is generated in the target. The high-harmonic generation is localized to sub-wavelength sized regions at the apexes of the cones due to a concentration of the incident laser light by total internal reflection and interference, leading to an at least 10-fold intensity enhancement. The localized emission leads to a 2-dimensional diffraction pattern in the far-field, which is shown for the 3rd (red spots) and 5th (blue) harmonic far-field intensity in the inset in Fig. 19.2a.

At this point, we employ two approaches to get information on the intensity distribution at the sample plane. Since the wavelength of the 3rd and 5th harmonics are in the visible range, direct imaging using a high-numerical-aperture objective in combination with a bandpass filter and a charge-coupled device camera is possible. Figure 19.2d shows the directly imaged 5th harmonic pattern at the sample plane.

In order to illustrate that such an imaging approach is also transferable to harmonics with higher photon energies, where refractive optics are not available, we reconstruct the exit amplitude from the far-field intensity by applying a phase retrieval algorithm as described in Chap. 18. The reconstructed amplitudes are shown in Fig. 19.2e. The field of view of the reconstruction is reduced due to limited spatial coherence and low signal-to-noise ratio of the diffraction pattern. However, the central spots clearly register to the arrangement of the cones in Fig. 19.2c, indicating the possibilities to image the nanoscale structures in solids.

In addition to structural modifications, which affect the driving laser field, local chemical changes also lead to enhanced high-harmonic generation in solid targets, as shown in Fig. 19.3 for the example of a gallium-ion-implanted silicon target. The

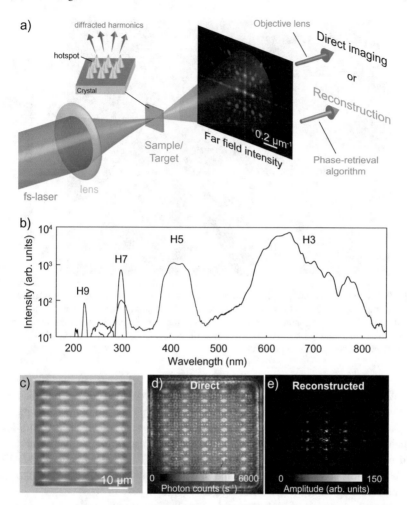

Fig. 19.2 High-harmonic generation in ZnO structures. **a** Schematic illustration of the experiment. The far-field intensity distribution, which is recorded with a 3-color CMOS sensor, shows diffracted third- (red) and fifth- (blue) harmonic signals emitted from a cone grating (see SEM image in **c**). The upper inset shows a 3-dimensional illustration of the microcones on the crystal's surface. **b** High-harmonic spectrum from a structured target illuminated with amplified 2.1-μm laser pulses. The harmonic orders are labeled H3–H9. Harmonics H7 and H9 were measured with a different spectrometer exhibiting a higher sensitivity in the spectral range below 350 nm. **c** Scanning electron micrograph of a 2D-grating of microcones on ZnO. **d**, **e** Comparison of the directly imaged fifth harmonic signal (Direct) using an objective and a phase-retrieval reconstruction of the exit amplitude (Reconstructed) using the far-field intensity

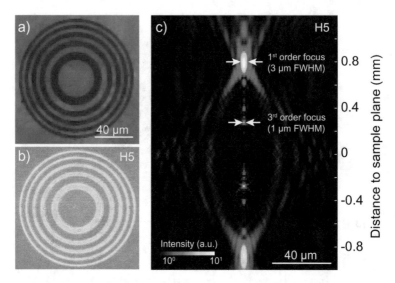

Fig. 19.3 Fresnel zone plate high harmonic source in silicon. **a** Scanning electron micrograph of a gallium-implanted Fresnel zone plate pattern (darker region) on a silicon wafer. **b** Intensity distribution of the fifth harmonic signal (H5, compare spectrum in Fig. 19.2b) at the sample plane. **c** Focus scan of imaging objective showing the azimuthally-integrated fifth-harmonic signal as a function of distance to the sample plane ($z = 0$)

scanning electron microscope image in Fig. 19.3a shows the Fresnel zone plate pattern written into the silicon surface (dark regions were exposed to low-dose gallium ions). The gallium ions create defects in the silicon matrix, which lead to a local enhancement of the high-harmonic generation process, most likely by inducting mid-gap state in the silicon band structure. Figure 19.3b is an image of the fifth-harmonic emission at the sample plane, recorded with the objective lens, showing an enhanced signal in the gallium-implanted regions.

Generally, imaging of high-harmonic signals from locally structured and chemically modified solid targets represents a novel means to investigate local strong-field phenomena in condensed matter systems at nanometer scales, with potential capabilities for temporally or spectroscopically resolved studies. In turn, such sources also allow for the control of the generated high-harmonic wave field, as the Fresnel zone target in Fig. 19.3 illustrates.

The source leads to a focusing of the generated harmonic radiation to diffraction limited spot sizes. Figure 19.3c shows a focus scan of the azimuthally integrated intensities at different distances to the sample plane along the optical axis. Further results show that also the phase and polarization of the generated emission can be modified in such schemes.

In conclusion, our study demonstrates that strong-field effects are generally excitable in localized laser fields, e.g. by employing plasmonic nanostructures for field enhancement. However, gas-phase high harmonic generation lacks conversion efficiency in such scenarios due to a small number of gas atoms in the localized generation volume and the resulting insufficient coherent build-up of the high-harmonic signal. Instead of coherent high-harmonic emission, we observed bright incoherent extreme-ultraviolet fluorescence, which stems from multiphoton and strong-field excited and ionized gas atoms.

In contrast to gas-phase targets, in solid media an efficient generation of high harmonics in localized fields is possible. In structured ZnO and silicon targets we concentrated the driving laser field in microscopic cones and wedges and observed enhanced coherent high-harmonic emission from sub-wavelength sized generation volumes. In a second approach, we demonstrated that chemical modifications also influence the high-harmonic generation in solids. Direct imaging or phase-retrieval reconstruction of the localized coherent emission will enable new means to study strong-field phenomena in solid-state systems. Additionally, tailored solids will enable the development of new types of sources for tailored high-harmonic wave fields.

References

1. Kim, S., et al.: High-harmonic generation by resonant plasmon field enhancement. Nature **453**(7196), 757–760 (2008). http://dx.doi.org/10.1038/nature07012
2. Park, I.Y., et al.: Plasmonic generation of ultrashort extreme-ultraviolet light pulses. Nat. Photonics **5**(11), 677–681 (2011). https://doi.org/10.1038/nphoton.2011.258, http://www.nature.com/doifinder/10.1038/nphoton.2011.258
3. Sivis, M., Ropers, C.: Generation and bistability of a waveguide nanoplasma observed by enhanced extreme-ultraviolet fluorescence. Phys. Rev. Lett. **111**(8), 085001 (2013). https://doi.org/10.1103/PhysRevLett.111.085001, http://link.aps.org/doi/10.1103/PhysRevLett.111.085001
4. Sivis, M., et al.: Nanostructure-enhanced atomic line emission. Nature **485**(7397), E1–E2 (2012). https://doi.org/10.1038/nature10978, http://www.nature.com/doifinder/10.1038/nature10978, http://www.ncbi.nlm.nih.gov/pubmed/22575967
5. Sivis, M., et al.: Extreme-ultraviolet light generation in plasmonic nanostructures. Nat. Phys. **9**(5), 304–309 (2013). https://doi.org/10.1038/nphys2590, http://www.nature.com/doifinder/10.1038/nphys2590
6. Raschke, M.B.: High-harmonic generation with plasmonics: feasible or unphysical? Ann. Phys. **525**(3), A40–A42 (2013). https://doi.org/10.1002/andp.201300721, http://doi.wiley.com/10.1002/andp.201300721
7. Han, S., et al.: High-harmonic generation by field enhanced femtosecond pulses in metal-sapphire nanostructure. Nat. Commun. **7**(May), 13105 (2016). https://doi.org/10.1038/ncomms13105, http://www.nature.com/doifinder/10.1038/ncomms13105
8. Vampa, G., et al.: Plasmon-enhanced high-harmonic generation from silicon. Nat. Phys. **13**(7), 659–662 (2017). https://doi.org/10.1038/nphys4087
9. Imasaka, K., et al.: Antenna-enhanced high harmonic generation in a wide-bandgap semiconductor ZnO. Opt. Express **26**(16), 21364 (2018). https://doi.org/10.1364/OE.26.021364, https://www.osapublishing.org/abstract.cfm?URI=oe-26-16-21364
10. Sivis, M., et al.: Tailored semiconductors for high-harmonic optoelectronics. Science **357**, 6348 (2017). https://doi.org/10.1126/science.aan2395

Chapter 20
Wavefront and Coherence Characteristics of Extreme UV and Soft X-ray Sources

Bernd Schäfer, Bernhard Flöter, Tobias Mey and Klaus Mann

Abstract The first part of this chapter comprises setups and results of the determination of wavefront and beam parameters for different EUV sources (free-electron lasers, HHG-sources, synchrotron radiation) by self supporting Hartmann-Sensors. We present here i.a. a sensor applied for alignment of the ellipsodial mirror at FLASH beamline 2, yielding a reduction of the rms-wavefront aberrations by more than a factor of 3. In the second part we report on the characterization of the Free-Electron-Laser FLASH at DESY by a quantitative determination of the Wigner distribution function. The setup, comprising an ellipsodial mirror and a moveable extreme UV sensitive CCD detector, enables the mapping of two-dimensional phase space corresponding to the horizontal and vertical coordinate axes, respectively. Furthermore, an extended setup utilizing a torodial mirror for complete 4D-Wigner reconstruction has been accomplished and tested using radiation from a multimode Nd:VO4 laser.

PACS Subject Classification: 42.15.Dp · 42.25.Kb · 42.55.Vc

20.1 Introduction

Electromagnetic radiation in the extreme UV and soft X-ray spectral range is of steadily increasing importance in fundamental research and industrial applications. For instance, the molecular structure of proteins and viruses has become accessible by coherent diffractive imaging techniques; currently, lithographic processes for the microchip production are being adapted to the extreme UV wavelength of 13.5 nm.

The original version of this chapter was revised: The Reference 14 replaced with correct reference. The correction to this chapter is available at https://doi.org/10.1007/978-3-030-34413-9_25

B. Schäfer (✉) · B. Flöter · T. Mey · K. Mann
Laser-Laboratorium Göttingen eV., Hans-Adolf-Krebs-Weg 1,
37077 Göttingen, Germany
e-mail: bernd.schaefer@llg-ev.de

K. Mann
e-mail: kmann@llg-ev.de

© The Author(s) 2020, corrected publication 2021
T. Salditt et al. (eds.), *Nanoscale Photonic Imaging*, Topics in Applied Physics 134,
https://doi.org/10.1007/978-3-030-34413-9_20

For both examples, a comprehensive beam characterization is an essential condition for an ideal use of the available photons, and only exact knowledge of the illuminating radiation field allows for further improvements of spatial resolution and reliability in nanoscale imaging and structuring. To this end, pioneering developments in large-scale and table-top light sources of extreme ultraviolet (EUV) radiation are necessarily complemented by implementing advanced beam characterization techniques. The Laser-Laboratorium Göttingen has developed metrological tools and analysis procedures for proper characterization of the propagation behaviour of short wavelength radiation. This contribution addresses wavefront measurements on free electron lasers (FELs) and high harmonic (HHG) sources emitting in the extreme UV and soft X-ray range. The diagnostics schemes based on Hartmann sensing accomplish, on the one hand, comprehensive beam analysis including prediction of focal distributions, on the other also fine-adjustment of beamline optics for optimization of peak intensities. Additionally, the coherence of laser beams is analyzed by measurements of the Wigner distribution function. This method is applied to the photon beam of the free-electron laser FLASH, resulting in the entire characterization of its propagation properties, including both global and local degrees of spatial coherence.

20.2 Wavefront Metrology and Beam Characterization with Hartmann Sensors

20.2.1 Hartmann Wavefront Sensing

The wavefront or phase distribution of a radiation field carries quantitative information over its directional distribution, and is therefore of utmost importance for the design of beam transport optics. On-line recording of the wavefront can also accomplish an optimization of the beam focusability by precision alignment of optical elements. Other relevant areas are the monitoring and possible reduction of thermal lensing effects, on-line resonator adjustment, or "at wavelength" testing of optics including Zernike analysis. The wavefront of a radiation source is defined as the surface $w(x, y)$ that is normal to the local direction of energy propagation in the electromagnetic field [1], i.e. normal to the Poynting vector (x, y) at the measurement plane (cf. Fig. 20.1, left). In case of highly coherent radiation, $w(x, y)$ is a surface of constant phase. The phase distribution $\Phi(x, y)$ is then related to the wavefront according to

$$\Phi(x, y) = \frac{2\pi}{\lambda} \cdot w(x, y), \qquad (20.1)$$

where λ is the mean wavelength of the light.

A variety of different techniques has been developed for wavefront sensing. Interferometric devices, as there are Twyman-Green, common path, lateral shear, Mach-Zehnder or Sagnac interferometers, can be applied over the full wavelength spectrum

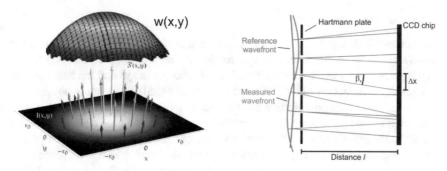

Fig. 20.1 left: Definition of the wavefront of a radiation field; right: Measurement principle of Hartmann-type wavefront sensors (cf. text)

for which detectors and optical materials are available, provided that the coherence is sufficient for detectable levels of interference. Alternatively, phase gradient measurement techniques, in particular Hartmann or Hartmann-Shack, can be used with both coherent and incoherent beams. In these instruments, the gradients of either wavefront or phase are measured, from which the two-dimensional phase distribution can be reconstructed. The Hartmann principle [1] is based on a subdivision of a beam into a number of beamlets (see Fig. 20.1, right). This is either accomplished by an opaque screen with pinholes placed on a regular grid (Hartmann sensor), or by a lenslet or micro-lens array (Hartmann-Shack sensor). The latter accomplishes a better radiation collection efficiency and a wider dynamic range. For this reason, Hartmann-Shack sensors are already widely used for Vis, NIR and UV radiation. However, since no transmissive optical materials for the fabrication of micro-lens arrays are available at extreme UV and soft X-ray wavelengths, only the Hartmann appoach using pinhole arrays is appropriate in this spectral region. The spot distribution produced by the segmenting array is recorded at a distance l by a position sensitive detector, most commonly a CCD camera. The position of the beamlet centroids is determined within each sub-aperture, both for the beam under test and a reference wavefront. The latter is provided preferably by a well collimated laser beam (plane wave), or a well defined spherical wave, using e.g. the output of a monomode fiber or the Airy pattern produced behind a diffracting pin-hole. The displacement of the spot centroid Δx divided by the distance l yields the local wavefront gradient β_x inside one subaperture relative to the reference wavefront (see Fig. 20.1, right). By direct integration or modal fitting techniques using Zernike or Legendre polynomials, the wavefront $w(x, y)$ is reconstructed from these local gradients [2, 3] and afterwards corrected for tip/tilt and defocus [1]. A detailed description of the wavefront reconstruction methods is given in the references. The main advantages of the Hartmann technique compared to interferometric devices are

- suitability for fully and partially coherent beams,
- no requirement of spectral purity,
- no ambiguity with respect to 2π increment in phase angle,

- compact and robust design.

Hartmann-Shack and Hartmann wavefront sensors can be successfully applied for real-time laser beam characterization, since they are recording simultaneously (i.e. in single pulses) the wavefront (directional distribution) $w(x, y)$ and the beam profile or intensity distribution $I(x, y)$ of a radiation field [4, 5]. The latter is obtained by summation over pixel data inside the individual subapertures, at a reduced spatial resolution given by the pitch of the segmenting array. As has been demonstrated for visible laser radiation, in case of coherent sources the knowledge of beam profile $I(x, y)$ and wavefront $w(x, y)$ allows for calculation of the relevant beam parameters [5–7]. For this purpose the moments method described in [5, 7, 8] is applied: The central second spatial (x, y) and angular (u, v) moments are computed from the intensity distribution and the local wavefront slopes $\beta_{x,y}$ according to

$$\langle x^2 \rangle = \frac{\sum_{i,j}(x_{ij} - \langle x \rangle)^2 I_{ij}}{\sum_{i,j} I_{ij}} \tag{20.2}$$

$$\langle xu \rangle = \frac{\sum_{i,j}(\beta_{xij} - \langle \beta_x \rangle)(x_{ij} - \langle x \rangle) I_{ij}}{\sum_{i,j} I_{ij}} \tag{20.3}$$

$$\langle u^2 \rangle = \frac{\sum_{i,j}(\beta_{xij} - \langle \beta_x \rangle)^2 I_{ij}}{\sum_{i,j} I_{ij}} + \left(\frac{\lambda}{2\pi}\right)^2 \frac{\sum_{i,j}\left(\frac{(\partial_x I)^2}{I}\right)_{ij}}{4\sum_{i,j} I_{ij}}, \tag{20.4}$$

where $\langle x \rangle$ and $\langle \beta_x \rangle$ are the first moments over x and β_x [7], respectively; the index (ij) denotes the subaperture. From the second moments the beam width d, divergence θ, beam propagation factor M^2, beam waist diameter d_θ, waist position z_θ and Rayleigh length z_R are computed according to the following equations [7]:

$$d = 4\sqrt{\langle x^2 \rangle}, \qquad\qquad \theta = 4\sqrt{\langle u^2 \rangle} \tag{20.5}$$

$$M^2 = \frac{4\pi}{\lambda}\sqrt{\langle x^2 \rangle \langle u^2 \rangle - \langle xu \rangle^2} \qquad\qquad d_\theta = \frac{4M^2\lambda}{\pi\theta} \tag{20.6}$$

$$z_\theta = \frac{z_R \langle xu \rangle}{|\langle xu \rangle|}\sqrt{\left(\frac{d}{d_\theta}\right)^2 - 1} \qquad\qquad z_R = \frac{d_\theta}{\theta}. \tag{20.7}$$

Moreover, once the intensity and the phase distributions are known from a Hartmann measurement, solving Fresnel-Kirchhoff's integral allows numerical propagation of the beam, i.e. computation of intensity distributions at different propagation distances z [9]:

$$I(x, y, z) = \left| \frac{ik}{2\pi z} \iint_\infty \sqrt{I(x', y')} e^{ikw(x',y')} e^{\frac{ik[(x-x')^2 + (y-y')^2]}{2z}} \, dx' \, dy' \right|$$

Here x, y and x', y' are the Cartesian coordinates in two coplanar planes separated by z. Thus, in particular the profile at the beam waist position can be predicted, which is in many cases hardly accessable for high power lasers, both due to the high intensity and the small size of the focal spot.

20.2.2   EUV Wavefront Sensor for FEL Characterization

Since currently operating (soft) X-ray FELs are based on the self-amplified spontaneous emission (SASE) process which builds up the laser emission from noise, their beam characteristics can signigicantly differ from pulse to pulse. Therefore, there is a strong requirement for single-pulse photon diagnostics and online characterization of the FEL beam propagation parameters [10, 11]. For this reason a Hartmann wavefront sensor for the extreme UV spectral range was developed and applied for photon diagnostics, beam propagation and optics alignment of the FLASH free electron laser in cooperation with DESY Photon Science/Hamburg [12]. The device was designed to operate from 4 to 40 nm, which is within the accessible FLASH wavelength range. It consists of a pinhole array (Hartmann plate) made of a 20 µm-thick nickel foil with orthogonally arranged electroformed holes (dia. 75 µm, pitch 250 µm) in front of a CCD camera at a distance of 200 mm behind the array (see Fig. 20.2). This distance as well as the dimensions of the Hartmann plate represent a compromise between attainable wavefront sensitivity at short wavelengths and spatial resolution at long wavelengths. For converting the soft X-rays into visible light the CCD chip is coated with a fluorescent coating (Gd_2O_2S:Tb, emission wavelength 545 nm). The Hartmann sensor is adjustable both laterally and with respect to tip and tilt. The device is self-supporting and compact (240 mm × 240 mm × 300 mm) and can be attached behind user experiments.

For absolute at-wavelength calibration of the Hartmann sensor a proper reference wavefront independent of the mentioned pulse-to-pulse fluctuations is essential. For this purpose a spherical wavefront is prepared by spatial filtering, placing a diffracting pinhole (dia. several µm) in the vicinity of the focal spot of the FEL beam. The

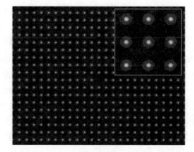

Fig. 20.2 Left: EUV Hartmann sensor with integrated tip/tilt and lateral adjustment; the inset shows a close-up of the Hartmann pinhole plate. Right: spot pattern of the reference wavefront ($\lambda = 13.5$ nm). The central pinhole is omitted for tip/tilt alignment

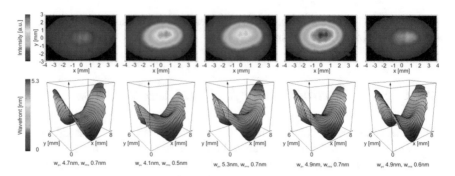

Fig. 20.3 Intensity profiles and wavefronts of single pulses at FLASH BL2 without focusing mirror ($\lambda = 7$ nm)

Hartmann sensor is positioned at a certain distance behind this pinhole, ensuring that its full field of view is illuminated by the central Airy disc. Thereafter, the reference spot distribution is registered (cf. Fig. 20.2, right). A temporally stable spherical wave as described above can also be utilized to assess the sensitivity of the Hartmann sensor. At FLASH beamline BL2 the single pulse wavefront repeatability was determined, recording a series of 100 single FEL pulses behind a $5\,\mu$m pinhole at an emission wavelength of 13.8 nm. A root-mean-square deviation $\Delta w_{\mathrm{rms}} = 0.12$ nm ($\lambda/116$) was evaluated, defining an upper limit for the achievable measurement precision of the wavefront sensor.

After these qualification tests the Hartmann sensor was used to analyze the FEL beam of FLASH, at first without focusing mirror [13]. Beam profiles and wavefronts recorded at BL2 for single pulses at a wavelength of 7 nm are displayed in Fig. 20.3, showing relatively strong pulse-to-pulse fluctuations as typical for the SASE process. The saddle-like shape of the wavefront indicates an astigmatism of the beam. Nevertheless, the peak-to-valley (w_{pv}) and root-mean-square (w_{rms}) wavefront aberrations computed after tip/tilt and defocus subtraction of the measured wavefronts are relatively low ($w_{\mathrm{rms}} \sim \lambda/10$).

Neglecting influences from partial coherence (cf. Sect. 20.3), the beam propagational parameters can be computed from the measured intensity and wavefront distributions according to the moments method described above (20.8)–(20.11). Corresponding beam characteristics are compiled in Table 20.1, taking the average over 20 single pulses. Despite the observed astigmatism (waist separation $x - y = 10$ m), the evaluated M^2 value of 1.15 is remarkably low, which can be explained by the small higher order wavefront expansion coefficients and the smooth intensity profile.

In contrast to these data, wavefront measurements performed behind beam line optics can lead to much less satisfactory results, caused by an insufficient fine-adjustment of the optics. An example is shown in Fig. 20.4 (left) for an ellipsoidal focusing mirror at BL 3 of FLASH: for this carbon-coated grazing incidence mirror with 2 m focal length the w_{pv} and w_{rms} values of the recorded wavefront are more than an order of magnitude higher than without focusing optics. However, by on-line wavefront diagnostics the EUV Hartmann sensor allows for fine-tuning the mirror

Table 20.1 Beam parameters of FLASH computed from Hartmann data (BL2, $\lambda = 7$ nm)

Beam parameters	X	Y
w_{pv} [nm]	5.3 ± 0.69	
w_{rms} [nm]	0.67 ± 0.09	
Beam propagation parameter M^2	1.15 ± 0.08	
Beam propagation parameter M_i^2	1.23 ± 0.1	1.1 ± 0.1
Beam width d [mm]	6 ± 0.2	4.4 ± 0.1
Waist position $z_{0,i}$ [m]	-109.2 ± 0.9	-99.2 ± 1.4
Rayleigh length z_R [mm]	3760 ± 484	5090 ± 731
Waist diameter $d_{0,i}$ [µm] 2nd moment	200 ± 20	223 ± 25
Divergence θ [µrad]	55 ± 2	44 ± 2

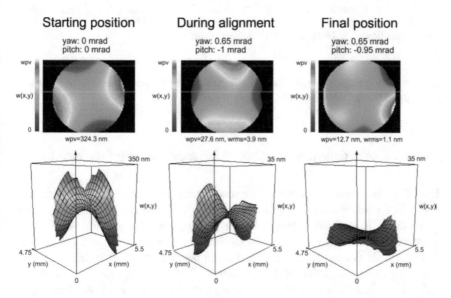

Fig. 20.4 Wavefronts measured at different steps of the alignment procedure of the ellipsoidal focusing mirror at FLASH beamline BL3 ($\lambda = 13.3$ nm). Note that the scale $w(x, y)$ is enlarged by a factor of ten for the starting position to account for the very large initial astigmatism

alignment. As seen from Fig. 20.4, the dominating strong astigmatism introduced by the optical element could be completely removed by real-time optimizing the pitch and yaw angles of the ellipsoidal mirror. After alignment a w_{pv} of 12 nm and a w_{rms} of 1.1 nm ($<\lambda/10$) could be achieved.

After optimized fine-adjustment of the focusing element, a Fresnel-Kirchhoff integration of the Hartmann data allows for propagation of the beam, as described in Sect. 20.2.1. Corresponding results recently obtained at FLASH II are displayed in Fig. 20.5 for three z-positions close to the beam waist [14]. The propagated profiles

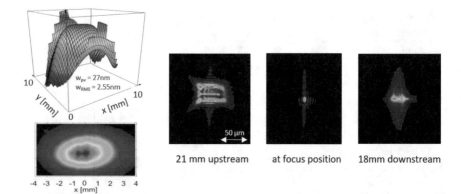

Fig. 20.5 Left: Wavefront and intensity distribution of FLASH 2 (FL24, $\lambda = 13.5\,\mathrm{nm}$) recorded $\sim$2 m behind KB optics with EUV Hartmann sensor; right: Profiles obtained by Fresnel-Kirchhoff back propagation of the Hartmann data to the beam waist

are currently compared with PMMA imprints taken at the respective positions by J. Chalupský et al. (Acad. of Sci., Czech Republic).

20.2.3 Beam Characterization of High-Harmonic Sources

The EUV Hartmann wavefront sensor successfully applied for beam characterization at FELs was also employed to investigate EUV radiation generated by High-Harmonic (HHG) sources. Especially for their use in CDI experiments, a proper alignment is crucial since successful reconstruction of phase objects can only be achieved if the phase distortions of the probe beam are negligible. In cooperation with Claus Ropers' group, the propagation of the 25th harmonic ($\lambda = 32$ nm) of a Ti:Saphire laser was studied after passing a toroidal grating that combines spectral filtering and focusing. The Hartmann sensor was positioned behind the focus, capturing the EUV wavefront while the angle of incidence of the harmonic on the grating was varied. As for the FEL beam line mirrors, the recorded wavefront initially shows a strong astigmatism (cf. Fig. 20.6, left), which can be minimized by real-time alignment. A description of the corresponding beam propagation by matrix methods [15] yields good agreement to the experimental data, especially the astigmatic waist difference (cf. Fig. 20.6 right, blue line). From the theoretical computations, the achievable beam intensity is estimated as a function of the incidence angle. As expected, the highest photon flux is obtained for an angle of incidence where the astigmatic aberration disappears. Apparently, already a slight misalignment of 0.5° leads to a decrease of the achievable intensity by 50% compared to its optimum. Thus, in order to achieve short exposure times and prevent reconstruction errors for following CDI experiments, this alignment procedure plays an essential role. With

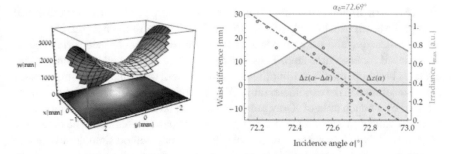

Fig. 20.6 Left: Wavefront (3D) and intensity profile (below) of a HHG beam (25th harmonic). Right: Astigmatic waist difference Δz and achievable irradiance I_{max} plotted as a function of the angle of incidence on toroidal grating. The theoretical curve $\Delta z(\alpha)$ (solid blue line) lies slightly above the experimental values (blue dots) [15]

the correspondingly optimized HHG source, it was possible to successfully image test samples at the diffraction limit [16].

20.2.4 Thermal Lensing of X-ray Optics

In addition to static aberrations of optical components given by figure errors or misalignment, beam propagation can also be deteriorated by transient distortions of the wavefront introduced by the beam itself due to local heating and surface deformation (thermal lensing). In order to investigate the influence of this effect on the performance of X-ray optics, in particular high power mirrors to be employed for the European XFEL/Hamburg, we have performed time-resolved wavefront measurements in pump-probe experiments at the ESRF/Grenoble [9]. In this investigation a Si mirror sample was exposed to an intense 15 keV beam, and its thermally induced

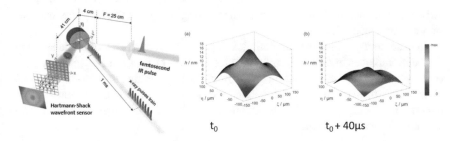

Fig. 20.7 Left: Schematic view of pump-probe setup to determine thermal wavefront distortions of high power X-ray optics at ESRF. Right: surface topology of a Si mirror reconstructed from wavefront measurements for different delays between the X-ray pump and infrared probe pulse. The decay of the heat bump on the mirror proceeds at a time scale of several ten microseconds

surface deformation was monitored by measuring the wavefront of a reflected optical laser probe beam with the help of a Hartmann-Shack wavefront sensor (cf. Fig. 20.7). By reconstructing and back propagating the wavefront, the deformed surface could be retrieved for each time step. Thus, the dynamics of the created heat bump, especially its build-up, maximum amplitude and relaxation, were analyzed with a surface height resolution in the nanometer range. For the investigated Si sample deformations induced by a bunch train of X-ray pulses were in the order of several ten nanometers (peak-to-valley); a relaxation time constant of $\sim 30\,\mu s$ was obtained. The data were interpreted taking into account results of finite element method simulations. Due to its robustness and simplicity this method can find further applications at new X-ray light sources (FEL), or to gain deeper understanding on thermo-dynamical behavior of highly excited materials under non-equilibrium conditions.

20.3 Wigner Distribution for Diagnostics of Spatial Coherence

Apart from the beam profile and shape of the wavefront, the degree of lateral coherence of a beam has a crucial impact on the minimum achievable focal spot size. Whereas both wavefront and irradiance distribution may be discovered from a single shot experiment, the latter is not true for the spatial degree of coherence γ, which is, like the mutual intensity J [17] defined on a four-dimenesional space of lateral position $\mathbf{x}$ and mutual distance $\mathbf{s}$. Earlier approaches for spatial coherence measurement at FELs utilize Young's double pinhole experiment [18] to derive the latter from fringe visibility in the corresponding interference patterns. However, any substantial mapping of $(\mathbf{x}, \mathbf{s})$-pairs requires a vast number of pinhole arrangements and image recordings to be evaluated, which appears to be very inefficient with respect to the experimental effort. Therefore, those measurements have only been carried out for a few selected points $\mathbf{x}$ within the beam profile and one or two perpendicular directions of the pinhole separation vector $\mathbf{s}$. An alternative approach is based on the investigation of lateral correlation of local intensity fluctuations in the beam profile [19]. Although this method is more efficient than Young's experiment, only the modulus of γ can be specified.

In order to determine the full complex degree of coherence, an alternate strategy has been employed to recover the mutual intensity $J(\mathbf{x}, \mathbf{s})$, i.e. through a measurement of the Wigner distribution function (WDF) $h(\mathbf{x}, \mathbf{u})$, representing the two-dimensional Fourier transform of the mutual coherence function [20]. Prior to the presentation of experimental setups and results the theoretical background of the applied formalism will be briefly summarized.

20.3.1 Theory

The Wigner distribution $h(\mathbf{x}, \mathbf{u})$ of a quasi-monochromatic paraxial beam is defined in terms of the mutual intensity $J(\mathbf{x}, \mathbf{s})$ as a two-dimensional Fourier transform of the latter [21]

$$h(\mathbf{x}, \mathbf{u}) = \left(\frac{k}{2\pi}\right)^2 \int J\left(\mathbf{x} - \frac{\mathbf{s}}{2}, \mathbf{x} + \frac{\mathbf{s}}{2}\right) e^{ik\mathbf{u}\cdot\mathbf{s}} \, ds_x \, ds_y,$$

where $\mathbf{x} = (x, y)$ and $\mathbf{s} = (s_x, s_y)$ denote spatial and $\mathbf{u} = (u, v)$ angular coordinates in a plane perpendicular to the direction of beam propagation, and k is the mean wave number of light. Propagation of the Wigner distribution h and its 4D Fourier transform $\tilde{h}$ through static and lossless paraxial systems from an input (index i) to an output (index o) plane, signified by a 4×4 optical ray propagation $ABCD$ matrix S, writes [22, 23]:

$$h_i(D\mathbf{x} - B\mathbf{u}, -C\mathbf{x}, +A\mathbf{u}) = h_o(\mathbf{x}, \mathbf{u}) \tag{20.8}$$

$$\tilde{h}_i(A^T\mathbf{w} + C^T\mathbf{t}, B^T\mathbf{w} + D^T\mathbf{t}) = \tilde{h}_o(\mathbf{w}, \mathbf{t}), \tag{20.9}$$

where $(\mathbf{w}, \mathbf{t})$ are the Fourier space coordinates corresponding to $(\mathbf{x}, \mathbf{u})$. Considering a set $\{p\}$ of parameters, defined by the optical system being employed to generate projections of the phase space, and a set of irradiance profiles $I_{\{p\}}$ recorded at positions which are connected to an arbitrary reference plane via corresponding ray transformation matrices $S_{\{p\}}$, one obtains:

$$\int h_{\{p\}}(\mathbf{x}, \mathbf{u}) \, du \, dv = I_{\{p\}}(\mathbf{x}) \overset{\text{FT}}{\longleftrightarrow} \tilde{h}_{\{p\}}(\mathbf{w}, \mathbf{t} = 0) = \tilde{I}_{\{p\}}(\mathbf{w}) \tag{20.10}$$

and from (20.9) and (20.10) [24]:

$$\tilde{h}_{\text{ref}}(A^T_{\{p\}}\mathbf{w}, B^T_{\{p\}}\mathbf{w}) = \tilde{I}_{\{p\}}(\mathbf{w}). \tag{20.11}$$

Propagation through free space in beam direction (z axis) is described by the $ABCD$ matrix

$$S_z = \begin{pmatrix} 1 & z \\ 0 & 1 \end{pmatrix} \tag{20.12}$$

corresponding to the detector position in the experimental arrangement described below. Thus, (20.10) becomes

$$\tilde{h}_{\text{ref}}(\mathbf{w}, z \cdot \mathbf{w}) = \tilde{I}_z(\mathbf{w}), \tag{20.13}$$

representing a four-dimensional mapping relation between Fourier transformed intensity distributions and the Wigner distribution of the beam (Projection Slice

Theorem). Following this equation, the phase space of $\tilde{h}$ is filled with data from intensity profiles measured at several z-positions, as for instance obtained from a caustic scan of the beam. A subsequent four-dimensional inverse Fourier transform of $\tilde{h}$ results in the Wigner distribution function.

The global degree of coherence K is calculated by

$$K = \frac{\lambda^2}{P^2} \int h(\mathbf{x}, \mathbf{u})^2 \, dx \, dy \, du \, dv, \qquad (20.14)$$

(P = total power of the beam) and the mutual coherence function is derived by a two-dimensional Fourier back-transform

$$J(\mathbf{x}, \mathbf{s}) = \int h(\mathbf{x}, \mathbf{u}) e^{-ik\mathbf{u} \cdot \mathbf{s}} \, du \, dv. \qquad (20.15)$$

The coherence lengths l_x and l_y are deduced as half width at half maximum of $J(0, 0, s_x, 0)$ and $J(0, 0, 0, s_y)$, respectively.

20.3.2 Experimental Results

As mentioned above, the Wigner distribution can be derived from intensity profiles along the propagation direction of a beam. Figure 20.8 shows the corresponding experimental setup employed for caustic measurements of the FEL FLASH. Here, focusing is achieved by a carbon-coated ellipsoidal mirror with a focal length of 2 m. The EUV sensor consists of a phosphorous screen imaged onto a CCD chip by a 10x magnifying microscope. A motorized translation stage allows for movement of the detector in z-direction within a range of 250 mm, covering up to ten Rayleigh lengths z_R in both directions around the beam waist. During the caustic measurement, FLASH was running in single bunch mode at a wavelength of $\lambda = 25$ nm. Typically, profiles are acquired at more than 100 different z-positions around the beam waist.

FEL beam profiles at three positions in the focal region behind the ellipsoidal mirror are displayed in Fig. 20.8, indicating pronounced vertical stripes which can be attributed to a residual ripple-like corrugation of the mirror surface [25], while for y-direction the profiles are distributed much smoother. In the focal position the structure vanishes into a uniform distribution. From the Wigner distribution function computed according to the previous section we reconstruct beam profiles at arbitrary positions z in the following fashion: $h(\mathbf{x}, \mathbf{u})$ is propagated via (20.8) applying propagation matrix $S_{\{p\}}$ from (20.12), subsequently the near field of the beam is generated by the integration $I_z(\mathbf{x}) = \int h_z(\mathbf{x}, \mathbf{u}) \, du \, dv$. The resulting reconstructed intensity distribution at average waist position is displayed in Fig. 20.8 (right) together with the corresponding experimental profile. Apparently, a good agreement between measured and reconstructed intensity distribution is achieved which confirms the validity of the obtained Wigner distribution. From the mutual coherence function J of the

Table 20.2 Beam propagation parameters of FLASH evaluated from Wigner measurements (beam waist diameter d_0, Rayleigh length z_R, beam quality factor M^2, coherence length l and global degree of coherence K)

	d_0 [µm]	z_R [mm]	M^2	l [µm]	K
x-direction	67	12.2	8.6	9.0	0.032
y-direction	53	4.3	4.6	11.6	0.032

beam, reconstructed at the average waist position by application of (20.15), the coherence lengths l_x in horizontal and l_y in vertical direction can be calculated as HWHM values of the 1D slices $J(0, 0, s_x, 0)$ and $J(0, 0, 0, s_y)$. The dotted curves in Fig. 20.8 show the corresponding ellipses $s_x^2/l_x^2 + s_y^2/l_y^2 = 1$ and $x^2/d_x^2 + y^2/d_y^2 = 1$, indicating the coherence area and total beam area, respectively. It is appearent, that the coherence length corresponds to a small fraction of the beam diameter only. The exact values are given in Fig. 20.2. The coherence for the vertical beam direction is found to be significantly larger than for horizontal direction. Furthermore, the global degree of coherence is calculated as $K = 0.032$, unveiling an apparently low coherence of the FLASH beam.

In comparison with existing coherence measurements based on Young's double slit good agreement is found for the coherence lengths, but the global degree of coherence is lower by one order of magnitude. This discrepancy can, at least partly, be explained by the fact that an ensemble of beam profiles is employed for the Wigner evaluation, resulting in beam properties in terms of mean values. In contrast, in Young's experiment individual pulses are analyzed which yield corresponding maximum values [18]. Another issue leading to underestimated values of the global coherence is the incomplete 3D mapping using only profiles from a standard caustic measurement (cf. next section).

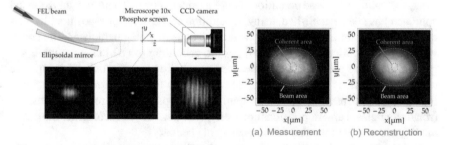

Fig. 20.8 Left: Experimental setup for the Wigner distribution measurement at FLASH and selected profiles close to the beam waist at $\lambda = 25$ nm. Right: Measurement of the beam profile at mean waist position together with reconstruction from the obtained Wigner distribution. Ellipses indicate the coherent fraction of the beam area

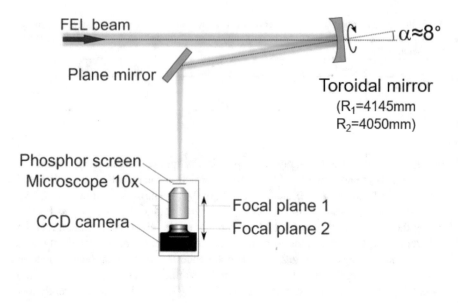

Fig. 20.9 Setup for a 4D measurement of the Wigner distribution

20.3.3 4D Wigner Measurements

The described reconstruction of the Wigner distribution based on beam profiles acquired from free space propagation according to (20.13) covers only a 3D subset of the phase space, as z is the only free parameter in the ray propagation matrix (20.12). Although such a reduced mapping is sufficient for some special cases covering e.g. separable or quasi-homogeneous beams and coherent beams with zero twist, the validity of these conditions is not a priori known. Therefore, an extended approach has been established, using, in addition to the detector z-position, the orientation angle ϕ of an astigmatic optical element as mapping parameter. The corresponding ray matrices $S_{\{z,\phi\}}$ [24, 26] permit, according to (20.11), a complete 4D map of the phase space. The extended setup including a non-rotational symmetric element is shown in Fig. 20.9. It applies a rotatable toroidal mirror, introducing a fourth degree of freedom into the system. Thus, choosing measurement parameters (rotation angles, camera positions) properly, it is possible to reconstruct the entire 4D Wigner distribution, also for non-separable beams.

In order to test and qualify the 4D approach, a diode-pumped Nd:YVO$_4$ laser operating at its fundamental wavelength $\lambda = 1064$ nm (continuous wave) was employed which accomplishes the selective excitation of Hermite-Gaussian modes by pumping the laser crystal at different lateral positions ("mode generator"). TEM$_{00}$, TEM$_{10}$, TEM$_{02}$, TEM$_{03}$ and an uncorrelated superposition of TEM$_{10}$ and TEM$_{01}$ were investigated in a setup similar to Fig. 20.9, using a polished aluminum toroidal mirror

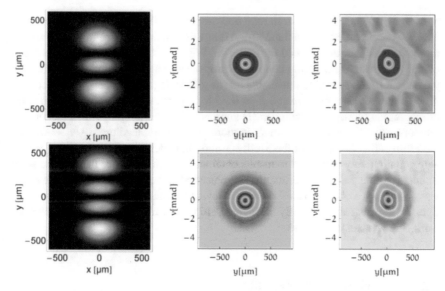

Fig. 20.10 Qualification of 4D Wigner formalism using near IR mode generator for different Hermite-Gaussian beams: Wigner distributions of TEM_{02} and TEM_{03} modes resulting from theory and experiment are shown for comparison

Table 20.3 Global degree of coherence K for different Hermite–Gaussian beams in theory and experiment

	TEM_{00}	TEM_{10}	TEM_{02}	TEM_{03}	$TEM_{10} + TEM_{01}$
Theory	1	1	1	1	0.5
Experiment	0.95	1.06	0.98	0.90	0.46

(radii 200 and 300 mm) and a standard CCD camera as detector. Rotation of the mirror and movement of the camera has been achieved by servo motors. The automated measurement for each laser mode consists of 410 beam profiles in total, corresponding to 10 rotation angles and 41 z-positions. In Fig. 20.10 the Wigner distribution functions reconstructed from the described measurement are depicted for the TEM_{02} and TEM_{03} modes, together with the expected theoretical WDFs of the analyzed beams. Obviously, the experimental results correspond nicely to the theory. Also quantitatively, the expected global degree of coherence $K = 1$ is reproduced with an accuracy better than 10%. As seen from Fig. 20.3, this holds also for the other investigated modes.

After successful qualification, the 4D Wigner method was adapted to the extreme UV wavelength of 13.5 nm by employing a toroidal MoSi multilayer mirror (curvature radii 4145 and 4050 mm, cf. Fig. 20.9), serving to further characterize FEL beams. Recently, first 4D measurements of the WDF have been performed at beamline FL24 of FLASH 2. The data were acquired for the unfocused beam, employing a tilt angle of 8° on the toroid. In total 500 profiles at 50 z-positions and 10 rotation angles were recorded within the 250 mm range of a motorized linear stage. Furthermore, various apertures placed upstream the Wigner setup were utilized in order to modify beam size and spatial coherence properties. The 4D Wigner reconstruction yielded preliminary evaluation results for the global degree of coherence K between $K = 0.1$ for a 5 mm aperture and $K = 0.25$ for a 3 mm aperture, which is one order of magnitude higher than with the 3D approach. For better comparison to Young's measurements the square root of K is more appropriate, leading to $\sqrt{K} = 0.3$ for the 5 mm and $\sqrt{K} = 0.5$ for the 3 mm aperture diameter, respectively. These values agree qualitatively quite well to results obtained from interference fringe contrast in Young's double pinhole experiments. The observed higher degree of coherence can at least in part be attributed to the focusing optics at FLASH II (high quality Kirkpatrick-Baez mirror instead of slightly corrugated ellipsoidal mirror at FLASH I). Further clarifying work is in progress.

20.4   Conclusion and Outlook

The feasibility of a compact Hartmann wavefront sensor to be employed for beam characterization of FELs and HHG sources emitting in the EUV spectral region has been demonstrated. The device accomplishes simultaneous recording of both wavefront and intensity distributions, allowing for an optimization of the beam transport by fine-tuning the focusing optics. For both FELs and HHG sources we demonstrated that Hartmann wavefront sensor assisted alignment can considerably reduce the astigmatic focal difference induced by grazing incidence mirrors and gratings. The resulting decrease of wavefront error leads to higher spot brightness resulting in an enhanced CDI performance. In case of the FEL FLASH a reduction of residual wavefront aberrations to $w_{rms} \sim \lambda/10$ could be achieved. Wavefronts and intensity profiles of single FLASH pulses were recorded, accomplishing an analysis of beam fluctuations of the SASE FEL. From these data characteristic propagational parameters of the FLASH beam were computed by applying the moments method. Fresnel-Kirchhoff integration allowed for numerical beam propagation, in particular for an analysis of profiles in the waist region which is hardly accessible for a direct measurement. Future activities will involve extension of the Hartmann approach to harder X-rays in cooperation with European XFEL/Hamburg, as well as an improved prediction of the propagation characteristics by employing a sensor with higher dynamics and spatial resolution.

In addition, it has been shown that the propagation of partially coherent radiation is successfully described by the formalism of the Wigner distribution function. In comparison to existing studies, this is achieved without the need of simplifying assumptions on the beam. It is expected that the obtained comprehensive beam characterization leads to further improvements in the field of CDI and related techniques. Inclusion of coherence information into the propagation formalism will enable a successful prediction of focal intensity distributions and spot sizes even for sources with relatively poor coherence properties.

Acknowledgements   The authors like to thank Barbara Keitel and Elke Plönjes (DESY Photon Science) for their support during the measurements at FLASH as well as for stimulating discussions.

References

1. Iso 15367: Lasers and laser-related equipment—test methods for determination of the shape of a laser beam wavefront (2003)
2. Cubalchini, R.: Modal wave-front estimation from phase derivative measurements. JOSA **69**(7), 972–977 (1979)
3. Noll, R.J.: Phase estimates from slope-type wave-front sensors. JOSA **68**(1), 139–140 (1978)
4. Mercère, P., Zeitoun, P., Idir, M., Le Pape, S., Douillet, D., Levecq, X., Dovillaire, G., Bucourt, S., Goldberg, K.A., Naulleau, P.P., et al.: Hartmann wave-front measurement at 13.4 nm with λ euv/120 accuracy. Opt. Lett. **28**(17), 1534–1536 (2003)
5. Schäfer, B., Lübbecke, M., Mann, K.: Hartmann-shack wave front measurements for real time determination of laser beam propagation parameters. Rev. Sci. Instrum. **77**(5), 053,103 (2006)
6. Neal, D.R., Alford, W.J., Gruetzner, J.K., Warren, M.E.: Amplitude and phase beam characterization using a two-dimensional wavefront sensor. In: Third International Workshop on Laser Beam and Optics Characterization, vol. 2870, pp. 72–83. International Society for Optics and Photonics (1996)
7. Schäfer, B., Mann, K.: Determination of beam parameters and coherence properties of laser radiation by use of an extended hartmann-shack wave-front sensor. Appl. opt. **41**(15), 2809–2817 (2002)
8. Iso 11146: Lasers and laser-related equipment—test methods for laser beam parameters—beam widths, divergence angle and beam propagation factor (1999)
9. Gaudin, J., Keitel, B., Jurgilaitis, A., Nüske, R., Guérin, L., Larsson, J., Mann, K., Schäfer, B., Tiedtke, K., Trapp, A., et al.: Time-resolved investigation of nanometer scale deformations induced by a high flux x-ray beam. Opt. Express **19**(16), 15516–15524 (2011)
10. Kuhlmann, M., Plönjes, E., Tiedtke, K., Toleikis, S., Zeitoun, P., Gautier, J., Lefrou, T., Douillet, D., Mercère, P., Dovillaire, S.B.G., et al.: Wave-front observations at flash. Proc. FEL **2006**, (2006)
11. Tiedtke, K., Azima, A., Von Bargen, N., Bittner, L., Bonfigt, S., Düsterer, S., Faatz, B., Frühling, U., Gensch, M., Gerth, C., et al.: The soft x-ray free-electron laser flash at desy: beamlines, diagnostics and end-stations. New J. Phys. **11**(2), 023,029 (2009)
12. Flöter, B., Juranić, P., Kapitzki, S., Keitel, B., Mann, K., Plönjes, E., Schäfer, B., Tiedtke, K.: Euv hartmann sensor for wavefront measurements at the free-electron laser in hamburg. New J. Phys. **12**(8), 083,015 (2010)
13. Flöter, B., Juranić, P., Großmann, P., Kapitzki, S., Keitel, B., Mann, K., Plönjes, E., Schäfer, B., Tiedtke, K.: Beam parameters of flash beamline bl1 from hartmann wavefront measurements. Nucl. Instrum. Methods Phys. Res. Sect. A: Accel. Spectrom. Detect. Assoc. Equip. **635**(1), S108–S112 (2011)
14. Private communication Barbara Keitel and Elke Plönjes (DESY), to be published in detail elsewhere
15. Mey, T., Zayko, S., Ropers, C., Schäfer, B., Mann, K.: Toroidal grating astigmatism of high-harmonics characterized by euv hartmann sensor. Opt. Express **23**(12), 15310–15315 (2015)
16. Zayko, S., Mönnich, E., Sivis, M., Mai, D.D., Saldtt, T., Schäfer, S., Ropers, C.: Coherent diffractive imaging beyond the projection approximation: waveguiding at extreme ultraviolet wavelengths. Opt. Express **23**(15), 19911–19921 (2015)

17. Born, M., Wolf, E.: Principles of Optics: Electromagnetic Theory of Propagation. Cambridge University Press, Interference and Diffraction of Light (2000)
18. Singer, A., Vartanyants, I., Kuhlmann, M., Duesterer, S., Treusch, R., Feldhaus, J.: Transverse-coherence properties of the free-electron-laser flash at desy. Phys. Rev. Lett. $101(25)$, 254,801 (2008)
19. Singer, A., Lorenz, U., Sorgenfrei, F., Gerasimova, N., Gulden, J., Yefanov, O., Kurta, R., Shabalin, A., Dronyak, R., Treusch, R., et al.: Hanbury brown–twiss interferometry at a free-electron laser. Phys. Rev. Lett. $111(3)$, 034,802 (2013)
20. Bastiaans, M.J.: The wigner distribution function of partially coherent light. Opt. Acta: Int. J. Opt. $28(9)$, 1215–1224 (1981)
21. Bastiaans, M.J.: Application of the wigner distribution function to partially coherent light. JOSA A $3(8)$, 1227–1238 (1986)
22. Eppich, B., Mann, G., Weber, H.: Measurement of the four-dimensional wigner distribution of paraxial light sources. In: Optical Design and Engineering II, vol. 5962, p. 59622D. International Society for Optics and Photonics (2005)
23. Tran, C., Williams, G., Roberts, A., Flewett, S., Peele, A., Paterson, D., de Jonge, M., Nugent, K.: Experimental measurement of the four-dimensional coherence function for an undulator x-ray source. Phys. Rev. Lett. $98(22)$, 224,801 (2007)
24. Schäfer, B., Mann, K.: Characterization of an arf excimer laser beam from measurements of the wigner distribution function. New J. Phys. $13(4)$, 043,013 (2011)
25. Mey, T., Schäfer, B., Mann, K., Keitel, B., Kuhlmann, M., Plönjes, E.: Wigner distribution measurements of the spatial coherence properties of the free-electron laser flash. Opt. Express $22(13)$, 16571–16584 (2014)
26. Mey, T., Schäfer, B., Mann, K.: Measurement of the wigner distribution function of non-separable laser beams employing a toroidal mirror. New J. Phys. $16(12)$, 123,042 (2014)

Chapter 21
Laboratory-Scale Soft X-ray Source for Microscopy and Absorption Spectroscopy

Matthias Müller and Klaus Mann

Abstract Soft X-ray microscopy and absorption spectroscopy are extremely useful tools for high-resolution imaging and chemical analysis of samples in various scientific fields. However, due to the required high photon flux of soft X-ray radiation, up to now, both methods are almost exclusively performed at synchrotron sources. Thus, great efforts have been made to develop table-top sources emitting in the soft X-ray spectral range ($\lambda = 1$–5 nm). Here, the development of a laser-produced plasma source from a pulsed gas jet is presented, enabling the construction of an almost debris-free, compact and long-term stable X-ray source. Based on this source a compact soft X-ray microscope (spatial resolution 50 nm) operating at a wavelength of $\lambda = 2.88$ nm was built and applied for imaging of various test and biological objects. In addition, a laboratory-scale NEXAFS spectrometer has been established, allowing for reliable analysis of different absorption edges at photon energies between 250 and 1250 eV.

PACS Subject Classification: 07.85.Fv · 07.85.Nc · 07.85.Tt · 32.30.Rj · 52.38.Ph · 52.50.Dg · 52.50.Jm. 61.10.Ht · 68.37.Yz. 78.70.Dm

21.1 Table-Top Soft X-ray Source Using a Pulsed Gas Jet

The laser-produced plasma source for the generation of intense soft X-ray radiation is based on a gas-puff target and a Nd:YAG laser system (wavelength $\lambda = 1064$ nm, pulse energy 700 mJ, pulse width 7 ns, repetition rate 5 Hz) as schematically depicted in Fig. 21.1. The radiation characteristics depend strongly on the gases used (see Fig. 21.3a, b): Photon emission spectra of low atomic number gases (e.g. oxygen and nitrogen) consist of individual, isolated lines, whereas those of gases

M. Müller · K. Mann (✉)
Laser-Laboratorium Göttingen e.V., Hans-Adolf-Krebs-Weg 1, 37077 Göttingen, Germany
e-mail: kmann@llg-ev.de

M. Müller
e-mail: matthias.mueller@llg-ev.de

© The Author(s) 2020
T. Salditt et al. (eds.), *Nanoscale Photonic Imaging*, Topics in Applied Physics 134,
https://doi.org/10.1007/978-3-030-34413-9_21

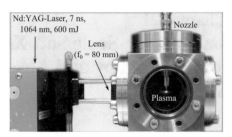

Fig. 21.1 Schematic drawing (left) and photograph (right) of the table-top laser-produced soft X-ray plasma source at Laser-Laboratorium Göttingen

with a high atomic number (argon, krypton, xenon) are quasi-continuous due to the higher number of possible electronic transitions.

The pulsed gas jet is created by a conical nozzle (diameters $d_1 \approx 550\ \mu$m, $d_2 \approx 300\ \mu$m, cone half angle $7°$) behind a fast valve. The latter is based on the Proch-Trickl setup [1], consisting of a piezo disk translator to generate short gas pulses ($t_{open} = 900\ \mu$s), allowing for a background pressure of about $5 \cdot 10^{-3}$ mbar during operation (gas pressure $p = 10$–20 bar).

As compared to alternative laser-produced plasma sources employing solid or liquid targets, the gas jet based source offers several advantages. In particular, these are:

- Low debris generation, i.e. cleanliness,
- Continuous supply of target material,
- Long-term stability.

However, due to the reduced particle density of the gas jet, the peak brilliance is definitely smaller as the plasma size increases to several hundreds of micrometer. Progress in the enhancement of gas density has been achieved by using cluster beam targets [2] or double-stream gas puff targets [3]. Another successful approach is the generation of a so-called "barrel shock", applying a small background pressure p_b to the supersonic flow emanating from the nozzle ([4], see Fig. 21.2a). On passing this barrel shock system, particles become locally concentrated, forming high-density regions. Focusing the laser beam into the high-density region behind one of these shocks, a higher number of gas atoms can be ionized, resulting in a brighter and smaller plasma (see Fig. 21.2b). The peak brilliance of the source is increased by one order of magnitude to $3.15 \cdot 10^{16}$ photons/(mm^2 mrad s) for the isolated nitrogen line at $\lambda = 2.88$ nm.

The emission characteristics of a soft X-ray plasma are not only affected by the target gas and its particle density, but also by the laser parameters. The influence of the laser pulse length was investigated, employing two lasers of about the same pulse energy (500 mJ, $\lambda = 1064$ nm), but with different pulse durations of 7 ns and 170 ps [5]. As seen from the compilation in Fig. 21.3, the spectral characteristics of the plasma emission are clearly affected by the pulse duration (i.e. power density): The

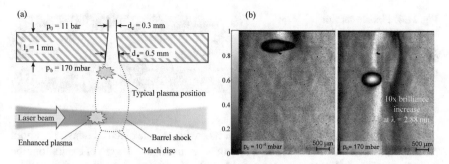

Fig. 21.2 a Schematic representation of the plasma generation in a "barrel shock": The local confinement of target gas atoms by an X-ray transparent background gas (He, $p_b > 10$ mbar) allows for a plasma ignition at a greater distance from the nozzle as compared to the standard expansion into vacuum ($p_b < 10^{-4}$ mbar). **b** Pinhole camera images indicating plasma intensities at $\lambda = 2.88$ nm, superimposed on Schlieren images of the gas jet under vacuum conditions and in an ambient He atmosphere, respectively [4]

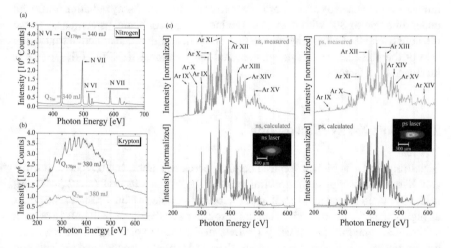

Fig. 21.3 Influence of laser pulse duration on emission spectra of **a** nitrogen and **b** krypton from both ns (red) and ps (blue) laser plasmas (target gas pressure 10 bar, 200 nm Al-filtered, average of 100 pulses). **c** Comparison of measured and calculated argon emission spectra for ns and ps laser and corresponding pinhole camera images [5]

brightness is strongly enhanced, for some emission lines by more than a factor of 10. The picosecond spectra are shifted considerably towards shorter wavelengths since the higher power density results in a higher degree of ionization.

By comparing the spectra with model calculations using a magneto-hydrodynamic code electron temperatures and densities were obtained, indicating the maximum achievable degree of ionization of the plasma. As an example, Fig. 21.3c shows measured and calculated emission spectra of the argon plasma for both ns and ps laser excitation. The average electron temperature resulting from the simulation is 66

Fig. 21.4 Pinhole camera
images of krypton plasma
(average 100 pulses) taken at
90° and 30° to the incident
laser beam

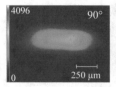
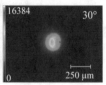

eV for the ps plasma (33% larger than for ns), whereas the electron density is about
3 times higher ($22.4 \cdot 10^{19}$ eV/cm³). Along with the higher degree of excitation, the
ps laser-produced plasmas are also considerably smaller, as could be monitored with
a soft X-ray sensitive pinhole camera (see Fig. 21.3c inset).

Recent investigations on the angular emission characteristics of the laser-produced
plasma have demonstrated that the emission of soft X-rays in backward direction is
strongly favored. This can be explained by a shorter path length and thus, a reduced
reabsorption of the radiation by the gas jet in direction of the laser beam. The plasma
intensity is enhanced by a factor of 1.6 when the emitted soft X-ray radiation is utilized
under an angle of 30° with respect to the incoming laser beam, as compared to the
previously chosen 90° geometry (see Fig. 21.4). In addition, the effective irradiation
area of the plasma becomes 4x smaller, and the positional stability of the source is
increased by a factor of 5.

21.2 Soft X-ray Microscopy

Benefiting from the high absorption contrast between carbon and oxygen, transmission X-ray microscopy in the spectral range of the "water window" ($\lambda = 2.3$–4.4 nm)
is ideally suited for the investigation of biological samples. Using Fresnel zone plates
as highly magnifying objectives spatial resolutions in the range of 10 nm have been
achieved at synchrotron sources [6]. However, in order to pave the way for a wider
dissemination of soft X-ray microscopy, there is definitely also the need for table-top
systems. Although considerable progress has already been achieved in this field [7–
10], a further compaction and simplification of these microscopes is still necessary.

Making use of the long-term stable and nearly debris-free laser-produced plasma
from a pulsed nitrogen gas jet target, an extremely compact soft X-ray microscope
operating in the "water window" region at the wavelength $\lambda = 2.88$ nm was installed
and tested [11, 12]. The setup of this table-top soft X-ray microscope is depicted
in Fig. 21.5a. It consists basically of the laser-produced plasma source described in
Sect. 21.1, an ellipsoidal condenser mirror, a Fresnel zone plate objective fabricated
by standard lithographic techniques, and a back-illuminated CCD camera sensitive
for soft X-ray radiation, all integrated into a vacuum system with a base pressure of
10^{-6} mbar.

The soft X-ray radiation emitted from the nitrogen plasma is filtered by a titanium
(Ti) filter to block out-of-band radiation, ensuring monochromatic irradiation of

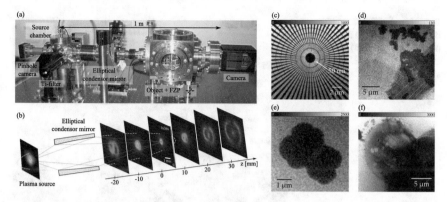

Fig. 21.5 **a** Photograph of the table-top soft X-ray microscope. **b** Spatial intensity profiles of soft X-ray radiation at $\lambda = 2.88$ nm for different positions along the optical axis behind the ellipsoidal condenser mirror; the minimum spot diameter at $z = 0$ mm (object plane) is about 400 μm (FWHM). **c–f** Soft X-ray micrographs of **c** Siemens star, **d** geo-colloids, **e** bacterium *Deinococcus Radiodurans* (DSM no. 20539) and **f** alga *Trachelomonas oblonga* (SAG 1283-11). Magnification and exposure time vary from 175 to 500 and 5 min to 60 min, respectively. **a, e, f** reproduced from [12], with permission of AIP Publishing. **d** reprinted with permission from [11], Optical Society of America

the sample at $\lambda = 2.88$ nm wavelength. The adjustment of the grazing incidence condenser mirror was optimized by the measurement of intensity profiles at various positions along the optical axis. Figure 21.5b displays corresponding distributions for different camera positions. A Gaussian-like spatial profile with a diameter of about 400 μm (FWHM) is measured at the focal plane, representing the object plane of the soft X-ray microscope.

In order to assess the performance of the microscope a Siemens star test pattern was imaged. Figure 21.5c shows a corresponding micrograph, indicating an almost uniform illumination over a field of view of about 10 μm. Structures with a size of about 50 nm are resolved in all directions [12]. Furthermore, various biological and geological samples were investigated (see Fig. 21.5d, e). The characteristic shape of these objects is clearly visible, but almost no internal structure is apparent due to the thickness of the samples. The signal-to-noise ratio of the micrographs is rather low due to the relatively low brilliance of the plasma. However, the presented system offers various opportunities for scalability of the photon flux (see Sect. 21.1), thereby maintaining the compactness of the microscope and especially the inherent cleanliness of the soft X-ray source.

21.3 X-ray Absorption Spectroscopy

Another prominent application of soft X-ray radiation is absorption spectroscopy, probing the fine structure of X-ray absorption edges (Near Edge X-ray Absorption Fine Structure—NEXAFS), by exciting core level electrons to higher lying unoccu-

Fig. 21.6 Schematic representation (top) and photograph (bottom) of the laboratory-scale NEXAFS spectrometer

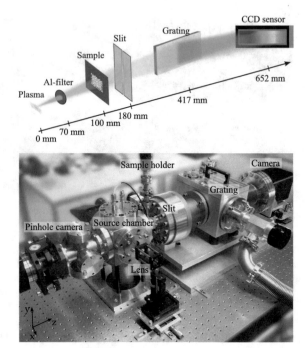

pied states. As the energy levels of both initial and final states depend on the involved molecular bonds and their chemical environment, the spectral features of the near-edge fine structure represent a "molecular fingerprint" of the sample. Thus, NEXAFS spectroscopy is a very common analytical method for compositional surface analysis, representing nowadays one of the most important applications of synchrotron radiation [13]. In contrast to soft X-ray microscopy, however, it has been conducted only a few times with laboratory-scale sources until now [14–17], although almost identical results as with radiation from storage rings are achievable.

At the Laser-Laboratorium Göttingen (LLG) a compact spectrometer based on the laser-produced soft X-ray plasma source was developed and utilized for NEXAFS measurements. Figure 21.6 shows the current setup of the table-top spectrometer, which makes use of a concave flat-field grating and a soft X-ray sensitive CCD camera. At synchrotrons absorption spectroscopy is conducted mainly by recording the total electron yield using monochromatic radiation for excitation. In contrast, the broadband radiation emitted from lab-scale sources allows for a polychromatic spectroscopic approach.

Up to now, a considerable number of NEXAFS investigations has been performed at the soft X-ray absorption edges of various elements. A brief overview of the spectra measured so far at the carbon K-edge is given in Fig. 21.7a, showing polyimide [16], humic substances and the organic matter of a Luvisol [18]. Differences in spectral features point out varying influences of e.g. aromatic, phenolic, and carbonyl functional groups.

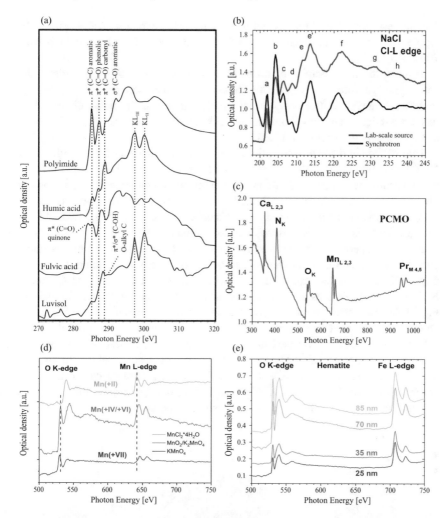

Fig. 21.7 NEXAFS spectra acquired at various absorption edges with the compact spectrometer at LLG: **a** C K-edge spectra of polyimide, humic acid, fulvic acid and a luvisol bulksoil [19], **b** Cl L-edge spectrum of NaCl [20] indicating EXAFS oscillations, in comparison with corresponding synchrotron results [21], **c** overview spectrum of PCMO [22], **d** O K-edge and Mn L-edge spectra of various Mn compounds [23] and **e** O K-edge and Fe L-edge spectra of hematite samples of different thickness

Furthermore, NEXAFS studies were conducted at the absorption edges of Si, S, Cl (see Fig. 21.7b, [20]), Ca [24], O [22, 23], and, for the first time with a laboratory-scale setup, at the L-edges of Mn, Fe and Cu [22, 23, 25] at photon energies >500 eV (Fig. 21.7c–e). In all cases, there is an excellent agreement of spectral features compared to spectra obtained with synchrotron radiation. The energy resolution of the spectrometer is about $E/\Delta E \approx 450$ at a photon energy of 430 eV. However, a higher spectral resolution is necessary, e.g. for the analysis of Fe oxides to resolve the underlying splitting of the pre-edge features around 530 eV and the $L_{3,2}$ peaks between 700 and 725 eV.

Besides the investigation of different absorption edges, the soft X-ray source was applied for time-resolved NEXAFS measurements in Perovskit-type manganite $Pr_{0.7}Ca_{0.3}MnO_3$ (PCMO). Pump-probe experiments reveal diminutive changes of the oxygen K-edge, stemming from an optically induced phase transition [22], which compare nicely to synchrotron data [26].

Due to the short mean free path of soft X-rays in air (only a few millimeters), NEXAFS experiments with the table-top spectrometer have so far been performed in a high vacuum system, excluding a number of interesting samples from spectroscopic investigations. To overcome this limitation, a new sample chamber was constructed using two silicon nitride membranes as vacuum windows (see Fig. 21.8a and b) to measure samples also under atmospheric conditions.

NEXAFS spectra have been recorded under different conditions in this sample chamber, i.e. vacuum, air and helium, respectively, on e.g. the calcium L-edge and oxygen K-edge of crystalline calcium chloride tetra- or hexahydrate [25]. Obviously, NEXAFS measurements are feasible for all conditions at the calcium L-edge (see Fig. 21.8c). However, the oxygen signal is absent in vacuum and air (see Fig. 21.8d). During measurement in vacuum, the crystalline structure is changing along with the decreasing atmospheric pressure from a tetra- or hexahydrate into a water-free distorted "rutile-structure" [27], thus no traces of water are detected. In air, these hydrate structures are still present, but could not be measured due to the strong absorption above the nitrogen K-edge. Therefore, helium purging turns out to be essential for the proper investigation of sensitive samples.

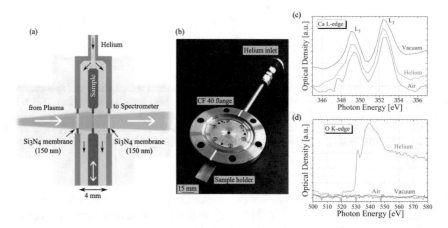

Fig. 21.8 **a** Drawing and **b** photograph of the helium purged sample chamber. **c** Ca L-edge and **d** O K-edge spectra of $CaCl_2 \cdot H_2O$ recorded for different conditions. Reproduced from [25], with the permission of the American Vacuum Society

21.4   Outlook

At present state, NEXAFS studies with the table-top spectrometer are conducted in transmission mode, requiring very thin samples (≈ 100 nm). Experiments in reflection geometry would also allow for investigation of thick samples, which cannot be prepared for measurements in transmission. Additionally, taking advantage of the short penetration depth of soft X-rays under grazing incidence (<30 nm), measurements in reflection would yield highly surface sensitive information. Thus, the setup shall be modified to accomplish NEXAFS experiments both in transmission and reflection mode.

The soft X-ray microscope shall be extended to a STED integrated X-ray nanoscope. Combining both methods allows for comprehensive studies of biological objects due to the complementary contrast of both imaging techniques. Thus, in addition to pure imaging of the sample also information about its chemical composition can be gained without transferring the sample between different setups.

Moreover, the table-top EUV source can be used for metrological applications at the microlithographic wavelength $\lambda = 13.5$ nm, e.g. for material testing or actinic sensor qualification.

References

1. Proch, D., Trickl, T.: A high-intensity multi-purpose piezoelectric pulsed molecular beam source. Rev. Sci. Instrum. **60**(4), 713–716 (1989)
2. Kubiak, G., Richardson, M.: US Patent 5,577,092 (1996)
3. Fiedorowicz, H., Bartnik, A., Daido, H., Choi, I., Suzuki, M., Yamagami, S.: Strong extreme ultraviolet emission from a double-stream xenon/helium gas puff target irradiated with a Nd:YAG laser. Opt. Commun. **184**(1–4), 161–167 (2000)
4. Mey, T., Rein, M., Großmann, P., Mann, K.: Brilliance improvement of laser-produced soft x-ray plasma by a barrel shock. New J. Phys. **14**(7), 073045 (2012)
5. Müller, M., Kühl, F.-C., Großmann, P., Vrba, P., Mann, K.: Emission properties of ns and ps laser-induced soft x-ray sources using pulsed gas jets. Opt. Express **21**(10), 12831 (2013)
6. Chao, W., Fischer, P., Tyliszczak, T., Rekawa, S., Anderson, E., Naulleua, P.: Real space soft x-ray imaging at 10 nm spatial resolution. Opt. Express **20**(9), 9777–9783 (2012)
7. Berglund, M., Rymell, L., Peuker, M., Wilhein, T., Hertz, H.: Compact water-window transmission x-ray microscopy. J. Microsc. **197**(3), 268–273 (2000)
8. Jansson, P., Vogt, U., Hertz, H.: Liquid-nitrogen-jet laser-plasma source for compact soft x-ray microscopy. Rev. Sci. Instrum. **76**, 043503 (2005)
9. Legall, H., Blobel, G., Stiel, H., Sandner, W., Seim, C., Takman, P., Martz, D., Selin, M., Vogt, U., Hertz, H., Esser, D., Sipma, H., Luttmann, J., Höfer, M., Hoffmann, H., Yulin, S., Feigl, T., Rehbein, S., Guttmann, P., Schneider, G., Wiesemann, U., Wirtz, M., Diete, W.: Compact x-ray microscope for the water window based on a high brightness laser plasma source. Opt. Express **20**(16), 18362–18369 (2012)

10. Takman, P.A., Stollberg, H., Johansson, G.A., Holmberg, A., Lindblom, M., Hertz, H.M.: High-resolution compact x-ray microscopy. J. Microsc. **226**(2), 175–181 (2007)
11. Müller, M., Mey, T., Niemeyer, J., Mann, K.: Table-top soft x-ray microscope using laser-induced plasma from a pulsed gas jet. Opt. Express **22**(19), 23489 (2014)
12. Müller, M., Mey, T., Niemeyer, J., Lorenz, M., Mann, K.: Table-top soft x-ray microscopy with a laser-induced plasma source based on a pulsed gas-jet. AIP Conf. Proc. **1764**, 030003 (2016)
13. Stöhr, J.: NEXAFS spectroscopy. Springer, Berlin (2003)
14. Przemyslaw, W., Duda, M., Bartnik, A., Sarzynski, A., Wegrzynski, L., Nowak, M., Jancarek, A., Fiedorowicz, H.: Compact system for near edge X-ray fine structure (NEXAFS) spectroscopy using a laser-plasma light source. Opt. Express **26**(7), 8260–8274 (2018)
15. Mantouvalou, I., Witte, K., Grötzsch, D., Neitzel, M., Günther, S., Baumann, J., Jung, R., Stiel, H., Kanngießer, B., Sandner, W.: High average power, highly brilliant laser-produced plasma source for soft X-ray spectroscopy. Rev. Sci. Instrum. **86**, 035116 (2015)
16. Peth, C., Barkusky, F., Mann, K.: Near-edge x-ray absorption fine structure measurements using a laboratory-scale XUV source. J. Phys. D Appl. Phys. **41**(10), 105202 (2008)
17. Vogt, U., Wilhein, T., Stiel, H., Legall, H.: High resolution x-ray absorption spectroscopy using a laser plasma radiation source. Rev. Sci. Instrum. **75**, 4606 (2004)
18. Sedlmair, J., Gleber, S.-C., Peth, C., Mann, K., Niemeyer, J., Thieme, J.: Characterization of refractory organic substances by NEXAFS using a compact X-ray source. J. Soils Sediments **12**, 24–34 (2012)
19. Sedlmair, J.: Soft x-ray spectromicroscopy of environmental and biological samples. Ph.D. thesis, Universität Göttingen (2011)
20. Olschewski, M.: Aufbau und Charakterisierung eines Absorptions- und Reflektionsspektrometers für den EUV-Bereich unter Verwendung eines laserinduzierten Plasmas. Master's thesis, TU Clausthal (2012)
21. Kasrai, M., Fleet, M., Bancroft, G., Tan, K., Chen, J.: X-ray-absorption near-edge structure of alkali halides: the interatomic-distance correlation. Phys. Rev. B **43**(2), 1763–1772 (1991)
22. Großmann, P., Rajkovic, I., More, R., Norpoth, J., Techert, S., Jooß, C., Mann, K.: Time resolved near-edge X-ray absorption fine structure spectroscopy on photo-induced phase transitions using a tabletop soft-X-ray spectrometer. Rev. Sci. Instrum. **83**, 053110 (2012)
23. Schellhorn, M.: Präparation von Probensystemen pedochemisch relevanter Metallverbindungen zur Untersuchung durch ein tabletop Röntgen-Nahkanten Spektrometer. Master's thesis, Universität Hohenheim (2015)
24. Kühl, F.-C.: Nah-Kanten-Absorptionsspektroskopie im weichen Röntgenbereich bei Atmosphärendruck. Bachelor's thesis, HAWK Göttingen (2013)
25. Kühl, F.-C., Müller, M., Schellhorn, M., Mann, K., Wieneke, S., Eusterhues, K.: Near edge X-ray absorption fine structure spectroscopy at atmospheric pressure with a table-top laser-induced soft X-ray source. J. Vac. Sci. Technol. A **34**, 041302 (2016)
26. Rini, M., Zhu, Y., Wall, S., Tobey, R.-I., Ehrke, H., Garl, T., Freeland, J.-W., Tomioka, Y., Tokura, Y., Cavalleri, A., Schoenlein, R.-W.: Transient electronic structure of the photoinduced phase of $Pr_{0.7}Ca_{0.3}MnO_3$ probed with soft X-ray pulses. Phys. Rev. B **80**, 155113 (2009)
27. Galasso, F.S.: Structure and properties of inorganic solids. In: Kurti, N., Smoluchowski, R. (eds.) International Series of Monographs in Solid State Physics. Elsevier, Amsterdam (2013)

Chapter 22
Multilayer Zone Plates for Hard X-ray Imaging

Markus Osterhoff and Hans-Ulrich Krebs

Abstract This chapter reviews progress both in the fabrication of multilayer zone plate optics for focusing X-rays, as well as in imaging experiments using these optics. The fabrication based on pulsed laser deposition is accompanied by analytical and numerical treatment of X-ray propagation to control volume diffraction effects. On the imaging side, different schemes are presented; these include scanning-scattering with focused X-rays, holography, as well as recent advances in lens-enhanced phase-reconstruction.

22.1 From Focusing to Imaging

In this chapter we review progress on hard X-ray imaging methods using Multilayer Zone Plate (MZP) optics. First, we have developed Multilayer Laue Lenses (MLLs), as a means to one-dimensionally focus soft X-rays produced at table-top laser-driven plasma sources, see also Chap. 21. Focusing experiments using a depth graded MLL consisting of Ti/ZrO$_2$ layer were performed with the table-top soft X-ray source at a wavelength of 2.88 nm, achieving a focal spot size of 280 nm [1, 2].

Then, first focusing experiments using MLLs and MZPs have demonstrated that two-dimensional focusing of hard X-rays (7.9 and 13.8 keV) is possible at 3rd generation synchrotron radiation sources. For technical reasons, only optics with rather small aperture sizes have been fabricated in Göttingen. To compensate, pre-focusing optics (Kirkpatrick-Baez mirrors, KBs, or Compound Refractive Lenses, CRLs) were used to increase the flux density on the zone plates. In the first experiment, an MLL with a height of 401 nm was illuminated by a pre-focused KB mirror beam. From

M. Osterhoff (✉)
Institute for X-ray Physics, Universität Göttingen, Friedrich-Hund-Platz 1,
37077 Göttingen, Germany
e-mail: mosterh1@gwdg.de

H.-U. Krebs
Institute for Material Physics, Universität Göttingen, Friedrich-Hund-Platz 1,
37077 Göttingen, Germany

© The Author(s) 2020   561
T. Salditt et al. (eds.), *Nanoscale Photonic Imaging*, Topics in Applied Physics 134,
https://doi.org/10.1007/978-3-030-34413-9_22

(a) reconstructed focus **(b)** reconstructed intensity profile in focal plane

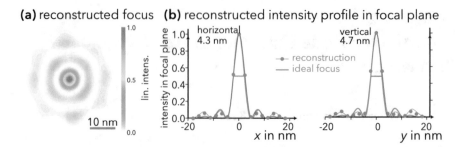

Fig. 22.1 Reconstructed intensity profile of an MZP focus: **a** two-dimensional rendering on linear colour scale, **b** one-dimensional intensity cuts with Gaussian fit. The full width at half maximum is determined to 4.3 nm × 4.7 nm. Adapted from [4]

far-field measurements, a focal spot size of 6.8 nm (FWHM) was reconstructed at 13.8 keV photon energy [3].

Later, a two-dimensional hard X-ray spot was achieved. While two MLL optics were successfully crossed, the fabrication of a round MZP optic succeeded, too: A depth-graded circular W/Si multilayer was deposited onto a rotating W wire. In contrast to lithographic fabrication, virtually unlimited aspect ratios of outer-most zone width (here: down to 5 nm) and optical thickness (here: length of straight sections of the wire, usually several hundred micrometres) are possible. In a first experiment— again with the nano-focusing optic placed in the KB beam—an unprecedented focal spot size of 4.3 nm × 4.7 nm was reconstructed (see Fig. 22.1) [4].

Backed by these achievements, the fabrication of round MZPs was improved significantly. In the following, we will concentrate on research dedicated to further optimise the lenses, and apply them for new hard X-ray imaging schemes [5]. First, we will cover the advanced design and improved fabrication; then we will review different experiments performed at synchrotron setups, and present results obtained during those imaging experiments.

Preliminary imaging experiments could be performed, see Fig. 22.2a for a scanning-SAXS measurement on a Siemens star [5]. In Fig. 22.2b, c, multi-order images of semiconductor nanowires are shown. Scanning experiments without an Order Sorting Aperture yield information both from the holographic −1st order and the focused +1st order. Figure 22.2b shows an overview hologram of nanowires; the information in the centre of the image is missing due to a beamstop. But a high-resolution scanning-SAXS measurement in differential phase contrast mode of a nanowire on an electrical contact is shown in Fig. 22.2c. Both images are extracted from the same dataset.

For a detailed description of Fresnel and Multilayer Zone Plates, see Chap. 3.

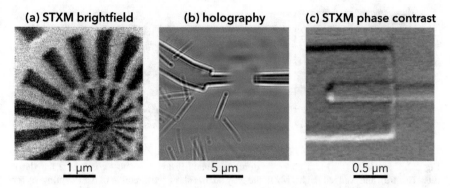

(a) STXM brightfield   **(b) holography**   **(c) STXM phase contrast**

1 µm   5 µm   0.5 µm

Fig. 22.2 a First MZP-based STXM brightfield scan of a Siemens star test pattern, adapted from [5]. The smallest features of 50 nm could not be resolved due to vibrations of the early experimental setups. **b** Holographic image obtained the diverging −1st order, after overlaying of shifted sample positions. The sample consists of semiconductor nano-wires [6] and electrical contacts. The centre is not visible in this image due to a beamstop; **c** but in scanning-SAXS mode, here differential phase contrast. The image has been extracted using a "software-OSA" from the same measurement as the hologram, and shows a single nano-wire lying on an electrical contact. The tip of the nano-wire consists of a Au sphere

22.2 Let There be an Ideal World

In this section we show simulation results for "perfect" zone plates to assess the best case scenario and to devise new measurement techniques.

Optically thin Fresnel Zone Plates (FZPs) are constructed via the Zone Plate Law, cf. Sect. 3.5. To model thin FZPs, an incoming wave-field can be multiplied pixel-wise with a complex-valued transmission function. For thick Multilayer Zone Plates, however, this approximation cannot be met. Usually, wave-optical propagation modelling through the structure has to be applied, if

$$F := \frac{(\Delta r_N)^2}{\lambda t} \lesssim 1, \tag{22.1}$$

with outer-most zone thickness Δr_N, wavelength λ, and optical thickness t. For $\Delta r_N = 5$ nm and $\lambda = 1$ Å, the geometrical approximation becomes invalid for $t \gtrsim 250$ nm.

The "imprint" of an infinitely thin zone plate on an impinging wave field $\psi(x, y)$ is usually described by a complex-valued transmission function $\tau(x, y)$; the outgoing field can then be propagated to e.g. the focal plane using the Fresnel-Kirchoff integral. For longitudinally extended objects, multi-slice approaches with various propagators are used; these alternate between numerical free-space propagation and transmission functions for "short" propagation distances with $F \gg 1$ instead of (22.1). In the general case, $\tau(x, y, z)$ can also change along the optical axis, so volume zone plates

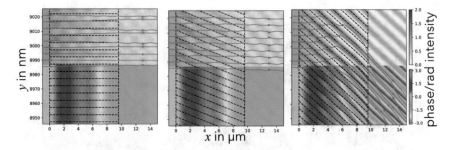

Fig. 22.3 Intensity (top) and phase (bottom) simulated inside outermost zones of a flat (left), tilted (centre), and wedged (right) MZP; the grey arrows show the direction of the Poynting vector field, i.e. the "flow" of intensity. In the flat geometry, a "beating mode structure" known from array waveguides emerges after only 1 μm, and the Poynting vectors are almost unaltered in direction, and also the phase fronts are almost perpendicular to the incoming beam direction; hence, the focusing efficiency is very limited. In the tilted case, the impinging field is coupled into the channels for about 2 μm, and a focusing effect becomes visible in the outgoing intensity pattern. But only for the wedged case also the Poynting vectors align in the direction of the desired focal spot

with tapering can be described. Here, we have mostly made use of the paraxial wave equation described in Chap. 2.

Such tapering becomes important at higher X-ray energies (i.e., long devices) and smaller zones (i.e., smaller focus). This is shown in the simulation shown in Fig. 22.3, where three layouts of MZPs are compared. The three columns show the intensity (top) and phase (bottom) for a flat MZP (left), a tilted MZP (centre) and a wedged MZP (right). Dashed lines show the nano-focusing zones on a scale of 5 nm, over an optical thickness of about 9.8 μm.

In both flat and tilted case, the phase fronts are almost flat and perpendicular to the optical axis. Also, a "mode beating" effect known from array waveguides can be seen in the intensity patterns. In the flat case, this beating starts at around 1 μm, while for the tilted case the incoming wave is coupled for about 2 μm into the "channels". In the flat case, the outgoing wave field reproduces the "checker board" pattern; the tilted geometry, on the other hand, shows stripes directed towards the focus, promising enhanced efficiency.

The grey arrows show the local direction of the Poynting vector field. In both cases, this field points almost parallel to the optical axis. Since the Poynting vectors locally measures the energy flow, this shows that the focusing efficiency into the +1st order is rather small.

For the wedged case, on the other hand, the Poynting vectors are bent towards the focal point.

22.3 Back to the Real World: Fabrication Challenges

Now we will leave the neat and clean simulations to "dive into the real world".

In this section we briefly review a particular fabrication technique for Multilayer Zone Plates. Firstly, by Pulsed Laser Deposition (PLD), a rotating glass fibre is coated with alternating layers; secondly, the final MZP is sliced out using a Focused Ion Beam (FIB) device.

First, a depth graded multilayer was grown on a wire according to the Zone Plate Law by PLD. Then, the MZP was fabricated by cutting a slice out of the multilayer with desired optical depth by FIB. The MZP was positioned onto a W tip, which can be used as sample holder during hard X-ray focusing. For the first focusing experiments, the MZP was illuminated by a pre-focused X-ray beam at the coherence beamline P10 of the PETRA III synchrotron. Later on, a sample is either put into the MZP focus (scanning-SAXS/scanning-WAXS) or into the MZP defocus (holography) for imaging experiments.

22.3.1 Pulsed Laser Deposition

As a thin film sputtering technique, PLD allows the deposition of multilayer structures with reliable layer thicknesses just in the range suitable for hard X-ray focusing optics. One advantage over other sputtering techniques originates in the high energetic bombardment: New particles enter the structure and increase the mobility. In the end, this "self-healing" property decreases roughness and tends towards cylindrical layers even on a slightly elliptically wire.

The thin film deposition was realized by a computer controlled KrF excimer laser (wavelength of 248 nm, pulse duration of 30 ns, repetition rate of 10 Hz). The laser beam was focused onto the different targets in ultrahigh vacuum of about 10^{-8} mbar. The targets were moved constantly following an algorithm that allows uniform ablation from different directions. The films were grown at room temperature at a target-to-substrate distance of 65 mm. By both changing the distance of the focusing lens to the target and the laser energy, the laser fluence was controlled in a range of $1-5$ J/cm^2 [7].

Different material combinations have been tested during this Collaborative Research Centre. We will present the results of W/Si, W/ZrO$_2$, and the final combination Ta$_2$O$_5$/ZrO$_2$ further [8–10].

22.3.2 FIB Processing

Within the FIB-facility (Nova NanoLab600, FEI), for further protection an additional layer of Pt is deposited onto the multilayer by electron beam deposition. Then, the

specimen is cut by a 30 kV Ga^+-ion beam (using a current of about 1 nA) close
to the region, where the wire had its desired diameter. A piece of the coated wire
is attached to a thin W-micromanipulator with electron and ion-beam deposited Pt
and transferred onto the final lens holder, e.g. a W tip prepared beforehand. Afterwards, the micromanipulator gets cut off and is drawn back, leaving the coated
wire on the tip. Finally, the lens is shaped and polished by less energetic Ga^+-ions
(5 kV, ca. 30 pA) down to the desired optical thickness, usually around 6 μm; for the
high-energy experiment, an MZP of 30 μm optical thickness has been prepared.

22.3.3   From MLL to MZP

For the table-top soft X-ray microscope developed in project C04, one-dimensional
Multilayer Laue Lenses (MLLs) have been manufactured. Other than round MZPs,
these consist of parallel films on a flat substrate. It was found that PLD also works on
curved substrates, and hence two-dimensionally focusing devices can be manufactured. To this end, a rotation motor was incorporated into the PLD vacuum chamber;
the alternating layers are then deposited onto a thin wire while the latter rotates
around its axis.

Geometrically, the layer thickness is reduced by a factor of π for the same number of pulses. Experimentally, however, transfer factors of $TF \approx 3.8$ (Ta_2O_5) and
$TF \approx 3.3$ (ZrO_2) were found. The deviation from the geometrical $TF_{geo} = \pi$ can
be explained by resputtering and reflection during the deposition on tilted substrates.
From experimental data it is found that the dependance of the deposition rate on the
angle φ to the substrate varies more strongly than the expected $\cos \varphi$. This behaviour
could be reproduced using SDTrimSP simulations [8].

Note that due to aging of the target materials the deposition rates also change in
time. This can be compensated by changing the number of pulses accordingly.

Glass wires for the MZP were prepared in a heat-and-pull process using standard
glass fibres [9, 11]. The core wire of the Ta_2O_5/ZrO_2 MZP was prepared using a Sutter
Instruments flaming/brown micropipette puller system P1000. With this technique
it has become possible to prepare round glass fibres of suitable diameters between
1 μm and 2 μm. Within our parameters, this corresponds to the first about ten zones
of the Zone Plate Law that are missing. In addition, it is possible to draw tapered
fibres with an opening angle of few degree, see Fig. 22.4a.

22.3.4   Material and Parameter Studies

During this Collaborative Research Centre, different material combinations (W/Si,
Ta_2O_5/Si, and Ta_2O_5/ZrO_2) have been tested, and the PLD parameters (mainly the
laser energy, or flux density) have been optimised.

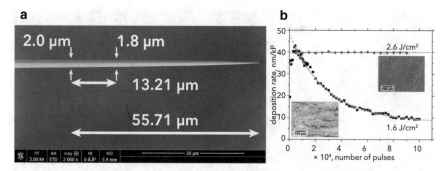

Fig. 22.4 **a** SEM image of a tapered glass fibre, prior to coating: both tapering angle (here about 0.9°) and diameter (here around 2 μm can be tweaked during the pulling process. **b** In situ deposition rate measurements during target aging for Ta$_2$O$_5$. For a laser fluence of 1.6 J/cm^2 (black), plate like structures are formed on the target surface leading to a strong decrease of the deposition rate at higher pulse numbers. In contrast, the deposition rate becomes extraordinary stable at 2.6 J/cm^2 (blue). The SEM pictures show the corresponding target morphologies after 10^5 pulses. From [8]

When calculating the number of pulses necessary to fulfil the Zone Plate Law, the deposition rate has to be known. The rate depends on the material and laser energy; it was also found that during the deposition process, the rate changes on short and long time scales due to target aging. Deposition rates have been obtained from reflectivity measurements on coated flat substrates. The time-dependent rates are then used by the control software to calculate the number of pulses for the individual layers.

Especially the W/Si system suffered from strongly varying deposition rates. With increasing number of laser pulses, in both cases the deposition rate first strongly rises and then monotonously decreases again. The very early stages of target aging depend on the initial surface topography. On longer time scales, the deposition rate decreases exponentially with the number of pulses; this is attributed to further surface roughening (for W) and to cone formation (for Si). Also, the W/Si system suffers from some limitations due to droplet formation during ablation of Si. The laser fluence was tuned to 1.7 J/cm^2 to reduce the number of droplets; still about 400 droplets per mm^2 are found on a 10 nm layer.

To circumvent droplet formation, Si was replaced by ZrO$_2$, which also offers a large phase shift for X-rays when combined with W. In contrast to Si, the surface of ZrO$_2$ targets remains relatively smooth at a laser fluence of 1.8 J/cm^2, and droplet formation is almost completely avoided. Furthermore, the deposition rate of ZrO$_2$ is about five times higher than for Si (about 45 nm per 10^3 pulses) and remains much more stable. In contrast to W/Si, no enhanced resputtering is observed at the interfaces of planar W/ZrO$_2$ multilayers by in situ rate measurements, even at increased laser fluence on the ZrO$_2$ target.

To increase the deposition rates, W was replaced by Ta$_2$O$_5$; it was found that for a laser fluence of 2.6 J/cm^2, the deposition rate becomes extraordinary stable and gives ideal conditions for the deposition of both thin and thick films. The development of the Ta$_2$O$_5$ deposition rates for two fluences is shown in Fig. 22.4b.

Fig. 22.5 SEM images (top) and TEM images (bottom) of "D13", an MZP with outermost zone width of 5 nm and a diameter of 16 μm. The optical thickness (top left) of 7 μm is designed for an X-ray energy of 14 keV. Adapted from [9]

Since the deposition rates of both materials were sufficiently high and did not change significantly over time, multilayers with precise layer thicknesses and larger overall thickness of 1.2 μm could be deposited onto the glass wire, while almost no droplet formation occurred at the same time.

Recently, MZPs with a diameter of 16 μm, outer-most zones down to 5 nm, and optical thicknesses of up to 7 μm have been built and successfully used in X-ray experiments; see Fig. 22.5 for SEM and TEM images.

22.3.5   Summary

Due to high and almost constant deposition rates, minimization of resputtering, sharp multilayer interfaces, and low transformation factors, Ta_2O_5/ZrO_2 is a very promising multilayer system for the fabrication of high quality lenses for hard X-rays by the combination of PLD and FIB.

22.4   The World of Synchrotron Instrumentation

Here we briefly describe the basic setup used at synchrotron radiation facilities for the imaging experiments presented later on. Most measurements were carried out using the GINIX end-station of the P10 beamline at the PETRA III synchrotron (DESY, Hamburg). The high-energy experiment at 60 keV up to above 100 keV was performed at the ID31 beamline of the European Synchrotron Radiation Facility (ESRF; Grenoble, France); recently, a ptychographic measurement was successfully carried out at the PtyNAMi setup of the P06 beamline at PETRA III.

22.4.1   Hard X-rays Near 14 keV

Most experiments were carried out using the versatile Göttingen Instrument for Nano-Imaging with X-rays (GINIX) at the P10 coherence beamline of PETRA III [12]. From the undulator radiation, a monochromatic beam around 8.0 or 13.8 keV is filtered by either a double-crystal Si(111) monochromator, or by a more stable channel-cut monochromator. Then, with Compound Refractive Lenses (CRLs), the X-ray beam is pre-shaped onto the MZP and the sample.

During first alignment of the zone plate, the CRLs are moved out of the beam. Alignment of the MZP-tip/tilt angles is eased when illuminated by a large beam of around 200 μm diameter. For the MZP-based measurements, the CRL beam is usually set to a beam size of approximately 30 μm, so an MZP with diameter of 16 μm is illuminated homogeneously. It was found during the scanning SAXS/WAXS measurements, that combining the ultra-small focal spot of the MZP with a 2 μm CRL beam and a lithographic FZP of 30 nm zone width enables a profitable zoom-in capability. Since all three kinds of optics are in-line (without lateral shifts), features of the sample can be measured at different sequential resolution levels.

For the success of these experiments, close collaboration with Michael Sprung, beam line responsible for the P10 at DESY, Hamburg, and Tim Salditt, project leader of C01, was crucial; we thank Michael, Tim, and their teams for outstanding support.

22.4.2 High Energies: From 60 to 101 keV

Opposed to Fresnel Zone Plates, which are traditionally fabricated using lithography and hence achieve only small aspect ratios, Multilayer Zone Plates are a promising optic to focus even higher X-ray energies. The regime above 30 keV is usually attributed to Compound Refractive Lenses; here, we show that also MZPs can be used to focus photon energies from 60 to 100 keV. The proof of principle experiments were carried out at the high energy beam line ID 31 at the European Synchrotron Radiation Facility (ESRF) in Grenoble, France.

The MZP setup (see next subsection for more details) was integrated into the HEMD setup (High Energy Microdiffraction Instrument). This X-ray diffractometer is used to study buried interfaces; micro-focused X-ray beams of high energies can penetrate even metallic specimens. Scattering from buried interfaces (Reflectivity), hidden nano-scaled particles (SAXS), or packaged crystallites can be investigated. As detector, a Pilatus 2M single-photon counting pixelated hybrid detector with CdTe sensor material is used.

For the experiment, an MZP with a diameter of 8 μm and smallest zones of 10 nm was fabricated using PLD; with the FIB, a thick slice of 30 μm was cut out. This length provides optimal phase shift at 60 keV.

We report on our experiment in the Imaging section further below [10].

22.4.3 Sampler Scanner

In an early phase of the project, we learned that the versatile approach of the GINIX setup poses severe limits on resolution once the 10 nm region is targeted. To minimise vibrations and drift, a new sample tower was designed and commissioned. Only the most essential degrees of freedom for the imaging experiments are included.

For the MZP, three translational stages based on the Piezo stick-slip principle by SmarAct GmbH (Oldenburg, Germany) are used for lateral alignment; the vertical z-movements are accomplished by an inclined linear positioner and an additional free-moving guideway to increase stability. On top, horizontal movements are facilitated by one positioner plus one guideway (lateral y-direction) and by one positioner of increased stability (longitudinal x-direction). Travel ranges are 10 mm in z (vertical), and >40 mm in x and y (horizontal). The x-motor then has an adjustable mount position for an FZP optic, and holds a tip/tilt motor for the MZP optic. This tip/tilt is adapted from a motorised optical mount with stick-slip motors in open-loop mode. The maximum angles are ±5°. A custom-built encoder based on a two-dimensional Position Sensitive Detector (SEEPOS PSD Signal Process System with 2L45_MH02 sensor; SiTek Electro Optics, Partille, Sweden) is being commissioned to improve the angular motorisation.

For the sample, three translations by Physik Instrumente (PI GmbH, Karlsruhe, Germany) are used for coarse alignment. For x and y, a PILine piezo ultrasonic drive

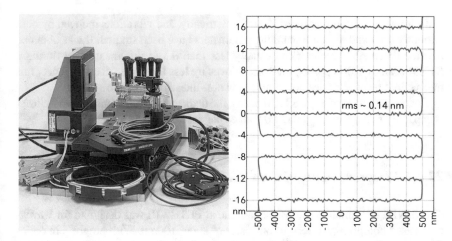

Fig. 22.6 Photograph of the Multilayer Zone Plate sample scanner (left) and encoder values recorded during a fast two-dimensional scan (right). The MZP is mounted on a tip-tilt stage with large translational stages by SmarAct; the sample is then coarsely aligned by three translations, and can be raster-scanned using a two-dimensional piezo stage by PI Physik Instrumente. According to encoder values, the positional accuracy of the continuous scan is better than 2 Å

of large area (model M-686) is used as base; a vertical NEXACT Piezo stepping drive (model N-765) with a high load capacity of 25 N is mounted on top. Travel ranges are 25 mm in x and y, and 6.5 mm in z. On top of the alignment motors is a vertically mounted Piezo scanner of high stiffness with a clear aperture of 50 mm edge length (model P-733). The built-in capacitive sensors show an r.m.s. noise of better than 0.2 nm. The maximum travel range is 30 μm; line scan speeds of 100 Hz and more are possible at reduced travel range.

All translational stages employ optical encoders with nanometre resolution. The non-linear, but reproducible movements of the tip/tilt stage have been characterised and adjusted in software by fourth order polynomials (Fig. 22.6).

The stage was designed for fast Scanning Transmission X-ray Microscopy (STXM) with new hybrid pixelated photon-counting detectors like the EigerX 4M (Dectris Inc., Baden-Daetwill, Switzerland). This particular detector is able to image at 750 Hz frame rate; each of the four million pixels has a full analog-digital processing chain to count single photon events. When configured for "fly scan mode" (continuous STXM), the Piezo moves a two-dimensional trajectory and starts a series of detector acquisitions via hardware trigger; a common scan consists of 255 lines with 255 detector frames each. In the fastest mode with 1.3 ms exposure times this is done in less than three minutes. This could be shortened further in bi-directional scanning mode [13, 14].

During fast scanning, inertial forces of the Piezo stage and sample holder have to be compensated by the mechanical system; otherwise, the MZP itself moves significantly, rendering the system unusable. We have measured the mechanical feedback from the sample part onto the optic part of the setup. For that, the movement of the

MZP holder has been measured interferometrically, and vibrations induced by both currently used and ultimate scanning parameters have been studied. It was obtained that the mechanical feedback is usually less than 0.1%; this means that during a STXM scan of 1,000 points, the MZP moves by less than one "pixel". Hence, this mechanical feedback does not induce new non-linearities into the system.

For data analysis of large STXM scans, the dedicated "Heinzelmännchen" cluster is capable of analysing up to 3,000 frames per second [15, 16].

22.4.4   Improvements of the GINIX Setup

The GINIX instrument at the P10 beamline at PETRA III was designed for waveguide based holography at resolution scales of about 100nm. With progress in optics fabrication (both for waveguides and for zone plates), this became a limiting factor in imaging. Together with project C01, several measures have been implemented to improve the stability. But first, the vibrations had to be quantified in a reliable manner. For that goal, two techniques have been applied. (i) Laser interferometry is able to measure the relative distance between two objects; this allows to quantify the absolute translational amplitude, but only relative between emitter and reflector. On the other hand, (ii) acceleration sensors (accelerometer 731-207 by Wilcoxon Research Inc., Meggitt PLC, Dorset, UK; including Dataq DI-155 data logger/Red Pitaya STEMlab) can measure absolute movements in space, but only the second time derivative. Numerical integration to extract the distance signal cannot reproduce slow and long-term drift-like movements. Note that both methods yield one-dimensional data.

After the implementation of vibration measurements, different strategies to improve the situation have been evaluated. We briefly discuss a few methods and share lessons learned.

Active vibration isolation using e.g. the *Nano Series* (Accurion GmbH, Göttingen) is a portable plug-in system that measures and actively damps vibrations in six degrees of freedom. We found that amplitudes of 50 nm and below are already "too good" for the system to work properly; instead, high frequencies were shifted to drift on the sub-second scale. Within the requirements of C01 and C12, the system did not work as desired.

Passive vibration isolation using a mechanical mass damper was implemented based on finite-element simulations modelling the vibrations. The principle is that an inert mass is accelerated by the vibrations, and the movements are subsequently dissipated by pushing a rubber absorber. During tuning of the resonance frequency, virtually no measurable effect could be seen; however, vibrations change significantly from night do day. It is assumed that the rubber absorber does not dissipate enough energy for very small amplitudes.

Further attempts include additional air springs, rubber mounts, and foams. Also, it has been tried to shift the resonance frequencies by additional masses on the breadboard. No measurable improvement of the vibrations could be achieved.

High-resolution platforms based on additional granite tables and combined sample towers with integrated mountings for optics. While the general flexible instrument of the GINIX is remained, high-resolution experiments are enabled by additional setups. For more details on the sample tower of project C12, see the previous subsection.

22.5 Imaging

In this section, we report on several imaging experiments. During the progress of project C12, many different imaging modalities have been implemented and tested for their suitability to study different samples in different geometries. Here, only a selection of results is presented.

22.5.1 Ptychography

During the second and at the beginning of the third funding period of our CRC, several attempts for a ptychographic reconstruction of the MZP focus were tried—unsuccessfully. Ptychography is a scanning technique that uses overlapping measurements to support both phase retrieval and the separation of wave-fields into illumination and sample scattering. Ultra-small beam sizes of MZPs on the order of sub-10 nm, however, introduce several problems, which were identified and subsequently resolved.

Vibrations on the order of 50 nm are a drawback of the flexibility of the GINIX setup. More details have been given in the previous section. First experiments suffered a lot; also because online-reconstruction was not possible during early stages of the project, visual feedback to steer the experiment was missing.

Sampling issues due to the diverging -1st diffractive orders rendered the reconstruction impossible. For ptychography to work correctly, the maximum beam size is bounded by $\lambda R/\Delta$, where R is the sample—detector distance and Δ the detector's pixel size. With Pilatus detectors and hence $\Delta = 172\,\mu m$ in a distance of $R = 5\,m$, the illuminated area has to be smaller than about $1\,\mu m$ in diameter. The negative order of our MZP optics, however, illuminates a circle of more than $30\,\mu m$.

22.5.1.1 Results at the PtyNAMi Instrument

To overcome the sampling issue of the negative order, we have successfully introduced on OSA close to the focal spot during a beamtime at the PtyNAMi instrument of the P06 beamline (PETRA III, DESY Hamburg). Due to the short focal length of $f \approx 1\,mm$, the OSA needs to be positioned about $200\,\mu m$ upstream of the focus. Then, a Siemens star test pattern was moved from the downstream side close to the

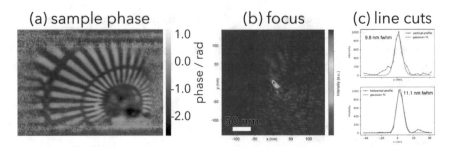

Fig. 22.7 **a** Ptychographic phase reconstruction of a Siemens star test pattern, **b** shows the intensity of the X-ray nano focus on a linear colour map in the best focal plane, **c** shows horizontal and vertical line cuts, with Gaussian fits of about 10 nm × 11 nm

focal region; after careful approaching, ptychographic scans could be performed. Results of the reconstruction are shown in Fig. 22.7. Figure 22.7a shows the reconstructed phase of the Siemens star; the smallest features of 50 nm are clearly resolved. Also damaged regions in the central part from an earlier experiment are visible. Figure 22.7b shows the intensity distribution in the best focal plane after numerical propagation; horizontal and vertical line cuts with Gaussian fits are shown in Fig. 22.7c. The FWHM of the focal spot size can be estimated to about 10 nm.

Note that the used MZP was not the "design cut" from the wire, and alignment could not be perfected during the ptychographic experiment during the allocated time slot. Nonetheless, this is the first successful ptychographic reconstruction of the MZP focus, indicating that resolution on the single-digit nanometre scale is in reach.

22.5.2 Holography and Scanning SAXS

Holography is a full-field imaging method, and the contrast is based on the local electron density inside the specimen and its imprint in the near-field intensity distribution. Scanning SAXS, on the other hand, is an imaging modality where sample or focused X-ray beam are scanned, and the contrast is extracted from far-field diffraction patterns and carries information about ordered structures inside the illuminated area [17–22]. For diffractive optics like Zone Plates, usually an order sorting aperture (OSA) is placed between sample and FZP. With a short focal length of ∼1 mm for the MZPs used here, the alignment of an OSA becomes impractical; hence, a "software OSA" has been used to extract multi-order holographic images from combined scanning-holography datasets. The scheme for each order is similar to the propagation-based imaging described in Chap. 2.

For sake of simplicity, we make use of the first diffractive orders, +1st and −1st; with the sample placed in a defocus plane $\Delta x \neq 0$, the propgation distances are

$x_{+1} = \Delta x$ and $x_{-1} = 2f + \Delta x$; with the detector placed at a position x_2 from the MZP, the effective distances become

$$\tilde{x}_{+1} = \frac{\Delta x \times x_2}{\Delta x + x_2} \approx 0.1\,\text{mm}, \qquad \tilde{x}_{-1} = \frac{(2f + \Delta x) \times x_2}{2f\,\Delta x + x_2} \approx 2.1\,\text{mm}.$$

For the numerical values, we have assumed a focal distance $f = 1$ mm, a defocal distance $\Delta x = 0.1$ mm, a detector distance $x_2 = 5$ m, an X-ray wavelength $\lambda = 1$ Å, and a detector pixel size of $p = 75\,\mu$m.

The corresponding Fresnel numbers and magnifications are

$$F_{+1} = \frac{(p/M_{+1})^2}{\lambda \times x_2} \approx 4.5 \times 10^{-9}, \qquad F_{-1} = \frac{(p/M_{-1})^2}{\lambda \times x_2} \approx 2.0 \times 10^{-6};$$

$$M_{+1} = \frac{\Delta x + x_2}{\Delta x} = 50{,}001\times, \qquad M_{-1} = \frac{2f + \Delta x + x_2}{2f + \Delta x} \approx 2382\,\times\,.$$

When the sample is laterally scanned in the defocus plane Δx, both holograms "move" with different velocities in the detector plane; by a simple re-arranging of the pixelated intensity values according to the magnifications $M_{\pm 1}$, two scaled holograms can be obtained. Also, an "average flat-field" can be extracted from the measurement.

For $\Delta x \to 0 \Rightarrow F_{+1} \to \infty$, and the $+1$st signal approaches a traditional Scanning X-ray Transmission Microscopy (STXM) contrast; in first order, differential phase contrast maps can be deduced from tracking the centred moment of the intensity distribution:

$$\varphi'(y) = k\,\sin\vartheta = \frac{2\pi}{\lambda}\,\sin\tan^{-1}\frac{\int I(y)\,y\,dy}{\int I(y)\,dy}.$$

Combing these imaging modalities of the 5 nm MZP with a "traditional" (i.e., lithographically produced) FZP with an outermost zone width of 30 nm, and a compound refractive lenses (CRL) optics with a focus size of about $2\,\mu$m, multi-scale imaging with a zoom-in capability has been explored. The different optics and field-of-views are shown in Table 22.1.

A phase-reconstruction using the relaxed averaged alternating reflection (RAAR) algorithm has been carried out on the -1st order hologram of semiconductor nanowires (for raw hologram, see Fig. 22.2). An overall view is shown in Fig. 22.8a, while Fig. 22.8b shows a zoom-in on a single nanowire. These reconstructions stem from the holographic dataset in Fig. 22.2b; Fig. 22.2 shows a close-up of the (here invisible) nanowire lying on an electrical contact in (vertical) differential phase contrast mode.

Table 22.1 For multi-scale imaging, different optical illuminations and measuring schemes are combined, with a decreasing field of view and increasing (ideal-case) resolution. The Compound Refractive Lenses (CRL) are from the P10 beamline, DESY Hamburg; the Fresnel Zone Plate (FZP) was generously provided by Christian David, PSI (CH); the Multilayer Zone Plate (MZP) is "D13" from the C12 project

Optical setup	Measuring scheme	Field of view	Resolution limit
Parallel beam	Holography, SAXS, WAXS	≥200 μm	Detector pixel size
MZP, −1st	Holography	30 μm	Demagnified pixel size
CRL pre-focus	Alignment scan	Overview scan	20 μm
CRL focus	Scanning SAXS/WAXS	Scan size	2 μm
FZP	Holography, scanning	Scan size	30 nm
MZP	Holography, scanning	Scan size	5 nm

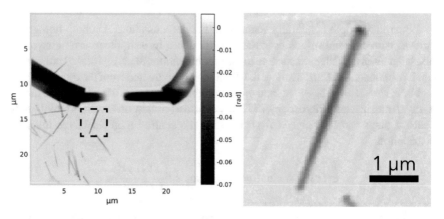

Fig. 22.8 Phase-reconstruction from a holographic measurement on semiconductor nanowires (see Fig. 22.2). Using the relaxed averaged alternating reflection (RAAR) algorithm, quantitative phase information of the nanowires and electrical contacts can be extracted

22.5.3   Scanning WAXS

High X-ray energies beyond 30 keV offer long penetration lengths into bulk material, so that even hidden and buried particles can be studied. Many composite materials are crystalline, with particle sizes in the sub-μm region; also, crystal domains within larger compounds of few hundred nm in size are common. So far, small structures can be studied ex-situ, i.e. as isolated or free-standing objects. But often, chemical and physical properties change when devices are assembled from individual components [10].

To demonstrate that MZP optics can be used in a scanning nano-WAXS setup, we have fabricated a focusing optic with unprecedented aspect ratios for hard X-ray

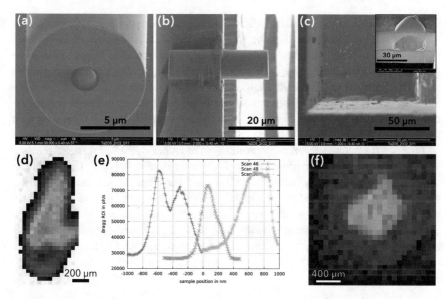

Fig. 22.9 **a** Front SEM view of the high-E MZP, showing an 8 μm aperture. **b** Side SEM view of the high-E MZP, showing the optical thickness of 30 μm. **c** An imprint of Ag droplets/nanocrystals, sandwiched between two layers of ZrO$_2$ (SEM top view). The inset shows an SEM image of one droplet after it has been exposed with FIB. **d** Spatially resolved Bragg peak intensity, measured at $E = 60$ keV. **e** 1D line scans of the same droplet (shifted for clarity), showing substructure in scan 46 (purple). In case of scan 50 (light blue), the droplet serves as a knife edge with a steepness of better than 50 nm. Note that the droplet was not actually in the best focal plane of the MZP. **f** Spatially resolved Bragg peak intensity, measured at $E = 101$ keV

energies. SEM images of the final MZP are shown in Fig. 22.9a, b. The lens has a diameter of 8 μm with outer-most zones of 10 nm, and an optical thickness of 30 μm. This length is optimised for a phase-shifting zone plate at an X-ray energy of 60 keV.

As a test sample, Ag droplets (nano crystallites developing during the PLD process for detuned parameters) were buried within 1.5 μm thick layers of ZrO$_2$ on a planar Si substrate. The typical droplet size is on the single μm scale. Figure 22.9c shows a top SEM view of the specimen. Although the droplets are buried, their shape and position can be seen as imprints on the surface. The inset shows an SEM image of a Ag droplet exposed with a FIB after the beamtime [10].

The experiment was performed at the high-energy station of ID31 at the ESRF (Grenoble, France). In the first user experiment with the new Laue-Laue bending monochromator, X-ray energies at 60 keV (bandwidth 0.44%) and 101 keV (bandwidth 0.57%) were selected from the 5th or 9th harmonic, respectively, of an in-vacuum cryo-cooled undulator with a 14.5 mm period. The X-ray beam was pre-focused by up to 288 Be lenses, in order to illuminate the MZP. The Bragg diffraction patterns were recorded with a Pilatus3 CdTe 2M detector (Dectris Inc., Baden-Daetwill, Switzerland).

The undiffracted order of the MZP illuminated basically a diluted powder of droplets and an amorphous thin Ag film. This resulted in a Debye-Scherrer ring at $0.4 \, Å^{-1}$ or a Bragg angle of $1.5°$ at $101 \, keV$ on the detector. While the sample was raster-scanned through the MZP focus, a single Bragg peak was heavily excited above background level. Figure 22.9d shows a real-space mapping of the intensity in a single Bragg peak, showing the outline of a particular crystal grain inside one droplet. The lateral size of the grain is about 500 nm, compatible with sizes seen with the FIB.

Line scans through the intensity profile are shown in Fig. 22.9e for measurements at 60 keV, and for a different droplet (Fig. 22.9f) measured at 101 keV. The resolution has been determined to be better than 50 nm at 60 keV and even better than 40 nm at 101 keV. The full potential of the MZP focus depends of course on properly aligning the sample in the focal plane, which has not been achieved in the presented experiment. Nonetheless, the results show the feasibility of nano-focusing three-digit energies that are able to penetrate into nano-sized particles buried deeply inside bulk material [10].

22.5.4   Correlative Scans

Recently, zone plate based holography was combined with scanning in both SAXS and WAXS regime. To this end, project C12 teamed up with projects B03 and B10 (Simone Techert). A sample made of amino acid crystallites (D- and L-Tryptophan) was raster scanned sequentially with a large WAXS detector (EigerX in "front position" at about 0.3 m) and then with a smaller SAXS detector (Pilatus in "rear position" at about 5.1 m). In addition, holographic datasets have been taken. Due to alignment difficulties, most scans were performed using a Fresnel Zone Plate (outermost zone width 30 nm) and Compound Refractive Lenses (focal spot size on the μm scale). A systematic measurement to study the growth process of such crystallites is envisioned; to spatially resolve the real-space sub-structure, a Multilayer Zone Plate will be used.

This setup combines certain strengths of the different modalities. Holography is—in principle—a full-field technique; here it is combined with small scans of the sample to implement the "software OSA". With the FZP, large overview images at moderate resolution can be obtained. Changing to the MZP optic, a zoom-in capability with a field of view slightly larger than the crystallites and at few nanometre spatial resolution become possible. The fundamental contrast mechanism is the phase shift due to the electron density of the specimens.

Scanning SAXS, on the other hand, is sensitive to surfaces and especially morphology; since length scales of up to 100 nm can be probed, it is beneficial to use the CRL beam. With scanning WAXS, then, it becomes possible to distinguish the crystalline grains within the sample.

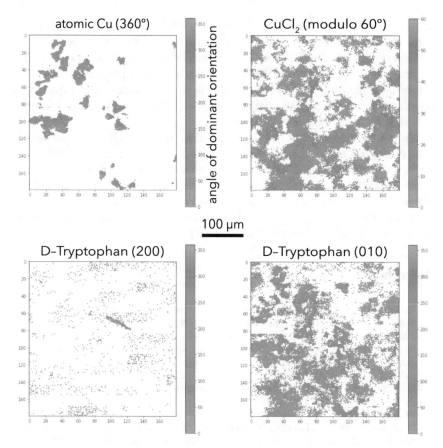

Fig. 22.10 Dominant crystallite orientations mapped into real-space, for **a** pure Cu, **b** CuCl, **c** D-Tryptophan (200) and **d** D-Tryptophan (010) lattices

From preliminary scanning-WAXS measurements, the dominant crystallite orientations can be mapped into real-space images. Figure 22.10 shows these dominant orientations for (b) CuCl, (c) D-Tryptophan (200), and (d) D-Tryptophan (010).

A detailed analysis to study correlations in reciprocal and real-space is still on-going.

22.6 Summary

During the second funding period, a collaboration of projects C01 and C04 have demonstrated the first two-dimensional 5 nm X-ray focus. Building on this, project C12 has further developed hard X-ray imaging in different areas. Together with C01, the GINIX setup at the P10 beamline (DESY Hamburg) has been optimised in terms of motorisation, reducing of vibrations / drift, automated data handling, and in the end image resolution. During a beamtime at the ESRF (Grenoble, France), high energy X-rays of 101 keV could be used to localise buried nano-crystallites at a spatial resolution better than 50 nm. In a joint work with projects B03/B10, the possibility of imaging and crystallite mapping has been successfully explored on organic matter. A full analysis on the complex structures of the Tryptophan sample is, however, beyond the scope of project C12.

With the retirement of one of the principal investigators (H.U.K.), the deposition process and optimisation thereof has been paused; however, new MZPs have successfully been cut out off existing coated wires for new experiments.

Due to rapidly increasing amounts of data from scanning experiments, processing and analysis had stagnated for a while; with the Heinzelännchen system, a dedicated analysis platform for scanning SAXS data was comissioned and has served different projects within this collaborative research centre.

References

1. Reese, M.: Fabrication of Soft X-ray Optics and their Characterisation with Labratory and Synchrotron Sources. Universitätsverlag Göttingen (2011)
2. Reese, M., Schäfer, B., Großmann, P., Bayer, A., Mann, K., Liese, T., Krebs, H.: Submicron focusing of xuv radiation from a laser plasma source using a multilayer laue lens. Appl. Phys. A Mater. Process. **102**, 85–90 (2010)
3. Ruhlandt, A., Liese, T., Radisch, V., Krüger, S.P., Osterhoff, M., Giewekemeyer, K., Krebs, H.U., Salditt, T.: A combined kirkpatrick-baez mirror and multilayer lens for sub-10 nm X-ray focusing. AIP Adv. **2**, 012,175 (2012). https://doi.org/10.1063/1.3698119
4. Döring, F., Robisch, A.L., Eberl, C., Osterhoff, M., Ruhlandt, A., Liese, T., Schlenkrich, F., Hoffmann, S., Bartels, M., Salditt, T., Krebs, H.U.: Sub-5 nm hard X-ray point focusing by a combined kirkpatrick-baez mirror and multilayer zone plate. Opt. Express **21**, 19311–19323 (2013). https://doi.org/10.1364/OE.21.019311
5. Osterhoff, M., Eberl, C., Döring, F., Wilke, R., Wallentin, J., Krebs, H., Sprung, M., Salditt, T.: Towards multi-order hard x-ray imaging with multilayer zone plates. J. Appl. Crystallogr. **48**, 116–124 (2015). https://doi.org/10.1107/S1600576714026016
6. Wallentin, J., Anttu, N., Asoli, D., Huffman, M., Åberg, I., Magnusson, M.H., Siefer, G., Fuss-Kailuweit, P., Dimroth, F., Witzigmann, B., Xu, H.Q., Samuelson, L., Deppert, K., Borgström, M.T.: Inp nanowire array solar cellsachieving 13.8% efficiency byexceeding the ray optics limit. Science **339**, 1057–1060 (2013). https://doi.org/10.1126/science.1230969
7. Eberl, C., Döring, F., Liese, T., Schlenkrich, F., Roos, B., Hahn, M., Hoinkes, T., Rauschenbeutel, A., Osterhoff, M., Salditt, T., Krebs, H.U.: Fabrication of laser deposited high-quality multilayer zone plates for hard x-ray nanofocusing. Appl. Surf. Sci. **307**, 638–644 (2014). https://doi.org/10.1016/j.apsusc.2014.04.089
8. Eberl, C., Liese, T., Schlenkrich, F., Döring, F., Hofsäss, H., Krebs, H.U.: Enhanced resputtering and asymmetric interface mixing in w/si multilayers. Appl. Phys. A **111**, 431 (2013)
9. Eberl, C., Osterhoff, M., Döring, F., Krebs, H.U.: Mzp design and fabrication for efficient hard x-ray nano-focusing and imaging. In: Proceedings of SPIE Advances in X-Ray/EUV Optics and Componoents X **958**, 808 (2015). https://doi.org/10.1117/12.2187788
10. Osterhoff, M., Soltau, J., Eberl, C., Krebs, H.U.: Ultra-high-aspect multilayer zone plates for even higher x-ray energies. In: Proceedings of SPIE Advances in X-Ray/EUV Optics and Componoents XII, vol. 10386, 1038,608 (2017). https://doi.org/10.1117/12.2271139
11. Warken, F., Rauschenbeutel, A., Bartholomäus, T.: Fiber pulling profits from precise positioning. Photonics Spectra **42**, 73–75 (2008)

12. Salditt, T., Osterhoff, M., Krenkel, M., Wilke, R., Priebe, M., Bartels, M., Kalbfleisch, S., Sprung, M.: Compound focusing mirror and x-ray waveguide optics for coherent imaging and nano-diffraction by Markus Osterhoff. J. Synchr. Radiat. **22**, 867–878 (2015). https://doi.org/10.1107/S1600577515007742
13. Osterhoff, M., Eberl, C., Soltau, J., Krebs, H.: Preparing for hard x-ray microscopy with multilayer zone plates. J Phys Conf Ser **849**, 012,049 (2016). https://doi.org/10.1088/1742-6596/849/1/012049
14. Osterhoff, M., Soltau, J., Eberl, C., Krebs, H.U.: Faster scanning and higher resolution: new setup for multilayer zone plate imaging. In: Proceedings of SPIE X-ray Nanoimaging: Instruments and Methods III, vol. 10389, 103,890T (2017). https://doi.org/10.1117/12.2271141
15. Dectris Award Winner Heinzelännchen. https://www.dectris.com/company/news/newsroom/news-details/dectris-award-winner-heinzelmaennchen-analyse-x-ray-images, 27 Mar (2018)
16. Osterhoff, M., Goeman, J., Salditt, T., Köster, S.: Stxm analysis: preparing to go live @ 750 hz. In: AIP Conference Proceedings, vol. 2054, 060,075 (2019). https://doi.org/10.1063/1.5084706
17. Bernhardt, M., Nicolas, J.D., Eckermann, M., Eltzner, B., Rehfeldt, F., Salditt, T.: Anisotropic x-ray scattering and orientation fields in cardiac tissue cells. New J. Phys. **19**, 013,012 (2016). https://doi.org/10.1088/1367-2630/19/1/013012
18. Feigin, L., Svergun, D.: Structure Analysis by Small-Angle X-Ray and Neutron Scattering. Springer (1987)
19. Hémmonot, C., Köster, S.: Imaging of biological materials and cells by x-ray scattering and diffractions. ACS Nano **11**, 8542–8559 (2017). https://doi.org/10.1021/acsnano.7b03447
20. Nicolas, J.D., Bernhardt, M., SChlick, S.F., Tiburcy, M., Zimmermann, W.H., Khan, A., Markus, A., Alves, F., Toischer, K., Salditt, T.: X-ray diffraction imaging of cardiac cells and tissue. Progress Biophys. Mol. Biol. (In press 2019). https://doi.org/10.1016/j.pbiomolbio.2018.05.012
21. Priebe, M., Bernhardt, M., Blum, C., Tarantola, M., Bodenschatz, E., Salditt, T.: Scanning x-ray nanodiffraction on dictyostelium discoideum. Biophys. J. **107**, 2662–2673 (2014). https://doi.org/10.1016/j.bpj.2014.10.027
22. Weinhausen, B., Nolting, J.F., Olendrowitz, C., Langfahl-Klabes, J., Reynolds, M., Salditt, T., Köster, S.: X-ray nano-diffraction on cytoskeletal networks. New J. Phys. **14**, 085,013 (2012). https://doi.org/10.1088/1367-2630/14/8/085013

Chapter 23
Convergence Analysis of Iterative Algorithms for Phase Retrieval

D. Russell Luke and Anna-Lena Martins

Abstract This chapter surveys the analysis of the phase retrieval problem as an inconsistent and nonconvex feasibility problem. We apply a convergence framework for iterative mappings developed by Luke, Tam and Thao in 2018 to the inconsistent and nonconvex phase retrieval problem and establish the convergence properties (with rates) of popular projection methods for this problem. Although our main purpose is to illustrate the convergence results and their underlying concepts, we demonstrate how our theoretical analysis aligns with practical numerical computation applied to laboratory data.

Mathematics Subject Classification: 65K10 · 49K40 · 49M05 · 65K05 · 90C26 · 49M20 · 49J53

23.1 Introduction

We highlight recent theoretical advances that have opened the door to a quantitative convergence analysis of well-known phase retrieval algorithms. As shown in Chap. 6, phase retrieval problems have a natural and easy characterization as feasibility problems, and issues like noise and model misspecification do not effect the abstract regularity of the problem formulation. This was also observed in studies by Bauschke et al. [1] and Marchesini [2] reviewing phase retrieval algorithms in the context of fixed point iterations, though in those works the theory only provided convex heuristics for understanding the most successful algorithms. A slow progression of the theory for nonconvex feasibility culminating in the work by Luke et al. in [3] now provides a firm theoretical basis for understanding most of the standard algorithms for phase retrieval.

D. R. Luke (✉) · A.-L. Martins
Institute for Numerical and Applied Mathematics, Universität Göttingen,
Lotzestr. 16-18, 37083 Göttingen, Germany
e-mail: r.luke@math.uni-goettingen.de

A.-L. Martins
e-mail: a.martins@math.uni-goettingen.de

© The Author(s) 2020
T. Salditt et al. (eds.), *Nanoscale Photonic Imaging*, Topics in Applied Physics 134,
https://doi.org/10.1007/978-3-030-34413-9_23

The approach is fix-point theoretic and is based on a framework introduced by Luke et al. in [3]. Given some (set-valued) mapping $T : \mathcal{E} \rightrightarrows \mathcal{E}$, where $\mathcal{E}$ is a finite-dimensional Euclidean space, the algorithms are studied as mere generators of sequences $(x^k)_{k \in \mathbb{N}}$ through the fixed point iteration $x^{k+1} \in T x^k$ ($\forall k \in \mathbb{N}$) with $x^k \to x^*$ where $x^* = T x^*$. We demonstrate the convergence framework of [3] on a few of the more prevalent iterative phase retrieval algorithms introduced in Chap. 6.

The analysis is based on two main properties. The first of these is the regularity of the mapping defining the fixed point iteration; the second property concerns the stability of the fixed points of the mapping. The first property is covered by the notion of *pointwise almost averagedness*, a generalization of regularity concepts like (firm) nonexpansiveness. Already in the 1960s Opial [4] showed that an iterative sequence defined by an averaged self-mapping with nonempty fixed point set converges to a fixed point. It is no surprise, then, that generalizations of averagedness should play a central role in convergence for more general fixed point mappings. In the setting of feasibility problems, i.e. finding a point in the intersection of a collection of sets, pointwise almost averagedness of the fixed point mapping is inherited from the regularity of the sets.

The other concept that is central to the analysis concerns stability of the fixed points. This is the characterized by the notion of *metric subregularity* as presented in Dontchev and Rockafellar [5], and Ioffe [6, 7]. Metric subregularity of the mapping at fixed points guarantees quantitative estimates for the rate of convergence of the iterates. This is closely related to the existence of error bounds, and weak-sharp minima, among other equivalent notions that provide a path to a quantitative convergence analysis.

In Sect. 23.2 we remind the reader of the phase retrieval problem. Section 23.3 and its subsections introduce basic notations and concepts. This is followed by a toolkit for convergence in Sect. 23.4 that describes the convergence framework we are working with. The use of this theoretical toolkit is demonstrated on two of the most prevalent algorithms for phase retrieval. We conclude this chapter with some numerical remarks in Sect. 23.8.

23.2 Phase Retrieval as a Feasibility Problem

The phase retrieval problem reviewed in Chaps. 2 and 6 involves reconstructing a complex valued field in a plane (the object plane) from measurements of its amplitude under a unitary mapping in a plane somewhere downstream from the object plane (the image plane). We use the notation for the phase retrieval problem already introduced in previous chapters. For a detailed description see Sect. 6.1.1. The measurements are represented by the sets

$$M_j := \left\{ u \in \mathbb{C}^n \mid \|(\mathcal{F}u)_i\| = \sqrt{I}_{j,i}, \ (i = 1, 2, \ldots, n) \right\} \quad (j = 1, 2, \ldots, m).$$
$$(23.1)$$

The problem of recovering the phase from just the modulus of unitary transformed measurements is impossible to solve uniquely. Usually nonuniqueness is associated with *ill-posedness*, but for feasibility problems it is rather *existence* that is the source of difficulty. In real-world problems measurement errors and model misspecification have profound implications for feasibility models, but not for the reasons that one might expect. The geometry of the individual measurement sets does not change in the presence of noise or model misspecification. The issue is that the measurements are not *consistent* with one another. In other words, there is no solution that satisfies the measurements and other model requirements (like nonnegativity, in the case of real objects). A solution from the provided information is then only an approximation to the actual signal. Mathematically these characteristics translate into an *inconsistent* feasibility problem. That is, the intersection of the sets in the feasibility model is empty. Inconsistency has been investigated in many works (see for instance [8–11]) but most of these studies consider convex sets. Unfortunately, the sets involved in the phase retrieval problem are mostly nonconvex and have empty intersecton. In [3] the authors provided a scheme to handle even this case. The following sections are devoted to their work and present the most important concepts.

To avoid ambiguities recovering the phase, one often uses a priori information about the model. Common examples are the knowledge of a support of the signal, real-valuedness, non-negativity, sparsity or the information about an amplitude:

$$
\begin{array}{ll}
\text{support constraint} & \mathfrak{S} := \{y \in \mathbb{C}^n \mid y_i = 0 \ \forall i \notin I\} \\
\text{real-valued support constraint} & \mathfrak{S}_r := \{y \in \mathbb{R}^n \mid y_i = 0 \ \forall i \notin I\} \\
\text{non-negative support constraint} & \mathfrak{S}_+ := \{y \in \mathbb{R}_+^n \mid y_i = 0 \ \forall i \notin I\} \\
\text{amplitude constraint} & A := \{y \in \mathbb{C}^n \mid |y_k| = a_k, \ 1 \le k \le n\} \\
\text{sparsity constraint} & A_s := \{y \in \mathbb{R}^n \mid \|x\|_0 \le s\}
\end{array}
$$

$$(23.2)$$

for a set of indices $I \subset \{1, 2, \ldots, n\}$, $a \in \mathbb{R}_+^n$ and $s \in \{1, 2, \ldots, n\}$, where $\mathbb{R}_+^n = \{x \in \mathbb{R}^n \mid x_i \ge 0, \ 1 \le i \le n\}$. In the following we focus on the (non-negative) support constraint.

23.3   Notation and Basic Concepts

Our setting throughout this chapter is a finite dimensional real Euclidean space $\mathcal{E}$ equipped with inner product $\langle \cdot, \cdot \rangle$ and induced norm $\| \cdot \|$. The open unit ball is denoted by $\mathbb{B}$, whereas $\mathbb{S}$ stands for the unit sphere in $\mathcal{E}$. The open ball with radius δ and center x is denoted by $\mathbb{B}_\delta(x)$. The iterative algorithms we analyze can be represented by mappings $T : \mathcal{E} \rightrightarrows \mathcal{E}$, where $\rightrightarrows$ indicates that T is a point-to-set mapping. $\mathbb{N}$ denotes the natural numbers. The *inverse mapping* T^{-1} at a point y in the range of T is defined as the set of all points x such that $y \in T(x)$.

23.3.1 Projectors

We follow in this section the definitions introduced in Chap. 6. As a reminder: the distance of a point x to a set $\Omega \subset \mathcal{E}$ is defined by dist $(x, \Omega) := \inf_{y \in \Omega} \{\|y - x\|\}$. The corresponding *projector* onto the set Ω is given by $\mathcal{P}_\Omega : \mathcal{E} \rightrightarrows \mathcal{E},\ x \mapsto \{y \in \Omega \mid$ dist$(x, \Omega) = \|y - x\|\}$. A single element of $\mathcal{P}_\Omega x$ is called a *projection*. Similarly to the projector, the *reflector* onto a set Ω is defined by $\mathcal{R}_\Omega : \mathcal{E} \rightrightarrows \mathcal{E},\ x \mapsto 2\mathcal{P}_C x - x$, which is again a set. A single element in $\mathcal{R}_\Omega$ is called a *reflection*.

The regularity of a set influences the properties of the corresponding projector onto the set. The best properties are generated by *convex* sets. A convex set Ω is defined as a set that contains the line segment between any two points $x, y \in \Omega$. The projector onto a convex set is not only single-valued, but can be characterized by a variational inequality (see for instance [12, Theorem 3.14]). As we see in Sect. 23.3.2 the algorithms considered here are all composed of projectors and reflectors. This leads to an analysis of the projectors onto the sets introduced in Sect. 23.2. The projector onto the measurement sets $\mathcal{M}_j$, defined in (23.1) was already discussed in Sect. 6.1.2. The projectors onto the support constraint sets are even simpler. The following statement is taken from [1, Example 3.14].

Lemma 23.1 (projectors onto support constraints) *Let* $y \in \mathbb{C}^n$, $I \subset \{1, 2, \ldots, n\}$. *Then the following hold.*

$$\mathcal{P}_{\mathbb{S}} y = z \quad \text{where } z_j = \begin{cases} y_j & \text{if } j \in I \\ 0 & \text{otherwise} \end{cases} \qquad \text{for } 1 \leq i \leq n, \quad (23.3)$$

$$\mathcal{P}_{\mathbb{S}_+} y = z \quad \text{where } z_j = \begin{cases} \max\{Re(y_j), 0\} & \text{if } j \in I \\ 0 & \text{otherwise} \end{cases} \qquad \text{for } 1 \leq i \leq n. \quad (23.4)$$

The projectors onto other constraint sets can be found, for instance, in [13] or [14] for a sparsity constraint, or in [1, Example 3.14] for an amplitude constraint or real-valued sparsity constraint. Except for the amplitude and sparsity constraint, all other mentioned constraint sets are closed and convex. The type of regularity of the constraint sets is later discussed in Remark 23.5.1.

Another concept closely related to that of projectors are normal cones.

Definition 23.2 (*Normal cones*) Let $\Omega \subseteq \mathcal{E}$. Define the cone containing Ω by

$$\text{cone}(\Omega) := \mathbb{R}_+ \cdot \Omega := \{\kappa s \mid \kappa \in \mathbb{R}_+, s \in \Omega\}.$$

Let $\Omega \subseteq \mathcal{E}$ and let $x \in \Omega$.

  (i) The *proximal normal cone* of Ω at $\bar{x}$ is defined by

$$N_\Omega^{\text{prox}}(x) = \text{cone}\left(\mathcal{P}_\Omega^{-1} x - x\right).$$

Equivalently, $x^* \in N_\Omega^{\text{prox}}(x)$ whenever there exists $\sigma \geq 0$ such that

$$\langle x^*, y - x \rangle \leq \sigma \|y - x\|^2 \quad (\forall y \in \Omega).$$

(ii) The *limiting (proximal) normal cone* of Ω at x is defined by

$$N_\Omega(x) = \operatorname{Lim\,sup}_{z \to x} N_\Omega^{\mathrm{prox}}(z),$$

where the limit superior is taken in the sense of *Painlevé-Kuratowski outer limit* (for more details on the outer limit see for instance [15, Chap. 4]).

When $x \notin \Omega$ all normal cones at x are empty (by definition). If the set Ω is convex, the given definitions of the normal cones coincide (see for instance [16]).

23.3.2 Algorithms

In the context of feasibility problems, a prominent class of iterative algorithms are projection algorithms. Under these, the most prominent and probably one of the easiest to compute is the *method of cyclic projections* as introduced in Sect. 6.2.1. Given a finite number of closed sets $\Omega_1, \Omega_2, \ldots, \Omega_m \subseteq \mathcal{E}$ and a point it generates the next iterate by consecutively projecting onto each of the individual sets. For only two sets the algorithm reduces to the method of alternating projections. In Sect. 6.2.3 the error reduction algorithm was identified with the method of alternating projections applied to a measurement and a support constraint. This connection was first made by Levi and Stark in [17]. Considering again only two sets, Sect. 6.1.2 introduced the well-known *Douglas-Rachford algorithm* as well as its relaxed version, the *relaxed averaged alternating reflection* algorithm introduced by Luke in [10]. For one magnitude constraint and a support constraint Douglas-Rachford yields *Fienup's hybrid input output method (HIO)* [18]. The connection of HIO and Douglas-Rachford was already observed by Bauschke et al. [1]. These three algorithms are the ones we want to focus on here. Nevertheless, we want to emphasize that the analysis shown below can be applied also to other projection methods.

Our survey is far from complete. Other approaches worthy of mention are several of the algorithms discussed in Chap. 5 and those in Chap. 6. Readers familiar with the physics literature will also miss the Hybrid Projection Reflection algorithm, [19], difference map, [20], solvent flipping algorithm, [21], and Fienup's Basic Input-Output algorithm (BIO). BIO is, in fact, nothing more than Dykstra's algorithm, see [1]. Like the BIO algorithm, most of the known approaches to phase retrieval fit into a concise scheme presented in [22].

23.3.3 Fixed Points and Regularities of Mappings

We refer to $\mathsf{Fix}\, T$ as the set of fixed points of the mapping T, i.e. $x \in \mathsf{Fix}\, T$ if and only if $x \in Tx$. The continuity of set-valued mappings is a well-developed concept and follows the familiar patterns of continuity for single-valued functions. One key property is *nonexpansiveness*, which nothing more than being Lipschitz continuous with constant 1. That is, given two points, their images under the mapping T are no further away from each other than the initial points. A slightly stronger notion than nonexpansiveness is *averagedness*. For set-valued mappings, a finer distinction of the types of continuity, whether pointwise, or uniform, for example, is necessary. The following definition captures the crucial types of continuity and regularity of set-valued mappings that lie at the heart of numerical analysis of algorithms for phase retrieval.

Definition 23.3 (*almost nonexpansive/averaged mappings*) Let $D \subseteq \mathcal{E}$ and $T : D \rightrightarrows \mathcal{E}$.

(i) T is said to be *pointwise almost nonexpansive* on D at $y \in D$ if there exists a constant $\epsilon \in [0, 1)$ such that

$$\|x^+ - y^+\| \le \sqrt{1 + \epsilon}\|x - y\| \quad (\forall\, y^+ \in Ty)(\forall x^+ \in Tx)(\forall x \in D). \quad (23.5)$$

If (23.5) holds with $\epsilon = 0$ then T is called *pointwise nonexpansive* at y on D.
If T is pointwise (almost) nonexpansive at every point on a neighborhood of y (with the same violation constant ϵ) on D, then T is said to be *(almost) nonexpansive at y (with violation ϵ) on D*.
If T is pointwise (almost) nonexpansive on D at every point $y \in D$ (with the same violation constant ϵ), then T is said to be *pointwise (almost) nonexpansive on D (with violation ϵ)*. If D is open and T is pointwise (almost) nonexpansive on D, then it is (almost) nonexpansive on D.

(ii) T is called *pointwise almost averaged on D at y* if there is an averaging constant $\alpha \in (0, 1)$ and a violation constant $\epsilon \in [0, 1)$ such that the mapping $\tilde{T}$ defined by $T = (1 - \alpha)\mathrm{Id} + \alpha\tilde{T}$ is pointwise almost nonexpansive at y with violation ϵ/α on D.
Similarly, if $\tilde{T}$ is (pointwise) (almost) nonexpansive on D (at y) (with violation ϵ), then T is said to be *(pointwise)(almost) averaged on D (at y) (with averaging constant α and violation $\alpha\epsilon$)*.
If the averaging constant $\alpha = \frac{1}{2}$, then T is said to be *(pointwise) (almost) firmly nonexpansive on D (with violation ϵ) (at y)*.

From the above definition it can easily be seen that if a set-valued mapping is non-expansive at a point, then it is single-valued there. This is a crucial property for our analytical framework, but should not be confused with uniqueness of fixed points: a multi-valued operator can be single-valued at its fixed points without having unique fixed points.

Proposition 23.4 (single-valuedness, Proposition 2.2 of [3]) *Let $T : \mathcal{E} \rightrightarrows \mathcal{E}$ be pointwise almost averaged on $D \subset \mathcal{E}$ at $\overline{x} \in D$ with violation $\epsilon \ge 0$. Then T is single-valued at $\overline{x}$. In particular, if $\overline{x} \in \mathsf{Fix}\, T$, then $T\overline{x} = \{\overline{x}\}$.*

Averaged mappings do not enjoy as nice a calculus as nonexpansive mappings, but the next proposition shows that averagedness of some sort is preserved under addition and composition.

Proposition 23.5 (compositions, Proposition 2.4 of [3]) *Let* $T_j : \mathcal{E} \rightrightarrows \mathcal{E}$ *for* $j = 1, 2, \ldots, m$ *be pointwise almost averaged on* U_j *at all* $y_j \in S_j \subset \mathcal{E}$ *with violation* ϵ_j *and averaging constant* $\alpha_j \in (0, 1)$ *where* $U_j \supset S_j$ *for* $j = 1, 2, \ldots, m$.

(i) *If* $U := U_1 = U_2 = \cdots = U_m$ *and* $S := S_1 = S_2 = \cdots = S_m$ *then the weighted mapping* $T := \sum_{j=1}^{m} w_j T_j$ *with weights* $w_j \in [0, 1]$, $\sum_{j=1}^{m} w_j = 1$, *is pointwise almost averaged at all* $y \in S$ *with violation* $\epsilon = \sum_{j=1}^{m} w_j \epsilon_j$ *with averaging constant* $\alpha = \max_{j=1,2,\ldots,m} \{\alpha_j\}$ *on* U.

(ii) *If* $T_j U_j \subseteq U_{j-1}$ *and* $T_j S_j \subseteq S_{j-1}$ *for* $j = 2, 3, \ldots, m$, *then the composite mapping* $T := T_1 \circ T_2 \circ \cdots \circ T_m$ *is pointwise almost averaged at all* $y \in S_m$ *on* U_m *with violation at most* $\epsilon = \prod_{j=1}^{m} (1 + \epsilon_j) - 1$. *and averaging constant at least*
$$\alpha = m \Big/ \left(m - 1 + \frac{1}{\max_{j=1,2,\ldots,m} \{\alpha_j\}} \right).$$

23.4 A Toolkit for Convergence

With the characterization of algorithms as simply self mappings with certain regularity properties, we show in this section how those properties come together to guarantee convergence of the algorithm iterations to fixed points. The fixed points need not be solutions to the feasibility problem (indeed, this does not exist for phase retrieval) but will in general be a point that allows one to compute *another point* that does have some physical significance, such as a *local best approximation point*.

It turns out that convergence itself is provided by regularity properties introduced in Sect. 23.3.3. The basic convergence idea goes back to Opial [4]. It says that averagedness of a single-valued mapping T and nonemptyness of the fixed point set imply convergence of the iterative sequence $(T^k x^0)_{k \in \mathbb{N}}$ to a point in Fix T for any $x^0 \in \mathcal{E}$. Henceforth, we will see that averagedness of T and a nonempty fixed point set is enough to get convergence. As one would expect, it can be difficult for a map to satisfy these properties globally. Nevertheless, this is often the case in nonconvex problem instances. Thus, we seek a statement that includes local properties. That is in our case pointwise almost averagedness as introduced in Definition 23.3.

But convergence alone for iterative procedures is not enough: eventually one has to stop the iteration and without knowing the *rate* of convergence it is impossible to estimate how far a given iterate must be to the solution. A quantitative convergence analysis is achieved with the second essential property: *metric (sub-)regularity*. This concept has been studied by many authors in the literature (see for instance [5–7, 15, 23, 24]). For the definition of metric regularity we need *gauge functions*. A function $\mu : [0, \infty) \to [0, \infty)$ is a gauge function if it is continuous and strictly increasing with $\mu(0) = 0$ and $\lim_{t \to \infty} \mu(t) = \infty$. The following definition is taken from [3, Definition 2.5].

Definition 23.6 (*metric regularity on a set*) Let $\Phi : \mathcal{E} \rightrightarrows \mathbb{Y}$, $U \subset \mathcal{E}$, $V \subset \mathbb{Y}$. The mapping Φ is called *metrically regular with gauge* μ *on* $U \times V$ *relative to* $\Lambda \subset \mathcal{E}$ if

$$\text{dist}\left(x, \Phi^{-1}(y) \cap \Lambda\right) \leq \mu\left(\text{dist}\left(y, \Phi(x)\right)\right) \tag{23.6}$$

. holds for all $x \in U \cap \Lambda$ and $y \in V$ with $0 < \mu\left(\text{dist}\left(y, \Phi(x)\right)\right)$. When the set V consists of a single point, $V = \{\bar{y}\}$, then Φ is said to be *metrically subregular for $\bar{y}$ on U with gauge μ relative to $\Lambda \subset \mathcal{E}$.*

When μ is a linear function (that is, $\mu(t) = \kappa t$, $\forall t \in [0, \infty)$) one says "with constant κ" instead of "with gauge $\mu(t) = \kappa t$". When $\Lambda = \mathcal{E}$, the quantifier "relative to" is dropped. When μ is linear, the smallest constant κ for which (23.6) holds is called *modulus* of metric regularity.

While this definition might seem abstract there are properties that directly imply metric regularity or reformulations that allow to prove metric regularity. One of these is *polyhedrality* (see [3, Proposition 2.6]). A mapping $T : \mathcal{E} \rightrightarrows \mathcal{E}$ is called polyhedral if its graph is the union of finitely many sets that can be expressed as the intersection of finitely many closed half-spaces and/or hyper-planes [5].

Collecting the concepts we have established so far, we present the following convergence result that goes back to Luke et al. in [3, Theorem 2.2] and was later refined in [25] by Luke et al. to convergence to a specific point.

Theorem 23.4.1 (basic convergence template with metric regularity) *Let $T : \Lambda \rightrightarrows \Lambda$ for $\Lambda \subset \mathcal{E}$, $\Phi := T - \text{Id}$ and let $S \subset \text{ri } \Lambda$ be closed and nonempty with $TS \subset \text{Fix } T \cap S$. Denote $(S + \delta \mathbb{B}) \cap \Lambda$ by S_δ for a nonnegative real δ. Suppose that, for all $\bar{\delta} > 0$ small enough, there are $\gamma \in (0, 1)$, a nonnegative scalar ϵ_i and a positive constant α bounded above by 1, such that,*

(i) *T is pointwise almost averaged at all $y \in S$ with averaging constant α and violation ϵ on $S_{\gamma^i \bar{\delta}}$, and*

(ii) *for $R_i := S_{\gamma^i \bar{\delta}} \setminus \left(\text{Fix } T \cap S + \gamma^{i+1} \bar{\delta} \mathbb{B}\right)$,*

    (i)

$$\text{dist}(x, S) \leq \text{dist}\left(x, \Phi^{-1}(\bar{y}) \cap \Lambda\right)$$

    for all $x \in R_i$ and $\bar{y} \in \Phi\left(\mathcal{P}_S x\right) \setminus \Phi(x)$,

    (ii) *Φ is metrically regular with gauge μ_i relative to Λ on $R_i \times \Phi\left(\mathcal{P}_S(R_i)\right)$, where μ_i satisfies*

$$\sup_{x \in R_i, \bar{y} \in \Phi(\mathcal{P}_S(R_i)), \bar{y} \notin \Phi(x)} \frac{\mu_i\left(\text{dist}\left(\bar{y}, \Phi(x)\right)\right)}{\text{dist}\left(\bar{y}, \Phi(x)\right)} \leq \kappa_i < \sqrt{\frac{1 - \alpha_i}{\epsilon_i \alpha_i}}. \tag{23.7}$$

Then, for any $x^0 \in \Lambda$ close enough to S, the iterates $x^{j+1} \in Tx^j$ satisfy $\text{dist}\left(x^j, \text{Fix } T \cap S\right) \to 0$ and

$$\text{dist}\left(x^{j+1}, \text{Fix } T \cap S\right) \leq c\,\text{dist}\left(x^j, S\right) \quad \left(\forall x^j \in R_i\right), \tag{23.8}$$

where $c_i := \sqrt{1 + \epsilon - \left(\frac{1-\alpha}{\kappa_i^2 \alpha}\right)} < 1$.

In particular, if $\kappa_i \leq \bar{\kappa} < \sqrt{\frac{1-\alpha}{\alpha\epsilon}}$ for all i large enough, then convergence is eventually at least R-linear with rate at most $\bar{c} := \sqrt{1 + \bar{\epsilon} - \left(\frac{1-\alpha}{\bar{\kappa}^2\alpha}\right)}$ to some point in Fix $T \cap S$. *If $S \cap \Lambda$ is a singleton, then (iii) is redundant and convergence is Q-linear.*

In both Opial's original statement as well as Theorem 23.4.1 averagedness is the essential property for convergence of iterative algorithms. Whereas assumption (ii) of Theorem 23.4.1 serves to quantify the convergence.

23.5 Regularities of Sets and Their Collection

In this section we connect the regularities of sets to regularities of the projectors on these, which effect the regularity of the mapping T. When dealing with nonconvex sets there are numerous set-regularity definitions available. A recent survey by Kruger et al. [26], sorted the different classes of nonconvex sets to highlight their dependencies and differences. Uniting several concepts of regularity, we propose to use the notion of ϵ-*set regularity* as introduced in [26] and refined in [27].

Definition 23.7 (ϵ-*set regularity*) Let $\Omega \subset \mathcal{E}$ be nonempty and let $\bar{x} \in \Omega$. The set Ω is said to be ϵ-*subregular relative to* Λ *at* $\bar{x}$ *for* $(\bar{y}, \bar{v}) \in$ gph (N_Ω) if it is locally closed at $\bar{x}$ and there exists an $\epsilon > 0$ together with a neighborhood U of $\bar{x}$ such that

$$\langle \bar{v} - (x - x^+), x^+ - \bar{y} \rangle \leq \epsilon \|\bar{v} - (x - x^+)\| \|x^+ - \bar{y}\|,$$
$$(\forall x \in \Lambda \cap U))(x^+ \in \mathcal{P}_\Omega x). \qquad (23.9)$$

if *for every* $\epsilon > 0$ there is a neighborhood (depending on ϵ) such that (23.9) holds, then Ω is said to be *subregular relative to* Λ *at* $\bar{x}$ *for* $(\bar{y}, \bar{v}) \in$ gph (N_Ω). If $\Lambda = \{\bar{x}\}$, then the qualifier "relative to" is dropped.

In the phase retrieval problem one type of nonconvexity, that is also covered by ϵ-subregularity, is *prox-regularity*.

Definition 23.8 (*prox-regular sets*) A closed set Ω is prox-regular at $\bar{x} \in \Omega$ if for $\bar{v} \in N_\Omega(\bar{x})$ there exist $\epsilon, \delta > 0$ such that

$$\frac{\epsilon}{2}\|x - c\|^2 \geq \langle v, x - c \rangle \quad (\forall x, c \in \Omega \cap \mathbb{B}_\delta(\bar{x}))(\forall v \in N_\Omega(c) \cap \mathbb{B}_\delta(\bar{v})).$$

This definition dates back to Federer [28] who called the property *sets with positive reach*. The definition presented here is taken from [29, Proposition 1.2]. The authors in [29] showed that their definition of prox-regularity at $\bar{x} \in C$ is equivalent to several statements. One of the most prominent might be local single-valuedness of the projector [29, Theorem 1.3] around $\bar{x}$. Kruger et al. showed that prox-regularity implies ϵ-subregularity in [26, Proposition 4(vi)]. As the next remark shows all constraint sets involved in the phase retrieval problem are, in fact, prox-regular.

Remark 23.5.1 (phase retrieval constraint sets are prox-regular) Of great importance for the convergence analysis of the introduced algorithms is the ϵ-subregularity of the measurement sets defined in (23.1). By [3, Example 3.1.b] circles are subregular at any of their points $\overline{x}$ for all $(\overline{x}, v)$ in the graph of the normal cone of the sets. As mentioned before ϵ-subregularity covers a divers range of regularity notions for sets. The measurement sets investigated here are in fact shown to be semi-algebraic [30, Proposition 3.5] and prox-regular by [29, Theorem 1.3] and (6.11).

The other sets that are involved in the phase retrieval problem are the qualitative constraints introduced in (23.2) or mentioned before. Except for the amplitude constraint and the sparsity constraint all of these sets are convex and thus by [3, Proposition 3.1 (vii)] subregular. Fortunately, the amplitude constraint describes coordinatewise circles when the other coordinates are fixed, like the measurement constraint. Hence, the amplitude constraint is ϵ-subregular as well (and additionally semi-algebraic and prox-regular). The sparsity constraint $\mathcal{A}_s$ is prox-regular at all points $\overline{x}$ satisfying $\|\overline{x}\|_0 = s$ (similar to the proof in [14, Proposition 4.4]).

By [12, Proposition 4.8] the projector onto a closed convex set is averaged with constant $\alpha = 1/2$. Allowing sets to have a more general regularity, here prox-regularity, yield regularity of the projectors as well.

Proposition 23.9 (projectors and reflectors onto prox-regular sets) *Let $\Omega \subset \mathcal{E}$ be nonempty closed, and let U be a neighborhood of $\overline{x} \in C$. Let $\Lambda \subset \Omega \cap U$. If Ω is prox-regular at $\overline{x}$ with constant ϵ on the neighborhood U, then the following hold.*

(i) *Let $\epsilon \in [0, 1)$. The projector $\mathcal{P}_\Omega$ is pointwise almost firmly nonexpansive at each $y \in \Lambda$ with violation $\epsilon_2 := 2\epsilon + 2\epsilon^2$ on U. That is, at each $y \in \Lambda$*

$$\|x - y\|^2 + \|x' - x\| \le (1 + \epsilon_2) \|x' - y'\|^2 \quad (\forall x' \in)\, (\forall x \in \mathcal{P}_\Omega x') .$$

(ii) *The reflector $\mathcal{R}_\Omega$ is pointwise almost nonexpansive at each $y \in \Lambda$ with violation $\epsilon_3 := 4\epsilon + 4\epsilon^2$ on U; that is, for all $y \in \Lambda$*

$$\|x - y\| \le \sqrt{1 + \epsilon_3}\|x' - y\| \quad (\forall x' \in U)\,(\forall x \in \mathcal{P}_\Omega x') .$$

Proof By [26, Proposition 4(vi)] prox-regularity of Ω at $\overline{x}$ implies that the set Ω is ϵ-subregular at $\overline{x}$ for all $(c, v) \in \mathrm{gph} N_\Omega$, where $c \in U$. The result follows then from [3, Theorem 3.1].

Note that Proposition 23.9 presents a special case of [3, Theorem 3.1], where the authors allowed their sets to be ϵ-subregular for certain normal vectors. By Proposition 23.5 compositions and convex combinations of averaged mappings are again averaged. Combining this with Proposition 23.9 implies that compositions of projectors are averaged. Thus, the algorithms presented in Sect. 23.3.2 are pointwise almost averaged as we see in Sect. 23.7.

Whereas the regularity of the individual sets imply almost averagedess of the mapping T, metric regularity relies on the regularity of the whole collection of sets

$\{\Omega_1, \Omega_2, \ldots, \Omega_m\}$. The idea of regularities of collections of sets traces back to [26, Theorem 3] by Kruger, Luke and Thao, but the analysis there covers only consistent feasibility problems, i.e. the intersection of sets is nonempty. A generalized notion of *subtransversality* proposed in [3, Definition 3.2] includes inconsistent settings too.

Definition 23.10 (*subtransversal collection of sets*) Let $\{\Omega_1, \ldots, \Omega_m\}$ be a collection of nonempty closed subsets of $\mathcal{E}$ and define $\Upsilon : \mathcal{E}^m \rightrightarrows \mathcal{E}^m$ by $\Upsilon(x) :=$ $P_\Omega (\Pi x) - \Pi x$ where $\Omega := \Omega_1 \times \Omega_2 \times \cdots \times \Omega_m$, the projection P_Ω is with respect to the Euclidean norm on $\mathcal{E}^m$ and $\Pi : x = (x_1, x_2, \ldots, x_m) \mapsto (x_2, x_3, \ldots, x_m, x_1)$ is the permutation mapping on the product space $\mathcal{E}^m$ for $x_j \in \mathcal{E}$ $(j = 1, 2, \ldots, m)$. Let $\overline{x} = (\overline{x}_1, \overline{x}_2, \ldots, \overline{x}_m) \in \mathcal{E}^m$ and $\overline{y} \in \Upsilon(\overline{x})$. The collection of sets is said to be *subtransversal with gauge μ relative to* $\Lambda \subset \mathcal{E}^m$ *at* $\overline{x}$ *for* $\overline{y}$ if Υ is metrically subregular at $\overline{x}$ for $\overline{y}$ on some neighborhood U of $\overline{x}$ (metrically regular on $U \times \{\overline{y}\}$) with gauge μ relative to Λ. As in Definition 23.6, when $\mu(t) = \kappa t$, $\forall t \in [0, \infty)$, one says "constant κ" instead of "gauge $\mu(t) = \kappa t$". When $\Lambda = \mathcal{E}$, the quantifier "relative to" is dropped.

In [3, Proposition 3.3] Luke et al. showed that for a *consistent* feasibility problem subtransversality of the collection of sets is equivalent to what is elsewhere recognized as *linear regularity* of the collection [31].

23.6 Analysis of Cyclic Projections

Having introduced the main tools for convergence, this section is devoted to an explicit demonstration of how this framework can be applied. In particular, we present the main steps of the convergence analysis of the cyclic projection mapping as done by Luke et al. in [3].

As introduced in Algorithm 6.2.1 the method of cyclic projections on a finite collection of closed subsets of $\mathcal{E}$, $\{\Omega_1, \Omega_2, \ldots, \Omega_m\}$ $(m \geq 2)$, is defined by the mapping

$$\mathcal{P}_0 := P_{\Omega_1} P_{\Omega_2} \cdots P_{\Omega_m} P_{\Omega_1}, \tag{23.10}$$

that we denote for notational simplicity by $\mathcal{P}_0$. For an initial point u^0 the algorithm generates a sequence $(u^k)_{k \in \mathbb{N}}$ by $u^{k+1} \in \mathcal{P}_0 u^k$. For the analysis of $\mathcal{P}_0$ it is convenient to introduce some auxiliary sets. We denote by Ω the product of the sets Ω_j on $\mathcal{E}^m$,

$$\Omega := \Omega_1 \times \Omega_2 \times \cdots \times \Omega_m.$$

Let $\overline{u} \in \text{Fix}\, \mathcal{P}_0$ and fix $\overline{\zeta} \in \mathcal{Z}(\overline{u})$ where

$$\mathcal{Z}(u) := \left\{ \zeta := z - \Pi z \,\middle|\, z \in W_0 \subset \mathcal{E}^m, \ z_1 = u \right\} \tag{23.11}$$

for

$$W_0 := \left\{ x \in \mathcal{E}^m \mid x_m \in \mathcal{P}_{\Omega_m} x_1, \ x_j \in \mathcal{P}_{\Omega_j} x_{j+1}, \ j = 1, 2, \ldots, m-1 \right\}. \quad (23.12)$$

Note that $\sum_{j=1}^m \bar{\zeta}_j = 0$. The elements of W_0 are all cycles of the cyclic projection method, where each coordinate of x corresponds to an inner iterate of $\mathcal{P}_0$. The first coordinate x_1 of $x \in W_0$ is, thus, a fixed point of $\mathcal{P}_0$. The vectors $\zeta \in \mathcal{Z}(u)$ are called *difference vectors*. Their coordinate entries provide information about the gaps between the inner iterates of a cycle of the mapping $\mathcal{P}_0$.

To monitor the inner iterations, we consider the cyclic projection algorithm lifted to the product space $\mathcal{E}^m$. That is, generate the sequence $(x^k)_{k \in \mathbb{N}}$ by $x^{k+1} \in T_{\bar{\zeta}} x^k$ with

$$T_{\bar{\zeta}} : \mathcal{E}^m \rightrightarrows \mathcal{E}^m \, x \mapsto \left\{ \left(x_1^+, x_1^+ - \bar{\zeta}_1, \ldots, x_1^+ - \sum_{j=1}^{m-1} \bar{\zeta}_j \right) \, \middle| \, x_1^+ \in \mathcal{P}_0 x_1 \right\} \quad (23.13)$$

for $\bar{\zeta} \in \mathcal{Z}(\bar{u})$ where $\bar{u} \in \mathsf{Fix}\,\mathcal{P}_0$. Thus, the first entry of $T_{\bar{\zeta}}$ belongs to the cyclic projection mapping $\mathcal{P}_0$. Whereas the other entries of $T_{\bar{\zeta}} x$ indicate how close or distant x_1^+ is from a certain cycle specified by $\bar{\zeta}$. In order to isolate cycles, we restrict our attention to relevant subsets of $\mathcal{E}^m$. These are

$$W(\bar{\zeta}) := \left\{ x \in \mathcal{E}^m \mid x - \Pi x = \bar{\zeta} \right\}, \quad (23.14)$$

$$L := \text{an affine subspace with } T_{\bar{\zeta}} : L \rightrightarrows L, \quad (23.15)$$

$$\Lambda := L \cap W(\bar{\zeta}). \quad (23.16)$$

The set $W(\bar{\zeta})$ contains all points whose entries have a certain distance to each other, namely $\bar{\zeta}_i$. In particular, $W(\bar{\zeta})$ contains all fixed points of $T_{\bar{\zeta}}$. The affine subspace L is used to restrict the analysis to an affine subspace that contains the iterates x^k of $T_{\bar{\zeta}}$.

To apply the convergence framework, Theorem 23.4.1, there are two major steps we have to take. First, we have to show that the mapping is averaged. Since the cyclic projection mapping is, as its name suggests, a composition of projectors averagedness, this not hard to show by the concepts presented in Sect. 23.5. Second, metric subregularity needs to be proven. For this, we state an auxiliary result that relates metric subregularity to subtransversality of the collection of sets (see [3, Proposition 3.4]).

Proposition 23.11 (metric subregularity of cyclic projections) *Let $\bar{u} \in \mathsf{Fix}\,\mathcal{P}_0$ and $\bar{\zeta} \in \mathcal{Z}(\bar{u})$ and let $\bar{x} = (\bar{x}_1, \bar{x}_2, \ldots, \bar{x}_m) \in W_0$ satisfy $\bar{\zeta} = \bar{x} - \Pi \bar{x}$ with $\bar{x}_1 = \bar{u}$. For L an affine subspace containing $\bar{x}$, let $T_{\bar{\zeta}} : L \rightrightarrows L$ and define the mappings for $\Phi_{\bar{\zeta}} := T_{\bar{\zeta}} - \mathrm{Id}$ and $\Upsilon := (\mathcal{P}_\Omega - \mathrm{Id}) \circ \Pi$. Suppose the following hold:*

(i) *the collection of sets $\{\Omega_1, \Omega_2, \ldots, \Omega_m\}$ is subtransversal at $\bar{x}$ for $\bar{\zeta}$ relative to $\Lambda := L \cap W(\bar{\zeta})$ with constant κ and neighborhood U of $\bar{x}$;*
(ii) *there exists a positive constant σ such that*

$$\text{dist}\left(\bar{\zeta}, \Upsilon(x)\right) \le \sigma \text{dist}\left(0, \Phi_{\bar{\zeta}}(x)\right), \quad \forall x \in \Lambda \cap U \text{ with } x_1 \in \Omega_1.$$

Then Φ is metrically subregular for 0 on U (metrically regular on $U \times \{0\}$) relative to Λ with constant $\bar{\kappa} = \kappa\sigma$.

Proposition 23.11 indicates that subtransversality plus the additional assumption (ii) are enough to deduce metric subregularity of $\Phi_{\bar{\zeta}} := T_{\bar{\zeta}} - \text{Id}$ as required in Theorem 23.4.1. Using this connection and the development in Sect. 23.5 about almost averagedness we can state the following convergence result which is an implication of Theorem 23.4.1.

Theorem 23.6.1 (convergence of cyclic projections) *Let $S_0 \subset \text{Fix} \, \mathcal{P}_0 \ne \emptyset$ and $Z := \cup_{u \in S_0} \mathcal{Z}(u)$. Define*

$$S_j := \cup_{\zeta \in Z} \left(S_0 - \sum_{i=1}^{j-1} \zeta_i \right) \quad j = 1, 2, \ldots, m. \tag{23.17}$$

Let $U := U_1 \times U_2 \times \cdots \times U_m$ be a neighborhood of $S := S_1 \times S_2 \times \cdots \times S_m$ and suppose that

$$\mathcal{P}_{\Omega_j}\left(u - \sum_{i=1}^{j} \zeta_i\right) \subseteq S_0 - \sum_{i=1}^{j-1} \zeta_i \quad \forall u \in S_0, \, \forall \zeta \in Z \text{ for each } j = 1, 2, \ldots, m,$$
$$\tag{23.18}$$

$$\mathcal{P}_{\Omega_j} U_{j+1} \subseteq U_j \text{ for each } j = 1, 2, \ldots, m \quad (U_{m+1} := U_1). \tag{23.19}$$

For fixed $\bar{\zeta} \in Z$ and $\bar{x} \in S$ with $\bar{\zeta} = \bar{x} - \Pi\bar{x}$, generate the sequence $(x^k)_{k \in \mathbb{N}}$ by $x^{k+1} \in T_{\bar{\zeta}} x^k$ for $T_{\bar{\zeta}}$ defined by (23.13), seeded by a point $x^0 \in W(\bar{\zeta}) \cap U$ for $W(\bar{\zeta})$ defined by (23.14) with $x_1^0 \in \Omega_1 \cap U_1$.

Suppose that, for $\Lambda := L \cap \text{aff}\left(\cup_{\zeta \in Z} W(\zeta)\right) \supset S$ such that $T_\zeta : \Lambda \rightrightarrows \Lambda$ for all $\zeta \in Z$ and an affine subspace $L \supset \text{aff}(x^k))k \in \mathbb{N}$, the following hold:

(i) *Ω_j is prox-regular at all $\hat{x}_j \in S_j$ with constant $\epsilon_j \in (0, 1)$ on the neighborhood U_j for $j = 1, 2, \ldots, m$;*

(ii) *for each $\hat{x} = (\hat{x}_1, \hat{x}_2, \ldots, \hat{x}_m) \in S$, the collection of sets $\{\Omega_1, \Omega_2, \ldots, \Omega_m\}$ is subtransversal at $\hat{x}$ for $\hat{\zeta} := \hat{x} - \Pi\hat{x}$ relative to Λ with constant κ on the neighborhood U;*

(iii) *for $\Upsilon_{\bar{\zeta}} := T_{\bar{\zeta}} - \text{Id}$ and $\Psi := (\mathcal{P}_\Omega - \text{Id}) \circ \Pi$ there exists a positive constant σ such that for all $\bar{\zeta} \in Z$*

$$\text{dist}\left(\hat{\zeta}, \Upsilon(x)\right) \le \sigma \, \text{dist}\left(0, \Phi_{\bar{\zeta}}(x)\right)$$

holds whenever $x \in \Lambda \cap U$ with $x_1 \in \Omega_1$;

(iv) *$\text{dist}(x, S) \le \text{dist}\left(x, \Phi_{\bar{\zeta}}^{-1}(0) \cap \Lambda\right)$ for all $x \in U \cap \Lambda$, for all $\hat{\zeta} \in Z$.*

Then the sequence $(x^k)_{k \in \mathbb{N}}$ *initialized by a point* $x^0 \in W(\bar{\zeta}) \cap U$ *with* $x_1^0 \in \Omega_1 \cap U_1$
satisfies

$$\text{dist} \left(x^{k+1}, \text{Fix } T_{\bar{\zeta}} \cap S \right) \leq c \, \text{dist} \left(x^k, S \right)$$

whenever $x^k \in U$ *with* $c := \sqrt{1 + \bar{\epsilon} - \frac{1-\alpha}{\alpha \bar{\kappa}^2}}$ *where* $\bar{\epsilon} := \prod_{j=1}^{m} \left(1 + \widetilde{\epsilon}_j \right) - 1$, $\widetilde{\epsilon}_j :=$
$4\epsilon_j \frac{1+\epsilon_j}{(1-\epsilon_j)^2}$, $\alpha := \frac{m}{m+1}$ *and* $\bar{\kappa} := \kappa \sigma$. *If, in addition,* $\bar{\kappa} < \sqrt{\frac{1-\alpha}{\bar{\epsilon}\alpha}}$, *then* dist
$\left(x^k, \text{Fix } T_{\bar{\zeta}} \cap S \right) \to 0$, *and hence* dist $\left(x_1^k, \text{Fix } \mathcal{P}_0 \cap S_1 \right) \to 0$ *at least linearly with*
rate $c < 1$.

Proof This is a special case of [3, Theorem 3.2] when the sets are prox-regular.

Remark 23.6.2 Theorem 23.6.1 is rather long and technical at first sight, though
the pieces are easily parsed. Equations (23.17)–(23.19) force the iterations to stay in
specific neighborhoods. This is needed to apply Proposition 23.9 with the help of
(i) to deduce pointwise almost averagedness of $\mathcal{P}_0$ and likewise of $T_{\bar{\zeta}}$. Assumptions
(ii) and (iii) then yield metric subregularity of $\Phi_{\bar{\zeta}} = T_{\bar{\zeta}} - \text{Id}$ by Proposition 23.11.
This is where the construction in the product space comes into play. Working on
$\mathcal{E}^m$, we were able to use subtransversality to show metric subregularity of $\Phi_{\bar{\zeta}}$. It is
worth mentioning that, until now, we were not able to show metric subregularity for
the mapping directly associated to $\mathcal{P}_0$. Adding assumption (iv) in Theorem 23.6.1
we can finally apply Theorem 23.4.1 and deduce convergence of $T_{\bar{\zeta}}$ with the given
constants. At this point the definition of $T_{\bar{\zeta}}$ becomes crucial. Since the first iterate
of the sequence x^k generated under the mapping $T_{\bar{\zeta}}$ is nothing more than applying
the method of cyclic projections $\mathcal{P}_0$, convergence of x^k implies convergence of x_1^k,
that is, the sequence generated by cyclic projections. In [25] Luke et al. discussed
the necessity of subtransversality for alternating projections to converge R-linearly.

23.7   Application to Phase Retrieval Algorithms

In Sect. 23.6 we have seen how to apply Theorem 23.4.1 on the method of cyclic
projections. This section is devoted to the analysis of other well known algorithms
which we introduced in Sect. 23.3.2. The analysis in Sect. 23.6 focuses on showing
how to satisfy the assumptions of Theorem 23.4.1 in the context of set-feasibility.
This section aims to provide a broad intuition of the convergence of projection based
algorithms used to solve the phase retrieval problem. This explains also why the
statements given next are presented in a cartoon-like manner. The statements include
only the most important parts that yield local convergence, but not how to construct
it nor at which rate. Nevertheless, these are verifiable by following the approach in
Sect. 23.6.

Corollary 23.12 (convergence of the error reduction algorithm) *Let* Fix $\mathcal{P}_{\mathfrak{S}} \mathcal{P}_{\mathcal{M}_1} \neq$
$\emptyset$. *The error reduction algorithm, that is alternating projections as discussed*

*in Sect. 6.2.3 on the sets $\mathfrak{S}$ and $\mathcal{M}_1$, converges locally linearly to a point $\tilde{x} \in$
Fix $\mathcal{P}_{\mathfrak{S}}\mathcal{P}_{\mathcal{M}_1}$ whenever the mapping $\Phi = \mathcal{P}_{\mathfrak{S}}\mathcal{P}_{\mathcal{M}_1} - \mathrm{Id}$ is locally metrically subregular at its zeros.*

Proof Following Luke et al. in [32, Sect. 3.2.2], we represent $\mathbb{C}$ as $\mathbb{R}^2$ and reformulate the phase retrieval problem as a feasibility problem with entrywise values in $\mathbb{R}^2$. Then this is an application of Theorem 23.4.1 using Remark 23.5.1.

Remark 23.7.1 In contrast to Theorem 23.6.1 metric subregularity is required directly in Theorem 23.12. Equivalently, we could demand subtransversality of the collection of sets $\{\mathfrak{S}, \mathcal{M}_1\}$ plus the additional assumption (iii) in Theorem 23.6.1. The problem here is, that, until now, it is not clear when and where these two assumptions are satisfied. Illustrative examples and numerical simulations indicate that they hold in many instances. Nevertheless, there are certain situations when at least one of the two assumptions is violated (see for instance [33]). Moreover, allowing metric subregularity under some gauge depicts the reality sometimes better than restricting the analysis to a linear setting. One example is the setting of alternating projections applied to the sphere $\mathbb{S}$ and a line tangent to $\mathbb{S}$ at $\bar{x} = (0, -1)$. In this instance the algorithm does not converge linearly to $\bar{x}$, although it converges depending on the initial point (see for instance [3]). This problem is not only interesting for the type of convergence, but also when it comes to the actual numerical implementation of algorithms. Although sets in real-life applications intersect tangentially on a set of measure zero, beyond a certain numerical accuracy the distinction between tangential intersection and linear convergence with a rate constant within 15 digits of 1 is rather academic. Having a relatively large gap between sets for inconsistent feasibility is in fact an advantage for the numerical performance of an algorithm.

Theorem 23.7.2 (convergence of Fienup's HIO method) *Let $\beta_n = 1$ for all n and* Fix $\frac{1}{2}\left(\mathcal{R}_{\mathfrak{S}}\mathcal{R}_{\mathcal{M}_1} + \mathrm{Id}\right) \neq \emptyset$. *The HIO algorithm, defined in (6.9) that is Douglas-Rachford as defined in (6.15) on the sets $\mathfrak{S}$ and $\mathcal{M}_1$, converges locally linearly to a point $\tilde{x} \in$* Fix $\frac{1}{2}\left(\mathcal{R}_{\mathfrak{S}}\mathcal{R}_{\mathcal{M}_1} + \mathrm{Id}\right)$ *whenever the mapping $\Phi = \frac{1}{2}\left(\mathcal{R}_{\mathfrak{S}}\mathcal{R}_{\mathcal{M}_1} + \mathrm{Id}\right) - \mathrm{Id}$ is locally metrically subregular at its zeros.*

Proof Since Fienup's HIO for $\beta_n = 1$ for all n can be identified with the Douglas-Rachford method the result follows from [3, Theorem 3.4]. □

Even if one had an infinite detector, noisy measurements make the phase retrieval problem almost always inconsistent. It is easy to prove [8, Theorem 3.13] that, in this case, Fix $\frac{1}{2}\left(\mathcal{R}_{\mathfrak{S}}\mathcal{R}_{\mathcal{M}_1} + \mathrm{Id}\right) = \emptyset$ and so Φ does not possess zeros. Consequently, Fienup's HIO algorithm *cannot* converge. To circumvent this problem, one can use a relaxed version of Douglas-Rachford, the relaxed averaged alternating reflections method (RAAR), that we introduced in Sect. 6.1.2 which is adapted to inconsistent feasibility.

Theorem 23.7.3 (convergence of RAAR) *relaxed averaged alternating reflections. Let $\bar{x} \in$* Fix T_{RAAR} *for T_{RAAR} defined in (6.22). The relaxed averaged alternating reflections applied to a phase retrieval problem converges locally linearly to a point*

$\tilde{x} \in \mathsf{Fix}\, T_{RAAR}$ *whenever the mapping* $\Phi = \frac{\lambda}{2}\left(\mathcal{R}_{\mathfrak{S}}\mathcal{R}_{\mathcal{M}_1} + \mathrm{Id}\right) + (1 - \lambda)\mathcal{P}_{\mathcal{M}_1} - \mathrm{Id}$
is locally metrically subregular at its zeros.

A detailed proof of the convergence analysis for the relaxed averaged alternating reflection algorithm can be found in [33] by the authors of this chapter. There we use subtransversality of the collections of sets in general feasibility problems to make the connection to metric subregularity of the algorithm in question. The analysis does not use prox-regularity as the desired type of regularity for sets yielding the almost averaging property, but rather the property of being *super-regular at a distance*. This extends notions of regularity of sets to their effect on points that are not in the sets. Their definition is in line with ϵ-subregularity and is thus connected to the analysis of [3].

Remark 23.7.4 In [33] we not only provided a convergence statement for the relaxed averaged alternating reflections method, but also gave a description of the fixed point set of the underlying mapping. For super-regular sets at a distance, the fixed points, if they exist, are either points in the intersection of both sets or relate to the local gap between these, if the intersection of the sets is empty. This result is in line with [11] where Luke studied the case of one set being convex and the other prox-regular. In contrast to the original Douglas-Rachford algorithm, the main advantage of the relaxed version is that existence of fixed points does not depend on whether the feasibility problem is consistent. Connecting this observation to the convergence analysis presented here, in practice the Douglas-Rachford/HIO is much less stable than the relaxed version.

Following the ideas above, it is not hard to show that most projection methods are pointwise almost averaged mappings when applied to the phase retrieval problem. Nonetheless, the property of metric subregularity is still an open problem in some important cases. Thus, local convergence can be easily verified, but it is hard to quantify.

23.8   Final Remarks

When it comes to computing (see Remark 23.7.1), whether a method converges, let alone determining the rate depends on the numerical precision. But also inconsistency has an impact on the numerical performance. Closely related to this, we want to stress another feature of the analysis surveyed here. That is, sometimes less information can lead to better performance of an algorithm. For a demonstration we analyze a data set recorded by undergraduates at the X-Ray Physics Institute at the University of Göttingen. It is an optical diffraction image with model constraints $\sqrt{I}_{j,i}$, $j = 1, 2, \ldots, m$, as in (23.1) with $m = 1$ and n the dimension of the image and additional support constraint. The full data set has dimension $n = 1392 \times 1040$, the cropped data set $n = 128^2$. The graphs shown in Figs. 23.1 and 23.2 are produced by applying the alternating projection algorithm, i.e. error reduction, on the data sets individually.

As it turns out alternating projections on the full data set (Fig. 23.2) shows a worse convergence behavior than the image with the limited data set (Fig. 23.1). Not only that the algorithm needs more iterations to reach a certain accuracy (9.8485×10^4 instead of 666), but also the rate of linear convergence when the iterates reach a suitable neighborhood is worse. Noteworthy is the observed gap in both problem instances. In the full data set version the gap is smaller than in the version with a limited data set. We conjecture that this behavior is closely related to the property of metric subregularity, or in the context of set feasibility, subtransversality. The more, and better, information one has, the closer the constraint sets come to intersect. But this can included cases in which the sets intersect transversally as well. In cases like these the method of alternating projections does not have to converge locally linearly but can show a sublinear convergence behavior (see for instance [3, Remark 3.2]). The take home message in this context is that more information does not have to yield a better image when applying numerical algorithms. This is good news and bad news for these algorithms. The good news is that one can profit from implicit regularization with smaller problem sizes. The bad news is that this indicates a type of *dimension dependence* of these methods: the higher the dimension, the worse the constants in the linear convergence rates. This is not surprising and points to the need for models that lead to algorithms whose performance (that is, regularity) is dimension independent. While our discussion here focuses on the theoretical analysis rather than the comparison of the presented algorithms we point the reader to a study by Luke et al. [22], where the authors present a thorough review of first-order proximal methods for phase retrieval algorithms.

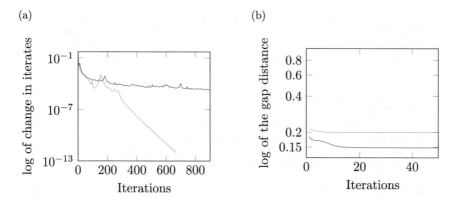

(a)

(b)

Fig. 23.1 Alternating projections applied to the data set "tasse" for full data (blue) and cropped data (orange). **a** Change in iterates until iteration 900. **b** Gap distance until iteration 50

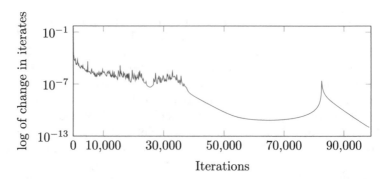

Fig. 23.2 Change in iterates for full data set "tasse"

References

1. Bauschke, H.H., Combettes, P.L., Luke, D.R.: Phase retrieval, error reduction algorithm, and Fienup variants: a view from convex optimization. J. Opt. Soc. Am. A **19**(7), 1334–1345 (2002)
2. Marchesini, S.: A unified evaluation of iterative projection algorithms for phase retrieval. Rev. Sci. Instrum. **78**(1), 011301 (2007)
3. Luke, D.R., Thao, N.H., Tam, M.K.: Quantitative convergence analysis of iterated expansive, set-valued mappings. Math. Oper. Res. **43**(4), 1143–1176 (2018). https://doi.org/10.1287/moor.2017.0898
4. Opial, Z.: Weak convergence of the sequence of successive approximations for nonexpansive mappings. Bull. Am. Math. Soc. **73**(4), 591–597 (1967)
5. Dontchev, A.L., Rockafellar, R.T.: Implicit Functions and Solution Mapppings, 2nd edn. Springer, Dordrecht (2014)
6. Ioffe, A.D.: Regularity on a fixed set. SIAM J. Optim. **21**(4), 1345–1370 (2011)
7. Ioffe, A.D.: Nonlinear regularity models. Math. Program. **139**(1–2), 223–242 (2013)
8. Bauschke, H.H., Combettes, P.L., Luke, D.R.: Finding best approximation pairs relative to two closed convex sets in Hilbert spaces. J. Approx. Theory **127**(2), 178–192 (2004)
9. Borwein, J.M., Tam, M.K.: The cyclic Douglas-Rachford method for inconsistent feasibility problems. J. Nonlinear Convex Anal. **16**, 537–584 (2015)
10. Luke, D.R.: Relaxed averaged alternating reflections for diffraction imaging. Inverse Probl. **21**(1), 37 (2005)
11. Luke, D.R.: Finding best approximation pairs relative to a convex and prox-regular set in a Hilbert space. SIAM J. Optim. **19**(2), 714–739 (2008)
12. Bauschke, H.H., Combettes, P.L.: Convex Analysis and Monotone Operator Theory in Hilbert Spaces. CMS Books in Mathematics/Ouvrages de Mathématiques de la SMC. Springer, New York (2011)
13. Hesse, R., Luke, D.R., Neumann, P.: Alternating projections and Douglas-Rachford for sparse affine feasibility. IEEE Trans. Signal Process. **62**(18), 4868–4881 (2014)
14. Tam, M.K.: Regularity properties of non-negative sparsity sets. J. Math. Anal. Appl. **447**(2), 758–777 (2017)
15. Rockafellar, R., Wets, R.: Variational Analysis. Springer, Berlin (1998)
16. Mordukhovich, B.S.: Variational Analysis and Applications. Springer Monographs in Mathematics. Springer International Publishing (2018). https://books.google.de/books?id=6DxnDwAAQBAJ
17. Levi, A., Stark, H.: Image restoration by the method of generalized projections with application to restoration from magnitude. JOSA A **1**(9), 932–943 (1984)
18. Fienup, J.R.: Phase retrieval algorithms: a comparison. Appl. Opt. **21**(15), 2758–2769 (1982)

19. Bauschke, H.H., Combettes, P.L., Luke, D.R.: Hybrid projection-reflection method for phase retrieval. J. Opt. Soc. Am. A **20**(6), 1025–1034 (2003)
20. Elser, V.: Phase retrieval by iterated projections. JOSA A **20**(1), 40–55 (2003)
21. Abrahams, J.P., Leslie, A.G.W.: Methods used in the structure determination of bovine mitochondrial F1 atpase. Acta Crystallogr. Sect. D: Biol. Crystallogr. **52**(1), 30–42 (1996)
22. Luke, D.R., Sabach, S., Teboulle, M.: Optimization on spheres: models and Proximal algorithms with computational performance comparisons. SIAM J. Math. Data Sci. **1**(3), 408–445 (2019)
23. Aze, D.: A unified theory for metric regularity of multifunctions. J. Convex Anal. **13**(2), 225 (2006)
24. Penot, J.P.: Metric regularity, openness and lipschitzian behavior of multifunctions. Nonlinear Anal.: Theory Methods Appl. **13**(6), 629–643 (1989)
25. Luke, D.R., Teboulle, M., Thao, N.H.: Necessary conditions for linear convergence of iterated expansive, set-valued mappings. Math. Program. 180:1–31 (2020). https://doi.org/10.1007/s10107-018-1343-8
26. Kruger, A.Y., Luke, D.R., Thao, N.H.: Set regularities and feasibility problems. Math. Program. **168**(1–2), 279–311 (2018)
27. Daniilidis, A., Luke, D.R., Tam, M.K.: Characterizations of super-regularity and its variants. In: Splitting Algorithms. Modern Operator Theory and Applications. Springer (2019). https://arxiv.org/abs/1808.04978
28. Federer, H.: Curvature measures. Trans. Am. Math. Soc. **93**(3), 418–491 (1959)
29. Poliquin, R.A., Rockafellar, R., Thibault, L.: Local differentiability of distance functions. Trans. Am. Math. Soc. **352**(11), 5231–5249 (2000)
30. Hesse, R., Luke, D.R., Sabach, S., Tam, M.K.: Proximal heterogeneous block implicit-explicit method and application to blind ptychographic diffraction imaging. SIAM J. Imaging Sci. **8**(1), 426–457 (2015)
31. Bauschke, H.H., Borwein, J.M.: On projection algorithms for solving convex feasibility problems. SIAM Rev. **38**(3), 367–426 (1996)
32. Luke, D.R., Burke, J.V., Lyon, R.G.: Optical wavefront reconstruction: theory and numerical methods. SIAM Rev. **44**(2), 169 (2002)
33. Luke, D.R., Martins, A.L.: Convergence analysis of the relaxed Douglas-Rachford algorithm. SIAM J. Optim. (to appear). https://arxiv.org/abs/1811.11590

Chapter 24
One-Dimensional Discrete-Time Phase Retrieval

Robert Beinert and Gerlind Plonka

Abstract The phase retrieval problem has a long and rich history with applications in physics and engineering such as crystallography, astronomy, and laser optics. Usually, the phase retrieval consists in recovering a real-valued or complex-valued signal from the intensity measurements of its Fourier transform. If the complete phase information in frequency domain is lost then the problem of signal reconstruction is severely ill-posed and possesses many non-trivial ambiguities. Therefore, it can only be solved using appropriate additional signal information. We restrict ourselves to one-dimensional discrete-time phase retrieval from Fourier intensities and particularly consider signals with finite support. In the first part of this section, we study the structure of the arising ambiguities of the phase retrieval problem and show how they can be characterized using the given Fourier intensity. Employing these observations, in the second part, we study different kinds of a priori assumptions on the signal, where we are especially interested in their ability to reduce the non-trivial ambiguities or even to ensure uniqueness of the solution. In particular, we consider the assumption of non-negativity of the solution signal, additional magnitudes or phases of some signal components in time domain, or additional intensities of interference measurements in frequency domain. Finally, we transfer our results to phase retrieval problems where the intensity measurements arise, for example, from the Fresnel or fractional Fourier transform.

2010 Mathematics Subject Classification: 42A05 · 94A08 · 94A12 · 94A20

R. Beinert (✉)
Institute for Mathematics and Scientific Computing, University of Graz,
Heinrichstraße 36, 8010 Graz, Austria
e-mail: robert.beinert@uni-graz.at

G. Plonka
Institute for Numerical and Applied Mathematics, Universität Göttingen,
Lotzestraße 16-18, 37083 Göttingen, Germany
e-mail: plonka@math.uni-goettingen.de

24.1 Introduction

In the classical phase retrieval problem, one is usually faced with the recovery of a complex-valued signal from intensity measurements of its Fourier transform. Recovery problems of this kind have many interesting applications in physics and engineering like crystallography [1–3], astronomy [4, 5], and laser optics [6, 7]. We particularly refer to Chap. 2. Without further information about the unknown signal, the phase retrieval problem is highly ambiguous such that the recovery of the true solution within the solution set is nearly hopeless.

In this chapter, we consider the one-dimensional discrete-time variant of the phase retrieval problem, where we restrict ourselves to the recovery of complex-valued signals with finite support length. The solution set of this problem can be characterized by investigating and factorizing the related autocorrelation function, which coincides with the squared given Fourier intensity, see [4, 8, 9]. As a consequence of this characterization, we show how ambiguities of the discrete-time phase retrieval problem are related to the true solution signal. Trivial ambiguities are caused by multiplication with a unimodular constant, time-shifts, and reflection and conjugation. Non-trivial ambiguities are essentially obtained by conjugation of linear factors of the algebraic polynomial being defined by the signal values [8]. The number of these non-trivial ambiguities essentially depends on the structure of the given intensities [10].

Depending on the application, one can incorporate different a priori conditions or further information about the unknown signal in order to get rid of the unwanted ambiguities. One approach to reduce the solution set is to assume that the unknown signal is real-valued and non-negative. The non-negativity condition is usually employed if the original signal represents an intensity or a probability density, see for instance [3, 4, 11]. The a priori non-negativity is, moreover, exploited by a variety of numerical methods like the alternating projection algorithm [5, 11–13] or adapted multilevel Gauß–Newton methods [6]. In the one-dimensional case considered here, the non-negativity constraint is, however, very erratic [14]. In special cases, the restricted phase retrieval problem can become uniquely solvable. However, in many situations, the non-negativity assumption may either not reduce the solution set at all or may lead to an empty solution set.

Sometimes, like in wave front sensing and laser optics [7], one has additionally access to the magnitudes of the unknown signal itself. The obtained restricted one-dimensional phase retrieval problem with a priori magnitude information in time domain can be efficiently solved by multilevel Gauß–Newton methods [6, 15, 16]. While these numerical methods work well in certain cases, their stability strongly depends on the given Fourier intensities [17]. Moreover, the algorithms can converge to approximate solutions which essentially differ from the true solution signal. On the basis of these numerical observations, we study the question whether the knowledge of magnitudes of the signal components can guarantee uniqueness. Our findings imply that the related phase retrieval problems are uniquely solvable for almost all finite-length signals [8, 18]. But there also exist instances of non-unique phase retrieval problems with given magnitudes of the signal components [10]. Our results on uniqueness of solutions can be transferred to phase retrieval problems with additional phase information in time domain [18].

A further approach to reduce the solution set or to ensure uniqueness is to exploit additional measurements in frequency domain, which arise from the interference of the unknown true signal with an appropriate reference signal. If the reference signal is known beforehand, the solution set of the discrete phase retrieval problem is reduced to at most two different signals [8, 19, 20]. Under mild assumptions, one can also use an unknown reference signal to guarantee uniqueness [8, 21–23]. Besides employing known or unknown reference signals that are not related to the unknown true signal, it is also possible to use a modulation of the unknown signal itself as a reference [24, 25].

One possible generalization of the classical phase retrieval problem is to replace the discrete-time Fourier transform by some other signal transform. If we restrict the one-dimensional discrete phase retrieval problem to signals with fixed support $\{0, \ldots, M - 1\}$, which can be represented as M-dimensional vectors, then the Fourier intensities $|\widehat{x}(\omega_k)|$ at different points $\omega_k \in [-\pi, \pi]$ can be written as magnitudes $|\langle x, v_k \rangle|$ with $v_k := (e^{i\omega_k m})_{m=0}^{M-1}$. If we now replace the the Fourier vectors v_k by elements of an arbitrary frame of $\mathbb{C}^M$, the question arises how to choose the frame vectors to ensure a unique recovery of the true signal. This question has been studied, for instance, in [26–29]. Further generalizations, where the Fourier transform is replaced by the signed Fourier transform or by the short-time Fourier transform, have been studied in [30] and [31–33] respectively, see also the references therein.

The generalization of the Fourier phase retrieval problem with respect to a suitable frame often goes hand in hand with the a priori assumption that the true signal possesses a sparse representation in the frame. Phase retrieval problems of this kind have been studied for the shearlet frame [34] and for the translation invariant Haar pyramid tight frame [35]. Certainly, the sparsity assumption is not restricted to frame representations. If the sparsity of the true signal is sufficiently strong, this a priori condition guarantees uniqueness in the classical phase retrieval problem too, see [31] and references therein. Moreover, the sparsity of the true signal plays a key role in the recovery of spike and spline functions [36, 37] as well as in the reconstruction of structured functions [38].

This chapter is organized as follows. In the first part, Sects. 24.2 and 24.3, we introduce the one-dimensional discrete-time phase retrieval problem in more detail and derive a characterization of the entire solution set by factorizing the autocorrelation function—the squared Fourier intensity—suitably. Using this characterization, we show that each ambiguity is caused by rotation, time-shift, and conjugation and reflection of the factors in a convolution representation of the true signal.

In the second part—Sects. 24.4–24.6—we exploit our findings on the solution set to investigate different a priori assumptions and additional information about the signal with respect to their capability to ensure a unique recovery of the unknown true signal. In particular, we study the three a priori assumptions: non-negativity, additionally known direct measurements or intensity measurements in time domain, and additional intensity measurements in the frequency domain. Here the measurements in the frequency domain arise from the interference of the true signal with another signal. We particularly study the interference with a known reference signal, the interference with an unknown reference signal, and interference with modulations of the unknown solution signal.

Finally, in Sect. 24.7, we briefly discuss a generalization of the discrete-time phase retrieval problem where the Fourier transform is replaced by a so-called linear canonical transform. The linear canonical transform covers an entire class of well-known transforms like the Fresnel and the fractional Fourier transform. Due to the structure of these transforms the characterization of the solution set and uniqueness guarantees can be easily transferred to the new setting.

24.2 The Discrete-Time Phase Retrieval Problem

The central task in phase retrieval is the recovery of an unknown complex-valued signal from the measured intensity of its Fourier transform. In other words, we have completely lost the phase information in the frequency domain. Although the Fourier transform itself is a well-understood isometric isomorphism, the missing phase significantly hampers the reconstruction process and turns the phase retrieval problem into an ill-posed, quadratic inverse problem.

In this chapter, we consider the discrete-time version of the phase retrieval problem that can be stated as follows: recover an unknown complex-valued signal $x := (x[n])_{n \in \mathbb{Z}}$ from its *Fourier intensity*

$$|\mathcal{F}[x](\omega)| := |\widehat{x}(\omega)| := \left| \sum_{n \in \mathbb{Z}} x[n] e^{-i\omega n} \right| \qquad (\omega \in \mathbb{R}). \tag{24.1}$$

Throughout the paper, we assume that the true signal x has a finite support, which means that only finitely many components $x[n]$ are non-zero. We say that the signal x has a support of length N if there exists an integer n_0 such that $x[n_0]$ and $x[n_0 + N - 1]$ are non-zero and $x[n] = 0$ for all $n \notin \{n_0, \ldots, n_0 + N - 1\}$. Since the exponential sum in (24.1) has only finitely many terms, the Fourier intensity $|\widehat{x}|$ is here always well-defined.

The Fourier intensity $|\widehat{x}|$ is closely related to the *autocorrelation signal* $a := (a[n])_{n \in \mathbb{Z}}$ of x given by

$$a[n] := \sum_{k \in \mathbb{Z}} \overline{x[k]} \, x[k+n] \qquad (n \in \mathbb{Z}).$$

The coefficients of the autocorrelation signal are conjugate symmetric, which means $a[n] = \overline{a[-n]}$ for all n in $\mathbb{Z}$. Further, the support of the autocorrelation signal is always $\{-N + 1, \ldots, N - 1\}$, where N again denotes the support length of the original signal x, and does not depend on the actual position of the non-zero elements of the true signal x.

Using the definition of the autocorrelation signal, we observe

$$|\widehat{x}(\omega)|^2 = \sum_{n\in\mathbb{Z}}\sum_{k\in\mathbb{Z}} x[n]\,\overline{x[k]}\,e^{-i\omega(n-k)} = \sum_{n\in\mathbb{Z}}\sum_{k\in\mathbb{Z}} \overline{x[k]}\,x[k+n]\,e^{-i\omega n} = \widehat{a}(\omega),$$

where $\widehat{a}$ is called the *autocorrelation function* of x. The phase retrieval problem is thus equivalent to the recovery of the true signal x from its autocorrelation signal a. Due to the support $\{-N+1, \ldots, N-1\}$ of the autocorrelation signal a, the squared intensity $|\widehat{x}|^2$ is here a trigonometric polynomial of degree $N-1$, which implies that the Fourier intensity $|\widehat{x}|$ is already completely determined by $2N-1$ measurements in $[-\pi, \pi)$. For convenience, we nevertheless assume that the entire Fourier intensity is given.

24.3 Trivial and Non-trivial Ambiguities

The unknown phase of $\widehat{x}$ in the frequency domain cannot be completely arbitrary since the squared Fourier intensity is a trigonometric polynomial. However, without further information, the phase retrieval problem is never uniquely solvable. The simplest occurring ambiguities are

1. rotated signals $(e^{-i\alpha}\,x[n])_{n\in\mathbb{Z}}$ with $\alpha \in \mathbb{R}$,
2. time-shifted signals $(x[n-n_0])_{n\in\mathbb{Z}}$ with $n_0 \in \mathbb{Z}$, and
3. the reflected and conjugated signal $(\overline{x[-n]})_{n\in\mathbb{Z}}$,

which obviously have the same Fourier intensity $|\widehat{x}|$ as the true signal x. Since these signals are, however, closely related to the true signal x, we call these ambiguities *trivial*.

In the following, we are interested in all non-trivial solutions of the discrete-time phase retrieval problem. Before we give an explicit characterization, let us start with the following observation. If our true signal x can be represented as a convolution $x = x_1 * x_2$ defined by

$$(x_1 * x_2)[n] := \sum_{k\in\mathbb{Z}} x_1[k]\,x_2[n-k] \qquad (n \in \mathbb{Z}),$$

where x_1 and x_2 are two signals with finite support, than the Fourier convolution theorem implies that the signal

$$\left(e^{-i\alpha}\,\overline{x_1[n]}\right)_{n\in\mathbb{Z}} * \left(x_2[n-n_0]\right)_{n\in\mathbb{Z}} \tag{24.2}$$

with $\alpha \in \mathbb{R}$ and $n_0 \in \mathbb{Z}$ has the same Fourier intensity $|\widehat{x}|$. Differently from the trivial ambiguities, the constructed signal in (24.2) can have a completely different structure than the original signal x. In this section, we will show that all ambiguities—trivial and non-trivial—in discrete-time phase retrieval can be written as in (24.2), which means that they are caused by rotation, time-shifts, and reflection and conjugation of the single factors with respect to an appropriate convolution.

For this purpose, we will derive a suitable factorization of the given autocorrelation function $\widehat{a}$ by exploiting that the trigonometric polynomial $\widehat{a}$ is closely related to the algebraic polynomial P_a of degree $2N - 2$ defined by

$$P_a(z) := \sum_{n=0}^{2N-2} a[n - N + 1] z^n \qquad (z \in \mathbb{C}). \tag{24.3}$$

More precisely, we have

$$|\widehat{x}(\omega)|^2 = \widehat{a}(\omega) = e^{i\omega(N-1)} P_a(e^{-i\omega}).$$

In the following, we call P_a the *algebraic polynomial associated* to $\widehat{a}$.

Due to the conjugate symmetry $a[n] = \overline{a[-n]}$ for $n = 0, \ldots, N - 1$, the polynomial P_a is here conjugate palindromic, which implies that all roots occur in pairs of the form $(\gamma, \overline{\gamma}^{-1})$, where γ and $\overline{\gamma}^{-1}$ have exactly the same multiplicity. Moreover, zeros on the unit circle have an even multiplicity. Hence, the associated polynomial can be written as

$$P_a(z) = a[N - 1] \prod_{j=1}^{N-1} (z - \gamma_j)(z - \overline{\gamma}_j^{-1}).$$

Using the identity

$$\begin{aligned}
|(e^{-i\omega} - \gamma_j)(e^{-i\omega} - \overline{\gamma}_j^{-1})| &= |\overline{\gamma}_j^{-1}| \, |e^{-i\omega} - \gamma_j| \, |\overline{\gamma}_j - e^{i\omega}| \\
&= |\gamma_j|^{-1} |e^{-i\omega} - \gamma_j|^2,
\end{aligned} \tag{24.4}$$

we obtain the factorization

$$\begin{aligned}
\widehat{a}(\omega) = |P_a(e^{-i\omega})| &= |a[N - 1]| \prod_{j=1}^{N-1} |(e^{-i\omega} - \gamma_j)(e^{-i\omega} - \overline{\gamma}_j^{-1})| \\
&= |a[N - 1]| \prod_{j=1}^{N-1} |\gamma_j|^{-1} \cdot \prod_{j=1}^{N-1} |e^{-i\omega} - \gamma_j|^2,
\end{aligned}$$

see for instance [8, 10, 39].

The square root of $\widehat{a}$ now yields the Fourier transform of a finitely supported signal, and hence a solution of the phase retrieval problem with respect to the autocorrelation function $\widehat{a}$. Interchanging the role of γ_j and $\overline{\gamma}_j^{-1}$ in (24.4), we can explicitly construct further non-trivial solutions of the problem. With this idea in mind, we can characterize all solutions x of the discrete-time phase retrieval problem with given squared Fourier intensity $|\widehat{x}|^2 = \widehat{a}$.

Theorem 24.1 ([8]) *Let $\widehat{a} \colon \mathbb{R} \to [0, \infty)$ be an arbitrary non-negative trigonometric polynomial of degree $N - 1$. The Fourier transform of every finitely supported signal x with $|\widehat{x}|^2 = \widehat{a}$ can be written in the form*

$$\widehat{x}(\omega) = e^{i(\alpha - n_0 \omega)} \sqrt{|a[N-1]| \prod_{j=1}^{N-1} |\beta_j|^{-1} \cdot \prod_{j=1}^{N-1} \left(e^{-i\omega} - \beta_j \right)}, \qquad (24.5)$$

where α is a real number, n_0 is an integer, and β_j is chosen from the zero pair $(\gamma_j, \overline{\gamma}_j^{-1})$ of the associated polynomial P_a.

In Theorem 24.1, the trivial rotation ambiguity is covered by the factor $e^{i\alpha}$, and the time-shift ambiguity by the factor $e^{in_0\omega}$. Further, if the true signal x corresponds to the zero set $\{\beta_1, \ldots, \beta_{N-1}\}$, then the reflected and conjugated signal $\overline{x[-\cdot]}$ corresponds to the zero set $\{\overline{\beta}_1^{-1}, \ldots, \overline{\beta}_{N-1}^{-1}\}$. Consequently, the trivial reflection and conjugation ambiguity is also covered.

Employing the representation (24.5) of all ambiguities in the frequency domain, we can finally show that every non-trivial solution x of the phase retrieval problem $|\widehat{x}|^2 = \widehat{a}$ can be described by a suitable convolution factorization of the true signal x.

Theorem 24.2 ([8]) *Let x and y be two discrete-time signals with finite support and the same Fourier intensity $|\widehat{x}|$. Then there exist two finitely supported signals x_1 and x_2 such that*

$$x = x_1 * x_2$$

and

$$y = \left(e^{i\alpha} \overline{x_1[-n]} \right)_{n \in \mathbb{Z}} * \left(x_2[n - n_0] \right)_{n \in \mathbb{Z}},$$

where α is a suitable real number and n_0 is a suitable integer.

Using the characterization of all solutions in Theorem 24.1, we can construct 2^{N-1} zero sets $\{\beta_1, \ldots, \beta_{N-1}\}$ by choosing either $\beta_j = \gamma_j$ or $\beta_j = \overline{\gamma}_j^{-1}$ for $j = 1, \ldots, N - 1$, and each zero set determines a solution $\widehat{x}$ of the phase retrieval problem $|\widehat{x}|^2 = \widehat{a}$. Remembering that the reflection of all zeros on the unit circle corresponds to the reflection and conjugation of the related signal, we can therefore have at most 2^{N-2} different non-trivial solutions.

The true number of non-trivially different solutions $\widehat{x}$ can, however, be much smaller and depends on the number of different zero sets $\{\beta_1, \ldots, \beta_{N-1}\}$ that can be constructed. If all zeros lie on the unit circle, then γ_j and $\overline{\gamma}_j^{-1}$ coincide for all $j = 1, \ldots, N - 1$, and the solution is thus unique. A similar observation holds if some zero pairs $(\gamma_\ell, \overline{\gamma}_\ell^{-1})$ have a higher multiplicity $m_\ell > 1$, where the number of pairwise different zero sets $\{\beta_1, \ldots, \beta_{N-1}\}$ is then reduced accordingly.

Theorem 24.3 ([10]) *Let x be a discrete-time signal with finite support. Further-more, let L be the number of distinct zero pairs $(\gamma_\ell, \overline{\gamma}_\ell^{-1})$ of the associated polynomial P_a to the autocorrelation function $\widehat{a}$ not lying on the unit circle, and let m_ℓ be the multiplicity of these zero pairs. The corresponding phase retrieval problem to recover the signal x exactly has*

$$\left\lceil \frac{1}{2} \prod_{\ell=1}^{L} (m_\ell + 1) \right\rceil$$

non-trivial ambiguities.

Example 24.1 The actual number of non-trivial ambiguities in phase retrieval strongly depends on the zeros of the autocorrelation function. For example, the phase retrieval problem related to the autocorrelation function given by

$$\widehat{a}(\omega) = |P_a(e^{-i\omega})| = \left|\left(e^{-i\omega} + \tfrac{1}{2}\right)\left(e^{-i\omega} + 2\right)\right|^4 \cdot \left|e^{-i\omega} + e^{i\frac{\pi}{10}}\right|^{10},$$

has exactly three non-trivially different solutions, namely

$$\widehat{x}_1(\omega) = \left(e^{-i\omega} - \tfrac{1}{2}\right)^4 \left(e^{-i\omega} + e^{i\frac{\pi}{10}}\right)^5,$$

$$\widehat{x}_2(\omega) = \tfrac{1}{2}\left(e^{-i\omega} - \tfrac{1}{2}\right)^3 \left(e^{-i\omega} - 2\right)\left(e^{-i\omega} + e^{i\frac{\pi}{10}}\right)^5,$$

and

$$\widehat{x}_3(\omega) = \tfrac{1}{4}\left(e^{-i\omega} - \tfrac{1}{2}\right)^2 \left(e^{-i\omega} - 2\right)^2 \left(e^{-i\omega} + e^{i\frac{\pi}{10}}\right)^5.$$

Reflecting more than two zeros at the unit circle from $1/2$ to 2 only produces further trivial ambiguities caused by conjugation of the linear factors. The absolute value and the coefficients of the three non-trivially different solutions x_1, x_2, and x_3 are shown in Fig. 24.1.   □

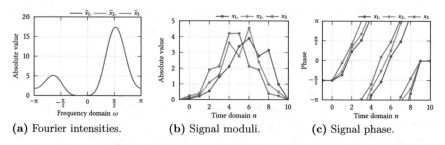

(a) Fourier intensities.   **(b)** Signal moduli.   **(c)** Signal phase.

Fig. 24.1 Phase retrieval problem $\widehat{a} = |\widehat{x}|^2$ with exactly three non-trivially different solutions as in Example 24.1, see [10]

24.4 Non-negative Signals

As we have seen in Theorem 24.3, the solution set of the discrete-time phase retrieval problem usually consists of a vast number of non-trivially different solutions that strongly differ in shape and form. To recover the true signal x within the solution set, we have to rely on further a priori knowledge on the desired signal. In many applications, we can assume that the unknown signal is real-valued and non-negative, see for instance [3, 4, 11]. Therefore, we study the problem: how many non-trivial real-valued non-negative signals exist satisfying $|\widehat{x}|^2 = \widehat{a}$ for a given autocorrelation function? In other words, can the a priori assumption that x is real-valued and non-negative help us to find a unique solution or at least essentially reduce the number of non-trivial ambiguities?

Let us now assume that x has a finite support of the form $\{0, \ldots, N-1\}$, and that all components $x[n]$ with $n \in \{0, \ldots, N-1\}$ are real and non-negative. The representation (24.5) in Theorem 24.1 without rotations and time-shifts—with $\alpha = 0$ and $n_0 = 0$—yields the solution

$$\widehat{x}(\omega) = |a[N-1]|^{\frac{1}{2}} \prod_{j=1}^{N-1} |\beta_j|^{-\frac{1}{2}} \left(e^{-i\omega} - \beta_j \right). \tag{24.6}$$

The non-negativity of x is now equivalent with the condition that the coefficients of the algebraic polynomial Q given by

$$Q(z) := \prod_{J=1}^{N-1} (z - \beta_j) \tag{24.7}$$

are non-negative. Since the zeros β_j are always non-zero, the leading coefficient and the absolute term of Q have even to be strictly positive. Algebraic polynomials of this kind are usually called *positive polynomials*.

A closer inspection shows that the non-negativity condition does not always reduce the number of non-trivial ambiguities of the phase retrieval problem.

Theorem 24.4 ([14]) *Let x be a real-valued discrete-time signal with finite support. If the zero set $\{\beta_1, \ldots, \beta_{N-1}\}$ corresponding to $\widehat{x}$ in (24.6) is contained in the left half plane, which means that $\mathrm{Re}\, \beta_j < 0$ for all $j = 1, \ldots, N-1$, then all occurring real-valued non-trivial ambiguities of the corresponding phase retrieval problem are non-negative.*

Proof Since the polynomial Q in (24.7) is real-valued, the zeros β_j have to be real or have to come in conjugated pairs $(\beta_j, \overline{\beta_j})$. The corresponding linear or quadratic factors have the form

$$(e^{-i\omega} - \beta_j) \quad \text{or} \quad (e^{-i\omega} - \beta_j)(e^{-i\omega} - \overline{\beta}_j) = e^{-2i\omega} - 2\,(\mathrm{Re}\,\beta_j)\,e^{-i\omega} + |\beta_j|^2.$$

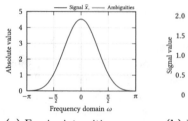

(a) Fourier intensities.

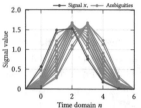

(b) Real-valued non-trivial solutions.

Fig. 24.2 Non-reduced non-negative solution set for the phase retrieval problem $\widehat{a} = |\widehat{x}|^2$, see [10]

By assumption all coefficients in these factors are non-negative. Therefore, also all coefficients of the polynomial Q are non-negative, and Q therefore always leads to a non-negative solution x of the phase retrieval problem. □

However, if the assumption of Theorem 24.4 is not satisfied, the number of non-negative non-trivial ambiguities of the discrete-time phase retrieval problem is usually reduced.

Example 24.2 Figures 24.2, 24.3 and 24.4 show some different cases which can occur under the restriction of non-negativity. Figure 24.2 presents all non-trivial solutions that can be constructed from the autocorrelation function $|\widehat{x}|^2$, where x is the marked signal of length 6 being determined by the zero set

$$\{\beta_1, \ldots, \beta_5\} := \left\{ -\frac{18}{5}, -\frac{5}{2}, -\frac{9}{5}, -\frac{6}{5}, -\frac{7}{8} \right\}.$$

As shown in Theorem 24.4, all non-trivial solutions being constructed via (24.6) are real and non-negative. The solution set is presented without reflected, conjugated signals. In this example, we have $2^4 = 16$ different solutions, which is the maximal number of non-trivial ambiguities by Theorem 24.3.

In the second example, see Fig. 24.3, the condition of nonnegativity is strong enough to ensure uniqueness of the phase retrieval problem. Here, the unique non-negative solution x_1 corresponds to the zero set

$$\{\beta_1, \ldots, \beta_5\} := \left\{ -\frac{3}{2}, -\frac{1}{2} + \frac{3}{2}i, -\frac{1}{2} - \frac{3}{2}i, 1 + 1i, 1 - 1i \right\}.$$

Note that the problem has only four non-trivial ambiguities since the complex zeros of the characterization (24.5) additionally have to be chosen as complex conjugated pairs.

In the last example, Fig. 24.4, the restriction of non-negativity is too strong since every solution of the phase retrieval problem possesses some negative coefficients,

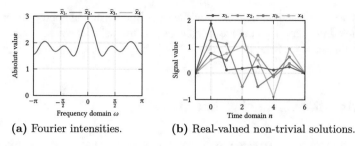

(a) Fourier intensities.   (b) Real-valued non-trivial solutions.

Fig. 24.3 Unique non-negative solution of the phase retrieval problem $\widehat{a} = |\widehat{x}|^2$, see [10]

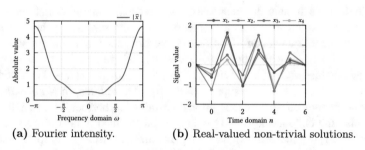

(a) Fourier intensity.   (b) Real-valued non-trivial solutions.

Fig. 24.4 Empty non-negative solution set for the phase retrieval problem $\widehat{a} = |\widehat{x}|^2$, see [10]

which means that the given phase retrieval problem cannot be solved by a real-valued non-negative signal. The signal x_1 in Fig. 24.4b here corresponds to the zero set

$$\{\beta_1, \ldots, \beta_5\} := \left\{ \tfrac{1}{2}, -\tfrac{1}{2} + \tfrac{3}{2}\,\mathrm{i}, -\tfrac{1}{2} - \tfrac{3}{2}\,\mathrm{i}, 1 + 1\mathrm{i}, 1 - 1\mathrm{i} \right\}. \qquad \square$$

Since the coefficients of an algebraic polynomial continuously depend on the zero set $\{\beta_1, \ldots, \beta_{N-1}\}$, the number of non-negative non-trivial ambiguities around a true signal x with non-zero components remains unchanged in a certain (small) neighbourhood of the zero set. On the basis of this observation, one can show that neither the signals that are uniquely defined by their Fourier intensity and the a priori known non-negativity nor the signals that are not uniquely defined by these conditions form negligible sets. Unfortunately, there is no simple way to decide from the given intensity data whether the a priori condition of non-negativity is helpful for reducing the number of ambiguities of the discrete-time phase retrieval problem. Mathematically, we can show the following result.

Theorem 24.5 ([14]) *The set of real-valued discrete-time signals with support $\{0, \ldots, N-1\}$ of length $N > 0$ that can be recovered uniquely up to reflection as well as the set of signals that cannot be recovered uniquely from their Fourier intensities employing the non-negativity constraint are both unbounded sets containing a cone of infinite Lebesgue measure.*

In other words, the non-negativity of the true signal is not an appropriate a priori assumption in order to guarantee uniqueness of the solution of the related one-dimensional phase retrieval problem.

24.5 Additional Data in Time-Domain

In certain applications like electron microscopy, wave front sensing and laser optics [6], we may have direct access to one or more signal values $x[n]$ or to magnitudes $|x[n]|$ in the time domain. In order to exploit these additional information, we need to know the position of the measurements within the support of the true signal. To simplify the following considerations, we restrict ourselves to finite discrete-time signals $x = (x[n])_{n \in \mathbb{Z}}$ with support $\{0, \ldots, N - 1\}$. These signals may be interpreted as complex-valued vectors $x = (x[n])_{n=0}^{N-1}$ in $\mathbb{C}^N$. If $x[0]$ and $x[N - 1]$ are additionally non-zero, then we call the support of the signal x with length N *normalized*.

Xu et al. already considered the a priori constraint that, besides $|\widehat{x}|$ for a real-valued signal, also the endpoint $x[N - 1]$ is known [40], which almost always enforces uniqueness. In [18], these ideas have been generalized to discrete-time phase retrieval problems with given magnitudes of the form $|x[n]|$ or partial phase information $\arg(x[n])$ in the time domain.

Again, the question arises whether a priori information of this type is sufficient to determine a unique solution of the discrete-time phase retrieval problem (up to trivial ambiguities).

24.5.1 Using an Additional Signal Value

In order to get a heuristic idea, we start with the following question: for a given non-negative trigonometric polynomial $\widehat{a}$ of degree $N - 1$ as in Theorem 24.1 and a given constant $C \in \mathbb{C}$, how many non-trivial solutions x with support $\{0, \ldots, N - 1\}$ exist for the constrained phase retrieval problem

$$|\widehat{x}|^2 = \widehat{a} \quad \text{and} \quad x[N - 1] = C ? \tag{24.8}$$

As we know already from Theorem 24.1, there exist at most 2^{N-2} non-trivially different signals $x = (x[n])_{n \in \mathbb{Z}}$ with Fourier intensity $|\widehat{x}|^2 = \widehat{a}$. But how many of these solutions also satisfy the side condition $x[N - 1] = C$?

To answer this question, we employ our knowledge about the structure of the solutions in (24.5). Recalling that $\widehat{x}(\omega) = \sum_{n=0}^{N-1} x[n] e^{-i\omega n}$, we notice that the coefficient $x[N - 1]$ in (24.5) with $n_0 = 0$ is given by

$$x[N - 1] = e^{i\alpha} \left[|a[N - 1]| \prod_{j=1}^{N-1} |\beta_j|^{-1} \right]^{1/2}.$$

We therefore derive the *consistency condition*

$$|C|^2 = |a[N-1]| \prod_{j=1}^{N-1} |\beta_j|^{-1}. \tag{24.9}$$

If this condition is not satisfied, there will be no solution signal satisfying the side condition $x[N-1] = C$. Assuming that (24.9) is satisfied for some zero set $\{\beta_1, \ldots, \beta_{N-1}\}$, we find at least one solution of (24.8), where we take α according to the phase of the complex value C. By Theorem 24.1, all further solutions with Fourier intensity $|\widehat{x}|^2 = \widehat{a}$ are obtained by reflecting zeros from β_j to $\overline{\beta}_j^{-1}$ at the unit circle for some indices $j \in \{1, \ldots, N-1\}$ in the representation (24.5).

Let us now assume that there is indeed a second solution $\tilde{x}$ of (24.8) satisfying (24.9), and let $\{\tilde{\beta}_1, \ldots, \tilde{\beta}_{N-1}\}$ be the corresponding zero set in (24.5). Then we can assume without loss of generality that the corresponding zeros are given by

$$\tilde{\beta}_j = \begin{cases} \overline{\beta}_j^{-1} & j = 0, \ldots, L, \\ \beta_j & \text{otherwise} \end{cases}$$

for some $L \in \{1, \ldots, N-1\}$. The consistency condition (24.9) now implies

$$\prod_{j=1}^{N-1} |\beta_j|^{-1} = \prod_{j=1}^{L} |\beta_j| \prod_{j=L+1}^{N-1} |\beta_j|^{-1}$$

and thus $\prod_{j=1}^{L} |\beta_j|^2 = 1$. We can therefore state the following theorem.

Theorem 24.6 ([8]) *Let x be a complex-valued discrete-time signal with normalized support of length N, i.e., $\widehat{x}$ is of the form (24.5) with $n_0 = 0$. Then the constrained phase retrieval problem to recover the signal x from its Fourier intensity $|\widehat{x}|$ and the signal value $x[N-1]$ is uniquely solvable if and only if*

$$\prod_{\beta_j \in \Lambda} |\beta_j|^2 \neq 1$$

for each non-empty subset Λ of B, where B denotes the set of values in the corresponding zero set of x not lying on the unit circle.

Since the support of x is normalized to $\{0, \ldots, N-1\}$, and since the rotation factor α in (24.5) is here fixed by the phase of the given constant C, we have no trivial ambiguities caused by rotation or shift. Moreover, the reflection and conjugation ambiguity cannot occur since we have here particularly assumed $\prod_{j=1}^{N-1} |\beta_j|^2 \neq 1$. The simplification of the set of zeros to those with modulus different from 1 can be done since the reflection of zeros on the unit circle does not lead to further non-trivial solutions.

24.5.2   Using Additional Magnitude Values of the Signal

Next, we generalize the problem considered in (24.8) and assume that, besides the Fourier intensity $|\widehat{x}|^2$, either all or at least some of the magnitudes $|x[n]|$ with $n = 0, \ldots, N - 1$ are given. Phase retrieval problems with these constraints have been considered for example in [6, 15, 16]. The numerical approaches to find the phase retrieval solution in [15] are based on multilevel Gauß–Newton methods. However, these algorithms are not always stable and sometimes reconstruct signals that are different from the desired solution.

We therefore study the uniqueness of solutions of the following constrained phase retrieval problem: for a given non-negative trigonometric polynomial $\widehat{a}$ of degree $N - 1$ as in Theorem 24.1 and a given $C > 0$, how many non-trivial solutions x with support $\{0, \ldots, N - 1\}$ exist to the constrained phase retrieval problem

$$|\widehat{x}|^2 = \widehat{a} \quad \text{and} \quad |x[N - 1]| = C \,? \tag{24.10}$$

To characterize the solutions of (24.10), we can proceed similarly as in Sect. 24.5.1. Doing so, we obtain the same consistency condition (24.9) but, obviously, the given absolute value $|x[N - 1]|$ will give us no information how to choose the rotation factor $e^{i\alpha}$; so we cannot get rid of the rotation ambiguity.

Corollary 24.1 ([8]) *Let x be a complex-valued discrete-time signal with normalized support of length N, i.e., $\widehat{x}$ is of the form (24.5) with $n_0 = 0$. Then the constrained phase retrieval problem to recover the signal x from its Fourier intensity $|\widehat{x}|$ and the absolute value $|x[N - 1]|$ is uniquely solvable up to rotations if and only if*

$$\prod_{\beta_j \in \Lambda} |\beta_j|^2 \neq 1$$

for each non-empty subset Λ of B, where B denotes the set of values in the corresponding zero set of x not lying on the unit circle.

In [18], we have generalized these observations to the constrained phase retrieval problem of the form

$$|\widehat{x}|^2 = \widehat{a} \quad \text{and} \quad |x[n]| = C \tag{24.11}$$

for some $n \in \{0, \ldots, N - 1\}$. Similarly as before, one can derive a consistency condition and a condition in terms of the zeros of the autocorrelation function $\widehat{a}$ such that uniqueness of a solution x of (24.11) is guaranteed up to trivial ambiguities. However, the corresponding conditions are more complex and require an extensive investigation of the $(N - 1)$-variate elementary symmetric polynomials, which are related to the components of the true signal x by Vieta's formulae. The important outcome of these investigations can be summarized as follows.

Theorem 24.7 ([18]) *Let x be a complex-valued discrete-time signal with normalized support of length N, and let ℓ be an arbitrary integer between 0 and N − 1. The phase retrieval problem to recover the signal x from its Fourier intensity $|\widehat{x}|$ and the absolute value $|x[N − 1 − \ell]|$ is almost always uniquely solvable up to rotations whenever $\ell \neq (N−1)/2$. In the special case that $\ell = (N−1)/2$, the reconstruction is almost always unique up to rotations and conjugate reflections.*

'Almost always' means here that the union of all signals with normalized support of length N, which permit a further non-trivial solution, corresponds to the union of finitely many algebraic varieties with Lebesgue measure zero in $\mathbb{R}^{2N}$, see [18]. In particular, we almost always obtain uniqueness if the magnitudes of all signal values $|x[n]|, n = 0, \ldots, N − 1$, are given.

Corollary 24.2 ([10]) *Let x be a complex-valued discrete-time signal x with normalized support of length N. The phase retrieval problem to recover the signal x from its Fourier intensity $|\widehat{x}|$ and its moduli $(|x[n]|)_{n\in\mathbb{Z}}$ is almost always uniquely solvable up to rotations.*

Obviously, Corollary 24.2 is a simple consequence of Theorem 24.7. But the following question remains: is the knowledge of all magnitudes $|x[n]|$ with $n = 0, \ldots, N − 1$ already sufficient to obtain a unique solution of the constrained problem

$$|\widehat{x}|^2 = \widehat{a} \quad \text{and} \quad |x[n]| = C_n \quad \text{for} \quad n = 0, \ldots, N − 1$$

up to rotation ambiguities? The following example shows that this is unfortunately not the case.

Example 24.3 We consider the complex-valued signal x determined by th corresponding zeros

$$\beta_1 := -\tfrac{1}{2} - \tfrac{1}{2}\,\mathrm{i}, \qquad \beta_2 := -\mathrm{e}^{-\frac{2\pi}{3}\mathrm{i}}\,(1 + \mathrm{i}), \qquad \text{and} \qquad \beta_3 := -\mathrm{e}^{\frac{2\pi}{3}\mathrm{i}}\,(1 + \mathrm{i})$$

and by $\alpha = 0$ and $n_0 = 0$ in the representation (24.5). Knowing the autocorrelation function $\widehat{a}(\omega) = |\widehat{x}(\omega)|^2$ for $\omega \in \mathbb{R}$ and the moduli of all components $|x[n]|$ for $n \in \mathbb{Z}$, we still cannot recover x uniquely. In this specific example, we find, up to rotations, three non-trivial solutions that are presented in Fig. 24.5.

It is possible to construct further examples with several non-trivial solutions for all dimensions N of the problem, see [10]. Hence, the a priori known moduli of the components strongly reduce the set of ambiguities, but we cannot ensure uniqueness (up to trivial ambiguities) for every signal. □

Using an analogous approach, one can study the restricted discrete-time phase retrieval problem where the knowledge of additional magnitudes in time domain is replaced by a priori phase information in time domain. Due to the trivial rotation ambiguity, we can only expect to reduce the non-trivial solution set if we have given the phase of at least two signal components.

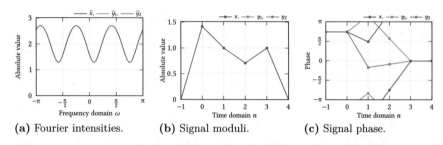

(a) Fourier intensities. **(b)** Signal moduli. **(c)** Signal phase.

Fig. 24.5 Three non-trivial solutions of phase retrieval problem $\widehat{a} = |\widehat{x}|^2$ with given moduli $|x[n]|$ for $n = 0, \ldots, 3$ as in Example 24.3, see [10]

Theorem 24.8 ([18]) *Let x be a complex-valued discrete-time signal with normalized support of length N, and let ℓ_1 and ℓ_2 be different integers between 0 and $N - 1$. The phase retrieval problem to recover the signal x from its Fourier intensity and the two phases $\arg x[N - 1 - \ell_1]$ and $\arg x[N - 1 - \ell_2]$ is almost always uniquely solvable whenever $\ell_1 + \ell_2 \neq N - 1$. If $\ell_1 + \ell_2 = N - 1$, then the reconstruction is only unique up to conjugate reflections, except for the special case where ℓ_1 and ℓ_2 correspond to the two endpoints.*

24.6 Interference Measurements

Another possibility to reduce the ambiguities in the considered discrete-time phase retrieval problem is to exploit additional reference measurements of the form $| \mathcal{F}[x + h]|$, where h is a suitable reference signal with finite support. We consider here two cases, either the reference signal h is known beforehand, or it is also unknown. In the first case, we will show that the corresponding phase retrieval problem with given intensities $|\mathcal{F}[x]|$ and $| \mathcal{F}[x + h]|$ has at most two non-trivial solutions. If h is unknown, we assume that the Fourier intensities $|\mathcal{F}[x]|$, $|\mathcal{F}[h]|$, and $| \mathcal{F}[x + h]|$ are given and show unique recovery results under suitable side conditions. Finally, we will examine the special case where the unknown reference signal is a modulated version of the true signal x itself.

24.6.1 Interference with a Known Reference Signal

Let us assume that the considered reference signal $h = (h[n])_{n \in \mathbb{Z}}$ has finite support and is completely known beforehand. The corresponding phase retrieval problem is then nearly unique solvable up to at most one ambiguity, see [8, 20].

Theorem 24.9 ([8]) *Let x and h be two discrete-time signals with finite support, where the non-vanishing reference signal h is known beforehand. Then the signal x can be recovered from the Fourier intensities*

$$|\mathcal{F}[x]| \quad and \quad |\mathcal{F}[x+h]|$$

except for at most one ambiguity. This ambiguity ist trivial if h possesses a linear phase.

Proof Let $y = x + h$ be the interference between the unknown signal x and the known reference signal h. Then y is a finite length signal with known Fourier intensity $|\mathcal{F}[y]| = |\mathcal{F}[x+h]|$. Further, with $\widehat{x}(\omega) = |\widehat{x}(\omega)|\, e^{i\phi(\omega)}$ and $\widehat{h}(\omega) = |\widehat{h}(\omega)|\, e^{i\psi(\omega)}$, it follows

$$|\widehat{y}(\omega)|^2 = |\widehat{x}(\omega)|^2 + |\widehat{h}(\omega)|^2 + 2\operatorname{Re}\left(\widehat{x}(\omega)\overline{\widehat{h}(\omega)}\right)$$
$$= |\widehat{x}(\omega)|^2 + |\widehat{h}(\omega)|^2 + 2\,|\widehat{x}(\omega)|\,|\widehat{h}(\omega)|\,\cos(\phi(\omega) - \psi(\omega))$$

such that we can extract the phase difference $\phi(\omega) - \psi(\omega)$ up to the sign and a multiple of 2π for every $\omega \in \mathbb{R}$. Due to the piecewise continuity of the phases ϕ and ψ, there is an open interval where the sign has to be either plus or minus everywhere. Since each trigonometric polynomial is completely determined by its values on an open set, we conclude that there can be at most two different solutions. If we write these solutions x and $\tilde{x}$ in the form $\widehat{x}(\omega) = |\widehat{x}(\omega)|\, e^{i\phi_1(\omega)}$ and $\widehat{\tilde{x}}(\omega) = |\widehat{x}(\omega)|\, e^{i\phi_2(\omega)}$, then the phases ϕ_1 and ϕ_2 are related by

$$\phi_1(\omega) - \psi(\omega) = -\phi_2(\omega) + \psi(\omega) + 2\pi\,\ell_\omega,$$

i.e., $\phi_2(\omega) = -\phi_1(\omega) + 2\psi(\omega) + 2\pi\,\ell_\omega$. If the reference h has linear phase, which means that the phase ψ is of the form $\psi(\omega) = n_0\omega + \alpha$ for some $n_0 \in \mathbb{Z}$ and $\alpha \in \mathbb{R}$, then $\tilde{x}$ is a trivial ambiguity of x obtained by support shift and rotation. □

If the signal h does not have linear phase, then the ambiguity $\tilde{x}$ can be non-trivially different as discussed in the next example.

Example 24.4 Let us consider the discrete-time phase retrieval problem to recover the true signal

$$x := \tfrac{1}{128}\left(\ldots, 0, \underline{55 - 15i}, -84 + 87i, 34 + 82i,\right.$$
$$\left. 204 - 120i, -16 + 16i, -96, 128, 0, \ldots\right)$$

from the Fourier intensities $|\mathcal{F}[x]|$ and $|\mathcal{F}[x+h]|$, where h is the known reference signal

$$h := \left(\ldots, 0, \underline{0}, 20 - 10i, 19 - 17i, -4 - 4i, 4 - 4i, 16, 0, \ldots\right).$$

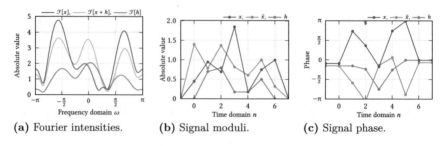

(a) Fourier intensities.   **(b)** Signal moduli.   **(c)** Signal phase.

Fig. 24.6 Discrete-time phase retrieval problem to recover x from the Fourier intensities $|\mathcal{F}[x]|$ and $|\mathcal{F}[x+h]|$, where h is a known reference signal as in Example 24.4. Besides the true solution x, the non-trivial ambiguity $\check{x}$ is the only further solution of the problem, see [10]

Here, we have underlined the entry with index 0. Since h does not possess a linear phase, cf. Fig. 24.6c, there may exist a further non-trivially different solution by Theorem 24.9, and, indeed, the signal

$$\check{x} := \tfrac{1}{128} \left(\ldots, 0, \underline{160 - 80i}, -28 - 96i, -173 + 31i, \right.$$
$$\left. 95 - 44i, 76 + 16i, -120 - 44i, 40 - 8i, 0, \ldots \right)$$

yields the same Fourier intensities. The signals and the given Fourier intensities are presented in Fig. 24.6.   □

24.6.2 Interference with an Unknown Reference Signal

Let us now consider the case where the finitely supported signal h is also unknown. For real signals, this problem has already been studied in [21]. For complex signals, we want to refer to the work of Raz et al. [22], where, besides the three Fourier intensities in the next theorem, a fourth intensity of the form $|\hat{x}(\omega) + i\hat{h}(\omega)|$ was used for the recovery of x. From a theoretical point of view, this intensity is not needed to ensure uniqueness, but this additional information allows the derivation of an explicit analytic solution.

Theorem 24.10 ([8]) *Let x and h be two discrete-time signals with finite support. If the corresponding zero sets of the signals x and h are disjoint, then the two signals x and h can be recovered from the Fourier intensities*

$$|\mathcal{F}[x]|, \quad |\mathcal{F}[h]|, \quad \text{and} \quad |\mathcal{F}[x+h]|$$

uniquely up to common trivial ambiguities.

'Common trivial ambiguities' means here that we can multiply the two signals x and h with the same unimodular constant $e^{i\alpha}$ or shift the two signals with the same integer n_0 or take the reflection and conjugation for both signals, and all these actions do not change the given Fourier intensities in Theorem 24.10. For a detailed proof of this theorem, we refer to [8]. The main part of the proof is heavily based on the result of Theorem 24.2, where we have shown that each further solution $(\tilde{x}, \tilde{h})$ with Fourier intensities $|\mathcal{F}[\tilde{x}]| = |\mathcal{F}x]|$ and $|\mathcal{F}[\tilde{h}]| = |\mathcal{F}[h]|$ is related to the true solution (x, h) by some factorization

$$\widehat{\tilde{x}}(\omega) = e^{i\alpha_1 + n_1\omega}\,\widehat{x_1}(\omega)\,\overline{\widehat{x_2}(\omega)} \quad \text{for} \quad \widehat{x}(\omega) = \widehat{x_1}(\omega)\,\widehat{x_2}(\omega)$$

and

$$\widehat{\tilde{h}}(\omega) = e^{i\alpha_2 + n_2\omega}\,\widehat{h_1}(\omega)\,\overline{\widehat{h_2}(\omega)} \quad \text{for} \quad \widehat{h}(\omega) = \widehat{h_1}(\omega)\,\widehat{h_2}(\omega)$$

with rotations α_1, $\alpha_2 \in [-\pi, \pi)$ and shifts n_1, $n_2 \in \mathbb{Z}$. The assertion then follows from a detailed comparison of the third Fourier intensity

$$|\widehat{x}(\omega) + \widehat{h}(\omega)|^2 = |\widehat{\tilde{x}}(\omega) + \widehat{\tilde{h}}(\omega)|^2$$

by incorporating the product representations of the signals.

If the assumption of Theorem 24.10 is violated and x and h have common zeros in the defining zero sets in representation (24.5), then we may find more non-trivial solutions, as shown in the following example.

Example 24.5 We want to recover the signal x in Example 24.4, which corresponds to the zero set

$$\{\beta_1, \ldots, \beta_6\} := \tfrac{1}{4}\left\{1 + i, 3 - 2i, -3 - i, -4 + 2i, 4 + 4i, 2 - 4i\right\}$$

in the representation (24.5) of $\widehat{x}$. Further, we choose the reference signal h with the corresponding zero set

$$\{\eta_1, \ldots, \eta_5\} := \tfrac{1}{4}\left\{1 + i, 4 + 4i, -4 - 3i, -4 + 2i, -4i\right\}.$$

If both signals are unknown, we have to recover x and h from the Fourier intensities $|\mathcal{F}[x]|$, $|\mathcal{F}[h]|$, and $|\mathcal{F}[x + h]|$. The intersection of the corresponding zero sets of x and h is here given by

$$\tfrac{1}{4}\left\{1 + i, 4 + 4i\right\}$$

such that the uniqueness of the solution is not covered by Theorem 24.10. Indeed, reflecting the zeros $1/4\,(1 + i)$ and $1/4\,(4 + 4i)$ in the representations (24.5) of both signals at the unit circle, we find a second non-trivial solution $(\check{x}, \check{h})$. Both solutions (x, h) and $(\check{x}, \check{h})$ are presented in Fig. 24.7. □

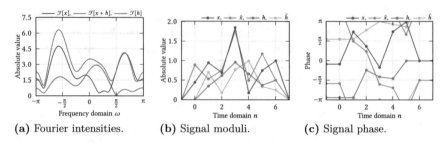

(a) Fourier intensities.   (b) Signal moduli.   (c) Signal phase.

Fig. 24.7 Discrete-time phase retrieval problem to recover x and h from the Fourier intensities $|\mathcal{F}[x]|$, $|\mathcal{F}[h]|$ and $|\mathcal{F}[x+h]|$, where h is an unknown reference signal as in Example 24.5. Besides the true solution (x, h), the non-trivial ambiguity $(\check{x}, \check{h})$ is also a solution of the problem, see [10]

24.6.3 Interference with the Modulated Signal

Finally we consider the model, where the unknown reference signal is a modulated version of the signal x itself. Similar approaches for the (periodic) discrete Fourier transform have already been studied in [25, 41]. We here especially rely on the results in [24]. The discrete-time phase retrieval problem can be now posed as follows: recover a finitely supported signal x from its Fourier intensity $|\widehat{x}|$ and a set of interference measurements

$$| \mathcal{F}[x + e^{i\alpha}\, e^{i\mu \cdot} x]|,$$

where the modulations and rotations are described by $\mu \in \mathbb{R}$ and $\alpha \in [0, 2\pi)$. In order to guarantee uniqueness, besides the Fourier intensity $|\widehat{x}|$, we merely need two additional interference signals.

Theorem 24.11 ([24]) *Let x be a discrete-time signal with finite support of length N. If μ satisfies the assumption that $k\mu \not\equiv 0 \bmod 2\pi$ for all $k = 1, \ldots, 2N - 1$, then the signal x can be uniquely recovered up to a rotation ambiguity from its Fourier intensity $|\widehat{x}|$ and the Fourier intensities of two interference signals*

$$\left| \mathcal{F}\left[x + e^{i\alpha_1}\, e^{i\mu \cdot}\, x\right]\right| \quad and \quad \left| \mathcal{F}\left[x + e^{i\alpha_2}\, e^{i\mu \cdot}\, x\right]\right|,$$

where α_1 and α_2 are two real numbers satisfying $\alpha_1 - \alpha_2 \neq \pi k$ for all $k \in \mathbb{Z}$.

Proof Writing the unknown Fourier-transformed signal in the form $\widehat{x}(\omega) = |\widehat{x}(\omega)|\, e^{i\phi(\omega)}$, we only need to recover the phase $\phi(\omega)$ to solve the given phase retrieval problem. The Fourier intensity measurements of the first interference signal yield

$$\left| \mathcal{F}\left[x + e^{i\alpha_1}\, e^{i\mu \cdot} x\right]\right| = |\widehat{x}(\omega) + e^{i\alpha_1}\widehat{x}(\omega - \mu)|^2$$
$$= |\widehat{x}(\omega)|^2 + |\widehat{x}(\omega - \mu)|^2$$
$$+ 2\,|\widehat{x}(\omega)|\,|\widehat{x}(\omega - \mu)|\cos(\phi(\omega - \mu) - \phi(\omega) + \alpha_1)$$

and thus the cosine of the relative phase

$$\cos(\phi(\omega - \mu) - \phi(\omega) + \alpha_1) = \cos(\alpha_1)\cos(\phi(\omega - \mu) - \phi(\omega))$$
$$- \sin(\alpha_1)\sin(\phi(\omega - \mu) - \phi(\omega)).$$

Analogously, we can extract $\cos(\phi(\omega - \mu) - \phi(\omega) + \alpha_2)$ from the Fourier intensity measurements of the second interference signal. Since $\alpha_1 - \alpha_2 \neq \pi k$ for all $k \in \mathbb{Z}$, we can therefore uniquely determine the phase difference $\phi(\omega - \mu) - \phi(\omega)$ for every $\omega \in \mathbb{R}$. Obviously, the solution can be only recovered up to rotations. Taking an arbitrary phase $\phi(\omega_0)$, we can compute the corresponding phases $\phi(\omega_0 + \mu k)$ for $k = 0, \ldots, 2N - 1$ and thus the Fourier values $\widehat{x}(\omega_0 + \mu k)$ for $k = 0, \ldots, 2N - 1$. It remains to recover the signal x and especially the unknown support from these Fourier values. Due to the support length N, the Fourier transform $\widehat{x}$ can be written in the form

$$\widehat{x}(\omega) = e^{-i\omega n_0} \sum_{n=0}^{N-1} c_n e^{-i\omega n}$$

with $c_n := x[n + n_0]$. Using the found Fourier values, we obtain the equation system

$$\widehat{x}(\omega_0 + \mu k) = \sum_{n=0}^{N-1} [c_n e^{-i\omega_0(n+n_0)}] e^{-ik\mu(n+n_0)} = \sum_{n=0}^{N-1} d_n z_n^k, \quad k = 0, \ldots, 2N - 1,$$

with $d_n := c_n e^{-i\omega_0(n+n_0)}$ and $z_n := e^{-i\mu(n+n_0)}$. This system can be solved by Prony's method if the values $\omega_0 + \mu k \mod 2\pi$ are pairwise different for $k = 0, \ldots, 2N - 1$, which means that $k\mu$ is not a multiple of 2π for all $k = 1, \ldots, 2N - 1$, see for instance [42].   □

24.7 Linear Canonical Phase Retrieval

Up to this point, we have assumed that the given measurements in the frequency domain arise from the Fourier-transformed true signal. These measurements can be seen as intensities in the so-called far field in Fourier optics. In this section, we briefly investigate the question: how do the established uniqueness guarantees change if we replace the far field intensity measurements for example by near field intensity measurements—what happens if we replace the Fourier transform by the Fresnel or fractional Fourier transform?

The discrete-time Fourier, fractional Fourier, and Fresnel transform are special cases of the so-called linear canonical transform. Referring to [43, 44], for the real parameters a, b, c, and d with $ad - bc = 1$ and $b \neq 0$, we define the discrete-time linear canonical transform of the signal $x := (x[n])_{n \in \mathbb{Z}}$ by

$$\mathcal{C}_{(a,b,c,d)}\,[x]\,(\omega) := \sum_{n\in\mathbb{Z}} x\,[n]\ K_{(a,b,c,d)}\,(\omega,n)\,,$$

where the kernel $K_{(a,b,c,d)}$ is given by

$$K_{(a,b,c,d)}\,(\omega,t) := \frac{1}{\sqrt{2\pi b}}\,e^{-i\frac{\pi}{4}}\,e^{\frac{i}{2}(\frac{a}{b}t^2 - \frac{2}{b}\omega t + \frac{d}{b}\omega^2)}.$$

The inverse discrete-time linear canonical transform is given by

$$\mathcal{C}^{-1}_{(a,b,c,d)}[\tilde{x}][n] = \int_{-\pi|b|}^{\pi|b|} \tilde{x}\,(\omega)\ \overline{K_{(a,b,c,d)}\,(\omega,n)}\,d\omega,$$

see [43, 44]. The classical discrete-time Fourier transform $\mathcal{F}$ coincides with the linear canonical transform $\mathcal{C}_{(0,1,-1,0)}$ up to a multiplicative constant $\theta := \theta_{(a,b,c,d)} := 1/\sqrt{2\pi b}\,e^{-i\pi/4}$. The discrete-time Fresnel transform [45] and the fractional Fourier transform [46] are covered by the linear canonical transforms $\mathcal{C}_{(1,1/2\alpha,0,1)}$ and $\mathcal{C}_{(\cos\alpha,\sin\alpha,-\sin\alpha,\cos\alpha)}$ with $\alpha \in \mathbb{R}$, respectively.

Since $b \neq 0$ by assumption, we can rewrite the linear canonical transform with respect to the discrete-time Fourier transform as

$$\mathcal{C}_{(a,b,c,d)}\,[x]\,(\omega) = \theta_{(a,b,c,d)}\,e^{i\frac{d}{2b}\omega^2}\,\mathcal{F}\!\left[x\,[\cdot]\,e^{i\frac{a}{2b}\cdot^2}\right]\!\left(\frac{\omega}{b}\right). \tag{24.12}$$

Let us now consider the linear canonical phase retrieval problem, where we wish to recover a complex-valued discrete-time signal $x := (x[n])_{n\in\mathbb{N}}$ with finite support from the intensity $|\mathcal{C}_{(a,b,c,d)}[x]|$ of its linear canonical transform. Similarly to the Fourier phase retrieval problem, we are particularly interested in the arising ambiguities and in uniqueness guarantees. Using the alternative formulation (24.12) that relates the linear canonical transform to the discrete-time Fourier transform, it can be simply seen that the linear canonical phase retrieval problem to recover the true signal x is also solved by

1. the rotated signal $e^{i\alpha}\,x$ with $\alpha \in \mathbb{R}$,
2. the shifted signal $e^{-ian_0\cdot/b}\,x[\cdot - n_0]$ with $n_0 \in \mathbb{Z}$, and
3. the conjugated and reflected signal $e^{-ia\cdot^2/b}\,\overline{x[-\cdot]}$.

Again, these trivial ambiguities are of minor interest.

The representation (24.12) implies that the linear canonical phase retrieval problem to recover x from $|\mathcal{C}_{(a,b,c,d)}[x]|$ is equivalent to the recovery of the signal $(x[n]\,e^{ian^2/2b})_{n\in\mathbb{N}}$ from the Fourier intensity $|\mathcal{F}[x[\cdot]\,e^{ia\cdot^2/2b}]|$, and we can immediately transfer the characterization of the complete solution set in Theorem 24.1 to the new setting.

Theorem 24.12 ([44]) *Let x be a discrete-time signal with finite support. Then each signal y with finite support satisfying*

$$|\mathcal{C}_{(a,b,c,d)}[y]| = |\mathcal{C}_{(a,b,c,d)}[x]|$$

is characterized by

$$\mathcal{F}\left[\theta\, e^{\frac{ia}{2b}\cdot^2}\, y\right](\omega) = e^{i(\alpha+\omega n_0)}\sqrt{|a[N-1]|\prod_{j=1}^{N-1}|\beta_j|^{-1}\cdot\prod_{j=1}^{N-1}\left(e^{-i\omega}-\beta_j\right)},$$

where α is a real number, n_0 is an integer, and β_j is chosen from the zero pair $(\gamma_j, \overline{\gamma}_j^{-1})$ of the associated polynomial P_a with respect to the autocorrelation signal of $\theta\, e^{ia\cdot^2/2b}x[\cdot]$.

With the characterization of trivial and non-trivial ambiguities, the linear canonical phase retrieval problem also inherits the uniqueness guarantees of the discrete-time Fourier phase retrieval problem; so the solutions in linear canonical phase retrieval are almost always unique (up to trivial ambiguities) if we have access to further magnitudes or phases in the time-domain, cf. Sect. 24.5, or if further interference measurements are available, cf. Sect. 24.6.

Acknowledgements The first author gratefully acknowledges the funding of this work by the Austrian Science Fund (FWF) within the project P28858. The Institute of Mathematics and Scientific Computing of the University of Graz, with which the first author is affiliated, is a member of NAWI Graz (http://www.nawigraz.at/).

References

1. Hauptman, H.A.: The phase problem of X-ray crystallography. Rep. Progr. Phys. **54**(11), 1427–1454 (1991)
2. Kim, W., Hayes, M.H.: The phase retrieval problem in X-ray crystallography. In: IEEE International Conference on Acoustics, Speech and Signal Processing. Proceedings: ICASSP 91, May 14–17, vol. 3, pp. 1765–1768. IEEE Signal Processing Society (1991)
3. Millane, R.P.: Phase retrieval in crystallography and optics. J. Opt. Soc. Am. A **7**(3), 394–411 (1990)
4. Bruck, Y.M., Sodin, L.G.: On the ambiguity of the image reconstruction problem. Opt. Commun. **30**(3), 304–308 (1979)
5. Dainty, J.C., Fienup, J.R.: Phase retrieval and image reconstruction for astronomy. In: Stark, H. (ed.) Image Recovery: Theory and Application, pp. 231–275. Academic Press, Orlando (Florida) (1987)
6. Seifert, B., Stolz, H., Donatelli, M., Langemann, D., Tasche, M.: Multilevel Gauss-Newton methods for phase retrieval problems. J. Phys. A, Math. Gen. **39**(16), 4191–4206 (2006)
7. Seifert, B., Stolz, H., Tasche, M.: Nontrivial ambiguities for blind frequency-resolved optical gating and the problem of uniqueness. J. Opt. Soc. Amer. B Opt. Phys. **21**(5), 1089–1097 (2004)
8. Beinert, R., Plonka, G.: Ambiguities in one-dimensional discrete phase retrieval from Fourier magnitudes. J. Fourier Anal. Appl. **21**(6), 169–1198 (2015)
9. Oppenheim, A.V., Schafer, R.W.: Discrete-Time Signal Processing. Prentice Hall Signal Processing Series. Prentice Hall, Englewood Cliffs, NJ (1989)
10. Beinert, R.: Ambiguities in one-dimensional phase retrieval from Fourier magnitudes. Dissertation, University of Göttingen (2015)

11. Fienup, J.R.: Reconstruction of an object from the modulus of its Fourier transform. Opt. Lett. **3**(1), 27–29 (1978)
12. Adams, D., Martin, L.S., Seaberg, M.D., Gardner, D., Kapteyn, H., Murnane, M.: A generalization for optimized phase retrieval algorithms. Opt. Express **20**(22), 24,778–24,790 (2012)
13. Bauschke, H.H., Combettes, P.L., Luke, D.: Hybrid projection-reflection method for phase retrieval. J. Opt. Soc. Am. A Opt. Image Sci. Vis. **20**(6), 1025–1034 (2003)
14. Beinert, R.: Non-negativity constraints in the one-dimensional discrete-time phase retrieval problem. Inf. Inference **6**(2), 213–224 (2017)
15. Langemann, D., Tasche, M.: Phase reconstruction by a multilevel iteratively regularized Gauss-Newton method. Inverse Probl. **24**(3), 035006(26) (2008)
16. Langemann, D., Tasche, M.: Multilevel phase reconstruction for a rapidly decreasing interpolating function. Results Math. **53**(3–4), 333–340 (2009)
17. Beinert, R.: Multilevel Gauss-Newton-Methoden zur Phasenrekonstruktion. Master thesis, University of Göttingen (2013)
18. Beinert, R., Plonka, G.: Enforcing uniqueness in one-dimensional phase retrieval by additional signal information in time domain. Appl. Comput. Harmon. Anal. **45**(3), 505–525 (2018)
19. Kim, W., Hayes, M.H.: Iterative phase retrieval using two Fourier transform intensities. In: IEEE International Conference on Acoustics, Speech and Signal Processing. Proceedings: ICASSP, 90, Apr 3–6, vol. 3, pp. 1563–1566. IEEE Signal Processing Society (1990)
20. Kim, W., Hayes, M.H.: Phase retrieval using two Fourier-transform intensities. J. Opt. Soc. Am. A **7**(3), 441–449 (1990)
21. Kim, W., Hayes, M.H.: Phase retrieval using a window function. IEEE Trans. Signal Process. **41**(3), 1409–1412 (1993)
22. Raz, O., Dudovich, N., Nadler, B.: Vectorial phase retrieval of 1-d signals. IEEE Trans. Signal Process. **61**(7), 1632–1643 (2013)
23. Raz, O., Schwartz, O., Austin, D., Wyatt, A.S., Schiavi, A., Smirnova, O., Nadler, B., Walmsley, I.A., Oron, D., Dudovich, N.: Vectorial phase retrieval for linear characterization of attosecond pulses. Phys. Rev. Lett. **107**(13), 133902(5) (2011)
24. Beinert, R.: One-dimensional phase retrieval with additional interference measurements. Results Math. **72**(1), 1–24 (2017)
25. Candès, E.J., Eldar, Y.C., Strohmer, T., Voroninski, V.: Phase retrieval via matrix completion. SIAM J. Imaging Sci. **6**(1), 199–225 (2013)
26. Balan, R., Bodmann, B.G., Casazza, P.G., Edidin, D.: Painless reconstruction from magnitudes of frame coefficients. J. Fourier Anal. Appl. **15**(4), 488–501 (2009)
27. Balan, R., Casazza, P.G., Edidin, D.: On signal reconstruction without phase. Appl. Comput. Harmon. Anal. **20**(3), 345–356 (2006)
28. Bandeira, A.S., Chen, Y., Mixon, D.G.: Phase retrieval from power spectra of masked signals. Inf. Inference **3**(2), 83–102 (2014)
29. Bodmann, B.G., Hammen, N.: Stable phase retrieval with low-redundancy frames. Adv. Comput. Math. **41**(2), 317–331 (2015)
30. van Hove, P., Hayes, M.H., Lim, J.S., Oppenheim, A.V.: Signal reconstruction from signed Fourier transform magnitude. IEEE Trans. Acoust. Speech Signal Process. ASSP **31**(5), 1286–1293 (1983)
31. Bendory, T., Beinert, R., Eldar, Y.: Fourier phase retrieval: uniqueness and algorithms. In: Boche, H., Caire, G., Calderbank, R., März, M., Kutyniok, G., Mathar, R. (eds.) Compressed Sensing and Its Applications, pp. 231–275. Birkhäuser (2017)
32. Nawab, S., Quatieri, T.F., Lim, J.S.: Algorithms for signal reconstruction from short-time Fourier transform magnitude. In: IEEE International Conference on Acoustics, Speech, and Signal. Proceedings: ICASSP 83, vol. 8, pp. 800–803. IEEE (1983)
33. Nawab, S., Quatieri, T.F., Lim, J.S.: Signal reconstruction from short-time Fourier transform magnitude. IEEE Trans. Acoust. Speech Signal Process. ASSP **31**(4), 986–998 (1983)
34. Loock, S., Plonka, G.: Phase retrieval for Fresnel measurements using a shearlet sparsity constraint. Inverse Probl. **30**(5), 055005(17) (2014)

35. Shi, B., Lian, Q., Chen, S.: Sparse representation utilizing tight frame for phase retrieval. EURASIP J. Adv. Signal Process. **96** (2015)
36. Beinert, R., Plonka, G.: Sparse phase retrieval of one-dimensional signals by Prony's method. Front. Appl. Math. Stat. **3**(5) (2017)
37. Ranieri, J., Chebira, A., Lu, Y.M., Vetterli, M.: Phase retrieval for sparse signals: uniqueness conditions (2013). Preprint, arXiv:1308.3058v2
38. Beinert, R., Plonka, G.: Sparse phase retrieval of structured signals by Prony's method. PAMM. Proc. Appl. Math. Mech. **17**(1), 829–830 (2017)
39. Fejér, L.: Über trigonometrische Polynome. J. Reine Angew. Math. **146**(2), 53–82 (1916)
40. Xu, L., Yan, P., Chang, T.: Almost unique specification of discrete finite length signal: from its end point and Fourier transform magnitude. In: IEEE International Conference on Acoustics, Speech, and Signal Processing. Proceedings: ICASSP 87, vol. 12, pp. 2097–2100. IEEE (1987)
41. Alexeev, B., Bandeira, A.S., Fickus, M., Mixon, D.G.: Phase retrieval with polarization. SIAM J. Imaging Sci. **7**(1), 235–66 (2014)
42. Plonka, G., Tasche, M.: Prony methods for recovery of structured functions. GAMM-Mitt. **37**(2), 239–258 (2014)
43. Wolf, K.B.: Integral Transforms in Science and Engineering. Plenum Press, New York (1979)
44. Beinert, R.: Ambiguities in one-dimensional phase retrieval from magnitudes of a linear canonical transform. ZAMM Z. Angew. Math. Mech. **97**(9), 1078–1082 (2017)
45. Gori, F.: Fresnel transform and sampling theorem. Opt. Commun. **39**(5), 293–297 (1981)
46. Pei, S.C., Ding, J.J.: Relations between fractional operations and time-frequency distributions, and their applications. IEEE Trans. Signal Process. **49**(8), 1638–1655 (2001)

Correction to: Wavefront and Coherence Characteristics of Extreme UV and Soft X-ray Sources

Bernd Schäfer, Bernhard Flöter, Tobias Mey, and Klaus Mann

Correction to:
Chapter 20 in: T. Salditt et al. (eds.), *Nanoscale Photonic Imaging*, Topics in Applied Physics 134,
https://doi.org/10.1007/978-3-030-34413-9_20

The book was inadvertently published with the error in the reference section of chapter 20. The correction detail is as follows:

[14] Ruiz-Lopez, M.: Wavefront propagation simulations supporting the design of a time delay compensating monochromator beamline at flash2 (2018) Photon Diag. 2018needs to be replaced by:

[14] Private communication Barbara Keitel and Elke Plönjes (DESY), to bepublished in detail elsewhere

The chapter and book has been updated with the change.

The updated version of this chapter can be found at
https://doi.org/10.1007/978-3-030-34413-9_20

Index

T. Salditt et al. (eds.), *Nanoscale Photonic Imaging*, Topics in Applied Physics 134,
https://doi.org/10.1007/978-3-030-34413-9

Printed in the United States
by Baker & Taylor Publisher Services